DATE DUE

The Structure of Sociological Theory

Sixth Edition

JONATHAN H. TURNER
University of California, Riverside

With Contributions by

P.R. Turner
Kenneth Allan
Charles H. Powers
Rebecca S.K. Li
Alexandra Maryanski
Paul Colomy
Stephan Fuchs
David G. Wagner

WADSWORTH PUBLISHING COMPANY
I(T)P® An International Thomson Publishing Company

Belmont, CA • Albany, NY • Bonn • Boston • Cincinnati • Detroit • Johannesburg • London • Madrid • Melbourne • Mexico City • New York • Paris • Singapore • Tokyo • Toronto • Washington

Sociology Editor: Denise Simon
Editorial Assistant: Angela Nava
Marketing Manager: Chuan Hightower
Project Manager: John Walker
Production: Robin Gold / Forbes Mill Press
Print Buyer: Karen Hunt
Permissions Editor: Peggy Meehan
Copyeditor: Robin Gold
Cover Designer: Ross Carron
Compositor: Wolf Creek Press / Forbes Mill Press
Printer: The Maple-Vail Book Manufacturing Group

Printed in the United States of America
1 2 3 4 5 6 7 8 9 10
For more information, contact Wadsworth Publishing Company, 10 Davis Drive, Belmont, CA 94002, or electronically at http://www.thomson.com/wadsworth.html

International Thomson Publishing Europe
Berkshire House 168-173
High Holborn
London, WC1V 7AA, England

Thomas Nelson Australia
102 Dodds Street
South Melbourne 3205
Victoria, Australia

Thomas Nelson Canada
1120 Birchmount Road
Scarborough, Ontario
Canada M1K 5G4

International Thomson Publishing GmbH
Königswinterer Strasse 418
53227 Bonn, Germany

International Thomson Editores
Campos Eliseos 385, Piso 7
Col. Polanco
11560 México D.F. México

International Thomson Publishing Asia
221 Henderson Road
#05-10 Henderson Building
Singapore 0315

International Thomson Publishing Japan
Hirakawacho Kyowa Building, 3F
2-2-1 Hirakawacho
Chiyoda-ku, Tokyo 102, Japan

International Thomson Publishing Southern Africa
Building 18, Constantia Park
240 Old Pretoria Road
Halfway House, 1685 South Africa

Library of Congress Cataloging-in-Publication Data

Turner, Jonathan H.
The structure of sociological theory / Jonathan H. Turner : with contributions by P. R. Turner . . . [et al.]. — 6th ed.
p. cm.
Includes index.
ISBN 0-534-51353-0
1. Sociology. I. Title.
HM24.T84 1997
301—dc21 97-37817

Brief Contents

Contents

Preface

The basic goal of this extensively revised edition remains the same: to analyze the *structure* of sociological theories. What has made a theoretical perspective unique? What are the assumptions, points of attack, and strategies employed by theorists working within a particular theoretical tradition? How are arguments constructed and presented? What kinds of models and propositions, if any, are generated or, at least, suggested by a theory? These kinds of questions have guided the presentation in the previous five editions of this book, and they still inform the analysis in this current edition. But, unlike the previous revisions, this edition is dramatically new. Not only have all existing chapters been rewritten, but, more significantly, twenty new chapters have been introduced.

That this sixth edition of *The Structure of Sociological Theory* will be published near the end of the twentieth century has given me pause to consider what to include. Sociologists still talk as if certain scholars are the cutting edge of theory, when these scholars are actually theorists of the century's midpoint. Some of their ideas will endure; others will fade away and collect dust on library shelves. As I though about rewriting this book, which I started in the late 1960s, I not only felt older and perhaps wiser, but, more important, I realized that the book must be changed in significant ways to reflect more closely the changed structure of theory in sociology.

WHAT IS NEW

I have retained the detailed discussion of the origins of various theoretical perspectives because it is important to understand the intellectual heritage of current theoretical arguments. Then, I have created a series of chapters for each of the major theoretical perspectives on what can be termed "the maturing tradition." The ideas of scholars in these chapters dominated when I first started

The Structure of Sociological Theory, but now these theorists are becoming historical figures of the field. They have passed the torch on; other theorists have been carrying their perspectives forward or developing new, alternative perspectives. A book on sociological theory at the close of the twentieth century must communicate this passage of time—even though we often talk about these mid-century figures as if they were still the advance guard of sociological theory.

Next, I have written a series of new chapters on "the continuing tradition" of theorizing within each broad theoretical perspective. In most cases, the coherent theoretical perspectives of the century's midpoint have fractured into a wide variety of more specialized theories, although in some cases, a particular perspective has not developed much since the 1960s. The chapters on continuing tradition roughly embrace work from the 1970s to the present, and they emphasize that an approach is continuing to produce theory or is passing into intellectual history.

I have redrawn somewhat the lines separating various broad theoretical approaches. I now group the chapters into sections on functional, evolutionary or bio-ecological, conflict, exchange, interactionist, structuralist, and critical theorizing. This diversity of perspectives as well as the varied theories within these perspectives reflect the practice of theory as we reach the close of the century.

Perhaps most visibly, I have added chapters and sections on many of the anti-science, or at least nonpositivistic, theories that currently occupy the discipline's attention. In presenting this new material I have sought to avoid the critical stance against nonscience approaches that I took in earlier editions.

In making these changes, the book has grown but not as much as one might think. There are many more chapters, to be sure, but many of the old chapters have been shortened or eliminated to make room for the new. And, I have removed the criticisms that I once made about theory. This time, I have let the theories and theorists speak for themselves.

I have also tried to make this edition a more modular book in which readers can choose for themselves topics to pursue. The organization of the book does not lock the reader into a particular pattern or order of reading. You can chose and select, and the book will remain coherent—or, at least as coherent as sociological theory is these days.

The end result of these changes is a more comprehensive and balanced book. In a sense, this is a new book, built on the foundation laid by the previous five editions. It is, I feel, a book that can be useful well into the twenty-first century.

ACKNOWLEDGEMENTS

I would like to thank the Academic Senate, University of California at Riverside, for support in preparing this edition. I also am grateful to my coauthors of various chapters who helped extend the breadth of the book and my own knowledge of sociological theory.

Jonathan H. Turner

1

Sociological Theory

Theories are stories about how and why events occur. Sociological theories are thus stories about how humans behave, interact, and organize themselves. When viewed in this way, there is little controversy about the nature of sociological theory. All sociologists would agree that theories seek to explain how and why social processes operate, but once we move beyond this generality, disagreements become evident.

At the center of this controversy is the question of whether or not sociological theories can be scientific. Some argue that they can, others argue the opposite, and many offer some moderating position.

To appreciate the dimensions of this controversy, let us begin by briefly outlining the nature of scientific theories. Then, we can return to what critics of scientific theory find objectionable.

THE NATURE OF SCIENTIFIC THEORY

Scientific theories begin with the assumption that the universe, including the social universe created by acting human beings, reveals certain basic and fundamental properties and processes that explain the ebb and flow of events in specific contexts. Because of this concern with discovering fundamental properties and processes, scientific theories are always stated abstractly, rising above specific empirical events and highlighting the underlying forces that drive these events. In the context of sociological inquiry, for example, theoretical explanations are not so much about the specifics of a particular economy as about the dynamics of production and distribution in general. Similarly, scientific theories are not so much about a particular form of government but about the nature of power as a basic social force. Or, to illustrate further, theories are not so much about particular behaviors and interactions among actual persons in a specific setting as about the nature of human interpersonal behavior in general. The goal, then, is always to see if the underlying forces that govern particulars of specific cases can be discovered. To realize this goal, theories must be about generic properties and processes transcending the unique characteristics of any one situation or case. Thus, scientific theories always seek to transcend the particular and the time bound. Theories are, instead, about the generic, the fundamental, the timeless, and the universal.

Another characteristic of scientific theories is that they are stated more formally than ordinary language. At the extreme, theories are couched in another language, such as mathematics, but more typically in the social sciences and particularly in sociology, theories are phrased in ordinary language. Still, even when using regular language, an effort is made to speak in neutral, objective, and unambiguous terms so that the theory means the same thing to all who examine it. Terms denoting properties of the world and their dynamics are defined clearly so that their referents are clear, and relationships among concepts denoting phenomena are stated in ways such that their interconnections are understood by all who examine the theory. At times, this attention to formalism can make theories seem stiff and dull, especially when these formalisms are couched at higher levels of abstraction. Yet, without attention to what terms and phrases denote and connote, a theory could mean very different things to diverse audiences.

A final characteristic of scientific theories is that they are designed to be systematically tested with replicable methods against the facts of particular empirical settings. Despite being stated abstractly and formally, then, scientific theories do not stand aloof and alone from the empirical. Useful theories all suggest ways that they can be assessed against empirical events.

All scientific fields develop theories. For in the end, science seeks (1) to develop abstract and formally stated theories and (2) to test these theories against empirical cases to see if they are plausible. If the theory seems plausible in light of empirical assessment, then it represents *for the present time* the explanation of events. If a theory is contradicted by empirical tests, then it must be discarded or revised. If competing theories exist to explain the same phenomena, they too must be empirically assessed, with the better explanation winning out.

Science is thus a rather slow process of developing theories, testing them, and then rejecting, modifying, or retaining them, at least until a better theory is proposed. Without attention to stating theories formally and objectively assessing them against the empirical world, theory would become self-justifying and self-contained, reflecting personal biases, ideological leanings, or religious convictions. The differences between scientific theory and other types of knowledge are presented in Figure 1-1.

The typology asks two basic questions: (1) Is the search for knowledge to be evaluative or neutral? (2) Is the knowledge developed to pertain to actual empirical events and processes, or is it to be about nonempirical realities? In other words, should knowledge tell us what *should be* or *what is?* And should it refer to the observable world or to other, less observable, realms? If knowledge is to tell us what should exist (and, by implication, what should not occur) in the empirical world, then it is ideological knowledge. If it informs us about what should be but does not pertain to observable events, then the knowledge is religious, or about forces and beings in another realm of existence. If knowledge is

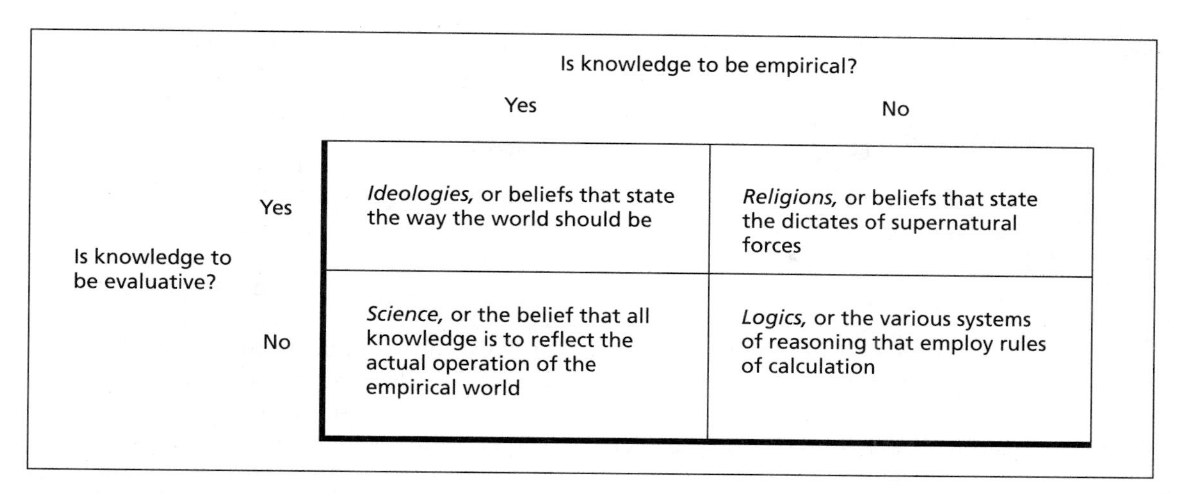

FIGURE 1-1 Types of Knowledge

neither empirical nor evaluative, then it is a formal system of logic, such as mathematics. And if it is about empirical events and is nonevaluative, then it is science.

This typology is crude, but it makes the essential point: there are different ways to look at, interpret, and develop knowledge about the world. Science is only one way. Science is based on the presumption that knowledge can be value free, that it can explain the actual workings of the empirical world, and that it can be revised as a result of careful observations of empirical events. These characteristics distinguish science from other beliefs about how we should generate understanding and insight.

In the "hard sciences," this view of theory and science is widely accepted, whereas in the social sciences, this view is often questioned and criticized. From the very beginning of sociology as a discipline, controversy about whether or not it could be a natural science emerged. The titular founder of sociology, Auguste Comte, recognized in the early 1800s that the status of a "science of society" was precarious.[1] To defend sociology from its many critics and to legitimate his claims that the emergence of sociology as a science was now possible, Comte posited a "law of the three stages." In the early stage, interpretations of events are provided by religious beliefs or by reference to the activities of sacred and supernatural forces. From religion comes a metaphysical stage in which logic, mathematics, and other formal systems of reason come to dominate how events are interpreted. And from these gains in formal reasoning in the metaphysical stage emerges the possibility for "positivism" or a scientific stage, during which formal statements are critically examined against carefully collected facts. Comte argued that the accumulation of knowledge about each domain of the universe—the physical, the chemical, the biological, and, finally, the social—passed successively through these three stages. Patterns of human organization were to be the last such domain to move into the positive stage, and in 1830, Comte trumpeted the call for the use of science to develop knowledge about human affairs.

Comte's advocacy raised issues that still haunt sociology. More than 165 years after Comte pronounced sociology to be in the positive phase, he is denounced in many quarters as naive and just plain wrong. True, his point of view has many supporters, but his faith that there could be a "natural science of society" is hardly shared by all. Hence the question of whether or not sociology can be a science is the basic issue over which sociologists find themselves in disagreement.

THE CRITIQUE OF SCIENTIFIC THEORY IN SOCIOLOGY

As we approach the study of sociological theory at the close of the twentieth century, Comte's advocacy seems remote. Many of the theories that we will examine are couched in purely scientific terms, being committed to sociology as a natural science. Other theories are less precise and formal, but still maintain a commitment to developing science. Yet, many of the theories are highly suspicious about, if not outright antagonistic toward, the view that human social relations can be studied scientifically. Indeed, this latter group of theorists argue that it is not only pretentious to view sociology as a science but also very naive. This critical view takes many forms: Some argue that humans' capacity for agency and creativity enables them to change the very nature of social reality, thereby making any search for underlying and fundamental social processes merely a chimera. Others assert that social theories will always, no matter how hard theorists try, contain ideological elements arising from vested interests of those who create theories, or fund their development. Still others argue that theories will inevitably support the status quo, generalizing from what is observable and assuming that this is the way the social world must operate instead of searching for more liberating alternatives. And so the criticisms go, as we will appreciate as we explore the structure of sociological theory. Thus, the optimism of sociology's titular founder is now tempered by a pessimism about the scientific prospects for sociology.

Theoretical sociology is often not scientific theory but merely abstract statements that are ideological pronouncements, that are untestable, and that are loose frameworks for interpreting social events. Theory is now the label that encompasses many diverse kinds of intellectual activity, from the history of ideas through biographies of major thinkers, from philosophical discourses to analyses of the great works of masters, from critiques of modern society to lambasting the prospects for a science of society. Only a portion of theoretical activity in sociology is devoted to the production of scientific theory.

SEVEN BASIC THEORETICAL ORIENTATIONS

How, then, do we get a handle on this diversity of activity that all goes by the label "theory?" One answer is to present a variety of theories that fall within broad perspectives or orientations. In the pages that follow, seven broad perspectives will be used to organize very diverse theories, some of which are scientific and others of which are less so. The first perspective is functionalism, which was sociology's first theoretical orientation. Functionalism is unique in analyzing social forces and structures to determine their consequences for meeting the needs and requisites of social wholes. The second theoretical perspective is bio-ecological theory, which emerged from functionalism. Here, ideas from biology are incorporated into social theory and used to analyze the dynamics of social processes. Some of these theories are ecological and emphasize resource niches, competition for resources, and selection processes, whereas others are more biological and draw on insights from genetics and evolutionary forces to explain human behavior and organization. The third perspective is conflict theory, which examines how systems of inequality systematically generate conflict among superordinates and subordinates. The fourth perspective is exchange theory, which views all social processes among individuals and collective actors as the exchange of valued resources. The fifth perspective is interactionism, which is a broad label that encompasses theories seeking to understand the dynamics of face-to-face encounters among individuals. The sixth perspective is structuralist theory, which emphasizes that the social universe is guided by underlying patterns in cultural symbols or forms of social relations. And finally, the last perspective is critical theory, in which the goal is to use theory to critique not only the way the social world is organized—and, by implication, to propose alternatives—but also the notion that science is a useful way to understand the social universe.

As will become evident, this eclectic mix of theories roughly grouped into these broad perspectives is the way that sociological theory is structured. There is now little consensus over what sociological theory is, or should be. And as we examine the structure of each of these theories, we will see that at the close of the twentieth century, theoretical sociology reveals little coherence or singularity of purpose.

NOTES

1. Auguste Comte, *System of Positive Philosophy,* vol. 1 (Paris: Bachelier, 1930). Subsequent portions were published between 1831 and 1842. For a more detailed analysis of Comte's thought, see Jonathan H. Turner, Leonard Beeghley, and Charles Powers, *The Emergence of Sociological Theory* (Belmont, CA: Wadsworth, 1998).

PART 1

Functional Theorizing

2

The Emerging Tradition:

The Rise of Functionalist Theorizing

August Comte (1798–1857) is usually credited as being the founder of sociology. Philosophizing about humans and society had, of course, long been a preoccupation of lay people and scholars alike, but Comte advocated a "science of society" and coined the term *sociology*. Although Comte's work soon fell into neglect and obscurity—leading him to live out his later years in frustration and bitterness—his work profoundly influenced social thought. It is regrettable that few recognize this influence, even today, but despite Comte's current obscurity, the emergence of the functionalist perspective began with his work.[1] Comte created functionalism by comparing societies to biological organisms—a point of emphasis that remained for the entire nineteenth century.

THE ORGANISMIC ANALOGY

Comte felt that human evolution in the nineteenth century had reached the "positive stage" in which empirical knowledge could be used to understand the social world and to create a better society. Comte thus advocated applying the scientific method to the study of society—a strategy that, in deference to Comte, is still termed *positivism* in the social sciences. This application of the scientific method gave birth to a new science, sociology.

Comte's entire intellectual life represented an attempt to legitimate sociology. His efforts on this score went so far as to construct a "hierarchy of the sciences," with sociology as the "queen" of the sciences. Although this hierarchy allowed Comte to assert the importance of sociology and thereby separate it from social philosophy, his most important tactic for legitimating sociology was to borrow terms and concepts from the highly respected biological sciences. Sociology was thus initiated and justified by appeals to the biological sciences—which helps explain why functionalism was sociology's first and, until the 1970s, most dominant theoretical orientation.

Seeing the affinity between sociology and biology as residing in their common concern with organic bodies, Comte divided sociology into social "statics," or morphology, and "dynamics," or social growth and progress. But Comte was convinced that, although "Biology has hitherto been the guide and preparation for Sociology . . . Sociology will in the future . . . (provide) the ultimate systematization of Biology." Comte visualized that, with initial borrowing of concepts from biology and later with the development of positivism in the social sciences, the principles of sociology would inform biology. Thus, sociology must first recognize the correspondence between the individual organism in biology and the social organism in sociology:

> We have thus established a true correspondence between the Statistical Analysis of the Social Organism in Sociology, and that of the Individual Organism in Biology. . . . If we take the best ascertained points in Biology, we may decompose structure anatomically into *elements, tissues,* and *organs*. We have the same things in the Social Organism; and may even use the same names.[2]

Comte then began to make clear analogies between specific types of social structures and the biological concepts:

> I shall treat the Social Organism as definitely composed of the Families which are the true elements or cells, next the Classes or Castes which are its proper tissues, and lastly, of the cities and Communes which are in real organs.[3]

In Comte's hands, the organismic analogy was rough and crude, but it provided a model for legitimating sociology under the mantra of the more respected biological sciences as well as a strategy for conducting sociological inquiry. Yet, Comte never followed through on his advocacy; it was left to the British sociologist, Herbert Spencer, to develop more fully the implications of the organismic analogy.

THE ANALYTICAL FUNCTIONALISM OF HERBERT SPENCER

Herbert Spencer (1820–1903) was a broad-based philosopher who, before writing on sociology, had produced multivolume treatises on ethics,[4] biology[5] and psychology.[6] All were a part of his Synthetic Philosophy, which was to unify the diverse realms of the universe under a common set of abstract principles.[7] In writing his major work on sociology, *The Principles of Sociology,*[8] Spencer developed an organismic analogy that systematically compared society to organisms[9]:

1. As organic and superorganic (societal) bodies increase in size, they increase in structure. That is, they become more complex and differentiated.
2. Such differentiation of structures is accompanied by differentiation of functions. Each differentiated structure serves distinctive functions for sustaining the "life" of the systemic whole.
3. Differentiated structures and functions require in both organic and superorganic bodies integration through mutual dependence. Each structure can be sustained only through its dependence on others for vital substances.
4. Each differentiated structure in both organic and superorganic bodies is, to a degree, a systemic whole by itself (that is, organs are composed of cells, and societies are composed of groupings of individuals); thus the larger whole is always influenced by the systemic processes of its constituent parts.
5. The structures of organic and superorganic bodies can "live on" for a while after the destruction of the systemic whole.

These points of similarity between organism and society, Spencer argued, must be qualified for their points of "extreme unlikeness"[10]:

1. There are great differences in the degree of connectedness of the parts, or structures, in organic and social wholes. In superorganic wholes, there is less direct and continuous physical contact and more dispersion of parts than in organic bodies.
2. There are differences in the modes of contact between organic and superorganic systems. In the superorganic there is much more reliance on symbols than in the organic.[11]
3. There are differences in the levels of consciousness and voluntarism of parts in organic and superorganic bodies. All units in society are conscious, goal seeking, and reflective, whereas only one unit can potentially be so in organic bodies.

As Spencer continued to analogize the points of similarity between organicism and societies, he began to develop what can be termed *requisite functionalism*. That is, organic and superorganic bodies reveal certain universal requisites that must be fulfilled for these bodies to adapt to an environment. Moreover, these same requisites exist for all organic and superorganic systems. To quote Spencer on this point:

> Close study of the facts shows us another striking parallelism. Organs in animals and organs in societies have internal arrangements framed on the same principle.
>
> Differing from one another as the viscera of a living creature do in many respects, they have several traits in common. Each viscus contains appliances for conveying nutriment to its parts, for bringing it materials on which to operate, for carrying away the product, for draining off waste matters; as also for regulating its activity.[12]

It is not hard to see the seeds of an argument for universal functional requisites in this passage. Indeed, on the next page from this quote, Spencer argued "it is the same for society" and proceeded to list the basic functional requisites of societies. For example, each superorganic body

> Has a set of agencies which bring the raw material . . . ; it has an apparatus of major and minor channels through which the necessities of life are drafted out of the general stocks circulating through the kingdom . . . ; it has appliances . . . for bringing those impulses by which the industry of the place is excited or checked; it has local controlling powers, political and ecclesiastical, by which order is maintained and healthful action furthered.

Even though these universal requisites are not so clearly separated as they were to become in modern functional approaches, the logic of the analysis is clear. First, there are certain universal needs or requisites that structures function to meet. These revolve around (a) securing and circulating resources, (b) producing usable substances, and (c) regulating and integrating internal activities through power and symbols. Second, each system level—group, community, region, or whole society—reveals a similar set of needs. Third, the important dynamics of any empirical system revolve around processes that function to meet these universal requisites. Fourth, the level of adaptation of a social unit to its environment is determined by the extent to which it meets these functional requisites.

Thus, by recognizing that certain basic or universal needs must be met, analysis of organic and superorganic systems is simplified. One examines processes to determine needs for integrating differentiated parts, needs for sustaining the parts of the system, needs for producing and distributing information and substances, and needs for political regulation and control. In simple systems these needs are met by each element of the system, but, when structures begin to grow and to become more complex, they are met by distinctive types of structures that specialize in meeting one of these general classes of functions. As societies become highly complex, structures become even more specialized and meet only specific subclasses of these general functional needs.

The logic behind this form of requisite functionalism guided much of Spencer's substantive analysis and is the essence of functional analysis today. The list of basic needs to be met varies among theorists, but the mode of the analysis remains the same: Examine specific types of social processes and structures for the needs or requisites that they meet.

FUNCTIONALISM AND ÉMILE DURKHEIM

We should not be surprised that, as the inheritor of a long French tradition of social thought, especially Comte's organicism, Émile Durkheim's (1858–1917) early works were heavily infused with organismic terminology. Although his major work, *The Division of Labor in Society,* was sharply critical of Herbert Spencer, many of Durkheim's formulations were clearly influenced by the nineteenth-century intellectual preoccupation with biology.[13] Aside from the extensive use of biologically inspired terms, Durkheim's basic assumptions reflected those of the organicists: (1) Society was to be viewed as an entity in itself that can be distinguished from and is not reducible to its constituent parts. In conceiving society as a reality, *sui generis,* Durkheim in effect gave analytical priority to the social whole. (2) Although such an emphasis by itself did not necessarily reflect organismic inclinations, Durkheim, in giving causal priority to the whole, viewed system parts as fulfilling basic functions, needs, or requisites of that whole. (3) The frequent use of the notion "functional needs" is buttressed by Durkheim's conceptualization of social systems as "normal" and "pathological" states. Such formulations, at the very least, connote the view that social systems have needs that must be fulfilled if "abnormal" states are to be avoided. (4) When we view systems as normal and pathological, as well as by functions, the additional implication is that systems have equilibrium points around which normal functioning occurs.

Durkheim recognized the dangers in this kind of analysis and explicitly tried to deal with several of them. First, he was clearly aware of the dangers of teleological analysis—of implying that some future consequence of an event causes that very event to occur. Thus he warned that the causes of a phenomenon must be distinguished from the ends it serves:

> When, then, the explanation of a social phenomenon is undertaken, we must seek separately the efficient cause which produces it and the function it fulfills. We use the word "function" in preference to "end" or "purpose," precisely because social phenomena do not generally exist for the useful results they produce.[14]

Thus, despite giving analytical priority to the whole and viewing parts as having consequences for certain normal states and hence meeting system requisites, Durkheim remained aware of the dangers of asserting that all systems have "purpose" and that the need to maintain the whole causes the existence of its constituent parts. Yet Durkheim's insistence that the function of a part for the social whole always be examined sometimes led him, and certainly many of his followers, into questionable teleological reasoning. For example, even when distinguishing *cause* and *function* in his major methodological statement, he leaves room for an illegitimate teleological interpretation: "Consequently, to explain a social fact, it is not enough to show the cause on which it depends; we must also, at least in most cases, show its function in the establishment of social order."[15] In this summary phrase, the words "in the establishment of" could connote that the existence of system parts can be explained only by the whole, or social order, that they function to maintain. From this view it is only a short step to outright teleology: The social fact in question is caused by the needs of the social order that the fact fulfills. Such theoretical statements do not necessarily have to be illegitimate, for it is conceivable that a social system could be programmed to meet certain needs or designated ends and thereby have the capacity to cause variations in cultural items or "social facts" to meet these needs or ends. But if such a system is being described by an analyst, it is necessary to document how the system is programmed and how it causes variations in social facts to meet needs or ends. As the previous quotation illustrates, Durkheim did not have this kind of system in mind when he formulated his particular brand of functional analysis; thus he did not want to state his arguments teleologically.

Despite his warnings to the contrary, Durkheim appears to have taken this short step into teleological reasoning in his substantive works. In his first major work on the division of labor, Durkheim went to great lengths to distinguish between cause (increased population and moral density) and function (integration of society). However, the causal statements often become fused with functional statements. The argument is, generally, like this: Population density increases moral density (rates of contact and interaction); moral density leads to competition, which threatens the social order; in turn, competition for resources results in the specialization of tasks as actions seek viable niches in which to secure resources; and specialization creates pressures for mutual interdependence and increased willingness to accept the morality of mutual obligation. This transition to a new social order is not made consciously, or by "unconscious wisdom"; yet the division of labor is necessary to restore the order that "unbridled competition might otherwise destroy."[16] Hence, the impression is left that the threat or the need for social order causes the division of labor. Such reasoning can be construed as an illegitimate teleology, because the consequence or result of the division of labor—social order—is the implied cause of it. At the very least, then, cause and function are not kept as analytically separate as Durkheim so often insisted.

In sum, then, despite Durkheim's warnings about illegitimate teleology, he often appears to waver on the brink of the very traps he wanted to avoid. The reason for this failing can probably be traced to the organismic assumptions built into this form of sociological analysis. In taking a strong sociologistic position on the question of emergent properties—that is, on the irreducibility of the whole to its individual parts—Durkheim separated sociology from the naive psychology and anthropology of his day.[17] However, in supplementing this emphasis on the social whole with organismic assumptions of function, requisite, need, and normality/pathology, Durkheim helped weld organismic principles to sociological theory for nearly three-quarters of a century. The brilliance of his analysis of substantive topics, as well as the suggestive features of his analytical work, made a functional mode of analysis highly appealing to subsequent generations of sociologists and anthropologists.

FUNCTIONALISM AND THE ANTHROPOLOGICAL TRADITION

Functionalism might have died with Durkheim except that anthropologists began to find it an appealing way to analyze simple societies. Indeed, functionalism as a well-articulated conceptual perspective was perpetuated in the first half of the twentieth century by the writings of two anthropologists, Bronislaw Malinowski and A. R. Radcliffe-Brown.[18] Each of these thinkers was heavily influenced by the organicism of Durkheim, as well as by their own field studies among primitive societies. Despite the similarities in their intellectual backgrounds, however, the conceptual perspectives developed by Malinowski and Radcliffe-Brown reveal many dissimilarities.

The Functionalism of A. R. Radcliffe-Brown

Recognizing that "the concept of function applied to human societies is based on an analogy between social life and organic life" and that "the first systematic formulation of the concept as applying to the strictly scientific study of society was performed by Durkheim," Radcliffe-Brown (1881–1955) tried to indicate how some problems of organismic analogizing might be overcome.[19] Radcliffe-Brown believed the most serious problem with functionalism was the tendency for analysis to appear teleological. Noting that Durkheim's definition of function pertained to the way in which a part fulfills system needs, Radcliffe-Brown emphasized that, to avoid the teleological implications of such analysis, it would be necessary to "substitute for the term 'needs' the term 'necessary condition of existence.'" In doing so, he felt that no universal human or societal needs would be postulated; rather, the question of which conditions were necessary for survival would be an empirical one, an issue that would have to be discovered for each given social system. Furthermore, in recognizing the diversity of conditions necessary for the survival of different systems, analysis would avoid asserting that every item of a culture must have a function and that items in different cultures must have the same function.

Once the dangers of illegitimate teleology were recognized, functional or (to use his term) structural analysis could legitimately proceed from several assumptions: (1) One necessary condition for survival of a society is minimal integration of its parts. (2) The term *function* refers to those processes that maintain this necessary integration or solidarity. (3) Thus, in each society, structural features can be shown to contribute to the maintenance of necessary solidarity. In such an analytical approach, social structure and the conditions necessary for its survival are irreducible. In a vein similar to that of Durkheim, Radcliffe-Brown saw society as a reality in and of itself. For this reason he usually visualized cultural items, such as kinship rules and religious rituals, as explicable through social structure—particularly social structure's need for solidarity and integration. For example, in analyzing a lineage system, Radcliffe-Brown would first assume that some minimal degree of solidarity must exist in the system. Processes associated with lineage systems would then be assessed to determine their consequences for maintaining this solidarity. The conclusion was that lineage systems provided a systematic way of adjudicating conflict in societies where families owned land because such a system specified who had the right to land and through which side of the family it would always pass. The integration of the economic system—landed "estates" owned by families—is thus explained.[20]

This form of analysis poses a number of problems that continue to haunt functional theorists. Although Radcliffe-Brown admitted that "functional unity [integration] of a social system is, of course, a hypothesis," he failed to specify the analytical criteria for assessing just how much or how little functional unity is necessary for testing this hypothesis. As subsequent commentators discovered, without some analytical criteria for determining what is and what is not minimal functional integration and societal survival, the hypothesis cannot be tested, even in principle. Thus, what it typically done is to assume that the existing system is minimally integrated and surviving because it exists and persists. Without carefully documenting how various cultural items promote instances of both integration and malintegration of the social whole, such a strategy can reduce the hypothesis of functional unity to a tautology: If one can find a system to study, then it must be minimally integrated; therefore, lineages that are a part of this system must promote its integration. To discover the contrary would be difficult, because the system, by virtue of being a surviving system, is already composed of integrated parts, such as a lineage system. There is a non sequitur in such reasoning, because it is quite possible to view a cultural item as a lineage system as having both integrative and malintegrative (and other) consequences for the social whole. In his actual ethnographic descriptions, Radcliffe-Brown often slips inadvertently into a pattern of circular reasoning: The fact of a system's existence requires that its existing parts, such as a lineage system, be viewed as contributing to the system's existence. Assuming integration and then assessing the contribution of individual parts to the integrated whole lead to an additional analytical problem. Such a mode of analysis implies that the causes of a particular structure—for example, lineages—lie in the system's needs for integration, which is most likely an illegitimate teleology.

Radcliffe-Brown would, of course, have denied this conclusion. His awareness of the dangers of illegitimate teleology would have seemingly eliminated the implication that the needs of a system cause the emergence of its parts. His repeated assertions that the notion of function "does not require the dogmatic assertion that everything in the life of every community has a function" should have led to a rejection of tautological reasoning.[21] However, much like Durkheim, what Radcliffe-Brown asserted analytically was frequently not practiced in the concrete empirical analysis of societies. Such lapses were not intended but appeared to be difficult to avoid with functional needs, functional integration, and equilibrium as operating assumptions.[22]

Thus, although Radcliffe-Brown displayed an admirable awareness of the dangers of organicism—especially of the problem of illegitimate teleology and the hypothetical nature of notions of solidarity—he all too often slipped into a pattern of questionable teleological reasoning. Forgetting that integration was only a working hypothesis, he opened his analysis to problems of tautology. Such problems were persistent in Durkheim's analysis, and, despite his attempts to the contrary, their specter haunted even Radcliffe-Brown's insightful essays and ethnographies.

The Functionalism of Bronislaw Malinowski

Functionalism might have ended with Radcliffe-Brown because it had very little to offer sociologists attempting to study complex societies. Both Durkheim and Radcliffe-Brown posited one basic societal need—integration—and then analyzed system parts to determine how they meet this need. For sociologists who are concerned with differentiated societies, this is likely to become a rather mechanical task. Moreover, it does not allow analysis of those aspects of a system part that are not involved in meeting the need for integration.

Bronislaw Malinowski's (1884–1942) functionalism removed these restrictions; by reintroducing Spencer's approach, Malinowski offered a way for modern sociologists to employ functional analysis.[23] Malinowski's scheme reintroduced two important ideas from Spencer: (1) the notion of system levels and (2) the concept of different and multiple system needs at each level. In making these two additions, Malinowski made functional analysis more appealing to twentieth-century sociological theorists.

Malinowski's scheme has three system levels: the biological, the social structural, and the symbolic.[24] At each level we can discern basic needs or survival requisites that must be met if biological health, social-structural

TABLE 2-1 Requisites of System Levels

Cultural (Symbolic) System Level	Structural (Instrumental) System Level
1. Requisites for systems of symbols that provide information necessary to adjust to the environment.	1. The requisite for production and distribution of consumer goods.
2. Requisites for systems of symbols that provide a sense of control over people's destiny and over chance events.	2. The requisite for social control of behavior and its regulation.
3. Requisites for systems of symbols that provide members of a society with a sense of a "communal rhythm" in their daily lives and activities.	3. The requisite for education of people in traditions and skills.
	4. The requisite for organization and execution of authority relations.

integrity, and cultural unity are to exist. Moreover, these system levels constitute a hierarchy, with biological systems at the bottom, social-structural arrangements next, and symbolic systems at the highest level. Malinowski stressed that the way in which needs are met at one system level sets constraints on how they are met at the next level in the hierarchy. Yet he did not advocate a reductionism of any sort; indeed, he thought that each system level reveals its own distinctive requisites and processes meeting these needs. In addition, he argued that the important system levels for sociological or anthropological analysis are the structural and symbolic. And in his actual discussion, the social-structural level receives the most attention. Table 2-1 lists the requisites or needs of the two most sociologically relevant system levels.

In analyzing the structural system level, Malinowski stressed that institutional analysis is necessary. For Malinowski, institutions are the general and relatively stable ways in which activities are organized to meet critical requisites. All institutions, he felt, have certain universal properties or "elements" that can be listed and then used as dimensions for comparing different institutions. These universal elements are

1. *Personnel:* Who and how many people will participate in the institution?
2. *Charter:* What is the purpose of the institution? What are its avowed goals?
3. *Norms:* What are the key norms that regulate and organize conduct?
4. *Material apparatus:* What is the nature of the tools and facilities used to organize and regulate conduct in pursuit of goals?
5. *Activity:* How are tasks and activities divided? Who does what?
6. *Function:* What requisite does a pattern of institutional activity meet?

By describing each institution along these six dimensions, Malinowski believed that he had provided a common analytical yardstick for comparing patterns of social organization within and between societies. He even constructed a list of universal institutions as they resolve not just structural but also biological and symbolic requisites.

In sum, Malinowski's functional approach opened new possibilities for sociologists who had long forgotten Spencer's similar arguments. Malinowski suggested to sociologists that attention to system levels is critical in analyzing requisites; he argued that there are universal requisites for each system level; he forcefully emphasized that the structural level is the essence of sociological analysis; and, much like Spencer before him and Talcott Parsons a decade later (see Chapter 4), Malinowski posited four universal functional needs at this level—economic adaptation, political authority, educational socialization, and social control—that were to be prominent in subsequent functional schemes. Moreover, he provided a clear method for analyzing institutions as they operate to meet functional requisites. It is fair to say, therefore, that Malinowski drew the rough contours for modern sociological functionalism.

FUNCTIONALISM AND THE GHOST OF MAX WEBER

During the latter nineteenth century and into the early part of this century, Max Weber (1864–1920) developed a particular approach for sociological analysis. His approach in a wide range of substantive areas—economic sociology, stratification, complex organizations, sociology of religion, authority, and social change, for example—still guides modern research and theory in these areas. In the development of general theoretical orientations, however, Weber's influence has been less direct. Although his impact on some perspectives is clear, his influence on functionalism is less evident. Yet, because several contemporary functionalists were so important in initially exposing American scholars to Weber's thought, it is likely that these functionalists' theorizing was influenced by the power of his approach.

What, then, has been Weber's impact on the emergence of functionalism? Generally, two aspects of Weber's work have had an important influence: (1) his substantive vision of "social action" and (2) his strategy for analyzing social structures. Weber argued that sociology must understand social phenomena on two levels—at the "level of meaning" of the actors themselves and at the level of collective action among groupings of actors. Weber's substantive view of the social world and his strategy for analyzing its features were thus influenced by these dual concerns. In many ways, Weber viewed two realities—that of the subjective meanings of actions and that of the emergent regularities of social institutions.[25] Much functionalism similarly addresses this dualism: How do the subjective states of actors influence emergent patterns of social organization, and vice versa?

Talcott Parsons, in particular, labeled his functionalism *action theory,* and his early theoretical scheme was devoted to analyzing the basic components and processes of the subjective processes of individual actors. But, much like Weber, Parsons and other functionalists moved to a more macroscopic concern with emergent patterns of collective action.

This shift from the micro to the macro represents only part of the Weberian analytical strategy. One of Weber's most enduring analytical legacies is his strategy for constructing "ideal types." Thus, an ideal type represents a category system for "analytically accentuating" the important features of social phenomena. Ideal types are abstractions from empirical reality, and their purpose is to highlight certain common features among similar processes and structures. Moreover, they can be used to compare and contrast empirical events in different contexts by providing a common analytical yardstick. By noting the respective deviations of two or more concrete, empirical situations from the ideal type, it is possible to compare these two situations and thus better understand them. For virtually all phenomena he studied—religion, organizations, power, and the like—Weber constructed an ideal type to visualize its structure and functioning. In many ways, the ideal-type strategy corresponds to taxonomic procedures for categorizing species and for describing somatic structures and processes in the biological sciences. It encourages a concern with conceptual schemes and categories rather than propositions and laws. So, although Weber's work is devoid of the extensive organismic imagery of Durkheim's or Spencer's, the concern with categorization of different social structures is highly compatible with the organismic reasoning of early functionalism. Thus it is not surprising that contemporary functionalists borrowed the substantive vision of the world implied by the concepts of structure and function as well as Weber's use of the taxonomic approach for studying structures and processes.

For functionalists in general and Talcott Parsons in particular, the construction of conceptual schemes remains an important activity. Functionalists elaborately categorize the social world to emphasize the importance of some structures and processes for maintaining the social system. For example, much like Weber before him, Parsons first developed a category system for individual social action and then elaborated this initial system of categories into an incredibly complex, analytical edifice of concepts. What is important to recognize is that this strategy of developing first category systems and then propositions about the relationships among categorized phenomena lies at the heart of contemporary functionalism. This emphasis on category systems is, no doubt, one subtle way that Weber's ideal-type strategy continues to influence functional theorizing in sociology.

THE EMERGENCE OF FUNCTIONALISM: AN OVERVIEW

With its roots in the organicism of the early nineteenth century, functionalism is the oldest and, until recent decades, the dominant conceptual perspective in sociology. The organicism of Comte and later that of Spencer and Durkheim clearly influenced the first functional anthropologists—Malinowski and Radcliffe-Brown—who in turn, with Durkheim's timeless analysis, helped shape the more modern functional perspectives.[26] Coupled with Weber's emphasis on social taxonomies, or ideal types, of both subjective meaning and social structure, a strategy for studying the properties of the "social organism" similarly began to shape contemporary functionalism.

In emphasizing the contribution of sociocultural items to the maintenance of a more inclusive systemic whole, early functional theorists often conceptualized social needs or requisites. The most extensive formulations of this position were those of Malinowski, in which institutional arrangements meet one of various levels of needs or requisites: biological, structural, and symbolic. Durkheim and Radcliffe-Brown believed it was important to analyze separately the causes and functions of a sociocultural item, because the causes of an item could be unrelated to its function in the systemic whole. In their analyses of actual phenomena, however, both Durkheim and Radcliffe-Brown lapsed into assertions that the need for integration caused a particular event—for example, the emergence of a particular type of lineage system or the division of labor.

By the midcentury, functionalism was sociology's dominant theoretical perspective. Some, such as Robert Merton, recognized the problems in functional analysis and advocated an empirical assessment of function requisites. Others, such as Talcott Parsons, emphasized general functional requisites that apply to all systems, and then began to build theoretical schemes to explain phenomena in terms of their consequences for meeting these requisites. Whatever the approach, the period between 1950 and 1970 was the zenith of functional theorizing. Today, prominent functional approaches persist but in more muted form and in more marginal intellectual niches.

NOTES

1. August Comte, *The Course of Positive Philosophy* (1830–1842). References are to the more commonly used edition that Harriet Martineau condensed and translated, *The Positive Philosophy of August Comte*, vols. 1, 2, and 3 (London: Bell & Sons, 1898; originally published in 1854).

2. August Comte, *System of Positive Polity or Treatise on Sociology* (London: Burt Franklin, 1875; originally published in 1851), pp. 239–240.

3. Ibid., pp. 241–242.

4. Herbert Spencer, *Social Statics* (New York: D. Appleton, 1870; originally published in 1850) was only one volume, but it became a much larger work near the end of Spencer's career when he wrote *Principles of Ethics* (New York: D. Appleton, 1879-1893).

5. Herbert Spencer, *The Principles of Biology* (New York: D. Appleton, 1864-1867).

6. Herbert Spencer, *The Principles of Psychology* (New York: D. Appleton, 1898; originally published 1855).

7. These are contained in his *First Principles* (New York: A.C. Burt, 1880; originally published in 1862).

8. Herbert Spencer, *The Principles of Sociology* (1874–1896). This work has been reissued in varying volume numbers. References in this chapter are to the three-volume edition (the third edition) issued by D. Appleton, New York, in 1898. When you read this long work, it is much more critical to note the parts (numbered I through VII) than the volumes, because pagination can vary with editions.

9. See Jonathan H. Turner, *Herbert Spencer: Toward a Renewed Appreciation* (Beverly Hills, CA: Sage, 1985), Chapter 4. See also Spencer's *Principles of Sociology*, vol. I, part II, pp. 449–457. The bulk of Spencer's organicism is on these few pages of a work that spans more than 2,000 pages, yet this is what we most remember about Spencer.

10. *Principles of Sociology*, part II, pp. 451–462.

11. Spencer's theory of symbolism is commonly ignored by contemporary sociologists who simply accept Durkheim's critique. Part I of *Principles* contains a very sophisticated analysis of symbols.

12. Principles of Sociology, part II, p. 477.

13. Émile Durkheim, *The Division of Labor in Society* (New York: Macmillan, 1933; originally published in 1893). Durkheim tended to ignore the similarity between Spencer's organismic analogy and his own organic formulations as well as the close correspondence between their theories of symbols. For more details on this line of argument, see Jonathan H. Turner, "Émile Durkheim's Theory of Social Organization," *Social Forces* 68 (3, 1990), pp. 1–15, and "Spencer's and Durkheim's Principles of Social Organization," *Sociological Perspectives* 27 (January 1984), pp. 21–32.

14. Émile Durkheim, *The Rules of the Sociological Method* (New York, Free Press, 1938, originally published in 1895), p. 96.

15. Ibid., p. 97.

16. Ibid., p. 35. For a more detailed analysis, see Jonathan H. Turner and Alexandra Maryanski, *Functionalism* (Menlo Park, CA: Benjamin/Cummings, 1979). See also Percy S. Cohen, *Modern Social Theory* (New York: Basic Books, 1968), pp. 35–37.

17. Robert A. Nisbet, *Émile Durkheim* (Englewood Cliffs, NJ: Prentice-Hall, 1965), pp. 9–102.

18. For basic references on Malinowski's functionalism, see his "Anthropology," *Encyclopedia Britannica*, supplementary vol. 1 (London and New York, 1936); *A Scientific Theory of Culture* (Chapel Hill: University of North Carolina Press, 1944); and *Magic, Science, and Religion and Other Essays* (Glencoe, IL: Free Press, 1948). For basic references on A.R. Radcliffe-Brown's functionalism, see his "Structure and Function in Primitive Society," *American Anthropologist* 37 (July–September, 1935), pp. 58–72; *Structure and Function in Primitive Society* (Glencoe, IL: Free Press, 1952); and *The Andaman Islanders* (Glencoe, IL: Free Press, 1948). See also Turner and Maryanski, *Functionalism* (cited in note 16).

19. Radcliffe-Brown, "Structure and Function in Primitive Society," *American Anthropologist* 37 (July–September, 1935), p. 68. This statement is, of course, incorrect, because the organismic analogy was far more developed in Spencer's work.

20. Radcliffe-Brown, *Structure and Function in Primitive Society* (Glencoe, IL: Free Press, 1952), pp. 31–50. For a secondary analysis of this example, see Arthur L. Stinchcombe, "Specious Generality and Functional Theory," *American Sociological Review* 26 (December, 1961), pp. 929–930.

21. See, for example, Radcliffe-Brown, *Structure and Function in Primitive Society* (cited in note 20).

22. A perceptive critic of an earlier edition of this manuscript provided an interesting way to visualize the problems of tautology:

When do you have a surviving social system?

When certain survival requisites are met.

How do you know when certain survival requisites are met?

When you have a surviving social system.

23. Don Martindale, *The Nature and Types of Sociological Theory* (Boston: Houghton Mifflin, 1960), p. 459.

24. Bronislaw Malinowski, *A Scientific Theory of Culture and Other Essays* (London: Oxford University

Press, 1964), pp. 71–125; see also Turner and Maryanski, *Functionalism* (cited in note 16), pp. 44–57.

25. For basic references on Weber, see his *The Theory of Social and Economic Organization* (New York: Free Press, 1947); "Social Action and Its Types" in *Theories of Society*, eds. Talcott Parsons et al. (New York: Free Press, 1961); *From Max Weber: Essays in Sociology*, eds. Hans Gerth and C. Wright Mills (New York: Oxford University Press, 1958).

26. For a more thorough analysis of the historical legacy of functionalism, see Don Martindale's *The Nature and Types of Sociological Theory* (cited in note 23) and his "Limits of and Alternatives to Functionalism in Sociology," in *Functionalism in the Social Sciences*, American Academy of Political and Social Science Monograph, no. 5 (Philadelphia, 1965), pp. 144–162; see also Ivan Whitaker, "The Nature and Value of Functionalism in Sociology," also in *Functionalism in the Social Sciences*, pp. 127–143.

3

The Maturing Tradition I: Robert K. Merton's Empirical Approach

In the late 1930s, a unique convergence of ideas and individuals occurred at Harvard University. Robert K. Merton, then a graduate student, conducted several seminars with fellow graduate students and young instructors, one of whom was Talcott Parsons. Although Parsons had studied briefly with Bronislaw Malinowski in England, it is likely that Parsons and others who became well-known functional theorists[1] received their first intense exposure to A.R. Radcliffe-Brown's and Bronislaw Malinowski's functionalism through Robert K. Merton. Both Merton and Parsons became the preeminent functionalists of the midcentury, but they took very different directions in their views about how to conduct functional analysis. Merton emphasized the importance of a functionalism that is empirical and that avoids the tendency to construct grand analytical schemes into which the flow of empirical events is pushed and shoved, whereas Parsons constructed a great analytical edifice—an edifice that Merton sought, at least implicitly, to tear down.

Because Merton's criticism of early functionalisms came before Parsons' grand scheme began to unfold in the 1950s, we should begin with his views about how to conduct functional analysis and what to avoid in so doing. Merton's views were invoked again and again as Parsons' more analytical functionalism emerged and proliferated in the 1950s and 1960s. Merton's goal was to keep functional assumptions to a minimum, whereas Parsons' intent was to build a functional analytical scheme that could explain all reality.

At the heart of Merton's criticism of Malinowski's analytical scheme building and, later, of Parsons' as well was the contention that concern for developing an all-encompassing system of concepts would prove both futile and sterile.[2] To Merton, the search for "a total system of sociological theory, in which observations about every aspect of social behavior, organization, and change promptly find their preordained place, has the same exhilarating challenge and the same small promise as those many all-encompassing philosophical systems which have fallen into deserved disuse."[3]

Merton believed grand theoretical schemes are premature because the theoretical and empirical groundwork necessary for their completion has not been performed. Just as Einsteinian theory did not emerge without a long cumulative research foundation and theoretical legacy, so sociological theory will have to wait for its Einstein, primarily because "it has not yet found its Kepler—to say nothing of its Newton, Laplace, Gibbs, Maxwell, or Planck."[4]

In the absence of this foundation, what passes for sociological theory, in Merton's critical eye, consists of "general orientations toward data, suggesting types of variables which theorists must somehow take into account, rather than clearly formulated, verifiable statements of relationships between specified variables."[5] Strategies advocated by those such as Parsons are not really theory but philosophical systems, with "their varied suggestiveness, their architectonic splendor, and their sterility."[6] However, to pursue the opposite strategy of constructing inventories of low-level empirical propositions will prove equally sterile, thus suggesting to Merton the need for "theories of the middle range" in sociology.

THEORIES OF THE MIDDLE RANGE

In Merton's view, theories of the middle range offer more theoretical promise than does grand functional theory. These middle-range theories are couched at a lower level of abstraction and reveal clearly defined and operationalized concepts that are incorporated into statements about a limited range of phenomena. Although middle-range theories are abstract, they are also connected to the empirical world, thus encouraging the research so necessary for the clarification of concepts and reformulation of theoretical generalizations. Without this interplay between theory and research, Merton contended, theoretical schemes will remain suggestive congeries of concepts, which are incapable of being refuted, whereas, on the other hand, empirical research will remain unsystematic, disjointed, and of little utility in expanding a body of sociological knowledge. Thus, by following a middle-range strategy, the concepts and propositions of sociological theory will become more tightly organized as theoretically focused empirical research forces clarification, elaboration, and reformulation of the concepts and propositions of each middle-range theory.

From this growing clarity in theories directed at a limited range of phenomena and supported by empirical research can eventually come the more encompassing theoretical schemes. Merton reasoned that, although it is necessary to concentrate energies on constructing limited theories that inspire research, theorists must also be concerned with "consolidating the special theories into a more general set of concepts and mutually consistent propositions."[7] The special theories of sociology must therefore be formulated with an eye toward what they can offer more general sociological theorizing. However, just how these middle-range theories should be formulated to facilitate their eventual consolidation into a more general theory poses a difficult analytical problem, for which Merton had a ready solution: A form of functionalism should be used in formulating middle-range theories. Such functional theorizing should take the form of a paradigm that allows easy specification and elaboration of relevant concepts, while encouraging systematic revision and reformulation as empirical findings would dictate. Conceived in this way, Merton believed that functionalism could build not only middle-range theories but also the grand theoretical schemes that would someday subsume such middle-range theories. Thus, for Merton, functionalism represents a strategy for ordering concepts and for sorting significant from insignificant social processes.[8]

MERTON'S PARADIGM FOR FUNCTIONAL ANALYSIS

As have most commentators on functional analysis, Merton began his discussion with a review of the mistakes of early functionalists, particularly the anthropologists Malinowski and Radcliffe-Brown.[9] Generally, Merton saw functional theorizing as potentially embracing three questionable postulates: (1) the functional unity of social systems, (2) the functional universality of social items, and (3) the indispensability of functional items for social systems.[10]

The Functional Unity Postulate

As we can recall from Chapter 2, Radcliffe-Brown, following Émile Durkheim's lead, frequently transformed the hypothesis that social systems reveal social integration into a necessary requisite or need for social survival. Although it is difficult to argue that human societies do not possess some degree of integration—otherwise they would not be systems—Merton viewed the degree of integration in a system as an issue to be empirically determined. To assume, however subtly, that a high degree of functional unity must exist in a social system is to define away the important theoretical and empirical questions: What levels of integration exist for different systems? What various types of integration can be discerned? Are varying degrees of integration evident for different segments of a system? And, most important, what variety of processes leads to different levels, forms, and types of integration for different spheres of social systems? Merton, believed that beginning analysis with the postulate of "functional unity" or integration of the social whole can divert attention away not only from these questions but also from the varied and "disparate consequences of a given social or cultural item (usage, belief, behavior pattern, institutions) for diverse social groups and for individual members of these groups."[11]

Instead of the postulate of functional unity, emphasis should be on varying types, forms, levels, and spheres of social integration and the varying consequences of the existence of items for specified segments of social systems. In this way, Merton sought to direct functional analysis away from concern with total systems and toward an emphasis on how different patterns of social organization within more inclusive social systems are created, maintained, and changed, not only by the requisites of the total system but also by interaction among sociocultural items within systemic wholes.

The Issue of Functional Universality

One result of a functional unity emphasis was for some early anthropologists to assume that if a social item existed in an ongoing system, it must therefore have positive consequences for the integration of the social system. This assumption tended to result in tautologous statements: A system exists; an item is a part of the system; therefore the item is positively functional for the maintenance of the system.

Merton reasoned that if an examination of empirical systems is undertaken, it is clear that there is a wider range of empirical possibilities. First, items can be not only positively functional for a system or another system item but also dysfunctional for either particular items or the systemic whole. Second, some consequences, whether functional or dysfunctional, are intended and recognized by system incumbents and are thus manifest, whereas other consequences are not intended or recognized and are therefore latent. Thus, in contrast with Malinowski and Radcliffe-Brown, Merton proposed the analysis of diverse consequences or functions of socio-cultural items—whether positive or negative, manifest or latent—"for individuals, for sub-groups, and for the more inclusive social structure and culture."[12] In turn, the analysis of varied consequences requires calculating a "net balance of consequences" of items for each other and more inclusive systems. In this way, Merton visualized contemporary functional thought as compensating for the excesses of earlier analysis forms by focusing on the crucial types of consequences of socio-cultural items for each other and, if the facts dictate, for the social whole.

The Issue of Indispensability

Somewhat out of context and unfairly, Merton quoted Malinowski's assertion that every cultural item "fulfills some vital function, has some task to accomplish, represents an indispensable part within a working whole"[13] as simply an extreme statement of two interrelated issues in functional analysis: (1) Do social systems have functional requisites or needs that must be fulfilled? (2) Are certain crucial structures indispensable for fulfilling these functions?

In response to the first question, Merton provided a tentative yes, but with an important qualification: The functional requisites must be established empirically for specific systems. For actual groups or whole societies, it is possible to ascertain the "conditions necessary for their survival," and it is of theoretical importance to determine which structures, through what specific processes, have consequences for these conditions. But to assume a system of universal requisites adds little to theoretical analysis because to stress that certain functions must be met in all systems simply leads observers to describe processes in social systems that meet these requisites. Such descriptions, Merton contended, can be done without the excess baggage of system requisites. It is more desirable to describe cultural patterns and then assess their various consequences in meeting the specific needs of different segments of concrete empirical systems.

Merton's answer to the second question is emphatic: Empirical evidence makes the assertion that only certain structures can fulfill system requisites obviously false. Examination of the empirical world reveals quite clearly that alternative structures can exist to fulfill basically the same requisites in both similar and diverse systems. This led Merton to postulate the importance in functional analysis of concern with various types of "functional alternatives," or "functional equivalents," and "functional substitutes" within social systems. In this way, functional analysis would not view the social items of a system as indispensable and thereby would avoid the tautologous trap of assuming that items must exist to assure the continued existence of a system. Furthermore, in looking for functional alternatives, analytical attention would be drawn to questions about the range of items that could serve as functional equivalents. If these questions are to be answered adequately, analysts should then determine why a particular item was selected from a range of possible alternatives, leading to questions about the "structural context" and "structural limits" that might circumscribe the range of alternatives and account for the emergence of one item over another. Merton believed examining these interrelated questions would thus facilitate the separate analysis of the causes and consequences of structural items. By asking why one particular structure, instead of various alternatives, had emerged, analysts would not forget to document the specific processes leading to an item's emergence as separate from its functional consequences. In this way the danger of assuming that items must exist to fulfill system needs would be avoided.

In reviewing Merton's criticisms of traditional anthropological reasoning and at some contemporary functionalists' implications, we can see much of his assessment of these three functional postulates involves the destruction of "straw men." Yet, in destroying these assumptions, Merton formulated alternative postulates that advocated a concern for the multiple consequences of sociocultural items for one another and for more inclusive social wholes, without assumptions of functional needs or imperatives. Rather, functional analysis must specify (1) the social patterns under consideration, whether a systemic whole or some subpart; (2) the various types of consequences of these patterns for empirically established survival requisites; and (3) the processes whereby some patterns rather than others exist and have the various consequences for one another and for systemic wholes.

A PROTOCOL FOR EXECUTING FUNCTIONAL ANALYSIS

To ascertain the causes and consequences of particular structures and processes, Merton insisted that functional analysis begin with "sheer description" of individual and group activities. In describing the patterns of interaction and activity among units under investigation, we can discern clearly the social items to be subjected to functional analysis. Such descriptions can also provide a major clue to the functions performed by such patterned activity. For these functions to become more evident, however, additional steps are necessary.

The first steps is for investigators to indicate the principal alternatives that are excluded by the dominance of a particular pattern. Such description of the excluded alternatives indicates the structural context from which an observed pattern first emerged and is now maintained—thereby offering further clues about the functions or consequences the item might have for other items and perhaps for the systemic whole. The second analytical step beyond sheer description involves an assessment of the meaning, or mental and emotional significance, of the activity for group members. Description of these meanings can offer some indication of the motives behind the activities of the individuals involved and thereby shed some tentative light on the manifest functions of an activity. These descriptions require a third analytical step of discerning some array of motives for conformity or for deviation among participants, but these motives must not be confused with either the objective description of the pattern or the subsequent assessment of the functions served by the pattern. Yet by understanding the configuration of motives for conformity and deviation among actors, an assessment of the psychological needs served (or not served) by a pattern can be understood—offering an additional clue to the various functions of the pattern under investigation.

But focusing on the meanings and motives of those involved in an activity can skew analysis away from unintended or latent consequences of the activity. Thus a final analytical step describing the patterns under investigation reveal regularities not recognized by participants but appearing to have consequences for both the individuals involved and other central patterns or regularities in the system. In this way, analysis will be attuned to the latent functions of an item.

Merton assumed that, by following each of these steps, it would be possible to assess the net balance of consequences of the pattern under investigation, as well as to determine some of the independent causes of the item. These steps assure that a proper functional inquiry will ensue because postulates of functional unity, assumptions of survival requisites, and convictions about indispensable parts do not precede the analysis of

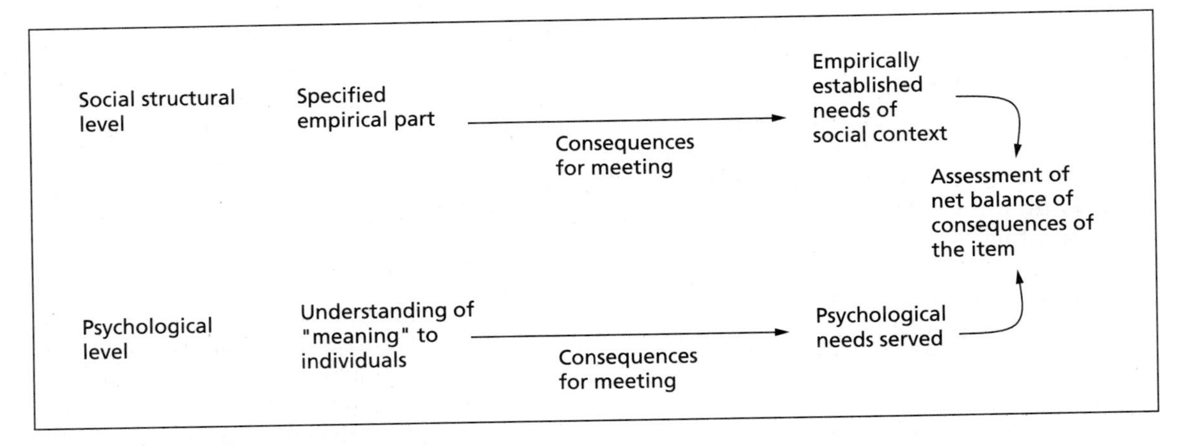

FIGURE 3-1 Merton's Net Functional Balance Analysis

social structures and processes. On the contrary, attention is drawn only to observable patterns of activity, the structural context in which the focal pattern emerged and persists despite potential alternatives, the meaning of these patterns for actors involved, the actors' motives for conformity and deviation, and the implications of the particular pattern for unrecognized needs of individuals and other items in the social system. Thus, with this kind of preliminary work, functional analysis will avoid the logical and empirical problems of previous forms of functionalism. And in this way functional analysis can provide an understanding of the causes and consequences of system parts for one another and for more inclusive system units.[14]

Figure 3-1 recapitulates the essential elements of Merton's strategy. First, only empirical units are to be analyzed, and the part and the social context of the part must be clearly specified. Then the task is to establish the particular survival requisites of the empirical system—that is, what is necessary for this particular empirical system to survive. By assessing the functions or consequences of an item's meeting or not meeting these needs, we can achieve insight into the nature of a part and its social contexts. In addition to this structural analysis must come an analysis of the meaning for participants, particularly the psychological needs served or not served by participation in a structure. In this way, we can assess the net balance of consequences of an item at diverse levels of social organization.

CONCLUSION

In the end, most functional theorists ignored Merton's protocol, although "middle range" functional theories are easy to find in the literature.[15] Instead, Merton's protocol was used to criticize all functionalist theory, and his pleas for "theories of the middle range" were taken as an anti-grand theory pronouncement. Although Merton's position was originally published before Parsons' functional scheme began to unfold, the republishing of these early essays in subsequent editions of Merton's *Social Theory and Social Structure* was interpreted as an attack on Parsons' functional approach. Parsons, in the analytical tradition of Malinowski and Herbert Spencer, took functionalism to its analytical extreme—as we will explore in the next chapter.

NOTES

1. For example, Wilbert Moore, Kingsley Davis, and Marion J. Levy were all exposed to functionalism at this time. Early functionalist theories, such as Davis' and Moore's famous "Some Principles of Stratification," *American Sociological Review* 10 (1945), pp. 242–247 was, no doubt, influenced by Merton's teachings, as was the functional slant of Kingsley Davis' *Human Societies* (New York: Macmillan, 1948). Other early functional statements, such as David F. Aberle, Albert K. Cohen, A.K. Davis, Marion J. Levy, and F.Y. Sutton's "The Functional Requisites of Society," *Ethics* 55 (1950), pp. 100–111, came from participants in these seminars. For further details, see Jonathan H. Turner and Alexandra Maryanski, *Functionalism* (Menlo Park, CA: Benjamin/Cummings, 1979).

2. Robert K. Merton, "Discussion of Parsons' 'The Position of Sociological Theory,'" *American Sociological Review* 13 (1948), pp. 164–168.

3. Ibid. Most of Merton's significant essays on functionalism have been included, and frequently expanded, in Robert K. Merton, *Social Theory and Social Structure* (Glencoe, IL: Free Press, 1949). Quotation taken from page 45 of the 1968 edition of this classic work. Most subsequent references will be made to the articles incorporated into this book. For more recent essays on Merton's work, see *The Idea of Social Structure*, ed. Lewis A. Coser (New York: Free Press, 1975). For excellent overviews of Merton's sociology, see Piotr Sztompka, *Robert K. Merton: An Intellectual Profile* (New York: St. Martin's, 1986); and Charles Crothers, *Robert K. Merton* (London: Tavistock, 1981).

4. Merton, *Social Theory and Social Structure* (cited in note 3), p. 47.

5. Ibid., p. 42.

6. Ibid., p. 51.

7. Merton, *Social Theory and Social Structure* (1957), p. 10.

8. See Turner and Maryanski, *Functionalism* (cited in note 1), pp. 65–68, for additional details.

9. Robert K. Merton, "Manifest and Latent Functions," in *Social Theory and Social Structure* (Glencoe, IL: Free Press, 1968), pp. 74–91.

10. See Merton, *Social Theory and Social Structure* (cited in note 3), pp. 45–61.

11. Merton, *Social Theory and Social Structure* (1968), (cited in note 3), pp. 81–82.

12. Ibid., p. 84.

13. This quote is from an encyclopedia article in which Malinowski argued against ethnocentrism. His more scholarly work (see Chapter 2) is much less extreme. See also Turner and Maryanski, *Functionalism* (cited in note 1), for a more balanced analysis of Malinowski's work.

14. Ibid., p. 136.

15. For samples of these, see *Neofunctionalist Sociology*, ed. Paul Colomy (Brookfield, VT: Edward Elgar, 1990).

4

The Maturing Tradition II: Talcott Parsons' Analytical Approach

Talcott Parsons was probably the most prominent theorist of his time, and it is unlikely that any one theoretical approach will so dominate sociological theory again. In the years between 1950 and the late 1970s, Parsonian functionalism was clearly the focal point around which theoretical controversy raged. Even those who despised Parsons' functional approach could not ignore it. Even now, years after his death and more than two decades since its period of dominance, Parsonian functionalism is still the subject of controversy.[1] To appreciate Parsons' achievement in bringing functionalism to the second half of the twentieth century, it is best to start at the beginning, in 1937, when he published his first major work, *The Structure of Social Action*.[2]

THE STRUCTURE OF SOCIAL ACTION

In *The Structure of Social Action,* Parsons advocated using an "analytical realism" to build sociological theory. Theory in sociology must use a limited number of important concepts that "adequately 'grasp' aspects of the external world. . . . These concepts do not correspond to concrete phenomena, but to elements in them that are analytically separable from other elements."[3] Thus, first, theory must involve the development of concepts that abstract from empirical reality, in all its diversity and confusion, common analytical elements. In this way, concepts will isolate phenomena from their embeddedness in the complex relations that constitute social reality.

The unique feature of Parson's analytical realism is the insistence about how these abstract concepts are to be employed in sociological analysis. Parsons did not advocate the immediate incorporation of these concepts into theoretical statements but rather advocated their use to develop a "generalized system of concepts." This use of abstract concepts would involve their ordering into a coherent whole that would reflect the important features of the "real world." What is sought is an organization of concepts into analytical systems that grasp the salient and systemic features of the universe without being overwhelmed by empirical details. This emphasis on systems of categories represents Parsons' application of Max Weber's ideal-type strategy for analytically accentuating salient features of the world. Thus, much like Weber, Parsons believed that theory should initially resemble an elaborate classification and categorization of social phenomena that reflects significant features in the organization of these social phenomena. This strategy was evident in Parsons' first major work, where he developed the "voluntaristic theory of action."[4]

Parsons believed that the "voluntaristic theory of action" represented a synthesis of the useful assumptions and concepts of utilitarianism, positivism, and idealism. In reviewing the thought of classical economists, Parsons noted the excessiveness of their utilitarianism: unregulated and atomistic actors in a free and competitive marketplace rationally attempting to choose those behaviors that will maximize their profits in their transactions with others. Parsons believed such a formulation of the social order presented several critical problems: Do humans always behave rationally? Are they indeed free and unregulated? How is order possible in an unregulated and competitive system? Yet Parsons saw as fruitful several features of utilitarian thought, especially the concern with actors as seeking goals and the emphasis on the choice-making capacities of human beings who weigh alternative lines of action. Stated in this minimal form, Parsons felt that the utilitarian heritage could indeed continue to inform sociological theorizing. In a similar critical stance, Parsons rejected the extreme formulations of radical positivists, who tended to view the social world in terms of observable cause-and-effect relationships among physical phenomena. In so doing, he felt, they ignored the complex symbolic functionings of the human mind. Furthermore, Parsons saw the emphasis on observable cause-and-effect relationships as too easily encouraging a sequence of infinite reductionism: groups were reduced to the causal relationships of their individual members; individuals were reducible to the cause-and-effect relationships of their physiological processes; these were reducible to physico-chemical relationships, and so on, down to the most basic cause-and-effect connections among particles of physical matter. Nevertheless, despite these extremes, radical positivism draws attention to the physical parameters of social life and to the deterministic impact of these parameters on much—but of course not all—social organization. Finally, in assessing idealism, Parsons saw the conceptions of "ideas" to circumscribe both individual and social processes as useful, although all too frequently these ideas are seen as detached from the ongoing social life they were supposed to regulate.

The depth of scholarship in Parsons' analysis of these traditions is impossible to communicate. More important than the details of his analysis is the weaving of selected concepts from each of these traditions into a voluntaristic theory of action. At this starting point, in accordance with his theory-building strategy, Parsons began to construct a functional theory of social organization. In this initial formulation, he conceptualized voluntarism as the subjective decision-making processes of individual actors, but he viewed such decisions as the partial outcome of certain kinds of constraints, both normative and situational. Voluntaristic action therefore involves these basic elements: (1) Actors, at this point in Parsons' thinking, are individual persons. (2) Actors are viewed as goal seeking. (3) Actors also possess alternative means to achieve the goals. (4) Actors are confronted with a variety of situational conditions, such as their own biological makeup and heredity as well as various external ecological constraints, that influence the selection of goals and means. (5) Actors are

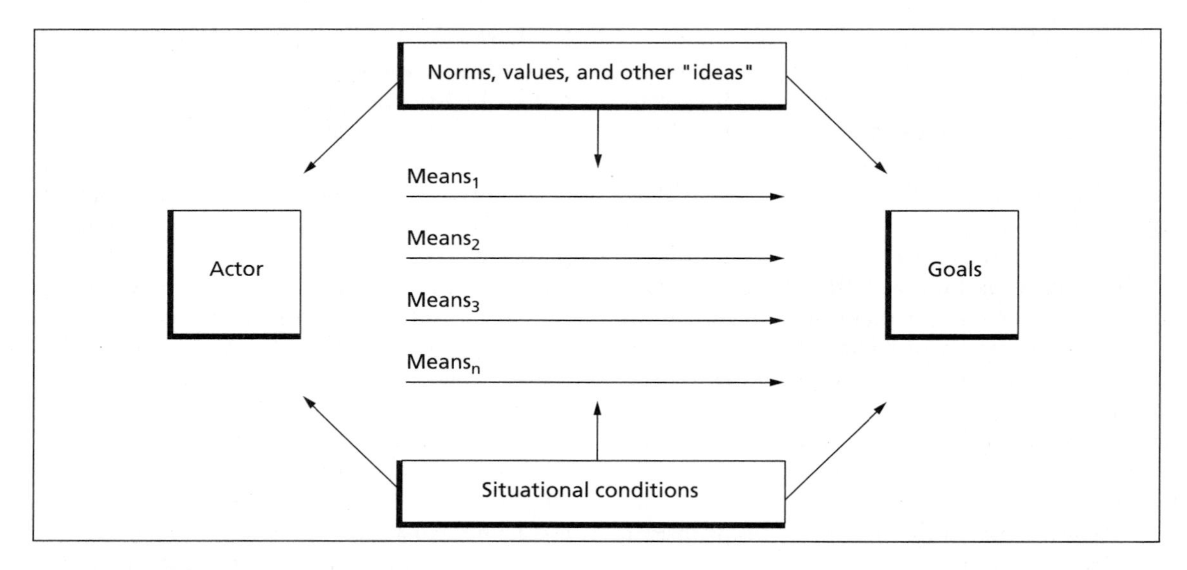

FIGURE 4-1 The Units of Voluntaristic Action

governed by values, norms, and other ideas such that these ideas influence what is considered a goal and what means are selected to achieve it. (6) Action involves actors making subjective decisions about the means to achieve goals, all of which are constrained by ideas and situational conditions.

Figure 4-1 represents this conceptualization of voluntarism. The processes diagrammed are often termed the *unit act,* with social action involving a succession of such unit acts by one or more actors. Parsons chose to focus on such basic units of action for at least two reasons. First, he felt it necessary to synthesize the historical legacy of social thought about the most basic social process and to dissect it into its most elementary components. Second, given his position on what theory should be, the first analytical task in the development of sociological theory is to isolate conceptually the systemic features of the most basic unit from which more complex processes and structures are built.

Once these basic tasks were completed, Parsons began to ask: How are unit acts connected to each other, and how can this connectedness be conceptually represented? Indeed, near the end of *The Structure of Social Action,* he recognized that "any atomistic system that deals only with properties identifiable in the unit act . . . will of necessity fail to treat these latter elements adequately and be indeterminate as applied to complex systems."[5] However, only the barest hints of what was to come were evident in those closing pages.

THE SOCIAL SYSTEM

Figure 4-2 summarizes the transition from unit acts to social system.[6] This transition occupies the early parts of Parsons' next significant work, *The Social System.*[7] Drawing inspiration from Weber's typological approach to this same topic,[8] Parsons viewed actors as "oriented" to situations in terms of motives (needs and readiness to mobilize energy) and values (conceptions about what is appropriate). There are three types of motives: (1) cognitive (need for information), (2) cathectic (need for emotional

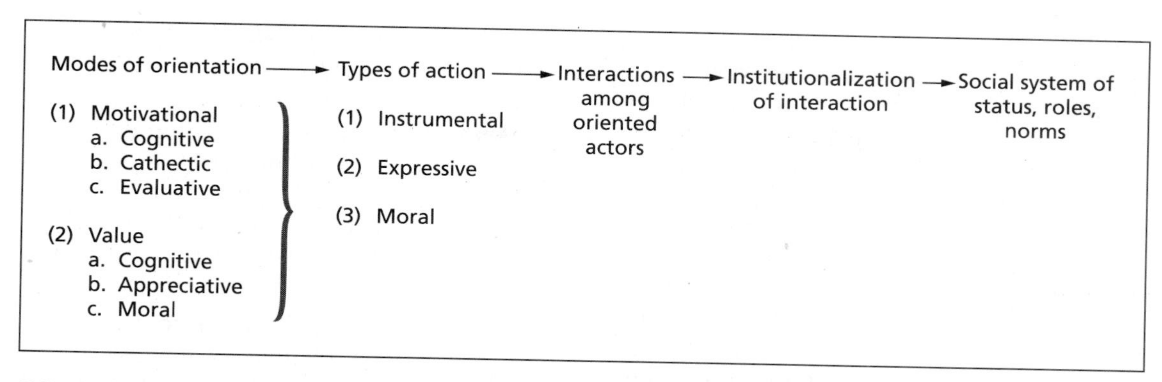

FIGURE 4-2 Parsons' Conception of Action, Interaction, and Institutionalization

attachment), and (3) evaluative (need for assessment). Also, there are three corresponding types of values: (1) cognitive (evaluation by objective standards), (2) appreciative (evaluation by aesthetic standards), and (3) moral (evaluation by absolute rightness and wrongness). Parsons called these *modes of orientation*. Although this discussion is somewhat vague, the general idea seems to be that the relative salience of these motives and values for any actor creates a composite type of action, which can be one of three types: (1) instrumental (action oriented to realize explicit goals efficiently), (2) expressive (action directed at realizing emotional satisfactions), and (3) moral (action concerned with realizing standards of right and wrong). That is, depending on which modes of motivational and value orientation are strongest, an actor will act in one of these basic ways. For example, if cognitive motives are strong and cognitive values most salient, then action will be primarily instrumental, although the action will also have expressive and moral content. Thus the various combinations and permutations of the modes of orientation—that is, motives and values—produce action geared in one of these general directions.

"Unit acts" therefore involve motivational and value orientations and have a general direction as a consequence of what combination of values and motives prevails for an actor. Thus far Parsons had elaborated only on his conceptualization of the unit act. The critical next step which was only hinted at in the closing pages of *The Structure of Social Action*: As variously oriented actors (in the configuration of motivational and value orientations) interact, they develop agreements and sustain patterns of interaction, which become "institutionalized." Such institutionalized patterns can be, in Parsons' view, conceptualized as a social system. Such a system represents an emergent phenomenon that requires its own conceptual edifice. The normative organization of status-roles becomes Parsons' key to this conceptualization; that is, the subject matter of sociology is the organization of status, roles, and norms. Yet, Parsons recognized that the actors who are incumbent in status-roles are motivationally and value oriented; thus, as with patterns of interaction, the task now becomes one of conceptualizing these dimensions of action in systemic terms. The result is the conceptualization of action as composed of three "interpenetrating action systems": the cultural, the social, and the personality. That is, the organization of unit acts into social systems requires a parallel conceptualization of motives and values that become, respectively, the personality and cultural systems. The goal of action theory now becomes understanding how institutionalized patterns of interaction (the social system) are circumscribed by complexes of values, beliefs, norms, and other ideas (the cultural system) and by configurations of motives and role-playing skills (the personality system). Later Parsons added the organismic (subsequently called behavioral) system, but let us not get ahead of the story. At this stage of conceptualization, analyzing social systems involves developing a system of concepts that, first, captures the systemic features of society at all its diverse levels and, second, points to the modes of articulation among personality systems, social systems, and cultural patterns.

In his commitment to developing concepts that reflect the properties of all action systems, Parsons was led to a set of concepts denoting some of the variable properties of these systems. Termed *pattern variables,* they simultaneously allow the categorization of the modes of orientation in personality systems, the value patterns of culture, and the normative requirements in social systems. The variables are phrased as polar dichotomies that, depending on the system under analysis, allow a rough categorization of decisions by actors, the value orientations of culture, or the normative demands on status roles.

1. *Affectivity/affective neutrality* concerns the amount of emotion or affect that is appropriate in a given interaction situation. Should a great deal or little affect be expressed?
2. *Diffuseness/specificity* denotes the issue of how far-reaching obligations in an interaction situation are to be. Should the obligations be narrow and specific, or should they be extensive and diffuse?
3. *Universalism/particularism* points to the problem of whether evaluation of others in an interaction situation is to apply to all actors, or should all actors be assessed by the same standards?
4. *Achievement/ascription* deals with the issue of how to assess an actor, whether by performance or by inborn qualities, such as sex, age, race, and family status. Should an actor treat another on the basis of achievements or ascriptive qualities that are unrelated to performance?
5. *Self-collectivity* denotes the extent to which action is to be oriented to self-interest and individual goals or to group interests and goals. Should actors consider their personal or self-related goals over those of the group or large collectivity in which they are involved?[9]

Some of these concepts, such as self-collectivity, were later dropped from the action scheme, but others, such as universalism-particularism, assumed greater importance. The intent of the pattern variables remained the same, however: to categorize dichotomies of decisions, normative demands, and value orientations. In *The Social System,* however, Parsons was inclined to view them as value orientations that circumscribe the norms of the social system and the decisions of the personality system. Thus the structure of the personality and social systems reflects the dominant patterns of value orientations in culture. This implicit emphasis on the impact of cultural patterns on regulating and controlling other systems of action became ever more explicit in his later work.

By 1951 Parsons had already woven a complex conceptual system that emphasizes the process of institutionalization of interaction into stabilized patterns called social systems, which are penetrated by personality and circumscribed by culture. The profile of institutionalized norms, of decisions by actors in roles, and of cultural value orientations can be typified by concepts—the pattern variables—that capture the variable properties in each of these action components.

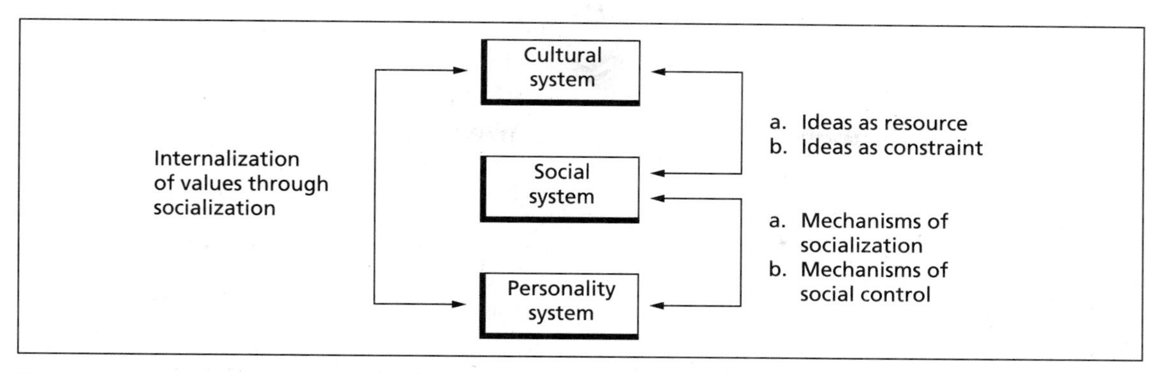

FIGURE 4-3 Parsons' Early Conception of Integration among Systems of Action

Having built this analytical edifice, Parsons then returned to a question, first raised in *The Structure of Social Action,* that guided all his subsequent theoretical formulations: How do social systems survive? More specifically, why do institutionalized patterns of interaction persist? Such questions raise the issue of system imperatives or requisites. Parsons was asking how systems resolve their integrative problems. The answer is provided by the elaboration of additional concepts that point to how personality systems and culture are integrated into the social system, thereby providing assurance of some degree of normative coherence and a minimal amount of commitment by actors to conform to norms and play roles. Figure 4-3 delineates the key ideas in Parsons' reasoning.

Just how are personality systems integrated into the social system, thereby promoting equilibrium? At the most abstract level, Parsons conceptualized two mechanisms that integrate the personality into the social system: (1) mechanisms of socialization and (2) mechanisms of social control.

1. *Mechanisms of socialization* are the means through which cultural patterns—values, beliefs, language, and other symbols—are internalized into the personality system, thereby circumscribing its need structure. Through this process actors are made willing to deposit motivational energy in roles (thereby willing to conform to norms) and are given the interpersonal and other skills necessary for playing roles. Another function of socialization mechanisms is to provide stable and secure interpersonal ties that alleviate much of the strain, anxiety, and tension associated with acquiring proper motives and skills.

2. *Mechanisms of social control* involve those ways in which status-roles are organized in social systems to reduce strain and deviance. There are numerous specific control mechanisms, including (a) institutionalization, which makes role expectations clear and unambiguous while segregating in time and space contradictory expectations; (b) interpersonal sanctions and gestures, which actors subtly employ to mutually sanction conformity; (c) ritual activities, in which actors act out symbolically sources of strain that could prove disruptive while they reinforce dominant cultural patterns; (d) safety-valve structures, in which pervasive deviant propensities are segregated in time and space from normal institutional patterns; (e) reintegration structures, which are specifically charged with bringing deviant tendencies back into line; and, finally, (f) institutionalizing the capacity to use force and coercion into some sectors of a system.

These two mechanisms resolve one of the most persistent integrative problems facing social systems. The other major integrative problem facing social systems concerns how cultural patterns contribute to the maintenance of social order and equilibrium. Again at the most abstract level, Parsons visualized two ways in which this occurs: (1) Some components of culture, such as language, are basic resources necessary for interaction to occur. Without symbolic resources, communication and, hence, interaction would not be possible. Thus, by providing common resources for all actors, interaction is made possible by culture. (2) A related but still separable influence of culture on interaction is exerted through the substance of ideas contained in cultural patterns (values, beliefs, ideology, and so forth). These ideas can provide actors with common viewpoints, personal ontologies, or, to borrow from W.I. Thomas, a common "definition of the situation." These common meanings allow interaction to proceed smoothly with minimal disruption.

Naturally, Parsons acknowledged that the mechanisms of socialization and social control are not always successful, hence allowing deviance and social change to occur. But clearly the concepts developed in *The Social System* weight analysis in the direction of looking for processes that maintain the integration and, by implication, the equilibrium of social systems.

THE TRANSITION TO FUNCTIONAL IMPERATIVISM

In collaboration with Robert Bales and Edward Shils, Parsons published *Working Papers in the Theory of Action* shortly after *The Social System*. In *Working Papers,* conceptions of functional imperatives dominated the general theory of action,[10] and by 1956, with Parsons' and Neil Smelser's publication of *Economy and Society,* the functions of structures for meeting system requisites were well institutionalized into action theory.[11]

During this period, systems of action were conceptualized to have four survival problems, or requisites: adaptation, goal attainment, integration, and latency. *Adaptation* involves securing sufficient resources from the environment and then distributing these throughout the system. *Goal attainment* refers to establishing priorities among system goals and mobilizing system resources for their attainment. *Integration* denotes coordinating and maintaining viable interrelationships among system units. Latency embraces two related problems: pattern maintenance and tension management. Pattern maintenance pertains to how to ensure that actors in the social system display the appropriate characteristics (motives, needs, role-playing, and so forth). Tension management concerns dealing with the internal tensions and strains of actors in the social system.

All these requisites were implicit in *The Social System,* but they tended to be viewed under the general problem of integration. In Parsons' discussion of integration within and between action systems, problems of securing facilities (adaptation), allocation and goal seeking (goal attainment), and socialization and social control (latency) were conspicuous. Thus, the development of the four functional requisites—abbreviated A, G, I, and L—was not so much a radical departure from earlier works as an elaboration of concepts implicit in *The Social System.*

With the introduction of A, G, I, and L, however, a subtle shift away from the analysis of structures to the analysis of functions occurs. Structures are now evaluated explicitly by their functional consequences for meeting the four requisites. Interrelationships among specific structures are now analyzed by how their interchanges affect the requisites that each must meet.

As Parsons' conceptual scheme became increasingly oriented to function, social

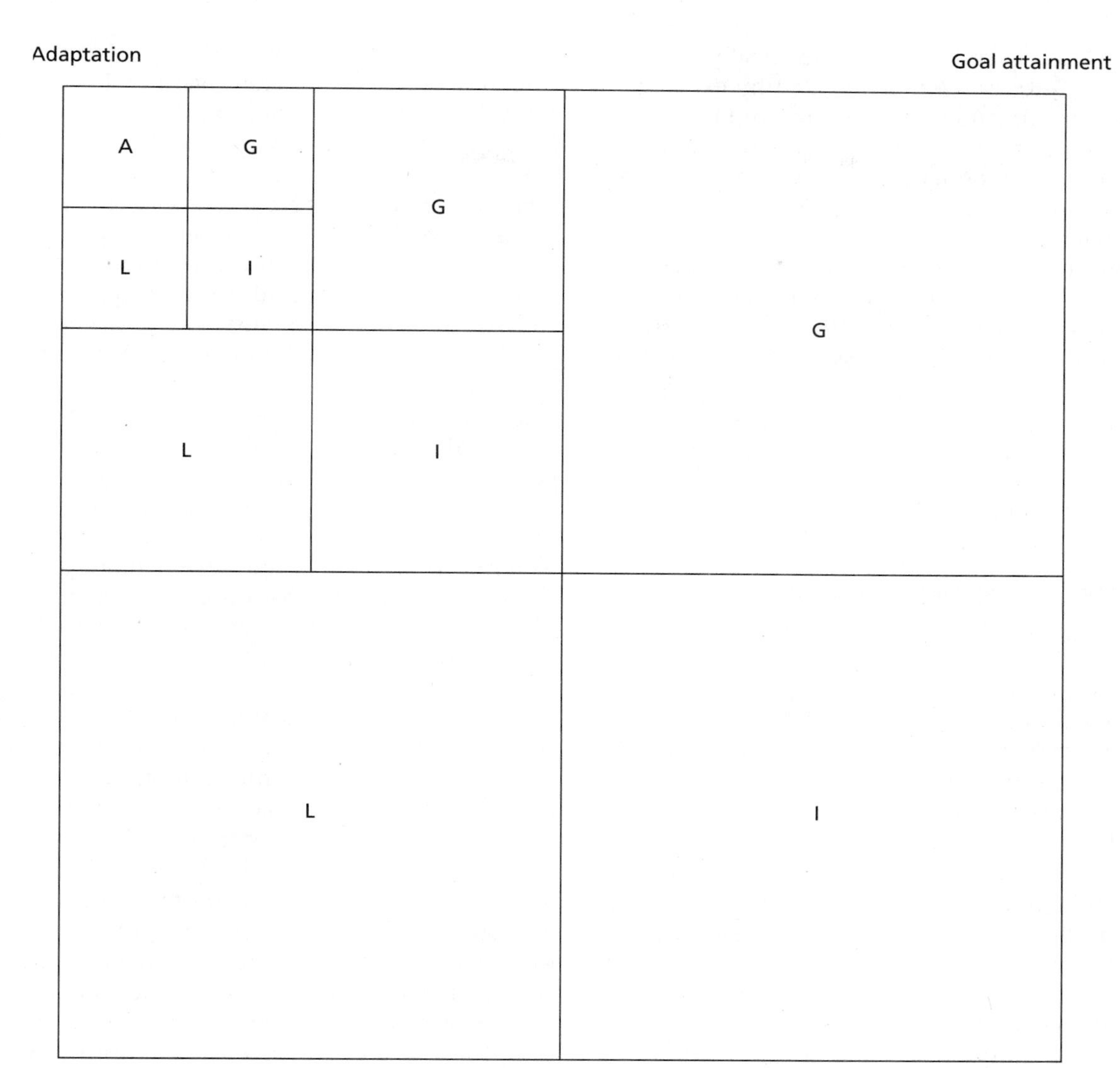

FIGURE 4-4 Parsons' Functional Imperativist View of Social Systems

systems are divided into sectors, each corresponding to a functional requisite—that is, A, G, I, or L. In turn, any subsystem can be divided into these four functional sectors. Then, each subsystem can be divided into four functional sectors, and so on. This process of "functional sectorization," to invent a word to describe it, is illustrated for the adaptive requisite in Figure 4-4.

Of critical analytical importance in this scheme are the interchanges among systems and subsystems. It is difficult to comprehend the functioning of a designated social system without examining the interchanges among its A, G, I, and L sectors, especially because these interchanges are affected by exchanges among constituent subsystems and other systems in the environment. In turn, the functioning of a designated subsystem cannot be understood without examining internal interchanges among its adaptive, goal attainment, integrative, and latency sectors, especially because these interchanges are influenced by exchanges with other subsystems and the more inclusive system of which it is a subsystem. Thus, at this juncture, as important interchanges among the functional sectors of systems and subsystems are outlined, the Parsonsian scheme begins to resemble an elaborate mapping operation.

THE INFORMATIONAL HIERARCHY OF CONTROL

Toward the end of the 1950s, Parsons turned his attention toward interrelationships *among* (rather than *within*) what were then four distinct action systems: culture, social structure, personality, and organism. In many ways, this concern represented an odyssey back to the analysis of the basic components of the unit act outlined in *The Structure of Social Action*. But now each element of the unit act is a full-fledged action system, each confronting four functional problems to resolve: adaptation, goal attainment, integration, and latency. Furthermore, although individual decision making is still a part of action as personalities adjust to the normative demands of status-roles in the social system, the analytical emphasis has shifted to the input/output connections among the four action systems.

At this juncture Parsons began to visualize an overall action system, with culture, social structure, personality, and organism composing its constituent subsystems.[12] Each of these subsystems is seen as fulfilling one of the four system requisites—A, G, I, L—of the overall action system. The organism is considered to be the subsystem having the most consequences for resolving adaptive problems because it is ultimately through this system that environmental resources are made available to the other action subsystems. As the goal-seeking and decision-making system, personality is considered to have primary consequences for resolving goal-attainment problems. As an organized network of status-norms integrating the patterns of the cultural system and the needs of personality systems, the social system is viewed as the major integrative subsystem of the general action system. As the repository of symbolic content of interaction, the cultural system is considered to have primary consequences for managing tensions of actors and assuring that the proper symbolic resources are available to ensure the maintenance of institutional patterns (latency).

After viewing each action system as a subsystem of a more inclusive, overall system, Parsons explored the interrelations among the four subsystems. What emerged is a hierarchy of informational controls, with culture informationally circumscribing the social system, social structure informationally regulating the personality system, and personality informationally regulating the organismic system. For example, cultural value orientations would be seen as circumscribing or limiting the range of variation in the norms of the social system; in turn, these norms, as translated into expectations for actors playing roles, would be viewed as limiting the kinds of motives and decision-making processes in personality systems; these features of the personality system would then be seen as circumscribing biochemical processes in the organism. Conversely, each system in the hierarchy is also viewed as providing the "energic conditions" necessary for action at the next higher system. That is, the organism

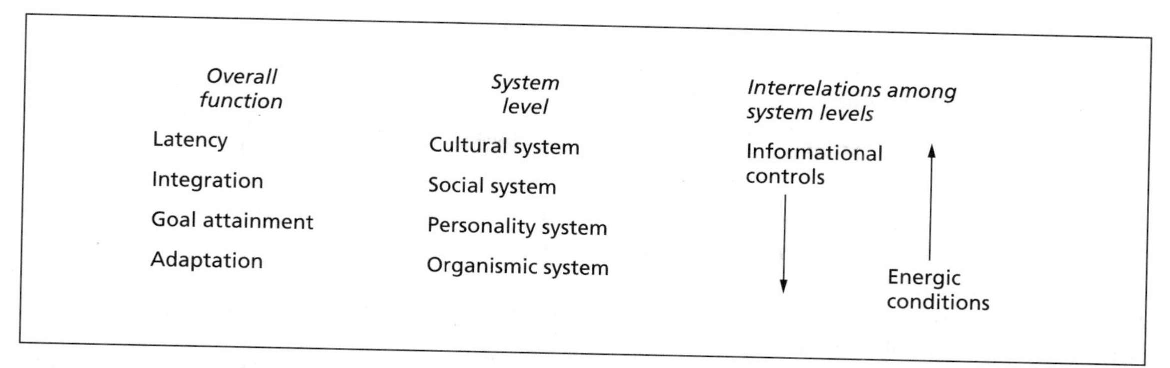

FIGURE 4-5 Parsons' Cybernetic Hierarchy of Control

provides the energy necessary for the personality system, the personality system provides the energic conditions for the social system, and the organization of personality systems into a social system provides the conditions necessary for a cultural system. Thus the input/output relations among action systems are reciprocal, with systems exchanging information and energy. Systems high in information circumscribe the utilization of energy at the next lower system level, and each lower system provides the conditions and facilities necessary for action in the next higher system. This scheme has been termed a *cybernetic hierarchy* of control and is diagrammed in Figure 4-5.

GENERALIZED MEDIA OF EXCHANGE

Until his death, Parsons maintained his interest in the intra- and intersystemic relationships of the four action systems. Although he never developed the concepts fully, he had begun to view these relationships as *generalized symbolic media of exchange.*[13] In any interchange, generalized media are employed—for example, money is used in the economy to facilitate the buying and selling of goods. What typifies these generalized media, such as money, is that they are really symbolic modes of communication. The money is not worth much by itself; its value is evident only for what it says symbolically in an exchange relationship.

Thus, what Parsons proposed is that the links among action components are ultimately informational. This means that transactions are mediated by symbols. Parsons' emphasis on information is consistent with a cybernetic hierarchy of control. Informational exchanges, or cybernetic controls, are seen as operating in at least three ways. First, the interchanges or exchanges *among* the four subsystems of the overall action system are carried out by different types of symbolic media; that is, money, power, influence, or commitments. Second, the interchanges *within* any of the four action systems are also carried out by distinctive symbolic media. Finally, the system requisites of adaptation (A), goal attainment (G), integration (I), and latency (L) determine the type of generalized symbolic media used in an inter- or intrasystemic exchange.

Within the social system, the adaptive sector uses money as the medium of exchange with the other three sectors; the goal-attainment sector employs power—the capacity to induce conformity—as its principal medium of exchange; the integrative sector of a social system relies on influence—the capacity to persuade; and the latency sector uses commitments—especially the capacity to be loyal. The analysis of interchanges of specific structures within social systems should thus focus on the input/output exchanges using different symbolic media.

Among the subsystems of the overall action system, a similar analysis of the symbolic media used in exchanges should be undertaken, but Parsons never clearly described the nature of these media.[14] What he appeared to be approaching was a conceptual scheme for analyzing the basic types of symbolic media, or information, linking systems in the cybernetic hierarchy of control.[15]

PARSONS ON SOCIAL CHANGE

In the last decade of his career, Parsons became increasingly concerned with social change. Built into the cybernetic hierarchy of control is a conceptual scheme for classifying the locus of such social change. What Parsons visualized was that the information and energic interchanges among action systems provide the potential for change within or between the action systems. One source of change can be excesses in either information or energy in the exchange among action systems. In turn, these excesses alter the informational or energic outputs across systems and within any system. For example, excesses of motivation (energy) would have consequences for the enactment of roles and perhaps ultimately for the reorganization of these roles or the normative structure and eventually of cultural value orientations. Another source of change comes from an insufficient supply of either energy or information, again causing external and internal readjustments in the structure of action systems. For example, value (informational) conflict would cause normative conflict (or anomie), which in turn would have consequences for the personality and organismic systems. Thus, inherent in the cybernetic hierarchy of control are concepts that point to the sources of both stasis and change.[16]

To augment this new macro emphasis on change, Parsons used the action scheme to analyze social evolution in historical societies. In this context, the first line of *The Structure of Social Action* is of interest: "Who now reads Spencer?" Parsons then answered the question by delineating some of the reasons why Spencer's evolutionary doctrine had been so thoroughly rejected by 1937. Yet, after some 40 years, Parsons chose to reexamine the issue of societal evolution that he had so easily dismissed in the beginning. And in so doing, he reintroduced Spencer's and Durkheim's evolutionary models back into functional theory.

In drawing heavily from Spencer's and Durkheim's insights into societal development, Parsons proposed that the processes of evolution display the following elements:

1. Increasing differentiation of system units into patterns of functional interdependence.
2. Establishment of new principles and mechanisms of integration in differentiating systems.
3. Increasing adaptive capacity of differentiated systems in their environments.

From the perspective of action theory, then, evolution involves (a) increasing differentiation of the personality, social, cultural, and organismic systems from one another; (b) increasing differentiation within each of these four action subsystems; (c) escalating problems of integration and the emergence of new integrative structures; and (d) the upgrading of the survival capacity of each action subsystem, as well as of the overall action system, to its environment.

Parsons then embarked on an ambitious effort in two short volumes to outline the pattern of evolution in historical systems through primitive, intermediate, and modern stages.[17] In contrast with *The Social System,* where he stressed the problem of integration between social systems and personality, Parsons drew attention in his evolutionary model to the inter- and intradifferentiation of the cultural and social systems and to the resulting integrative problems. Each stage of evolution is seen as reflecting a new set of integrative problems between society and culture as each of these systems has become more internally differentiated as well as differentiated from the other. Thus, the concern with the issues of integration within and among action systems, so evident in earlier works, was not abandoned but applied to the analysis of specific historical processes.

Even though Parsons was vague about the causes of evolutionary change, he saw evolution as guided by the cybernetic hierarchy of controls, especially the informational component. In his documenting of how integrative problems of the differentiating social and cultural systems have been resolved in the evolution of historical systems, the informational hierarchy is regarded as crucial because the regulation of societal processes of differentiation must be accompanied by legitimation from cultural patterns (information). Without such informational control, movement to the next stage of development in an evolutionary sequence will be inhibited.

Thus, the analysis of social change is an attempt to use the analytical tools of the general theory of action to examine a specific process, the historical development of human societies. What is of interest in this effort is that Parsons developed many propositions about the sequences of change and the processes that will inhibit or accelerate the unfolding of these evolutionary sequences. It is of more than passing interest that tests of these propositions indicate that, on the whole, they have a great deal of empirical support.[18]

PARSONS ON "THE HUMAN CONDITION"

Again in a way reminiscent of Spencer's grand theory, Parsons attempted to extend his analytical scheme to all aspects of the universe.[19] In this last conceptual addition, it is ironic that, Parsons' work increasingly resembled Spencer's. Except for the opening line in *The Structure of Social Action*—"Who now reads Spencer?"—Parsons ignored Spencer. Indeed, he may not have realized how closely his analyses of societal evolution and his conceptualization of the "human condition" resembled Spencer's effort of 100 years earlier. At any rate, this last effort was more philosophy than sociology. Yet it represents the culmination of Parsons' thought. Parsons began in 1937 with an analysis of the smallest and most elementary social unit, the act. He then developed a requisite functionalism that embraced four action systems: the social, cultural, personality, and what he called "the behavioral" in later years (he had earlier called this the organismic). Finally, in this desire to understand basic parameters of the human condition, he viewed these four action systems as only one subsystem within the larger system of the universe. This vision is portrayed in Figure 4-6.

As can be seen in Figure 4-6, the universe is divided into four subsystems, each meeting one of the four requisites—that is, A, G, I, or L. The four action systems resolve integrative problems, the organic system handles goal-attainment problems, the physicochemical copes with adaptation problems, and the telic ("ultimate" problems of meaning and cognition) deals with latency problems.

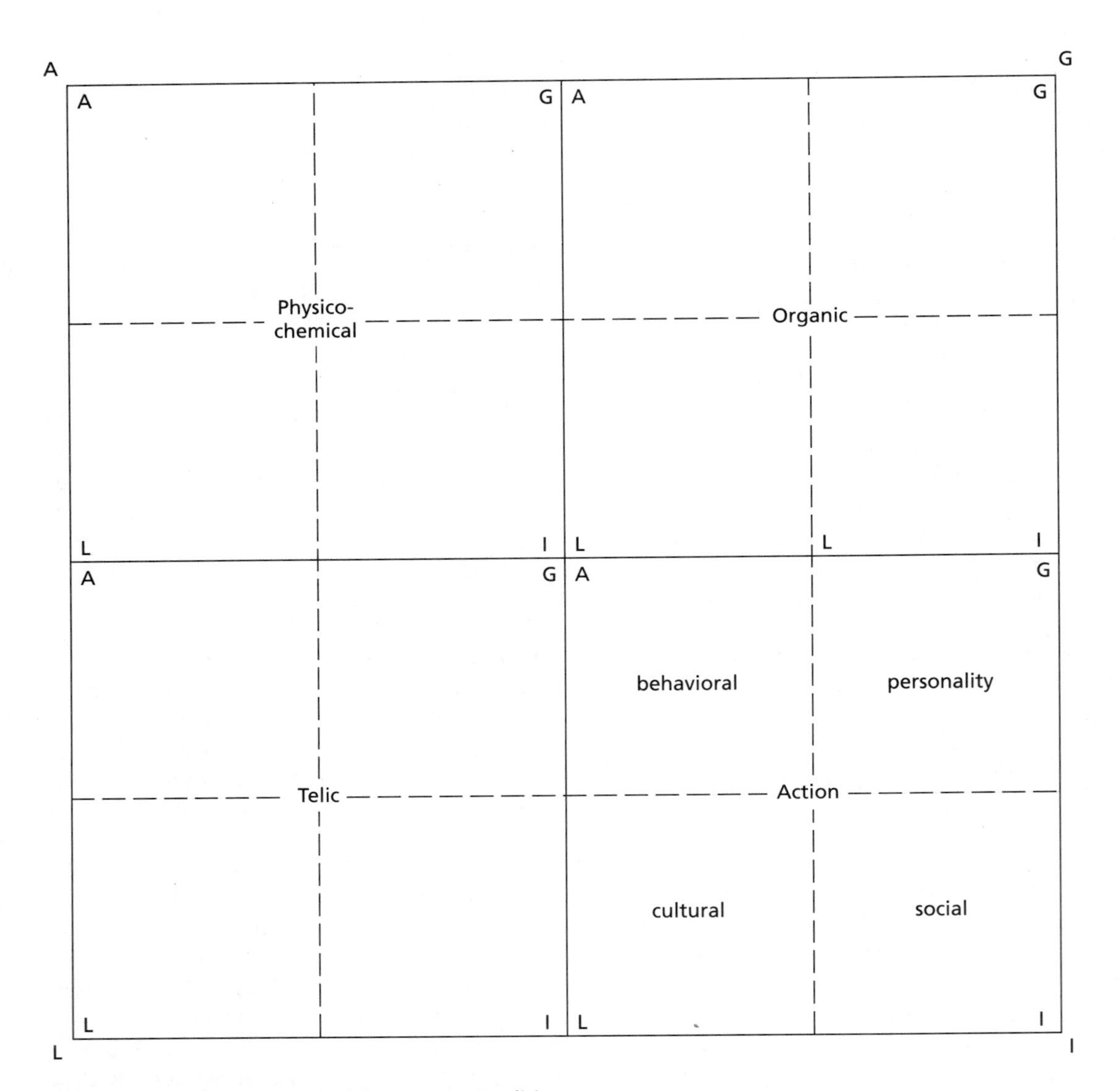

FIGURE 4-6 The Subsystems of the Human Condition

Each subsystem employs its own media for intra-and intersubsystem activity. For the action subsystem, the distinctive medium is symbolic meanings; for the telic, it is transcendental ordering; for the organic, it is health; and for the physicochemical, it is empirical ordering (lawlike relations of matter, energy, and so forth). There are double interchanges of these media among the four A, G, I, L sectors, with "products" and "factors" being reciprocally exchanged. That is, each subsystem of the universe transmits a product to the others, while it also provides a factor necessary for the operation of other subsystems. This can be illustrated with the L (telic) and I (action) interchange. At the product level, the telic system provides "definitions of human responsibility" to the action subsystems and receives "sentiments of justification" from the action subsystem. At the factor level, the telic provides "categorical imperatives" and receives "acceptance of moral obligations." These double interchanges are, of course, carried out with the distinctive media of the I and A subsystems—that is, transcendental ordering and symbolic meaning, respectively.

The end result of this analysis was a grand metaphysical vision of the universe as it impinges on human existence. Parsons' analysis represents an effort to categorize the universe into systems, subsystems, system requisites, generalized media, and exchanges involving these media. As such, this analysis is no longer sociology but philosophy or, at best, a grand meta-theoretical vision. Parsons had indeed come a long way since the humble unit act made its entrance in 1937.

NOTES

1. Although few appear to agree with all aspects of Parsonian theory, rarely has anyone quarreled with the assertion that he has been the dominant sociological figure of this century. For documentation of Parsons' influence, see Robert W. Friedrichs, *A Sociology of Sociology* (New York: Free Press, 1970), and Alvin W. Gouldner, *The Coming Crisis of Western Sociology* (New York: Basic Books, 1970).

2. Talcott Parsons, *The Structure of Social Action* (New York: McGraw-Hill, 1937); the most recent paperback edition (New York: Free Press, 1968) will be used in subsequent footnotes.

3. Ibid., p. 730.

4. For useful analyses of Parsons' work in relation to the issues he raised in *The Structure of Social Action*, see Leon Mayhew, "In Defense of Modernity: Talcott Parsons and the Utilitarian Tradition," *American Journal of Sociology* 89 (1984), pp. 1273–1306, and Jeffrey C. Alexander, "Formal and Substantive Voluntarism in the Work of Talcott Parsons: A Theoretical Reinterpretation," *American Sociological Review* 13 (1978), pp. 177–198.

5. Parsons, *Structure of Social Action* (cited in note 2), pp. 748–749.

6. See also, Jonathan H. Turner, "The Concept of 'Action' in Sociological Analysis," in *Analytical and Sociological Theories of Action*, eds. G. Seeba and Raimo Toumea (Dordrecht, Holland: Reidel, 1985).

7. Talcott Parsons, *The Social System* (New York: Free Press, 1951).

8. Max Weber, *Economy and Society*, vol. I (Totowa, NJ: Bedminster, 1968), pp. 1–95.

9. These pattern variables were developed in collaboration with Edward Shils and were elaborated in *Toward a General Theory of Action* (New York: Harper & Row, 1951), pp. 76–98, 203–204, 183–189. Again, Parsons' debt to Max Weber's concern with constructing ideal types can be seen in his presentation of the pattern variables.

10. Talcott Parsons, Robert F. Bales, and Edward A. Shils, *Working Papers in the Theory of Action* (Glencoe, IL: Free Press, 1953).

11. Talcott Parsons and Neil J. Smelser, *Economy and Society* (New York: Free Press, 1956). These requisites are the same as those enumerated by Malinowski. See Chapter 2 in this work, Table 2–1.

12. Talcott Parsons, "An Approach to Psychological Theory in Terms of the Theory of Action," in *Psychology: A Science*, ed. S. Koch, vol. 3 (New York: McGraw-Hill, 1958), pp. 612–711. By 1961 these ideas were even more clearly formulated; see Talcott Parsons, "An Outline of the Social System," in *Theories*

of Society, eds. T. Parsons, E. Shils, K.D. Naegele, and J.R. Pitts (New York: Free Press, 1961), pp. 30–38. See also Jackson Toby, "Parsons' Theory of Social Evolution," *Contemporary Sociology* 1 (1972), pp. 395–401.

13. Parsons' writings on this topic are incomplete, but see "On the Concept of Political Power," *Proceedings of the American Philosophical Society* 107 (1963), pp. 232–262; "On the Concept of Influence," *Public Opinion Quarterly* 27 (Spring 1963), pp. 37–62; and "Some Problems of General Theory," in *Theoretical Sociology: Perspectives and Developments*, eds. J.C. McKinney and E.A. Tiryakian (New York: Appleton-Century-Crofts, 1970), pp. 28–68. See also Talcott Parsons and Gerald M. Platt, *The American University* (Cambridge, MA: Harvard University Press, 1975).

14. For his first attempt at a statement, see Parsons, "Some Problems of General Theory" (cited in note 3), pp. 61–68.

15. For a more readable discussion of these generalized media, see T.S. Turner, "Parsons' Concept of Generalized Media of Social Interaction and Its Relevance for Social Anthropology," *Sociological Inquiry* 38 (Spring 1968), pp. 121–134.

16. For a fuller discussion, see Alvin L. Jacobson, "Talcott Parsons: A Theoretical and Empirical Analysis of Social Change and Conflict," in *Institutions and Social Exchange: The Sociologies of Talcott Parsons and George C. Homans*, eds. H. Turk and R.L. Simpson (Indianapolis: Bobbs-Merrill, 1970).

17. Talcott Parsons, *Societies: Evolutionary and Comparative Perspectives* and *The System of Modern Societies* (Englewood Cliffs, NJ: Prentice-Hall, 1966, 1971, respectively). The general stages of development were first outlined in Talcott Parsons, "Evolutionary Universals in Society," *American Sociological Review* 29 (1964), pp. 339–357.

18. See Gary L. Buck and Alvin L. Jacobson, "Social Evolution and Structural-Functional Analysis: An Empirical Test," *American Sociological Review* 33 (June 1968), pp. 343–355; A.L. Jacobson, "Talcott Parsons: Theoretical and Empirical Analysis" (cited in note 16).

19. Talcott Parsons, *Action Theory and the Human Condition* (New York: Free Press, 1978). See the last chapter in this book and my analysis in "Parsons on the Human Condition," *Contemporary Sociology* 9 (1980), pp. 380–383.

5

The Continuing Tradition I: The Neofunctionalism of Jeffrey C. Alexander*

At a time when most sociological theorists were highly critical of functional theorizing, Jeffrey C. Alexander began to argue for a neofunctionalism.[1] This neofunctionalism involved "an intellectual tendency" among a diverse range of scholars to retain the useful elements of Talcott Parsons' analytical approach, while downplaying the notion of functional requisites. Thus, for Alexander and others, such as Paul Colomy,[2] who have been instrumental in the rise of neofunctionalism, sociological theorizing should visualize society as a system of interrelated parts, retain the distinctions among cultural, social, and personality systems, visualize culture as a distinctive realm of social reality, examine the forces causing both the integration and disintegration of social orders, and recognize that social differentiation is a central feature of social change. All these had been prominent pieces in the complex analytical system built by Parsons, and Alexander felt that these were still central to sociological analysis. However, the most distinctive feature of functional theory—that is, analyzing sociocultural forces by their functions for sustaining a more inclusive system—was to be downplayed. Indeed, in Alexander's eye, the idea of systems having needs and requisites cannot fully explain social processes that operate within these systems; at best, functional requisites establish broad limits for the way that human social action and patterns of social organization operate.

ALEXANDER'S ADVOCACY FOR POSTPOSITIVISM[3]

Alexander's neofunctionalism is situated in a *postpositivist* conception of social science. He argues that sociological work can be understood as falling along a continuum stretching from the very abstract, general, and metaphysical,

*This chapter is coauthored with Paul Colomy.

on the one end, to the concrete, empirical, and factual on the other.[4] Other elements of social scientific activity, including ideologies, models, methodological assumptions, concepts, laws, propositions, and observational statements, fall between these endpoints. Even though its overall form can be characterized more by one element than another, every social scientific statement contains implicit or explicit commitments about the nature of every other element on the continuum. Figure 5-1 represents Alexander's formulation of this continuum.

Alexander maintains that various forms of sociology are carried forward by traditions or schools of thought. According to this postpositivist model of social science, a key impetus to advancing sociological knowledge is conflict and competition between these traditions and schools.[5] The primary reference points for determining whether a particular social scientific statement advances knowledge are comparisons with parallel statements offered by competing traditions or with previous statements formulated within a given tradition itself. Thus, instead of speaking about theoretical or empirical progress, per se, Alexander speaks of a social scientific statement's relative explanatory and theoretical success compared with a scholar's own tradition or a competing one.[6]

This conception of social science implies that a comprehensive discussion of any tradition would examine developments at every point along the scientific continuum presented in Figure 5-1. Alexander proposes the more economical, and less tedious, strategy of examining traditions for what he refers to as two discursive genres: generalized discourse and research programs.[7] *Generalized discourse* refers to discussion of issues listed at the left side of the continuum portrayed in Figure 5-1: presuppositions, general models purporting to describe and explain social processes and systems, and explicit ideological commitments or the ideological implications of a particular formulation. By contrast, within the context of *research programs* portrayed on the right side of the continuum in Figure 5-1, these generalized issues are assumed to be relatively unproblematic, because this mode of sociological inquiry is devoted to explaining or interpreting specific empirical structures and processes.

Both generalized discourse and research programs are problem-solving activities,[8] but the two genres differ in dramatic ways. Research programs are organized around relatively concrete problems with readily apparent empirical referents. Research programs investigate specific empirical issues, such as recruitment to social movements or the amount and sources of intergenerational mobility in the contemporary United States. In contrast, generalized discourse is preoccupied with relatively abstract issues (like the nature of social action or social organization) whose empirical referents are not always clear. The solutions that generalized discourse offers to these highly abstract issues tend to be very remote from the researchers' concerns, sometimes bordering on the philosophical. The essential point is that in formulating general problems and endeavoring to solve them, generalized discourse establishes a context of relevance and plausibility. Generalized discourse suggests what there is to see, how to approach what can be seen, and the types of questions that might be asked about what is seen. Within this frame of reference supplied by generalized discourse, research programs identify and attempt to solve specific empirical problems. From a postpositivist perspective, then, generalized discourse or what is sometimes called "general theory," plays a crucial, if often unrecognized, role in sociology.

Competition occurs within both generalized discourse and research programs.[9] At the generalized discourse level, competition proceeds through debates about a tradition's categories, its analytic and empirical breadth, its ability to provide accurate and incisive interpretations of classic sociological work, its

FIGURE 5-1 Alexander's Continuum of Social Scientific Thought

Adapted from Jeffrey Alexander, *Positivism, Pressupositions, and Current Controversies,* p. 40.

avowed or implied ideological stance, its resonance with the era's reigning sensibilities, issues, and social movements, its logical coherence, and its utility for empirical investigation. At the research program level, competition is organized around rival attempts to explain empirical structures and processes regarded as important by the discipline at a given time. In either case, a social scientific tradition advances when its proponents formulate statements deemed superior relative to comparable work produced by other schools or earlier work within the same school.

At any given time, the field on which traditions compete is organized hierarchically.[10] Traditions are invidiously compared, and a small subset is accorded high levels of recognition and prestige. The disciplinary judgment that a particular tradition is especially meritorious encourages competing schools to frame their discussions as critical alternatives to the most highly regarded approaches. Proponents of less esteemed or newly minted traditions are constrained to demonstrate their perspective's relative merit by highlighting its conceptual and empirical strengths relative to more hegemonic paradigms.

Over time, this competition fosters significant changes in a disciplinary field.[11] Once established and highly regarded, traditions are discredited and sometimes disappear, lowly ranked schools gain prominence, and new paradigms flourish. A given tradition's ability to survive partly reflects its capacity to respond to the intellectual critiques issued by rival traditions and its ability to remain germane to contemporaneous societal and global transformations. Because of the difficulties in responding satisfactorily to these recurring challenges, most traditions experience periods of crisis or are completely discredited shortly after they are initiated. To persist, traditions must change, and those that last for more than a generation are almost always substantially revised and reconstructed.

Alexander uses this postpositivist model to describe both the decline of Talcott Parsons' functionalism and the emergence of neofunctionalism.[12] Parsons' general theorizing, which commenced with the publication of *The Structure of Social Action* in 1937,[13] established a discursive context that fueled the development of several research programs across a wide array of substantive areas. From World War II until the early 1960s, Parsons and his colleagues presented functionalism as a conceptually sophisticated and empirically supported paradigm. By the mid-1960s and throughout the 1970s, however, functionalism's generalized discourse and its research programs were subject to increasingly critical assessments by proponents of competing schools. Some of these critiques were partially inspired by the social and political questions posed by the insurgent social movements of the 1960s and early 1970s, and at least one observer has noted that the decline of Parsonian functionalism "must be at least partly explained by the observation that Parsonian system theory did not seem to offer fruitful lines of research in period marked by a strong demand for critical social theory."[14] Emerging in the mid-1980s, contemporary neofunctionalism can be understood as a response to three developments. First, it represents an effort to address the legitimate objections raised by critics of Parsonian functionalism, an effort that has revised the older tradition in significant ways while retaining the core elements identified at the beginning of this chapter. Second, contemporary neofunctionalism is partly fueled by the renewed interest in synthesizing ostensibly irreconcilable perspectives, both within sociology and across disciplinary boundaries. Third, contemporary neofunctionalism has emerged in an historic climate in which Marxism and many orthodox critical theories that flowered in the 1960s and 1970s have been morally delegitimated and in which new types of questions are being asked. Neofunctionalism attempts to make this new world intelligible by addressing these compelling questions.

NEOFUNCTIONALISM AS GENERALIZED DISCOURSE

A substantial portion of Alexander's theoretical work has revolved around generalized discourse, or discussion about the modes of scientific activity portrayed on the left side of the continuum presented in Figure 5-1. Thus, Alexander believes that it is necessary to examine the presuppositions, the ideological content, and the models of system processes in a postpositivist social science. Only with this kind of activity, Alexander appears to argue, can other modes of scientific activity presented in the middle and right portions of the continuum in Figure 5-1 be fully developed. For in Alexander's view, it is difficult to develop definitions of concepts, classification schemes, laws of fundamental processes, propositions, methodologies, and observational statements without recognizing the underlying presuppositions, implicit and explicit ideologies, and models guiding these forms of scientific activity.

The Multidimensionality of the Social World

At the core of Alexander's generalized discourse is the notion of multidimensionality. He maintains that different points on the continuum presented in Figure 5-1 are interdependent but partially autonomous. These points are interdependent in the sense that every scientific statement contains implicit references to every other level on the continuum. At the same time, however, every point listed on the continuum has its own partial autonomy because each possesses distinctive criteria for assessing scientific merit.

The Problems of Reductionism and Conflation. If the partial autonomy of different modes of social science is ignored, two related errors result: (1) reductionism, and (2) conflation. *Reductionism* occurs when the inherently multileveled character of sociological work is reduced to a single level that is arbitrarily deemed the most significant. *Conflation* exists when the differentiated modes of scientific activity arrayed in Figure 5-1 are ignored. Conflation occurs, for example, when presuppositional positions are treated as expressions of ideological commitments or when descriptions of sociological models are taken for exact observational statements. By failing to distinguish consistently among the different points along the continuum of science, conflation confuses sociological discourse, promotes sterile debates, and generates false dilemmas.[15]

An important source of conflation and reductionism, and one especially salient in American sociology, is the failure to recognize the partial autonomy of generalized discourse as well as the impact of general theoretical commitments on other types of sociological work. The positivist persuasion embraced by most sociologists (as distinct from the postpositivist position advanced by Alexander) is essentially reductionist in that it "reduces theory to fact."[16] What Alexander means by this bold assertion is that, in traditional views of science, facts from data collected by some methodological procedure are what determine the plausibility and utility of theory. In positivistic science, theories are produced to explain facts; the collection of data occurs to assess theories. But if each mode of scientific activity along the continuum portrayed in Figure 5-1 is both autonomous and interdependent, then theories cannot be "reduced to facts" because the way facts are collected and interpreted is related to the generalized discourse or presuppositions, ideologies, and models that have guided the definition of concepts, the formulation of laws and propositions, the construction of methodologies, and the collection of data. A postpositivist view of science, then, introduces additional criteria for assessing theories. Theories should be evaluated not only by recourse to facts, but also by their underlying discursive commitments, particularly their presuppositions, ideological stances, and conceptions of systems.

Action and Order. The general problems of "action" and "order," Alexander maintains, are the two most important presuppositions in sociological theory, and hence, they should be essential topics of generalized discourse.[17] Historically, sociologists have typically addressed the presuppositional problem of *action* by selecting one of two options: rational or nonrational. A *rational* approach presumes that action should be understood as an instrumental, calculating, and efficient adaptation to external, material conditions. This is the conception of action adopted by many economists and by proponents of rational choice theory in sociology, as we will see in Chapters 23 and 24, or as we saw for Talcott Parsons' portrayal of utilitarianism in Chapter 4. These social scientists portray social action as resulting from calculations of efficiency, with people selecting behaviors that involve the fewest costs and produce the greatest rewards.

The alternative approach emphasizes the *nonrational* dimensions of action and asserts that action can only be understood by reference to actors' subjective motivations. These subjective motivations include values, ideals, intentions, purposes, impulses, and unconscious needs, and although the particular subjective motivations differ, they share one common attribute as mediators of the relationship between a person and the environment. Hence, an individual's (or group's) conduct cannot be accurately predicted without sufficient knowledge of the actor's subjective dispositions.

The second presuppositional issue is *order*, or the presumption made by most practicing social scientists that the social world exhibits recurring patterns, regularities, and uniformities. A concern with order, however, does not only embrace forces of consensus, harmony, and integration because nonrandom patterns and uniformities are no less common in conflictual than in cooperative social relations. So, whether sociologists examine class conflict or wedding ceremonies, their studies presuppose that order—that is, nonrandomness—exists in all types and forms of social settings. The presuppositional problem of order, then, is concerned with how individual units of whatever motivation (that is, whether they are presumed to act rationally or nonrationally) are arranged in nonrandom social patterns. Individualist and collectivist approaches are the two most common solutions to the problem of order.

Individualist approaches explain order as the result of individual negotiations and choice. Individualist theories assert or imply that social structures and patterned activities are produced by individual actors' perceptions and conduct in particular settings. Because individuals' choices, negotiations, perceptions, and behaviors are the ultimate sources of order, actors can alter established structures and patterns at any time.

A *collectivistic approach*, by contrast, explains order by reference to the emergent properties of social organization itself—properties that cannot be reduced to concrete individuals. Collectivistic theories frequently explain recurring social patterns by introducing traditional sociological concepts such as values, norms, class structures, or gender regimes.

Alexander insists that the conventional options to the dual problems of action and order generate serious conceptual and empirical difficulties in every theory or study that embraces them. Theories with an individualistic conception of order are unable to explain recurring social patterns without recourse to collectivistic concepts such as culture, social organization, or normative principles,[18] but because of their presuppositional commitment to individualism, these theories wind up smuggling collectivistic notions in through the back door, a dubious strategy that renders their theories internally contradictory. Similar difficulties are evident in theories that adapt a one-dimensional approach to action. Karl Marx's theory of revolution, for example, adopts a highly rational approach to action, predicting that working class revolutions will emerge as rational responses to deteriorating, objective economic and political conditions. Marx's empirical studies of actual revolutions, however, indicate that several "nonrational" considerations—for example, charismatic leaders, prevailing traditions, and participants' political convictions—intervene between objective economic conditions and workers' revolutionary struggles. Although such considerations are important to understanding many revolutions, in light of Marx's rationalist presuppositions about action they represent residual categories and are inconsistent with the core assumptions of his theory.[19]

The traditional, analytically one-sided conceptions of action and order, Alexander maintains, invariably engender sterile polemics in sociology. To move beyond the warring schools and the unproductive debates that polarize proponents, Alexander presents a synthesis, which he calls *multidimensionality*. In effect, multidimensionality involves two distinct syntheses: (1) A "strong synthesis" maintains that action should be viewed, at the presuppositional level, as the product of *both* rational and nonrational elements. In essence, a multidimensional approach to this problem insists that action is shaped by rational adaptations to external conditions *and* by internal, subjective commitments and perceptions. (2) A "weaker synthesis" recommends a basically collectivistic stance to order but one that explicitly acknowledges that individualistic theories supply useful empirical insights into how social patterns and structures are reproduced and transformed by individual actors. In a multidimensional framework, then, rational and nonrational determinants of action are given roughly equal analytic weight, and a collectivistic approach to order is preferred, though individualistic theories' insights into empirical processes are recognized as crucial to a fully satisfactory understanding of social life.

It should be clearly understood that multidimensionality is not a simple classificatory scheme contrived to sift and sort theories into mutually exclusive categories. Rather, its primary purpose is evaluative. Alexander's postpositivist position presumes that the development of social science is a "two-tiered process, propelled as much by theoretical as by empirical argument."[20] Conventional scientific approaches assert that reliable data are the final arbiter of theoretical disputes, and that the facts must be consulted to determine which of two competing theories has more scientific credibility. By contrast, Alexander's "two-tiered" conception of scientific knowledge implies that sociological theory and research should be evaluated not only by reference to data but also by their respective presuppositional commitments. Facts alone are not the final arbiter of sociological statements; theories must also be assessed in light of their theoretical logic.

The Micro-Macro Linkage

One of the most debated issues during the last two decades in sociological theory has been the connection between micro-level interpersonal processes, on the one hand, and macro-level collective or organization processes, on the other. Given Alexander's commitment to synthesizing competing approaches, it is not surprising that he has been a leading contributor to this debate.[21] Alexander began with Talcott Parsons' original formulation of the "unit act" (see Figure 4-1 on p. 30).[22] Parsons' original action frame of reference included the following components: effort, means, ends, norms, and conditions. Alexander treats norms and conditions as macrosociological elements and treats ends and means as situationally specified products of individual action. Effort carves means and ends from conditions and norms, respectively. Every end is viewed as a compromise between individuals' efforts, their objective possibilities, and their normative standards of reference; every means represents an aspect of individuals' conditional world that they have succeeded in using according to their objective possibilities and internalized needs. Whereas Parsons had said little about action as effort and failed to indicate how ends and means might be conceptualized as micro-translations of norms and conditions, Alexander argues that effort "is the contingent element of action . . ., the motor, the micro process that drives the combination of the other elements."[23] This reconceptualization of the unit act, which accords effort—the unexamined black box in Parsons' earlier formulation—a more central position, provides a foundation for synthesizing key elements of microsociological theories. In Alexander's view, each micro perspective emphasizes one analytic dimension of effort and aids our understanding of ends and means as well as contributes to the micro explanation of norms and conditions.

Alexander's reconstructed unit act underscores the complementary features of nonrationalistic and rationalistic micro theories. From a multidimensional perspective, action can be conceived as moving along the dimensions of (1) interpretation and (2) strategic calculations. Borrowing from interactionist traditions (see Part V), Alexander argues that *interpretation* consists of two distinct processes: (1) *typification* occurs when each new impression is interpreted using an established frame of reference, and (2) *invention* accompanies each cognition as the actor discovers understanding in slightly new ways. Thus, in Alexander's approach, invention almost always accompanies typification as either minor or major alterations in classification systems.

Action is guided not only by interpretive understandings, but also by *strategic calculations* of rewards and costs. With this insight, Alexander incorporates many elements of rational choice theory's characterization of action (see Chapters 23 and 24) as well as the general notion that strategic considerations are formulated in terms of available time and energy that, in turn, are allocated according to the principle of least expense.

These two general dimensions of action—interpretation and calculation—interact. Strategic calculations depend on relevant but almost always imperfect knowledge about the future, and interpretive understanding provides both an environment for formulating strategies and, hence, profoundly affects the calculus of strategic action itself. Conversely, interpretive efforts are usually expended on phenomena in which actors have some interest and over which they sense some control, and thus, strategic calculations influence interpretations.

The claims of micro theorists notwithstanding, action cannot be completely understood solely as a product of interpretation and strategic calculation. This limitation can be overcome by drawing on macro theories that examine the constraining facets of social order. The product of this theoretical union is what might be called a "conditional voluntarism" that conceptualizes contingent action as occurring within macro environments, which inform and constrain action, whereas action shapes environments, either reproducing and re-creating them or perhaps changing them.

Alexander adopts a modified version of Parsons' tripartite systemic model of action systems—social, culture, and personality—to analyze the pertinent collective environments. *Social systems* provide actors with real objects on which they can act: (1) a division of labor and institutions of political authority that constitute crucial settings for interpretation and strategization; (2) a solidarity component, or those with whom actors establish a sense of community and the qualitative nature of these solidarity bonds that affects how actors interpret and strategize; and (3) a social role component of norms and sanctions that shape action. *Cultural systems* constitute a second environment and enter into action by naming reality, articulating the sacred and profane, and establishing the universe of potentially institutionalized values. By establishing the grid of cognitive, moral, and valuational classifications, symbol systems inform both interpretation and strategization. The third action system—the *personality system* includes the two micro dimensions of action, or interpretation and strategic calculation. These capacities vary developmentally over the life cycle and across social systems and thus condition the employment of interpretation and strategization.

Alexander's general theory is an important attempt to synthesize micro and macro perspectives. Because action is partially a product of interpretation and strategic calculation, it possesses a contingent quality. These dimensions of action—that is, interpretation and strategic—carve a space between agents and their macro environments that enable actors to formulate new courses of action and re-create their environments. At the same time, however, systemic environments of society, culture, and personality within which actors' efforts are contemplated and carried out establish limits on contingent action.

The Method of Specification

Alexander's postpositivist conception of science argues that generalized discourse is a legitimate and vitally important activity in sociology. This view maintains that the points along the continuum of science portrayed in Figure 5-1 are interdependent. As a consequence, advances in general theory, such as Alexander's statements on multidimensionality and the micro-macro link, ought to have crucial ramifications for more empirical work. How can these ramifications be determined? Though Alexander does not directly address this issue, an answer is implicit in much of his own work and that of his colleagues. That answer revolves around what might be called the *method of specification*.[24]

Specification involves identifying the importance and relevance of abstract social scientific statements for more empirical work. (As shown in Figure 5-1, specification means moving from left to right on the social scientific continuum.) In principle, specification can occur between any two points on the continuum, provided that the statement being specified is more general than the level at which the specification is being carried out. Specifying general theory for more empirical levels involves two interrelated steps: The first is a "critical reading," and the second introduces "correctives" designed to fill in theoretical and factual inconsistencies or anomalies uncovered through the critical reading.

(1) In the context of specification, a fruitful *critical reading* uses a general theoretical statement to assess and evaluate the usually tacit assumptions underlying more empirically oriented sociological studies. Although sociological research commonly emphasizes observational statements or the complex relations

between carefully measured variables, such research also rests on, even if only implicitly, more abstract assumptions. Moreover, Alexander's postpositivism maintains that sociological statements, even the seemingly most factual, should be evaluated not only for their empirical accuracy but also for their general theoretical logic. Critical readings identify how deficiencies in a study's underlying discursive commitments produce conceptual and factual anomalies.

(2) If critical readings reveal significant analytic and empirical deficiencies, the next step is to develop suitable remedies. This *corrective phase* must be responsive to the interdependence and partial autonomy of different points on the social scientific continuum. Specification presumes that general theories have ramifications for more concrete sociological work but recognizes that the partial autonomy of the latter is an essential protection against conflating general theory with empirical observation.

The remedies or correctives extracted from general theory are applied to a particular substantive specialty area or field of research. This corrective phase also uses innovative, middle-range conceptualizations and empirical observations already developed in the specialty area. The task of specifying general theory, therefore, typically falls to specialists, for only those well acquainted with all the subtleties, nuances, and history of a particular substantive area can adapt highly general statements to the existing corpus of work in that field. When carried out successfully, specification introduces more comprehensive, middle-range concepts and redresses empirical anomalies in the specialist's particular area of expertise.

Ideology

Talcott Parsons insisted on a sharp separation between sociological theory and ideology. In contrast, Alexander[25] argues that every social scientific tradition contains an ideological dimension. In more recent work, Alexander adopts Geertz's[26] cultural conception of ideology as a meaning system to argue that the ideological dimension of sociological theory helps to make the social world intelligible to contemporary audiences.[27] Major social and global developments (for example, the Civil Rights struggle, the anti-Vietnam War movement, the collapse of the Soviet Union, and the repudiation of communist regimes across Eastern Europe) affect the course of sociological theory by posing different questions and placing a premium on theories that can make sense of these portentous developments. A successful sociological theory, Alexander holds, must be responsive to shifts in ideological sensibility that follow in the wake of unexpected and initially disorienting social transformations.[28]

Systems Models

Alexander finds considerable merit in Parsons' general notion that society can be modeled as a system. Indeed, he maintains that Parsons' interchange model with its divisions into the functional requisites of adaptation, goal attainment, integration, and latency (AGIL) and with its notion of the generalized media of exchange for each division is an illustration of a systems model. Alexander believes this aspect of Parsons' work illustrated specification of multidimensionality at the level of models (see page 37 of Chapter 4).[29] At the same time, Alexander maintains that Parsons tended to reify systems and functionalist reasoning and conflate the distinctions in AGIL with the actual structure and operation of contemporary societies. In response, Alexander has stressed the importance of examining the historical and institutional content of systems, the dynamic and conflictual elements within and between systems, and the tensions between systems and their environments.

NEOFUNCTIONALIST RESEARCH PROGRAMS: TWO ILLUSTRATIONS

Complementing Alexander's generalized discourse, neofunctionalism has produced several research programs in a variety of different substantive areas. We will focus on only two, Alexander's analysis of social change and his more recent approach to the nature of culture.

The Analysis of Change

Alexander's work on social change must be understood in the context of earlier functionalist treatments of this problem. Partly in response to critics who charged that functionalism could not account for social change, Parsons[30] and his colleagues initiated in the mid-1950s a broad-ranging research program, which its proponents call *differentiation theory*, to address this issue.[31] Drawing on the insight Émile Durkheim presented in *The Division of Labor in Society*, this program first postulated a "master trend" of differentiation, arguing that long-term social change has involved the emergence of ever more distinct institutional subsystems. Second, differentiation theorists invoked systemic needs or requisites to explain the transition from simple and undifferentiated to more specialized structures, arguing that when functional requisites were not being fulfilled effectively, disequilibrium would spur the development of more efficient, differentiated subsystems. Third, the institutionalization of more specialized units purportedly increased the effectiveness and efficiency of a social system and its constituent subsystems. In addition, high levels of differentiation were correlated with value generalization and greater inclusion, the latter processes ostensibly contributing to the reintegration under common, though highly general, values in a system grown more complex through differentiation.

Parsons' theory of social change provoked a second round of criticism. Constructively engaging these criticisms, Alexander's work on social change revises Parsons' research program in several ways. First, Alexander contends that a more comprehensive characterization of structural change is essential. Parsons' general description of the master trend must be combined with accounts of more historically delimited phases of change. Moreover, differentiation theory must also be applied to particular, historically significant events to demonstrate its applicability to meaningful occurrences in everyday life. In addition, Alexander supplements Parsons' evolutionary ordering of total societies by their relative degree of structural differentiation by specifying how differentiation works itself out in particular institutional spheres such as the mass media and systems of solidarity.[32] Supplementing these institutional specifications of differentiation is the need to identify more variable patterns of structural change, such as unequal, uneven, incomplete, and blunted differentiation.[33]

Faulting Parsons for treating adaptation as both the cause and effect of differentiation, Alexander insists that a viable explanation of differentiation must investigate the processual dynamics and structural parameters of change in much greater depth than conventional functionalists did. Micro dynamics, he argues, are frequently the proximate impetus for differentiation. The analysis of these dynamics is essential because differentiation cannot be fully understood as a natural, inevitable response to structural strain or as an immanent, systemic impulse toward greater adaptive capacity. The analysis of micro dynamics highlights the role of social groups, underscores the impact of competition and conflict, and examines how processes of group mobilization, coalition formation, control over pertinent resources, and group struggle affect the course of differentiation. In this vein, Alexander applauds the pioneering work of

S.N. Eisenstadt, giving particular attention to his discussion of institutional entrepreneurs.[34] Institutional entrepreneurs refer to small groups of individuals who crystallize broad symbolic orientations, articulate specific and innovative goals, establish new normative and organizational frameworks for the pursuit of these goals, and mobilize resources necessary to achieve them. The very occurrence of differentiation as well as the specific directions it takes are shaped by these innovating groups. These movers and shakers are not disinterested, altruistic agents of greater systemic effectiveness or efficiency. Rather, their advocacy of differentiation is partially informed by their own particular material and ideal interests. Nevertheless, these particular interests are typically blunted and compromised by the frequently countervailing interests of their coalition partners, by the larger cultural patterns by which they seek to legitimate their project, by the environing social structure that conditions their activities, resources, and sources of potential support, and by the conflicting interests of their opponents.

Alexander emphasizes this conflictual dimension of differentiation, insisting that conflict is a ubiquitous feature of social life and central to a viable explanation of differentiation. Entrepreneurs' efforts to institutionalize new levels of differentiation, for example, are commonly rebuffed or significantly compromised by groups opposed to their innovative project. Even successfully differentiated institutions are continually typified by conflict.

Alexander also regards orthodox functionalism's emphasis on two consequences of structural differentiation—increased adaptive capacity and reintegration—as much too constricting. By confusing differentiation with adaptive success, for example, the Parsonian formulation was unable to recognize that highly differentiated societies can fail. Further, in Parsons' mature work there is no account of the anxiety and pathologies associated with high levels of differentiation. Alexander treats increased efficiency and reintegration as possible but not inevitable products of differentiation. Differentiated institutions frequently establish circles of material and ideal interests around which groups rally when they perceive illegitimate intrusion on their "turf," even if that intrusion represents an effort to increase efficiency by introducing still higher levels of differentiation. Structural innovations produce new organizational niches and social roles, which constitute the bases for collectivities that can become politicized constituencies intent on maintaining or advancing their interests.

Besides reintegration, other consequences associated with differentiation include new forms of conflict as well as strains within and between differentiated subsystems. Alexander hypothesizes that the emergence of functionally differentiated and partly autonomous subsystems actually increase the amount of conflict in modern societies. Discordant specifications of general values, strain-inducing changes in a system's or subsystems' environments, advancement of parochial subsystem interests at the expense of a more encompassing public good, and the unresolved tensions within each subsystem are likely to engender conflict. At the same time, however, these conflicts in highly differentiated systems are usually contained by other subsystems specializing in "tension management," and in this vein Alexander suggests that although the sheer amount of conflict rises directly with differentiation, the scope of conflict is rarely generalized to the entire society.

Alexander points out that orthodox differentiation theory was always more than a social scientific attempt to explain the world. Like every viable sociological theory, it was also "a meaning structure, a form of existential truth"[35] that offered a compelling interpretation of the world. The ideological thrust of conventional differentiation theory valorized Western societies, particularly the

United States, as highly differentiated social systems that, unlike the fused structures of the traditional societies that they ostensibly superseded, embodied democracy, universalism, free markets, science, secularism, stability, autonomy, and achievement. The theory's facile equation of differentiation with progress and the good society was discredited by popular movements of the 1960s and 1970s as a cover for the essentially repressive, exploitive, bureaucratic, and morally bankrupt character of modern society. The theory was further discredited by postmodernists (see Chapter 44) who effectively portrayed a society that in the 1980s and 1990s was fatalistic, private, particularistic, fragmented, and local.

The Analysis of Culture

Alexander's research program on cultural sociology maintains that culture represents a partially autonomous dimension of social life.[36] He rejects competing programs that either reduce culture to a mere expression of class, power, or other purportedly more substantial social structures and interests or that treat culture as synonymous with values. Both of these alternative approaches effectively omit the analysis of "purely symbolic phenomena like ritual, sacralization, pollution, metaphor, myth, narrative, metaphysics, and code."[37] Alexander's cultural sociology, in contrast, directly addresses these issues, examining "how people make their lives meaningful" and "the ways in which social actors invest their worlds with sentiment and significance."[38]

Drawing from several intellectual and social scientific traditions, Alexander characterizes culture as "a structure composed of symbolic sets."[39] Symbols are generalized signs that supply "categories for understanding the elements of social, individual, and organic life."[40] Symbols define and interrelate these elements in an "arbitrary" fashion—that is, in a way that does not simply reflect the nature or exigencies of social, individual, and organic life. These interrelated symbols thus constitute a nonmaterial structure that patterns action as surely as more visible material structures.

There are different types of cultural structures. Narratives, for instance, refer to stories individuals, groups, and societies articulate to understand their progress through time. Narratives consist of plots with beginnings, middles, and ends; heroes and antiheroes; epiphanies and denouements; and dramatic, comic, and tragic forms. Alexander's primary interest, however, resides in the more fundamental cultural structures that "organize concepts and objects into symbolic patterns and convert them into signs. Complex cultural logics of analogy and metaphor, feeding on differences, enable extended codes to be built up from simple binary structures."[41] Cultural structures are autonomous because meaning derives not from the concrete referent signified by the symbol but from *the interrelations of symbols themselves*. Symbols are situated in sets of binary relations, so the cultural life of society can be portrayed as "web of intertwining sets of binary relations."[42]

Signs sets are organized into discourses that structure reality cognitively, affectively, and evaluatively. The antinomy between sacred and profane is central to the affective and evaluative dimensions of culture. Sacred symbols supply images of purity and oblige those committed to them to protect sacred objects from harm, whereas profane symbols embody this harm, providing images of pollution and danger and identifying groups and behaviors that must be defended against. The Watergate episode in American political life was, for example, not simply a political event but a redolent symbol of pollution and embodies a sense of evil and impurity.[43] The Watergate hearings in the 1970s can be understood as a ritualized event that reinterpreted banal political machinations as "higher" antitheses between the pure and impure elements of American civil religion.

As the notion of profane symbols implies, culture must not be equated with positive idealized imagery—with symbols and myths that command virtually unanimous, uncritical assent. Negativity, Alexander insists, is an essential component of culture and is symbolized "every bit as elaborately as the good."[44] Positive codes must be understood in relation to negative ones, and the conflict between good and bad operates inside culture "as an internal dynamic. Conflict and negation are coded and expected; repression, exclusion, and domination are part of the very core of the evaluative system itself."[45] Conflict, in other words, cannot be fully explained by reference to antagonistic material interests; it is also patterned by autonomous cultural structures.

Alexander specifies these general notions through an analysis of the discourse of American civil society. This discourse occurs within more encompassing semiotic codes that are built from elementary, binary oppositions and that are organized around contrasting conceptions of actors, social relationships, and institutions. For the last two hundred years, Alexander asserts, American civil society has been partially structured around a discourse of liberty and a contrasting discourse of repression. According to this argument, the striking continuity in the thematic content of public debate across a wide array of substantively divergent issues during the last two centuries reflects the cultural patterning of these opposed discourses.

Complementing his theory of cultural sociology is a method Alexander calls "reflexive hermeneutics."[46] This method begins with the social scientist's own emotional, moral, and behavior responses to meaningful events and symbols (for example, the fall of the Berlin Wall). Continuities and patterns in the feelings evoked by these symbols are identified by eliciting others' responses to the same events through casual conversation and investigating mass media reports. Next, comparisons are made between reactions from different periods of the same longitudinal event, across opinion groupings, and from responses to similar events (for example, comparing public responses to Watergate with responses to other presidential scandals). When these data are then analyzed from the vantage point of a theory of culture, the social scientist gains distance from his or her own experience and the particular experience of others, understanding the responses to highly charged symbolic events in a more dispassionate, decentered way. At the same time, understanding these data frequently requires revisions in the cultural theory with which the investigation began.

CONCLUSION

Alexander has engaged in other research programs, but his enduring contribution is the effort to rethink the logic of sociological theory. Substantively, he begins with Parsons' famous categories, minus the emphasis on functional requisites, but epistemologically, he seeks to examine theories for their presuppositions and how these influence explanations of particular empirical phenomena. As his work on differentiation illustrates, the suppositions of Parsons' scheme, including its ideological elements, initiates a research program where Alexander's views on action and order can be used to develop a more nuanced and fine-grained analysis of differentiation. The same is true for his work in cultural sociology.

Theorizing in Alexander's hands thus becomes a highly complex activity of analyzing all the elements presented in Figure 5-1 as we seek to explain specific empirical and historical cases. The facts of the cases are not determinative, however, because the logics of theorizing at each point on the continuum and against potential competitors become equally important. The end result is that theorizing becomes heavily infused with "discourse" about itself, and then somewhat ad hoc interpretations of empirical cases in light of the problems and issues raised in this discourse. Many find this approach appealing, but it wanders very far from Parsons' effort to construct a universal abstract scheme for analyzing all realms of the empirical world in terms of functional imperatives and requisites.

NOTES

1. Jeffrey C. Alexander, ed., *Neofunctionalism* (Beverly Hills, CA: Sage, 1985); Jeffrey C. Alexander and Paul Colomy, "Toward Neofunctionalism," *Sociological Theory* 3 (2, Fall 1985), pp. 11–23; "Neofunctionalism Today: Restructuring a Theoretical Tradition" in *Frontiers of Social Theory*, ed. George Ritzer (New York: Columbia University Press, 1990).

2. Paul Colomy, ed., *Functionalist Sociology: Classic Statements* (London: Edward Elgar, 1990), and *Neofunctionalist Sociology: Contemporary Statements* (London: Edward Elgar, 1990); Jeffrey C. Alexander and Paul Colomy, "Funzionalismo e Neofunzionalismo," *Enciclopedia Italiana*, volume 4 (1994), p. 199.

3. This section draws extensively from Alexander and Colomy, "Neofunctionalism Today" (cited in note 1); Jeffrey C. Alexander and Paul Colomy, "Traditions and Competition: Preface to a Postpositivist Approach to Knowledge Cumulation" in *Metatheorizing*, ed. George Ritzer (Newbury Park, CA: Sage, 1992); Colomy, *Neofunctionalist Sociology* (cited in note 2); and Paul Colomy, "Metatheorizing in a Postpositivist Frame," *Sociological Perspectives* 34 (1991), pp. 269–286.

4. Jeffrey C. Alexander, *Positivism, Presuppositions, and Current Controversies* (Berkeley: University of California Press, 1982).

5. Alexander and Colomy, "Neofunctionalism Today" (cited in note 1), p. 42.

6. Ibid.; and Alexander and Colomy, "Traditions and Competition" (cited in note 3).

7. Alexander and Colomy, "Neofunctionalism Today" (cited in note 1), p. 41.

8. Colomy, "Metatheorizing in a Postpositivist Frame," pp. 272–273.

9. Alexander and Colomy, "Traditions and Competition" (cited in note 3).

10. Ibid.

11. Ibid.

12. Alexander and Colomy, "Neofunctionalism Today" (cited in note 1); Jeffrey C. Alexander, *The Modern Reconstruction of Classical Thought: Talcott Parsons* (Berkeley: University of California Press, 1983).

13. Talcott Parsons, *The Structure of Social Action* (New York: Free Press, 1937), pp. 27–42.

14. Peter Hamilton, ed., *Readings from Talcott Parsons* (London: Tavistock, 1985), p. 14.

15. Jeffrey C. Alexander, *The Antinomies of Classical Thought: Marx and Durkheim* (Berkeley: University of California Press, 1982), p. 300.

16. Alexander, Positivism, Presuppositions, and Current Controversies (cited in note 3), p. 5.

17. Ibid.

18. Jeffrey C. Alexander, *Twenty Lectures* (New York: Columbia University Press, 1987), pp. 156–194, 215–237; "The Individualist Dilemma in Phenomenology and Interactionism" in Jeffrey C. Alexander, *Actions and Its Environments* (New York: Columbia University Press, 1988), pp. 222–265.

19. Alexander, *The Antinomies of Classical Thought* (cited in note 15), pp. 330–343.

20. Alexander, *Positivism, Presuppositions, and Current Controversies* (cited in note 4), p. 30.

21. Jeffrey C. Alexander, "Action and Its Environments" *in The Micro-Macro Link,* eds. Jeffrey C. Alexander, Bernhard Giesen, Richard Münch, and Neil J. Smelser (Berkeley: University of California Press), pp. 289–318; "Some Remarks on 'Agency' in Recent Sociological Theory," *Perspectives* 15 (1992), pp. 1–4; "Recent Sociology between Agency and Social Structure," *Schweiz. Z. Soziol/Rev. Suisse Sociol.* 18 (1992), pp. 7–11; "More Notes on the Problem of Agency: A Reply," *Schweiz. Z. Soziol/Rev. Suisse Sociol.* 19 (1993), pp. 501–506; Jeffrey C. Alexander and Bernhard Giesen, "From Reduction to Linkage: The Long View of the Micro-Macro Debate" in *The Micro-Macro Link,* pp. 1–42.

22. This overview of Alexander's statement on the micro-macro link draws extensively from Paul Colomy and Gary Rhoades, "Toward a Micro Corrective of Structural Differentiation Theory," *Sociological Perspectives* 37 (1994), pp. 547–583.

23. Alexander, "Action and Its Environments" (cited in note 21), p. 296.

24. We infer this method from a careful reading of Alexander's own work and the more empirical studies of his students and colleagues that appear to have used this method in an almost intuitive way to draw out the implications of Alexander's generalized discourse for their research. See, for example, Edward W. Lehman, "The Theory of the State versus the State of Theory," *American Sociological Review* 53 (1988), pp. 807–823; Duane Champagne, "Transocietal Cultural Exchange, World-System Incorporation, and Geopolitical Competition: Explaining Institutional Change in Eastern Native North America" in *The Dynamics of Social Systems,* ed. Paul Colomy (London: Sage International, 1992), pp. 120–153; Paul Colomy and Gary Rhoades, "Toward a Micro Corrective of Structural Differentiation Theory" (cited in note 22); Laura Desfor Edles, "Rethinking Democratic Transition: A Culturalist Critique and the Spanish Case," *Theory and Society* 24 (1995), pp. 355–384.

25. Jeffrey C. Alexander, "Formal and Substantive Voluntarism in the Work of Talcott Parsons: A Theoretical and Ideological Reinterpretation," *American Sociological Review* 43 (1978), pp. 177–198; "Sociology for Liberals," *New Republic* (June 2, 1979), pp. 10–12; "Paradigm Revision and Parsonianism," *Canadian Journal of Sociology* 4 (1979), pp. 343–357; "Revolution, Reaction, and Reform: The Change Theory of Parsons's Middle Period," *Sociological Inquiry* 51 (1981), pp. 267–280; *The Modern Reconstruction of Classical Thought: Talcott Parsons* (cited in note 12), pp. 128–150.

26. Clifford Geertz, "Ideology as a Cultural System" in *The Interpretation of Cultures* (New York: Basic Books, 1973), pp. 193–233.

27. Jeffrey C. Alexander, "Modern, Anti, Post, and Neo: How Intellectuals Have Coded, Narrated, and Explained the 'New World of Out Time'" in Jeffrey C. Alexander, *Fin de Siecle Social Theory: Relativism, Reduction and the Problem of Reason* (New York: Verso, 1995), pp. 6–64.

28. Jeffrey C. Alexander, "Science, Sense, and Sensibility," *Theory and Society* 15 (1986), pp. 443–463.

29. Jeffrey C. Alexander, *The Modern Reconstruction of Classical Thought: Talcott Parsons* (cited in note 12), pp. 73-118; *Twenty Lectures* (cited in note 18), pp. 89–110.

30. See, for example, Talcott Parsons, *Societies: Evolutionary and Comparative Perspectives* (New York: Free Press, 1966); *The System of Modern Societies* (Englewood Cliffs, NJ: Prentice-Hall, 1971). See also discussion on pages 38 to 39 in Chapter 4.

31. Jeffrey C. Alexander, "Differentiation Theory: Problems and Prospects" in *Differentiation Theory and Social Change: Comparative and Historical Perspectives,* eds. Jeffrey C. Alexander and Paul Colomy (New York: Columbia University Press, 1990), pp. 1–15; "Durkheim's Problem and Differentiation Theory Today" in Alexander, *Action and Its Environments* (New York: Columbia University Press, 1988), pp. 49–77.

32. Jeffrey C. Alexander, "Core Solidarity, Ethnic Out-Groups, and Social Differentiation" in *Differentiation Theory and Social Change,* pp. 267–293; "The Mass News Media in Systemic, Historical, and Comparative Perspective" in *Differentiation Theory and Social Change* (cited in note 31), pp. 323-66.

33. These patterns are discussed at length in the papers assembled in Alexander, *Neofunctionalism* (cited in note 1); in Alexander and Colomy, *Differentiation Theory and Social Change* (cited in note 31); and in Colomy, *The Dynamics of Social Systems* cited in note 24). This argument was initially made by S.N. Eisenstadt, "Social Change, Differentiation, and Evolution," *American Sociological Review* 29 (1964), pp. 235–247.

34. Jeffrey C. Alexander and Paul Colomy, "Institutionalization and Collective Behavior: Points of Contact between Eisenstadt's Functionalism and Symbolic Interactionism" in *Comparative Social Dynamics,* eds. E. Cohen, M. Lissak, and U. Almagor (Boulder, CO: Westview, 1985), pp. 337–345; Jeffrey C. Alexander and Paul Colomy, "Toward Neofunctionalism: Eisenstadt's Change Theory and Symbolic Interactionism," *Sociological Theory* 2 (1985), pp. 11–23. Also see Colomy and Rhoades, "Toward a Micro Corrective of Structural Differentiation Theory" (cited in note 22).

35. Jeffrey C. Alexander, "Modern, Anti, Post, and Neo" (cited in note 27), p. 13.

36. Jeffrey C. Alexander, "Analytic Debates: Understanding the Relative Autonomy of Culture" in *Culture and Society: Contemporary Debates,* eds. Jeffrey C. Alexander and Steven Seidman (Cambridge: Cambridge University Press, 1990), pp. 1–27.

37. Ibid., p. 6.

38. Jeffrey C. Alexander, Philip Smith, and Steven Jay Sherwood, "Risking Enchantment: Theory and Method in Cultural Studies." Forthcoming, p. 10.

39. Jeffrey C. Alexander and Philip Smith, "The Discourse of American Civil Society: A New Proposal for Cultural Studies," *Theory and Society* 22, p. 156.

40. Ibid.

41. Ibid., pp. 156–157.

42. Ibid., p. 157.

43. Jeffrey C. Alexander, "Culture and Political Crisis: Watergate and Durkheimian Sociology" in *Durkheimian Sociology: Cultural Studies,* ed. Jeffrey C. Alexander (Cambridge: Cambridge University Press, 1988), pp. 187–224.

44. Alexander and Smith, "The Discourse of American Civil Society" (cited in note 39), p. 158.

45. Ibid.

46. Alexander, Smith, and Sherwood, "Risking Enchantment" (cited in note 38).

6

The Continuing Tradition II:

The Systems Functionalism of Niklas Luhmann

The grand architecture of the Parsonian functional scheme, as it evolved over a forty-year period at the midcentury, inspired a great amount of criticism, both inside and outside the functionalist perspective. Within the functionalist camp, Robert Merton criticized the Parsonian conceptual edifice for being too abstract and detached from empirical reality—as we saw in Chapter 4. Unlike Merton, however, other functionalists were not so willing to abandon high-level abstractions in favor of empirical assessments of functional needs. Rather, the alternative was to reduce the complexity and rigidity of the conceptual edifice, while maintaining a broad scheme at a high level of abstraction. As we saw in the last chapter, neofunctionalists such as Jeffrey C. Alexander downplayed the notion of functional requisites, although he sustained an interest in the substantive problems raised by Parsons. Yet, others continued to build on Parsons' approach, focusing on the same substantive questions as Parsons, and sustaining a commitment to analyzing these problems within the four-functions paradigm.[1]

In this chapter, we will examine the work of Niklas Luhmann, who took the criticisms about the analytical complexity of Parsons' four functions approach seriously and dropped many particulars of this scheme itself but not the goal of producing abstract frameworks for analyzing social reality. Luhmann studied for a time with Parsons but he eventually criticized Parsonian action theory for being "overly concerned with its own architecture." Luhmann considers himself a systems theorist more than a functionalist, and in recent years the systems aspects of his scheme have been given increasing emphasis.[2] But he is still a functionalist because he tends to analyze system processes by how they meet one master functional requisite: reduction of environmental complexity.[3]

LUHMANN'S GENERAL SYSTEMS APPROACH

System and Environment

Luhmann employs a *general systems* approach to stress that human action becomes organized and structured into systems. When the actions of several people become interrelated, a social system can be said to exist. The basic mechanism by which actions become interrelated to create social systems is communication via symbolic codes, such as words and other media. All social systems exist in multidimensional environments, posing potentially endless complexity with which a system must deal. To exist in a complex environment, therefore, a social system must develop mechanisms for reducing complexity, lest the system simply merge with its environment. These mechanisms involve selecting ways and means for reducing complexity. Such selection creates a boundary between a system and its environment, thereby allowing the system to sustain patterns of interrelated actions.

The basic functional requisite in Luhmann's analysis is thus the need to reduce the complexity of the environment in relation to a system of interrelated actions. All social processes are analyzed with respect to their functions for reducing complexity in relation to an environment. Processes that function in this way are typically defined as *mechanisms* in a manner reminiscent of Talcott Parsons' early discussion in *The Social System*[4] (see pp. 30–34 in Chapter 4). Indeed, the bulk of Luhmann's sociology revolves around discussions of such mechanisms—differentiation, ideology, law, symbolic media, and other critical elements of his scheme.

Dimensions of the Environment

There are three basic dimensions along which the complexity of the environment is reduced by these mechanisms: (1) a temporal dimension, (2) a material dimension, and (3) a symbolic dimension. More than most social theorists, Luhmann is concerned with time as a dimension of the social universe. Time always presents complexity for a system because it reaches into the past, because it embodies complex configurations of acts in the present, and because it involves the vast horizons of the future. Thus a social system must develop mechanisms for reducing the complexity of time. The system must find a way to order this dimension by developing procedures to orient actions to the past, present, and future.[5]

Luhmann is also concerned with the material dimension of the environment—that is, with all the possible relations among actions in potentially limitless physical space. Luhmann always asks these questions: What mechanisms are developed to order interrelated actions in physical space? What is the structure and form of such ordering of relations?

Luhmann visualizes the third dimension of human systems as the symbolic. Of all the complex symbols and their combinations that humans can conceivably generate, what mechanisms select some symbols over others and organize them in some ways as opposed to the vast number of potential alternatives? What kinds of symbolic media are selected and used by a social system to organize social actions?

Thus the mechanisms of a social system that reduce complexity and thereby maintain a boundary between the system and the environment function along three dimensions, the temporal, material, and symbolic. The nature of a social system—its size, form, and differentiations—will be reflected in the mechanisms that the system uses to reduce complexity along these dimensions.

Types of Social Systems

A social system exists any time the actions of individuals are "meaningfully interrelated and interconnected," thereby setting them off from the temporal, material, and symbolic environment by virtue of the selection of functional mechanisms. From such processes come three basic types of social systems: (1) interaction systems, (2) organization systems, and (3) societal systems.[6]

Interaction Systems. An interaction system emerges when individuals are co-present and perceive each other. The very act of perception is a selection mechanism that sorts from a much more complex environment, creating a boundary and setting people off as a system. Such systems are elaborated by the use of language in face-to-face communication, thereby reducing complexity even further along the temporal, material, and symbolic dimensions. For example, Luhmann would ask: How does the language and its organization into codes shape people's perceptions of time? Who is included in the conversation? And what codes and agreements guide conversation and other actions?

Interaction systems reveal certain inherent limitations and vulnerabilities, however. First, only one topic can be discussed at a time, lest the system collapse as everyone tries to talk at once (which of course frequently occurs). Second, the varying conversational resources of participants often lead to competition over who is to talk, creating inequalities and tensions that can potentially lead to conflict and system disintegration. Third, talk and conversation are time-consuming because they are sequential; as a result, an interaction system can never be very complex.

Thus, interaction systems are simple because they involve only those who can be co-present, perceived, and talked to; they are vulnerable to conflict and tension; and they consume a great deal of time. For a social system to be larger and more complex, additional organizing principles beyond perceptions of co-presence and sequential talk are essential.

Organizational Systems. These systems coordinate the actions of individuals with respect to specific conditions, such as work on a specific task in exchange for a specific amount of money. Organizational systems typically have entry and exit rules (for example, come to work for this period of time and leave with this much money), and their main function is to "stabilize highly 'artificial' modes of behavior over a long stretch of time." They resolve the basic problem of reconciling the motivations and dispositions of individuals and the need to get certain tasks done. An organization does not depend on the moral commitment of individuals; nor does it require normative consensus. Rather, the entrance and exit rules specify tasks in ways that allow individuals to do what is required without wholly identifying with the organization.

Organization systems are thus essential to a complex social order. They reduce environmental complexity by organizing people (1) *in time* by generating entrance and exit rules and by ordering activities in the present and future; (2) *in space* by creating a division of labor, which authority coordinates; and (3) *in symbolic terms* by indicating what is appropriate, what rules apply, and what media, such as money or pay, are to guide action. In his delineation of organization systems, Luhmann stresses that complex social orders do not require consensus over values, beliefs, or norms to be sustained; they can operate quite effectively without motivational commitments of actors. Their very strength—flexibility and adaptability to changing environmental conditions—depends on delimited and situational commitments of actors, along with neutral media of communication, such as money.[7]

Societal Systems. These systems cut across interaction and organization systems. A societal system is a "comprehensive system of all reciprocally accessible communication actions."[8] Historically, societal systems have been limited by geopolitical considerations, but today Luhmann sees a trend toward one world society. Luhmann's discussion on the societal system is rather vague, but the general idea can be inferred from his analysis of more specific topics: Societal systems use highly generalized communication codes, such as money and power, to reduce the complexity of the environment. In so doing, they set broad limits on how and where actions are to be interrelated into interaction and organization systems. These systems also organize how time is perceived and how actions are oriented to the past, present, and future.

System Differentiation, Integration, and Conflict

These three systems—interaction, organization, and societal—cannot be totally separated because "all social action obviously takes place in society and is ultimately possible only in the form of interaction."[9] Indeed, in very simple societies they are fused together, but, as societies become larger and more complex, these systems become clearly differentiated from and irreducible to one another. Organizations become distinctive with respect to (1) their functional domains (government, law, education, economy, religion, science), (2) their entrance and exit rules, and (3) their reliance on distinctive media of communication (money, truth, power, love, and so on). As a consequence, they cannot be reduced to a societal system. Interaction systems follow their own laws, for rarely do people strictly follow the guidelines of organizations and society in their conversations.

The differentiation of these systems poses several problems for the more inclusive system. First is the problem of what Luhmann calls "bottlenecks." Interaction systems are slow, sequentially organized patterns of talk, and they follow their own dynamics as people use their resources in conversations. As a result, interaction systems often prevent organizations from operating at high levels of efficiency. As people interact, they develop informal agreements and take their time, with the specific tasks of the organization going unperformed or underperformed. Similarly, as organization systems develop their own structure and programs, their interests often collide, and they become "bottlenecks" to action requirements at the societal level. Second is the problem of conflict in differentiated systems. Interactants may disagree on topics; they may become jealous or envious of those with conversational resources. And because interaction systems are small, they cannot become sufficiently complex "to consign marginals to their borders or to otherwise segregate them." At the organizational level, diverse organizations can pursue their interests in ways that are disruptive to both the organization and the more inclusive societal system.

Yet, countervailing these disruptive tendencies are processes that maintain social integration. One critical set of processes is the "nesting" of system levels inside each other. Actions within an interactive system are often nested within an organization system, and organizational actions are conducted within a societal system. Hence the broader, more inclusive system can promote integration in two ways: (1) It provides the temporal, material, and social premises for the selection of actions; and (2) it imposes an order or structure on the proximate environment around any of its subsystems. For example, an organizational system distributes people in space and in an authority hierarchy; it orients them to time; it specifies the relevant communication codes; and it orders the proximate environment (other people, groupings, offices, and so on) of any interaction system. Similarly, the functional division of a society into politics, education, law, economy, family, religion, and science determines the substance of an organization's action, while it orders the proximate environment of any particular organization. For example, societal differentiation of a distinctive economy delimits what any economic organization can do. Thus a corporation in a capitalist economy will use money as its distinctive communications media; it will articulate with other organizations concerning market relations; it will organize its workers into bureaucratic organizations with distinctive entrance and exit rules ("work for money"); and it will be oriented to the future, with the past as only a collapsed framework to guide present activity in the pursuit of future outcomes (such as profits and promotions).

In addition to these nesting processes, integration is promoted by the deflection of people's activities across different organizations in diverse functional domains. When many organizations exist in a society, none consumes an individual's sense of identity and self because people's energies are dispersed across several organization systems. As a consequence of their piecemeal involvement, members are unlikely to be emotionally drawn into conflict among organization systems, and, when individual members cannot be pulled emotionally into a conflict, its intensity and potential for social disruption are lessened. Moreover, because interaction systems are distinct from the more inclusive organization, any conflict between organizations is often seen by the rank and file as distant and remote to their interests and concerns; it is something "out there" in the environment of their interaction systems, and hence it is not very involving.

Yet another source of conflict mitigation are the entrance and exit rules of an organization. As these become elaborated into hierarchies, offices, established procedures, salary scales, and the like, these rules reduce the relevance of members' conflicts outside the organization—for example, their race and religion. Such outside conflicts are separated from those within the organization, and as a result their salience in the broader societal system is reduced.

Finally, once differentiation of organizations is an established mechanism in a society, specific social control organizations—law, police, courts—can be easily created to mitigate and resolve conflicts. That is, the generation of distinct organizations that are functionally specific represents a new "social technology"; once this technology has been used in one context, it can be applied to additional contexts. Thus the integrative problems created by the differentiation and proliferation of organizations create the very conditions that can resolve these problems—the capacity to create organizations to mediate among organizations.

And so, although differentiation of three system levels creates problems of integration and conditions conducive to conflict, it also produces countervailing forces for integration. In making this argument, Luhmann emphasizes that, in complex systems, order is not sustained by consensus on common values, beliefs, and norms. On the contrary, there is likely to be considerable disagreement about these, except perhaps at the most abstract level. This point of emphasis is an important contribution of Luhmann's sociology, for it distinguishes his theoretical approach from that of Talcott Parsons, who overstressed the need for value consensus in complex social systems. In addition, Luhmann stresses that individuals' moral and emotional attachment to the social fabric is not essential for social integration. To seek a romantic return to a cohesive community, as Émile Durkheim, Karl Marx, and others have, is impossible in most spheres of a complex society. And, rather than viewing this as a pathological state—as concepts like alienation, egoism, and anomie connote—the impersonality and neutrality of many encounters in complex systems can be seen as normal and analyzed less evaluatively. Moreover, people's lack of emotional embeddedness in complex systems gives them more freedom, more options, and more flexibility.[10] This also liberates them from the constraints of tradition, the restrictions of dependency on others, and the indignities of surveillance by the powerful that are so typical of less complex societies.

Communications Media, Reflexivity, and Self-Thematization

Luhmann's system theory stresses the relation of a system to its environment and the mechanisms used to reduce its complexity. All social systems are based on communication among actors as they align their respective modes of conduct. Because action systems are built from communication, Luhmann devotes considerable attention to *communications theory,* as he defines it. He stresses that human communications become reflexive and that this reflexiveness leads to self-thematization. Luhmann thus develops a communications theory revolving around communication codes and media as well as reflexiveness and self-thematization. Each of these elements in his theory will be explored briefly.

Communication and Codes. Luhmann waxes philosophically and metaphorically about these concepts, but in the end he concludes that communication occurs through symbols that signal actors' lines of behavior, and such symbols constitute a code with several properties.[11] First, the organization of symbols into a code guides the selection of alternatives that reduce the complexity of the environment. For example, when someone in an interaction system says that he or she wants to talk about a particular topic, these symbols operate as a code that reduces the complexity of the system in an environment (its members will now discuss this topic and not all the potential alternatives). Second, codes are binary and dialectical in that their symbols imply their opposite. For example, the linguistic code "be a good boy" also implicitly signals its opposite—that is, what is not good and what is not male. As Luhmann notes, "language makes negative copies available" by its very nature. Third, in implying their opposite, codes create the potential for the opposite action—for instance, "to be a bad boy." In human codes, then, the very process of selecting lines of action and reducing complexity with a code also expands potential options (to do just the opposite or some variant of the opposite). This makes the human system highly flexible because the communications codes used to organize the system and reduce complexity also contain implicit messages about alternatives.

Communications Media. Communication codes stabilize system responses to the environment (while implying alternative responses). Codes can organize communication into distinctive media that further order system responses. As a society differentiates into functional domains, distinctive media are used to organize the resources of systems in each domain.[12] For example, the economy uses money as its medium of communication, which guides interactions within and among economic organizations. Thus, in an economy, relations among organizations are conducted in money terms (buying and selling in markets), and intraorganizational relations among workers are guided by entrance and exit rules structured by money (pay for work at specified times and places). Similarly, power is the distinctive communications medium of the political domain; love is the medium of the family; truth, the medium of science; and so on for other functional domains.[13]

Several critical generalizations are implicit in Luhmann's analysis of communications media. First, the differentiation of social systems into functional domains cannot occur without the development of a distinctive medium of communication for that domain. Second, media reduce complexity because they limit the range of action in a system. (For example, love as a medium limits the kinds of relations that are possible in a family system.)[14] Third, even in reducing complexity, media imply their opposite and thus expand potential options, giving systems flexibility (for instance, money for work implies its opposite, work without pay; the use of power implies its opposite, lack of compliance to political decisions).

Reflexivity and Self-thematization. The use of media allows for *reflexivity,* or the capacity to examine the process of action as a part of the action itself. With communications media structuring action, we can use these media to think about or reflect on action. Social units can use money to make money; they employ power to decide how power is to be exercised; they can analyze love to decide what is true love; they can use truth to specify the procedures to get at truth; and so on. Luhmann sees this reflexivity as a mechanism that facilitates adaptation of a system to its environment. Reflexivity does so by ordering responses and reducing complexity, while providing actors in a system with the capacity to think about new options for action. For example, it becomes possible to mobilize power to think about new and more adaptive ways to exercise power in political decisions, as is the case when a society's political elite create a constitutional system based on a separation of powers.

As communications media are used reflexively, they allow for what Luhmann terms *self-thematization.* Using media, a system can conceptualize itself and relations with the environment as a "perspective" or "theme." Such self-thematization reduces complexity by providing guidelines about how to deal with the temporal, material, and symbolic dimensions of the environment. It becomes possible to have a guiding perspective about how to orient to time, to organize people in space, and to order symbols into codes. For example, money and its reflexive use for self-thematization in a capitalist economy create a concern with the future, an emphasis on rational organization of people, and a set of codes that emphasize impersonal exchanges of services and commodities. The consequence of these self-thematizations is that economic organizations reduce the complexity of their environments and, thereby, coordinate social action more effectively.

Luhmann's Basic Approach

In sum, Luhmann's general systems approach revolves around the system versus environment distinction. Systems need to reduce the complexity of their environments in their perceptions about time, their organization of actors in space, and their use of symbols. Processes that reduce complexity are conceptualized as functional mechanisms. There are three types of systems: interaction, organization, and societal. All system processes occur through communications that can develop into distinctive media and allow reflexivity and self-thematization in a system.

LUHMANN'S CONCEPTION OF SOCIAL EVOLUTION

Because Luhmann's substantive discussions are cast into an evolutionary framework, it is wise to begin by extracting from his diverse writings the key elements of this evolutionary approach. Like other evolutionary theorists, Luhmann views evolution as the process of increasing differentiation of a system in relation to its environment.[15] Such increased differentiation allows a system to develop more flexible relations to its environment and, as a result, to increase its level of adaptation. As systems differentiate, however, there is the problem of integrating diverse subsystems; as a consequence, new kinds of mechanisms emerge to sustain the integration of the overall system. But, unlike most evolutionary theorists, Luhmann uses this general image of evolution in a way that adds several new twists to previous evolutionary approaches.

The Underlying Mechanisms of Evolution

Luhmann is highly critical of the way traditional theory has analyzed the process of social differentiation.[16] First, traditional theories—from Marx and Durkheim to Parsons—all imply that there are limits to how divided a system can be, so they all postulate an end to the process, which, in Luhmann's view, is little more than an evaluative utopia. Second, traditional theories overstress the importance of value consensus as an integrating mechanism in differentiated systems. Third, these theories see many processes, such as crime, conflict, dissensus about values, and impersonality, as deviant or pathological; however, they are inevitable in differentiated systems. Fourth, previous theories have great difficulty handling the persistence of social stratification, viewing it as a source of evil or as a perpetual conflict-producing mechanism.

Luhmann's alternative to these evolutionary models is to use his systems theory to redirect the analysis of social differentiation. Like most functionalists, he analogizes to biology, but not to the physiology of an organism; rather, his analogies are to the processes delineated in the theory of evolution. Thus he argues for an emphasis of those processes that produce (1) variation, (2) selection, and (3) stabilization of traits in societal systems.[17] The reasoning here is that sociocultural evolution is like other forms of biological evolution. Social systems have mechanisms that are the functional equivalents of those in biological evolution. These mechanisms generate variation in the structure of social systems, select those variations that facilitate adaptation of a system, and stabilize these adaptive structures.[18]

Luhmann argues that the "mechanism for variation" inheres in the process of communication and in the formation of codes and media. All symbols imply their opposite, so there is always the opportunity to act in new ways (a kind of "symbolic mutation"). The very nature of communication permits alternatives, and at times people act on these alternatives, thereby producing new variations. Indeed, compared with the process of biological mutation, the capacity of human systems for variation is much greater than in biological systems.

The "mechanism for selection" can be found in what Luhmann terms *communicative success*. The general idea behind this concept is that certain new forms of communication facilitate increased adjustment to an environment by reducing its complexity while allowing more flexible responses to the environment. For example, the creation of money as a medium greatly facilitated adaptation of systems and subsystems to the environment, as did the development of centralized power to coordinate activity in systems. And, because they facilitated survival and adaptation, they were retained in the structure of the social organism.

The "stabilization mechanism" resides in the very process of system formation. That is, new communication codes and media are used to order social actions among subsystems, and, in so doing, they create structures, such as political systems and economic orders, that regularize for a time the use of the new communications media. For example, once money is used, it creates an economic order revolving around markets and exchange that, in turn, feeds back and encourages the extension of money as a medium of communication. From this reciprocity ensues some degree of continuity and stability in the economic system.

Evolution and Social Differentiation

Luhmann believes that sociocultural evolution involves differentiation in seven senses.

1. *Evolution is the increasing differentiation of interaction, organization, and societal systems from one another.* That is, interaction systems increasingly become distinct from organization systems, which in turn are more clearly separated from societal systems. Although these system levels are nested in each other, they also have their unique dynamics.
2. *Evolution involves the internal differentiation of these three types of systems.* Diverse interaction systems multiply and become different from one another (for example, compare conversations at work, at a party, at home, and at a funeral). Organization systems increase in number and specialize in different activities (compare economic with political organizations, or contrast different types of economic organizations, such as manufacturing and retail organizations). And the societal system becomes differentiated from the organization and interaction systems that it comprises. Moreover, there is an evolutionary trend, Luhmann claims, toward a one world society.
3. *Evolution involves the increasing differentiation of societal systems into functional domains, such as economy, polity, law, religion, family, science, and education.* Organization subsystems within these domains are specialized to deal with a limited range of environmental contingencies, and, in being specialized, subsystems can better deal with contingencies. The overall result for a societal system is increased adaptability and flexibility in its environment.
4. *Functional differentiation is accompanied by (and is the result of) the increasing use of distinctive media of communication.* For example, organization systems in the economy employ money, those in the polity or government exercise power, those in science depend on truth, and those in the family domain use love.
5. *There is a clear differentiation during evolution among the persons, roles, programs, and values.* Individuals are entities separated from the roles and organizations in which they participate. One plays many roles, and each involves only a segment or part of a person's personality and sense of self; many roles are played with little or no investment of oneself in them. Moreover, most roles persist whether or not any one individual plays them, thereby emphasizing their separation from the person. Such roles are increasingly grouped together into an ever-increasing diversity of what Luhmann calls programs (work, family, play, politics, consumption, and so on) that typically exist inside different kinds of organization systems operating in a distinctive functional domain. In addition, these roles can be shuffled around into new programs, emphasizing

the separation of roles and programs. Finally, societal values become increasingly abstract and general, with the result that they do not pertain to any one functional domain, program, role, or individual.[19] They exist as very general criteria that can be selectively invoked to help organize roles into programs or to mobilize individuals to play roles; however, their application to roles and programs is made possible by additional mechanisms such as ideologies, laws, technologies, and norms. For, by themselves, societal values are too general and abstract for individuals to use in concrete situations. Indeed, one of the most conspicuous features of highly differentiated systems is the evolution of mechanisms to attach abstract values to concrete roles and programs.

6. *Evolution involves the movement through three distinctive forms of differentiation: (a) segmentation, (b) stratification, and (c) functional differentiation.*[20] That is, the five processes outlined earlier have historically created, Luhmann believes, only three distinctive forms of differentiation. When the simplest societies initially differentiate, they do so segmentally in that they create like and equal subsystems that operate very much like the ones from which they emerged. For example, as it initially differentiates, a traditional society will create new lineages, or new villages, that duplicate previous lineages and villages. But segmentation limits a society's complexity and, hence, its capacity to adapt to its environment. And so, alternative forms of differentiation are selected during sociocultural evolution. Further differentiation creates stratified systems in which subsystems vary in their power, wealth, and other resources. These subsystems are ordered hierarchically, and this new form of structure allows more complex relations with an environment but imposes limitations on how complex the system can become. As long as the hierarchical order must be maintained, the options of any subsystem are limited by its place in the hierarchy.[21] Thus pressures build for a third form of differentiation, the functional. Here communication processes are organized around the specific function to be performed for the societal system. Such a system creates inequalities because some functions have more priority for the system (for example, economics over religion). This inequality is, however, fundamentally different from that in hierarchically ordered or stratified systems. In a functionally differentiated society, the other subsystems are part of the environment of any given subsystem—for example, organizations in the polity, law, education, religions, science, and family domains are part of the environment of the economy. And although the economy might have functional priority in the society, it treats and responds to the other subsystems in its environment as equals. Thus inequality in functionally differentiated societies does not create a rigid hierarchy of subsystems; as a consequence, it allows more autonomy of each subsystem, which in turn gives them more flexibility in dealing with their respective environments. The overall consequence of such subsystem autonomy is increased flexibility of the societal system to adjust and adapt to its environment.

7. *Evolutionary differentiation increases the complexity of a system and its relationship with the environment.* In so doing, it escalates the risks, as Luhmann terms the matter, of making incorrect and maladaptive decisions about how to relate to an environment. With increased complexity comes an expanded set of options for a system, but there is a corresponding chance that the selection of options will

be dysfunctional for a system's relationship to an environment. For example, any organization in the economy must make decisions about its actions, but there are increased alternatives and escalated unknowns, resulting in expanded risks. In Luhmann's view, the ever-increasing risk level that accompanies evolutionary differentiation must be accompanied by mechanisms to reduce risk, or at least by the perception or sense that risk has been reduced. Thus evolution always involves an increase in the number and complexity of risk-reducing mechanisms. Such mechanisms also decrease the complexity of a system's environment because they select some options over others. For example, a conservative political ideology is a risk-reducing mechanism because it selects some options from more general values and ignores others. In essence, an ideology assures decision makers that the risks are reduced by accepting the goals of the ideology.[22]

Before proceeding further, we should review these elements of Luhmann's view of evolution and how they change a society's and its constituent subsystem's relation to the temporal, material, and symbolic dimensions of the environment. Temporally, Luhmann argues that social evolution and differentiation lead to efforts at developing a chronological metric, or a standardized way to measure time (clocks, for example). Equally fundamental is a shift in people's perspective from the past to the future. The past becomes highly generalized and lacks specific dictates of what should be done in the present and the future. For, as systems become more complex, the past cannot serve as a guide to the present or future because there are too many potential new contingencies and options. The present sees time as ever more scarce and in short supply; thus people become more oriented to the future and to the consequences of their present actions. Materially, social differentiation involves (1) the increasing separation of interaction, organization, and societal systems; (2) the compartmentalization of organization systems into functional domains; (3) the growing separation of person, role, program, and values; and (4) the movement toward functional differentiation and away from segmentation and stratification. And, symbolically, communication codes become more complex and organized as distinctive media for a particular functional domain. Moreover, they increasingly function as risk-reducing mechanisms for a universe filled with contingency and uncertainty.

Luhmann has approached the study of specific organizational systems from this overall view of sociocultural evolution and differentiation. As he has consistently argued, an analytical framework is only as good as the insights into empirical processes that it can generate. Luhmann's framework is much more complex than he contends such a framework should be, and it is often more metaphorical than analytical. Yet, it allows him to analyze political, legal, and economic processes in functionally differentiated societies in very intriguing ways.

THE FUNCTIONAL DIFFERENTIATION OF SOCIETY

Politics as a Social System

As societies grow more complex, new structures emerge for reducing complexity. Old processes, such as appeals to traditional truths, mutual sympathy, exchange, and barter, become ever more inadequate. A system that reaches this point of differentiation, Luhmann argues, must develop the "capacity to make binding decisions." Such capacity is generated from the problems of increased complexity, but this capacity also becomes an important condition for further differentiation.

To make binding decisions, the system must use a distinctive medium of communication: power.[23] Power is defined by Luhmann as "the possibility of having one's own decisions select alternatives or reduce complexity for others." Thus, whenever one social unit selects alternatives of action for other units, power is being employed as the medium of communication.

The use of power to make binding decisions functions to resolve conflicts, to mitigate tensions, and to coordinate activities in complex systems. Societies that can develop political systems capable of performing these functions can better deal with their environments. Several conditions, Luhmann believes, facilitate the development of this functional capacity. First, there must be time to make decisions; the less time an emerging political system is allowed, the more difficulty it will have in becoming autonomous. Second, the emerging political system must not confront a single power block in its environment, such as a powerful church. Rather, it requires an environment of multiple subsystems whose power is more equally balanced. So, the more the power in the political subsystem's environment is concentrated, the more difficult is its emergence as an autonomous subsystem. Third, the political system must stabilize its relations with other subsystems in the environment in two distinctive ways: (1) at the level of diffuse legitimacy, so that its decisions are accepted as its proper function; and (2) at the level of daily transactions among individuals and subsystems.[24] That is, the greater the problems of a political system are in gaining diffuse support for its right to make decisions for other subsystems and the less salient the decisions of the political system are for the day-to-day activities, transactions, and routines of system units, then the greater will be its problems in developing into an autonomous subsystem.

Thus, to the extent that a political system has time to develop procedures for making decisions, confront multiple sources of mitigated power, and achieve diffuse legitimacy as well as relevance for specific transactions, then the more it can develop into an autonomous system and the greater will be a society's capacity to adjust to its environment. In so developing, the political system must achieve what Luhmann calls *structural abstraction*, or the capacity to (1) absorb multiple problems, dilemmas, and issues from a wide range of system units and (2) make binding decisions for each of these. Luhmann sees the political system as "absorbing" the problems of its environment and making them internal to the political system. Several variables, he argues, determine the extent to which the political system can perform this function: (1) the degree to which conflicts are defined as political (instead of moral, personal, and so forth) and therefore in need of a binding decision, (2) the degree of administrative capacity of the political system to coordinate activities of system units, and (3) the degree of structural differentiation within the political system itself.

This last variable is the most crucial in Luhmann's view. In response to environmental complexity and the need to absorb and deal with problems in this environment, the political system must differentiate along three lines: (1) the creation of a stable bureaucratic administration that executes decisions, (2) the evolution of a separate arena for politics and the emergence of political parties, and (3) the designation of the public as a relevant concern in making binding decisions. Such internal differentiation increases the capacity of the political system to absorb and deal with a wide variety of problems; as a consequence, it allows greater complexity in the societal system.

This increased complexity of the political and societal systems also increases the risks of making binding decisions that are maladaptive. As complexity increases, there are always unknown contingencies. Therefore, not only do political systems develop mechanisms such as internal differentiation for dealing with complexity, but they also develop mechanisms for reducing risk or the perception of risk. One mechanism is the growing reflexiveness of the political process—that is, its increased reflection on itself. Such reflection is built into the nature of party politics where the manner and substance of political decisions are analyzed and debated. Another mechanism is what Luhmann calls the *positivation of law,* or the creation of a separate legal system that makes "laws about how to make laws" (more on this in the next section). Yet another mechanism is ideology or symbolic codes that select which values are relevant for a particular set of decisions. A related mechanism is the development of a political code that typifies and categorizes political decisions into a simple typology.[25] For example, the distinction between progressive and conservative politics is, Luhmann argues, an important political code in differentiated societies. Such a code is obviously very general, but this is its virtue because it allows very diverse political acts and decisions to be categorized and interpreted with a simple dichotomy, thereby giving political action a sense of order and reducing perceptions of risk. Luhmann even indicates that it is a system's capacity to develop a political code, more than consensus of values, that leads to social order. For in interpreting actions in terms of the code, a common perspective is maintained, but it is a perspective based on differences—progressive versus conservative—rather than on commonality and consensus. Thus, complex social orders are sustained by their very capacity to create generalized and binary categories for interpreting events rather than by value consensus.

Still another mechanism for reducing risks is arbitrary decision making by elites. However, although such a solution achieves order, it undermines the legitimacy of the political system in the long run because system units start to resent and to resist arbitrary decision making. And a final mechanism is invocation of a traditional moral code (for example, fundamentalistic religious values) that, in Luhmann's terms, "remoralizes" the political process. But when such remoralization occurs, the political system must de-differentiate because strict adherence to a simple moral code precludes the capacity to deal with complexity (an example of this process would be Iran's return to a theocracy from its previously more complex political system).

In sum, then, it is fair to say that Luhmann uses his conceptual metaphor to analyze insightfully specific institutional processes, such as government. Yet he does not use his scheme in a rigorous deductive sense; much like Parsons before him, he employs the framework as a means for denoting and highlighting particular social phenomena. Although much of his analysis of political system differentiation is "old wine in new bottles," there is a shift in emphasis and, as a result, some intriguing but imprecise insights. In a similar vein, Luhmann analyzes the differentiation of the legal system and the economy.

The Autonomy of the Legal System

As discussed earlier, Luhmann visualizes social evolution as involving a separation of persons, roles, programs, and values. For him, differentiation of structure occurs at the level of roles and programs. Consequently, there is the problem of how to integrate values and persons into roles organized into programs within organization systems. The functional mechanism for mobilizing and coordinating individuals to play roles is law, whereas the mechanism for making values relevant to programs is ideology.[26] Thus, because law regulates and coordinates people's participation in roles and programs and because social differentiation must always occur at the roles level, it becomes a critical subsystem if a society is to differentiate and evolve. That is, a society cannot become complex without the emergence of an autonomous legal system to specify rights, duties, and obligations of people playing roles.[27]

A certain degree of political differentiation must precede legal differentiation because there must be a set of structures to make decisions and enforce them. But political processes often impede legal autonomy, as is the case when political elites have used the law for their own narrow purposes. For legal autonomy to emerge, therefore, political development is not enough. Two additional conditions are necessary: (1) "the invocation of sovereignty," or references by system units to legal codes that justify their communications and actions; and (2) "lawmaking sovereignty," or the capacity of organizations in the legal system to decide just what the law will be.

If these two conditions are met, then the legal system can become increasingly reflexive. It can become a topic unto itself, creating bodies of procedural and administrative law to regulate the enactment and enforcement of law. In turn, such procedural laws can themselves be the subject of scrutiny. Without this reflexive quality, the legal system cannot be sufficiently flexible to change in accordance with shifting event in its environment. Such flexibility is essential because only through the law can people's actions be tied to the roles that are being differentiated. For example, without what Luhmann calls the "positivization of law," or its capacity to change itself in response to altered circumstances, new laws and agencies (for example, workers' compensation, binding arbitration of labor and management disputes, minimum wages, health and safety) could not be created to regulate people's involvement in roles (in this case, work roles in a differentiating economy).

Thus, positivization of the law is a critical condition for societal differentiation. It reduces complexity by specifying relations of actors to roles and relegating cooperation among social system units. But it reduces complexity in a manner that presents options for change under new circumstances; thus it becomes a condition for the further differentiation of other functional domains, such as the economic.

The Economy as a Social System

Luhmann defines the economy as "deferring a decision about the satisfaction of needs while providing a guarantee that they will be satisfied and so utilizing the time thus acquired."[28] The general idea seems to be that economic activity—production and distribution of goods and services—functions to satisfy basic or primary needs for food, clothing, and shelter as well as derived or secondary needs for less basic goods and services. But this happens in a way not fully appreciated in economic analysis: The economic activity restructures humans' orientation to time because economic action is oriented to the satisfaction of future needs. Present economic activity is typically directed at future consumption; so, when a person works and a corporation acts in a market, they are doing so to guarantee that their future will be unproblematic.

Luhmann's definition of the economic subsystem is less critical than his analysis of the processes leading to the creation of an autonomous economic system in society. In traditional and undifferentiated societies, Luhmann argues, only small-scale solutions are possible with respect to doing something in the present to satisfy future needs. One solution is stockpiling of goods, with provisions for the redistribution of stocks to societal members or trade with other societies.[29] Another solution is mutual assistance agreements among individuals, kin groups, or villages. But such patterns of economic organization are very limited because they merge familial, political, religious, and community activity. Only with the differentiation of distinctly economic roles can more complexity and flexibility be structured into economic action. The first key differentiation along these lines is the development of markets with distinctive roles for buyers and sellers.

A market performs several crucial functions. First, it sets equivalences or the respective values of goods and services. Second, it neutralizes the relevance of other roles—for instance, the familial, religious, and political roles of parties in an exchange. Value is established by the qualities of respective goods, not by the positions or characteristics of the buyers and sellers.[30] Third, markets inevitably generate pressures for a new medium of communication that is not tied to other functional subsystems. This medium is money, and it allows quick assessments of equivalences and value in an agreed-on metric. In sum, then, markets create the conditions for the differentiation of distinctly economic roles, for their separation and insulation from other societal roles, and for the creation of a uniquely economic medium of communication.

Money is a very unusual medium, Luhmann believes, because it "transfers complexity." Unlike other media, money is distinctive because it does not reduce complexity in the environment. For example, the medium of power is used to make decisions that direct activity, thereby reducing the complexity of the environment. The medium of truth in science is designed to simplify the understanding of a complex universe. And the medium of love in the family circumscribes the actions and types of relations among kindred and, in so doing, reduces complexity. In contrast, money is a neutral vehicle that can always be used to buy and sell many different things. It does not limit; it opens options and creates new opportunities. For example, to accept money for a good or for one's work does not reduce the seller's or worker's options. The money can be used in many different ways, thereby preserving and even increasing the complexity of the environment. Money thus sets the stage for—indeed it encourages—further internal differentiation in the economic subsystem of a society.

In addition to transferring complexity, Luhmann sees money as dramatically altering the time dimension of the environment. Money is a liquid resource that is always "usable in the future." When we have money, it can be used at some future date—whether the next minute or the following year. Money thus collapses time, because it is to be used in the future, hence making the past irrelevant; and the present is defined by what will be done with money in the future. However, this collapsing of time can come about only if (1) money does not inflate over time, and (2) it is universally used as the medium of exchange (that is, barter, mutual assistance, and other traditional forms of exchange do not still prevail).[31]

Like all media of communication, money is reflexive. It becomes a goal of reflection, debate, and action itself. We can buy and sell money in markets; we can invest money to make more money; we can condemn money as the root of evil or praise it as a goal that is worth pursuing; we can hoard it in banks or spread it around in consumptive activity. This reflexive quality of money, coupled with its capacity to transfer complexity and reorient actors to time, is what allows money to become an ever more dominant medium of communication in complex societies. Indeed, the economy becomes the primary subsystem of complex societies because its medium encourages constant increases in complexity and growth in the economic system. As a consequence, the economy becomes a prominent subsystem in the environment of other functional subsystems—that is, science, polity, family, religion, and education. The economy becomes something that must always be dealt with by these other subsystems.

This growing complexity of the economic subsystem increases the risks in human conduct. The potential for making a mistake in providing for a person's future needs or a corporation's profits increases because the number of unknown contingencies dramatically multiplies. Such escalated risks generate pressures, Luhmann argues, for their reduction through the emergence of specific mechanisms. The most important of these mechanisms is the tripart internal differentiation of the economy around (1) households, (2) firms, and (3) markets.[32] There is a "structural selection" for this division, Luhmann believes, because these are structurally and functionally different. Households are segmental systems (structurally the same) and are the primary consumption units. Firms are structurally diverse and the primary productive units. And markets are not as much a unit as a set of processes for distributing goods and services. Luhmann is a bit vague on this point, but it seems that there is strength in this correspondence of basically different structures with major economic functions. Households are segmented structurally and are functionally oriented to consumption; firms are highly differentiated structurally and are functionally geared to production; and markets are processually differentiated by their function to distribute different types of goods and services. Such differentiation reduces complexity, but at the same time it allows flexibility: Households can change consumption patterns, firms can alter production, and markets can expand or contract. And, because they are separated from one another, each has the capacity to change and redirect its actions, independent of the others. This flexibility is what allows the economic system to become so prominent in modern industrial societies.

Yet, Luhmann warns, the very complexity of the economy and its importance for other subsystems create pressures for other risk-reducing mechanisms. One of these is intervention by government so that power is used to make binding decisions on production, consumption, and distribution as well as on the availability of money as a medium of communication. The extensive use of this mechanism, Luhmann believes, reduces risk and complexity in the economy at the expense of its capacity to meet needs in the future and to make flexible adjustments to the environment.

NOTES

1. For the best of these efforts, see Richard Münch's work: *Theory of Action: Towards a New Synthesis Going Beyond Parsons* (London: Routledge, 1988); *Die Struktur der Modene* (Frankfurt am Main: Suhrkamp, 1984).

2. See, for example, Niklas Luhmann, *Systems Theory* (Stanford, CA: Stanford University Press, 1995).

3. Luhmann has published extensively, but most of his work is in German. The best sample of his work in English is *Systems Theory* and *The Differentiation of Society*, trans. S. Holmes and C. Larmore (New York: Columbia University Press, 1982).

4. Talcott Parsons, *The Social System* (New York: Free Press, 1951).

5. Luhmann, *The Differentiation* (cited in note 3), Chap. 12.

6. Ibid., pp. 71–89.

7. In making this assertion, Luhmann directly attacks Parsons. Luhmann, *The Differentiation*, Chapter 3 (cited in note 3).

8. Ibid., p. 73.

9. Ibid., p. 79.

10. Here Luhmann takes a page from Georg Simmel's *The Philosophy of Money*, trans. T. Bottomore and D. Frisby (Boston: Routledge & Kegan Paul, 1978).

11. Luhmann, *The Differentiation* (cited in note 3), p. 169.

12. Obviously Luhmann is borrowing Parsons' idea about generalized media. See Chapter 4.

13. Much like Parsons, this analysis of communications media is never fully explicated or systematically discussed for all functional domains.

14. We will see shortly, however, that money is the sole exception here.

15. This is essentially Parsons' definition (see Chapter 4). It was Spencer's and Durkheim's as well (see Chapter 2).

16. Luhmann, *The Differentiation* (cited in note 3), pp. 256–257.

17. Luhmann's interpretation of the synthetic theory of evolution in biology is, at best, loose and inexact.

18. Luhmann, *The Differentiation* (cited in note 3), p. 265. Luhmann seems completely unaware that Herbert Spencer in his *The Principles of Sociology* (New York: D. Appleton, 1885; originally published in 1874) performed a similar, and more detailed, analysis 100 years ago.

19. Luhmann is borrowing here from Émile Durkheim's analysis in *The Division of Labor in Society* (New York: Free Press, 1949; originally published in 1893) as well as from Talcott Parsons' discussion of "value generalization." See Chapter 4.

20. Luhmann, *The Differentiation* (cited in note 3), pp. 229–254.

21. Ibid., p. 235.

22. Ibid., p. 151.

23. Ibid., p. 151.

24. Ibid., pp. 143–144.

25. Ibid., pp. 168–189.

26. Ibid., pp. 90–137.

27. This is essentially the same conclusion Parsons reached in his description of evolution in *Societies: Evolutionary and Comparative Perspectives* and *The System of Modern Societies* (Englewood Cliffs, NJ: Prentice-Hall, 1966 and 1971, respectively).

28. Luhmann, *The Differentiation* (cited in note 3), p. 194.

29. Ibid., p. 197.

30. Luhmann fails to cite the earlier work of Georg Simmel on these matters. See Simmel's *The Philosophy of Money* (cited in note 10).

31. Luhmann, *The Differentiation* (cited in note 3), p. 207.

32. Ibid., p. 216.

PART II

Evolutionary Theorizing

7

The Emerging Tradition: The Rise of Evolutionary Theorizing*

Sociology was born in the nineteenth century when biology was fast becoming the dominant science. The idea of evolution was in the air, and despite the power of religious orthodoxy to repress nonbiblical speculation about the origins of humans, the Age of Science had its own powerful momentum. Thoughts about the mechanisms accounting for the great diversity of species on earth, including humans, were not going to be repressed, whatever the risks. Indeed, one key mechanism was staring scholars in the face because, after all, animal and plant breeders had been engaged in unnatural selection for millennia.

When Charles Darwin, under pressure of being scooped by Alfred Wallace, finally published his decades-old conclusions in *On the Origin of Species*,[1] biology was changed forever. The idea of evolution by *natural selection* not only provided a simple explanation for speciation, it also led to a search for other mechanisms creating the variations on which natural selection works—mechanisms already discovered by Gregor Mendel[2] but that had to be rediscovered decades later. Equally significant, theorizing in sociology began to borrow ideas from biology and, relatedly, what became known later as ecology.

This borrowing still occurs, and indeed, there is even more excitement at the close of the twentieth century about the prospects for a biologically informed evolutionary sociology than there was at the end of the nineteenth century.[3] The key figure in the emergence of an evolutionary theory was, of course, Darwin, but closer to home, Herbert Spencer and Émile Durkheim more directly inspired evolutionary sociology. Let us see what each contributed to what is now reemerging as an important new form of theorizing in sociology.

*This chapter is coauthored with Alexandra Maryanski.

HERBERT SPENCER AS THE FIRST SOCIOLOGICAL BIO-ECOLOGIST

Almost a decade before Darwin published *On the Origin of Species,* Spencer coined the phrase "survival of the fittest."[4] He used this phrase in a moral and philosophical sense, arguing that the best forms of social organization emerge with unregulated competition among humans, which allows the most fit to survive, thereby elevating the level of society. Obviously, this moral philosophy is highly flawed, but it became a significant argument in the nineteenth and early twentieth centuries; indeed, what became known as Social Darwinism[5] was actually more Spencerian in tone. Indeed, Darwin would have turned over in his grave if he knew the extremes to which Spencer's moral philosophy had gone under his name.

Darwin acknowledged Spencer in the preface to his famous book, in which the basic idea of natural selection as the driving force of speciation is developed, but this acknowledgment was perhaps only a courtesy to a well-known social philosopher who was friends with many of Darwin's supporters. Thus, even though Spencer's original phrasing of the argument of evolution as survival of the fittest preceded Darwin's formulation, at least in print, it had little direct impact on the reemergence of biologically inspired thinking in the second half of the twentieth century. Yet, Spencer had made the connection between biology and sociology.

Of course, we must remember that twenty years before Spencer's publication of his moral philosophy, and more than forty years before Spencer published his first sociological treatises, Auguste Comte[6] had allied sociology with biology, arguing that in the *hierarchy of the sciences,* sociology would emerge from biology and become the "queen science." Comte was the key figure to reintroduce organismic analogies to sociology, by seeking to find the counterparts of cells, tissues, and organs of biological organisms in the structure of society. As we saw in Chapter 2, this argument evolved into functionalism, where theory focuses on the consequences of a social phenomenon for the larger "body social" in which it operates. Much later in the 1870s, when Spencer finally turned to sociology[7] (after writing treatises on morals,[8] physics,[9] biology,[10] and psychology[11]), he made the comparison between social and biological organisms more explicit (see page 10 for a review of these comparisons). But Spencer did more than make superficial analogies between biological and social bodies, he proclaimed that *sociology* was to be the study of *superorganic* organisms—that is, relations among living organisms—and he included more than human organisms in this definition. All species that organize, or create the *superorganic,* were to be the topic of sociology—an idea that has been rediscovered just recently. In making these analogies and assertions about sociology's subject matter, Spencer also developed several modes of analysis that influenced sociological theory in the nineteenth and early twentieth centuries. In turn, these lines of influence were, indirectly and often unacknowledged, to shape the reemergence of biologically oriented sociological theory.

One point of emphasis was functional, as we saw in Chapter 2, and need only be briefly mentioned here[12]: A social system can be analyzed for how its respective parts sustain the system in its physical, organic, and social environment. There were, in Spencer's view, several basic functions that all systems, whether biological or social, had to meet if they are to survive: the *production* of life-sustaining substances, the *reproduction* of system parts, the *regulation* and control of actions by system parts, and the *distribution* of information and materials among system units. As this kind of argument evolved into functionalism, it tended to bring other analogies from biology.

Among these analogies, which social thinkers had maintained for more than a century, was the idea of social evolution as moving in a certain direction—an idea that early figures of The Enlightenment had phrased as "progress." Society was seen progressing to an ever better state—a view held not only by eighthteenth century moral philosophers but also by virtually all early sociologists, except perhaps for Max Weber. But Spencer gave this moralistic argument a more sophisticated ring: The evolution of society involves increasing complexity of social structure and associated cultural symbols, and this complexity increases the capacity of the human species to adapt and survive in its environment. This argument was picked up by early and modern-day functionalists, as we saw in Chapters 4 through 6, but it was carried forward in many alternative approaches analyzing the forces guiding the adaptation of human populations to their respective environments. Some of these alternative approaches carry elements of functional analysis, but most are more purely ecological and borrow from biology proper, as we will see in this section of chapters on biologically oriented evolutionary theorizing.

Another point of emphasis in Spencer's work, but one that was more fully developed by Émile Durkheim twenty years later in 1893, is the Darwinian analogy: Social differentiation, or "social speciation," is the result of competition among actors for resources; from such competition comes differentiation as those most fit to secure resources in a niche win out, whereas those actors who are less fit change (or die) and seek resources in other resource niches.[13] This process of seeking resource niches is, then, the driving force of social differentiation and, hence, of societal evolution. For Spencer, growth in size of a population increases competition for resources—an idea that Spencer borrowed from Thomas Malthus' famous essay on population,[14]—and sets into motion the selection processes that cause social differentiation and societal evolution.

A related point in Spencer's sociology blends this evolutionary argument into a theory of geopolitics.[15] More complex societies generally win wars, Spencer argued, because they are better organized; as they vanquish or colonize their adversaries, the overall efficiency of social organization and the complexity of the social universe increase. For this reason Spencer saw war as an important force in societal evolution, because war had historically pitted societies against each other in a kind of Darwinian struggle, but at the same time, Spencer argued that war had now (that is, at the turn of the nineteenth into twentieth century) served its purpose and would, he believed, be a hindrance to further evolution. Instead, competition in markets should be allowed to increase social complexity and societal fitness rather than war, which, he argued, discouraged economic activity, biased production toward purely military ends, generated concentrations of power that overregulated society, and diminished human freedom and innovation.

What we see in Spencer's work, then, is a concern with biological modes of thinking that were emerging throughout the nineteenth century. Spencer himself had, of course, written a multivolume treatise on biology,[16] but far more important for sociological theory was his incorporation of biological analogies and metaphors into thinking about the dynamics of society. Thus, where Comte had been a bit vague in his pronouncements about the connection between biology and sociology, Spencer added considerably more substance and detail to how sociological and biological theories could be blended together.

ÉMILE DURKHEIM'S BIO-ECOLOGICAL ANALOGY

Today, we tend to forget how important a figure Spencer was, not just in the nineteenth century but in America well into the twentieth century. Today, we see Durkheim as a far more imposing figure, and we often give him credit for ideas that, in reality, come from Spencer.[17] But Durkheim made the connection between the Darwinian idea of evolution by natural selection and societal evolution more explicit than Spencer. Thus, as bio-ecological theorizing and evolutionary theory were developing in the early twentieth century, Durkheim became as significant a contributor as Spencer was.

Durkheim's first major work was *The Division of Labor in Society*[18] and, as we saw in Chapter 2, it served as an inspiration for functional theorizing. In analyzing the causes of the division of labor—that is, what forces increase specialization of activities (or social "speciation")—Durkheim began to draw from his reading of Darwin:[19]

> Thus, Darwin says that in a small area, opened to immigration, and where, consequently, the conflict of individuals must be acute, there is always to be seen a very great diversity in the species inhabiting it.

Durkheim then noted that such diversity made the survival of each species less problematic and, indeed, contributed to the well-being of each species. And so, Durkheim posited,[20]

> Men submit to the same law. In the same city, different occupations can coexist without being obliged mutually to destroy one another, for they pursue different objects. The soldier seeks military glory, the priest moral authority, the statesman power, the businessman riches, the scholar scientific renown. Each of them can attain this end without preventing the others from attaining theirs.

Durkheim was seeking an answer to a question that had vexed all social theory since the time of Adam Smith (who, in the eighteenth century, created the basic utilitarian ideas that so dominate economics today): If societies are differentiating, what force is to hold them together? Durkheim's answer became part of his functionalism, but in the previous passage, we can see his effort to paint the division of labor in a benign light. For our purposes, Durkheim's ecological model on the causes of the division of labor is more important than his ultimate answer about what holds society together. Figure 7-1 presents his basic model. In this model, Durkheim argues that those forces that increase the *material density* of a population—forces such as immigration, population growth, and ecological barriers—and those that reduce the "social space" between individuals—forces like improved transportation and communication technologies—all increase *competition*. Such competition, in turn, leads to *social speciation* or the division of labor, which, Durkheim felt, reduced competition and increased cooperation as individuals in different social niches went their own way while exchanging resources with each other.

In a manner more explicit than Spencer, then, Durkheim argued that population density increases competition for resources, which then leads to social differentiation, but like Spencer, Durkheim argued that this mechanism is ultimately what is responsible for social evolution of society from simple to more complex forms. That is, as populations grow or the social space among members is reduced by new technologies, competition escalates; from competition comes social differentiation and increased societal complexity.

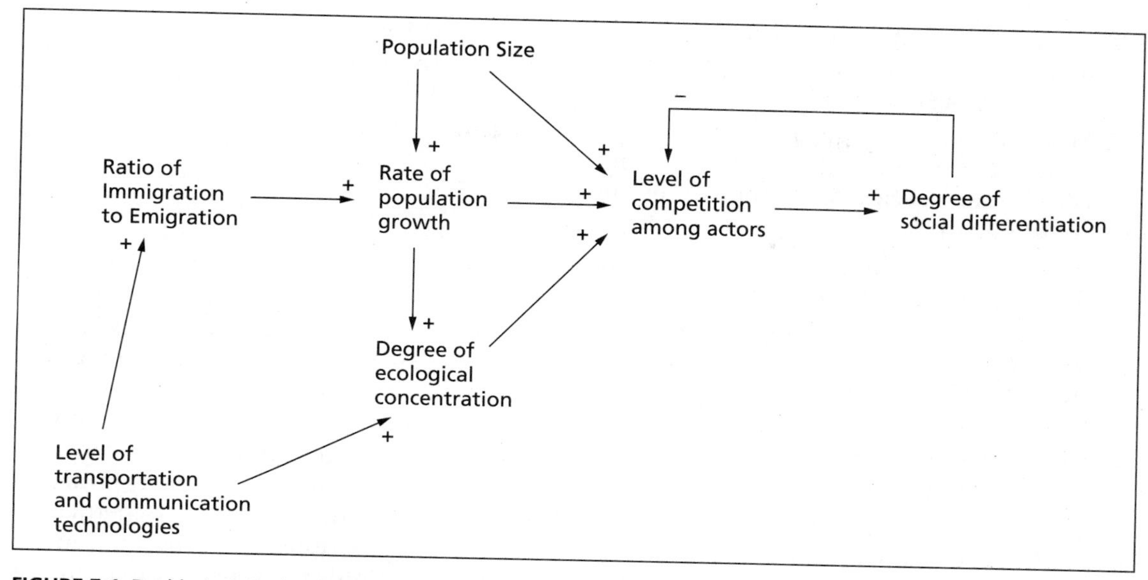

FIGURE 7-1 Durkheim's Ecological Model of Social Differentiation

CHARLES DARWIN AND NATURAL SELECTION

We have alluded to Darwin, but let us review how his ideas set the framework not only for nineteenth century evolutionary theorizing but, more significantly, for the revival of evolutionary theory in the twentieth century. Like Spencer, Darwin was greatly influenced by Malthus' essay on population, especially Malthus' "notion of a natural elimination mechanism." In his early notebooks,[21] for example, Darwin considered the power of the Malthusian force when he wrote that "no structure will last . . . without it(s) . . . adaptation to *whole* life . . . it will decrease and be driven outwards in the grand crush of population," and ". . . there is a contest . . . and a grain of sand turns the balance." Coupled with Darwin's observations as the naturalist on the famous voyage of the *Beagle*, where it became clear to him that somewhat different environments create new variants of species, Darwin was led to the concept of *natural selection*. This was to be the title of his book, but he later settled on the more controversial title *On the Origin of Species*.[22] Darwin held the idea of natural selection for twenty years before he published his treatise; only when Wallace,[23] who had also been inspired by Malthus, came up with a similar idea to explain evolution did Darwin begin to publish his ideas. Darwin's and Wallace's initial papers were read jointly at the Royal Society, and both argued that the natural world reveals a "struggle" among species and that this competition "selects" for survival those organisms that are better equipped to adapt to the conditions imposed by an environment. As a result, these organisms are more likely than less fit organisms to produce offspring. Thus, the basic assumptions of Darwin's (and Wallace's) model are as follows:

1. Members of any given species reveal variations in their physical and behavioral traits.

2. Members of any given species tend to produce more offspring than can be supported by an environment.
3. Members of any given species must therefore compete with one another and with other species for resources in an environment.
4. Members of a given species revealing those traits that enable them to compete and secure resources will be more likely to survive and produce offspring, whereas those members evidencing traits that are unsuited to competition or ability to secure resources will be less likely to survive and produce offspring.

Thus the environment "selects" those traits of organisms that enable them to compete, secure resources, survive, and reproduce. Evolution is driven by this process of natural selection, as those varying traits of organisms are successively selected by the environment. In this line of argument, an important distinction must be emphasized: Selection operates on an *individual organism,* whereas evolution involves a *population of organisms*. That is, the individual organism survives and reproduces, or fails to do so, but the population as a whole evolves. This population consists of all those individual organisms that possess those traits that enabled them to survive and reproduce. This distinction is critical for understanding modern evolutionary theory. We must remember, then, that as different individual organisms revealing varying traits are "selected for" by the conditions of the environment, the overall composition of the population of organisms changes, or evolves.

We can see an affinity of these ideas to Spencer's famous phrase "survival of the fittest," but the most direct source of inspiration for modern evolutionary theory in the social sciences was the blending of Darwin's discovery of the mechanism of natural selection with Mendel's overlooked insights into the mechanisms of inheritance. For, despite the power of the concept of natural selection to explain evolution, Darwin's portrayal could not answer some basic questions: What is the source of those variations in organisms that are the objects of natural selection? How are traits passed on from one organism to another? What is the mechanism of inheritance? The answers to these questions sat for thirty-five years in dusty academic journals in the form of Mendel's short manuscripts on the inheritance of characteristics in garden peas.[24] But not until 1900 was the science of genetics born with the rediscovery of Mendel's work and its independent confirmation by Karl Correns in Germany, Hugo de Vries in Holland, and Erich Tschermak in Austria. This synthesis of the notions of natural selection and genetic variation was coupled with additional insights into the forces of evolution in what is called the "modern synthesis." This initial synthesis was, as we will see, crucial to reconnecting biology and sociology in the twentieth century.

CONCLUSION

Biological reasoning in sociology has come and gone in several distinct waves, each time resurfacing in a somewhat new guise. There have, in essence, been three basic theoretical approaches in sociology that have derived inspiration from biology. One is ecological and emphasizes the process of competition among social actors for limited resources. Another is based more on genetics and emphasizes the effects of genic selection on the nature of human behavior and social organization. The third overlaps with functional theories of evolution and stresses that the long-term trend in human societies has been for increased complexity and differentiation.

More significant, as we will examine in the last chapter of this section, a fourth direction that blends these three approaches together into a more unified framework. This last effort at synthesis is the future of evolutionary sociology as theoretical sociology approaches the twenty-first century, where, as in the nineteenth century, a revolution in biology—in this case, the biotechnologies that come with genetic engineering—will make biology once again the dominant science. Perhaps, this time around, Comte's rather fanciful belief that sociology would come to dominate biology can be made more realistic: Biological thinking can expand the explanatory power of sociological theories. In a sense, this is just what Spencer and Durkheim sought to do, and, as we will see in the next chapter, what others in the early to mid-decades of this century also attempted to accomplish. They had the basic insight, but they lacked the tools to execute their insights fully. Today, after a century of grappling with bio-ecological processes and their implications for patterns of social organization, sociology is poised to take these earlier efforts much further.

NOTES

1. Charles Darwin, *On the Origin of Species* (New York: New American Library, 1958; originally published 1859).

2. Gregor Mendel, "Versuche über pflanzenhybriden," translated into English in the *Journal of the Rural Horticulture Society* 26 (1901), originally published in 1865.

3. Alexandra Maryanski, "The Pursuit of Human Nature in Sociobiology and Evolutionary Sociology," *Sociological Perspectives* 37 (1994), pp. 375–390; see also Jonathan H. Turner and Alexandra Maryanski, "The Biology of Human Organization," *Advances in Human Ecology* 2 (1993), pp. 1–33.

4. Herbert Spencer, *Social Statics; or the Conditions Necessary for Human Happiness Specified, and First of Them Developed* (New York: Appleton, 1888; originally published in 1852). Spencer had used the phrase earlier when writing articles for newspapers.

5. Richard Hofstader, *Social Darwinism in American Thought* (Boston: Beacon, 1955).

6. Auguste Comte, *The Course of Positive Philosophy* (originally published in serial form between 1830–1842). More accessible is Harriet Martineau's translation and condensation, published under the title, *The Positive Philosophy of August Comte,* 3 volumes (London: Bell and Sons, 1898; originally published 1854).

7. Herbert Spencer, *The Principles of Sociology,* 3 volumes (New York: Appleton, 1898; originally published in serial form between 1874 and 1896).

8. Spencer, *Social Statics*, cited in note 4

9. Herbert Spencer, *First Principles* (New York: A.L. Burt, 1880; originally published in 1862).

10. Herbert Spencer, *The Principles of Biology,* 2 volumes (New York: Appleton, 1897; originally published in serial form between 1864 and 1867).

11. Herbert Spencer, *The Principles of Psychology,* 2 volumes (New York: Appleton, 1898; originally published in 1855).

12. For a review of Spencer's theoretical principles, see Jonathan H. Turner, *Herbert Spencer: A Renewed Appreciation* (Beverly Hills, CA, and London: Sage, 1985).

13. Spencer, *The Principles of Biology* and *The Principles of Sociology*, cited in notes 7 and 10.

14. Thomas R. Malthus, *An Essay on the Principle of Population as It Affects the Future Improvement of Society* (London: Oxford University Press, 1798).

15. Spencer, *The Principles of Sociology,* Part V of volume 2, cited in note 7.

16. Spencer, *The Principles of Biology,* cited in note 10.

17. For the details of this argument, see Jonathan H. Turner, "Durkheim's and Spencer's Principles of Social Organization," *Sociological Perspectives* 27 (1984), pp. 21–32.

18. Émile Durkheim, *The Division of Labor in Society* (New York: Free Press, 1933; originally published in 1893).

19. Ibid., p. 266.

20. Ibid., p. 267.

21. *Charles Darwin's Notebooks, 1836–1844: Geology, Transmutation of Species, Metaphysical Enquiries.* Notebook E, transcribed and edited by David Kohn (British Museum, New York: Cornell University Press), p. 395.

22. In a letter to Asa Gray on September 5, 1857, Darwin wrote the following: "I think it can be shown that there is such an unerring power at work in *Natural Selection* (the title of my book) which selects exclusively for the good of each organic being." See George Gaylord Simpson, *The Book of Darwin* (New York: Washington Square, 1982), p. 24. Also see Charles Darwin, *On the Origin of Species* (New York: New American Library, Mentor Books, 1958; originally published in 1859).

23. See Charles Darwin and Alfred Russell Wallace, *Evolution by Natural Selection* (Cambridge, England: Cambridge University Press, 1958).

24. See note 2, Mendel, "Versuche über pflanzenhybriden," but for the still definitive account of Mendel's discoveries, see W. Bateson, *Mendel's Principles of Heredity* (Cambridge: Cambridge University Press, 1909).

8

The Maturing Tradition: Ecological and Biological Theorizing*

Functional approaches carried into the present era the theorizing about long-term evolutionary trends toward increasing societal complexity, as we saw in Chapters 4, 5, and 6. The other two branches or fronts of biological reasoning that influenced sociological theorizing—that is, ecology and genetics—were extended in new and creative ways between World Wars I and II, then in the middle decades of the century.

Our goal in this chapter is to examine the basic nature of these two branches of biologically inspired social thought, providing an overview of the arguments that guided evolutionary sociology in the last decades of the century.

URBAN ECOLOGY

The Chicago School

The period between World Wars I and II cannot be noted for great theoretical progress. Even as the last works of the early masters began to appear or be translated into English, theoretical sociology did not develop in the same way as it had during sociology's first one hundred years. Part of the reason for this dearth of theoretical activity resides in the efforts to make sociology a more rigorous research discipline, especially in America where sociologists concentrated on ethnographic studies or on developing quantitative survey

*This chapter is coauthored with Alexandra Maryanski.

techniques.[1] As a consequence, theoretical ideas developed during this interwar period were almost always connected to empirical work on particular topics. One of the most famous efforts to blend theory and research was in what is often termed the "Chicago School," which derives its name from the activities of a group of sociologists at the University of Chicago.

One facet of the Chicago School was its study of urban problems; the city of Chicago became the laboratory for conducting research on these problems. Louis Wirth was perhaps the key figure in developing a more theoretically informed analysis of urban Chicago, and he reintroduced ideas that had been at the center of Durkheim's and Spencer's analyses of social differentiation. But in Wirth's approach, these big ideas about society and human evolution were downsized to study the dynamics of urban areas. Wirth believed urban development could be understood by studying the size and density of urban populations because these influence the diversity and heterogeneity of the population. For example, in his famous essay "Urbanism as a Way of Life,"[2] Wirth argued that a dense population of a certain size inevitably leads to the proliferation of secondary groups, a lessening of the intensity of personal interaction, and increased cultural heterogeneity. These events, he continued, lead to a weakening of family ties and to a subversion of traditional bases of social control, such as religion, common folklore, and shared cultural heritage. There was, then, a social problem bias in Chicago School research, but, within this tradition, Wirth and others brought ecological theorizing into the twentieth century and, in essence, kept it alive and available to succeeding generations.

Borrowing from the Science of Ecology

Thinkers in the Chicago School[3]—Ernest Burgess, Chauncy Harris, Homer Hoyt, Roderick McKenzie, Robert Park, Edward Ullman, Louis Wirth, and their students—borrowed quite self-consciously from the emerging subfield of ecology in biology. As Amos Hawley noted:[4]

> In their search for order in the turbulent urban centers of America . . . sociologists were stimulated by work then being done by bioecologists. . . . Those researchers showed that plant species adapt to their environment by distributing themselves over a localized area in a pattern which enables them to engage in contemporary uses of habitat and resources. That idea opened a vista to an understanding of what was occurring in the burgeoning industrial city. For then it was apparent that various subpopulations were jostling for spatial positions from which they could perform their diverse functions in an unfolding division of labor.

Chicago School ecologists thus began to view urban areas as a kind of sociocultural ecosystem in which different zones, sectors, and nuclei became differentiated by virtue of competition for resources.

The General Chicago-School Model of Urban Ecology

A variety of models on the growth of urban areas were proposed, but the general model underlying Chicago school thinking is better represented in Figure 8-1. For the Chicago school, urban growth was related to production and population growth because these aggregated people and the various corporate units, such as family housing and business structures, that people used to survive in space. As aggregation intensified, population density increased, which, in turn, led to a competition for resources, including urban space, governmental resources, retail markets, and virtually any resource that could be used to facilitate the survival of individuals and

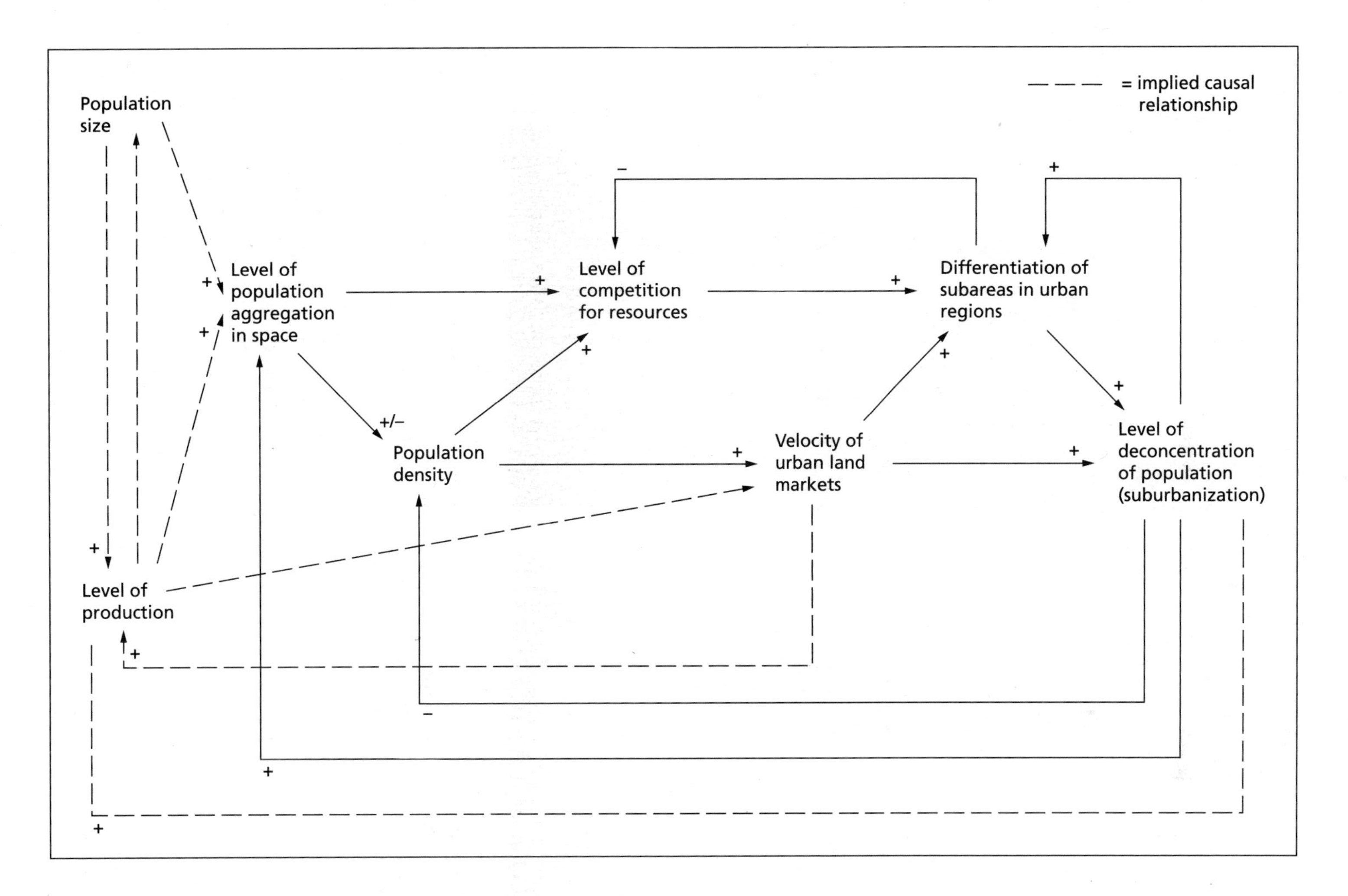

FIGURE 8-1 How Early Urban Ecologists Saw Urban Growth

corporate units. Real estate markets greatly accelerated this competition for resources. For once a market exists, price becomes the gateway to individuals' and corporate actors' ability to gain access to the resources in an urban area and, thereby, to survive in a particular spatial niche that, to Chicago School theorists, was the counterpart of a niche in a biotic system. From this competition came the differentiation of urban areas by the kinds of residents they had, the nature of the productive and business activity evident, and the cultural symbols associated with these. As competition increased, urban areas expanded as actors sought resource niches further from those areas where the competition was too intense, and the market-driven price for living or doing business was simply too high.

Because they relied too heavily on data from Chicago, the models that these early urban ecologists developed were too simplistic and, indeed, rather parochial But these models made an important connection between biological thinking about "selection" processes arising out of competition for urban space and resources, on the one hand, and sociocultural dynamics, on the other. Because of the dominance of the University of Chicago's department of sociology, this connection could not be ignored, and, as we will see in the next chapter, during the last sixty years it has become transformed into one of the more sophisticated theoretical approaches in sociology. Thus, taking a cue from Spencer and Durkheim but, equally so, from biology proper, the most important orientation to emerge from the theoretically fallow years between World Wars I and II was human ecology.

GENETICS AND NATURAL SELECTION

Within biology, Gregor Mendel's insights into the mechanism of inheritance were being developed at about the same time that ecological ideas were being downsized into analyses of urban areas by sociologists. The word *genetics* was coined in 1906 to denote the new discoveries about how inheritance operates. Although these discoveries occurred over a number of decades, they are discussed together here. What is critical is the recognition that an increasing understanding of the mechanisms of inheritance would provide the basis for the modern synthesis of evolutionary theory and recent efforts to bring biology back into sociological theory. What, then, were the new insights from genetics? Let us divide them into two types: (1) those pertaining to individual organisms and (2) those dealing with populations of organisms.

The Genetics of the Individual

Genes, or what Mendel had termed "merkmals," are the basic units of inheritance.[5] The information regarding the transmission of characteristics is stored here. Genes can be dominant or recessive, and the various combinations of these—as worked out by Mendel—determine what traits will be visible for an organism. Yet, even when not visible, information can still be in the genes and, as a result, emerge in subsequent generations. Genes are strung along thread-shaped bodies, or chromosomes, within the nuclei of cells. *Alleles* are the alternative or variant forms of a gene that affect the same trait in different ways—for example, potential variations in eye color are alleles (because they affect the variation of the same basic trait). *Genotype* is the sum of all the alleles making up an individual, including those that are visible and those that are stored in the genes but not manifest. *Phenotype* refers to the visible traits of an organism that are regulated by genes.

Thus, at the level of the individual, inheritance is regulated by genes. Those genes of an individual that are visibly expressed are its phenotype, whereas the complete collection (both expressed and invisible) of genes and alleles of an individual are its genotype. Differences in the characteristics of individuals result from the information on discrete genes that provide the fund of variation; hence, natural selection is the mechanism by which genetic material is preserved or lost.

The Genetics of the Population

Selection works on the individual and its phenotype and, as a result, an organism's more inclusive genotype, but the breeding population as a whole evolves. From a genetic point of view, however, it is not so much a population of individual organisms that evolves but, rather, it is a cluster of genotypes. Individual organisms come and go, but what is passed on and remains is their genetic information. Those genes that produce traits facilitating individual organisms' survival in an environment will increase the likelihood that such organisms will survive and produce offspring. From a genetic perspective, then, the genes that survive with the living organisms are simply temporary vessels, carrying a most interesting cargo—genotypes.

As this perspective developed, the concept of *gene pool* was introduced to characterize this shift in emphasis from the individuals who carry the genes to the sum total of their different alleles. A gene pool is thus the pooled sum of all the genotypes in a population of organisms. A species as a whole is the most inclusive gene pool because it contains all the genetic information for those genotypes that have survived, whereas a less inclusive gene pool would be a breeding population in a particular area or region.

What causes changes in the composition of genes in the pool? One obvious force—which, surprisingly, was not initially recognized—is *natural selection*. As some individual organisms (and their phenotypes and genotypes) survive and reproduce, others are selected out, leading to a shift in the composition of the gene pool. Another force changing the gene pool composition is random *mutations*, which add new DNA (the actual codes of information on genes) to the gene pool. Another force is *gene flow*, which results from the movement of individuals to and from different breeding populations (say, for example, intermarriage and offspring among Asians and Caucasians). Yet another force is *genetic drift*, or random changes in gene frequencies in the pool stemming from the fact that a smaller breeding population will be less likely to evidence over time as many duplicate alleles as a larger ones. The overall degree of variation in the gene pool is related to the interactions among these forces. Mutations and gene flow increase variation within a breeding population because they add new genetic material. Natural selection normally reduces variation because it weeds out those phenotypes (and genotypes) less suited to an environment. Similarly, genetic drift reduces variation through the random loss of genetic variance available to a small breeding population.

This shift to conceptualizing populations as pools of genes rather than individuals emerged in the twentieth century, although the actual term *gene pool* was a midcentury formulation.[6] As early as 1907, for instance, G.H. Hardy and W. Weinberg applied the new discoveries about particulate inheritance to the population level.[7] Hardy and Weinberg believed evolution could be conceptualized as "changes in gene frequencies" (later conceptualized as "the gene pool"). In what became known as the Hardy-Weinberg equilibrium model for assessing the degree of evolution, Hardy and Weinberg presented a decisive argument. If we know the allele frequencies in a population (that is, all the dominant and recessive genes relevant to various traits), we can predict the genotypic frequencies in the next generation, but, to do so, we must assume the following: No natural selection is operating; no mutations are produced; no gene flow resulting from migrations of breeding populations is evident; no genetic drift is to be found; no bias in mating can be observed (it is random); and no limits on population size exist (it is infinite). Obviously such assumptions do not correspond to natural populations in real environments, but they allow us to compute the expected distribution of genotypes in the next generation if *none* of these forces of change in the gene pool is operating. Then, by comparing this idealized computation of gene frequencies with the actual frequencies, we get some indication of how much change or evolution is occurring. If not much difference occurs between the idealized prediction and the actual genotypes in the population, then not much evolution is evident, but, if there is a large difference, then the degree of evolution can be measured (by comparing the predicted and actual gene frequencies). What is important about the Hardy-Weinberg Law is that it states that evolution (or a change in gene frequencies) can occur only when variations exist in a population on which selection forces can act. These selective differences among genotypes are what change the gene frequencies.

In the first decade of this century, this revolution in population genetics was not well integrated with Darwinian views of natural selection as the mechanism of evolution. Curiously, as concepts in genetics emerged, they were often considered to represent an alternative explanation to Darwinian natural selection. Indeed, during the first part of this century, a major anti-Darwinian movement rejecting natural selection as the force behind evolution emerged with opposition so negative that one noted scholar could write, "We are now standing at the death bed of Darwinism. . . ." By the 1920s, even major textbooks on evolution carried forth an antiselectionist position. And, for many scholars it was already a foregone conclusion that the Darwinian revolution had passed and that ". . . a new generation has grown up that knows not Darwin."[8]

The modern synthetic theory of evolution began to emerge from the revival of Darwinian notions of natural selection and their eventual coupling with genetics And at this point, the basic ideas of sociobiology were first articulated.

The Codification of Sociobiology

In *The Genetical Theory of Natural Selection*, R. A. Fisher was the first to recognize the implications of synthesizing genetics and Darwinian selection.[9] Fisher's first task was to refute the main competitor to natural selection as the force behind evolution. This competitor was "mutation theory," which argued that large mutations are the driving force behind evolution. Fisher demonstrated with elegant mathematical equations that the vast majority of mutations are harmful and doomed to extinction by natural selection. In particular, Fisher demonstrated that those large mutations that were posited as the key force behind evolution would be harmful and, hence, selected out of the gene pool. Instead, only small mutations offering slight advantages in promoting the fitness of organisms to an environment could be involved in evolution, but these modest mutations could not by themselves alter the gene pool. Rather, the power of natural selection to favor such mutations drives evolution.

With this argument, Fisher welded population genetics and natural selection, while introducing an important concept: *fitness*. This concept was as old as Malthus, Spencer, and Darwin, but it took on new meaning in Fisher's hands. The *mean fitness* of a population, Fisher argued, will usually be proportional to some component of its genetic variance. That is, gene pools revealing considerable variation provide a greater range of options for selection to work on, thereby increasing the mean fitness of that population to survive. Moreover, even with small variations in a gene pool, selection will still remain the significant force in determining gene frequencies.

In this way, the concepts of selection, genetic variation, and fitness were linked together. Others in the 1930s extended this argument, but what is crucial for our purposes is the view of evolution as revolving around the power of natural selection to promote fitness by "selecting" those variations in genes that promote adaptation.

In the 1940s, all these leads were crystallized in the "modern synthesis," with natural selection hailed as the only directional force in evolution. But before this synthesis had fully emerged, Fisher had set the stage for modern human sociobiology. In the last one-third of *The Genetical Theory of Natural Selection*, Fisher quoted Herbert Spencer and turned to an analysis of "Man and Society," stressing the need for forms of eugenics to promote the "survival of the fittest genes." And although Fisher's ideas on eugenics are not now important, they firmly planted in the minds of some biologists that human behavior and organization might be understood through the same natural processes affecting other species—that is, variation in genes, natural selection, and fitness as the adaptive value of genes.

The First True Sociobiologists

Early Instinct Theories. The first "sociobiologists" were, curiously, not those working in the tradition established by Fisher and the synthesis of population genetics and natural selection. Rather, the initial thrust for a biological view of humans and society was a group of scholars who were throwbacks to the older "instinct theories" that had always existed in sociology and social philosophy. Perhaps the most influential figure was the father of ethology, Konrad Lorenz. In particular, Lorenz investigated the "aggression instinct," which he saw as becoming channeled by rituals (presumably through natural selection) to space members of an animal species for food gathering, mating, and mitigation of violent encounters.[10] Lorenz viewed human aggression as maladaptive, however, because natural selection had not equipped humans' early ancestors with ritualized mechanisms for inhibiting aggression in high-density situations. Other speculation on human instincts for aggression and domination followed—for example, Robert Ardrey's portrayal of the "killing instinct" in *African Genesis* and *The Territorial Imperative*,[11] Desmond Morris' "naked ape,"[12] Lionel Tiger's and Robin Fox's "imperial animal,"[13] and the early sociobiological work of Pierre van den Berghe on age, sex, and domination.[14]

Like their predecessors in the nineteenth century, these instinct speculations passed into obscurity although newer approaches along these lines have emerged in the last decades of the twentieth century (see pp. 143–148 of Chapter 10). In the place of crude instinct approaches came sociobiology. Sociobiology also emerged partly in response to what is known as the "group selectionist" argument. Perhaps V.C. Wynne-Edwards is the scholar most closely associated with this group selection perspective.[15] Wynne-Edwards saw the basic problem as this: How is altruistic behavior in animals to be explained? That is, how does natural selection, with its emphasis on the individual organism's effort to survive and reproduce, explain cooperative behavior in which organisms sacrifice their own fitness for the good of the group? Despite being a competitive world, Wynne-Edwards argued in 1962, "the members of social groups cooperate in civilizing it and, so far as the competition is concerned, they act according to rules. Everything the social code decrees is done for the common good. . . ."[16] Thus it might be that, in higher animals, the group is often the unit of selection rather than the individual, as Spencer had recognized a hundred years earlier.[17] Variations in group structures are selected for because of their capacity to adapt and survive in an environment. Such an argument was not terribly different from Spencer's view that social groups, especially whole societies, often compete for existence in a given area, with the better organized society surviving (usually as a result of its superior military ability); so, evolution had, in Spencer's view, involved the successive competition and survival of ever more "fit" societies.[18] Whether the author was a Spencer or a Wynne-Edwards, the idea was that the individual organism is not the only unit of selection; groups or, in Spencer's terms, "super-organic" units composed of interacting and mutually dependent organisms can also constitute a "body" or "social organism" subject to selection pressures.

Sociobiology emerged as a reaction against such group selection arguments.[19] In particular, George C. Williams launched a critique of group selection, arguing that the real unit of selection is not the group or even the individual organism.[20] Rather, the unit of selection is "the gene," leading Williams to posit the concept of *genic selection*. Those genes temporarily housed in individuals and groups that promote survival and reproduction in an environment—that is, "fitness"—will be retained. Whatever the effects of selection for promoting "groupness," selection at the gene level is the operative mechanism. For "group-related adaptations do not, in fact, exist"; instead, the characteristics of groups—altruism, reciprocity, and exchange, for example—are the result of natural selection on individuals because "simply stated, an individual who maximizes his friendships and minimizes his antagonisms will have an evolutionary advantage, and selection should favor those characters that promote the optimization of personal friendship."[21] Thus, particular genes that promote those traits in individuals facilitating "groupness" will, in certain environments, promote fitness—that is, survival and reproduction. It is not necessary, Williams argued, to explain group processes using group selection; "genic selection" can explain such group processes, for "the fitness of a group will be high as a result of (the) summation of the adaptations of its members."[22]

W.D. Hamilton took this kind of reasoning a step further by introducing the important concept of *inclusive fitness*.[23] This concept was intended to account for cooperation among relatives, and the argument goes something like this: Natural selection promotes "kin selection" in the sense that those who share genes will interact and cooperate to promote one another's fitness—or capacity to pass on their genes. Self-sacrifice for a biological relative is, in reality, not altruism at all but the selfish pursuit of fitness because, in helping a relative to survive and reproduce, one is passing on one's own genetic material (as stored in relatives' genotypes). Thus, from this point of view, self-sacrifice for, and cooperation with, relatives will be greater as the amount of shared genetic material increases. So, altruistic behaviors among parents and offspring or among siblings can be understood as behaviors that were "selected for" as a way to pass on one's genetic material, or to keep it in the gene pool. This is the process of inclusive fitness—"inclusive" in the sense that one shares identical genes with others, and "fitness" in the sense that, in helping these others, one is also assuring that shared genetic material will remain in the gene pool. This kind of argument takes "the altruism out of altruism" among family members and sees such behaviors as simple matters of self-interest: to maximize the amount of one's genetic material that stays in the gene pool. Hence the "goal" of genes is to preserve themselves, and it is "rational" for them to help preserve the bodies of those individuals who carry common genetic material. Of course, genes do not "think," but blind natural selection has operated in the distant past to promote behaviors in organisms, such as altruism among relatives, that increased "fitness" in ways that *maximize* the passing on of particular sets of genes.

Although Hamilton's notions of "kin selection" and "inclusive fitness" might be seen to account for cooperation among relatives, the question was soon raised: How can such arguments explain altruism and cooperation among nonrelatives who do not share genetic material? Robert Trivers sought to overcome this objection with the concept of *reciprocal altruism*.[24] In a series of modeling procedures, he presented the following scenario: Natural selection can produce organisms that will incur the "costs" of helping a nonrelative because at some later time this nonrelative can "reciprocate" and help "altruistic" organisms (thereby increasing the latter's fitness, or ability to survive and pass on genes). Thus, for species that live a long time and congregate, natural selection can promote reciprocal altruism and increase all individuals' fitness, whereas those that would "cheat" and fail to reciprocate others' altruism will be selected out (because eventually, without signs of reciprocity, others would not come to their aid). And so, once again what seems like altruism is, in reality, "selfishness" by the individual organisms, each of which is trying to maximize its capacity to keep genes in the pool.

The last major conceptual development in sociobiology has been the use of "game theory" to describe the process of fitness. Here the key figure has been J. Maynard-Smith,[25] although Trivers had started his analysis with the classic "Prisoner's Dilemma" game to show how selfish individuals can cooperate to increase their fitness beyond what it would be without cooperation (see Chapter 23 for discussion of these game-theoretic models).[26] In game theory, it is assumed that, under particular conditions imposed by the game, rational decision-making actors seek the best possible payoff by adopting a particular behavioral strategy. The payoffs in game theory are typically some unit of subjective value. In contrast, unlike classical utilitarianism, in which actors are assumed to be conscious, rational, and payoff maximizing, game theory as it is applied to evolutionary theory cannot assume rational consciousness of its players (genes), and the payoffs are always a measure of fitness (capacity to pass on genes). The process of selection is presumed to have "decided" (unconsciously) the strategy that maximizes fitness in a given environment; the investigator's task is then to determine what behavioral strategy for a particular species in an environment would best assure maximal payoffs of fitness, or the capacity to survive, reproduce, and pass on genes. This is what Hamilton did with the concept of inclusive fitness: Helping one's biological relatives is the best strategy, as "decided" by the forces of natural selection, for passing on one's genetic material.

Maynard-Smith went a step further and developed the concept of an *evolutionary stable strategy*, or ESS, to describe the stabilization of behavioral strategies among the individuals of a population. Without outlining the mathematical and statistical details, the ESS enabled Maynard-Smith to calculate an equilibrium point, around which the relative amounts of various behaviors, or strategies, for fitness will stabilize. In this way, it is possible to show that all members of a population do not have to adopt the same strategy; rather, each potential strategy affects the payoffs of the others, and over time the relative frequencies of various strategies for survival will, as a result of natural selection, reach equilibrium, or the ESS.

This application of game theory gave sociobiologists a well-developed and powerful set of mathematical tools for making predictions about how natural selection will produce behavioral strategies maximizing fitness and how varying configurations of such strategies can reach equilibrium. Such configurations can, sociobiologists argued, explain patterns of social organization.

Richard Dawkins popularized this emerging sociobiological approach in his well-known work *The Selfish Gene*.[27] His argument is that genes are "replicator" or "copy" machines that try to reproduce themselves. Natural selection favored those replicators that could find a "survival machine" to live in—initially, in the distant past, a cell wall, then a grouping of cells, then an organism, and eventually a grouping of organisms. As Dawkins notes:[28]

> What weird machines of self-preservation would the millennia bring forth. . . . They (replicators) did not die out, for they are past masters of the survival arts. . . . Now they swarm in huge colonies, safe inside gigantic lumbering robots, sealed off from the outside world, communicating with it by tortuous indirect routes, manipulating it by remote control. They are in you and me; they created us, body and mind; and their preservation is the ultimate rationale for our existence. They have come a long way, those replicators. Now they go by the name of *genes*, and we are their survival machines.

Such a rich metaphor captures the essence of modern sociobiology because the unit of selection becomes the gene, and evolution is the result of genes competing and adopting strategies that allow them to leave their DNA in the gene pool. Evolution is not an effort of individuals or species to survive; these are only vehicles for the real driving force of evolution: genes that are "ruthlessly selfish" in adopting strategies to maximize their fitness. At times it serves the genes' interest to foster limited forms of altruism and other social behaviors that are often considered to be the exclusive domain of the social sciences. But from a sociobiological perspective, many of the behaviors, strategies, and organizational traits of animals, including humans, are simply the genes' way of coping with an unpredictable environment. Indeed, even the human capacity for thinking and learning, Dawkins avers, can be viewed as the genes' way to construct a better survival machine; cooperation can similarly be seen as one survival machine making use of another survival machine in an effort to further assure its fitness; and various patterns of social organization can thus be conceptualized as nothing more than more complex and inclusive survival machines for genes.

Yet, Dawkins hedged in his last chapter, as have many contemporary sociobiologists in recent years. Dawkins posits a "new replicator," which he terms *memes*. The basic tenets of sociobiology—genic selection, inclusive fitness, and reciprocal altruism, all producing strategies and "survival machines" for genes—can explain how humans came to exist, but culture begins to supplement and supplant biology as the major replicating mechanism. Memes are those new cultural units that exist inside brains and that, via socialization, are passed on and preserved in a "meme pool." Meme evolution will now begin to accelerate, for "once genes have provided their survival machines with brains which are capable of rapid imitation, the memes will automatically take over." And it might even be possible for memes to rebel against their creators, the selfish genes. Similarly, other sociobiologists now talk of "co-evolution," operating at both the genetic and the cultural levels.

Although these metaphors are colorful, sociobiology is highly technical—involving extensive use of mathematics, game theory, and computer simulations. It changes the image of natural selection as a process working on passive individuals and posits, instead, active actors driven by their genes to maximize their reproductive success by any strategy available (which can be modeled and simulated with various game-theoretic approaches). The challenge of this perspective is that many processes considered by sociologists to be explicable only by sociological laws are seen by sociobiologists to be understandable as biological processes derived from the laws of the synthetic theory of evolution.

CONCLUSION

By the middle 1970s, then, biological ideas had begun to penetrate the social sciences along two fronts: (1) the analysis of ecosystems and how concepts from this branch of biology could be used to understand competition and selection processes in the units of sociocultural systems; and (2) the analysis of genetics and how notions of fitness, inclusive fitness, and reciprocal altruism could be employed to explain human behavior and social organization as survival machines for the real driving force of society, genes. The third branch of biological thinking about the long-term evolution of complexity sustained itself within functional theories of evolution (see Chapters 3 through 5), and perhaps other more conflict-oriented approaches, such as world systems theory (see Chapter 17).

The ecological front of biologically inspired theory, as we explore in the next chapter, became the more prominent, at least to the last decade of the century. As we will see, Amos Hawley not only continued the urban ecology approach (of his early work), but, more significantly, moved ecological analysis back to the macro level where both Spencer and Durkheim had originally employed ecological ideas. Hawley's students and others influenced by the Chicago School, however, worked at the more meso level, continuing to expand urban ecological theory and developing an entirely new ecological analysis of organizations.

The genetic front of theory became translated into rather extreme sociobiology arguments about behavior and social organization as driven by genes trying to maximize their fitness—arguments which became, to say the least, highly controversial. Yet, as we will see in Chapter 10, the reaction against these extreme arguments has proven to be healthy because it led to efforts to tone down the extremes of sociobiology and, more important, to reintegrate biological arguments in genetics, ecology, and even social evolution into more general theoretical arguments.

NOTES

1. See Stephen Park Turner and Jonathan H. Turner, *The Impossible Science: An Institutional Analysis of American Sociology* (Newbury Park, CA: Sage, 1990).

2. Louis Wirth, "Urbanism as a Way of Life," *American Journal of Sociology* 44 (1938), pp. 46–63.

3. See, for examples, Ernest W. Burgess, "The Growth of the City" in *An Institution to Sociology*, ed. R.E. Park and E.W. Burgess (Chicago: University of Chicago Press, 1921); Robert E. Park, "Human Ecology," *American Journal of Sociology* 42 (1936), pp. 1–15; Homer Hoyt, *The Structure and Growth of Residential Neighborhoods in American Cities* (Washington, D.C., Federal Housing Authority, 1939); Robert E. Park, Ernest Burgess, and Roderick D. McKenzie, *The City* (Chicago: University of Chicago Press, 1925); Chauncy D. Harris and Edward L. Ullman, "The Nature of Cities," *Annals of the American Academy of Political and Social Science* (1945), pp. 789–796.

4. Amos H. Hawley, "Human Ecology: Persistence and Change," *American Behavioral Science* 24 (January 1981), p. 423.

5. Gregor Mendel, "Versuche über pflanzenhybriden," English translation in *Journal of the Rural Horticultural Society* 26 (1901), originally published in 1865.

6. Although the term *genetics* was coined by William Bateson in 1906 as the basic construct to describe individual heredity and variation, the term *gene pool* was coined by Dobzhansky in 1950 and became the fundamental construct of population genetics. See Theodosius Dobzhansky, "Mendelian Populations and Their Evolution," *American Naturalist* 14 (1950), pp. 401–418. For further readings on the history of genetics, see Theodosius Dobzhansky, *Genetics and the Origin of Species* (New York: Columbia University Press, 3rd rev. ed., 1951), and *Mankind Evolving* (New York: Bantam, 1962). See also Mark B. Adams, "From 'Gene Fund' to 'Gene Pool': On the Evolution of Evolutionary Language," *History of Biology* 3 (1979), pp. 241–285, and "The Founding of Population Genetics: Contributions of the Chetvevikov School 1924–1934," *Journal of the History of Biology* 1 (1968), pp. 23–39; Alfred Sturtevant, *A History of Genetics* (New York: Harper & Row, 1965); and James Crow, "Population Genetics History: A Personal View," *Annual Review of Genetics* 21 (1987), pp. 1–22.

7. Extending Mendel's Laws by deducing the mathematical consequences of a nonblending system of heredity (that is, genes are discrete and do not blend), G.H. Hardy and W. Weinberg laid the cornerstone for population genetics; in turn, the modern science of statistics was born in the study of quantitative genetics. G.H. Hardy, "Mendelian Proportions in Mixed Populations," *Science* 28 (1908), pp. 49–50; W. Weinberg, "Über den Nachweis der Vererbung beim Menschen," *Jh. Ver. Vaterl. Naturk. Wurttemb.* 64 (1908), pp. 368–382.

8. See Eberhart Dennert, *At the Deathbed of Darwinism*, trans. E. G. O'Hara and John Peschges (Burlington, Iowa: German Literary Board, 1904), p. 4. Also see J.H. Bennett, *Natural Selection, Heredity, and Eugenics* (Oxford: Clarendon, 1983), p. 1; Garland Allen, "Hugo de Vries and the Reception of the Mutation Theory," *Journal of the History of Biology* 2 (1969), pp. 56–87; and Sewall Wright, "Genetics and Twentieth-Century Darwinism," *American Journal of Human Genetics* 12 (1960), pp. 24-38.

9. R.A. Fisher, *The Genetical Theory of Natural Selection* (Oxford: Clarendon, 1930). For an overview of Fisher's contribution, see J.H. Bennett, *Natural Selection, Heredity, and Eugenics* (Oxford: Clarendon, 1983). J.B.S. Haldane and Sewall Wright also laid the foundation for the reconciliation of Mendelian heredity and Darwinian selection, but Fisher's work triggered this revitalization. Fisher was primarily interested in how an organism can increase its fitness—however it is achieved. This emphasis on fitness is summarized in his fundamental theorem: "The rate of increase in fitness of any organism at any time is equal to its genetic variance in fitness at that time." Fisher, *The Genetical Theory of Natural Selection*, p. 35.

10. Konrad Lorenz, *On Aggression* (New York: Harcourt Brace Jovanovich, 1960).

11. Robert Ardrey, *African Genesis* (New York: Delta, 1961) and *The Territorial Imperative* (New York: Atheneum, 1966).

12. Desmond Morris, *The Naked Ape* (New York: Dell, 1967).

13. Lionel Tiger and Robin Fox, *The Imperial Animal* (New York: Holt, Rinehart & Winston, 1971).

14. Pierre van den Berghe, *Age and Sex in Human Societies: A Biosocial Perspective* (Belmont, CA: Wadsworth, 1973). A comparison of this work with later works hints at the changes that van den Berghe was to make. See *Human Family Systems: An Evolutionary View* (New York: Elsevier, 1979).

15. V.C. Wynne-Edwards, *Evolution through Group Selection* (Oxford: Blackwell, 1986), and *Animal Dispersion in Relation to Social Behavior* (New York: Hafner, 1962). For a review of the controversy surrounding group selection arguments, see David Sloan Wilson, "The Group Selection Controversy: History and Current Status," *Annual Review of Ecological Systems* 14 (1983), pp. 159–187.

16. Wynne-Edwards, *Evolution through Group Selection* (cited in note 15), p. 9, outlining his original ideas in the course of writing *Animal Dispersion*.

17. Herbert Spencer, *The Principles of Sociology*, 3 volumes (New York: Appleton, 1898; originally published in serial form between 1874 and 1896).

18. Ibid.

19. For an effort to extend group selection arguments, especially those developed by Spencer but also Durkheim, see Jonathan H. Turner, *Macrodynamics: Toward a Theory on the Organization of Human Populations* (New Brunswick, NJ: Rutgers University Press for Rose Book Series, 1995).

20. George C. Williams, *Adaptation and Natural Selection: A Critique of Some Current Evolutionary Thought* (Princeton, NJ: Princeton University Press, 1966). For his defense of reductionism away from the group level, see "A Defense of Reductionism in Evolutionary Biology," in *Oxford Surveys in Evolutionary Biology* 2, eds. R. Dawkins and M. Ridley (Oxford: Oxford University Press, 1985), pp. 1–27.

21. Williams, *Adaptation and Natural Selection*, p. 95.

22. Ibid.

23 W.D. Hamilton, "The Evolution of Altruistic Behavior," *American Naturalist* 97 (1963), pp. 354–356; "The Genetical Theory of Social Behavior I and II," *Journal of Theoretical Biology* 7 (1964), pp. 1–52; "Innate Social Aptitudes of Man: An Approach from Evolutionary Genetics," in *Biosocial Anthropology*, ed. R. Fox (New York: Wiley, 1984), pp. 135–155; "Geometry for the Selfish Herd," *Journal of Theoretical Biology* 31 (1971), pp. 295–311.

24. Robert L. Trivers, "The Evolution of Reciprocal Altruism," *Quarterly Review of Biology* 46 (4, 1971), pp. 35–57; "Parental Investment and Sexual Selection," in *Sexual Selection and the Descent of Man, 1871–1971*, ed. B. Campbell (Chicago: Aldine, 1972); and "Parent-Offspring Conflict," *American Zoologist* 14 (1974), pp. 249–264.

25. J. Maynard-Smith, "The Theory of Games and the Evolution of Animal Conflicts," *Journal of Theoretical Biology* 47 (1974), pp. 209–221; "Optimization Theory in Evolution," *Annual Review of Ecological Systems* 9 (1978), pp. 31–56; *Evolution and the Theory of Games* (London: University of Cambridge Press, 1982). See also Susan E. Riechert and Peter Hammerstein, "Game Theory in the Ecological Context," *Annual Review of Ecological Systems* 14 (1983), pp. 377–409.

26. The basic format for the "Prisoner's Dilemma" is this: Two criminals are caught together and accused of a crime; they are taken to separate rooms for questioning, with each being offered leniency in prosecution for telling on the other. If both refuse to talk, the police have no real evidence; yet, if one talks and the other does not, then the latter is at a disadvantage. Thus the dilemma is to "talk" or "keep quiet" under conditions in which each partner in crime does not know what the other will do. The maximizing strategy is for both to "keep quiet," but each can get less than the maximum benefit (and far more than the worst outcome) by telling on the other.

27. Richard Dawkins, *The Selfish Gene* (Oxford: Oxford University Press, 1976).

28. Ibid., p. 21.

9

The Continuing Tradition I: Ecological Theories

The macro-level ideas of Herbert Spencer and Émile Durkheim about the ecology of human social organization were downsized in the first half of the twentieth century to the meso-level analysis of urban social processes, as we reviewed in Chapter 8. Amos Hawley, who was a direct descendant of the Chicago School tradition, continued this emphasis on the differentiation of urban space in his early work in the late 1940s and early 1950s,[1] and yet, he was becoming "increasingly disenchanted with the then received conception of human ecology. The prevailing preoccupation with spatial distributions, which had attracted me at first, seem to me a theoretical cul-de-sac."[2] By the 1980s, he had pushed ecological analysis back to the macro or societal level.[3] During this transition in Hawley's thinking, urban ecology continued as a viable theoretical tradition. Equally significant, a new ecological approach focusing on the ecology of complex organizations, such as populations of business firms or voluntary associations, emerged from some of Hawley's students and other creative scholars. In this chapter, we will review this range of theories, beginning first with Hawley's macro-level theory, then move to urban ecology, and finally to organizational ecology.

AMOS H. HAWLEY'S MACRO-LEVEL ECOLOGICAL THEORY

Production, Transportation, and Communication

Hawley's theory of ecological processes begins with three basic assumptions:[4]

1. Adaptation to environment proceeds through the formation of a system of interdependencies among the members of a population.
2. System development continues, other things being equal, to the maximum

complexity afforded by the existing facilities for transportation and communication.

3. System development is resumed with the introduction of new information that increases the capacity for movement of materials, people, and messages and continues until that capacity is fully used.

Hawley terms these assumptions, respectively, the *adaptive*, *growth*, and *evolution* "propositions." These assumptions resurrect in altered form ideas developed by Spencer[5] and Durkheim.[6] To survive and adapt in an environment, human populations become differentiated and integrated by a system of mutual interdependencies. The size of a population and the complexity of social organization for that population are limited by its knowledge base, particularly with respect to transportation and communication technologies. Populations cannot increase in size, nor elaborate the complexity of their patterns of organization, without expansion of knowledge about (1) communication and (2) movement of people and materials. Hawley conceptualized the combined effects of transportation and communication technologies as *mobility costs*.

Linked to transportation and communication technologies is another variable, *productivity*. Curiously, in his most recent theoretical essay,[7] this variable is somewhat subordinate, whereas it is highlighted in earlier statements. No great contradiction or dramatic change in conceptualization occurs in this more recent statement, so we can merely reintroduce the productivity variable in more explicit terms. Basically, a reciprocal set of relations exists between production of materials, information, and services, on the one side, and the capacity of a system to move these products to other system units, on the other side. The development of new transportation and communication technologies encourages expanded production, whereas the expansion of production burdens existing capacities for mobility and thereby stimulates a search for new technologies. There is also a more indirect linkage among productivity, growth, and evolution, because productivity "constitutes the principal limiting condition on the extent to which a system can be elaborated, on the size of the population that can be sustained in the system, and on the area or space that the system can occupy."[8] Thus, to support a larger, more differentiated, population in a more extended territory requires the capacity to (1) produce more goods and services and (2) distribute these goods and services through transportation and communication technologies. If productivity cannot be increased or if the mobility costs of transportation and communication cannot be reduced, then there is an upper limit on the size, scale, and complexity of the system.

The Environment

An ecosystem is "an arrangement of mutual dependencies in a population by which the whole operates as a unit and maintains a viable environmental relationship."[9] The environment is the source of energy and materials for productivity, but the environment reveals more than a biophysical dimension. There is also an "ecumenic" dimension composed of the "ecosystems or cultures possessed by peoples in adjacent areas and beyond."[10] Moreover, in Hawley's view, ecological analysis "posits an external origin of change" because "a thing cannot cause itself"[11]; and thus, in examining a population as a whole in its physical, social, or biological environment, Hawley's approach emphasizes that change comes more from these environmental systems than from processes internal to organization of a population.

Functions and Key Functions

In Hawley's approach, the arrangement of mutual dependencies of a population in an environment is conceptualized as classes or types of *units* that form *relations* with one another with respect to *functions*.[12] Functions are defined as "repetitive activity that is reciprocated by another or other repetitive activities." Of particular importance are *key functions,* which are repetitive activities "directly engaged with the environment." As such, key functions transmit environmental inputs (materials and information) to other "contingent functions" (or repetitive activities joining units in a relation).[13] Hawley visualizes that a relatively small number of key functions exists, and "to the extent that the principle of key functions does not obtain, the system will be tenuous and incoherent."[14] A system is thus composed of functional units, a few of which have direct relations with the environment and perform key functions. Most other units, therefore, must "secure access to the environment indirectly through the agency of the key function."[15]

For example, production is a key function, and in earlier essays Hawley seemed to see productivity as *the* primary key function. Yet there are obviously other key functions—political, military, and perhaps ideological—that also influence the flow of resources to and from the environment. As a result, other functional units gain access to the environment only through their interconnections with those units engaged in these various key functions. Thus the relations of units that form the structure of a population are conceptualized as functions and key functions, or classes of reciprocated repetitive activity that join units together. This is, of course, another way of denoting specialization and differentiation of various types in clusters of activity. Just why Hawley proposes this particular terminology is unclear, but, in doing so, Hawley transforms the ecology perspective into a more functional form of analysis.

Indeed, slipping into Hawley's analysis is a stronger version of the term *function*. In part, this stronger notion of function is implied by the concept of "key function." A key function regulates inputs of energy, materials, and information into the system, and it is not hard to see how the next step on the road to functionalism is made: Certain key functions are necessary for adaptation and survival. Hawley himself takes this step when he notes,[16]

> We might suppose, for purposes of illustration, that every instance of collective life is sustained by a mix of activities that produce sustenance and related materials, distribute the production among the participants, maintain the number of units required to produce and distribute the products, and exercise the controls needed to assure an uninterrupted performance of all tasks with a minimum of friction.

Indeed, these requisites look very much like those proposed by Herbert Spencer—production, regulation, distribution, and sustenance.

Whatever the merits or defects in such functionalism, Hawley translated his ideas into a series of "hypotheses." Table 9-1 restates in somewhat modified form some of the most critical propositions that can be pulled from his analysis thus far.[17] These propositions represent abstract "laws" from which Hawley's many hypotheses can be derived. The basic ideas in the propositions of Table 9-1 are these: key functions, or those that mediate exchanges with the environment, disproportionately influence other functions and hold power over these other functions (for example, units involved in key economic or political functions in a society usually hold more power and influence than others, because they are engaged in interchanges, respectively, with the physical and social environment); the more proximate a function is to a key function, the greater this influence is; conversely, the more remote a function is from a key function, the less

TABLE 9-1 General Propositions on Functions in Ecosystems

I. The more a function (recurrent and reciprocated activity) mediates critical environmental relationships (key function), the more it determines the conditions under which all other functions are performed.

II. The more proximate is a function to a key function, the more the latter constrains the other, and vice versa.

III. The more a function is a key function, the greater is the power of those actors and units involved in this function, and vice versa.

IV. The more differentiated are functions, the greater is the proportion of all functions indirectly related to the environment.

V. The greater the number of units using the products of a function and the less the costs of the skills used in the function, the greater is the number of units in the population engaged in this function.

VI. The greater the mobility costs (for communication and transportation) associated with a function, the more stable are the number of, and the interrelations among, units implicated in this function.

VII. The more stable the number of, and interrelations among, units implicated in functions, the more a normative order corresponds to the functional order.

influence this function has on a key function. Differentiation of key functions decreases other functions' direct access to the environment, because such access is now mediated by units involved in key functions (for example, most people do not grow their own food or provide their own military defense in highly differentiated societies). As mobility costs for personnel and materials needed for functions increase, the functions' number and relations stabilize; under these conditions, a normative order can develop to regulate the internal relations of functions, as well as their interrelations.

Equilibrium and Change

An ecosystem is thus a population organized to adapt to an environment, with change in this system being defined by Hawley as "a shift in the number and kinds of functions or as a rearrangement of functions in different combinations."[18] In contrast with change, growth is "the maturation of a system through the maximization of the potential for complexity and integration implicit in the technology for movement and communication possessed at a given point in time,"[19] whereas evolution is "the occurrence of new structural elements from environmental inputs that lead to synthesis of new with old information and a consequent increase in the scope of the accessible environment."[20] This series of definitions presents a picture of ecosystem dynamics as revolving around (1) the internal rearrangement of functions, (2) the increase of complexity to the maximum allowed by an existent level of communication and transportation technologies, and (3) the receipt of environmental inputs, especially new information that expands transportation and communication (or capacity for mobility) and that, as a consequence, increases the scale and complexity of the ecosystem (to the limits imposed by the new technologies).

There is an image, then, of a system in equilibrium that is then placed into disequilibrium by new knowledge about production as it influences mobility (of people, materials, and information). Somewhat less clear is where this new information comes from. Must it be totally exogenous (from other societies, migrants, changes in biophysical forces that generate new knowledge)? Or can the system itself generate the new information through a particular array of functions? It would appear that both can be the source of change, yet, the imagery of

TABLE 9-2 General Propositions, Change, Growth, and Evolution in Ecosystems

I. The greater the exposure of an ecosystem to the ecumenic environment (other societies or cultures of other societies), the greater is the probability of new information and knowledge penetrating the system and, hence, the greater is the probability of change, growth, and evolution.

II. The more new information increases the mobility of people, materials, and information, as well as production, the more likely that change will be cumulative, or evolutionary, to the limits of complexity allowed by the new information as it is translated into technologies for production, transportation, and communication.

III. The more new information improves various mobility and productive processes at differential rates, the more the slower rate of technology will impose limits on the faster-changing technology.

IV. The more a system approaches the scale and complexity allowed by technologies, the slower the rate of change, growth, and evolution is and the more likely the system is to achieve a state of closure (equilibrium) in its ecumenic environment.

Hawley's model[21] connotes a system that must be disrupted from the outside if it is to evolve and develop new levels of structural complexity. Internal dialectical processes, or self-transforming processes that increase technology, seem to be underemphasized as crucial ecosystem dynamics. Table 9-2 summarizes the more abstract "laws" that can be culled from Hawley's hypotheses on these dynamics.[22]

These propositions reinforce the emphasis in human ecology that the source of change is exogenous, residing particularly in the "ecumenic" environment. From this environment new knowledge will come and then become "synthesized" with the existing knowledge base. Such synthesized knowledge will then change production, transportation, and communication in ways that allow the system to increase its complexity, size, and territory. Yet, new knowledge can introduce change only to a point. If some technologies lag behind others, the rate of change will be pulled down by the lower technology. And eventually the maximal size, scale, and complexity of the system will be reached, unless new technologies are inserted into the system from the environment. Thus systems that have grown to the maximum size, scope, and complexity allowed by production, transportation, and communication technologies will achieve an *equilibrium*. New knowledge from the environment can disrupt the equilibrium when such knowledge is used to achieve increases in productivity and mobility. But each technology has limits on how much growth and evolution it can facilitate; when this limit is reached, the system will tend to reequilibrate.

The concept of *equilibrium* is most problematic, although Hawley employs it only as a heuristic device. Hawley means the notion of equilibrium to connote "the balance of nature, denoting a tendency toward stabilization of the relative numbers of diverse organisms within the web of life and their several claims on the environment. . . ."[23] Yet Hawley recognizes that "equilibrium . . . is a logical construct"[24] and connotes that ecological systems tend toward stability, although Hawley also employs terms like "partial equilibrium" to connote only a tendency toward some degree of instability.

Growth and Evolution

The most interesting portions of ecological theory are those dealing with growth and evolution—that is, increasing size, scale, scope, and complexity of the systematic whole in its environment. This analysis builds on the propositions in Tables 9-1 and 9-2, but it extends them in creative ways and, as a result, goes considerably beyond the early formulations of Spencer and Durkheim.

In Figure 9-1, Hawley's model is redrawn in a way that makes the causal dynamics more explicit. Starting on the far left of the model, Hawley believes that an expanded knowledge base must come from the ecumenical environment. As the model stresses, new knowledge causes growth and change when it increases the level of communication and transportation technologies, either directly or indirectly, through increasing production (which then causes expansion of these technologies). A critical variable in Hawley's scheme is *mobility costs;* for any given technology, a cost (time, energy, money, materials) is associated with the movement of information, materials, and people. As these costs reach their maximum—that is, the system cannot "pay" for them without degenerating—they impose a limit on the scale of the system: the size of its population, the extent of its territory, the level of its productivity, and the level of complexity. Conversely, as the feedback arrows in the model indicate, the size of territory and population will, as they expand and grow, begin to impose higher mobility costs. Eventually these costs will increase to a point at which the population cannot grow or expand its territory—*unless* new communication and transportation technologies that reduce costs are discovered.

Much as Spencer and Durkheim argued, population and territorial size, because these are influenced by mobility costs, cause specialization of functions—what is termed *differentiation* in the model. As Hawley notes, however, the relationship between population size and differentiation is not unambiguous, but it does create "conditions that foster, if not necessitate, increases in the sizes of subsystems" and the number of such subsystems serving various functions. Thus, for Hawley, "the greater the size, the greater the probable support for units with degrees of specialization." And, as he adds, "Other pertinent conditions are the rate or volume of intersystem communications, scope of a market, and amount of stability in intersystem relations."[25]

These causal connections are not clearly delineated by Hawley, so the model involves making many causal inferences. The causal paths moving from level of productivity through extensiveness of markets and level of competition to selection pressures and differentiation of functions represent the old Spencerian and Durkheimian argument: Expansion of markets increases the level of competitiveness among units and, at the same time, increases the capacity to distribute goods and services because these are constrained by mobility costs (note arrows connecting markets and mobility costs); competition under conditions of increased production and population size allows—indeed encourages—specialization as actors seek their most viable niche. Similarly, the causal arrows moving from communication and transportation technologies through mobility costs, size of territory, and size of population to level of differentiation of functions restate Spencer's, but more particularly Durkheim's, argument in a more sophisticated form: Changes in communication and transportation technologies reduce mobility costs and allow population growth and territorial expansion; all these forces together create selective pressures to adjust and adapt varying attributes and competencies, especially under conditions of intense competition for resources.

Hawley believed that differentiation of subunits engaged in various functions occurs along two axes: (1) corporate and (2) categoric.[26] Corporate units are constructed from "symbiotic relations" of mutual dependence among differentiated actors, whereas

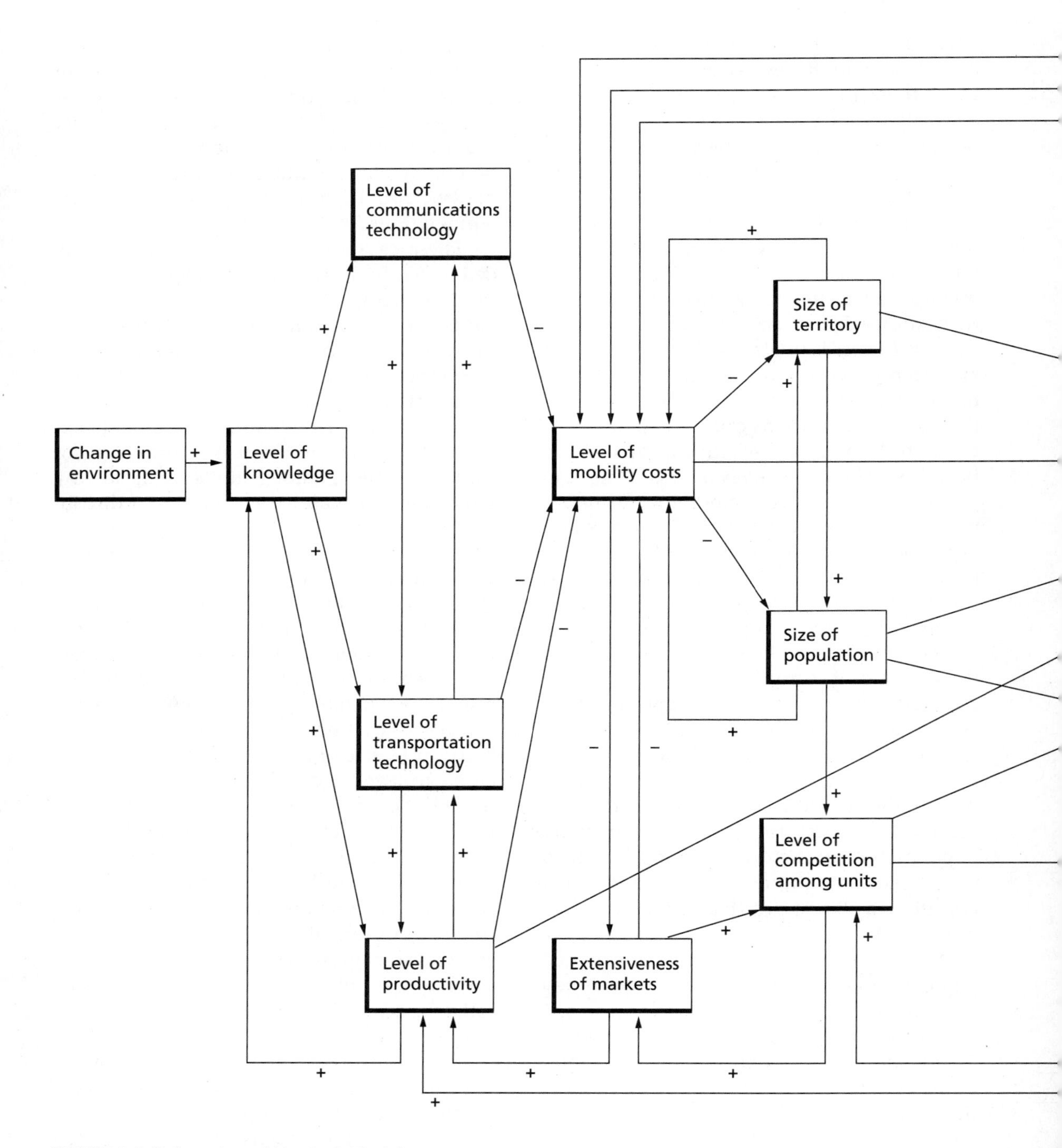

FIGURE 9-1 Elaboration of Hawley's Model

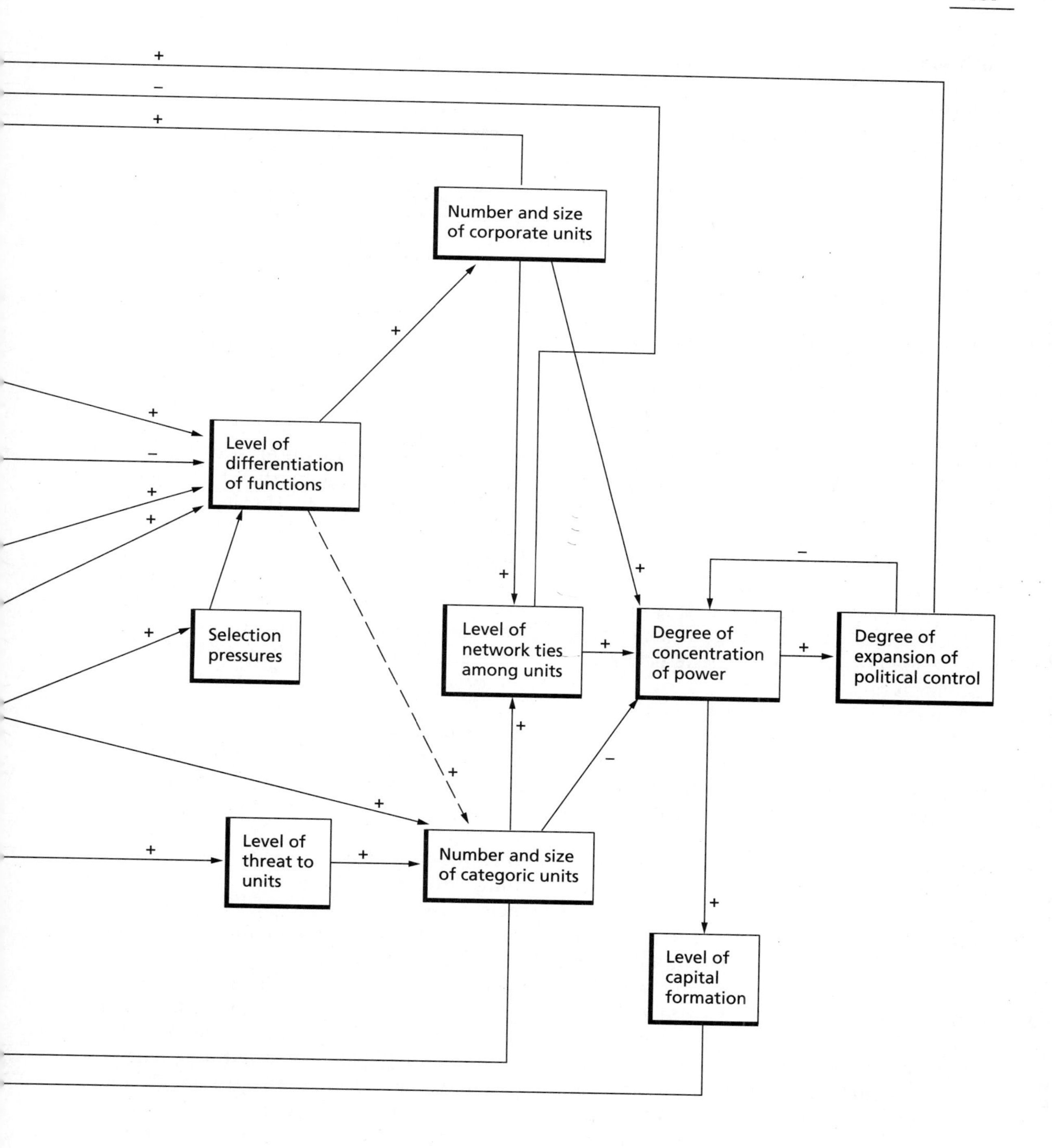
Number and size of corporate units
Level of differentiation of functions
Selection pressures
Level of network ties among units
Degree of concentration of power
Degree of expansion of political control
Level of threat to units
Number and size of categoric units
Level of capital formation

TABLE 9-3 A Typology of System Units

UNIFYING PRINCIPLE	RELATIONAL STRUCTURE	
	Corporate	**Categoric**
Familial	Household-producing and personal service unit	Kin, clan, and tribe
Territorial	Village, city, and ecumene	Polity, neighborhood, ethnic, enclave, ghetto
Associational	Industry, retail, store, school, government	Caste, class, sect, guild, club, union, professional organization

categoric units are composed of "commensalistic" relations among actors who reveal common interests and who pool their activities to adapt more effectively to their environments. Table 9-3 delineates these types of units along an additional dimension—their "unifying principle."[27] Thus, as differentiation increases, an ecosystem will represent a complex configuration of corporate and categoric units along various "unifying principles": familial, territorial, associational. It is not clear if Hawley meant this typology to be exhaustive or merely illustrative. Nonetheless, the typology is provocative.

The dynamics of these two types of units are very different. Corporate units form around functions, or sets of related activities, and are engaged in interchanges with other corporate units. As a consequence of this contact, corporate units tend to resemble one another, especially those that engage in frequent interchanges or are closely linked to corporate units engaged in key functions (interchanges with the environment). Moreover, as they engage in interchanges, corporate units tend toward closure of structure and establishment of clear boundaries. The size and number of such corporate units depend, of course, on the size of the population, the inter- and intra-unit mobility costs associated with communication and transportation technologies, the capacity for production and distribution in markets (as constrained by mobility costs), and the level of competition among units.

In contrast, categoric units involve interdependencies that develop "on the basis of similarities among the members of a population"[28]; their number and size are related to the size of the population and territory as well as to the level of threat imposed by their environment. As Hawley noted, the nature of the threat can vary—a "task too large for the individual to accomplish in a limited time, such as the harvesting of a crop. . ."; "losing land to an invader"; "possible destruction of a road or other amenity"; "a technological shift that might render an occupation obsolete"; and so on. If a threat is persistent, the actors in a categoric unit will form a more "lasting association," and, if similar units are in competition (say, rival labor unions or ideologically similar political parties), the costs and destructiveness of such competition will eventually lead to their consolidation into a larger categoric unit. Moreover, those categoric units that persist will develop a corporate core to sustain the flow and coordination of resources necessary to deal with the persisting or recurring threat. Categoric units can also get much larger than corporate units because their membership criteria—mere possession of certain characteristics (ethnic, religious, occupational, ideological)—are much more lax than those of corporate units, which recruit members to perform certain specialized and interdependent activities or functions. Of course, the size and number of categoric units are still circumscribed by the size and complexity of the ecosystem and, to a lesser extent than with corporate units, by the mobility costs associated with communication and transportation technologies.

As is indicated in the right portion of Figure 9-1, differentiation of categoric and corporate units leads to the consolidation of units into networks, which create larger subsystems. Such regularization of ties among units engaged in similar and symbiotic activities or functions reduces mobility costs, thereby facilitating growth of the ecosystem (note the feedback arrow to mobility costs).

Differentiation of units and their consolidation into larger networks also have consequences for the concentration of power. Categoric unit formation, and the consolidation of such units into larger networks and subsystems, tend to reduce concentrations of power because various confederations of categoric units will pose a check on one another. In contrast, corporate units are more likely to cause the concentration of power. This concentration is directly related to the capacity of some units to perform key functions and thereby dictate the conditions under which interrelated functions must operate. This control is facilitated by consolidation of networks into subsystems because such networks connect outlying and remote corporate units, via configurations of successive network ties, to those engaged in key functions. Such connections among corporate units, as they facilitate the concentration of power, enable political control to expand to the far reaches of the ecosystem; political and territorial boundaries tend to become coterminous in ecosystems. Yet, centralization of power and extension of control can increase mobility costs as rules and regulations associated with efforts at control escalate, setting limits on how complex the ecosystem can become, without a change in communication and transportation technologies (hence the long feedback arrow at the top of Figure 9-1).

We can add a variable to the model that is implicit but crucial: capital formation. Hawley does not address this issue extensively, so we are clearly adding it to his model. Nonetheless, it is important to recognize that concentration of power also consolidates the flow of resources, facilitating capital formation. If not squandered on maintenance of control, defense, or offensive efforts at military expansion, this capital can be used to expand productivity and, indirectly, to change the knowledge and technological base of the system (note the long feedback arrows at the bottom of Figure 9-1).

As with Tables 9-1 and 9-2, Table 9-4 extracts the most crucial hypothesis from Hawley's many hypotheses. As territory and population size increase, because of new knowledge about production, transportation, and communication, it is possible and perhaps necessary to differentiate units around specific functions. This is particularly true for corporate units, which represent clusters of interdependencies revolving around a particular function. Categoric units form in response to threats, which ultimately stem from the competition that results from increases in population size, productivity, and markets. As the number of differentiated units in a system increases, the number of relations increases at an exponential rate, increasing mobility costs. Corporate and categoric units both tend to consolidate into larger networks, forming subsystems and reducing mobility costs. But the effects of corporate and categoric unit differentiation and consolidation on the concentration of power vary. Corporate units consolidate, centralize, and extend power and regulation, whereas categoric units form power blocks that diffuse power in a system of checks and balances.

As with the other propositions, the many specific hypotheses in Hawley's scheme can be deduced from these and from the scenario

TABLE 9-4 Basic Propositions of Patterns of Ecosystem Differentiation

I. The greater the size of a populations and its territory and the greater the selection pressure stemming from competition among members of the population are, the greater will be the differentiation of functions and the number and size of corporate units, to the maximum allowed by mobility costs.

II. The greater the size of a population and the greater the threats posed by competition and environmental change to actors in similar situations are, the greater will be the number and size of categoric units, to the maximum allowed by mobility costs.

III. The greater the number and size of categoric and corporate units are, the more the potential number of relations increases as a geometric rate and the greater is the amount of time and energy allocated to mobility.

IV. The greater the number of relations among units and the higher the costs of mobility are, the more likely are differentiated units to establish networks and combine into more inclusive subsystems, thereby reducing mobility costs.

V. The more differentiated corporate units are, the more centralized is power around those units and subsystems performing key functions, whereas, the more differentiated categoric units are and the greater is their size, the less centralized is power.

VI. The more concentrated power is, the the more prominent networks and subsystems are, the more extensive is political regulation of units in the ecosystem.

delineated.[29] Thus the propositions in Tables 9-1, 9-2, and 9-4 do not do full justice to the depth and extent of Hawley's scheme. They are intended as more abstract statements rather than as the many "hypotheses" that punctuate Hawley's theory.

In sum, then, Hawley's ecological theory retains some important ideas of early sociology. One of these ideas is the obvious but often ignored view that "society" represents an adaptation of the human species to its environment. Another related idea is that it is not possible to understand human social organization without reference to the interchanges between environment and internal social structure. Yet another crucial idea is that the basic dynamics of a society revolve around (1) aggregation of actors in physical space, competition, and differentiation; and (2) integration through subsystem formation and centralization of power. Still another useful point is the emphasis on population size, territory, productivity, communication and transportation technologies, and competition as important causes of those macro structural processes—differentiation, conflict, class formation, consolidation of power, and the like—that have long interested sociologists. Finally, a significant, though problematic, idea is that the altered flow of resources—energy, information, materials—into the system is the ultimate source of social system growth and evolution.

THEORIES OF URBAN ECOLOGY

Before and during the time that Hawley was raising urban ecology to the macro-level, urban ecology remained a viable and vibrant meso-level theoretical approach. Recast as the study of spatial processes, theorists sought to explain such variables as the size of settlements, the concentration of populations within these settlements, the rate and form of geographical expansion of settlements, and the nature of connections among settlements. Much of what is termed "urban sociology" has sought to examine specific cases empirically, just as the original Chicago School once used the city of Chicago for its laboratory. But, at a more purely theoretical level, an effort was made to conceptualize urban processes generically as fundamental processes influencing patterns of organizing a population in space. This latter theoretical thrust can properly be seen as ecological, and the kinds of models developed by these spatial theorists owe a great deal to early Chicago School ecologists.

This debt can be best appreciated by examining Figure 9-2, which presents a composite and abstracted model of various approaches in urban ecology.[30] This model does not represent any one theorist's ideas but, rather, communicates the general thrust of various approaches combined.[31] As can be seen on the left of the model, technology and demographics influence two important variables, evident in Spencer's and Durkheim's as well as in Hawley's approaches. These two variables are (1) the level of development in communication and transportation technologies and (2) the level of production of goods and services. Population size and technology both determine the level of production directly, and, as can be seen by the arrows flowing from right to left into production, other forces made possible by expanded production feed back and increase production even more. Similarly, transportation and communication technologies set into motion many urban processes, and these also feed back, especially via production, to increase the level of these technologies and, in turn, the store of a population's technology in general.

Both technology and production increase the scale of the material infrastructure of a population—that is, its roadways, canals, ports, railroads, airports, subways, buildings, and all other physical structures built in space. The scale of this infrastructure gets an extra boost as the distributive capacities of a population increase—that is, its capacity to move information, materials, goods, and services about space. As can be seen along the bottom of the model in Figure 9-2, the volume and velocity of markets is important in this process; among populations with well-developed market systems, distributive activities increase. They do so because markets create new kinds of administrative and authority systems—banks, governmental agencies, insurance, sales, advertising, services, wholesale and retail outlets, and all the organizational structures required to sustain high volume markets. These are labeled, respectively, scale of administrative infrastructure and centralization of administrative authority in the model, and they have important effects on not only increasing the level of distribution but also on settlement patterns.

These latter effects are most noticeable on the size and density of settlement patterns. Administrative infrastructures and authority systems directly influence the size and density of settlements by concentrating activities and, thereby, pulling a population to urban areas. These administrative variables also operate indirectly on density by increasing distribution and production which, in turn, expands transportation and communication technologies and the scale of the material infrastructure that have their own effects on increases in the size and density of settlements.

Immigration patterns are influenced not only by transportation and technology but also by the level of production. When these are high, existing dense settlements become magnates for new immigrants, especially in settlements with dynamic market systems and administrative structures that offer opportunities to secure a living and other resources. Immigration, in turn, increases the size and density of settlements, particularly when previous waves of similar immigrants have already settled and can provide friends, relatives, and others of similar origins a place to house themselves and, perhaps, job opportunities.

As the size and density of settlements increase, these forces concentrate populations in ways that discourage geographical expansion of the urban area. But eventually, the population in urban cores must begin to

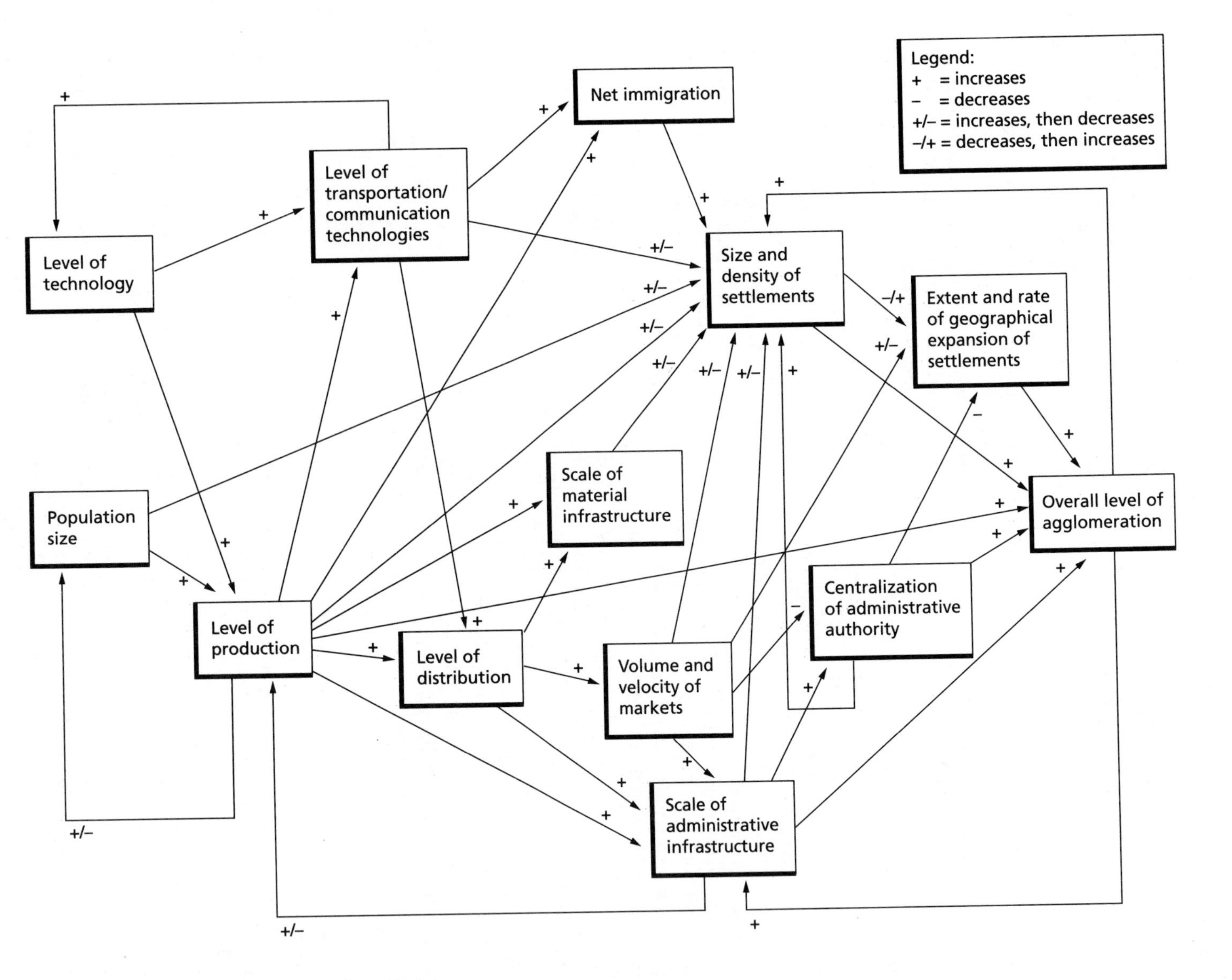

FIGURE 9-2 The Abstracted Urban Ecology Model

move outward. Such movement is facilitated by markets, especially real estate markets but also markets that can distribute goods and services in new locations. Centralized authority systems, such as governmental functions, can for a time discourage movement too far from this center of administrative control, but in the end, the growing size and density of the population, coupled with real estate and other markets, enable or force some of the population to extend the boundaries of the settlement or to create new settlements.

The term on the far right of the model in Figure 9-2—agglomeration—is meant to connote what is evident all over the world today: Outward movement of settlements, even when this movement involves creating separate settlements, eventually leads to relative contiguous systems of dense settlements—often termed urban or suburban "sprawl." That is, as movement out from the original urban core occurs, new settlements typically remain within physical proximity of this core; as these new areas attract migrants from the older core or new immigrants from other urban areas, these areas also become large and more dense. Eventually, they begin to bump into each other, creating high levels of agglomeration or the proximity of contiguous and relatively dense settlements spread across a comparatively large geographical space. Such agglomeration increases the scale of the administrative infrastructure, which, in turn, increases governmental functions and its administrative structure across the urban space. Via the reverse causal arrows in the model, agglomeration increases, indirectly, the level of distribution of goods, services, materials, and information around the agglomerated urban space, which, in turn, affects the level of production and transportation and communication technologies.

Much urban sociology describes specific empirical cases within these general ecological dynamics. Yet, from these more specific empirical studies that began with the Chicago School (see Figure 7-1 on page 89) have emerged interesting generalizations that can supplement the processes outlined in the model presented in Figure 9-2. One older principle is that the density of settlements in an urban area declines exponentially (that is, at an accelerating rate) as the distance from the center of an urban area increases.[32] This idea follows from early Chicago School observations that high demand for space in the core of urban areas will raise market prices and force out those who cannot afford to live or do business in the core; these actors must now assume the additional mobility costs of settling in lower-cost, outlying areas.[33] Yet, as more recent studies indicate,[34] the recent technological and organizational changes, especially those associated with information technologies, have tended to attenuate the connection between the central urban core and outlying areas. In fact, there is a movement of material and administrative infrastructures from the core to less densely settled outlying regions, creating a more polycentric system across agglomerated settlement patterns. Thus, the densities of settlements in outlying areas can increase as the distance from the old settlement core increases, but eventually, the principle that settlement density decreases with distance from urban core or cores in a polycentric system will become operative.

A related principle is that the relative size of settlements or cities decreases as movement from the urban core or cores increases.[35] That is, as movement from the large cities occurs, the size and density of settlements will decrease in a pattern: Large urban cores will be surrounded by mid-sized cities which, in turn, will be connected to smaller settlements.

Another related principle is that the flow of resources across settlements will reflect the degree to which they constitute an integrated system, especially in their markets and hierarchies of governmental structures.[36] When settlements are connected by markets and governmental agencies, the flow of resources—information, goods, and services—will be more rapid and efficient.

These and other principles specify in more detail what is subsumed in the model in Figure 9-2 under the label of agglomeration. These kinds of principles, in essence, indicate the ways that settlements become connected to each other and form ever-larger settlement patterns in physical space. Many of these generalizations are time-bound and relevant to particular empirical cases, but they do point to several more generic and fundamental forces that organize a population in physical space. Thus, the general intent of the early Chicago School of urban ecology has been retained in more recent work: to see the patterns of settlement of populations in physical space and, then, to develop more abstract generalization describing these patterns.

THEORIES OF ORGANIZATIONAL ECOLOGY

One creative extension of theory during the last twenty years has been the analysis of organizational dynamics from an ecological perspective. In these theories, *populations of organizations* of a given type are viewed as competing for resources, with selection favoring those most fit in a given environment. Thus, the rise and fall in the numbers and proportions of various kinds of organizational forms in a society can be seen as a kind of Darwinian struggle in which organizations compete with each other in resource niches, dying out if they are unsuccessful or, if they can, moving to find a new resource niche in which they can survive. The first well-developed theory about the ecology of complex organizations was presented by Michael Hannan and John Freeman in the late 1970s[37]; later others have extended their approach, typically by analyzing empirically specific populations of organizations. We will first examine Hannan and Freeman's general theory, then review a creative addition to theorizing on organizational ecology by Miller McPherson and various collaborators.

Michael T. Hannan and John Freeman's Ecological Theory

Hannan and Freeman had an important insight[38]: Populations of organizations of various kinds can be viewed as competing for resources. For example, automobile companies, clothing outlets, newspapers, governmental agencies, service clubs, and just about any organized corporate unit depend on particular kinds and levels of resources from their respective environments. Thus, a population of organizations, such as automobile companies, can be seen as competing in the same resource niche; for automobile companies, the resource environment consists of those who can afford to buy cars. This basic situation is analogous to evolutionary processes in that organizations must compete with each other to secure resources, particularly as the number of organizations occupying a given niche increases; from such competition comes selection of those organizational forms that are most fit. With this basic insight, Hannan and Freeman extended the analogy, and Figure 9-3[39] attempts to summarize all the key variables in their theory, as it has developed during the last twenty years.[40]

Hannan and Freeman's basic question focused on why organizations of a given type die out and others increase in frequency. The key dynamic is shown at the center of the model in Figure 9-3: *competition within a population of organizations* for resources. High levels of competition increase the *selection pressures* on organizations; those that can secure resources in this competition survive, and those that cannot will fail or move to another resource niche. The theory then examines the forces that increase competition

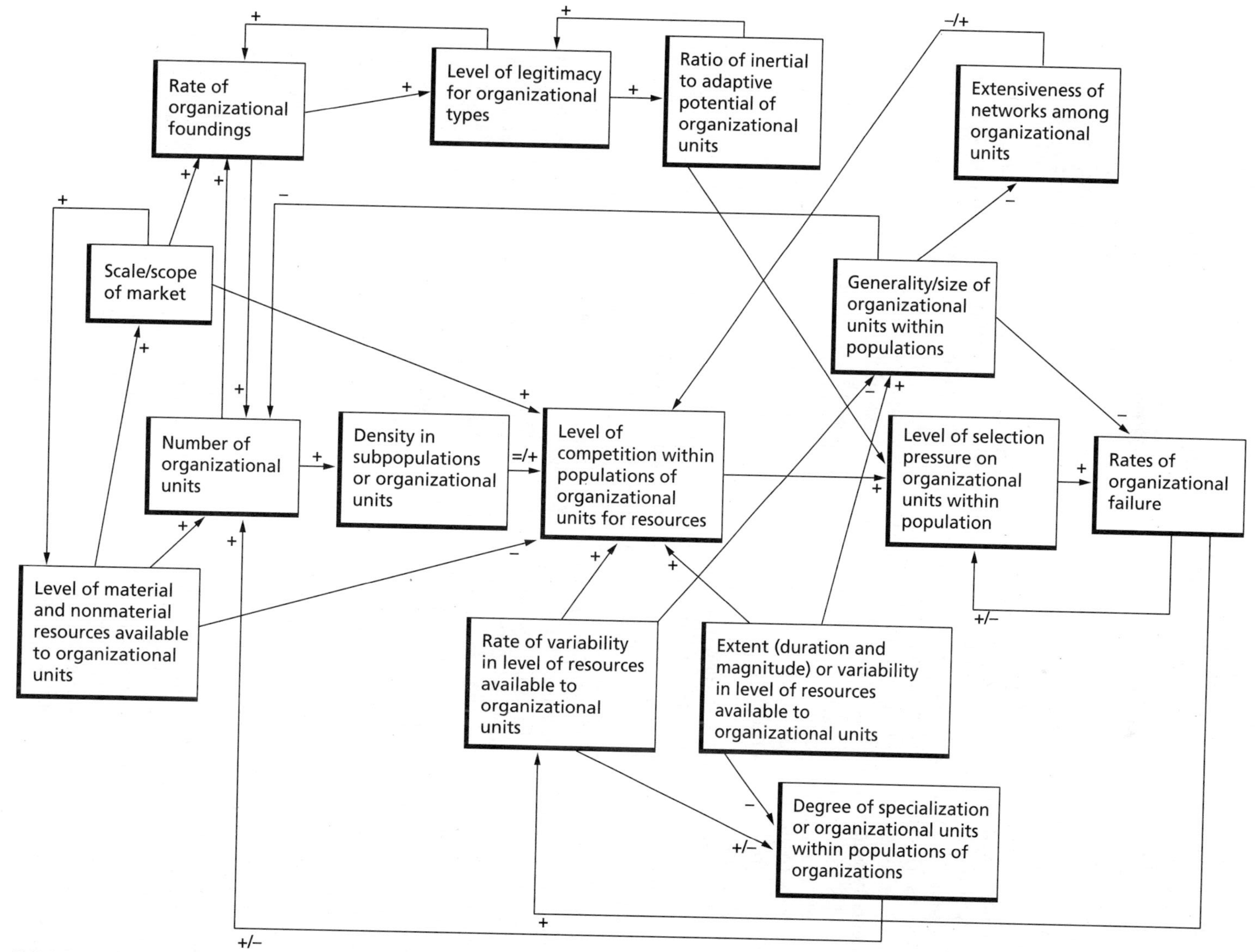

FIGURE 9-3 Hannan and Freeman's Ecological Model

and selection. One critical set of organizational forces is presented in the middle of the model, moving from left to right: the *number of organizations* of a given type increases the *density of organizations* in a niche, thereby increasing *competition, selection*, and *rates of organizational failure*.

Another force increasing competition is open and free markets. Such markets institutionalize competition, forcing ever more organizations to compete with each other for customers, members, or any other resource. Thus, as the *scale and scope of markets* increase, the level of competition increases, especially when niche density is high. If monopolies can emerge or government regulates markets extensively, however, then the level of competition is reduced, thereby lowering selection pressures and rates of organizational failure.

Still another set of variables moves across the top of the model, from left to right. When an organization of a given type first emerges in a niche, it must legitimate itself by surviving, and once it enjoys success, then the *rate of organizational foundings*, or the creation of new organizations of this type, will increase. These new foundings escalate niche density, competition, and selection, but they also do something else: They make organizations of a given type *legitimate*, which only encourages more foundings. With legitimation also comes what is phrased in the model as the *ratio of inertial to adaptive* tendencies in organizations. When organizations have structured themselves successfully in a particular manner and thereby achieved legitimacy, they can also develop structural rigidities or inertial tendencies. They become conservative, locked into the old ways of performing activity. These inertial tendencies give selection processes something to work on. As density in a resource niche increases or the level of resources declines, then those organizations that are too rigid or inertial are likely to be selected out of the population of organizations, whereas those that reveal flexibility or new and creative ways of organizing themselves in the pursuit of resources will be more likely to survive.

A third set of variables moves across the bottom portion of the model in Figure 9-3. The resources available to organizations will vary in several respects. One source of variation in resources is the *rate of variability*, or how often resources increase and decline. Are resources constantly shifting, or is the fluctuation gradual and slow? Another source of variation in the available resources is the *magnitude and duration of variability*, or the degree and length of fluctuation between high and low periods of resource availability. When there is rapid fluctuation in resources, specialized types of organizations are likely to emerge and be able to out-compete larger and more generalized organizational structures that, because of their inertial tendencies, cannot move fast enough to respond to rapid shifts in the resources available. When the magnitude of shifts is great and prolonged, however, the specialization of organizations is discouraged because larger and more generalized organizations can ride out the dramatic drop in the level of resources available more effectively than can smaller and highly specialized organizations; these larger organizations have other resource niches that they can pursue, and they typically have bigger resource *reserves*, whereas the more specialized organizations are likely to have too few reserves to survive large drops in the resources available.

As can be seen from the arrows in the model going into the competition variable, environmental change, whether a rapid or severe drop in resources, or both, will increase the struggle among organizations. As Darwin noted, when the environment changes, the resource niches of species are disrupted, escalating competition and natural selection. When change occurs over longer periods and is of high magnitudes, selection favors *larger, more generalized organizations* that draw from more than one niche and can ride out fluctuations of high magnitude in any one niche. As organizations become large and general, they often create *extensive networks* of ties and agreements to reduce competition that could potentially select them out. Examples of these networks can include cartels, trade agreements, interlocking boards of directors in private corporations, liaisons with government, joint production agreements, price fixing among oligopolies, and many other mechanisms by which organizations seek to reduce competition. These networks in effect decrease the density among organizations and hence their competition, which, in turn, reduces their rates of organizational failure.

Hannan and Freeman's theory has thus taken Spencer's and Durkheim's down to a more meso-level of analysis, but more directly, the theory adapts Darwinian ideas to the analysis of organizations. Thus, Hannan and Freeman inject a new meso-level phenomenon into ecological analysis: the dynamics within populations of complex organizations. Their approach has stimulated an entirely new branch of research and theory in sociology, and this branch has dominated ecological theorizing during the last two decades, although new, more macro approaches to ecology have also begun to rival the preeminence of theory and research on the ecology of organizations, as we will explore in the next chapter. First, however, we should examine another creative use of Darwinian ideas in the analysis of organizations.

J. MILLER MCPHERSON'S ECOLOGY THEORY

J. Miller McPherson and various collaborators have developed a variant on Hannan and Freeman's model of organizational ecology.[41] McPherson's empirical work has been primarily on voluntary associations and organizations, and this emphasis has led to several additional insights into the dynamics of organizational ecology. McPherson begins with an idea that he adapted from Peter M. Blau's theory of macrostructure, which is summarized in Chapter 39: The environment of organizations consists of members of a population who reveal a diversity of characteristics, such as age, sex, ethnicity, income, education, recreational interests, and so on. These characteristics distinguish individuals from each other and, often, become important markers of categorization (as is the case with sex and ethnicity) and inequality (as with income and years of education). These characteristics are also potential resource niches for organizations seeking members and clients. Thus, McPherson conceptualizes the diversity of characteristics among members of a population as *Blau-space* in deference to the theorist whose work gave him this idea (again, consult Chapter 39 for the details of Blau's theory). Blau-space is the environment of organizations, and the greater diversity of characteristics that differentiate members of a population, the greater is the number of resource niches in Blau-space available for organizations to recruit members and clients.

Figure 9-4 summarizes the model developed by McPherson in more general terms, giving us a way to visualize the causal relations among the variables in the theory. The *size of the population* is, as Spencer and Durkheim both recognized, an important determinant of the level of *diversity of characteristics* of individuals in Blau-space, as is indicated on the left of the model. The larger the population is, the more likely the characteristics of its members will be differentiated. Moreover, population size, per se, generates resources in the niches of Blau-space; that is, the more people there are, the more resources are available for organizational systems.

Population size also reduces the *density of networks* among members of a population;

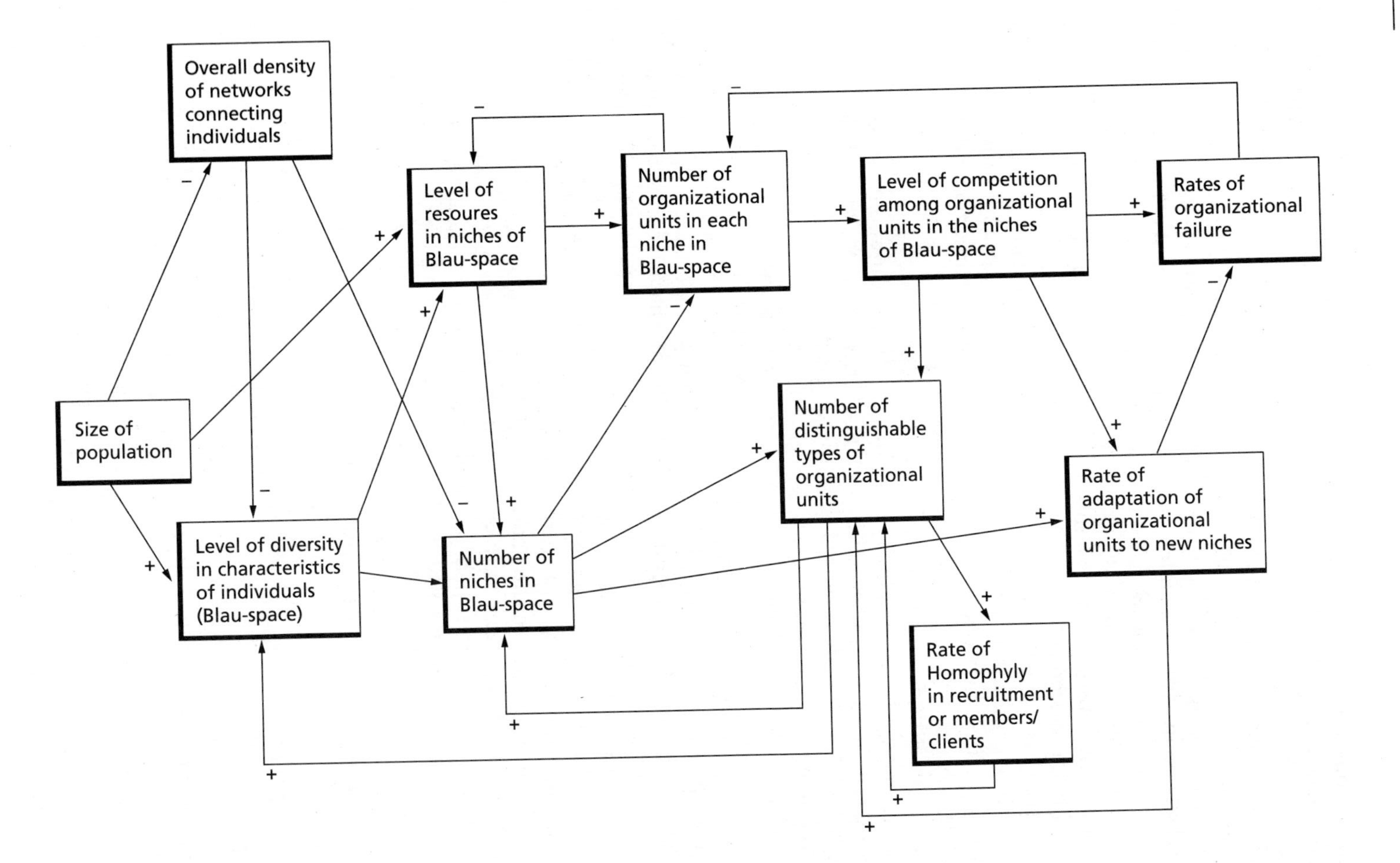

FIGURE 9-4 McPerson's Ecological Model

the more people there are to organize, the less likely individuals are to be connected to each other directly or indirectly (as summarized in Chapter 38 on networks, "network density" is a concept denoting the degree of connectedness among actors). With low density, or low rates of connectedness among members of a population, these members are more likely to develop distinctive characteristics because they do not have direct contact and the informal social control and conformity that such contacts generate. Thus, *low network density* among members of a large population increases the niches in Blau-space available that organizations can exploit.

As the number of niches in Blau-space increases, the number of organizational units in each niche will also tend to increase, and as their numbers grow, the *level of competition* among organizational units in a niche will begin to escalate. In turn, as niches become densely populated with organizational units, *rates of organizational failure* will increase, thereby lowering the number of units competing for resources in a particular niche in Blau-space.

Competition among organizations will increase the *number of distinguishable organizational units* for two reasons: First, each organization seeks to distinguish itself from competitors, thereby increasing the diversity of organizations in a niche. Second, as organizational units distinguish themselves, they create more niches in Blau-space because the members of organizations can reveal somewhat different characteristics. Indeed, as the model portrays, there is a mutually reinforcing cycle between number of *niches* and number of *distinguishable types of organizations* in Blau space. Competition only accelerates these forces.

Organizations in Blau-space also become distinctive because they tend to recruit members with similar characteristics, or what is termed rate of *homophyly* in the model. Thus, for example, service organizations such as the Lions, Kiwanis, Optimists, American Legion, and the like will seek members whose characteristics converge; as these organizations do so, they sustain their distinctiveness and, hence, the diversity of characteristics among members in a population and the corresponding niches in Blau-space. Competition for members or clients, however, places selection pressure on organizations, forcing them to adapt and change if they find themselves less able to compete in a niche. For example, in recent decades in America, service organizations have had great difficulty sustaining their memberships because the number of individuals in this niche has declined as the demographics and structure of the society have changed. Such competition has led to a decline in membership and some organizational failures, but it has also done something else: Some organizations have been forced to seek new niches in Blau-space. For example, a service organization might shift from a middle-income and high-education pool of members to lower-income and less-educated members because the competition is less intense. *Rates of adaptation* to new niches are influenced not only by the level of competition, but also by the number of niches in Blau-space. If there are many niches, then an organization that is having trouble recruiting members and clients has options that would not be available if there were only a few niches in Blau-space. Adaptation to new niches is particularly likely when there are adjacent niches that do not require a complete restructuring of the organization. For example, when the polio vaccine was created, the March of Dimes lost its resource base because its cause for recruiting donations was obviated; to survive, the March of Dimes moved to a new but adjacent charity niche that still involved the basic structure of soliciting charitable contributions.

As is evident, the key idea of McPherson's model is much the same as other organizational ecological models[42]: competition and selection among organizations, because these lead to organizational failure or movement of organizations to new niches. McPherson's most important addition is expansion of what constitutes the resource environment of organizations. Hannan and Freeman's model connoted a more money- and market-driven image of the resource environment, whereas McPherson's model expands the notion of what constitutes resources. Virtually any set of characteristics that distinguishes people in a population can become a resource niche for organizations that seek members, clients, or customers. The more varied the Blau-space is, the more diversity of organizational forms the environment can support and the more likely are less successful organizations in one niche to move to new, adjacent niches in efforts to survive.

CONCLUSION

Ecological models in sociological theory have all represented an analogy to the forces of evolution, particularly natural selection through competition. Whether operating at the macro- or meso-level, these theories seek to demonstrate that certain parallel processes operate in populations of collective actors within societies and populations of species in the biotic world.

At the same time that these ecological theories were developing, the genetic side of evolutionary theory was becoming codified into sociobiology, as we examined in the last chapter. The reaction against sociobiology created a great deal of acrimony, but emerging from this controversy came a softening of the extremes of sociobiology as well as a new kind of biologically oriented sociology. Here, old questions about human evolution, human nature, and societal development have become blended with ecological models from biology, producing what we will term for convenience a new type of "evolutionary sociology." This new evolutionary theorizing is an eclectic mix of theoretical ideas from biology, genetics, developmental notions about increasing complexity from functionalism, bio-ecology, and more sociologically oriented ecology models such as those examined in this chapter. There is little coherence to this new evolutionary theory, but it goes beyond mere analogies and uses biological and ecological concepts to explain patterns of human behavior and social organization.

In a very real sense, this newer approach has little in common with the models presented in this chapter, despite the common view that a science of human ecology should draw concepts from its biological counterpart. Yet, as we will come to appreciate, this newer approach is not nearly as developed or coherent as the ecological models examined in this chapter. Thus, ecologically oriented evolutionary sociology represents one of the emerging trends in evolutionary theorizing.

NOTES

1. Amos H. Hawley, *Human Ecology: A Theory of Community Structure* (New York: Ronald, 1950).

2. Amos H. Hawley, "The Logic of Macrosociology," *Annual Review of Sociology* 18 (1992), pp. 1–14.

3. The following list of titles from Hawley's work reviews this progression of thinking that culminated in the last reference at the end of this note. "Human Ecology," in *International Encyclopedia of the Social Sciences*, ed. D.C. Sills (New York: Crowell, Collier and Macmillan, 1968); *Urban Society: An Ecological Approach* (New York: Ronald, 1971 and 1981); "Human Ecology: Persistence and Change," *American Behavioral Scientist* 24 (3, January 1981), pp. 423–444; "Human Ecological and Marxian Theories," *American Journal of Sociology* 89 (1984), pp. 904–917; "Ecology and Population," *Science* 179 (March 1973), pp. 1196–1201; "Cumulative Change in Theory and History," *American Sociological Review* 43 (1978), pp. 787–797; "Spatial Aspects of Populations: An Overview," in *Social Demography*, eds. K.W. Taueber, L.L. Bumpass, and J.A. Sweet (New York: Academic, 1978); "Sociological Human Ecology: Past, Present and Future," in *Sociological Human Ecology*, eds. M. Micklin and H.M. Choldin (Boulder, CO: Westview, 1980); and, most significantly, *Human Ecology: A Theoretical Essay* (Chicago: University of Chicago Press, 1986).

4. Hawley, "Human Ecological and Marxian Theories" (cited in note 3), p. 905.

5. Herbert Spencer, *The Principles of Sociology*, 3 volumes (New York: Appleton, 1898; originally published in serial form between 1874 and 1896).

6. Émile Durkheim, *The Division of Labor in Society* (New York: Free Press, 1933; originally published in 1893).

7. Hawley, *Human Ecology: A Theoretical Essay* (listed in Note 3).

8. Hawley, "Human Ecology" (cited in note 3).

9. Hawley, *Human Ecology: A Theoretical Essay* (cited in note 3).

10. Ibid., p. 13.

11. Hawley, "Human Ecological and Marxian Theories," cited in note 3.

12. Hawley, "Human Ecology," and *Human Ecology: A Theoretical Essay*, p. 32, both cited in note 3.

13. Hawley, *Human Ecology: A Theoretical Essay* (cited in note 3), p. 34.

14. Hawley, "Human Ecology," p. 332, cited in note 3.

15. Ibid.

16. Hawley, *Human Ecology: A Theoretical Essay* (cited in note 3), p. 32.

17. Ibid., pp. 43–44.

18. Ibid., p. 46.

19. Ibid., p. 52.

20. Ibid.

21. Ibid., p. 59. Hawley "boxes" lists of variables and then draws arrows among the boxes, but not among the variables within each box. Hence, detailed causal arguments need to be inferred.

22. Ibid., pp. 85–87.

23. Hawley, "Human Ecology" (cited in note 3), p. 329.

24. Ibid., p. 334.

25. Hawley, *Human Ecology: A Theoretical Essay* (cited in note 3), pp. 80–81.

26. Ibid., pp. 68–73, and "Human Ecology," pp. 331–332, cited in note 3.

27. Hawley, *Human Ecology: A Theoretical Essay* (cited in note 3), p. 74.

28. Ibid., p. 70.

29. Ibid., pp. 106–108, 123–124.

30. For a more detailed analysis, see Jonathan H. Turner, "The Assembling of Human Populations: Toward a Synthesis of Ecological and Geopolitical Theories," *Advances in Human Ecology* 3 (1994), pp. 65–91 and *Macrodynamics: Toward Theory on the Organization of Human Populations* (New Brunswick, NJ: Rutgers University Press for Rose Book Series, 1995), Chapter 6.

31. In particular, the model summarizes ideas from Parker W. Frisbie, "Theory and Research in Urban Ecology," in *Sociological Theory and Research: A Critical Approach*, ed. H.M. Blalock (New York: Free Press, 1980); Parker W. Frisbie and John D. Kasarda, "Spatial Processes," in *Handbook of Sociology*, ed. N.J. Smelser (Newbury Park, CA: Sage, 1988); Mark Gottdiener, *The Social Production of Urban Space* (Austin: University of Texas Press, 1985); Amos H. Hawley, *Urban Society: An Ecological Approach* (New York: Ronald, 1981); John D. Kasarda, "The Theory of Ecological Expansion: An Empirical Test," *Social Forces* 51 (1972), pp. 165–175.

32. C. Clark, "Urban Population Densities," *Journal of the Royal Statistical Society*, series A, 114 (1951), pp. 490–496.

33. B.J.L. Berry and John D. Kasarda, *Contemporary Urban Ecology* (New York: Macmillan, 1977).

34. Frisbie and Kasarda, "Spatial Processes" (cited in note 3).

35. This idea was originally formulated by George Zipf, *Human Behavior and the Principle of Least Effort*

(Reading, MA: Addison-Wesley, 1949), and expanded on in other studies: Hawley, *Urban Society* (see note 3); E.G. Stephan, "Variation in County Size: A Theory of Segmental Growth," *American Sociological Review* 36 (1979), pp. 451–461 and "Derivation of Some Socio-Demographic Regularities from the Theory of Time Minimization," *Social Forces* 57 (1979), pp. 812–823.

36. See note 3: Frisbie and Kasarda, "Spatial Processes."

37. Michael T. Hannan and John Freeman, "The Population Ecology of Organizations," *American Journal of Sociology* 82 (1977), pp. 929–964.

38. Ibid.

39. For an analysis of the more macrostructural implications of the variables delineated in Figure 9-3, see Jonathan H. Turner, *Macrodynamics* (see note 3), Chapter 7, and "The Ecology of Macrostructure," *Advances in Human Ecology* 3 (1994), pp. 113–137.

40. For representative works by Hannan and Freeman, see "Structural Inertia and Organizational Change," *American Sociological Review* 49 (1984), pp. 149–164; "The Ecology of Organizational Founding: American Labor Unions 1836–1985," *American Journal of Sociology* 92 (1987), pp. 910–943; "The Ecology of Organizational Mortality: American Labor Unions," *American Journal of Sociology* 94 (1988), pp. 25–52; *Organizational Ecology* (Cambridge, MA: Harvard University Press, 1989).

41. See J. Miller McPherson, "A Dynamic Model of Voluntary Affiliation," *Social Forces* 59 (1981), pp. 705–728; "An Ecology of Affiliation," *American Sociological Review* 48 (1983), pp. 519–532; "The Size of Voluntary Organizations," *Social Forces* 61 (1983), pp. 1044–1064; "A Theory of Voluntary Organization," in *Community Organizations*, ed. C. Milofsky (New York: Oxford University Press, 1988), pp. 42–76; "Evolution in Communities of Voluntary Organization," in *Organizational Evolution*, ed. J. Singh (Newbury Park, CA: Sage, 1990); J.M. McPherson, P.A. Popielarz, and S. Drobnic, "Social Networks and Organizational Dynamics," *American Sociological Review* 57 (1992), pp. 153–170; J.M. McPherson, J. Ranger-Moore, "Evolution on a Dancing Landscape: Organizations and Networks in Dynamic Blau-Space," *Social Forces* 70 (1991), pp. 19–42; and J. M. McPherson and T. Rotolo, "Testing a Dynamic Model of Social Composition: Diversity and Change in Voluntary Groups," *American Sociological Review* 61 (1996), pp. 179–202.

42. For some general overviews of research and theory on organizational ecology, see Glenn R. Carroll, ed., *Ecological Models of Organizations* (Cambridge, MA: Ballinger, 1988) and "Organizational Ecology," *Annual Review of Sociology* 10 (1984), pp. 71–93; Jitendra V. Singh and Charles J. Lumsden, "Theory and Research in Organizational Ecology," *Annual Review of Sociology* 16 (1990), pp. 161–195.

10

The Continuing Tradition II:

New Evolutionary Theories*

As we saw in the last chapter, the most prominent form of biologically inspired theorizing in sociology has been ecological, where competition and selection processes are seen as the force behind social differentiation of whole societies, spatial arrangements in urban areas, and distributions among populations of complex organizations. Alongside ecological approaches, developmental theories of societal evolution from simple to ever more complex forms have also persisted, arising with Herbert Spencer and Émile Durkheim and moving forward within a number of theoretical traditions in the second half of the twentieth century. And, as we saw in Chapter 9 on the maturing tradition, with the modern synthesis in biology, a new form of biologically inspired theory has emerged: sociobiology, where emphasis is on genes as the driving force of not only organismic forms but social forms as well.

In this chapter, we will review some of the more promising types of evolutionary sociology. None of these approaches dominates, but each attempts to reconnect biology and sociology in creative ways. Moreover, each of the theories draws inspiration from different aspects of the modern evolutionary synthesis in biology. What emerges is a rather eclectic mix of innovative theoretical leads rather than mature theories. Yet, as the twenty-first century arrives, these leads might well develop into more coherent theoretical orientations that could become more prominent, perhaps rivaling the ecological theories that have dominated biologically inspired theorizing in sociology for most of the twentieth century.

Although all theories in evolutionary sociology rest on the modern synthesis, these theories can be sorted and grouped into two distinctive, though overlapping traditions:

*This chapter is coauthored with Alexandra Maryanski.

(1) theories working within the sociobiology framework, (2) theories of cross species comparisons, (3) theories employing biological ideas in evolutionary stage models, and (4) theories on the biological nature of humans. We will begin with sociobiology and then examine the rather eclectic mix of theories within a comprehensive evolutionary-ecological framework.

SOCIOBIOLOGICAL THEORIZING

Pierre van den Berghe's Approach

Pierre van den Berghe has consistently been among the most prolific advocates of a biological perspective on human affairs. Van den Berghe believes, it "is high time that we seriously look at ourselves as merely one biological species among many. . . ." Until this shift away from what he calls "species-wide anthropocentrism" is accomplished, sociology will remain stagnant, for there can be no doubt that biological forces shape and constrain patterns of human organization.[1] In his early advocacy, van den Berghe proposed a "biological" approach that corresponded to the instinct approach of some early sociologists and several contemporary thinkers, whereas in his later works, van den Berghe fully adopted the sociobiological approach and, since then, has consistently used sociobiology theory to interpret a variety of social behaviors among humans. Van den Berghe began his turn to sociobiology by posing the question: Why are humans "social" in the first place? His answer is that the banding together of animals in cooperative groups increases their reproductive fitness by (1) protecting them against predators and (2) providing advantages in locating, gathering, and exploiting resources. Such fitness allows the alleles of those who are social to stay in the gene pool.

Three sociobiological mechanisms produce the sociality that promotes reproductive fitness:[2] (1) kin selection (or what was termed in Chapter 7 "inclusive fitness"), (2) reciprocity (or what was labeled "reciprocal altruism in Chapter 7), and (3) coercion. Because each of these mechanisms promotes reproductive fitness of individuals, they lie at the base of sociocultural phenomena. Let us examine each mechanism separately and then see how van den Berghe generates theoretical explanations with them.

Kin Selection. In van den Berghe's view, *kin selection* (that is, a propensity to favor kin) is the oldest mechanism behind sociality. He sees this mechanism, as with all modern human sociobiological arguments, as operating at the genetic level, for "the gene is the ultimate unit of natural selection, (although) each gene's reproduction can be modified by cultural forces and, overall, is dependent on what Robert Dawkins terms its 'survival machine,' the organism in which it happens to be at any given time."[3]

This "survivor machine" carrying genetic material in its genotype need not be consciously aware of how natural selection in the distant past has created "replicators" or genes in its body that seek to survive and to be immortal. But selection clearly favored those replicators that could pass themselves on to new and better survival machines. One early survivor machine was the body of individual organisms, and a later one has been the development of cooperative arrangements among bodies sharing alleles in their genotypes. In van den Berghe's words,

> Since organisms are survival machines for genes, by definition those genes that program organisms for successful reproduction will spread. To maximize their reproduction, genes program organisms to do two things: successfully compete against . . . organisms that carry

> alternative alleles . . . , and successfully cooperate with (and thereby contribute to the reproduction of) organisms that share the same alleles of the genes.[4]

So, in van den Berghe's terms, individuals are *nepotistic* and favor kin over nonkin, and close kin over distant kin, for the simple reason that close kin will share genetic material (in their genotypes):

> Each individual reproduces its genes directly through its own reproduction and indirectly through the reproduction of its relatives, to the extent that it shares genes with them. In simple terms, each organism may be said to have a 100 percent genetic interest in itself, a 50 percent interest in its parents, offspring, and full-siblings, a 25 percent interest in half-siblings, grandparents, grandchildren, uncles, aunts, nephews, and nieces, a 12½ percent interest in first cousins, half-nephews, great-grandchildren, and so on.[5]

The degree of "altruism" for kin or "nepotism" will vary with the amount of genetic material shared with a relative *and* with the ability of that relative to reproduce this material. For van den Berghe, then, "blood is thicker than water" for a simple reason: reproductive fitness, or the capacity to help those "machines" carrying one's genetic material to survive and reproduce.

Van den Berghe is careful, however, to emphasize that these biologically based propensities for nepotism, or kin selection, are elaborated and modified by both environmental and cultural variations. Yet clearly a biological factor operates in the overwhelming tendency of humans to be nepotistic. One cannot, van den Berghe insists, visualize such nepotism as a purely cultural process.

Reciprocity. When individuals exchange assistance, they create a bond of *reciprocity*. Such reciprocity increases the "fitness" of genes carried by the organisms (and organizations of organisms) involved; that is, if organisms help each other or can be relied upon to reciprocate for past assistance, the survival of genetic material is increased for both organisms. Such exchange, or what sociobiologists sometimes term "reciprocal altruism,"[6] greatly extends cooperation beyond nepotism, or "inclusive fitness" and "kin selection" in sociobiological jargon. Thus, reciprocal exchange is not a purely social product; it is a behavioral tendency programmed by genes. Such programming occurred in the distant past as natural selection favored those genes that could create new kinds of survivor machines beyond an individual's physical body and social groupings of close kin. Those genes that could lodge themselves in nonkin groupings organized around bonds of exchange and reciprocity were more likely to survive; today the descendants of these genes provide a biological push for creating and sustaining such bonds of reciprocity.

At this point, van den Berghe reveals the affinity of such arguments with utilitarianism by introducing the problem of "free-riding" (see pages 303–304 of Chapter 24), or, as he terms it, "free-loading." That is, what guarantee does an individual (and its genotype) have that others will indeed reciprocate favors and assistance? Here van den Berghe appears to argue that this problem of free-riding created selection pressures for greater intelligence so that our protohuman ancestors could remember and monitor whether or not others reciprocated past favors. But, ironically, intelligence also generated greater capacities for sophisticated deceit, cheating, and free-riding, which in turn escalated selection pressures for the extended intelligence to "catch" and "detect" such concealed acts of nonreciprocation.

There is, then, a kind of selection cycle operating: Reciprocity increases fitness, but it also leads to cheating and free-riding. Hence, once reciprocity is a mechanism of cooperation, it can generate its own selection pressures for greater intelligence to monitor free-riding, but, ironically, increased intelligence enables individuals to engage in more subtle and sophisticated deceptions to hide their free-riding; to combat this tendency, there is increased pressure for greater intelligence; and so on. At some point culture and structure supplant this cycle by creating organizational mechanisms and cultural ideas to limit free-riding (see Chapter 24 on "rational choice" theory for a discussion of what these sociocultural mechanisms might be).

Van den Berghe does not pursue this discussion, however, except to point out that this cycle eventually produces "self-deceit." Van den Berghe believes that the best way to deceive is to believe one's own lies and deceptions, and in this way one can sincerely make and believe verbal pronouncements that contradict, at least in some ways, one's actual behavior. Religion and ideology, van den Berghe posits, are "the ultimate forms of self-deceit," because religion "denies mortality" and ideology facilitates "the transmission of credible, self-serving lies."[7] This conceptual leap from free-riding to religion and ideology is, to say the least, rather vague, but it does provide a sense of what sociobiology tries to do: connect what might be considered purely "cultural" processes—religion and ideology—to a fundamental biological process—in this case, reproductive fitness through reciprocity.

Coercion. There are limits to social organization for reciprocal exchanges, van den Berghe argues, because each party has to perceive that it receives benefits in the relationship.[8] Such perceptions of benefit can, of course, be manipulated by ideologies and other forms of deceit that hide the asymmetry of the relationship, but there are probably limits to the manipulation of perceptions. Power is an alternative mechanism to both kin selection and reciprocal exchanges because its mobilization allows some organisms to dominate others in their access to resources that promote fitness. *Coercion* thus enables some to increase their fitness at a cost to others. Although this mechanism is hardly unique to our species, humans "hold pride of place in their ability to use to good effect conscious, collective, organized, premeditated coercion in order to establish, maintain, and perpetuate systems of intraspecific parasitism."[9] Coercion allows the elaboration of the size and scale of human organization (states, classes, armies, courts, and so forth), but it is nonetheless tied to human biology as this was molded by selection. That is, genes that could use coercion to create larger and more elaborate social structures as their survival machines were more likely to survive and reproduce.

In sum, then, sociality or cooperation and organization are the result of natural selection because it preserved genes that could produce better survival machines through nepotism (kin selection and inclusive fitness), reciprocity (or reciprocal altruism), and coercion (or territorial and hierarchical patterns of dominance). The linkages between biology, on the one side, and culture and society, on the other, are complex and often indirect, even after we recognize that these linkages occur along the three basic dimensions or axes of nepotism, reciprocity, and coercion. For there have certainly been complex interactions between ecological and genetic factors to produce patterns of human (and many other animal) organization; once initiated, selection processes producing greater intelligence allow culture, as "an impressive bag of tricks," to operate as a force of human evolution.

Conceptualizing Cultural Processes. From van den Berghe's vantage point, culture is created and transmitted in humans through mechanisms fundamentally different from those involved in genetic natural selection.

Actually, van den Berghe portrays cultural evolution in Lamarckian rather than Darwinian terms: "Acquired cultural characteristics, unlike . . . genetic evolution, can be transmitted, modified, transformed, or eliminated through social learning." In recent years, van den Berghe has come to see culture in some ways as "an emergent phenomenon" that provides humans with another method of adaptation. For van den Berghe, the issue is not the genes versus culture but their interface, for both are intimately intertwined. Although culture is an outgrowth of biological evolution, it now has an autonomy, he says, that provides humans with the ability to modify their own genotypes, making the intricate feedbacks between nature and nurture more complex than ever.[10] So culture is not a *separate* entity; it is yet one more product and process of biological evolution driven by natural selection as it produced genes trying to maximize their fitness by nesting themselves in better and better survival machines.

Explanations of Social Phenomena with Sociobiology. At various times van den Berghe has used the concepts of sociobiology to explain such empirical phenomena as kinship systems, incest taboos, ethnicity, skin color, sexual selection, and classes. Probably his two most detailed empirical analyses are on (1) kinship systems and (2) ethnicity. Each of these will be briefly summarized.

1. *Kinship*. In one of his early sociobiological articles, van den Berghe and David Barash endeavored to explain various features of kinship systems by the behavioral strategies males and females pursue to maximize their fitness (that is, keep their alleles in the gene pool). Because the details of their suggestive arguments can be complex, just one example of the kinds of arguments that they make will be summarized.[11]

The widespread preference in human societies for polygyny (males with the option for multiple wives), hypergamy (females marrying males of higher-ranking kin groups), and double standards of sexual morality (males being given more latitude than females) can be explained in terms of reproductive fitness strategies. Women have comparatively fewer eggs to offer in their lifetime, have intervals when they are infertile (for example, during lactation), and, even in the most liberal or egalitarian societies, have to spend more time than men raising children. Hence a female will seek a reproductive strategy that will assure the survival of her less abundant genetical material, and the most maximizing strategy is to marry a male with the resources and capacity to assure the survival of her offspring (and, thereby, one-half of her genotype). Thus a woman will seek to "marry up" (hypergamy) in the sense of securing a man who has more resources than her kinship group.

On the male side, men produce an uninterrupted and large supply of sperm and have fewer child-care responsibilities, so they can afford to be more promiscuous with little cost (and they will derive some benefit from the fitness that comes with spreading their genes around). Although men have an interest in assuring that as many women as possible bear children with one-half their genes, a male cannot know that a female is bearing his child if she is promiscuous, so men also have an interest in restricting female sexual activity through polygyny to assure that their genes are indeed contained in the genotype of the children born to a female (creating limits of female sexuality outside marriage). Thus, what are commonly viewed as purely cultural phenomena—the preference for polygyny in human societies (or monogamy with promiscuity), hypergamy ("marrying up"), and sexual double standards (favoring male promiscuity)—can be explained as the result of varying fitness strategies for males and females.

2. *Ethnicity*. Turning to another empirical example, van den Berghe has sought to apply

sociobiology to what has always been his main area of research, ethnicity.[12] Again, kin selection is his starting point, but he extends this idea beyond family relatives helping one another (to maximize their fitness) to a larger subpopulation. Historically, larger kin groups (composed of lineages) constituted a breeding population of close and distant kin who would sustain trust and solidarity with one another while mistrusting other breeding populations. Van den Berghe coined the term *ethny* for "ethnic group" and views an ethny as an extended nepotism of these more primordial breeding populations. An ethny is a cluster of kinship circles that is created by endogamy (intermarriage of its members) and territoriality (physical proximity of its members and relative isolation from nonmembers). An ethny represents a reproductive strategy for maximizing fitness beyond the narrower confines of kinship, because, by forming an ethny—even a very large one of millions of people—individuals create bonds with those who can help preserve their fitness, whether by actually sharing genes or, more typically, by reciprocal acts of altruism to fellow "ethnys." An ethny is, therefore, a manifestation of more basic "urges" for helping "those like oneself." Whereas ethnys become genetically diluted as their size increases and become subject to social and cultural definitions, the very tendency to form and sustain ethnys is the result of natural selection as it produced nepotism. Van den Berghe's argument is, of course, much more complicated and sophisticated, but we can get at least a general sense for how a supposedly emergent phenomenon—ethnic groups—is explained by reduction to a theoretical perspective built on the principles of genetic evolution.

In summary, the sociobiology models developed by van den Berghe emphasize the interactive effects among genes, culture, and environment. Individuals are viewed as "selfish maximizers" who seek to maximize their own inclusive fitness. To accomplish this goal, different reproductive strategies are employed with variations in reproductive strategies among human societies mostly the result of different cultural adaptations to particular environmental conditions. In recent years, however, van den Berghe and collaborators have highlighted the power of culture in overriding the reproductive consequences of human actions, thereby subverting the genes' game of reproducing themselves. Van den Berghe[13] now says that this subversion of the genes is especially evident in industrial societies where many individuals "are no longer maximizing fitness" because the contraceptive technology and the comforts and security of material affluence have separated the hedonistic rewards of the "good life" from reproductive efforts. Thus, in modern societies, van den Berghe suggests that a more complex model is necessary to assess the quality versus quantity of offspring and the balance between human needs for luxury goods versus reproductive investment.

Joseph Lopreato's Approach

In the "hard-core" days of sociobiology (see Chapter 7), scholars followed the evolutionary logic that, in the past, natural selection favored those behavioral adaptations that increased the fitness of individuals in a given environment. All nonadaptive behaviors were weeded out and adaptive behaviors were preserved and transmitted for the ultimate function of maximizing fitness. The research objective, then, was to discover how particular social behaviors were adaptive for their fitness payoff.[14] In the words of R.D. Alexander, who underscored the importance of viewing humans as organisms whose behaviors evolved to maximize reproductive fitness, "*all* organisms should have logically evolved to avoid every instance of beneficence or altruism unlikely to bring returns greater than the expenditure it entails."[15]

Hence, through the application of such concepts as "kin selection" and "reciprocal altruism," even acts of cooperation or kindness were viewed as "selfish genes."

Joseph Lopreato[16] believes the "maximization principle" is still the fountainhead of sociobiology, but if we want to keep "with the logic of natural selection," its strict application is unwarranted. For this reason, Lopreato has undertaken the task of overhauling the maximization principle.

In revamping the maximization principle, Lopreato begins by rejecting the assumption that *all* adaptations are either related to reproduction or organized around the ultimate goal of reproductive success. Instead, he proposes that for all organisms there is only a *tendency* for individuals to behave in a way that maximizes their reproductive success.[17] This "could hardly be otherwise" he maintains, because many individual behaviors are clearly neutral or outright maladaptive, and there are good evolutionary reasons for this variability. As he points out, genotypic variation in every generation must exist for selection to act on. This pool of variability must include both neutral traits and maladaptive traits to provide a deviation of fitness from the maximization principle. This variability is retained when genes are recombined into new genotypes, virtually guaranteeing a differential in the adaptive quality of organisms. Over time this differential will, in turn, produce "a more or less adaptive fit between variations and environmental pressures."[18]

The second step taken by Lopreato is acknowledging that cultural evolution has placed heavy constraints on the maximization principle. In particular, the trappings of human culture have greatly augmented the basic mammalian "pleasure principle," which is now skewed in humans toward creature comforts, rather than the maximization of reproductive success. This requires a readjustment of the maximization principle: "Organisms are predisposed to behave so as to maximize their inclusive fitness, but this predisposition is conditioned by the quest for creature comforts."[19]

The third step taken by Lopreato is suggesting that complex causal agents in both evolutionary and cultural phenomena have generated a widespread predisposition for "self-deception," or the ability to engage in one form of behavior in the belief that it is actually another form of behavior.[20] Once self-deceptive behaviors evolved, Lopreato says, the necessary conditions were created for true ascetic altruism to evolve where individuals in varying degrees evidence a type of "Mother Theresa complex" or self-sacrifice that can, in some instances, redirect the sex drive away from physical satisfaction to "spiritual satisfaction" through altruistic good works. Lopreato submits that, ironically, self-deceptive behaviors must have evolved through natural selection making it possible to have "true altruism in the absence of altruistic genes."[21] For this reason he recommends a third modification of the maximization principle: "Organisms are predisposed to behave so as to maximize their inclusive fitness, but this predisposition is conditioned by the quest for creature comforts and by self-deception."[22]

A fourth constraint on the maximization principle is sociopolitical revolution, which Lopreato defines as "forcible action in a dominance order by individuals who desire to replace those who have organizational . . . control of their group's resources."[23] Lopreato suggests that the "ultimate cause of revolution is the quest for fitness maximization," because in traditional societies, highly ranked individuals with resources are more likely to have access to multiple females through polygyny, thereby increasing their reproductive success. However, to the degree that resource accumulation becomes an end in itself in modern industrial societies, resource acquisition and maximization of fitness become detached. Here, culture once again disrupts the maximization as behaviors directed at resource acquisition become separated from considerations of genetic fitness. For this reason, Lopreato adds a final variable in his restatement of the maximization principle: "Organicisms are predisposed to behave so as to maximize their inclusive fitness, but this predisposition is conditioned by the quest for creature comforts, by self-deception, and by autonomization of phenotype from genotype."[24]

In the last few years, Lopreato has used his theoretical model to enhance traditional sociological explanations of social phenomena. For example, in a recent paper with Arlen Carey,[25] Lopreato examines the relationship between fertility and mortality in human societies. Essentially, Carey and Lopreato argue that variations in human fertility have typically been explained by such fertility-reducing mechanisms as abortion, use of contraceptives, sterility, age at marriage, economic value of children, social status of women, and the like. Yet, despite these obvious and important influences on fertility, these explanations cannot provide by themselves an adequate account of one fact: Until recently, rates of reproduction have changed very little in humans' evolutionary history, with human females on average producing only two viable adult offspring. Why should this have historically been the case?

In considering this demographic puzzle, Carey and Lopreato submit that population dynamics do not fluctuate randomly, but fertility rates roughly correspond with mortality rates, with these two rates tending to behave like systems in equilibrium. In traditional societies there is a quasi-equilibrium between high mortality rates and high fertility rates. In addition, as Charles Darwin first realized (drawing his inspiration from Thomas Malthus), it is characteristic of most species, notwithstanding the tendency of populations to outpace their resources, to evidence a countertendency toward population stability, suggesting that "the fertility of individuals displays a vigorous tendency to track mortality—a tendency . . . toward a coupled-replacement reproductive strategy."[26] Carey and Lopreato ask this: In humans, what factors might be responsible for maintaining this "fertility-mortality quasi-equilibrium?"

Carey and Lopreato submit that although all humans are motivated in principle to maximize their reproductive success, they are constrained first by limited resources and the human need for "creature comforts." Instead, what the "demographic quasi-equilibrium" pattern suggests is a stabilizing reproductive strategy with natural selection tending to favor fertilities that meet and only slightly exceed corresponding mortalities.[27]

A second mechanism used to regulate the fertility-mortality relationship in human populations, they maintain, is collectively called "life history characteristics."[28] Basically, life history theory holds that natural selection guides the life characteristic of a species to optimize and regulate fertility.[29] Evolutionary ecologists have documented life-history characteristics in other species, which include distinct reproductive periods, inter-birth intervals, litter size, parental investment, age-specific probabilities of survival, and size and maturity of newborns. Hence, animal populations with low rates of mortality will also have lower fertility rates that are conditioned by smaller litters, delayed births, and bigger offspring because they represent a great investment of parental resources. Carey and Lopreato argue that the fertility-mortality relationship in humans also corresponds to life history characteristics. Members of the human species have a relatively high probability of survival through the reproductive period that is matched historically by a low fertility rate. To regulate this process, puberty begins late in humankind, conception is relatively difficult, fetuses take nine months to develop, single births are the norm, newborns are large, and births are difficult. Carey and Lopreato suggest that "these life history characteristics contribute to a relatively low level of fertility that is associated with a relatively low probability of mortality."[30] Although these facts are well known to demographers, they are rarely viewed as phenomenon forged by natural selection.

Still, Carey and Lopreato do not consider these factors sufficient to explain the relationship between fertility and mortality. Instead, they seek a third explanation for the regulation of human fertility, suggesting that psychological attributes "most closely associated with survival and reproduction may be expected to be especially prone to fitness maximization tendencies."[31] In short, there are perhaps psychologically regulated reproductive behaviors toward an optimal level of reproductive behavior that is activated by environmental cues. These cues are used to gauge the relative probabilities of survival of the young.[32] Thus, in a society where infant and child mortality are high, a higher fertility rate would prevail.

Identifying some psychological mechanisms that help regulate human fertility rates is now possible: Belsky, Steinberg, and Draper[33] suggest, for example, that reproductive strategies for individuals in a given population are largely decided during the first five to seven years of life. The crucial determining factor here is resource availability. The amount of resources then influences family life and childrearing and these shape both the psychological and behavioral development of a child.[34] In turn, these affect the age of puberty and the lifelong reproductive strategy. For instance, Chisholm[35] maintains that this developmental process offers the proximate mechanisms by which individuals internalize the mortality characteristics of their population.

Thus, Carey and Lopreato believe that human neurobiology has reproductively based predispositions, which are activated by the environment, charting the best reproductive strategy. As they note, "the evolution of the relationship between human fertility and human mortality has forged a tendency toward a reproductive psychology that *revolves closely around a near-average two-child family.*"[36] This can be called the "two-child psychology" tendency. It charts that in human history an average of two offspring (or a little more) is typically the best strategy given the costs of motherhood in comfort, health, and overall quality of life. In supporting their thesis, they use historical data that suggest population growth was minimal—at least, until recently.

In concluding, they offer these propositions:[37] (1) The greater the perceived probability of offspring survival is within a population, the more intense the two-child psychology is; and (2) the greater the tendency toward creature comforts is among members of a society, the more widespread is the two-child psychology.

CROSS-SPECIES COMPARISONS OF SOCIAL FORMS

Richard Machalek's Approach

Richard Machalek has applied modern evolutionary theory to traditional sociological problems.[38] Machalek would like to see a truly comparative sociology or one that crosses species lines. His approach is to search for the foundations and development of "sociality" wherever it is found, in both human and nonhuman species. By identifying the elementary forms of social life among human and nonhuman organisms, information can be gleaned about how the organizational features among species are assembled. In this effort to create a truly comparative sociology, Machalek outlines a four-step protocol for conducting a sociological analysis of generic social forms "with a priority on sociality, not the organism."[39]

1. Identify and describe a social form that is distributed across two or more species lines.
2. Identify the "design problems" that might constrain the evolution of this social form. In other words, what prerequisites are necessary for a particular social form to come into existence?
3. Identify the processes that generate a social form.
4. Identify those benefits and beneficiaries of a social form that will help explain the persistence and proliferation of certain social forms over other forms.

In applying this protocol, Machalek focused on the evolution of macro societies, a social form that first appeared in human social evolution about 5,000 years ago. He asked this: What makes human macro societies possible? Machalek suggests that we cannot just look at agrarian and industrial societies to answer this question, but, rather, we must subordinate the study of human macro-level societies to the study of macro societies as a general social form. If we take a cross-species comparative approach, it is evident that "macro sociality" is rare and exists in only two taxonomic orders: insects and primates.

Machalek describes a macro society as a society with hundreds of millions of members with distinct social classes and a complex division of labor. Among social insects, this social form is very old, but in humans it is very recent, beginning about 5,000 years ago with the emergence of agrarian societies. Obviously, humans and insects are remote species, separated by at least 600 million years of divergent evolution; hence, they cannot be compared by individual biological characteristics. Indeed, humans and insects are separated by major anatomical differences that include "six orders of magnitude in brain size," so, intelligence did not play a role in the evolution of insect macro societies. Instead, insect and human macro-societal social forms must be compared strictly for their "social structural design" features in what appears to be a case of convergent evolution.

In considering the fundamental similarities between the organization of human and insect macro societies, Machalek maintains that "whatever the species, all social organisms confront the same basic problems of organizational design and regulation if they are to succeed in evolving a macro society."[40] When looked at this way, the existence of this social form in two such distinctive and biologically remote taxa allows us to address such questions as this: What constraints must be surmounted before a species can evolve a macro society?

Machalek suggests that macro societies are rare because the evolution of this social form requires successful solutions to a series of difficult and complex problems. Machalek suggests that only insects and humans have managed to push aside or overcome (1) organismic constraints, (2) ecological constraints, (3) cost-benefit constraints, and (4) sociological constraints. Each of these will be briefly examined.

Organismic Constraints. In detailing the organismic constraints that must be overcome before complex cooperative behavior can evolve, Machalek highlights the morphology of a species as an important factor that can either promote or inhibit the ability of a species to evolve a macro society. For example, aquatic social species such as whales, who are extremely intelligent and who clearly enjoy a "social life," are hopelessly constrained by their enormous "body plans," a constraint that makes it difficult for them to engage in "diverse forms of productive behavior."[41] And, when a body plan constrains the variety of cooperative behaviors possible, it "also constrains the evolution of a complex and extensive division of labor."[42]

Ecological Constraints. In addition to organismic constraints, the ecological niche of a species sets limits on both the population size and complexity of a society. An ecosystem's physical properties can vary in the number of predators, competition for resources like food and shelter, diversity of other species, and mortality rates because of disease. All these can become factors in limiting population size for a given species. Social insects are more likely to find a habitat with ample resources to support their macro societies because they are such very small creatures.

Cost-Benefit Constraints. In addition to organismic and ecological constraints, the evolution of a macro society will depend on economic factors or various "costs and benefits" that accompany any macro society. Although the evolution of a macro society would seem to be beneficial to any social species, a society with complex and extensive cooperation has both costs and benefits. Using the logic of cost-benefit analysis, a particular evolved trait can be analyzed for the ratio of its costs to benefits. Among social insects like ants, costs (which include such problems as social parasitism where alien species expropriate labor or food from unsuspecting ants) do not exceed benefits. This is because social insects greatly benefit from a complex division of labor that allows them to compensate for the small size of each individual "and thus increase their ergonomic efficiency and effectiveness."[43]

Sociological Constraints. Of all the constraints, this is the most important. Even if all other constraints are overcome, the evolution of a macro society requires a unique form of social interaction that is rare in nature and beyond the capacity of most organisms. Essentially, an organism must overcome three large sociological problems to evolve a macro society:[44]

1. The individuals must be able to engage in impersonal cooperation.
2. The labor of members must be divided among distinct social categories.
3. The division of labor among members must be integrated and coordinated.

In considering these critical "design problems" that must be surmounted before a macro society can evolve, we should ask why it is that only the social insects and humans have been able to generate a rare and complex form of sociality. If we turn to other social species for clues, we find that the fundamental mechanism underlying social organization in most animals is kinship or genetic relatedness. Machalek argues that kinship bonds effectively restrict the number of individuals within a particular cooperative group, making it very difficult for most species to evolve a macro society. Machalek notes that the general principle that links kinship to social behavior among animals can be stated as follows: "The greater the degree of genetic relatedness among individuals, the higher the probability that they will interact cooperatively."[45] In other words, natural selection has seemingly favored social species with the basic capacity to distinguish individual kin from nonkin, and this makes kinship networks possible. Thus, kinship connections based on individual recognition of relatives are the basis for social cooperation for most social species.

In social insects, however, kin are distinguished from nonkin largely through remote chemical communication, for there is no evidence that "blood relatives" recognize each other as individuals. Thus, in ant societies, members interact with five or six *types* of ants—not millions of individual ants. Ants treat each other as members of distinct categories or castes. In turn, social categories or castes are occupationally specialized, allowing task specialization (that is, foraging, brood-tending, nest-repair, defense, and so on) and leading to a complex division of labor. Caste types are recognized by olfactory cues, the dominant mechanism behind the organization of ants. Machalek notes that humans often link a complex division of labor to human intelligence, culture, and technological development, but this social form clearly exists outside the range of human intelligence.

In contrast, despite selection for language and culture, human societies were small and based on face-to-face individualized kinship relations for most of human evolutionary history. Yet, in agrarian times, full-blown hierarchical stratification evolved, leading to the question: How were humans able to escape the constraining influences of personalized kinship relations and their highly evolved capacity for individual recognition? Following Machalek,

> Humans have evolved macro societies because they are empowered by culture to form highly cooperative patterns of behavior with "anonymous others." Thus, for the social insects, a state of permanent personal anonymity enables them to form large, complex societies comprising purely impersonal cooperation among members of different castes. Humans, on the other hand, are capable of forming cooperative social systems based either upon personal relationship or impersonal status-role attributes.[46]

Thus, chemical communication allows insects to convert individuals into social types, whereas humans employ "cognitive culture" and socially constructed typifications. This capacity allows humans to interact cooperatively not as individuals but as personal strangers, dividing individuals into types of social categories. Machalek believes "it is impersonal cooperation that lies at the very foundation of macrosociality."[47] Social insects and humans have thus "used different but functionally analogous strategies to achieve the capacity for close cooperation among anonymous others, thereby facilitating the evolution of macrosociety."[48] In addition, this impersonality specifies and limits the rights and obligations between (or among) parties to an interaction, for as Machalek notes, status-role constructs "are the evolutionary convergent human analogue to the chemical and tactile typification processes among social insects upon which caste systems are built."[49] Essentially, status-role constructs allow humans to ignore the unique and distinctive qualities of persons, thereby increasing the economy of a cooperative interaction. Unlike social insects, however, humans can also move between personal and impersonal attributes in organizing their social lives.

In sum, then, only in insects and humans has macro sociality become a reality, primarily because of the design problems in creating macro societies. Machalek emphasizes that sociologists have long struggled to understand the elementary forms of social behavior, but this quest has been limited because of a general reluctance by sociologists to expand their perspective to include inquiry into nonhuman social species. It is important, Machalek argues, to see how particular social traits are spread across species. The ability to research questions such as the emergence of a complex division of labor and why it is found in only a few societies can help us discover how it evolved in human societies. In addition, if we compare sociality forms across species by consequences, "we can enhance our understanding of the adaptive value of sociality as a response to ecological challenges."[50] Finally, beginning with the social form and then selecting for observation those species in which that form appears would also allow us to better understand the emergent properties of social systems, the adaptive value and processes that generate particular social forms, and the essential design features that might represent a solution to common problems facing diverse species.

STAGE THEORIES OF EVOLUTION

In the 1960s, sociologists turned once again to a view of societal evolution as involving movement from simple to more complex forms of social organization. Earlier theorists, such as Herbert Spencer and Émile Durkheim, had been the first to conceptualize long-term historical processes as revolving around increasing differentiation of the structure and culture organizing a population.[51] Much of the spirit of these earlier approaches was recaptured in the 1960s, but with considerably more anthropological detail and conceptual sophistication. As we saw in chapters on functional theorizing, most macro-level theorists in this tradition—theorists such as Talcott Parsons,[52] Niklas Luhmann,[53] and Jeffrey Alexander[54]—addressed the question of differentiation. Yet, by far the most sophisticated conceptualization in the 1960s came from outside functionalism in the form of Gerhard Lenski's more conflict view of societal stratification.

In *Power and Privilege: A Theory of Social Stratification,*[55] Lenski examined what he termed "distributive systems" or the mechanisms by which power, prestige, and material wealth are dispersed among the members of a population. What made this analysis evolutionary was Lenski's emphasis on distribution of power and privilege in distinct societal types that were seen as stages in humans' long-term evolutionary history: (1) hunting and gathering societies, (2) simple horticultural societies, (3) advanced horticultural societies, (4) agrarian societies, and (5) industrial societies. Other types of societies, such as those revolving around herding and fishing were seen as variants of horticultural types, but the critical point was that human societies have moved through conspicuous stages and that different types of societies have revealed varying degrees of inequality. The driving force behind each type of society was the level and nature of technology for economic production, which, in turn, generated varying degrees of economic surplus that was then distributed unequally. The basic finding in this early work by Lenski was that technology, production, and surplus were curvilineally related to inequality; that is, low technology-production-surplus in hunting and gathering societies displayed the most equality; then through horticultural and agrarian societies inequality increased, but with high technology-production-surplus in industrial societies, redistribution lowered inequality somewhat but not to the level of hunter-gatherers.

Later in work[56] with his wife, Jean Lenski,[57] and then in productive collaboration with Patrick Nolan,[58] Lenski's scheme moved away from a concern with stratification processes, per se, toward a macro-level theory of social organization in general. During the course of this transition to a more macro-level theory of societal organization, concepts from evolutionary theory in biology became increasingly prominent, as did the use of field studies from the Human Area Relations Files[59] (which represent an effort to code and catalogue data gathered from the many ethnographies that anthropologists and others have conducted on diverse types of societies). Other theorists, such as Stephen K. Sanderson[60] and Lee Freese,[61] similarly began to employ concepts from evolutionary biology and ecology to examine the path of human history through various societal types. In this section, we will examine these three stage-model theories of evolution—the Lenski, Nolan, and Lenski approach, the

Sanderson framework, and finally, the Freese ecological model.

Gerhard Lenski, Patrick Nolan, and Jean Lenski's Evolutionary Theory

As Gerhard Lenski, Patrick Nolan, and Jean Lenski added biological theorizing to their stage model of societal development from hunting and gathering to modern industrial systems, they sought to highlight the similarities and differences between biological and social evolution. Both biological and social evolution are, first, "based on records of experience that are preserved and transmitted from generation to generation in the form of coded systems of information" and, second, on "processes that involve random variation and selection" of those traits that promote adaptation to the environment.[62] Yet, there are some important differences between biological and social evolution. One is that in organic evolution the genes are the preservers of the informational codes, whereas in social evolution, cultural "symbol systems are the functional equivalents of the genetic alphabet."[63] Another difference revolves around the way that information is transmitted. In biological evolution, genetic information can be transmitted only through the reproduction of new organisms; moreover, diverse species cannot interbreed, so the transmission of information is limited to one species. In contrast, cultural information is more readily and broadly transmitted, moving from one type of society to another. The end result is that in biological evolution, speciation leads to ever new patterns of differentiation and diversification, whereas in social evolution the movement of information across societal types "is likely to eventuate in ever fewer and less dissimilar societies than exist today."[64] A related difference is that in biological evolution, both simple and complex species can continue to exist in their respective resource niches, whereas in social evolution, simpler societal types tend to be extinguished by more complex types. Still another difference is that acquired traits can be transmitted through socialization, whereas in biological evolution, such Lamarckian processes do not occur. An outcome of this difference is that genetic change in biological evolution is slow (because selection processes have to sort out genes across many generations), whereas cultural evolution can be very rapid (because new traits can be created, learned, transmitted, and diffused within one generation).

These similarities and differences lead to the recognition that human societies are part of the natural world and subject to selection forces from both their biophysical and sociocultural environments, that humans like any other animal are influenced by their genetic heritage, and that only humans are the creators of their cultural heritage or the informational codes that guide behavior and social organization. The basic model of human social evolution developed by Lenski, Nolan, and Lenski is summarized in Figure 10-1. A given society has social structural and cultural (symbolic) characteristics that, for analytical purposes, can be divided into (1) its population size and characteristics, (2) its culture or systems of symbols, particularly technologies, (3) its material products generated by the application of its technology to productive processes, (4) its organizational forms that structure activities, and (5) its institutional systems that combine (1) through (4) into systems addressing basic problems of survival and adaptation for individuals and the society as a whole. These five components of a society influence, while being influenced by, the other forces delineated in Figure 10-1: (1) a society's biophysical environment, (2) its social environment of other societies and their respective cultures, (3) the genetic heritage of humans as a species, namely an evolved ape, and (4) the prior social and cultural characteristics of a

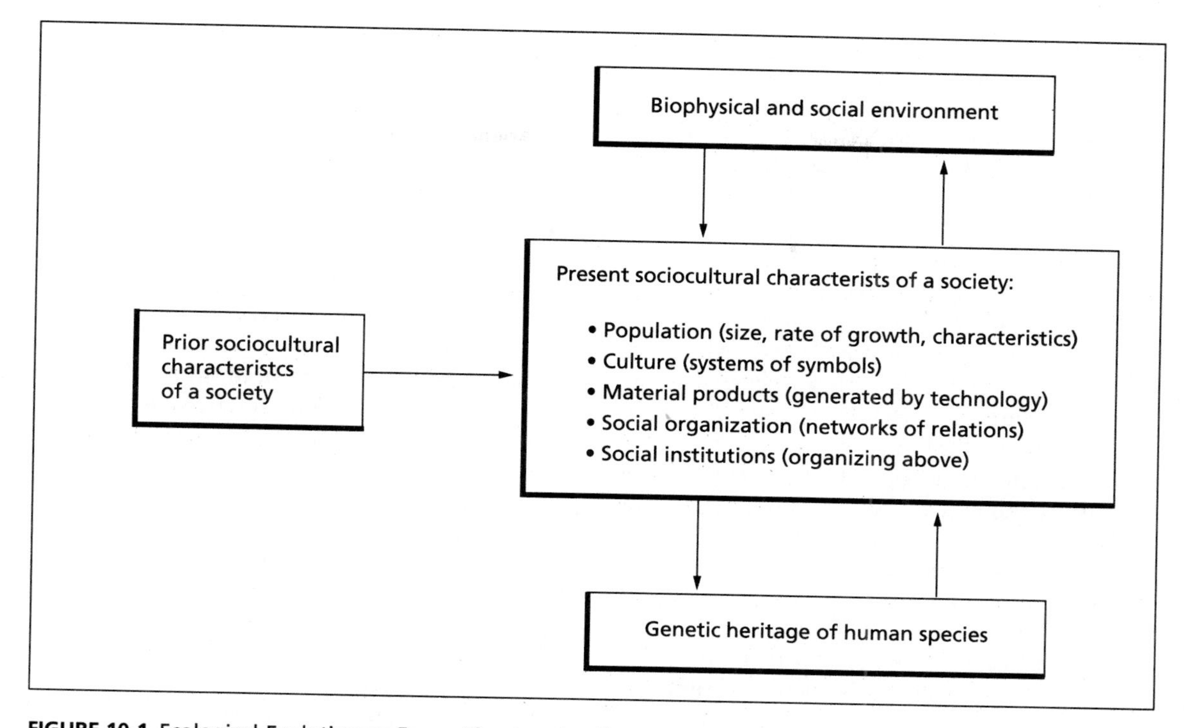

FIGURE 10-1 Ecological-Evolutionary Forces Shaping the Characteristics of a Society

Adapted from G. Lenski, P. Nolan, and J. Lenski, *Human Societies: An Introduction to Macrosociology,* 6th edition (New York: McGraw-Hill, 1995), pp. 21, 23–55.

society as these continue to influence its internal operation and its adaptation to the external environment.

When we approach the analysis with the orientation diagrammed in Figure 10-1, we can see that Gerhard Lenski's earlier emphasis on technologies as the driving force of social evolution is retained, but the argument is recast into a more evolutionary framework. As Lenski recently remarked, "It seems no exaggeration to say that advances in subsistence technology are functionally equivalent to adaptive changes in a population's gene pool; new energy resources and new materials enable populations to do things that they could not do before."[65]

Social evolution is a cumulative process in the sense that new technologies that prove more adaptive alter the pattern of social organization, generally toward larger and more complex forms of organization. Two basic forces drive change in human societies:[66] (1) *innovation* where new information and social structural patterns are created, whether by chance or conscious intent; and (2) *extinction* where old cultural and structural patterns are abandoned. Innovations in sociocultural evolution cause more rapid change than forces in biological evolution, because (a) humans have conscious capacities to develop new informational codes; (b) humans have "needs and desires" that are potentially "limitless" and, under certain conditions, drive them to make new discoveries as old needs are satisfied and new ones emerge; (c) humans can adopt the information of other societies through diffusion; (d) humans can force another society to adopt their informational codes through conquest and repression of older cultural and structural patterns, especially when larger and more complex societies conquer or co-opt smaller and less complex ones; (e) humans can institutionalize innovation in such structural forms as science, thereby creating a set of cultural codes and social structures specifically geared to constant innovation; and (f) humans can create complex interconnections among systems of information that force changes in other elements as changes in another occur.

Yet, Lenski, Nolan, and Lenski stress that there are also forces operating to sustain continuity in the cultural systems that guide the organization of a population.[67] One force for continuity is socialization, in which older patterns are transmitted to each new generation. Another force is ideology, which preserves cultural systems and guides the transmission of culture from one generation to another. Still another force is the systemic nature of human sociocultural systems, which resist change in one element because so many other elements will be forced to change (although, as noted earlier, once change in one element does occur, it has a cascading effect and actually accelerates change). Another force is vested interests, especially of the powerful in stratified societies who have the power to suppress innovations when changes threaten status quo interests. Yet another force is inertia, where past practices appear to promote adaptation and sufficient satisfaction for individuals to resist adopting new practices whose impact cannot be fully known.

Yet, despite these forces promoting continuity, the long-term historical record confirms that societal evolution has involved change, fueled by technological innovations, toward larger and more complex societies. Societies vary, of course, in their rates of innovation; for Lenski, Nolan, and Lenski, these rates vary because of several important forces. First, the amount of information already possessed by a society greatly influences its capacity to create and adopt more information. Second, the size of a population is another important factor because larger populations have more individuals who hold ideas and who can potentially generate new ideas. Third, the stability and nature of a society's environment, both social and biophysical, is another force of change; the more the environment changes, the more likely a society is to be innovative or adopt the innovations of another. Fourth, the nature of the innovations, per se, is a very significant factor; some innovations are fundamental and pave the way for additional innovations (for example, the discovery of metallurgy or new sources of energy stimulated even more innovations). And fifth, the ideology of a society greatly circumscribes the creation or adoption of innovations; powerful and conservative ideologies make it difficult for individuals to be innovative, while discouraging the diffusion of innovations from other societies.

Over the long course of societal development, however, productive technologies are the most important driving force of evolution.[68] In the end, technological innovations can overcome the forces promoting continuity, even the ideologies and the vested interests of the powerful. The reason for this significance of technology is that those societies that can better gather, produce, and distribute resources will generate an economic surplus that can support a larger population and its differentiation into new organizational forms and institutional systems. Eventually, their technologies diffuse to other societies, and particularly so when larger, more complex societies conquer, co-opt, or out-compete smaller and less complex societies. Thus, a kind of "group selection" operates in the history of human societies, as more powerful societies (with better technologies, productive capacities, and organizational forms) impose their cultural systems and structural patterns on others through conquest, provide models and incentives for less developed societies to adopt their cultural and structural systems, or take the resources on which less developed societies depend for their survival.[69] Indeed, selection processes have favored an emerging world system of societies.

Stephen K. Sanderson's Evolutionary Approach to Social Transformations

During the last decade, Stephen K. Sanderson has sought to bring biological concepts into sociological theory.[70] In his most recent efforts, Sanderson has produced a description of long-term historical *transformations* that is similar to Lenski, Nolan, and Lenski's stage model, but with rather different points of emphasis and with a very different presentation style.[71] The goal, however, remains much the same: to portray long-term historical development as a series of societal stages analyzed in terms of concepts adopted from evolutionary theory in biology.

Sanderson posits that the major point of similarity between biological and social evolution is that both involve adaptational processes that produce sequences of change.[72] However, unlike Lenski, Nolan, and Lenski, Sanderson does not argue for group selection; instead, he believes the basic unit of selection is the individual person. Indeed "adaptation is a process pertaining to individuals and never to any social unit larger than the individual."[73] Social structures cannot be the units of adaptation because they are "only abstractions," and only "concrete, flesh-and-blood individuals can be adaptational units, because only they have needs and wants."[74] Indeed, individuals are viewed as "egoistic" beings who are highly motivated to satisfy their own needs and wants, and they "seek to behave adaptively by maximizing the benefits and minimizing the costs of any course of action (or at least generating more benefits than costs)."[75] This egoistic and adaptive behavior must be the focus of evolutionary analysis, although it should also be stressed that "individuals acting in their own interests create social structures and systems that are the sum total and product of these socially oriented individual actions."[76] Moreover, there can also be a kind of "superstructural feedback" in which, once created, systems of symbols can become causal agents, forcing individuals to adopt and adapt to them. But still, in social evolutionary analysis, a social pattern must be viewed as adaptational only because it promotes adaptation for all, or nearly all, individuals within various social structures or the society as a whole.

Although the individual is the unit of adaptation and, hence, selection, only societies evolve. This line of reasoning draws from evolutionary biology stressing individual organisms as the unit of selection (and underlying genotype) but the population of organisms (and the underlying gene pool) as the unit of evolution. Thus, individual humans are the units of adaptation and selection, creating or adopting sociocultural traits because these meet their needs; in this process, the structure and culture of societies can be transformed. Such adaptations by individuals represent adaptive responses to "either the physical environment, the social environment, or both."[77]

Still, much like Lenski, Nolan, and Lenski, Sanderson stresses that "the principle causal factors in social evolution are the material conditions of human existence, that is demographic, ecological, technological, and economic forces at work in social life."[78] Demographic forces are variations in human populations, particularly their size and rates of growth as these put pressure on resources. Ecological factors are all those dimensions of the natural or physical environment that force members of a population to adapt, particularly as these factors interact with demographics and technology. Technological factors are those revolving around knowledge, tools, and techniques available to members of a society as means for adapting to their environment. And economic factors are the modes of social organization that enable people to produce, distribute, and exchange goods and services.[79] These basic "material conditions" of human life have causal significance because they influence "human needs concerning the production of subsistence and the reproduction of human life."[80] How these basic conditions play out is, however, highly variable depending on the empirical conditions in which individuals find themselves. Yet, "different types of social systems in different historical epochs and at different evolutionary stages embody different 'evolutionary logics'."[81]

Also, much like Lenski, Nolan, and Lenski, Sanderson recognizes that these "material conditions of human existence" point to differences between biological and social evolution.[82] First, in biological evolution, divergence of species is more likely than in human evolution where there is much more convergence of societal types. Second, in biological evolution, genetic variations are a random process on which selection can work, whereas in social evolution the variation on which selection operates is often deliberately and purposefully created by conscious thought and action by individuals. Third, social evolution has a much more predictive direction than biological evolution, for if human evolution started over again, the same sequence of societal types that have appeared during the last 100,000 years would again emerge, whereas such would not be the case for the biological evolution of species. Fourth, because humans can consciously create the variations on which selection works, social evolution is much more rapid than biological evolution, even when we factor in punctuated evolutionary leaps in the process of speciation. Fifth, in social evolution, traits diffuse across societies in a manner that has no counterpart in biological evolution through natural selection on genetic variations. And sixth, natural selection operates in social evolution, but it is not the actual cause of evolution because causality ultimately resides in the behavioral adaptations of motivated and conscious individuals who create cultural and social structural traits. So, Sanderson concludes, "the differences between social and biological evolution are great enough to require that social evolution be studied as a process in its own right, and not merely along the lines of an analogy with biological evolution."[83]

Social evolution, when viewed from the perspective of world history, evidences "social transformation and directional trends of sufficient generality such that topologies of social forms can be fruitfully constructed."[84] Yet, despite these patterns and directionality to social forms or societal types like hunting and gathering, horticulture, agrarianism, industrialism, and world system development, other forces such as social stasis, de-evolution back to simpler forms, and extinction of social structures are basic facts of world history; these too must be part of any evolutionary analysis of world history. Still, there is a direction trend to human history, but not one with a grand teleological design; rather, as individuals have adapted to their environments, they have created cultural and social systems that are similar. For example, Sanderson notes that the "Neolithic Revolutions," where individuals began to plant and harvest food, occurred independently in at least eight major regions of the world—Southwest Asia, Southeast Asia, Africa, China, Europe, Mesoamerica, South America, and North America—as well as in subregions within these major regions of the world. And these transformations all occurred within relatively similar times, within a few thousand years of each other. Similarly, the rise of the state is another case of such parallel evolution, emerging independently in several parts of the world. And even a third great evolutionary transformation—the emergence of capitalism—was not unique to Europe, as is generally argued, but occurred in Japan as well. Only an evolutionary analysis can explain these great transformations, Sanderson argues, because they did not emerge because of diffusion but rather as independent adaptations of individuals creating social structures and culture.

Of course, once created these adaptations can spread, or force others to adopt them to survive. The spread of capitalism across the world, creating a world system, is an excellent example of how a major social transformation can force members of other societies to adopt the structures and culture of capitalism to remain viable in their environment. Indeed, there is always the potential for a "superstructural feedback" in which, once created, systems like the culture of capitalism become one of the conditions to which individuals must adapt, thereby accelerating the spread of particular cultural systems and the social forms that these symbols regulate and legitimate.

Lee Freese's Model of "Biosociocultural Regimes"

Lee Freese has, in recent years, sought to develop a more ecologically oriented approach to long-term development of human societies.[85] Freese believes traditional accounts of social evolution are flawed because they make analogies to biological theory, while employing stage models of human development from simple to complex forms. The problem with this approach, Freese argues, is that it tends to focus almost exclusively on sociocultural phenomena as the basic evolving unit, whether this unit is a pattern of organization, a system of symbols, or a whole society. Analogies are then drawn to biological evolution, although, as we have seen for the stage theories examined thus far, there is always a list of differences between biological and sociocultural evolution. But Freese argues even more strongly that such differences are fundamental to how we theorize about patterns of human organization.

In biological evolution, genes are the unit of inheritance and are subject to selection forces that determine whether or not they will be transmitted to the next generation. In human social systems, symbolic information is not the analogue of genes, Freese says, for several reasons. First, cultural information and social structural patterns regulated by culture cannot make exact copies of themselves, as can genes. Second, unlike genetic material, cultural symbols are not discrete or particulate but, rather, are diffuse, often blending, fusing, differentiating, and disappearing in ways not typical of genetic codes. Third, as all stage theories also emphasize, inheritance is not Darwinian in the sense of being guided by the transmission of genetically coded information across generations, but instead, sociocultural evolution is Lamarckian, with innovations in one generation being used and then passed on, and perhaps modified or expanded on by subsequent generations. And fourth, because of the Lamarckian character of sociocultural evolution, change in human systems can become much more rapid than biological evolution. Thus, Freese contends, we should not even consider sociocultural systems or their respective parts as undergoing any type of evolution that is analogous to biological evolution. Instead, stage models of evolution like those developed by Lenski, Nolan, and Lenski and Sanderson are, in reality, developmental models describing the increasing complexity of societal social systems. How, then, do we account for these changes over the long-run of human history?

Freese argues that to answer this question, it is necessary to reemphasize the effects of ecological parameters. In particular, we need to view sociocultural phenomena as responding to environmental pressures at both the individual person and group level. Sociocultural phenomena are, in the end, adaptations to two primary habitat conditions: (1) the rate and extent of human mortality, and (2) the resource potential of a habitat. These two habitat conditions exert a "push and pull" on human culture and patterns of social organization, with the resource potential of a habitat generally moving upward (as new technologies for exploiting resources are developed) and with rates of mortality waxing and waning over the course of human history.

As sociocultural systems seek to extract resources from the environment and reduce mortality, the complex interplay between the biophysical and sociocultural becomes crucial to understanding the nature, types, and extent of social organization. Social systems are always seeking to extract resources and control mortality; thus the nature of such sociocultural systems will reflect the degree of success in adapting to the biophysical parameters of human existence. Reciprocally, the biophysical environment is changed as a result of sociocultural adaptations used to seek resources and regulate death rates. Essentially, Freese views a sociocultural system as resting on and adapting to a biophysical system. However, once these systems are in interaction, several internal self-generating dynamics are set into motion as biophysical and sociocultural assemblies connect, interact, organize, and reorganize in response to the challenges initiated by the other. This self-organizing and self-renewing process with its complex and mutual interdependencies, in turn, gives rise to a "biosociocultural system" or, when viewed from a point in time, a "biosociocultural regime." The unit of evolution for understanding societal change and complexity, then, is the biosociocultural regime, although Freese maintains that Darwinian processes of "transmission" and "inheritance" do not operate in this perspective. Instead, Freese entertains the theoretical possibility of a "descent with modification" as regimes—through self-generating, cumulative, and normally irreversible processes—proceed through time fashioning new material structures with a greater potential for energy capture and transfer, which modify and in some cases totally replace the existing assemblies involved in a biosociocultural regime.

Thus, biosociocultural regimes are the evolving unit. Here, natural selection is still operative but it is not the causal force of change. Instead, matter-energy forms and flow cycles initiate the complex interactions between biophysical-sociocultural assemblies.[86] In turn, the degree of capacity for greater energy capture and resource transfer reflects and influences the size of a population, its level of technology, and its division of labor, shaping in turn the nature and evolution of the biosociocultural regime. Over time as regimes are assembled for energy capture to sustain activities through self-organization, the result is larger and more complex regimes, as sociocultural development takes on a cumulative character toward ever more complexity and, ultimately, toward macro societies. Freese states "selection will tend to favor populations whose energy capture and resource exploitation are maximized relative to alternative populations that are competing for resources derived from the same matter-energy cycle."[87]

With this shift in emphasis from analogies to biological evolution toward a more ecological approach, revolving around the assembling of biosociocultural regimes that self organize to capture energy from the environment for economic production, Freese hopes to broaden the very nature of sociological theory. Rather than adopting biological concepts and theories, then modifying them because of the obvious differences between sociocultural and biological evolution, it is better to begin anew with a broad ecological framework in which sociocultural systems seeking to extract resources and control mortality are seen as parts of biosociocultural regimes. Then, once again a connection can be made between culture, society, and the ecological necessities humans had to confront in the development of complex human societies.

THE RETURN TO THEORIZING ABOUT HUMAN NATURE

Early sociology, indeed all early social thought, had a conception about the basic nature of humans. Most of this theorizing disappeared by the turn of the century, only to be revived again in midcentury—as noted at the beginning of this chapter. Sociobiology represented one effort to get back to the question of the biological forces underlying human behavior, and it did so with more biological sophistication than earlier "instinct" approaches. More recently, sociologists have made new efforts to bring considerations of humans' biological propensities into the mainstream of sociological theory. These efforts have stayed outside sociobiology, but they also evidence a level of biological sophistication that early "instinct" approaches lacked. Among the many potential theoretical approaches,[88] we will examine two in this closing section of the chapter: the respective approaches of Richard Udry and Alexandra Maryanski.

J. Richard Udry's Gender Theory

Richard Udry is a strong proponent for bringing biology back into sociology, arguing that humans have basic predispositions that can be divided into two broad categories: (1) predispositions unique to individuals, and (2) predispositions shared by all Homo sapiens. Udry maintains that it is now well accepted (even by sociologists) that human bodies and brains have evolved through natural selection; yet, understanding of behavior is still considered by most sociologists to reside outside the realm of evolutionary forces.[89] To document his advocacy for a biologically informed sociology, Udry has proposed a theory of "gendered behaviors" within and between the sexes.[90] Udry argues that most sociologists assume that all gendered behaviors are human inventions, the combined result of (1) gender roles and norms assigned to each sex in a given society, (2) diverse gender-specific socialization experiences, and (3) differential social opportunities for males and females. Although everyone recognizes that males and females differ in many ways, all notions of biological influences are rejected or considered trivial in sociological models and theories.[91] Yet, as Udry points out, there now exists outside sociology an active area of research on "sexual dimorphisms," or those characteristics that distinguish males and females. In addition, the hormones that control dimorphism are linked in all mammals with "sex-dimorphic body structures" and "sex-dimorphic reproductive behaviors," which, says Udry, lie "at the heart of gender."[92] Humans, and higher primates, have the same hormonal structure. In nonhuman primates, sex-dimorphic (or gendered behaviors) have been linked to two stages in the life cycle of primates: (1) a stage in mid-pregnancy when the testicles of a male fetus produce large amounts of testosterone for the genitals as well as for the neural structure of the brain; and (2) a stage of puberty when the sex hormones for all individuals trigger those changes in body morphology and neurology of the brain that produce adult sex-dimorphic behaviors.[93] If the sex hormones influence higher primate behavior, it is plausible that these sex hormones in humans create a gendered blueprint for human behaviors as well. Indeed, these hormonal and reproductive systems operate in humans much the same way they do in other primates, and particularly so for those reproductive mechanisms linked to infant survival, thereby making them likely candidates for preservation during human evolution.[94]

In an exploratory test of his hypothesis, Udry applied his primate-hormonal model to predict within-sex patterns of gendered behaviors among a sample of 250 adult women. From this study, Udry noted that "we were able to confirm several very specific hypotheses concerning the specific hormones involved prenatally, the trimester of effects of prenatal hormones, the specific hormones involved in adulthood, and the interaction of adult with prenatal hormones."[95] His successful predictions in this early study led him to conclude that "individual women differ in their biological propensity to sex-typed behavior."

On the basis of this early research, Udry proposed: "Those processes which affect within-sex variance in gendered behaviors are the same processes as cause between-sex differences . . . We can also infer that males and females differ from one another in their average biological propensity to the same behaviors . . . Gender has biological foundations."[96] In an effort to support this conclusion, Udry performed a thought experiment: Imagine a hypothetical society where both males and females *have gendered predispositions that are pan-human and universal to all human societies.* Let us also assume that this hypothetical society sequentially adopts three distinct socially constructed gender ideologies: (1) a traditional gender-segregated society with sex-typed normative rules for males and females, followed by (2) a single gender society supported by a "unisex" gender ideology and, (3) with institutional changes over time, a permissive society where males and females could follow their own gender proclivities. In a longitudinal examination of this society, what would the behavioral distributions for males and females look like for each structure?

Udry's theoretical model would predict the following patterns over time: In the traditional society, the strong normative sanctions for strictly segregated gender roles would push aside any individual predispositions that conflicted with socially constructed gender norms. In the "antigender" phase, the restrictive norms against divergent gender norms would also push males and females away from their biological inclinations toward conformity to a unisex standard. Finally, in the gender-permissive phase, where everyone could choose their gender preferences, individuals' gendered roles would mirror their biologically based predispositions. The important point here is that for all three ideologies, a "gender structure" would be clearly evident, but only in the gender-permissive society would gendered behaviors prevail "without the gender norms supporting the gendered structure."[97]

Yet, in the absence of a biosocial theory of gender, social scientists would conclude that all three ideologies, including the one advocating a gender-permissive society, were socially constructed. The reason for this conclusion is that sociologists think it impossible for a behavior to have biological foundations, in part because of the social change in ideologies guiding behavior. As this thought experiment demonstrates, however, ideological changes over time are always driven by purely social forces. What is constant in each phase is biologically based gendered predispositions. But only in the gender-permissive phase would gendered proclivities be allowed the potential of their fullest expression because of the absence of socially constructed gender roles.

In addition to this example, Udry maintains that his gender theory will help us better understand a variety of sexual differences between males and females that cannot be explained by purely sociological theories. These include such questions as the following:

1. Why do male and female behaviors vary so dramatically in some instances and not in others?
2. Why do these variations in behaviors show such cross-cultural commonalities, despite major societal and cultural differences?
3. Why do some males act much more masculine than other males?
4. Why do some females act much more feminine than other females?

Udry feels the challenge for the social sciences is to learn how to integrate biological variables into normal social sciences. He emphasizes that only "social scientists imagine that if a behavior is under biological influence, there is nothing we can do about it. . . ." The nature of social arrangements can override or alter any human predispositions.[98]

Finally, Udry offers the following propositions: The higher the level of social constraints in a society, the less is the variance in the behavior that is controlled by biological differences. In contrast, the more choices individuals are permitted, the more their behavior will be controlled by biological forces. Thus, biological forces influence behavior most when individual choice is broad. In complex societies, some people have more choice than others; hence, biological forces will exert more influence on those portions of the population that are less constrained by social forces. For Udry, an advanced industrial society represents a natural experiment in which to study the range of women's biologically based behaviors because modern women have more choice than they did previously about such matters as childbearing, occupation, contraception, marriage, and abortion. But the biological factors, Udry argues, will be invisible to most sociologists because in sociological models nobody is looking for biological factors. When societies provide gender-neutral opportunities, the naturally occurring variation in gender predispositions among individuals allows them to take advantage of these opportunities. The result will not be a "degendered" society, but rather one in which the range variation in biologically driven behaviors will be more evident in the actual role behaviors of males and females. He notes that work on the biology of genes has just begun, with his work as just one small step. He encourages replication of his exploratory work and also continued research on both males and females to see which behaviors are sensitive to hormonal influences.

Alexandra Maryanski's Cross-Species Comparative Analysis on the Origins of Human Sociality

In recent years, Alexandra Maryanski, in conjunction with sometime collaborator Jonathan Turner, has approached the question of human nature by examining the social network ties of humans' closest living relatives, the apes.[99] As is well known, humans share well over 98 percent of their genetic material with chimpanzees (Pan); indeed, chimpanzees might be closer to humans than they are to gorillas (Gorilla). And both chimpanzees and gorillas who are African apes are certainly closer to humans than they are to orangutans (Pongo) or gibbons (Hylobates), the other two genera who are Asian apes. In fact, humans and chimpanzees came from the same animal only about five million years ago, according to the latest fossil and molecular data.[100]

Long-term field studies have documented that primates are highly intelligent, slow to mature, undergo a long period of socialization, and live a long time, and the majority of primates are organized into year-round societies that require the integration of a wide variety of age and sex classes, not just adult males and females. In addition, primates have clear-cut social bonding patterns that vary widely among the 187 species of primates.

Using a historical comparative technique, which is termed cladistic analysis in biology, Maryanski began by examining the social relational data on present-day ape genera—that is, chimpanzees, gorillas, gibbons, and orangutans.[101] Following this procedure, Maryanski first identified a limited group of entities—in this case one crucial property of ape social structure, the strength of social bonds between and among age and sex classes in all ape genera—to see if there were structural regularities in the patterning of relations. If phyletically close species living in different environments reveal characteristic traits in common, then it can be assumed that their Last Common Ancestor (LCA) also had similar relational features. For this exercise, Maryanski undertook a comprehensive review of bonding propensities for apes living under natural field conditions in an effort to profile their social network structures with the idea of uncovering a blueprint of the LCA population to present-day apes and humans.

To assess the validity of these relational patterns, she followed the procedures of cladistic analysis by including an "outgroup lineage"—a sample of Old World monkey social networks—for comparison. She also subjected her data set to two fundamental assumptions associated with this comparative technique: (1) the Relatedness Hypothesis, which indirectly assesses whether or not the shared patterns of social relations are caused by chance, and (2) the Regularity Hypothesis which indirectly assesses whether the modifications from the ancestral to descendant forms evidence a systematic bias and are not randomly acquired. Both hypotheses provided strong empirical support for her reconstruction of hominoid (that is, apes and humans) Last Common Ancestral patterns.[102]

Her analysis led to a striking conclusion: Like the contemporary apes phyletically closest to humans, the Last Common Ancestral population evidenced a fluid organizational structure, consisting of a relatively low level of sociality and a lack of intergenerational group continuity over time. The proximal reasons for this structure are a combination of several forces that are still found in all living ape social networks: (a) a systematic bias toward female (and usually male) transfer from the natal unit at puberty, which is the opposite trend from monkeys where only males transfer and females stay to form intergenerational matrilines; (b) a shifting mating pattern that makes paternity difficult (the gibbon being the exception); and (c) an abundance of weak social ties and few strong ties among most adults. In addition, the Regularity Hypothesis that was used to assess the modifications from the LCA social structure suggested that after descendants separated from the ancestral population, the future trend in hominoid evolution (that is, ape and human) involved selection pressures for heightened sociality, seemingly to increase hominoid survival and reproductive success. Indeed, it is an established fact in the fossil record that about 18 million years ago, a huge number of ape species underwent a dramatic decline and extinction, just when species of monkeys suddenly proliferated and, according to the fossil record, moved into the former ape niches, perhaps because monkeys developed a competitive, dietary edge over apes. Whatever the explanation, the fossil record confirms that, when ape niches were being usurped by monkeys, apes began to undergo anatomical modifications to a marginal niche that required a peculiar locomotion pattern that involves hand-over-hand movement in the trees through space along with other novel skeletal features that characterize the anatomy of both apes and humans today.[103] Today, monkeys remain the dominant primates, and apes are a distinct tiny minority; moreover, with the exception of humans, the few remaining nonhuman hominoids—that is, chimpanzee, gorilla, orangutan, and gibbon are now considered "evolutionary failures" and "evolutionary leftovers" because of their small numbers and specialized and restricted niches.

The significance of this finding is important for thinking about human nature. If humans' closest relatives reveal a tendency for relatively weak social ties, then humans are also likely to have this social tendency as part of their genetic coding. What, however, is meant by weak and fluid ties?[104] Maryanski confirmed in her review of the data that monkeys have lots of strong ties and high-density matri-focal networks. In monkey societies, males at puberty disperse to other groups, whereas females remain behind, forming as many as four generations of strongly tied matrilines (that is, composed of grandmothers, mothers, sisters, and daughters). These extended female-bonds provide intergenerational continuity and are the backbone of most monkey societies. In contrast, females in ape societies evidence the rare pattern of dispersal at puberty with migration typically within a larger regional or community population. In addition, males in ape societies (with the exception of the chimpanzee) also depart. Thus, with both sexes dispersing at puberty, most kinship ties are broken (certainly matri-focal ties in all apes), intergenerational continuity is lost, and the result is a relatively fluid social structure with adult individuals moving about as a shifting collection of individuals within a larger regional population. In Asia, adult orangutans are nearly solitary, rarely interacting with others purely for social reasons. A mother with her dependent young are the only stable social unit. In Africa, chimpanzees and gorillas are more socially inclined, with gorillas living together peacefully in small groups, but individuals are so self-contained that it is uncommon to observe any overt social interactions between adults. Among our closest chimpanzee relatives, adult females are also self-contained, spending most of their days traveling about alone with their dependent offspring. Adult chimpanzee males, in contrast, are relatively more social and are likely to have a few individual "friendships" with other males. A mother and son also form strong ties. But, except for mother and her young offspring, there are no stable groupings in chimpanzee societies. Thus, chimpanzee males are still highly individualistic and self-reliant, preferring to move about independently in space within a large and fluid regional population.

Thus, if humans' closest African ape relatives evidence behavioral propensities for individualism, autonomy, mobility, and weak social ties, Maryanski argues that these genetically coded propensities are probably part of human nature as well. Indeed, if we examine the societal type within which humans as a species evolved—that is, hunting and gathering—it is clear that it approximates the pattern among the Great Apes, especially African apes: There is considerable mobility within a larger home range of bands; there is a high degree of individualism and personal autonomy; and, except for married couples, relatively loose and fluid social ties are evident. At a biological level, then, Maryanski argues that humans might not have the powerful biological urges for great sociality and collectivist-style social bonding that sociologists, and indeed social philosophy in general, frequently impute to our nature.[105]

In collaborative work with Jonathan Turner, Maryanski has described the implications in a review of the stages of societal development. Hunting and gathering is the stage in which human's basic human biological coding evolved; these societies of small, wandering bands within a territory evidence rather loose and fluid social ties, high individual autonomy, self-reliance, and mobility from band to band.[106] Yet, as human populations grew in size and were forced to adopt first horticulture and then agriculture to sustain themselves, they settled down to cultivate land, and, in the process, they "caged" themselves in sociocultural forms that violated basic needs for freedom, some degree of individual autonomy, and fluid ties within a larger community of local groups.

Thus, sociocultural evolution began to override the basic nature of humans. As Maryanski and Turner conclude, market-driven systems of the present industrial and post-industrial era are, despite their many obvious problems, closer than horticulture and agrarianism to the original societal type in which humans evolved biologically, at least in this sense: They offer more choices; they allow and indeed encourage individualism; they are structured in ways that make most social ties fluid and transitory; and they limit strong ties beyond family for many. Maryanski and Turner note that, for many sociologists of the past and today, the very features of human behavior required by market-driven societies are viewed as pathologies that violate humans' basic nature. For Maryanski and Turner, societal evolution has, since hunting and gathering was left behind, just begun to create conditions more compatible with humans' basic hominoid nature as an evolved ape.

Although many of these conclusions are obviously somewhat speculative, the point of Maryanski's analysis is clear: If we use evolutionary approaches from biology, such as cladistic analysis and cross-species comparison with humans' close evolutionary relatives, we can make informed inferences about human nature. Then, we can use these inferences to determine whether sociocultural evolution has been compatible or incompatible with humans' primate legacy. From this analysis, it is possible to examine basic institutional systems, such as kinship, polity, religion, and economy to determine how and why they evolved in the first societal type—that is, hunting and gathering—and how they have interacted with humans' basic nature as an evolved ape during the various stages of societal development.

CONCLUSION

In this long chapter, we have reviewed a variety of approaches to developing a brand of sociological theorizing that is informed by biology, or at least biologically inspired modes of thinking. In many ways, these new evolutionary sociologies represent an odyssey back to sociology's very beginnings in the early nineteenth century, where the connection between sociology and biology was first made explicit. As is evident, the strategy for developing evolutionary sociology is highly varied, and perhaps with the exception of stage models of societal development, none of these approaches rivals the influence of the ecological perspectives examined in Chapter 9.

Yet, as sociology enters the twenty-first century, these newer perspectives might rival the meso-level human ecology approaches that have so dominated biologically inspired theorizing for most of this century. For it is now clear, and should have been all through the development of sociology, that humans are an animal with an evolutionary history; we evolved like any animal and we must, like all species, adapt to the environment. Concepts from bio-ecology might prove increasingly useful. But, perhaps more important, sociological theory requires a mental set that forces a simple recognition: Human beings and their sociocultural creations are still subject to the basic problem of how to adjust and adapt to the environment. Theory in sociology should reflect this simple recognition.

NOTES

1. Pierre van den Berghe, *Age and Sex in Human Societies: A Biosocial Perspective* (Belmont, CA: Wadsworth, 1973), p. 2. See also his *Man in Society: A Biosocial View* (New York: Elsevier, 1975); "Territorial Behavior in a Natural Human Group," *Social Science Information* 16 (1977), pp. 421–430; "Bringing Beasts Back In: Toward a Biosocial Theory of Aggression," *American Sociological Review* 39 (1974), pp. 777–788; "Why Most Sociologists Don't (And Won't) Think Evolutionarily," *Sociological Forum* 5 (1990), pp. 173–185; "Genes, Mind and Culture," *Behavioral and Brain Sciences* 14 (1991), pp. 317–318.

2. van den Berghe, "Bridging the Paradigms," *Society* 15 (1977–1978), pp. 42–49; *The Ethnic Phenomenon* (New York: Elsevier, 1981); and *Human Family Systems* (Prospect Heights, IL: Waveland, 1990), pp. 14ff.

3. van den Berghe, "Bridging the Paradigms" (cited in note 2), p. 46, and *Human Family Systems*, p. 15.

4. van den Berghe, *The Ethnic Phenomenon* (cited in note 2), p. 7. See also Pierre van den Berghe and Joseph Whitmeyer, "Social Class and Reproductive Success," *International Journal of Contemporary Sociology* 27 (1990), pp. 29–48.

5. van den Berghe, "Bridging the Paradigms" (cited in note 2), pp. 46–47, and *Human Family Systems*, pp. 19–20.

6. See earlier discussion of Trivers in Chapter 7 and related references in footnotes.

7. van den Berghe, *The Ethnic Phenomenon* (cited in note 2), p. 9, and "Bridging the Paradigms," p. 48.

8. van den Berghe, "Bridging the Paradigms" (cited in note 2).

9. van den Berghe, *The Ethnic Phenomenon*, p. 10. See in note 4: van den Berghe and Whitmeyer, "Social Class and Reproductive Success," pp. 31–32.

10. van den Berghe, *The Ethnic Phenomenon* (cited in note 2), p. 6, and *Human Family Systems*, p. 220.

11. Pierre van den Berghe and David Barash, "Inclusive Fitness and Family Structure," *American Anthropologist* 79 (1977), pp. 809–823.

12. van den Berghe, *The Ethnic Phenomenon* (cited in note 2), and also see van den Berghe, "Heritable Phenotypes and Ethnicity," *Behavioral and Brain Sciences* 12 (1989), pp. 544–545.

13. van den Berghe, "Once More with Feeling: Genes, Mind and Culture," *Behavioral and Brain Sciences* 14 (1991), pp. 317–318. See note 4: van den Berghe and Whitmeyer, "Social Class and Reproductive Success," pp. 41–44.

14. J. Maynard-Smith, "The Theory of Games and the Evolution of Animal Conflicts," *Journal of Theoretical Biology* 47 (1974), pp. 209–221, and see Alexandra Maryanski, "The Pursuit of Human Nature in Sociobiology and Evolutionary Sociology," *Sociological Perspectives* 37 (Fall 1994), pp. 115–127.

15. R.D. Alexander, "The Search for a General Theory of Behavior," *Behavior Science* 20 (1975), pp. 77–100.

16. Joseph Lopreato, "The Maximization Principle: A Cause in Search of Conditions" in *Sociobiology and the Social Sciences*, eds. Robert and Nancy Bell (Lubbock: Texas Tech University Press, 1989), pp. 119–130.

17. Ibid., p. 121.

18. Ibid., p. 120–121.

19. Ibid., p. 125.

20. Ibid., p. 126.

21. Ibid., p. 127.

22. Ibid., p. 127.

23. Ibid., p. 127.

24. Ibid., p. 129.

25. Arlen Carey and Joseph Lopreato, "The Evolutionary Demography of the Fertility-Mortality Quasi-Equilibrium," *Population and Development Review* 21 (1995), pp. 613–630.

26. Ibid., p. 616.

27. Ibid., p. 617.

28. Ibid., p. 617–619.

29. Ibid., p. 619.

30. Ibid., p, 620.

31. Ibid.

32. Ibid.

33. Jay Belsky, Laurence Steinberg, and Patricia Draper, "Childhood Experience, Interpersonal Development, and Reproductive Strategy: An Evolutionary Theory of Socialization," *Child Development* 62 (1991), pp. 647–670.

34. Ibid.

35. James Chisholm, "Death, Hope, and Life: Life History Theory and the Development of Reproductive Strategies," *Current Anthropology* 34 (1993), pp. 1–12.

36. Carey and Lopreato, "The Evolutionary Demography of the Fertility-Mortality Quasi-Equilibrium," p. 621; Joseph Lopreato and Mei-Yu Yu, "Human Fertility and Fitness Optimization," *Ethnology and Sociobiology* 9 (1988), pp. 269–289.

37. Ibid., pp. 625–626. Carey and Lopreato emphasize that rapid sociocultural changes can temporarily disturb the two-offspring quasi-equilibrium. A case in point is the recent high fertility and low death frequencies now evident in some developing countries because of accelerated medical technology that caused an abrupt decline in mortality. They suggest that a quasi-equilibrium between fertility and mortality will be reestablished once such proximate cues as

an adequate food supply are linked with mortality reduction.

38. Richard Machalek, "Why Are Large Societies Rare?" *Advances in Human Ecology* 1 (1992), pp. 33–64.

39. Richard Machalek, "Crossing Species Boundaries: Comparing Basic Properties of Human and Non-human Societies." (Unpublished manuscript.)

40. Machalek, "Why Are Large Societies Rare?" (cited in note 38), p. 35.

41. Ibid., p. 42.

42. Ibid.

43. Ibid., P. 44.

44. Ibid., p. 45.

45. Ibid., p. 46.

46. Ibid., p. 47.

47. Ibid., p. 48.

48. Ibid., p. 50.

49. Ibid.

50. Ibid., p. 61.

51. See Chapter 7.

52. See Chapter 4.

53. See Chapter 6.

54. See Chapter 5.

55. Gerhard Lenski, *Power and Privilege: A Theory of Social Stratification* (New York: McGraw-Hill, 1966).

56. Gerhard Lenski, *Human Societies: An Introduction to Macrosociology* (New York: McGraw-Hill, 1970).

57. In subsequent editions of *Human Societies,* Lenski began to collaborate with Jean Lenski.

58. More recent editions of *Human Societies* are co-authored with Patrick Nolan. See, for the most recent edition, Gerhard Lenski, Patrick Nolan, and Jean Lenski, *Human Societies: An Introduction to Macrosociology,* 7th ed. (New York: McGraw-Hill, 1995).

59. See, for a recent discussion on their use, Patrick Nolan, *Annual Review,* forthcoming (1997).

60. Stephen K. Sanderson, *Social Transformations: A General History of Historical Development* (Cambridge, MA: Blackwell, 1995).

61. Lee Freese, *Evolutionary Connections,* in press (Greenwich, CT: JAI, 1997).

62. Lenski, Nolan, and Lenski, *Human Societies* (cited in note 58), p. 75.

63. Ibid.

64. Ibid., pp. 75–76.

65. Gerhard Lenski, "Societal Taxonomies: Mapping the Social Universe," *Annual Review of Sociology* 20 (1994), p. 23.

66. Lenski, Nolan, and Lenski, *Human Societies* (cited in note 58), pp. 57–58.

67. Ibid.

68. Ibid., p. 84.

69. Ibid., p. 54. This is an idea originally proposed by Herbert Spencer, as was emphasized in Chapter 7.

70. See, for example, Stephen K. Sanderson, *Social Evolutionism: A Critical History* (Oxford: Blackwell, 1990).

71. Stephen K. Sanderson, *Social Transformations* (cited in note 60). See also his *Macrosociology: An Introduction to Human Societies* (New York: HarperCollins, 1991).

72. Sanderson, *Social Transformations* (cited in note 60).

73. Ibid., pp. 10–11.

74. Ibid., p. 11.

75. Ibid., pp. 12–13.

76. Ibid., p. 13.

77. Ibid., p. 11.

78. Stephen Sanderson, "Evolutionary Materialism: A Theoretical Strategy for the Study of Social Evolution," *Sociological Perspectives* 37 (1994), pp. 47–73; see page 52 for quote.

79. Ibid., p. 53.

80. Ibid., p. 53.

81. Ibid.

82. Ibid., pp. 50–53.

83. Ibid., p. 52.

84. Ibid., p. 50.

85. Lee Freese, *Evolutionary Connections* (cited in note 61). See also Lee Freese, "Evolution and Sociogenesis. Part I: Ecological Origins; Part II: Social Continuities" in *Advances in Group Processes,* vol. 5, ed. E.J. Lawler and B. Markovsky (Greenwich, CT: JAI, 1988), pp. 53–118.

86. In assuming evolution by self-organization, Freese adopts Alfred Lotka's law as the force driving evolution in biosociocultural regimes. This principle states that "Natural selection will so operate as to increase the total flux through the system, so long as there is presented an unutilized residue of matter and available energy" (quoted in Freese, *Evolutionary Connections,* cited in note 61). The underlying assumption here is that acting entitles always generate change.

87. Freese, *Evolutionary Connections* (cited in note 61).

88. For instance, see William Catton, "Separation versus Unification in Sociological Human Ecology" in *Advances in Human Ecology,* VI, ed. L. Freese (1992), pp. 65–99; William Catton, "What Have We Done to Carrying Capacity?," *Human Ecology: Progress through*

Integrative Perspectives (1995), pp. 162–170; Walter Gove, "Why We Do What We Do: A Biopsychosocial Theory of Human Motivation," *Social Forces* 73 (1994), pp. 363–394; Walter Gove and G. Carpenter, *The Fundamental Connection Between Nature and Nurture* (Toronto: Lexington, 1982); Warren TenHouten, "Cerebral-Lateralization Theory and the Sociology of Knowledge," *The Dual Brain,* eds. D.F. Benson and E. Zaidel (New York: Guilford, 1985), pp. 341–355; Warren TenHouten, "Into the Wild Blue Yonder: On the Emergence of the Ethnoneurologies—the Social Science-Based Neurologies and the Philosophy-Based Neurologies," *Journal of Social and Biological Structures* 14 (1991), pp. 381–408; John Baldwin and Janice Baldwin, *Beyond Sociobiology* (New York: Elsevier, 1981); Marvin Olsen, "A Socioecological Perspective on Social Evolution" in *Advances in Human Ecology,* ed. L. Freese (London: JAI, 1993), vol. 2, pp. 69–92; Thomas Dietz and Tom Burns, "Human Agency and the Evolutionary Dynamics of Culture," *Acta Sociologica* 35 (1992), pp. 187–200; Thomas Dietz, Tom Burns, and F. Buttel, "Evolutionary Theory in Sociology: An Examination of Current Thinking," *Sociological Forum* 5 (1990), pp. 155–185; Theodore Kemper, *Social Structure and Testosterone* (New Brunswick, NJ: Rutgers University Press, 1990); Eugene Rosa, "Sociology, Biosociology or Vulgar Biologizing?" *Sociological Symposium* 27 (1979), pp. 28–45. And see Chapter 32.

89. J. Richard Udry, "Sociology and Biology: What Biology Do Sociologists Need to Know?," *Social Forces* 73 (1995), pp. 1267–1278.

90. J. Richard Udry, "The Nature of Gender," *Demography* 31 (1994), pp. 561–573.

91. Ibid., p. 562.

92. Ibid.

93. Ibid., p. 568.

94. Udry, "Sociology and Biology" (cited in note 89), p. 1274.

95. Udry, "The Nature of Gender" (cited in note 90), p. 570.

96. Ibid., p. 571.

97. Ibid., p. 565.

98. Ibid., p. 572.

99. A. Maryanski, "The Last Ancestor: An Ecological Network Model on the Origins of Human Sociality," *Advances in Human Ecology,* ed. L. Freese, vol. 1 (1992), pp. 1–32; A. Maryanski and Jonathan Turner, *The Social Cage* (Stanford, CA: Stanford University Press, 1992); and A. Maryanski, "African Ape Social Structure: Is There Strength in Weak Ties?" *Social Networks* 9 (1987), pp. 191–215.

100. See Charles G. Sibley, John A. Comstock, and Jon E. Ahlquist, "DNA Hybridization Evidence of Hominoid Phylogeny: A Reanalysis of the Data," *Journal of Molecular Evolution* 30 (1990), pp. 202–236; M. Goodman, D.A. Tagle, D.H.A. Fitch, W. Bailey, J. Czelusnak, B.F. Koop, P. Benson, and J.L. Slightom, "Primate Evolution at the DNA Level and a Classification of Hominids," *Journal of Molecular Evolution* 30 (1990), pp. 260–266.

101. This technique is a standard tool for reconstruction in such fields as comparative biology, historical linguistics, and textual criticism. Essentially, the basic procedure is to identify a set of characters believed to be the end points or descendants of an evolutionary or developmental process, with the idea that an "original" or common ancestor can be reconstructed through the detection of shared diagnostic characters.

102. For discussions of this methodology, see R. Jeffers and I. Lehiste, *Principles and Methods for Historical Linguistics* (Cambridge, MA: MIT Press, 1979); M. Hass, "Historical Linguistics and the Genetic Relationship of Languages," *Current Trends in Linguistics* 3 (1966), pp. 113-53; and N. Platnick and H.D. Cameron, "Cladistic Methods in Textual, Linguistic, and Phylogenetic Analysis," *Systematic Zoology* 26 (1977), pp. 380–385.

103. For discussions, see P. Andrews, "Species Diversity and Diet in Monkeys and Apes during the Miocene" in *Aspects of Human Evolution,* ed. C.B. Stringer (London: Taylor and Francis, 1981), pp. 25–61; J. Temerin and J. Cant, "The Evolutionary Divergence of Old World Monkeys and Apes," *American Naturalist* 122 (1983), pp. 335–351; and R. Ciochon and R. Corruccini, eds., *New Interpretations of Ape and Human Ancestry* (New York: Plenum Press, 1983).

104. To array social tie patterns, affective ties were assessed on the basis of mutually reinforcing and friendly interactions. Degrees of attachment were described along a simple scale of tie strength: null ties, weak ties, moderate ties, strong ties. Individuals without ties (for example, father-daughter where paternity cannot be known) or who rarely, if ever, interact have null ties; those who interact in a positive manner on an occasional basis have weak ties; those who affiliate closely for a time but without endurance over time (at least for adults) have moderate ties; and those who exhibit extensive nonsexual physical contact with much observable affect (for example, reciprocal grooming), have very high interactional rates, and show mutual support with stable and long-term relations over time have strong ties. Scaling tie strength for primates is a straightforward procedure because age and sex classes have clear-cut social tendencies that have been documented by field researchers during the last fifty years. Also see Maryanski, "The Last Ancestor" (note 99), and Maryanski and Turner, *The Social Cage,* for detailed discussions about how the network analysis of primate social ties was conducted. For general references on network analysis, see Chapter 38 on social network analysis in this book.

105. See for example M.G. Bicchieri, *Hunters and Gatherers Today* (New York: Holt, Rinehart & Winston, 1972) for a study of eleven food collecting societies. Also see Margaret Power, *The Egalitarians—Human and Chimpanzee: An Anthropological View of Social Organization* (Cambridge: Cambridge University Press, 1991), pp. xviii, 290; Robert C. Bailey and Robert Aunger, "Humans as Primates: The Social Relationships of Efe Pygmy Men in Comparative Perspective," *International Journal of Primatology,* Vol. 11, No. 2 (1990), pp. 127–145, which details some of the similarities between chimpanzee and human hunter-gatherer societies.

106. Maryanski and Turner, *The Social Cage* (cited in note 99), Chapter 4, pp. 69–90.

PART III

Conflict Theorizing

11

The Emerging Tradition:
The Rise of Conflict Theorizing

Conflict theory was one of sociology's first theoretical orientations, developing along with functionalism and the implicit bio-ecological theory contained within functionalism. Even some early functional theorists, such as Herbert Spencer,[1] developed conceptualizations of conflict; yet, over the years, these functional approaches increasingly came under attack for underemphasizing conflict and change. In seeking "the function" of sociocultural forces for meeting needs for integration and other requisites, functionalists tended to underemphasize the effects of inequality in systematically generating conflict, disintegration, and change. Talcott Parsons' functionalism, in particular, was viciously attacked during the late 1950s and 1960s for failing to conceptualize conflict processes adequately.[2]

Thus, one of sociology's first theoretical orientations was given new life in the second half of the twentieth century. This new life came at the expense of functionalism, which was replaced for a time by conflict theories. Today, the conflict theory reaction to functionalism—indeed, some might say an overreaction—is well entrenched within a broad array of sociological theories, as is explained in this part as well as in part seven on critical theories.

Conflict theory in sociology began with Karl Marx (1818–1883), but the development of the approach in the mid-twentieth century owes a debt to two other early German sociologists, Max Weber (1864–1920) and Georg Simmel (1858–1918). Weber and Simmel also articulated conflict theories, but they were suspicious of Marx's polemics. Taken together, however, Marx, Weber, and Simmel provided the core ideas that still inspire contemporary conflict approaches—as we will see in later chapters. For the present, we will examine how each approached the analysis of conflict.

KARL MARX AND CONFLICT THEORY

We will encounter Marx's work in analyzing several theoretical perspectives in present-day sociology, so, it is not necessary to present his entire theoretical corpus here. For the present, the goal is to describe the more general, abstracted model of conflict that is packaged between the polemics of the Marxian scheme. The reason for this concern here is that the reemergence of conflict theory as a reaction to functionalism took this form: scholars seeking to use Marx to develop general conflict theories that could serve as an alternative to the perceived inadequacies of functionalism.

Table 11-1 summarizes Marx's assumptions about the social world and the key forces behind conflict and change in societies.[3] As shown in Proposition I, Marx argued that the degree of inequality in the distribution of resources generates inherent conflicts of interest. Proposition II then emphasizes that when members of subordinate segments of the society become aware of their true interests in redistributing resources and, thereby, reducing inequality, they will begin to question the legitimacy of the system. Next, Proposition III specifies the conditions that facilitate subordinates' awareness of their true conflict of interest. Propositions III-A, B, C, and D deal, respectively, with the disruption in the social situation of deprived populations, the amount of alienation people feel as a result of their situation, the capacity of members of deprived segments to communicate with one another, and their ability to develop a unifying ideology that codifies their true interests. Marx saw these conditions as factors that increase and heighten awareness of subordinates' collective interests and, hence, decrease their willingness to accept as legitimate the right of superordinates to command a disproportionate share of resources.

In turn, some of these forces heightening awareness are influenced by such structural conditions as ecological concentration (III-C-1), educational opportunities (III-C-2), the availability of ideological spokespeople (III-D-1), and the control of socialization processes and communication networks by superordinates (III-D-2). Marx hypothesized (shown in Proposition IV), that the increasing awareness by deprived classes of their true interests and the resulting questioning of the legitimacy of the distribution of resources increases the likelihood that the disadvantaged strata will begin to organize collectively their opposition against the dominant segments of a system. This organization is seen as especially likely under several conditions: disorganization among the dominant segments with respect to organizing to protect their true interests (IV-A), sudden escalation of subordinates' sense of deprivation as they begin to compare their situation with that of the privileged (IV-B), and mobilization of political leadership to carry out the organizational tasks of pursuing conflict (IV-C). Marx emphasized (shown in Proposition V) that, once deprived groups possess a unifying ideology and political leadership, their true interests begin to take on clear focus and their opposition to superordinates begins to increase—polarizing the interests and goals of superordinates and subordinates. As polarization increases, the possibilities for reconciliation, compromise, or mild conflict decrease, because the deprived are sufficiently alienated, organized, and unified to press for a complete change in the pattern of resource distribution. As Proposition VI underscores, subordinates begin to see violent confrontation as the only way to overcome the inevitable resistance of superordinates. Finally, Marx noted (shown in Proposition VII) that violent conflict will cause great changes in patterns of social organization, especially its distribution of scarce resources. The propositions in Table 11-1 are stated much more

TABLE 11-1 Marx's Abstracted Propositions on Conflict Processes

I. The more unequal is the distribution of scarce resources in a society, the greater is the basic conflict of interest between its dominant and subordinate segments.

II. The more subordinate segments become aware of their true collective interests, the more likely they are to question the legitimacy of the existing pattern of distribution of scarce resources.

III. Subordinates are more likely to become aware of their true collective interests when
   A. Changes wrought by dominant segments disrupt existing relations among subordinates.
   B. Practices of dominant segments create alienative dispositions.
   C. Members of subordinate segments can communicate their grievances to one another, which, in turn, is facilitated by
      1. The ecological concentration among members of subordinate groups.
      2. The expansion of educational opportunities for members of subordinate groups.
   D. Subordinate segments can develop unifying ideologies, which, in turn, is facilitated by
      1. The capacity to recruit or generate ideological spokespeople.
      2. The inability of dominant groups to regulate socialization processes and communication networks among subordinates.

IV. The more that subordinate segments of a system become aware of their collective interests and question the legitimacy of the distribution of scarce resources, the more likely they are to join in overt conflict against dominant segments of a system, especially when
   A. Dominant groups cannot clearly articulate, nor act in, their collective interests.
   B. Deprivations of subordinates move from an absolute to a relative basis, or escalate rapidly.
   C. Subordinate groups can develop a political leadership structure.

V. The greater is the ideological unification of members of subordinate segments of a system and the more developed is their political leadership structure, the more likely are the interests and relations between dominant and subjugated segments of a society to become polarized and irreconcilable.

VI. The more polarized are the dominant and subjugated, the more will the conflict be violent.

VII. The more violent is the conflict, the greater is the amount of the structural change within a society and the greater is the redistribution of scarce resources.

abstractly than Marx would have considered appropriate,[4] but his ideas began to filter back into contemporary sociology in this form. As theorists sought explanations for the forces generating conflict and change, they implicitly drew from Marx this image of society as filled with conflicts of interests in the distribution of scarce resources, with inequality in the distribution of valued resources setting into motion the mobilization of the subordinates to pursue conflict against superordinates. Yet, few borrowed only from Marx; as conflict theories came to the forefront in sociology during the 1950s and 1960s, both Max Weber and Georg Simmel were also consulted.

MAX WEBER AND CONFLICT THEORY

Max Weber was implicitly critical of Marx's theory of conflict, arguing that the unfolding of history is contingent on specific empirical conditions. Revolutionary conflict is not, Weber believed, inevitable, coming to the revolutionary crescendo described by Marx. Yet, like Marx, Weber developed a theory of conflict, and despite a convergence in their theories, Weber saw conflict as highly contingent on the emergence of "charismatic leaders" who could mobilize subordinates. Unlike Marx, Weber saw the emergence of such leaders as far from inevitable, and hence,

TABLE 11-2 Weber's Abstracted Propositions on Conflict Processes

I. Subordinates are more likely to pursue conflict with superordinates when they withdraw legitimacy from political authority.

II. Subordinates are more likely to withdraw legitimacy from political authority when
   A. The correlation among memberships in class, status group, and political hierarchies is high.
   B. The discontinuity or degrees of inequality in the resource distributions within social hierarchies is high.
   C. Rates of social mobility up social hierarchies of power, prestige, and wealth are low.

III. Conflict between superordinates and subordinates becomes more likely when charismatic leaders can mobilize resentments of subordinates.

IV. When charismatic leaders are successful in conflict, pressures mount to routinize authority through new systems of rules and administration.

V. As a system of rules and administrative authority is imposed, the more likely are conditions II-A, II-B, and II-C to be met, and hence, the more likely are new subordinates to withdraw legitimacy from political authority and to pursue conflict with the new superordinates, especially when new traditional and ascriptive forms of political domination are imposed by elites.

revolutionary conflict would not always be produced in systems of inequality. Nonetheless, when Weber's implicit propositions shown in Table 11-2 are compared with those of Marx in Table 11-1, considerable overlap is evident.

Most of the principles in Table 11-2 can be found in Weber's discussion of the transition from societies based on traditional authority to those organized around rational-legal authority.[5] In societies where the sanctity of tradition legitimates political and social activity, the withdrawal of legitimacy from these traditions is a crucial condition of conflict, as is emphasized in Proposition I of Table 11-2. What, then, causes subordinates to withdraw legitimacy? As indicated in Proposition II-A, one cause is a high degree of correlation among power, wealth, and prestige or, in Weber's terms, among positions of political power (party), occupancy in advantaged economic positions (class), and membership in high-ranking social circles (status groups). When economic elites, for example, are also social and political elites, and vice versa, then those who are excluded from power, wealth, and prestige become resentful and receptive to conflict alternatives. Another condition (Proposition II-B) is dramatic discontinuity in the distribution of rewards, or the existence of large gaps in social hierarchies that give great privilege to some and very little to others. When only a few hold power, wealth, and prestige and the rest are denied these rewards, tensions and resentments exist. Such resentments become a further inducement for those without power, prestige, and wealth to withdraw legitimacy from those who hoard these resources. A final condition (Proposition II-C) is low rates of social mobility. When those of low rank have little chance to move up social hierarchies or to enter a new class, party, or status group, then resentments accumulate. Those denied opportunities to increase their access to resources become restive and unwilling to accept the system of traditional authority.

As stressed in Proposition III in Table 11-2, the critical force that galvanizes the resentments inhering in these three conditions is charisma. Weber felt that whether or not charismatic leaders emerge is, to a great extent, a matter of historical chance, but if such leaders do emerge to challenge traditional authority and to mobilize resentments caused by the hoarding of resources by elites and the lack of opportunities to gain access to wealth, power, or prestige, then conflict and structural change can occur.

When successful, however, such leaders confront organizational problems of consolidating their gains. As stated in Proposition IV, one result is that charisma becomes routinized as leaders create formal rules, procedures, and structures for organizing followers after their successful mobilization to pursue conflict. And as is emphasized in Proposition V, if routinization creates new patterns of ascription-based inequalities, thus erecting a new system of traditional authority, renewed conflict can be expected as membership in class, status, and party becomes highly correlated, as the new elites hoard resources, and as social mobility up hierarchies is blocked. If rational-legal routinization occurs, however, authority is based on equally applied laws and rules, and performance and ability become the basis for recruitment and promotion in bureaucratic structures. Under these conditions, conflict potential will be mitigated.

Unlike Marx, who tended to overemphasize the economic basis of inequality and to argue for a simple polarization of societies into propertied and nonpropertied (exploited) classes, Weber's Propositions I and II show more theoretical options. Weber believed that variations in the distribution of power, wealth, and prestige and the extent to which holders of one resource control the other resources become critical. Unlike Marx, who saw this correlation as inevitable, Weber saw more diverse relations among class, status, and party. Moreover, the degree of discontinuity in the distribution of these resources—in other words, the extent to which there are clear gaps and lines demarking privilege and nonprivilege—can also vary. Unlike Marx, Weber did not see the complete polarization of super- and subordinates as inexorable. Finally, the degree of mobility—the chance to gain access to power, wealth, and prestige—becomes a crucial variable in generating the resentments and tensions that make people prone to conflict; unlike Marx, Weber did not see a drop in mobility rates as always accompanying inequality.

In addition to the propositions in Table 11-2, which pertain primarily to *intra*societal conflict processes, Weber developed theoretical ideas on *inter*societal processes.[6] Because conflict between societies is, as Herbert Spencer recognized early in his work, a basic condition of human societies that have settled in territories and developed political leadership, it is not surprising that Weber also analyzed intersocietal conflict, or the "geopolitics" between societies. This emphasis has been a prominent theme in the dramatic revival of historical sociology in both its neo-Marxian[7] and neo-Weberian[8] forms and will be explored in later chapters. Weber believed the degree of legitimacy accorded political authority within a system very much depends on that authority's capacity to generate prestige in the wider geopolitical system, or what today we might term "world system." Thus, withdrawal of legitimacy is not just the result of conditions II-A, II-B, and II-C in Table 11-2; legitimacy also depends on the "success" and "prestige" of a state in relation to other states.[9]

Political legitimacy is a precarious situation because it relies on the capacity of political authority to meet the needs among system members for defense and attack against external enemies, even during periods of relative peace. Without this sense of "threat" and a corresponding "success" in dealing with this threat, legitimacy lessens. Weber did not argue that legitimacy is always necessary for superordinates to dominate—there are periods of apathy among members of a population, supported by tradition and routine. And there can also be periods of coercive force by superordinates to quell potential rebellion. Nor did Weber argue that "external enemies" must always be present to keep legitimacy revved up; rather, internal

TABLE 11-3 Weber's Abstracted Propositions on Geopolitics and Conflict

I. The capacity of political authority to dominate a society depends on its legitimacy.

II. The more those with power can sustain a sense of prestige and success in relations with external societies, the greater will be the capacity of leaders to be viewed as legitimate.

III. When productive sectors of a society depend on political authority for their viability, they encourage political authority to engage in military expansion to augment their interests. When successful, such expansion increases the prestige and, hence, the legitimacy of political authority.

IV. When productive sectors do not depend on the state for their viability, they encourage political authority to rely on co-optation rather than on military expansion, and when successful, such co-optation increases prestige and, hence, the legitimacy of political authority.

V. The more those with power can create a sense of threat from external forces, the greater is their capacity to be viewed as legitimate.

VI. The more those with power can create a sense of threat among the majority by internal conflict with a minority, the greater is their capacity to be viewed as legitimate.

VII. When political authority cannot sustain a sense of legitimacy, it becomes vulnerable to outbreaks of internal conflict, and when political authority loses prestige in the external system, it loses legitimacy and becomes more vulnerable to internal conflict.

conflicts that pose threats can also give legitimacy to political authority. Thus, the very processes that might lead some to withdraw legitimacy and initiate conflict under charismatic leadership can sometimes bolster the legitimacy of political authority, *if* enough other groupings in a society feel threatened. Indeed, Weber argued, political authorities often stir up internal or external "enemies" as a ploy for increasing their legitimacy and power to control the distribution of resources.

But the attention of those with political authority to the external system is not always political. Prestige, per se, can motivate some groupings to encourage military and other forms of contact with other societies. More important, however, are economic interests. Those economic interests—colonial and booty capitalists, privileged traders, financial dealers, arms exporters, and the like—who rely on the state to sustain their viability encourage foreign military expansion, whereas those economic interests that rely on market dynamics and free trade will usually resist military expansionism because it can hurt domestic productivity, or profits in external markets. Instead, these interests will encourage co-optive efforts through trade relations and market dependencies of external populations on commodities and services provided by these interests.

Table 11-3 presents Weber's argument in more abstract terms; these propositions supplement those in Table 11-2 where the loss of legitimacy is seen by Weber as increasing the likelihood of conflict. The essential point is not so much that Weber developed a mature theory but, rather, that he stimulated a conflict approach that examined the relationship between internal and external conflict processes.

GEORG SIMMEL AND CONFLICT THEORY

Georg Simmel was committed to developing theoretical statements that captured the *form of basic social processes*, an approach he labeled *formal sociology*. Primarily on the basis of his own observations, Simmel sought to extract the essential properties from processes and events in a wide variety of empirical contexts. In turn, abstract statements about these essential properties could be articulated.

TABLE 11-4 Simmel's Abstracted Propositions on Conflict Processes

I. The level of violence in conflict increases when
   A. The parties to the conflict have a high degree of emotional involvement, which, in turn, is related to the respective levels of solidarity among parties to the conflict.
   B. The membership of each conflict party perceives the conflict to transcend their individual self-interests, which, in turn, is related to the extent to which the conflict is about value-infused issues.

II. The level of violence in conflict is reduced when the conflict is instrumental and perceived by the conflict parties be a means to clear-cut and delimited goals.

III. Conflict will generate the following among the parties to a conflict:
   A. Clear group boundaries.
   B. Centralization of authority and power.
   C. Decreased tolerance of deviance and dissent.
   D. Increased internal solidarity among memberships of each party, but particularly for members of minority parties and for groups engaged in self defense.

IV. Conflict will have integrative consequences for the social whole when
   A. Conflict is frequent, low in intensity, and low in violence, which, in turn, allows disputants to release hostilities.
   B. Conflict occurs in a system whose members and subunits reveal high levels of functional interdependence, which, in turn, encourages the creation of normative agreements to regulate the conflict so that the exchange of resources is not disrupted.
   C. Conflict produces coalitions among various conflicting parties.

Much like Marx, Simmel viewed conflict as ubiquitous and, hence, subject to analysis in formal terms.[10] In his most famous essay on conflict, Simmel devoted considerable effort to analyzing the positive consequences of conflict for the maintenance of social wholes and their subunits. Simmel recognized, of course, that an overly cooperative, consensual, and integrated society would show "no life process," but his analysis of conflict is still loaded in the direction of how conflict promotes solidarity and unification. Thus unlike Marx, who saw conflict as ultimately becoming violent and revolutionary and leading to the structural change of the system, Simmel quite often analyzed the opposite phenomena—less intense and violent conflicts that promote the solidarity, integration, and orderly change.[11]

Simmel's key ideas on conflict are summarized in Table 11-4. Proposition I-A overlaps somewhat with those developed by Marx. Similarly to Marx, Simmel emphasized that violent conflict is the result of emotional arousal. Such arousal is particularly likely when conflict groups possess a great deal of internal solidarity. As shown in Proposition I-B, Simmel indicated that, coupled with emotional arousal, the extent to which members see the conflict as transcending their personal aims and self-interests increases the likelihood of violent conflict. Proposition II is Simmel's most important because it contradicts Marx's hypothesis that objective consciousness of interests will lead to organization for violent conflict. Simmel argued that the more clearly articulated are the interests of conflict parties, the more clear-cut and focused are their goals; with clearly articulated goals, less combative means, such as bargaining and compromise, are more likely to be used to meet the specific objectives of the group. Thus, for Simmel, consciousness of common interests can, under unspecified conditions, lead to highly instrumental and nonviolent conflict. In the context of labor-management relations, for example, Simmel's proposition is more accurate than Marx's prediction because violence has more often accompanied labor-management disputes in the initial formation of unions, when interests and goals are not well articulated. As interests become clarified, violent conflict has been increasingly replaced by less violent forms of social negotiation.[12]

The consequences of conflict for (1) the conflict parties and (2) the systemic whole in which the conflict occurs are summarized in Propositions III and IV. Propositions III-A, III-B, III-C, and III-D summarize Simmel's ideas about the functions of conflict for the respective parties to the conflict. Conflict increases the formation of clear-cut group boundaries, the centralization of authority, the control of deviance and dissent, and the enhancement of social solidarity within conflict parties.

Proposition IV, on the consequences of conflict for the social whole, provides an important qualification to Marx's analysis. Marx visualized initially mild conflicts as intensifying as the combatants become increasingly polarized, ultimately resulting in violent conflict that would lead to radical social change in the system. In contrast, Simmel argued that conflicts of low intensity and high frequency in systems of high degrees of interdependence do not necessarily intensify or lead to radical social change. On the contrary, these conflicts release tensions and become normatively regulated, thereby promoting stability in social systems. Further, with the increasing organization of the conflicting groups, and the formation of coalitions among conflict groups, violence will decrease as their goals become better articulated. The consequence of such organization and articulation of interests will be a greater disposition to initiate milder forms of conflict, involving competition, bargaining, and compromise.

THE PROLIFERATION OF CONFLICT THEORIES

Marx's, Simmel's, and Weber's work represented the beginnings of conflict theory in sociology, and these works continue to inform and inspire this orientation today. The initial wave of conflict theories in the late 1950s through the 1960s and into the early 1970s adopted and adapted the ideas of these theorists explicitly, as we will see in the next chapters on the maturing conflict tradition. But, as will become evident in the chapters on the current direction of conflict theory, the approach has become much more eclectic and varied during the last two decades. The ideas of Marx, Simmel, and Weber still inform, but their influence is not always so clearly evident. Indeed, as conflict theory has proliferated, it has quietly become, much like many functionalist ideas before it, absorbed into the sociological canon and not so easily distinguished as a unique perspective as it was in the 1960s and early 1970s. Perhaps the early conflict theorists had to establish a beachhead by engaging in rather exaggerated polemics, typically against functionalism, which then held the theoretical mainland. Only then could conflict theorists move ashore with more confidence; as they did so, the ideas of conflict theory penetrated ever more arenas of sociological inquiry.[13] We might even say that, outside the critical tradition that is examined in the chapters of Part VII of this book, a clear and distinct conflict approach is no longer so evident in sociology. Where it is found, as we will come to appreciate, it is in narrower substantive fields focusing on particular types of conflict.

NOTES

1. Herbert Spencer, *The Principles of Sociology* (New York: D. Appleton, 1898; originally published in serial form between 1874–1896). Spencer argued that war between populations had been an important evolutionary force, because the better organized society usually won and "selected out" the weaker one or at least brought it to the level of organization of the victor. This kind of geopolitical analysis became, in the latter half of the twentieth century, an ever more prominent form of conflict theorizing.

2. See, for example, David Lockwood, "Some Remarks on 'The Social System,'" *British Journal of Sociology* 7 (1956), pp. 134–146; Ralf Dahrendorf, "Out of Utopia: Toward a Reorientation of Sociological Analysis," *American Journal of Sociology* 744 (1958), pp. 115–127.

3. In Karl Marx's and Friedrich Engel's *The Communist Manifesto* (New York: International, 1971; originally published in 1847) are the key ideas in these propositions. Further amplification of these ideas came with Marx's *Capital: A Critical Analysis of Capitalist Production*, vol. 1 (New York: International, 1967; originally published in 1867). The critical and polemical substance of these and other works will be explored in Chapter 40 on the emergence of critical theory.

4. For criticism of efforts of such abstract renderings of Marx, see Richard P. Appelbaum, "Marx's Theory of the Falling Rate of Profit: Towards a Dialectical Analysis of Structural Social Change," *American Sociological Review* 43 (February 1978), pp. 64–73.

5. For a fuller discussion, see Jonathan H. Turner, Leonard Beeghley, and Charles Powers, *The Emergence of Sociological Theory* (Belmont, CA: Wadsworth, 1998), pp. 213–229. For original sources, see Max Weber, *Economy and Society* (New York: Bedminster, 1968).

6. Weber, *Economy and Society*, pp. 901–1372, especially pp. 901–920.

7. For example, see Immanuel Wallerstein, *The Modern World System*, 3 volumes (New York: Academic, 1974–1989).

8. For example, see Randall Collins, *Weberian Sociological Theory* (Cambridge, England: Cambridge University Press, 1986); Michael Mann, *The Sources of Social Power*, vol. 1 (New York: Cambridge University Press, 1986); Theda Skocpol, *States and Social Revolutions* (New York: Cambridge University Press, 1979). The last work combines the analysis of internal revolution and geopolitics, describing the former as a potential consequence of failed policies in the latter.

9. Ibid. (all references in note 8).

10. All subsequent references to this work are taken from Georg Simmel, *Conflict and the Web of Group Affiliation*, trans. K.H. Wolff (Glencoe, IL: Free Press, 1956).

11. Pierre van den Berghe has argued that a dialectical model of conflict is ultimately one in which unification, albeit temporary, emerges out of conflict. Yet, the differences between Marx and Simmel have inspired vastly different theoretical perspectives in contemporary sociology. See Pierre van den Berghe, "Dialectic and Functionalism: Toward a Theoretical Synthesis," *American Sociological Review* 28 (1963), pp. 695–705.

12. Admittedly, Marx's late awareness of the union movement in the United States forced him to begin pondering this possibility, but he did not incorporate this insight into his theoretical scheme.

13. For a diverse sampling of these efforts, see William Gamson, *The Strategy of Social Protest* (Belmont, CA: Wadsworth, 1990); Louis Kriesberg, *The Sociology of Social Conflict* (Englewood Cliffs, NJ: Prentice-Hall, 1973); Anthony Oberschall, *Social Conflict and Social Movements* (Englewood Cliffs, NJ: Prentice-Hall, 1973); Robin M. Williams, Jr., *Mutual Accommodation: Ethnic Conflict and Cooperation* (Minneapolis: University of Minnesota Press, 1977) and *The Reduction of Intergroup Tensions* (New York: Social Science Research Council, 1947); Randall Collins, *Conflict Sociology* (New York: Academic, 1975); Ted Gurr, "Sources of Rebellion in Western Societies: Some Quantitative Evidence," *Annals* 391 (1970), pp. 128–144; A.L. Jacobson, "Intrasocietal Conflict: A Preliminary Test of a Structural Level Theory," *Comparative Political Studies* 6 (1973), pp. 62–83; David Snyder, "Institutional Setting and Industrial Conflict," *American Sociological Review* 40 (1975), pp. 259–278; David Britt and Omer R. Galle, "Industrial Conflict and Unionization," *American Sociological Review* 37 (1972), pp. 46–57; E. McNeil, ed., *The Nature of Human Conflict* (Englewood Cliffs, NJ: Prentice-Hall, 1965); Jessie Bernard, "Where Is the Modern Sociology of Conflict?" *American Journal of Sociology* 56 (1950), pp. 111–116; and "Parties and Issues in Conflict," *Journal of Conflict Resolution* 1 (1957), pp. 111–121; Kenneth Boulding, *Conflict and Defense: A General Theory* (New York: Harper & Row, 1962); Thomas Carver, "The Basis of Social Conflict," *American Journal of Sociology* 13 (1908), pp. 628–637; James Coleman, *Community Conflict* (Glencoe, IL: Free Press, 1957); James C. Davies, "Toward a Theory of Revolution," *American Journal of Sociology* 27 (1962), pp. 5–19; Charles P. Loomis, "In Praise of Conflict and Its Resolution," *American Sociological Review* 32 (1967),

pp. 875–890; Raymond Mack and Richard C. Snyder, "The Analysis of Social Conflict," *Journal of Conflict Resolution* 1 (1957), pp. 388–397; John S. Patterson, *Conflict in Nature and Life* (New York: Appleton-Century-Crofts, 1883); Anatol Rapoport, *Fights, Games and Debates* (Ann Arbor: University of Michigan Press, 1960); Thomas C. Schelling, *The Strategy of Conflict* (Cambridge, MA: Harvard University Press, 1960); Pitirim Sorokin, "Solitary, Antagonistic, and Mixed Systems of Interaction," in *Society, Culture, and Personality* (New York: Harper & Row, 1947); Nicholas S. Timasheff, *War and Revolution* (New York: Sheed and Ward, 1965); and Clinton F. Fink, "Some Conceptual Difficulties in the Theory of Social Conflict," *Journal of Conflict Resolution* 12 (1968), pp. 429–431.

12

The Maturing Tradition I: Ralf Dahrendorf's Dialectical Theory

For the first half of the twentieth century, conflict theorizing remained somewhat dormant. There were, of course, analyses of conflict for specific empirical phenomena, such as ethnic strife, class tensions, inter-societal war, colonialism, and other dissociative processes. But, the conflict ideas contained in the German masters had not been explicitly incorporated into the mainstream of sociological theory, especially in America. Marxist scholarship was particularly recessive, being repressed in America by the anti-communism of the Cold War era.

With the 1960s, however, came a broad social and intellectual movement that, confronted then current institutional practices in western societies; in this new environment, conflict sociology was reborn, soon becoming an important part of the theoretical canon.

DAHRENDORF'S CRITIQUE OF FUNCTIONALISM

Much of this conflict approach reinvigoration was couched as a critique of the excesses of functionalism, which was often accused of being a conservative and supportive ideology of the status quo. As a corrective to these ideological implications of functionalism, a conflict approach was required; beginning in the late 1950s with the German sociologist, Ralf Dahrendorf, theoretical sociology took a decidedly conflict-oriented turn. By the decade of the 1970s, conflict theory had replaced functionalism as sociology's most dominant theoretical approach—although this dominance was not as strong as that enjoyed by functionalism in the 1950s and early 1960s. In this chapter, the dialectical approach of Ralf Dahrendorf will be examined.

Dahrendorf's was the first of the modern conflict theories to gain a wide audience, so it is appropriate to begin with his work.

In the late 1950s, Ralf Dahrendorf persistently argued that the Parsonian scheme, and functionalism in general, presented an overly consensual, integrated, and static vision of society—in his words a "utopia." In Dahrendorf's view, society has two faces—one of consensus, the other of conflict. It was time, Dahrendorf asserted, to begin analysis of society's ugly face and abandon the utopian image created by functionalism. To leave utopia, Dahrendorf offered the following advice:

> Concentrate in the future not only on concrete problems but on such problems as involve explanations in terms of constraint, conflict, and change. This second face of society may aesthetically be rather less pleasing than the social system—but, if all sociology had to offer were an easy escape to Utopian tranquillity, it would hardly be worth our efforts.[1]

To escape utopia, therefore, requires that a one-sided conflict model be substituted for the one-sided functional model. Although this conflict perspective was not considered by Dahrendorf to be the only face of society, it was seen as a necessary supplement that will make amends for the past inadequacies of functional theory.[2]

DAHRENDORF'S DIALECTICAL ASSUMPTIONS

The model that emerged from this theoretical calling is a dialectical conflict perspective, which still represents one of the best efforts to incorporate the insights of Marx and (to a lesser extent) Weber and Simmel into a coherent set of theoretical propositions. Dahrendorf believed that the process of institutionalization involves the creation of "imperatively coordinated associations" (hereafter referred to as *ICAs*) that, in terms of criteria not specified, represent a distinguishable organization of roles. This organization is characterized by power relationships, with some clusters of roles having power to extract conformity from others. Dahrendorf was somewhat vague on this point, but it appears that any social unit—from a small group or formal organization to a community or an entire society—could be considered an ICA for analytical purposes if an organization of roles displaying power differentials exists. Furthermore, although power denotes the coercion of some by others, these power relations in ICAs tend to become legitimated and can therefore be viewed as authority relations in which some positions have the "accepted" or "normative right" to dominate others. Dahrendorf thus conceived of the social order as maintained by processes creating authority relations in the various types of ICAs existing throughout all layers of social systems.[3]

At the same time, however, power and authority are the scarce resources over which subgroups within a designated ICA compete and fight. They are thus the major sources of conflict and change in these institutionalized patterns. This conflict is ultimately a reflection of where clusters of roles in an ICA stand in relation to authority, because the "objective interests" inherent in any role are a direct function of whether that role possesses authority and power over other roles. However, even though roles in ICAs possess varying degrees of authority, any particular ICA can be typified as just two basic types of roles, ruling and ruled. The ruling cluster of roles has an interest in preserving the status quo, and the ruled cluster has an interest in redistributing power, or authority. Under certain specified conditions, awareness of these contradictory interests increases, with the result that ICAs polarize into two conflict groups, each now aware of its objective interests, which then engage in a contest for authority. The resolution of this contest or conflict involves the redistribution of authority in the ICA, thus making conflict the source of change in social systems. In turn, the redistribution of authority represents the institutionalization of a new cluster of ruling and ruled roles that, under certain conditions, polarize into two interest groups that initiate another contest for authority. Social reality is thus typified by this unending cycle of conflict over authority within the various types of ICAs that constitute the social world.

Much like Marx, this image of institutionalization as a cyclical or dialectic process led Dahrendorf into the analysis of only certain key causal relations: (1) Conflict is assumed to be an inexorable process arising from opposing forces within social and structural arrangements; (2) such conflict is accelerated or retarded by a series of intervening structural conditions or variables; (3) conflict resolution at one point in time creates a structural situation that, under specifiable conditions, inevitably leads to further conflict among opposed forces. Moreover, Dahrendorf's and Marx's models reveal similar causal chains of events leading to conflict and the reorganization of social structure: Relations of domination and subjugation create an "objective" opposition of interests; awareness or consciousness by the subjugated of this inherent opposition of interests occurs under certain specifiable conditions; under other conditions this newfound awareness leads to the political organization and, then, to polarization of subjugated groups, which join in conflict with the dominant group; the outcome of the conflict will usher in a new pattern of social organization; this new pattern of social organization will have within it relations of domination and subjugation that set off another sequence of events leading to conflict and then change in patterns of social organization.

The intervening conditions affecting these processes are outlined by both Marx and Dahrendorf only with respect to the formation of awareness of opposed interests by the subjugated, the politicization and polarization of the subjugated into a conflict group, and the outcome of the conflict. The intervening conditions under which institutionalized patterns generate dominant and subjugated groups and the conditions under which these can be typified as having opposed interests remain unspecified—apparently because they are in the nature of institutionalization, or ICAs, and do not have to be explained.

Figure 12-1 outlines the causal imagery of Marx and Dahrendorf. The top row of the figure contains Marx's analytical categories, stated in their most abstract form. The other two rows specify the empirical categories of Marx and Dahrendorf, respectively. Separate analytical categories for the Dahrendorf model are not enumerated because they are the same as those in the Marxian model. The empirical categories of the Dahrendorf scheme differ greatly from those of Marx, but the form of analysis is much the same because each considers as nonproblematic and not in need of causal analysis the empirical conditions of social organization, the transformation of this organization into relations of domination and subjugation, and the creation of opposed interests. The causal analysis for both begins with an elaboration of the conditions leading to growing class consciousness (Marx) or awareness among quasi groups (Dahrendorf) of their objective interests; then analysis shifts to the creation of a politicized class "for itself" (Marx) or a true "conflict group" (Dahrendorf); finally, emphasis focuses on the emergence of conflict between polarized and politicized classes (Marx) or conflict groups (Dahrendorf).

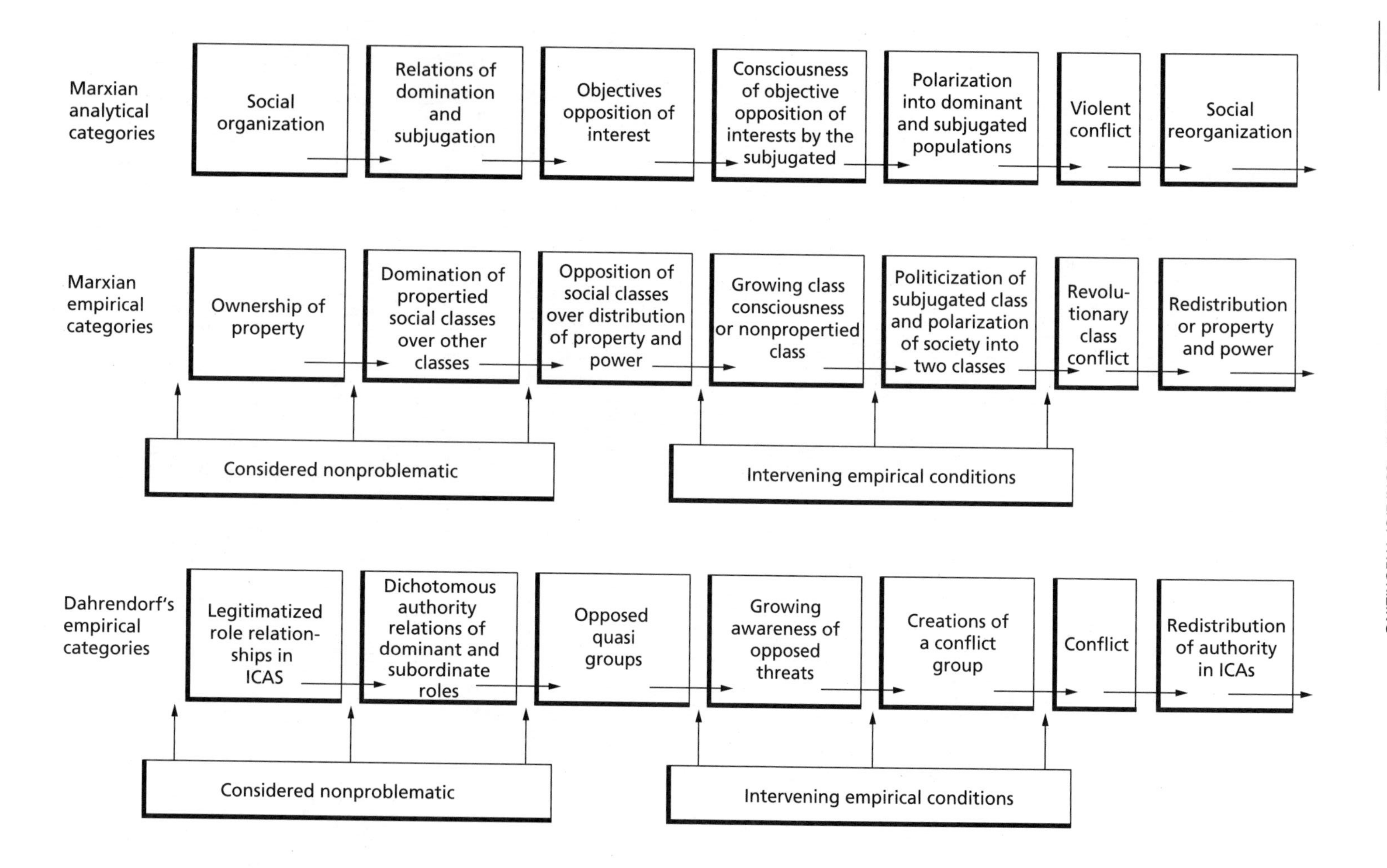

FIGURE 12-1 The Dialectical Causal Imagery

TABLE 12-1 Dahrendorf's Abstract Propositions

I. Conflict is likely to occur as members of quasi groups in ICAs can become aware of their objective interests and form a conflict group, which, in turn, are related to
   A. The "technical" conditions of organization which, in turn, depend on
      1. The formation of a leadership cadre among quasi groups.
      2. The codification of an idea system, or charter.
   B. The "political" conditions of organization, which are dependent on dominant groups permitting organization of opposed interests.
   C. The "social" conditions of organization, which, in turn, are related to
      1. Opportunities for members of quasi groups to communicate.
      2. Opportunities for recruiting members.

II. The less the technical, political, and social conditions of organization are met, the more intense the conflict will be.

III. The more the distribution of authority and other rewards are associated with each other (superimposed), the more intense the conflict will be.

IV. The less the mobility between super- and subordinate groups, the more intense the conflict will be.

V. The less the technical, political, and social conditions of organization are met, the more violent the conflict will be.

VI. The more the deprivation of the subjugated in the distribution of rewards shifts from an absolute to a relative basis, the more violent the conflict will be.

VII. The less is the ability of conflict groups to develop regulatory agreements, the more violent the conflict will be.

VIII. The more intense is the conflict, the more will be the degree of structural change and reorganization.

IX. The more violent is the conflict, the greater will be the rate of structural change and reorganization.

DAHRENDORF'S THEORETICAL PROPOSITIONS

More formally, Dahrendorf outlined three types of intervening empirical conditions: conditions of organization that affect the transformation of latent quasi groups into manifest conflict groups; conditions of conflict that determine the form and intensity of conflict; and conditions of structural change that influence the kind, speed, and depth of the changes in social structure. Thus the variables in the theoretical scheme are (1) the degree of conflict-group formation, (2) the degree of intensity of the conflict, (3) the degree of violence of the conflict, (4) the degree of change of social structure, and (5) the rate of such change. As is evident in Table 12-1,[4] Dahrendorf's propositions appear to be an elaboration of those developed by Marx.

Like Marx, Dahrendorf saw conflict as related to subordinates' growing awareness of their interests and formation into conflict groups (Proposition I). Such awareness and group formation are a positive function of the degree to which (a) the technical conditions (leadership and unifying ideology), (b) the political conditions (capacity to organize), and (c) the social conditions (ability to communicate) are met. These ideas clearly come from Marx's discussion (see Table 11-1 on page 157). However, as shown in Proposition II, Dahrendorf borrows from Simmel and contradicts Marx, emphasizing that if groups are *not* well organized—that is, if the technical, political, and social conditions are not met—then conflict is likely to be emotionally involving. Then Dahrendorf borrowed from Weber (Proposition III) by stressing that the superimposition of rewards—that is, the degree of correlation among those who enjoy privilege (power, wealth, and prestige)—also increases the emotional involvement of subordinates who pursue conflict. Proposition IV shows that Dahrendorf also takes as much from Weber as from Marx: Dahrendorf believed that the lack of mobility into positions of authority escalates the emotional involvement of subordinates. Proposition V is clearly from Simmel and contradicts Marx, in that the violence of conflict is related to the lack of organization and clear articulation of interests. But in Proposition VI, Dahrendorf returns to Marx's emphasis that sudden escalation in people's perception of deprivation—that is, relative deprivation—increases the likelihood of violent conflict. In Proposition VII, however, Dahrendorf returns to Simmel and argues that violence is very much related to the capacity of a system to develop regulatory procedures for dealing with grievances and releasing tensions. And in Propositions VIII and IX, Dahrendorf moves again to Marx's emphasis on how conflict produces varying rates and degrees of structural change in a social system.

CONCLUSION

Dahrendorf was not the first conflict theorist of the midcentury, but he soon became the most influential. Primarily because of his attack on Parsonian functionalism, he gained wide notoriety and, hence, a receptive audience for his reanalysis of Marx, which incorporated important qualifications from Simmel and Weber. Orthodox Marxists were very critical of Dahrendorf's abstract and analytical approach, because in their view Dahrendorf had taken the substance out of Marx's analysis. He had stripped Marx, in their view, of the very concepts that made Marx's approach so important. But Dahrendorf did liberate conflict theory from Marxian parochialism, making it appealing to a wider sociological audience. At the same time that Marxists were criticizing Dahrendorf, another conflict theorist with more sympathy for functional sociology—Lewis Coser—was proposing a less extreme alternative to dialectical approaches. Drawing more from Simmel than Marx, Coser proposed a functional theory of conflict, as is explored in the next chapter.

NOTES

1. Ralf Dahrendorf, "Out of Utopia: Toward a Reorientation of Sociological Analysis," *American Journal of Sociology* 64 (1958), p. 127.

2. As Dahrendorf emphasizes, "I do not intend to fall victim to the mistake of many structural-functional theorists and advance for the conflict model a claim to comprehensive and exclusive applicability. . . . it may well be that in a philosophical sense, society has two faces of equal reality; one of stability, harmony, and consensus and one of change, conflict and constraint" (ibid.). Such disclaimers are, in reality, justifications for arguing for the primacy of conflict in society. By claiming that functionalists are one-sided, it becomes fair game to be equally one-sided to "balance" past one-sidedness.

3. Ralf Dahrendorf, "Toward a Theory of Social Conflict," *Journal of Conflict Resolution* 2 (1958), pp. 170–183; *Class and Class Conflict in Industrial Society* (Stanford, CA: Stanford University Press, 1959), pp. 168–169; *Gesellschaft un Freiheit* (Munich: R. Piper, 1961); *Essays in the Theory of Society* (Stanford, CA: Stanford University Press, 1967).

4. The propositions listed in the table differ from those in a list provided by Dahrendorf, *Class and Class Conflict*, pp. 239–240, in two respects: (1) They are phrased consistently as statements of covariance, and (2) they are phrased somewhat more abstractly without reference to "class," which in this particular work was Dahrendorf's primary concern.

13

The Maturing Tradition II:

Lewis Coser's Conflict Functionalism

Lewis Coser was one of the first modern conflict theorists, and he published a major work on conflict before Ralf Dahrendorf. Yet, because this work had a functional flavor and borrowed from Simmel more than Marx, it was not initially seen as a devastating critique of functionalism in quite the same way as Dahrendorf's early polemic. Still, in his more functional version of conflict theory, Coser launched what became the standard polemic against functionalism: Conflict is not given sufficient attention, and related phenomena such as deviance and dissent are too easily viewed as "pathological" for the equilibrium of the social system.[1] Yet, although Coser consistently maintained that functional theorizing "has too often neglected the dimensions of power and interest," he did not follow either Marx's or Dahrendorf's emphasis on the disruptive consequences of violent conflict. Rather, Coser sought to correct Dahrendorf's analytical excesses by emphasizing the integrative and "adaptability" functions of conflict for social systems.[2] Thus, Coser justified his efforts by criticizing functionalism[3] for ignoring conflict and by criticizing conflict theory for underemphasizing the functions of conflict. In so doing, he turned to Georg Simmel's view of conflict as promoting social integration of the social systems, or at least of some of its critical parts.

Coser's analysis then proceeded as follows: (1) Imbalances in the integration of system parts lead to (2) the outbreak of varying types of conflict among these parts, which in turn causes (3) temporary reintegration of the system, which leads to (4) increased flexibility in the system's structure, increased capability to resolve future imbalances through conflict, and increased capacity to adapt to changing conditions. Coser executed this approach by developing, at least implicitly in his discursive argument, a variety of propositions that are extracted and presented in Tables 13-1 through 13-5.

He began with the causes of conflict (Table 13-1), turned to the issue of violence in conflict (Table 13-2), moved next to the duration of conflict (Table 13-3), and finally explored the functions of conflict (Tables 13-4 and 13-5).

THE CAUSES OF CONFLICT

Much like Weber, Coser emphasized (shown in Proposition I of Table 13-1) that the withdrawal of legitimacy from an existing system of inequality is a critical precondition for conflict.[4] In contrast, dialectical theorists such as Dahrendorf tended to view the causes of conflict as residing in "contradictions" or "conflicts of interest." In such dialectical theories, as subordinates become aware of their interests, they pursue conflict; hence the major theoretical task is to specify the conditions raising levels of awareness. But Coser argued that conflicts of interest are likely to be exposed only after the deprived withdraw legitimacy. Coser emphasized that the social order is maintained by some degree of consensus over existing sociocultural arrangements and that "disorder" through conflict occurs only when conditions decrease this consensus. Two such conditions are specified in Propositions I-A and I-B of Table 13-1, both of which owe their inspiration more to Weber than to Marx. When channels for expressing grievances do not exist and when the deprived's desire for membership in higher ranks is thwarted, the withdrawal of legitimacy becomes more likely.

TABLE 13-1 Coser's Propositions on the Causes of Conflict

I. Subordinate members in a system of inequality are more likely to initiate conflict as they question the legitimacy of the existing distribution of scarce resources, which, in turn, is caused by
   A. Few channels for redressing grievances.
   B. Low rates of mobility to more privileged positions.

II. Subordinates are most likely to initiate conflict with superordinates as their sense of relative deprivation and, hence, injustice increases, which, in turn, is related to
   A. The extent to which socialization experiences of subordinates do not generate internal ego constraints.
   B. The failure of superordinates to apply external constraints on subordinates.

As Proposition II in Table 13-1 indicates, the withdrawal of legitimacy, in itself, is not likely to result in conflict. People must first become emotionally aroused. The theoretical task then becomes one of specifying the conditions that translate the withdrawal of legitimacy into emotional arousal, instead of some other emotional state such as apathy and resignation. Here Coser drew inspiration from Marx's notion of relative deprivation. For, as Marx observed and as a number of empirical studies has documented, absolute deprivation does not always foster revolt.[5] When people's expectations for a better future suddenly begin to exceed perceived avenues for realizing these expectations, only then do they become sufficiently aroused to pursue conflict. The level of arousal will, in turn, be influenced by their commitments to the existing system, by the degree to which they have developed strong internal constraints, and by the nature and amount of social control in a system. Such propositions, for example, lead to predictions that, in systems with absolute dictators who ruthlessly repress the masses, revolt by the masses is less likely than in systems where some freedoms have been granted and where the deprived have been led to believe that things will be getting better. Under these conditions the withdrawal of legitimacy can be accompanied by released passions and emotions.

TABLE 13-2 Coser's Propositions on the Violence of Conflict

I. When groups engage in conflict over realistic issues (obtainable goals), they are more likely to seek compromises over the means to realize their interests, and hence, the less violent the conflict will be.

II. When groups engage in conflict over nonrealistic issues, the greater is the level of emotional arousal and involvement in the conflict, and hence, the more violent the conflict will be, especially when

  A. Conflict occurs over core values.

  B. Conflict endures over time.

III. When functional interdependence among social units is low, the less available are the institutional means for absorbing conflicts and tensions, and hence, the more violent the conflict will be.

THE VIOLENCE OF CONFLICT

Coser's most important propositions on the level of violence in a conflict are presented in Table 13-2.[6] As most functional theorists emphasized, Coser's Proposition I in Table 13-2 is directed at specifying the conditions under which conflict will be less violent. In contrast, dialectical theorists, such as Marx, often pursued just the opposite: specifying the conditions under which conflict will be more violent. Yet the inverse of Coser's first proposition can indicate a condition under which conflict will be violent. The key concept in this proposition is "realistic issues." Coser reasoned that realistic conflict involves the pursuit of specific aims against real sources of hostility, with some estimation of the costs to be incurred in such pursuit. As noted in Chapter 11, Simmel recognized that, when clear goals are sought, compromise and conciliation are likely alternatives to violence. Coser restated this proposition (shown in Proposition II in Table 13-2) on conflict over "nonrealistic issues," such as ultimate values, beliefs, ideology, and vaguely defined class interests. When nonrealistic, the conflict will be violent. Such nonrealism is particularly likely when conflict is about core values, which emotionally mobilize participants and make them unwilling to compromise (Proposition II-A). Moreover, if conflict endures for a long period of time, it becomes increasingly nonrealistic as parties become emotionally involved, as ideologies become codified, and as "the enemy" is portrayed in increasingly negative terms (Proposition II-B). Proposition III shows a more structural variable to the analysis of conflict violence. In systems in which there are high degrees of functional interdependence among actors—that is, where there are mutual exchanges and cooperation—conflict is less likely to be violent.

THE DURATION OF CONFLICT

As shown in the propositions of Table 13-3, Coser underscored that conflicts with a broad range of goals or with vague ones will be prolonged.[7] When goals are limited and articulated, it is possible to know when they have been attained. With perception of attainment, the conflict can be terminated. Conversely, with a wide variety or long list of goals, a sense of attainment is less likely to occur—thus prolonging the conflict. Coser also emphasized that knowledge of what would symbolically constitute victory and defeat will influence the length of conflict. If the parties do not have the ability to recognize defeat or victory, conflict is likely to be prolonged to a point where one party destroys the other. Leadership has important effects on conflict processes; the more leaders can perceive that complete attainment of goals is not possible and the greater their ability is to convince followers to terminate conflict, the less prolonged the conflict will be.

TABLE 13-3 Coser's Propositions on the Duration of Conflict

I. Conflict will be prolonged when
   A. The goals of the opposing parties to a conflict are expansive.
   B. The degree of consensus over the goals of conflict is low.
   C. The parties in a conflict cannot easily interpret their adversary's symbolic points of victory and defeat.
II. Conflict will be shortened when
   A. Leaders of conflicting parties perceive that complete attainment of goals is possible only at very high costs, which, in turn, is related to
      1. The equality of the power between conflicting groups.
      2. The clarity of indexes of defeat or victory in a conflict.
   B. Leaders' capacity to persuade followers to terminate conflict, which, in turn, is related to
      1. Centralization of power in conflict parties.
      2. Integration within conflict parties

THE FUNCTIONS OF SOCIAL CONFLICT

The concept of "function" presents several problems. If some process or structure has functions for some other feature of a system, there is often an implicit assumption about what is good and bad for a system. If this implicit evaluation is not operative, how do we assess when an item is functional or dysfunctional? Even seemingly neutral concepts, such as survival or adaptability, merely mask the implicit evaluation that is taking place. Sociologists are usually not in a position to determine what is survival and adaptation. To say that an item has more survival value or increases adaptation is frequently a way to mask an evaluation of what is "good."

This problem exists in Coser's propositions on the functions of conflict. Conflict is good when it promotes integration based on solidarity, clear authority, functional interdependence, and normative control. In Coser's terms, it is more *adaptive*. Other conflict theorists might argue that conflict in such a system is bad because integration and adaptability in this specific context could be exploitive. Nonetheless, Coser divided his analysis of the functions of conflict along lines similar to those by Simmel: the functions of conflict for (1) the respective parties to the conflict, and (2) the systemic whole in which the conflict occurs.

In the propositions listed in Table 13-4, the intensity of conflict—that is, people's involvement in and commitment to pursue the conflict—and its level of violence increase the demarcation of boundaries (Proposition I-A), centralization of authority (Proposition I-B), structural and ideological solidarity (Proposition I-C), and suppression of dissent and deviance (Proposition I-D) within each of the conflict parties.[8] Conflict intensity is presumably functional because it increases integration, although centralization of power as well as the suppression of deviance and dissent create malintegrative pressures in the long run (see Proposition II). Thus, there appears to be an inherent dialectic in conflict-group unification—one that creates pressures toward disunification. Unfortunately, Coser did not specify the conditions under which these malintegrative pressures are likely to surface. In focusing on positive functions—that is, forces promoting integration—the analysis ignored a promising area of inquiry. This bias becomes even more evident when Coser shifts attention to the functions of conflict for the systemic whole within which the conflict occurs. These propositions are listed in Table 13-5.[9]

Coser's propositions are not presented in their full complexity in Table 13-5, but the essentials of his analysis are clear. In Proposition I, complex systems that have a large

TABLE 13-4 Coser's Propositions on the Functions of Conflict for the Respective Parties

I. The more violent or intense is the conflict, the more the conflict will generate
   - A. Clear-cut boundaries for each conflict party.
   - B. Centralized decision-making structures for each conflict party, especially when these parties are structurally differentiated.
   - C. Structural and ideological solidarity among members of each conflict party, especially when the conflict is perceived to affect the welfare of all segments of the conflict parties.
   - D. Suppression of dissent and deviance within each conflict party as well as forced conformity to norms and values.

II. The more conflict between parties leads centers of power to force conformity within conflict groups, the greater is the accumulation of hostilities and the more likely is internal group conflict to surface in the long run.

TABLE 13-5 Coser's Propositions on the Functions of Conflict for the Social Whole

I. The more differentiated and functionally interdependent are the units in a system, the more likely is conflict to be frequent but of low degrees of intensity and violence.

II. The lower are the intensity and violence of conflicts, the more likely are conflicts to
   - A. Increase the level of innovation and creativity of system units.
   - B. Release hostilities before they polarize system units.
   - C. Promote normative regulation of conflict relations.
   - D. Increase awareness of realistic issues.
   - E. Increase the number of associative coalitions among social units.

III. The more conflict promotes II A through II E, the greater will be the level of internal social integration of the system whole and the greater will be its capacity to adapt to its external environment.

number of interdependencies and exchanges are more likely to have frequent conflicts that are less emotionally involving and violent than conflicts in those systems that are less complex and in which tensions accumulate. The nature of interdependence, Coser argued, causes conflicts to erupt frequently, but, because they emerge periodically, emotions do not build to the point that violence is inevitable. Conversely, systems in which there are low degrees of functional interdependence will often polarize into hostile camps; when conflict does erupt, it will be intense and violent. In Proposition II, frequent conflicts of low intensity and violence are seen to have certain positive functions. First, such frequent and low-intensity conflicts will force those in conflict to reassess and reorganize their actions (Proposition II-A). Second, these conflicts will release tensions and hostilities before they build to a point where adversaries become polarized around nonrealistic issues (Proposition II-B). Third, frequent conflicts of low intensity and violence encourage the development of normative procedures—laws, courts, mediating agencies, and the like—to regulate tensions (Proposition II-C). Fourth, these kinds of conflicts also increase a sense of realism over what the conflict is about. That is, frequent conflicts in which intensity and violence are kept under control allow conflict parties to articulate their interests and goals, thereby allowing them to bargain and compromise (Proposition II-D). Fifth, conflicts promote coalitions among units that are threatened by the action of one party or another. If conflicts are frequent and of low intensity and violence, such coalitions come and go, thereby promoting flexible alliances (Proposition II-E). If conflicts are infrequent and emotions accumulate, however, coalitions often polarize threatened parties into ever more hostile camps, with the result that, when conflict does occur, it is violent. And Proposition III simply states Coser's functional conclusion that, when conflicts are frequent and when violence and intensity are reduced, conflict will promote flexible coordination within the system and increased capacity to adjust and adapt to environmental circumstances. This increase in flexibility and adaptation is possible because of the processes listed in Proposition II-A through II-E.

CONCLUSION

Coser borrowed and extended Georg Simmel's initial insights. In its time, Coser's functionalism represented an important corrective to more Marxian-inspired dialectical approaches. None of Coser's ideas needed to be expressed in the language of functionalism, but the tactic was successful and enabled Coser to postulate an alternative to Marxian sociology. Today, Coser's and Simmel's ideas are incorporated into views of conflict processes that have abandoned the language of functionalism, and it is difficult to see an explicitly functional emphasis in any contemporary conflict theory.

NOTES

1. Lewis A. Coser, *The Functions of Social Conflict* (London: Free Press, 1956).

2. A listing of some of Coser's prominent works, to be used in subsequent analysis, reveals the functional flavor of his conflict perspective: *Functions of Social Conflict* (cited in note 1); "Some Social Functions of Violence" *Annals of the American Academy of Political and Social Science* 364, 1960; "Some Functions of Deviant Behavior and Normative Flexibility," *American Journal of Sociology* 68 (1962), pp. 172–181; and "The Functions of Dissent," in *The Dynamics of Dissent* (New York: Grune & Stratton, 1968), pp. 158–170. Other prominent works with less revealing titles but critical substance include "Social Conflict and the Theory of Social Change," *British Journal of Sociology* 8 (1957), pp. 197–207; "Violence and the Social Structure," in *Science and Psychoanalysis*, ed. J. Masserman, vol. 7 (New York: Grune & Stratton, 1963), pp. 30–42. These and other essays are collected in Coser's *Continuities in the Study of Social Conflict* (New York, Free Press, 1967). One should also consult his *Masters of Sociological Thought* (New York: Harcourt Brace Jovanovich, 1977).

3. Lewis Coser, "Durkheim's Conservatism and Its Implications for His Sociological Theory," in *Émile Durkheim, 1858–1917: A Collection of Essays,* ed. K.H. Wolff (Columbus: Ohio State University Press, 1960); also reprinted in Coser's *Continuities in the Study of Social Conflict.*

4. The propositions in Table 13-1 are extracted from *Functions of Social Conflict* (cited in note 2), pp. 8–385; "Social Conflict and the Theory of Social Change" (cited in note 2), pp. 197–207; and "Violence and Social Structure" (cited in note 2).

5. James Davies, "Toward a Theory of Revolution," *American Journal of Sociology* 27 (1962), pp. 5–19; Ted Robert Gurr, *Why Men Rebel* (Princeton, NJ: Princeton University Press, 1970), and "Sources of Rebellion in Western Societies: Some Quantitative Evidence," *Annals* 38 (1973), pp. 495–501.

6. These propositions are taken from Coser's *Functions of Social Conflict* (cited in note 1), pp. 45–50. Again, they have been made more formal than Coser's more discursive text.

7. These propositions come from Coser, "The Termination of Conflict," in *Continuities*, pp. 37–52; and *Functions of Social Conflict* (cited note 1), pp. 20, 48–55, 59, 128–133.

8. These propositions are taken from Coser, *Functions of Social Conflict* (cited in note 1), pp. 37–38, 45, 69–72, 92–95.

9. Ibid., pp. 45–48. See the following works cited in note 2: "Social Conflict and the Theory of Social Change"; "Some Social Functions of Violence"; and "The Functions of Dissent."

14

The Maturing Tradition III: Jonathan Turner's Synthetic Conflict Theory

By the mid 1970s, Dahrendorf's and Coser's theories had been the subject of considerable analysis. The problems of each theory had been frequently discussed, and most important, an entirely new generation of conflict theories were beginning to emerge—theories that were, as we will see shortly, to carry the conflict tradition to the end of the twentieth century.

In one last effort to resolve the points of dispute among conflict theorists, Jonathan Turner sought to synthesize the arguments of both Dahrendorf's and Coser's theories into one general approach that could articulate the conditions under which conflict will erupt in systems of inequality.[1]

This kind of effort took these early contemporary conflict theories about as far as they could go, thus bringing to a close the maturing conflict tradition and marking, at least approximately, the beginning to a new era of conflict theorizing.

A PROCESS MODEL OF CONFLICT

Turner began his synthesis by modeling Dahrendorf's and Coser's theories as presented in Figure 14-1. The numbers above each box in the diagram emphasize the steps in the processes leading to overt conflict. The arrows connecting each box mark the direction of this sequence. The arrows pointing upward between boxes represent propositions that each theorist develops for specifying the conditions under which the state of affairs described in each box will be realized. As can be seen, however, neither Dahrendorf nor Coser presented propositions to explain why social systems reveal interdependence (Box 1) or why an unequal distribution of scarce resources (Box 2) should exist. These propositions are simply boundary conditions of their theories. That is, in social systems that reveal the unequal

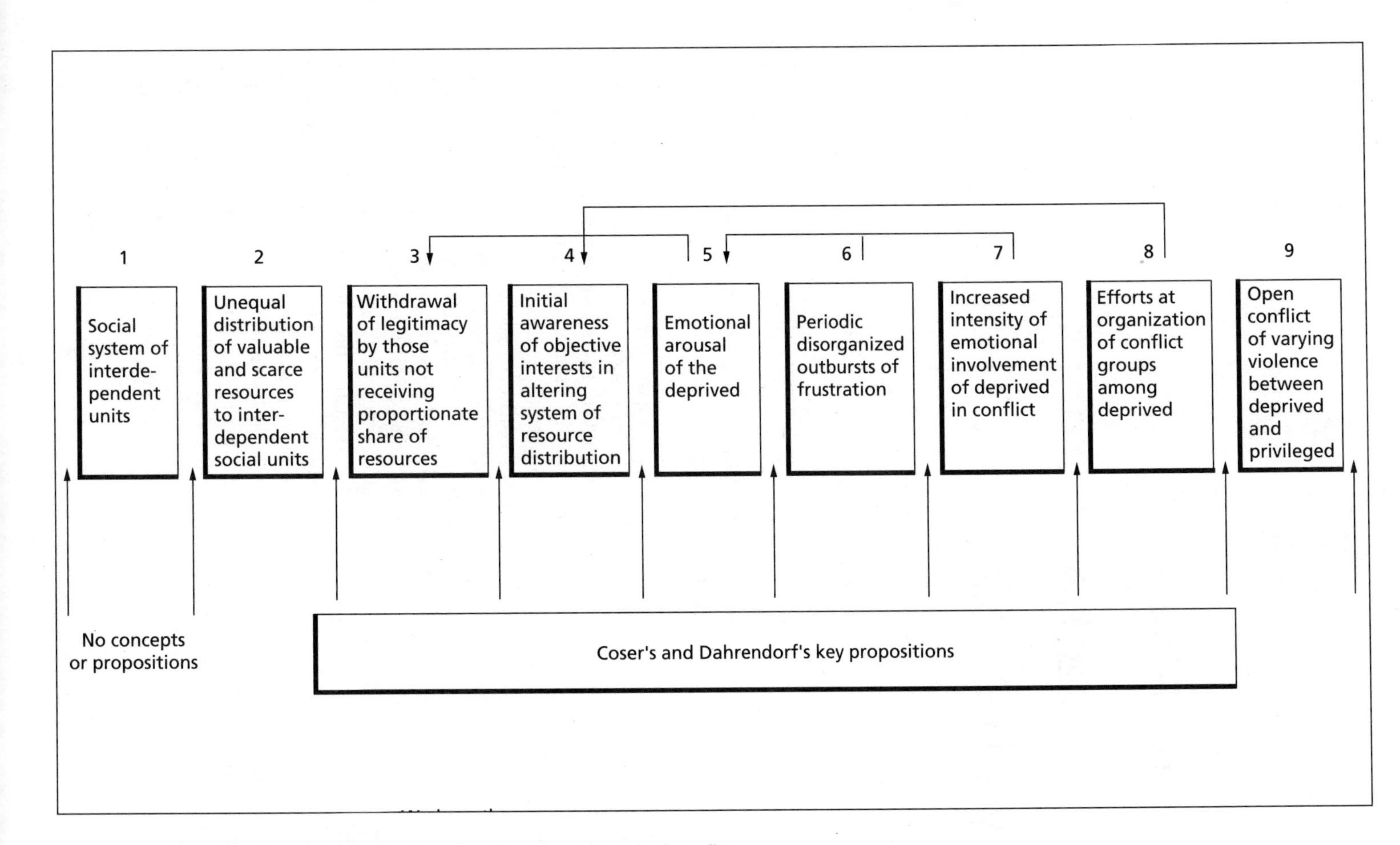

FIGURE 14-1 Coser's and Dahrendorf's Composite Causal Model of Conflict

distribution of scarce resources among interdependent units, Stages 3, 4, 5, 6, 7, 8, and 9 will be activated if certain conditions are met at each stage. The theory is thus devoted to outlining the conditions under which sequences of events will occur, ultimately resulting in overt conflict.

The reverse causal arrows at the top of the diagram connecting various boxes emphasize that the causal model is more complex than just indicated. Events at various stages feed back and influence the weights of variables at earlier stages, making the conflict process a sequence with many built-in cycles. For example, efforts at "organization of conflict groups" (Box 8) will feed back and influence "awareness of objective interests" (Box 4). If the reverse causal loop is positive—that is, organization is successful and thus increases awareness—then the weights of variables in Steps 5, 6, and 7 are altered. In turn, other stages, such as the increased emotional arousal (Box 5), the occurrence of collective outbursts (Box 6), and the escalation of emotional involvement (Box 7) will shape subsequent stages, such as the degree of organization of conflict groups (Box 8). Similar feedback cycles for other stages within the overall causal sequence were also postulated by Coser and Dahrendorf, as Turner indicated by the feedback arrows.

To synthesize the theories, Turner argued, it is best to begin at Box 3, where Coser's and Dahrendorf's theories begin, and then to specify the values for the variables at each point in the conflict process.

Stage 3. Withdrawal of Legitimacy

Coser and Dahrendorf differed in their conceptualizations of how inequalities initiate the conflict process. Dahrendorf emphasized awareness, Coser the withdrawal of legitimacy. In Figure 14-1, the combined model hypothesized that an initial withdrawal of legitimacy with respect to inequality is the first step in the conflict process. Such withdrawal is likely when (a) channels of upward mobility are insufficient to accommodate people's aspirations, thus creating a sense of blockage among deprived segments of the population; (b) channels for redressing grievances against the system of inequality are insufficient relative to the demand for expressing grievances; and (c) rewards and deprivations are superimposed on each other—that is, having (or not having) access to one resource is highly correlated with access (or lack of access) to other scarce resources. Hence, those with money also enjoy power, prestige, health, and other rewards, whereas the reverse is true for the deprived. These propositions are borrowed from Coser's analysis. In placing these propositions first in the causal sequence, Turner presumed that people must begin to question the system before they will begin to perceive their objective interests in altering the system of resource distribution.

Stage 4. Initial Awareness of Objective Interests

In Dahrendorf's theory, a group's awareness is influenced by the technical (leaders, ideology, and so forth), political (creation of opposition organizations), and social (opportunities to communicate, to recruit members) conditions. The more these technical, political, and social conditions can be met, the more likely are the deprived to be aware of their objective interests in altering the present system of resource distribution. However, Coser's theory emphasized the inadequacy of this formulation. As people withdraw legitimacy from a system, they do not suddenly become aware of their interests. Only an *initial* awareness is likely. Thus, Dahrendorf's technical, political, and social conditions are premature in that they do not exert their full impact until later in the causal sequence, when actors have become disillusioned, initially aware, and emotionally aroused. Only then do people begin to seek leaders, organization, unifying beliefs, and means for communication.

Stage 5. Emotional Arousal of Deprived

The major failing of Dahrendorf's scheme was that it is too mechanical. Actors do not seem to have emotion, and, actually, Dahrendorf stayed clear of the psychology of the deprived. Coser's emphasis on emotional arousal of the deprived was thus an important supplement to Dahrendorf's analysis. Coser appeared to recognize that withdrawal of legitimacy and initial awareness of interests lead to emotional arousal, which, under other conditions, drives actors to pursue conflict. Coser postulated two conditions influencing arousal: (a) the degree to which socialization practices among the deprived, and socialization agents in the broader system, create internal psychological controls in actors; and (b) the extent to which social control mechanisms can suppress, channel, or deflect emotional arousal. Thus, the greater are the internal psychological constraints and the more effective is the external social control, the less likely is overt emotional arousal among the deprived. The reverse is true under conditions of weak social and psychological control.

Stage 6. Periodic Collective Outbursts

The conflict process is often marked by individual and collective outbursts of emotion and frustration. These often result in conflict as the agencies of social control in a system seek to suppress these outbursts. Such outbursts are, of course, a form of conflict in themselves, but they are also a stage in a process leading to other forms of conflict, such as a societywide revolution or serious collective bargaining relations among conflict parties. Collective outbursts occur, as Simmel initially emphasized, when the technical, political, and social conditions postulated by Dahrendorf have not been realized. The impact of critical feedback loops outlined in Figure 14-1 must be recognized in this process. Aroused emotions (Box 5) feed back on questions about legitimacy. Aroused emotions will decrease commitments to the system and foster a sense of increased awareness of interests. In turn, increased withdrawal of legitimacy and awareness escalates emotions to a point where collective outbursts are more likely. Another critical feedback loop comes from Stages 6, 7, and 8 in the conflict process. When outbursts occur, they release frustrations, but if social control is harsh and highly repressive, outbursts also increase the level of emotional arousal (hence, the feedback loop between Boxes 6 and 5). Moreover, as actors become more motivated to channel emotions into conflict activities (Box 7), this too increases emotional arousal (Box 5). And, finally, if highly motivated actors can become organized—in accordance with technical, political, and social conditions (Box 8)—then this will influence awareness of objective interests (Box 4), which in turn will arouse emotions, but as Simmel understood more than Marx did, this arousal is now focused and less likely to lead to collective outbursts. Rather, deprived actors will become motivated to increase their organization and bargain with superordinates over resource redistribution.

Stage 7. Increased Intensity

Intensity is the degree to which actors are motivated to pursue their interests and engage in conflict. Intensity involves emotional arousal, but it denotes the channeling of emotional energies and the willingness to sustain these energies in the pursuit of objective interests and to incur the costs in doing so. One condition increasing intensity is the failure of collective outbursts. In the wake of an outburst—a ghetto riot or a wildcat strike, for example—some people become more committed to pursue conflict once they recognize that others would be prepared to join them. Moreover, the use of social control agents—police and troops, for instance—to suppress outbursts often helps solidify emotional commitments and bring into sharper perspective the targets of conflict-oriented activity.

Stage 8. Efforts at Organization

Once the deprived have withdrawn legitimacy, become somewhat aware of their interests, been emotionally aroused, participated in, or observed, outbursts of their fellows, and become committed to realizing their interests, then people are likely to become receptive to organization. Their ability to organize, as Dahrendorf emphasized, is a function of (a) the availability of leaders and unifying beliefs (technical conditions), (b) the tolerance of political organization and the resources to organize (political conditions), and (c) the capacity to communicate grievances and to recruit members into organizations (social conditions). With this increased recruitment into organizations, articulation of objective interests becomes more explicit (thus, feeding back to Box 4), and hence, the emotional arousal of actors (Box 5) will be less likely to result in spontaneous outbursts (Box 6), but instead to a growing commitment (or intensity) to use organizations to pursue objective interests (Box 7).

Stages 3 to 8 have thus set the stage for open conflict. Although conflict in the form of collective or individual outbursts might have preceded Stage 9, these outbursts can also be viewed as steps in a more inclusive conflict process. Moreover, several feedback cycles in the overall process will influence not only the probability of outbursts but also subsequent forms of conflict. Before presenting Coser's and Dahrendorf's propositions influencing the nature of conflict at Stage 9, however, it is wise to summarize and assess those presented thus far. In Figure 14-2, the basic causal scheme is presented again, but this time, the key propositions are inserted at the appropriate point in this causal sequence. As can be seen, certain stages lack propositions. For example, propositions specifying the conditions under which awareness of objective interests follows the withdrawal of legitimacy still need to be specified (Stage 4). Perhaps some incipient level of technical, political, and social conditions must exist, as Dahrendorf's theory emphasized. Another propositional gap can be seen for Step 6. The conditions under which emotional arousal leads to collective outbursts have not been specified. The literature on collective behavior emphasizes such variables as (a) a precipitating incident that symbolizes the situation of the deprived and suddenly escalates their emotions to a point where internal psychological inhibitions and external agents of social control are temporarily ineffective in preventing an outburst; (b) a high degree of propinquity among the deprived who witness the precipitating event, thus increasing mutual communication of hostilities and frustrations; and (c) the availability of objects—persons, organizations, or symbols—that can serve as targets of frustrations.

Stage 9. The Degree of Violence in the Conflict

Coser's and Dahrendorf's propositional inventories offered three propositions on the conditions influencing the degree of violence of a conflict: (a) The extent to which the technical, political, and social conditions are met is negatively related to violence; that is, the less the conditions are met, the more likely is conflict to be violent. (b) The failure to define true interests, independently of core values, is negatively related to the violence of conflict. Thus, if

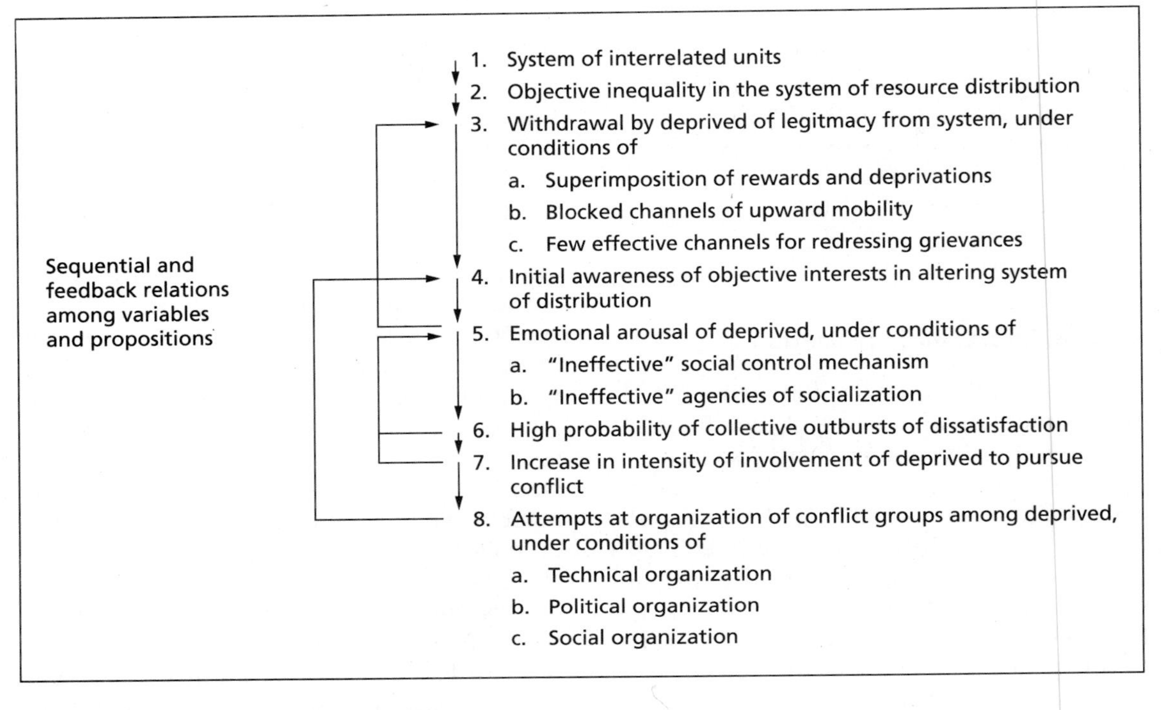

FIGURE 14-2 Propositions on the Conflict Process

conflict parties cannot distinguish between their core values and specific goals for realizing their interests, then the conflict is likely to be "moral" rather than "instrumental," with the result that compromises over moral issues become difficult to make. In contrast, compromises over specific goals are easier to make because they do not involve a moral issue that evokes great emotion. (c) A system that does not have a means for regularizing conflict interaction through legal norms and agencies of mediation is likely to reveal high rates of violent conflict. If a system cannot regulate conflicts between parties with laws, courts, mediating agencies, and other structures, it is difficult for conflict parties to bargain, compromise, and trust each other, because no mechanism exists for mediating conflicts of interest and for enforcing agreed-on compromises.

These three propositions have important interrelationships, Turner argued. A conflict group that realizes the technical, political, and social conditions of organization is likely to be able to articulate its interests, independently of values and beliefs. A system revealing well-organized conflict groups is also likely to have developed regulatory mechanisms, and if it has not, the potential for conflict might force the emergence of such regulatory mechanisms. Or, the existence of regulating mechanisms may actually facilitate the organization of conflict groups (constituting another political condition of conflict group organization). Of course, it is indeed possible for well-organized groups to mix values and goals, or to view them as inseparable. And it is possible for well-organized conflict groups to confront one another in the absence of any regulation. Under these conditions, then,

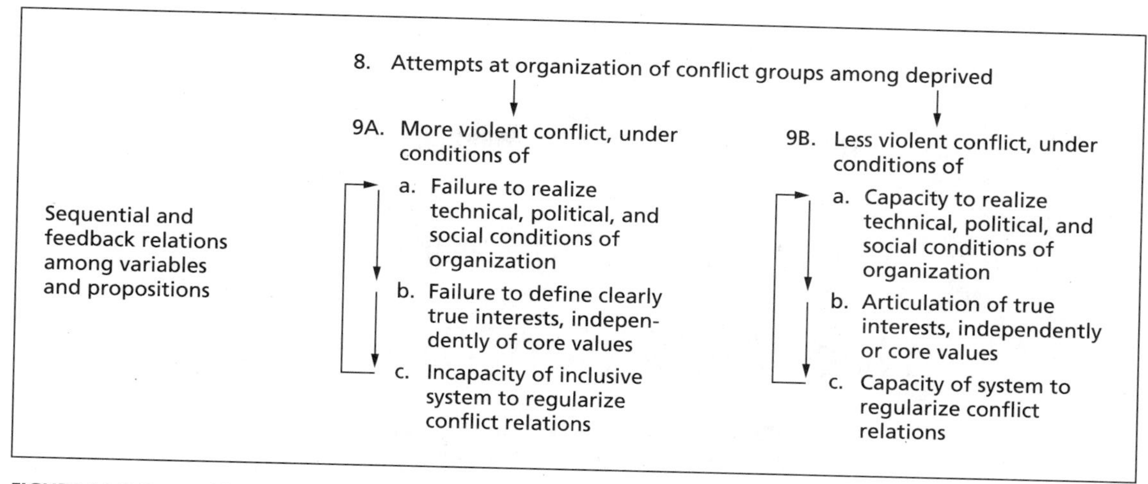

FIGURE 14-3 Propositions on the Degree of Conflict Violence

conflict violence is likely to increase. These propositions and their interrelations are summarized in Figure 14-3.

THE EMERGING CONTEMPORARY ERA IN CONFLICT THEORY

At about the time that Turner made this argument, new kinds of conflict theories were emerging. Some were highly general and abstract theories, whereas others focused on more specific phenomena. And today, at the close of the twentieth century, a rather eclectic mix of theories constitutes the current conflict orientation in sociology. In the next chapter, we will first examine Randall Collins' still evolving neo-Weberian conflict theory, which blends many of Weber's key insights with those from other theoretical traditions. Next, we will look at the Marxian-inspired theories that have sought to analyze class, state, and world-system processes. Then, we will turn to a variety of theories emerging from historical analysis of revolutions. Finally, in the last chapter of this section, we will explore how conflict ideas have been incorporated into theories to explain the conflict dynamics inhering in gender inequalities.

As is evident, then, conflict theory has gone in many different directions, once it became reestablished as one of sociology's main theoretical orientations. Conflict theory has, perhaps, lost some of its coherence as a distinctive perspective, but it has nonetheless exerted an enormous influence on virtually all aspects of sociological theory during the last twenty-five years.

NOTES

1. Jonathan Turner, "A Strategy for Reformulating the Dialectical and Functional Theories of Conflict," *Social Forces* 53 (1975), pp. 433–444.

15

The Continuing Tradition I:

Neo-Weberian Theories: Randall Collins' Analytical Approach

During the last thirty years, Randall Collins has consistently employed a conflict approach, emphasizing that inequalities inevitably set into motion conflict processes, some of which are relatively mild and routinized, but many of which can become more violent. As with any theorist committed to science, Collins sees the goal of a sociological theory of conflict as using a few key ideas to generate propositions explaining the full range of social processes in human interaction and organization. In this chapter, we will emphasize Collins' early Weberian thrust; in later chapters, his use of other theoretical traditions will be highlighted.

THE MICRO BASIS OF SOCIAL ORGANIZATION

Randall Collins has been one of the most forceful advocates of a sociology grounded in conceptualizations of face-to-face interaction. In Chapter 33, the details of Collins' micro-sociology are outlined; in this chapter, these ideas will be summarized in abbreviated form because our goal is to see how Collins developed a conflict approach[1] that, at its core, is Weberian but that adds elements from Émile Durkheim's analysis of rituals (see Chapter 26), Erving Goffman's dramaturgy (Chapter 30), conversation analysis within ethnomethodology (Chapter 31), and other micro-level theoretical perspectives. Collins' argument is that macro-level phenomena are, ultimately, created and sustained by micro encounters among individuals.[2] In essence, large and long-term social structures are built from what he terms *interaction rituals* that have been strung together over time in complex patterns. If true understanding of social reality is to be achieved by sociological theorizing, what transpires in face-to-face interaction must be examined, even if this examination only involves sampling of interaction rituals within a macrostructure.[3]

Interaction rituals occur when individuals are physically co-present, when these individuals reveal a common focus of attention, when they develop a common emotional mood, and when they represent their common focus and mood with symbols (words, objects, phrases, speech styles, and the like) and develop a sense of moral righteousness about these symbols. The dynamics of these interaction rituals revolve around several elements. First, individuals bring to a face-to-face encounter *cultural capital*, or resources that they command in the broader society (for example, power and authority, knowledge, education, network ties and alliances, experiences) or that they accumulated in past interactions of a particular type (for example, memories, information, knowledge, other resources that they can use again when an interaction is reconstituted). Second, individuals bring a level of *emotional energy* to the interaction, which, in turn, is related to (a) the level of cultural capital they possess, (b) the power and prestige or status that they enjoy in the interaction situation, and (c) their memories about the levels of positive emotions or enhanced cultural capital received from the previous time the interaction occurred. Third, individuals monitor situations along several lines: (a) the respective resources of other actors relative to self; (b) the number of others present in a situation; (c) the number of alternative options to the present interaction that are available to self and others; (d) the amount of work-practical, ceremonial, and social content of the interaction; and most important, (e) the payoffs in the amount of positive emotional energy and augmentation of cultural capital likely to be gained from a person's assessment of the inequalities in resources, the alternatives that might be pursued, the number of others presently monitoring the situation, the nature of the situation (as social, work-practical, or ceremonial), and the experiences (emotional energy and cultural capital received) in previous interactions of this nature.

These micro-level concepts have been extended during the last several decades into a more robust theory of emotions and interaction (see Chapter 32), but in his first major work, *Conflict Sociology*, Collins' approach was Weberian in several senses. First, he moved rapidly from the analysis of micro social processes to meso-level social forces, such as stratification and organizations, and then to truly macro-level processes operating at the societal and inter-societal level. Second, although his intent was to retain elements of the micro sociology, the key concepts Collins used to explain the dynamics of interaction rituals become ever more recessive as analysis became ever more macro. The result is a creative neo-Weberian theory of conflict processes.

CONFLICT SOCIOLOGY

In *Conflict Sociology*, Collins proposed the following steps for building social theory. First, examine typical real-life situations in which people encounter one another. Second, focus on the material arrangements that affect interaction—the physical layout of situations, the means and modes of communication, the available tools, weapons, and goods. Third, assess the relative resources that people bring to, use in, or extract from encounters. Fourth, entertain the general hypotheses that those with resources press their advantage, that those without resources seek the best deal they can get under the circumstances, and that stability and change are to be explained through the lineups and shifts in the distribution of resources. Fifth, assume that cultural symbols—ideas, beliefs, norms, values, and the like—are used to represent the interests of those parties who have the resources to make their views prevail. Sixth, look for the general and generic features of particular cases so that more abstract propositions can be extracted from the empirical particulars of a situation.

Collins is particularly concerned with the encounter, with the distribution of individuals in physical space, with their respective capital or resources to use in exchanges, and with inequalities in resources. Also as noted earlier, the respective resources of individuals are critical: *Power* is the capacity to coerce or to have others do so on one's behalf; *material resources* are wealth and the control of money as well as property or the capacity to control the physical setting and people's place in it; and *symbolic resources* are the respective levels of linguistic and conversational resources as well as the capacity to use cultural ideas, such as ideologies, values, and beliefs, for one's purposes.

A central consideration in all Collins' propositions is *social density,* or the number of people co-present in a situation where an encounter takes place. Social density is, of course, part of the macrostructure because it is typically the result of past chains of interaction. But it can also be a "material resource" that some individuals can use to their advantage. Thus the interaction in an encounter will be most affected by the participants' relative resources and the density or number of individuals co-present. These variables influence the two underlying micro dynamics in Collins' scheme, talk and ritual.

Talk and Ritual

Collins sees talk as the emission of verbal and nonverbal gestures that carry meaning and that are used to communicate with others and to sustain (or create) a common sense of reality.[4] Talk is one of the key symbolic resources of individuals in encounters, and much of what transpires among interacting individuals is talk and the use of this cultural capital to develop their respective lines of conduct. As can be seen in Proposition I in Table 15-1, the likelihood that people will talk is related to their sheer co-presence: If others are near, a person is likely to strike up a conversation. More important sociologically are conversations that are part of a "chain" of previous encounters. If people felt good about a past conversation, they will usually make efforts to have another; if they perceive each other's resources, especially symbolic or cultural but also material ones, as desirable, then they will seek to talk again. And if they have developed ritualized interaction that affirms their common group membership, they will be likely to enact those rituals again. As Proposition II indicates, conversations among equals who share common levels of resources will be more personal, flexible, and long-term because people feel comfortable with such conversations. As a result, the encounter raises their levels of emotional energy and increases their cultural capital. That is, they are eager to talk again and to pick up where they left off. However, the nature of talk in an encounter changes dramatically when there is inequality in the resources of the participants. As shown in Proposition III, subordinates will try to avoid wasting or losing emotional energy and spending their cultural capital by keeping the interaction brief, formal, and highly ritualized with trite and inexpensive words. Yet, as Proposition IV indicates, even under conditions of inequality and even more when equality exists, people who interact and talk in repeated encounters will tend, over time, to develop positive sentiments and will have positive emotional feelings. Moreover, they will also converge in their definitions of situations and develop common moods, outlooks, beliefs, and ideas. And, finally, they will be likely to develop strong attachments and a sense of group solidarity, which is sustained through rituals.

Thus the essence of interaction is talk and ritual as mediated by an exchange dynamic; as chains of encounters are linked together over time, conversations take on a more personal and also a ritualized character that results from and at the same time reinforces

TABLE 15-1 Key Propositions on the Conditions Producing Talk and Conversation

I. The likelihood of talk and conversational exchanges among individuals is a positive and additive function of (a) the degree of their physical co-presence, (b) the emotional gratifications retained from their previous conversational exchanges, (c) the perceived attractiveness of their respective resources, and (d) their level of previous ritual activity.

II. The greater the degree of equality and similarity that exists in the resources of individuals, the more likely conversational exchanges are to be (a) personal, (b) flexible, and (c) long-term.

III. The greater the level of inequality that exists in the resources of individuals, the more likely conversational exchanges are to be (a) impersonal, (b) highly routinized, and (c) short-term.

IV. The greater the amount of talk among individuals, especially among equals, the more likely are (a) strong, positive emotions; (b) sentiments of liking; (c) common agreements, moods, outlooks, and beliefs; and (d) strong social attachments sustained by rituals.

the growing sense of group solidarity among individuals. Such is the case because the individuals have "invested" their cultural capital (conversational resources) and have derived positive feelings from being defined as group members. Collins' intent is thus clear: to view social structure as the linking together of encounters through talk and ritual. This basic view of the micro reality of social life pervades all Collins' sociological theory, especially as he began the analysis of inequalities in social life.

Deferences and Demeanor

Inequality and stratification are structures only in the sense of being temporal chains of interaction rituals and exchanges among varying numbers of people with different levels of resources. Thus, to understand these structures, we must examine what people actually do across time and in space. One thing that people do in interaction is exhibit deference and demeanor. Collins and co-author Joan Annett define *deference* as the process of manipulating gestures to show respect to others, or, if one is in a position to command respect, the process of gesture manipulation is to elicit respect from others.[5] The actual manipulation of gestures is termed *demeanor*. Deference and demeanor are, therefore, intimately connected to each other. They are also tied to talk and rituals, because talk involves the use of gestures and because deference and demeanor tend to become ritualized. Hence, deference and demeanor can be visualized as one form of talk and ritual activity—a form that is most evident in those interactions that create and sustain inequalities among people.

As would be expected, Collins visualizes several variables as central to understanding deference and demeanor:

1. Inequality in resources, particularly wealth and power.
2. Social density variables revolving around the degree to which behaviors are under the "surveillance of others" in a situation.
3. Social diversity variables revolving around the degree to which communications networks are "cosmopolitan" (that is, not restricted to only those who are co-present in a situation).

In Table 15-2 these variables are incorporated into a few abstract propositions that capture the essence of Collins' and Annett's numerous propositions and descriptions of the history of deference and demeanor.[6] In these propositions, Collins and Annett argue that rituals and talk revealing deference and

TABLE 15-2 Key Propositions on Deference and Demeanor

I. The visibility, explicitness, and predictability of deference and demeanor rituals and talk among individuals increase with
  A. Inequality in resources among individuals, especially with respect to
    1. Material wealth.
    2. Power.
  B. Surveillance by others of behaviors, and surveillance increases with
    1. Co-presence of others.
    2. Homogeneity in outlook of others.
  C. Restrictiveness of communication networks (low cosmopolitanism), and restrictiveness decreases with
    1. Complexity in communications technologies.
    2. Mobility of individuals.

II. The greater is the degree of inequality among individuals and the lower is the level of surveillance, the more likely behaviors are to be directed toward
  A. Avoidance of contact and emission of deference and demeanor by individuals.
  B. Perfunctory performance of deference and demeanor by individuals when avoidance is not possible.

III. The greater the degree of inequality among individuals and the lower the level of cosmopolitanism among individuals, the more likely behaviors are to be directed toward simplified but highly visible deference and demeanor.

IV. The greater is the degree of inequality among individuals, and the less is the degree of mobility among groups with varying levels of resources, the more visible, explicit, and predictable are deference and demeanor rituals and talk within these groups.

V. The greater is the equality among individuals, and the greater is the degree of cosmopolitanism and/or the less is the level of surveillance, the less compelling are deference and demeanor talk and rituals.

demeanor are most pronounced between people of unequal status, especially when their actions are observable and when communication outside the situation is restricted. Such density and surveillance are, of course, properties of the macrostructure as it distributes varying numbers of people in space. As surveillance decreases, however, unequals avoid contact or perform deference and demeanor rituals in a perfunctory manner. For example, military protocol will be much more pronounced between an officer and enlisted personnel in public on a military base than in situations where surveillance is lacking (for example, off the base). Moreover, Collins and Annett stress that inequalities and low mobility between unequal groups create pressures for intragroup deference and demeanor rituals, especially when communications outside the group are low (for example, between new army recruits and their officers or between prison inmates and guards). But as communication outside the group increases or as surveillance by group members decreases, then deference and demeanor will decrease.

Class Cultures

These exchange processes revolving around talk, ritual, deference, and demeanor explain what are often seen as more macro processes in societies. One such process is variation in the class cultures. That is, people in different social classes tend to exhibit diverging behaviors, outlooks, and interpersonal styles. These differences can be seen in two main variables:

1. The degree to which one possesses and uses the capacity to coerce, to materially bestow, and to symbolically manipulate others so that one can give orders in an encounter and have these orders followed.

TABLE 15-3 Key Propositions on Class Cultures

I. Giving orders to others in a situation increases with the capacity to mobilize and use coercive, material, and symbolic resources.

II. The behavioral attributes of self-assuredness, the initiation of talk, positive self-feelings, and identification with the goals of a situation are positively related to the capacity to give orders to others in that situation.

III. The behavioral attributes of toughness and courage increase as the degree of physical exertion and danger in that situation escalates.

IV. The degree of behavioral conformity exhibited in a situation is positively related to the degree to which people can communicate only with others who are physically co-present in that situation and is negatively related to the degree to which people can communicate with a diversity of others who are not physically co-present.

V. The outlook and behavioral tendencies of an individual are an additive function of those spheres of life—work, politics, home, recreation, community—where varying degrees of giving or receiving orders, physical exertion, danger, and communication occur.

2. The degree to which communication is confined to others who are physically co-present in a situation or, conversely, the degree to which communication is diverse, involving the use of multiple modes of contact with many others in different situations.

Using these two variables, as well as several less central variables such as wealth and physical exertion on the job, Collins describes the class cultures of American society. More significantly for theory building, he also offers several abstract propositions that stipulate certain important relationships among power, order-giving, communication networks, and behavioral tendencies among individuals. These relationships are restated in somewhat altered form in Table 15-3.[7] With these principles, Collins explains variations in the behaviors, outlooks, and interpersonal styles of individuals in different occupations and status groups. For example, those occupations that require order giving, that reveal high co-presence of others, and that involve little physical exertion will generate behaviors that are distinctive and that circumscribe other activities, such as whom one marries, where one lives, what one values, and what activities one pursues in various spheres of life. Different weights to these variables would cause varying behavioral tendencies in individuals. Thus from the processes delineated in the propositions of Table 15-3, understanding of such variables as class culture, ethnic cultures, lifestyles, and other concerns of investigators of stratification is achieved. But such understanding is anchored in the recognition that these class cultures are built and sustained by interaction chains in which deference and demeanor rituals have figured prominently. Thus, a class culture is not mere internalization of values and beliefs or simple socialization (although this is no doubt involved); rather, a class culture is the result of repeated encounters among unequals under varying conditions imposed by the macrostructure as it has been built from past chains of interaction.

Organizational Processes

Like Weber before him, Collins also uses an extensive analysis of organizations and develops a rather long inventory of propositions on organizations' properties and dynamics.[8] These propositions overlap, to some degree, with those on stratification, because an organization is typically a stratified system with a comparatively clear hierarchy of authority.

TABLE 15-4 Key Propositions on Organizations

Processes of Organizational Control

I. Control in patterns of organizations is a positive and additive function of the concentration among individuals of (a) coercive resources, (b) material resources, and (c) symbolic resources.

II. The form of control in organizations depends on the configuration of resources held by those individuals seeking to control others.

III. The more control is sought through the use of coercive resources, the more likely those subject to the application of these resources are to (a) seek escape, (b) fight back, if escape is impossible, (c) comply if (a) and (b) are impossible and if material incentives exist, and (d) sluggishly comply if (a), (b), and (c) do not apply.

IV. The more control is sought through the use of material resources, the more likely those subject to the manipulation of material incentives are to (a) develop acquisitive orientations and (b) develop a strategy of self-interested manipulation.

V. The more control is sought through the use of symbolic resources, the more likely those subject to the application of such resources are to (a) experience indoctrination into values and beliefs, (b) be members of homogeneous cohorts of recruits, (c) be subject to efforts to encourage intra-organizational contact, (d) be subject to efforts to discourage extra-organizational contact, (e) participate in ritual activities, especially those involving rites of passage, and (f) be rewarded for conformity with upward mobility.

Administration of Control

VI. The more those in authority employ coercive and material incentives to control others, the greater is the reliance on surveillance as an administrative device to control.

VII. The more those in authority use surveillance to control, the greater are (a) the level of alienation by those subject to surveillance, (b) the level of conformity in only higher visible behaviors, and (c) the ratio of supervisory to nonsupervisory individuals.

VIII. The more those in authority employ symbolic resources to control others, the greater is their reliance on systems of standardized rules to achieve control.

IX. The greater is the reliance on systems of standardized rules, the greater are (a) the impersonality of interactions, (b) the standardization of behaviors, and (c) the dispersion of authority.

Organizational Structure

X. Centralization of authority is a positive and additive function of (a) the concentration of resources, (b) the capacity to mobilize the administration of control through surveillance, material incentives, and systems of rules; (c) the capacity to control the flow of information; (d) the capacity to control contingencies of the environment; and (e) the degree to which the tasks to be performed are routine.

XI. The bureaucratization of authority and social relations is a positive and additive function of (a) record-keeping technologies, (b) nonkinship agents of socialization of potential incumbents, (c) money markets, (d) transportation facilities, (e) nonpersonal centers of power, and (f) diverse centers of power and authority.

Table 15-4 lists three groups of propositions from Collins' analysis. These revolve around processes of organizational control, the administration of control, and the general organizational structure.

In the propositions shown in Table 15-4, control within an organization increases with the concentration of coercive, material, and symbolic resources. The pattern of control varies, however, with the particular type of resource—whether coercive, material, or symbolic—that is controlled and with the configuration among these resources, as is summarized in Propositions III, IV, and V. Control within an organization must be administered, and the pattern of such administrative control varies with the nature of the resources used to gain control. Collins extends Weber's analysis of organizations, and Propositions VI through IX summarize various patterns in the administration of control. In the end, Collins sees the profile of an organization's structure as reflecting the nature and concentration of resources, as well as

TABLE 15-5 Key Propositions on the State, Economy, and Ideology

I. The size and scale of political organization are a positive function of the productive capacity of the economy.

II. The productive capacity of the economy is a positive and additive function of (a) level of technology, (b) level of natural resources, (c) population size, and (d) efficiency in the organization of labor.

III. The form of political organization is related to the levels of and interactive effects among (a) size of territories to be governed, (b) the absolute numbers of people to be governed, (c) the distribution and diversity of people in a territory, (d) the organization of coercive force (armies), (e) the distribution (dispersion or concentration) of power and other resources among a population, and (f) the degree of symbolic unification within and among social units.

IV. The stability of the state is a negative and additive function of
   A. The capacity for political mobilization by other groups, which is a positive function of
      1. The level of wealth.
      2. The capacity for organization as a status group.
   B. The incapacity of the state to resolve periodic crises.

how these are used to administer control. Propositions X and XI review Collins' basic argument.

The State and Economy

As did Max Weber, Collins eventually moves to the analysis of the state which, though a type of complex organization, still controls and regulates the entire society. As the propositions in Table 15-5 summarize,[9] the size and scale of the state depend on the productive capacity of the economy; in the end, the state can only be supported by a large economic surplus. In turn, as is summarized in Proposition II, the productive capacity of the economy is related to technologies, natural resources, the number of people who must be supported, and the efficiency with which the division of labor is organized. The particular form of state power varies enormously, but these forms vary under the impact of basic forces summarized in Proposition III. The stability of the state is also a crucial variable, especially for a conflict theory. As Proposition IV summarizes, the state must be able to prevent mobilization by groups pursuing counter-power, and it must be able to resolve periodic crises. When it cannot, the state becomes unstable. Like Weber before him, Collins recognizes that much of the state's viability depends on the relation of the state to surrounding societies. No society exists in isolation; it almost always finds itself in competition with other societies. And, the ability of the state to prevail in this world of geopolitics often determines its form, viability, and stability.

Geopolitics

Borrowing from Weber but adding his own ideas, Collins argues that there are sociological reasons for the historical facts that only certain societies can form stable empires and that societies can extend their empires only to a maximal size of about three to four million square miles.[34] When a society has a resource (money, technology, population base) and marchland advantage (no enemies on most of its borders), it can win wars, but eventually it will (a) extend itself beyond its logistical capacities, (b) bump up against another empire, (c) lose its marchland advantage as it extends its borders and becomes ever more surrounded by enemies, and (d) lose its technological advantages as enemies adopt them. The result of these forces is that empires begin to stall at a certain size, as each of these points of resistance is activated.

TABLE 15-6 Key Propositions on Geopolitics

I. The possibility of winning a war between nation-states is a positive and additive function of
   A. The level of resource advantage of one nation-state over another, which is a positive function of
      1. The level of technology.
      2. The level of productivity.
      3. The size of the population.
      4. The level of wealth formation.
   B. The degree of "marchland advantage" of one nation-state over another, which is a positive and additive function of
      1. The extent to which the borders of a nation-state are peripheral to those of other nation-states.
      2. The extent to which a nation-state has enemies on only one border.
      3. The extent to which a nation-state has natural buffers (mountains, oceans, large lakes, and so on) on most of its borders.

II. The likelihood of an empire is a positive function to the extent to which a marchland state has resource advantages over neighbors and uses these advantages to wage war.

III. The size of an empire is a positive and additive function of the dominant nation-states' capacity to
   A. Avoid showdown war with the empire of other marchland states.
   B. Sustain a marchland advantage.
   C. Maintain territories with standing armies.
   D. Maintain logistical capacity for communications and transportation, which is a positive function of levels of communication, transportation, and military technologies and a negative and additive function of
      1. The size of a territory.
      2. The distances of borders from the home base.
   E. Diffusion of technologies to potential enemies.

IV. The collapse of an empire is a positive and additive function of
   A. The initiation of war between two empires.
   B. The overextension of an empire beyond its logistical capacity.
   C. The adoption of its superior technologies by enemy nation-states.

These processes indicate that internal nation-states will not build long-term or extensive empires because they are surrounded by enemies, increasingly so as they extend territory. Rather, marchland states, with oceans, mountains, or unthreatening neighbors at their back, can move out and conquer others, because they have to fight a war on only one front. But eventually they overextend, confront another marchland empire, lose their technological advantage, and acquire enemies on a greater proportion of their borders (thereby losing the marchland advantage and, in effect, becoming an internal state that must now fight on several borders). Sea and air powers can provide a kind of marchland advantage, but the logistical loads of distance from home bases and maintenance of sophisticated technologies make such empires vulnerable. Only when it encounters little resistance can an empire be maintained across oceans and at great distances by air; as resistance mounts, the empire collapses quickly as its supply lines are disrupted. Table 15-6 summarizes these ideas more formally.

CONCLUSION

It is not hard to see what makes Collins' approach to conflict processes neo-Weberian. Like Weber, Collins begins with a conceptualization of micro processes—in Weber's case, types of meaningful action, and in Collins' theory, interaction rituals. Then, their analysis shifts to the meso-level, examining patterns of stratification and forms of complex organizations. Finally, both Weber and

Collins moves to the analysis of the state and geopolitics. In all these levels of analysis, their concern is with inequalities of resources and how these inequalities generate tension and potential conflict.

Yet Collins is perhaps best known for what we have not emphasized in this chapter: the theory of interaction rituals and emotions. In Chapter 32, this aspect of his work will be examined in more detail, but if this micro-level theory is to be placed within its original Weberian showcase, pages 432 to 435 should be consulted now. Moreover, Collins has also developed a more elaborate theory of stratification that examines both gender and age as important dimensions. In Chapter 18 we will encounter the theory of gender stratification as it has been synthesized with other theories in its most recent treatment. Thus, Collins is much more than a strict adherent to the Weberian tradition; indeed, Weber is only a starting point from which Collins has theorized about many social processes.

NOTES

1. Randall Collins, Conflict Sociology: Toward an Explanatory Science (New York: Academic, 1975).
2. Randall Collins, "On the Micro-Foundation of Macro-Sociology," *American Journal of Sociology* 86 (1981), pp. 984–1014.
3. Randall Collins, "Micro-Translation as a Theory of Building Strategy," in *Advances in Social Theory and Methodology: Toward an Integration of Micro- and Macro-Sociology*, eds. K. Knorr-Cetina and A.V. Cicourel (London: Routledge, 1981), pp. 84–96.
4. Collins, *Conflict Sociology*, pp. 114–131.
5. Randall Collins and Joan Annett, "A Short History of Deference and Demeanor," in *Conflict Sociology*, pp. 161–224.
6. Ibid., pp. 216–219.
7. Collins, *Conflict Sociology*, pp. 49–88.
8. Ibid., pp. 286–347.
9. Ibid., pp. 348–413.
10. See Randall Collins, *Weberian Sociological Theory* (Cambridge, England: Cambridge University Press, 1986), pp. 167–212; "Long-Term Social Change and the Territorial Power of States," in his *Sociology Since Midcentury: Essays in Theory Cumulation* (New York: Academic, 1981).

16

The Continuing Tradition II:

Conflict Theories in Historical-Comparative Sociology*

Both Karl Marx and Max Weber saw societal revolutions as emerging from conditions of high inequality that, in turn, produce mass mobilization of subordinates to pursue conflict against superordinates. Their general analytical arguments are summarized in Tables 11-1 on page 157 and 11-2 on page 158, and although Marx's and Weber's abstract theories converged, important differences can be found in their respective formulations. One difference is that Marx saw revolution by the proletariat as inevitably arising from the contradictions of capitalism and the exploitation that capitalism systematically generates, whereas Weber saw revolutions as historically contingent and far from inevitable. Another difference is that Weber saw internal societal conflict as related to external geopolitical processes as these influence the legitimacy of political regimes (see the propositions in Table 11-3 on page 160). A related difference is that Weber saw power and its manifestation in the state as a distinct actor and, indeed, as a separate basis for stratification, whereas Marx tended to view the state as simply a tool of the dominant social class and as a "superstructure" of the underlying economic relations. Yet, despite their differences, both theorists sought to develop historical interpretations about when revolutions are likely to occur, and although Weber was certainly the more devoted historian, their efforts to analyze revolutionary conflict by marshaling historical data stimulated a distinctive branch of conflict theory in the modern era.

This branch of theorizing involves historical descriptions of conflicts in agrarian societies making the transition to modernity. Moreover, key figures in this branch of theorizing tend to use comparative analyses of case histories to make more analytical, abstract, and generalizable statements about the

*This chapter is coauthored with Rebecca S. K. Li.

generic conditions producing social conflict. In this sense, these approaches are Weberian because they systematically seek comparisons among societies and develop generalizations from historical cases, but there is almost always a Marxian strain in these works, emphasizing how the exploitive actions of dominant classes create the conditions that lead to mass mobilizations of subordinates. Indeed, the first efforts to develop a comparative-historical theory of conflict began with Marxian assumptions about inequality and class conflict, but over time, as comparative-historical sociology developed, it has taken on a more Weberian tone, seeking to document how certain conditions, when converging at a point or points in history, increase the likelihood of societal revolutions.

In discussing several prominent theories in this tradition, we will emphasize the more abstract theoretical statements that emerge from the comparative analyses of specific historical cases. In a sense, such an exercise violates the empirical intent of these Weber-inspired works, but this exercise is also true to Weber's goal of developing more analytical models and statements. Weber himself evidenced considerable ambivalence about the generalizability of models from one historical case to others; most comparative-historical analyses of conflict reveal a similar ambiguity about the appropriateness of more abstract generalizations. In this chapter, we will ignore this uncertainty and extract the theoretical statements of researchers and convert them into more abstract theoretical arguments.

MASS MOBILIZATION AND STATE BREAKDOWN

Theories of social conflict emerging from comparative-historical sociology tend to emphasize two related factors. One factor is the conditions that lead the masses to mobilize ideologically, politically, and organizationally to pursue conflict against those who dominate them. The other factor is the processes leading to the breakdown of the state and its capacity to control the population. Obviously, these two factors are interrelated, but the theories tend to give more emphasis to one or the other. Some emphasize the forces leading to mobilization, whereas others recognize that the power of the state must also erode before mobilizations can be successful. Indeed, the historical record reveals that revolts and other forms of civil disruption are fairly common in human societies, but true society-wide revolutions where the distribution of power and other resources are dramatically changed are rather rare. Because of their rarity, revolutions have fascinated historical sociologists. To explain why they occur, or do not occur, the various theories revolve around specifying the forces (1) leading to mobilization of the masses to pursue conflict with the dominant sectors of the society and (2) causing the state to lose legitimacy as well as the capacity to regulate and control members of the population. Figure 16-1 summarizes the two basic factors in various theories.

BARRINGTON MOORE'S STUDY OF THE ORIGINS OF DICTATORSHIP AND DEMOCRACY

One of the earliest analyses of conflict in societies undergoing the transition from an agrarian to an industrial base of social organization was Barrington Moore's comparative study of the conditions producing dictatorships or democracies.[1] In his analysis of these conditions, which can be considered the maturing tradition building on Marx and Weber, the theoretical leads to be developed by more contemporary historical sociologists were first exposed. Moore argued that there are three routes to modernization. One is where feudal landowners become capitalists, replacing the

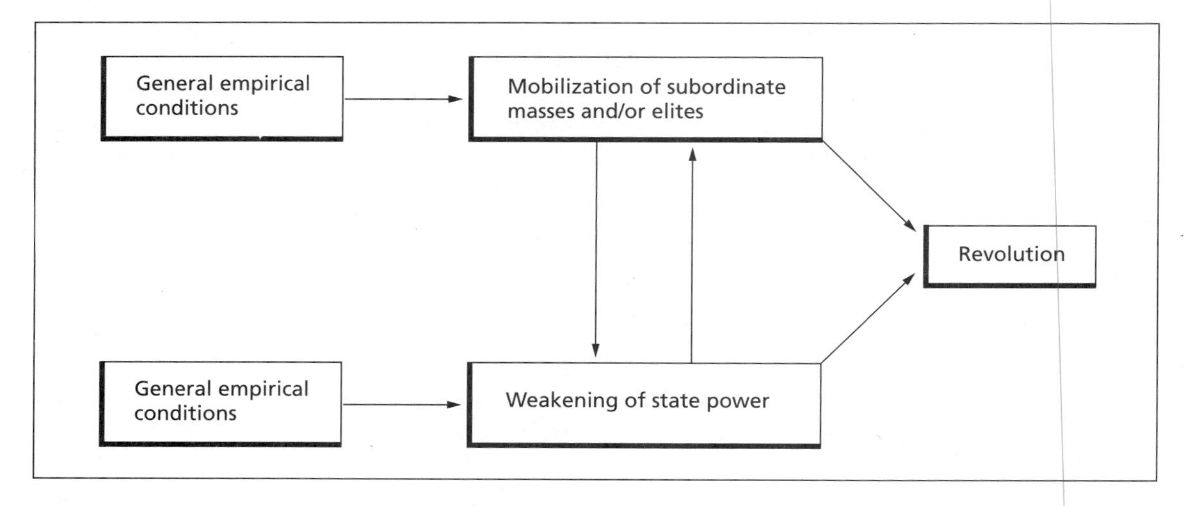

FIGURE 16-1 The Two Lines of Emphasis in Historical-Comparative Theories of Revolutions

peasants who once lived on and worked the land in tenancy agreements with hired labor, then selling the products of this hired labor in markets for profits. This route converts the old landed aristocracy into capitalists and, in the end, leads them to become reluctant allies with the bourgeoisie. This route is the most likely to produce democracy, and it is the path that both Britain and the United States (after the elimination of slavery) pursued.

A second route to modernization is where the landowners enter capitalist markets to sell their products but where they keep the peasants on the land, forcing them to increase productivity through the traditional controls of tenancy and patronage that inhere in feudalism. Under these conditions the landowners must ally themselves with the bureaucrats in the state to keep control of peasants who are increasingly exploited as capitalist landowners seek to extract ever more profit from the peasants' labor. In Moore's view, this path leads to a fascist state, and this was the route that Japan and Germany/ Prussia took during their initial transition to modernity.

The third and final route to modernization is where the landowners become absentee owners, collecting rents from the peasants who produce directly for markets in which prices can vary enormously but who must typically pay constant or even escalating rents to landlords. Decreases in the prices of goods sold in markets, coupled with constant or escalating rents, make peasants aware of their exploitation and the larger political-economy in which they must operate. This awareness and the emotion that it arouses result in peasant revolts, which can lead to mass mobilizations and societal revolutions. In the wake of revolution, whether successful or not, state power is increased to consolidate the power of the victorious faction in the revolution and to control potential dissidents. The exact form of the state can vary, as is evident with the difference between France after its major revolution and Russia or China in the wake of their respective revolutions. In either case, a strong state emerges.

In this descriptive scenario are more general propositions about the conditions increasing or decreasing mass mobilization of

TABLE 16-1 Abstracted Propositions from Barrington Moore's Analysis

I. The potential for mass mobilization of the subordinates to pursue conflict with superordinates in a system of inequality will increase when
   A. The subordinates constitute a coherent whole in their physical location, their daily routines, and their lived experiences.
   B. The subordinates experience collective solidarity. This sense of solidarity will increase when
      1. Subordinates experience a sense of threat from either those who dominate them or those from outside the society who might dominate them through conquest.
      2. Subordinates can avoid divisive competition with each other over
         a. Land and tenancy rights to farmland.
         b. Markets buying their labor.
         c. Commodity markets selling their goods.
      3. Subordinates experience the conditions listed under I-C and I-D.
   C. The traditional interconnections between the communities of subordinates and superordinates are weakened. This weakening will increase when
      1. Subordinates and superordinates are placed in more direct competition with each other for resources, or at least are perceived by subordinates to be so.
      2. Superordinates are no longer perceived as providing indispensable and otherwise unavailable resources to subordinates.
      3. Subordinates' relations to superordinates have moved from a paternalistic to market-driven form, but this effect of market forces operates only if subordinates do not compete directly with each other for resources in a market controlled by external sources of power.
      4. Superordinates become increasingly removed and absent from the daily routines of subordinates.
   D. The subordinates perceive that superordinates are exploiting them. This sense of exploitation will increase when
      1. Subordinates are forced to provide ever more of their resources to external powers, whether to the state or to those who own the means of their production.
      2. Subordinates experience the conditions listed in I-C, delineating the forces that weaken the connection between communities of subordinates and superordinates.

peasants for conflict. These more abstract propositions are summarized in Table 16-1, and we will discuss the implications of each. For mobilization of the subordinate masses to occur, they must remain a coherent whole performing similar tasks and having similar life experiences (Propositions I-A). In the historical cases studied by Moore, the absentee landlord extracting rents from peasants who remained on their land and produced for the market was most likely to meet this condition because the traditional residence patterns and productive work routines of the feudal peasantry were retained, but under conditions of exploitation by landowners and fluctuation in market prices for the goods produced by peasants.[2]

Proposition I-B in Table 16-1 stresses that subordinates must experience collective solidarity if they are to become mobilized to pursue conflict. Yet, other conditions in agrarian societies undergoing the transition to modernity will increase solidarity. For instance, solidarity increases with subordinates' sense of threat, and as Moore stressed, this sense of threat can come from many sources. One source is those who dominate them and engage in practices that threaten subordinates' livelihood and well being. Another source of threat can reside in the state's efforts to repress peasants. Still another source of threat can come from outside the society when armies invade and threaten the lives and livelihood of subordinates.[3] A final source of threat comes from the conditions leading to a breakdown of the traditional interconnections between superordinates and subordinates. Solidarity is not, however, inevitable. For example, as Moore emphasizes, commercialization of agriculture where estate labor is hired under highly competitive labor market conditions (that is, there is a surplus of labor relative to demand), will reduce the potential for mass mobilization—a proposition that goes against Marx's prediction about the effects of labor markets on workers.

This last source of threat is particularly important; the degree of connectedness and contact between the subordinate masses and those above them in the stratification system is critical to the mobilization of the masses. When there is a strong interconnection between the communities of superordinates and subordinates, or at least a sense of a strong connection by subordinates, the potential for mass mobilization is lessened, as was historically the case with traditional feudalism where the lord, his squire or overseer, and peasantry all lived on the manorial estate in nonmarket relations guided by paternalism and traditional tenancy understandings. Conversely, forces that weaken this interconnection will increase the likelihood of mass mobilization, as occurs when the landlord becomes absent and extracts rents regardless of the prices that peasants directly marketing their goods can secure. More generally, the connection between subordinates' and superordinates' respective communities is also weakened when they must compete with each other for resources, or think that they must, when superordinates are no longer perceived to provide indispensable and vital resources, when subordinates' relations with superordinates become market-driven, and when superordinates become absent and removed from daily work routines of subordinates. Again, the route to modernity most likely to meet these conditions is the absentee landlord extracting rents from peasants who work the land but who must also be subject to the vicissitudes of price fluctuations in markets.

As Marx argued, Moore also sees the rate and degree of exploitation experienced by the subordinate masses as crucial. Exploitation or at least the sense of being exploited increases as subordinates must give up more resources to superordinates, especially when the traditional interconnections between the communities of dominated and dominant are weakened. When this sense of exploitation is at its maximum—as is the case where peasants are forced to pay constant or escalating rents to absentee landlords but must endure price fluctuations in the markets where they sell their products—exploitation is laid bare, and hence, mass mobilization potential increases. The state can also play an important role in this sense of exploitation because the more peasants are restive, the more landlords depend on the state's coercive powers, which, in turn, means that the state must extract ever more resources from the lords and, ultimately, their peasants to finance repression. If the state taxes lords too severely, the lords become a potential source of revolt, but in the end the increased taxes are taken from the peasants who must increase their output or suffer a diminished standard of living as landowners extract ever more of the peasants' surplus product to pay for the coercive activities of the state. Again, when this additional extraction comes from increased rents by absentee landlords trying to repress their restive peasants who must endure market fluctuations, the sense of exploitation is likely to be maximal.

In sum, then, Moore's approach is Marxian in its assumptions—conflict emerges from high levels of exploitation—but it adds Weberian qualifications. The most obvious is that revolution is most likely in systems that have not gone completely capitalist; instead, the landed aristocracy has perhaps become more capitalistic but in a way that exposes their exploitation of peasants. Indeed, revolution is the least likely in the very industrial systems that Marx thought would experience a revolution where labor, even agrarian labor, had been pushed into exploitive labor markets. Moore's work thus marks an important beginning for a new form of conflict theory. Along with others working in the historical-comparative arena, Moore's efforts inspired not only his own students but also others to examine the forces producing mass mobilization of peasants and revolutionary conflict in agrarian societies.

JEFFREY PAIGE'S THEORY OF AGRARIAN REVOLUTION

Jeffrey Paige's book on *Agrarian Revolution*[4] was one of the first works in the more contemporary era to take Marx's basic ideas and adapt them to the analysis of revolutions and other forms of mass mobilization among peasants in agrarian societies. Paige argues that Marx's ideas about the conditions producing revolution and mass mobilization are, in essence, correct but that these conditions are not met in industrial societies.[5] Rather, they are much more likely to exist in agrarian economies engaged in exporting their agricultural products. The peasants who work the agricultural estates represent a more volatile pool of exploited labor than Marx's proletarian work force in factories. Thus, Paige believes the conditions producing mobilization and conflict are more likely in agrarian than industrial societies. Moreover, the nature of the relation between the cultivators (actual workers in the fields) and the noncultivators (who own the land and control the activities of cultivators) is crucial in determining the nature of the mobilization by cultivators and the intensity of the conflict between these subordinates and their superordinate masters.

Like Marx, Paige sees the conflict between owners and workers as flowing from their respective relations to the means of production in agrarian societies, but unlike Marx, he argues that revolution is unlikely when the conflict is about purely economic matters, such as wages. Rather, revolutionary conflict will occur only when the economic conflict moves into the political arena and attacks the system of control and authority in a society. Moreover, the likelihood that conflict will enter this broader arena depends more on the actions of the dominant than of the subordinate classes. Hence, the actions of the dominant class determine just how mobilized the subordinate class will be, and just how far their mobilization will proceed.

Some of Paige's key propositions have been extracted and presented in abstract terms in Table 16-2. These propositions lay out the conditions under which, first, the conflicts of interest between cultivators (peasants, farm labor) and noncultivators (owners and managers of the land) move from conflicts over narrow economic issues, escalate into violence, and thereby move into the political arena. The propositions listed under I in Table 16-2 apply Marx's essential argument to agrarian economies engaged in exporting agricultural products. Paige sees cultivators' receptiveness to radical, change-oriented political ideologies as crucial. When cultivators' ties and connections to the land on which they work are tenuous and unstable, they are more likely to be receptive to radical ideologies. But, when cultivators can barely survive by their work on the land, when they have few alternative means for making a living, and when they are part of traditional agrarian communities, they become less receptive to radical ideologies. Only when the production processes encourage high levels of interdependence among workers can solidarity be generated, and this solidarity-producing effect can be

TABLE 16-2 Key Propositions in Paige's Model of Agrarian Revolutions

I. Conflicts of interest between cultivators and noncultivators lead to the mass mobilization of cultivators when
   A. Cultivators are receptive to radical ideologies.
      1. This receptiveness to radical ideologies will increase when cultivators' ties to the land are tenuous and unstable.
      2. This receptiveness to radical ideologies is decreased, however, under the following conditions:
         a. When cultivators' survival margins are so thin and precarious that they fear the risks of accepting the tenets of a radical ideology.
         b. When cultivators have so few alternative means for making a living that the potential costs of accepting a radical ideology become too high.
         c. When cultivators live in traditional communities in which paternalistic traditions and tight social control make radical ideologies either unappealing or too risky.
   B. Cultivators can experience collective solidarity.
      1. This sense of solidarity will increase when agricultural production encourages high levels of interdependence among cultivators.
      2. This sense of solidarity will decrease, however, when cultivators are highly dependent on noncultivators for crucial resources and services.
   C. Cultivators have been able to engage in successful collective actions.
      1. This capacity to engage in collective actions is increased when
         a. Noncultivators' direct control of the cultivators' actions is weak.
         b. Cultivators constitute a homogenous population.
         c. Cultivators are required by the process of production to cooperate.
         d. Cultivators have access to organizational resources provided by outside political actors.
         e. Cultivators perceive that they can gain real economic benefits from collective action.
      2. This capacity to engage in collective actions is, however, decreased when
         a. Cultivators have opportunities for upward mobility from subordinate positions.
         b. Cultivators see themselves as engaged in competition among themselves for resources, especially opportunities for upward mobility.

II. The likelihood that mobilization of cultivators will lead to revolutionary movements is related to the actions of noncultivators. The reactions of noncultivators to mobilization by cultivators increase the potential for violence when
   A. The conditions increasing the mass mobilization of cultivators listed under I are met.
   B. Noncultivators do not possess great economic advantages over cultivators and, consequently, are more likely to resort to repression using the state's coercive power rather than pursuing economic manipulation of, and bargaining with, cultivators. Such repression will increase when
      1. Noncultivators have little capital beyond the export crops produced on their land, giving them relatively little economic advantage over cultivators.
      2. Noncultivators do not engage in capital-intensive processing of the crops produced on their land, thereby making them dependent on exports of unfinished crops and, hence, giving them relatively little economic advantage over cultivators.
      3. Noncultivators cannot afford to hire free labor, thereby making them unresponsive to the economic demands of cultivators and increasing their use of forced labor while employing tight and rigid control of labor.
         a. The ability of noncultivators to afford to yield to cultivators' economic demands is reduced when production is inefficient and less profitable.
         b. The ability of noncultivators to afford to yield to cultivators' demands is improved, however, when production can be increased through mechanization that expands production at lower costs per worker.
      4. Noncultivators' repressive control is less likely, however, when production necessitates a constant supply of disciplined labor, leading noncultivators to perceive the benefits of labor organizations and the need to make compromises in managing cultivators.

undermined when cultivators reveal very high degrees of dependence on noncultivators for crucial resources and services.

Receptiveness to radical ideologies and solidarity among cultivators is not, however, sufficient for mass mobilization of violence and movement into the political arena. Cultivators must also be capable of engaging in effective collective actions; without the ability to engage in such actions, revolutionary movements and changes in the politics of a society are not possible. This capacity to engage in collective actions is increased when noncultivators cannot directly control the actions of cultivators. The ability to pursue collective actions is further enhanced when cultivators constitute a homogenous population, when they cooperate in the production processes and hence have an organizational base for collective action built into the work setting, and when they perceive that it is possible to gain real benefits over potential losses from collective actions. Yet, the ability to engage in collective actions is undermined when cultivators have opportunities for individual upward mobility and when they perceive themselves in competition with each other for resources, particularly those revolving around chances for mobility.

Paige emphasizes, however, that these conditions facilitating mass mobilization—that is, ideology, solidarity, and collective action—are influenced by the options and reactions of noncultivators. The form of conflict will, Paige argues, ultimately reflect the way in which the dominant sectors of the society respond to the conflicts of interests as well as the conflicts over narrow economic issues between cultivators and noncultivators. Proposition II in Table 16-2 lists some conditions that increase the level of violence and extension of conflict beyond the economic to the political arena.

One condition is, of course, those forces listed under Proposition I increasing mass mobilization potential among cultivators, but this potential is more likely to be realized when noncultivators are forced to act in certain ways. When noncultivators depend on the state and military to maintain control of their land and the labor working the land, they are more likely to engage in rigid and repressive actions to control cultivators—a tactic that will increase the anger and dispositions of cultivators to engage in mass mobilization, if they can. Noncultivators engaged primarily in exporting raw agricultural products, especially products in which additional production processes are not performed before export, are particularly likely to depend on the external state and military to assure the viability of their operations. Repressive and rigid control is also likely when noncultivators cannot afford, or at least perceive that they cannot afford, free labor that makes them disposed to employ forced labor under rigid external control. Such perceptions by noncultivators are increased when production is inefficient and yields only low to modest profits. However, if noncultivators can increase production through capital investments, especially investments in equipment that reduce reliance on labor and that increase each cultivator's productivity, then noncultivators are more likely to perceive that they can yield to cultivators' economic demands and resolve the conflict as an economic rather than political matter. Furthermore, the need to repress labor is lessened even more when noncultivators depend on a constant supply of disciplined and skilled labor, leading noncultivators to perceive the benefits of labor organizations and compromises with them in exchange for maintaining a constant supply of labor.

CHARLES TILLY'S THEORY OF RESOURCE MOBILIZATION

In *From Mobilization to Revolution,*[6] Charles Tilly described the basic ideas that became

TABLE 16-3 Tilly's Core Theory of Resource Mobilization

I. The more evident are multiple contenders to the state's power, the more likely is a revolutionary situation to exist. The likelihood of multiple contenders increases when
   A. Segments of the population have been kept out of the political arena, or denied access to power.
   B. Older power elites feel that they are being excluded from centers of power and, hence, are losing some of the traditional access to power.

II. The more that significant segments of the population are willing to support or follow one of the competing contenders to the state's power, the more likely is a revolutionary situation to exist. The likelihood of an important segment following contenders to state power increases when
   A. The state is perceived to have failed in its responsibilities to the population, and the more rapidly escalating is the perception of failure to meet its responsibilities, the greater is the willingness of some segments to follow contenders to the state's power.
   B. The state escalates resource extraction from members of a population, with such rapid escalation typically occurring when the state needs resources to finance war.

III. The less willing, or able, is the state to suppress coercively competing contenders to its power, the more likely is a revolutionary situation to exist. The likelihood of such reluctance to use its power increases when
   A. The state is in fiscal crises and does not have the resources to engage in repression.
   B. The state has ineffective mechanisms for resource (tax) collection and inefficient ways of using the resources that it does collect.
   C. The state is inhibited from using its powers for repression because its military, or a segment of its coercive arm, is associated or allied with one of the contenders to its power.

important in all subsequent theorizing on social revolutions. Tilly distinguished in this early work between a revolutionary situation and a revolutionary outcome. *A revolutionary situation* exists when some kind of collective action against centers of power is evident—whether these actions be demonstrations, riots, social movements, revolts, civil wars, or other manifestations of antagonism toward the state. A *revolutionary outcome* is when there is an actual transfer of power. Ultimately, Tilly argues, revolutionary situations emerge when contenders to power can mobilize financial, organizational, and coercive resources, and a revolutionary outcome is likely when this mobilization is greater than the capacity of the state to mobilize its coercive, material, and administrative resources.

Table 16-3 delineates some general conditions of a revolutionary situation.[7] The first condition is multiple contenders to power; two of the most likely contenders are those segments of the population who are consistently denied access to power or the political arena and those segments of the older power elite that are left out, or feel that they are excluded, from their traditional access to centers of power in the state. Another general condition increasing the potential for a revolutionary situation is that a significant part of the population is prepared to follow or support one of the competing contenders to the state's power. Such support becomes more likely as the state fails to meet obligations to the population and as it makes efforts to extract rapidly more resources from the population. This latter condition is most likely when the state is engaged in war and, as a result, runs short of money. The third basic condition leading to a revolutionary situation is that the state cannot, or is unwilling to, repress with coercive force the activities of those who challenge its power. This inability to use its own power against competing sources of power is likely when the state has few resources; when it has made ineffective and inefficient use of its resources through support of privilege, patronage, military adventurism, corruption, and archaic system of tax collection, and when it is inhibited from using power because portions of its military have close associations with at least one of the contenders to its power.

Tilly's argument in this first important work was highly theoretical, but in his more empirical analyses of the history of societies in Europe, he has tended to stay closer to the details of particular cases. Nonetheless, the basic elements of the theory summarized in Table 16-3 remain, and in his more recent work on *European Revolution, 1492-1992,*[8] Tilly blends these theoretical ideas into his historical descriptions; in so doing, he adds considerable detail to the core theoretical ideas presented in Table 16-3.

Table 16-4 summarizes some of these refinements in more detail; later, we present only the most important lessons to be derived from the propositions in Table 16-4. One very important condition affecting the capacity of the state to mobilize its resources and, thereby, resist potential contenders to its power is its involvement in international military competition. The more the state engages in such competition with other state powers, the more resources it drains away from the domestic arena, and hence, the more vulnerable it becomes, particularly so when it loses a war. The loss of a war not only drains material resources from a society but also erodes the symbolic resources that the state uses to legitimate itself. The likelihood that a state will engage in war, and eventually lose a war, is related to the military strength of its neighbors; the stronger the neighbors are and the more frequent conflict is with them, the more likely is the state to lose a war at some point in time.

Mobilization for military conflict has two contradictory effects on state power. On the one hand, the stronger the military power of the state, the greater is its coercive power to suppress mobilization by contenders to its power. This coercive strength increases to the extent that the military is organized professionally and maintains "at the ready" a standing army that can be efficiently and rapidly used to repress contenders to state power. On the other hand, if the military is engaged in external military competition, its coercive capacity is not available domestically, and if military activity outside a society's core borders is expensive, then the capacity to finance alternative domestic sources of coercion is correspondingly reduced.

Whether inside or outside the state's borders, another dilemma of having a strong military is its loyalty to leaders in the state. If the military is not loyal to rulers, or exists as a separate entity somewhat detached from the administrative guidance of the leaders in the state, it can then become a source of contending power. Moreover, this type of military is more likely to ally itself with other sources of contending power.[9] Thus, if the military allies itself with a contender to the state's power, or itself seeks to take state power, then a revolutionary situation very likely will be transformed into a revolutionary outcome in which there is a transfer of power.

Thus, the state's ability to maintain high degrees of coercive power relative to potential contenders is a crucial condition in avoiding a revolutionary situation leading to a revolutionary outcome. This capacity is not only related to absolute military power but also to the capacity of the state to (a) deflect grievances away from itself and (b) co-opt into its structure potential contenders to power. The more the state can do either or both, the more likely is it to avoid both a revolutionary situation and outcome. Thus, if the state can make relatively inexpensive concessions to contenders to power, without hurting either its financial or coercive basis of power, the state might stave off a revolutionary

TABLE 16-4 Refinements to Tilly's Theory

I. The capacity of a state to avoid a revolutionary situation or outcome is positively related to its ability to mobilize coercive resources and to suppress potential contenders to power.
   A. This capacity to mobilize coercive resources is reduced when the state is engaged in international military competition that, in turn, causes
      1. Financial resources to flow away from the state.
      2. Coercive resources to be deployed away from geographical centers of power.
   B. This capacity to mobilize coercive resources is also reduced when the military is well organized but somewhat independent of the administrative centers of power, thereby making it more likely that
      1. The military will be less loyal to the key decision makers in the state.
      2. The military, itself, will become a contender for state power.
      3. The military will ally itself with at least one of the contenders for state power.
   C. This capacity to mobilize coercive resources increases, however, when the military meets all of the following conditions:
      1. The military is professional and well organized.
      2. The military maintains a standing army and does not need to recruit an army before it can act.
      3. The military is not engaged in extensive geopolitical conflict.

II. The capacity of the state to avoid a revolutionary situation or outcome is positively related to its ability to make strategic concessions to potential contenders to power. This capacity increases when
   A. The concessions to be made are not expensive.
   B. The concessions to be made do not erode the state's coercive basis of power.
   C. The concessions increase the strength of the state's symbolic basis of power, or its legitimacy in the eyes of key segments of the population.

III. The capacity of the state to avoid a revolutionary situation or outcome is positively related to the state's fiscal situation. The ability to remain fiscally sound is reduced when
   A. The state is engaged in expensive military activity against other states.
   B. The state must engage in excessive patronage of elites.
   C. The state does not have efficient or effective means for tax collection and other forms of resource extraction.

situation, and if it can avoid becoming the target for grievances by either elites or the masses, the state will be in a better position to hold onto its power and avoid a revolutionary situation.

The state is always in a precarious situation, however. Concessions to one segment of a population often arouse the hostility of others. Frequently, it is not possible to deflect hostility away from the state. This inability to deflect hostility becomes particularly likely when the state is engaged in expensive forms of patronage to satisfy demands of elite segments of the population and when it incurs the costs of military activity with other societies. Under these conditions, the state must often increase its demands (taxes) on the population for extra resources, and when it does so, resentments against the state increase because old agreements are now seen as violated and new indignities are being imposed. When mass discontent is coupled with resentments from portions of the nobility, a revolutionary situation exists and, moreover, a revolutionary outcome becomes more possible. These events are particularly likely when the state's mechanisms for collecting and distributing material resources are inefficient and ineffective, and when the state's political institutions are geographically concentrated, thereby making them an easier target for seizure by contenders to state power.

THEDA SKOCPOL'S ANALYSIS OF STATES AND SOCIAL REVOLUTIONS

Theda Skocpol's comparative analysis of social revolutions in various agrarian societies[10] builds on Moore's and Tilly's approaches, but adds some additional refinements. Like most historical analysts of revolutions, she sees the process of mass mobilization of peasants as crucial, and she adopts a key idea from Weber and Tilly: Internal revolutionary processes are related to the state's activities in the international arena. Skocpol's basic argument is that "revolutionary situations have developed due to the emergence of politico-military crises of the state and class domination."[11] Thus, although class inequalities provide much of the drive for the mass mobilization of peasants, such mobilization is not likely to be successful unless the state is experiencing a crisis of legitimacy stemming from military defeats in the international arena.

Skocpol's main examples of successful revolutions are the French Revolution of 1789, the Russian Revolution of 1917, and the Communist Chinese Revolution of 1949, which are compared with the relative stability in societies such as Japan, Germany/Prussia, and England where full-scale social revolutions did not occur during the agrarian era nor in the transition to commercial capitalism. All of the societies in Skocpol's pool of case studies possessed certain common characteristics: (1) They all evidenced administrative and military control under a monarch. (2) The monarch could not, however, directly control agrarian socioeconomic relations. (3) The major surplus-extracting class was the landed aristocracy who passed some of their wealth to the monarch. (4) All of the aristocracy depended on the work of a large peasant population. (5) Market relations, commercial classes, and even early industrial classes (in later periods) could exist within these societies, but these classes were subordinate to the monarchial state and landed aristocracy. (6) The aristocracy in all these societies had become dependent on the state to provide positions of patronage in the court and administrative bureaucracy. (7) Yet, although the aristocracy and state were generally aligned, their interests diverged over the aristocracy's desire, on the one side, to create personal wealth and privilege through peasant labor or commercial activities and the monarch's desire, on the other side, to finance military adventures and state-sponsored economic development.

The scenario for a full-scale social revolution begins to unfold when the state is defeated in military activities with other states, thereby unleashing the mass mobilization potential of the peasants and arousing the hostility of the aristocracy. Such hostility from elites is particularly likely when the state seeks reforms in the aftermath of a military defeat and, in the process, threatens the old landed aristocracy. As a result, the state becomes vulnerable to a revolution from below by the masses or above by the upper classes who often pursue their own narrow interests (for patronage and privilege) and weaken the state's capacity to respond to revolutionary mass mobilization by peasants.

But peasants do not automatically mobilize, despite their grievances. Peasants can successfully revolt only under certain conditions: First, they must be capable of developing solidarity with each other. Second, they must have some autonomy from direct day-to-day supervision and control by landlords and their agents. Third, the state must be so weakened by its military defeats that it cannot exert effective coercive control over periodic peasant revolts (which are quite common in all agrarian societies).

The scholarly detail of Skocpol's analysis is, of course, typical of historical-comparative

TABLE 16-5 Skocpol's Propositions on the State and Social Revolution

I. The mass mobilization of subordinates in a system of inequality increases when
   A. Subordinates can develop a sense of solidarity.
      1. Solidarity is increased when subordinates can perceive that they have a common enemy.
      2. Solidarity is reduced, however, when subordinates must compete with each other in a commercialized economy.
   B. Subordinates have autonomy from direct supervision by superordinates.
      1. Autonomy from direct supervision increases when
         a. Subordinates have control of the process of production.
         b. Subordinates have organizational forms that shield them from direct supervision and control.
         c. Subordinates can retain local communities insulating them from direct sanctioning. This is more likely when sanctioning systems are highly centralized in a state structure removed from the communities of subordinates.
      2. Autonomy from direct supervision decreases, however, when subordinates are structurally tied to, and hence dependent on, those who dominate them. This is related to the converse of the conditions listed under 1a, 1b, and 1c under I-B.
   C. Subordinates perform economic activities crucial to the well being of superordinates, thereby giving subordinates the capacity to disrupt the power of superordinates through collective actions.
   D. Subordinates have organizational resources with which to pursue conflict with superordinates.

II. The likelihood that the mass mobilization of subordinates will escalate into a full-scale and successful social revolution increases when
   A. The central coercive apparatus of a society is weak and cannot, therefore, suppress revolts by subordinates or power-plays by elites.
      1. The weakness of the central coercive apparatus increases when the state is defeated in a war. Defeat becomes ever more likely when
         a. The military structure of the state is disintegrated and poorly organized.
         b. The military has not been able to sustain its autonomy from domestic conflict and domestic intrusion.

work in sociology, but for our purposes, we need to lay out the underlying theory. Table 16-5 reviews the key propositions. Proposition I in Table 16-5 examines the conditions that facilitate mass mobilization, and Proposition II reviews the causes of state breakdown that, in turn, allow mass mobilization to flower into a full-scale revolution.

As Proposition I underscores, mass mobilization of subordinates requires solidarity among them, and such solidarity increases when subordinates perceive that they have a common enemy,[12] but solidarity is reduced to the extent that subordinates must compete with each other in commercialized labor markets and other competitive mechanisms for allocating resources. Mass mobilization also requires that subordinates have some degree of autonomy from those who are supposed to regulate their activities, and this autonomy is enhanced when subordinates exert control over the processes of production, when subordinates evidence organizational structures that insulate them somewhat from direct monitoring,[13] and when subordinates can retain community structures providing additional insulation from the central sanctioning system in a society. But, this autonomy will be reduced to the extent that subordinates are structurally interdependent with, and even more so if they are dependent on, those who dominate them. Conversely, mass mobilization is also facilitated when subordinates perform productive activities

TABLE 16-5 Skocpol's Propositions on the State and Social Revolution *(continued)*

c. The ratio of career and professional officers to noncareer and nonprofessional in the military is low.

d. The productive capacity of the economy supporting military activities is low relative to the productive capacity of enemies.

B. The state experiences fiscal crises and cannot, therefore, finance reforms or suppress revolts by subordinates and power plays by elites.

1. The state's fiscal crises will tend to worsen when it loses a war, or remains overextended in external military activities.
2. The state's fiscal crises will deepen when the economy's productivity remains low relative to the population size to be supported.
3. The state's fiscal crises will be aggravated when the mechanisms for extracting revenues (taxes) from the members of the society are indirect and inefficient.

C. The state's power relative to the dominant segments of the society decreases.

1. The state's relative power declines when the networks among elites are dense and strong.
2. The state's relative power declines when its control over the military is weakened.
3. The state's relative power declines if elite segments of the population have both short-term interests and sufficient organizational strength to prevent the state from instituting reforms that would lessen the fiscal crisis or appease subordinate segments of the population.
4. The state's relative power declines when the elite segments of the population are threatened by the state's activities, that is, when the elites fear erosion of their privileges and wealth. This sense of threat by elites increases as

a. Elites remain dependent on the central state for their wealth, prestige, and power.

b. Elites feel that their opportunities for mobility are restricted by the state and the economic system supported by the state.

c. Elites see efforts at social reform by the state as undermining their traditional sources of power, prestige, and wealth.

---

crucial to the well being of superordinates, and this dependence of superordinates can give subordinates a sense of their own power and capacity to mobilize. Finally, following Tilly's theory of resource mobilization (see Table 16-3), the organizational resources of subordinates will influence just how extensive mass mobilization can be. When there are existing organizational forms, or the capacity to use the organizational systems of superordinates, mass mobilization becomes more likely.

Mass mobilization cannot occur, or be effective, without a weakening state power, for without a weak state, mass mobilization cannot be transformed into full-scale social revolution. One critical force is a decline in the state's coercive capacities to repress revolts by the masses and power plays by elites. This weakening of the state's coercive power follows from a defeat in an external war. Factors influencing the likelihood of defeat include the level of disintegration of the military's organizational structure,[14] the degree of the military's autonomy from resource-draining domestic conflict, the ratio of professional to nonprofessional military personnel, and most important, lower productive capacity in the economy than a state's military adversaries (ultimately the economy supports the military, so if this capacity is less than an adversary has, a war is likely to be lost).

Another force weakening the state is fiscal crisis. When the state experiences a fiscal crisis, its legitimacy is undermined and its capacity to mobilize coercive and administrative control is diminished. These financial problems will escalate when a state loses a war or remains overextended in military commitments, draining resources away from the state and its ability to control the domestic population. Fiscal crises are also aggravated when the productivity of the economy is low relative to the size of the population to be employed and supported; thus, if a population grows without a corresponding increase in productivity, a fiscal crisis will eventually ensue. Fiscal crises are further aggravated when the state's tax collection systems are indirect (such as when the monarch extracts peasants' "surplus labor" through elites or delegates tax collection to local governments or elites), or when the tax collection system is inefficient (when accurate records are not kept, or adequate monitoring of tax collection does not occur). Thus, if the state cannot efficiently and effectively appropriate wealth, it will soon experience a fiscal crisis.

Yet another critical force weakening the state is the relative power of dominant segments of the population, especially elites. As the power of elites increases, there is a corresponding decline in the power of the state. This balance of power works against the state to the extent that elites have dense and strong network ties independent of the central authority of the state, to the degree that the state's control over its military becomes weakened, to the extent that elites' short-term interests in maintaining their privilege prevent the state from making reforms that might quell the masses, and to the degree that elites feel threatened by the activities of the state.

In sum, then, the propositions in Table 16-5 might have broader applicability than the particular historical cases examined by Skocpol. These propositions might also help explain why revolutions are so rare, especially once the agrarian phase of a society has passed and most societies have moved into more commercial, market-driven, and industrial forms of organization. When subordinates cannot easily mobilize (because the converse of the conditions listed in Proposition I prevail) and when the state is strong, periodic revolts and riots do not as easily become transformed into full-scale social revolutions.

JACK GOLDSTONE'S THEORY OF STATE BREAKDOWN

Unlike other theorists examined in this chapter, Jack Goldstone did not begin with Marxian assumptions about social class relations and their inherent potential for revolutionary conflict. In his *Revolution and Rebellion in the Early Modern World,*[15] Goldstone argued that revolutions in modernizing agrarian societies between 1640 and 1840 were caused, ultimately, by the effects of population growth. As populations grow, pressures are exerted on (1) the economy to increase productivity and (2) the polity to stimulate production, expand administrative control of the population, and maintain order through coercive force. When these key institutional systems cannot make the necessary adaptations to population growth, the potential for mass mobilization of peasants and revolt by elites against the state escalates, eventually causing state breakdown. Population growth does not, however, immediately throw a society into chaos and state breakdown; rather, it can take many decades for the pressures exerted by population growth to become evident and for the masses and elites to become sufficiently disenchanted to initiate conflict.

The fundamental problem with all agrarian societies is that they typically reveal rigid institutional structures. A hereditary monarchy is usually incapable of making the reforms that are necessary to accommodate a growing population's need for economic opportunities, and the land-based aristocracy generally resists changes in the economic system that would undermine its power and privilege. This rigidity can, in the end, lead to a situation where three forces converge: (1) a fiscal crisis in which the state simply does not have enough income to sustain its activities, to initiate economic reforms, or to control the restive population; (2) severe divisions and anger among elites, many of whom cannot secure the traditional spoils of state patronage and are, as a result, downwardly mobile; and (3) mass mobilization of peasants who cannot secure work or income. Goldstone's analysis of various empirical cases is detailed and complex, but the general model can be summarized in more abstract terms, as is done in the following discussion and in Table 16-5.

The driving force behind state breakdown through mass mobilization of peasants and revolt by at least some segments of traditional elites is population growth. As the rate of population growth increases, several demands are placed on economic institutions. First, the economy must expand to provide a living for the growing population; this expansion is often very difficult because of the nature of feudal agrarian systems where only so many tenants can be accommodated. Second, if the economy does not expand and keep pace with population growth, then resources become scarce. As goods become scarce, price inflation ensues in accordance with the laws of supply and demand. Third, as price inflation escalates, real incomes for all workers decline, increasing the level of misery in rural areas. Fourth, as conditions in rural areas deteriorate, many peasants leave the land and migrate to urban areas in search of opportunities that, if the economy is not expanding, cannot be fully realized, thereby creating a pool of disgruntled and unemployed urban workers. Fifth, population growth also assures that the proportion of younger age cohorts relative to cohorts of middle-aged and elderly will increase, and as their proportion increases, the young are more disposed to revolt and commit acts of violence than are older members of the society.

Sixth, population growth also affects the life chances of elites in several important ways: (a) the number of aspirants for elite positions increases; (b) price inflation dramatically affects mobility of elites, some of whom become downwardly mobile whereas others (who take advantage of commercial opportunities when demand for goods is high) become upwardly mobile; (c) the net affect of such mobility is to increase the pool of aspirants for state patronage as upwardly mobile elites aspire to new positions of privilege and influence, and downwardly mobile elites see patronage as their only hope against a worsening of their economic situation; (d) these demands for patronage increase the level of competition among elites for patronage positions, often at precisely the time the state itself is experiencing fiscal crises (ultimately caused by price inflation), thereby limiting the state's ability to meet the demand for elite patronage and, as a result, creating dissatisfaction among some segments of the elite population.

Seventh, the state's fiscal crisis will continue to mount for a number of interrelated reasons: (a) price inflation increases the costs to the state for administrators, goods and services, military personnel, and virtually all activities of the state; (b) the state might have engaged in military warfare in an effort to increase its resources because of inflation and growing demands of elites, or simply because of leaders' competition with other societies, but such military adventurism is expensive and can aggravate the fiscal crisis of the state; (c) the state often begins to borrow from elites, especially those upwardly mobile elites who have resources garnered from commercial activity, though this debt only increases the fiscal problems of the state; and (d) the state often increases tax revenues to stave off crisis, but increased taxation will antagonize

TABLE 16-6 Goldstone's Theory of State Breakdown

I. The long-term potential for state breakdown through mass mobilization of non-elites increases when
   A. Population growth generates demands among nonelites for goods and income that exceed the productive capacity of the economy.
   B. Population growth exceeding the productive capacity of the economy causes rapid price inflation.
   C. Population growth increases the proportion of the population in younger age cohorts whose members are potentially more violent and more easily mobilized to pursue conflict than those in older age cohorts.
   D. Population growth, as it causes price inflation and overburdens the productive capacities of the economy, escalates rural misery and forces many to migrate to urban areas in search of limited or nonexistent economic opportunities.

II. The long-term potential for state breakdown through mobilization of elites increases when
   A. Population growth creates a larger pool of elites who seek the state's patronage for privilege and positions.
   B. Population growth, as it causes price inflation, leads to a situation where
      1. Some traditional landed elites become financially strapped and, hence, desirous of state patronage and positions to prevent their downward mobility.
      2. Some upwardly mobile elites, often gaining wealth through commercial activity, seek state patronage and positions as a confirmation of their new-found station in life.
   C. Population growth, because it causes price inflation coupled with other fiscal pressures on the state, makes it impossible for the state to accommodate all of the demand for elite patronage and positions.

III. The long-term potential for state breakdown through fiscal crises increases when
   A. Population growth exceeds the economy's capacity to absorb and provide goods for the population, thereby creating a shortage of resources and causing price inflation.
   B. The state's mechanisms for revenue collection are inflexible and inefficient.
   C. The state's efforts to seek new formulas for revenue collection arouse the hostility of elites as well as that of nonelites.
   D. The state's expenditures on military activities exceed its ability to finance wars, especially when
      1. The state expands military activities during periods of rapid price inflation.
      2. The state engages in military activities in an effort to secure more resources to overcome the effects of price inflation, thereby incurring expenses for which it cannot afford to pay.
   E. The state is forced to borrow funds to sustain its military and administrative activities.

IV. The likelihood of state breakdown increases when
   A. Mass mobilization of nonelites to pursue conflict is increasing.
   B. Mobilization of segments of elites is increasing.
   C. Fiscal crises within the state are escalating to the point where the state's administrative and coercive control of the masses and elites is dramatically weakened.

some elites who are already downwardly mobile as well as many wage-earning workers who have little disposable income. If new tax revenues can be secured, this flow of money reduces the fiscal crisis even as it increases the potential for elite and mass mobilization against the state. Moreover, if the state can secure resources in the international arena via conquest or trade, it can similarly reduce the fiscal crisis and, thereby, avoid internal conflict.

The picture that Goldstone paints, then, is a series of converging forces that have been initiated by population growth. When they all come together, state breakdown becomes likely. That is, (1) if the fiscal crisis of the state is acute, (2) if elites are in intense competition with each other for position and patronage,

(3) if some elites are mobilized against the state because of their downward mobility, failure to receive patronage, and increased tax burden, and (4) if the younger masses in rural estates and migrants to urban areas cannot secure sufficient income and, hence, are restive and potentially mobilized for conflict, then the convergence of these four forces leads to state breakdown as the masses and segments of the elite population mobilize to overthrow the government and as the fiscal problems of the state make it incapable of responding with sufficient administrative or coercive power to resist this two-pronged revolt by elites and masses. Table 16-6 states Goldstone's argument more abstractly so that its potential for explaining revolt in a broader range of societies can be appreciated.

CONCLUSION

Historical-comparative sociology has, during the last three decades, produced some of the most scholarly and important works in sociology. Most of this work has had a theoretical bent, though always tempered by the particulars of specific empirical cases. Of particular interest to many historical-comparative sociologists has been the dynamics of power, especially as these dynamics are set into motion by conflict-producing inequalities and external conflicts with other societies. Historical-comparative sociology has, of course, examined many other substantive topics, but the most theoretical wing of this approach to sociology has tended to be most interested in conflict processes, particularly those generating social revolutions that redistribute power.

Reading the propositions in Tables 16-1 through 16-6, it is clear that the theories overlap, although each adds something new or provides a refinement to the others. The conditions producing a social revolution—the kind Marx thought to be inevitable and Weber saw as only probable—rarely converge. Some of these conditions always exist in agrarian societies, but the necessary mix of conditions producing a true revolution does not seem to have happened very often in human history. Moreover, the very conditions that generate high revolutionary potential in agrarian societies appear, on net balance, to be mitigated or eliminated in capitalist-industrial societies. Paige makes this point explicit, but it is one of the inescapable conclusions from this line of work in historical-comparative sociology. Thus, even though a good many historical-comparative sociologists have started with Marxian assumptions, they have tended to come to Weberian conclusions: Revolutions take a unique convergence of forces, even in agrarian societies where there is so much potential for conflict stemming from the high degrees of inequality. This convergence appears to be less likely in mature capitalist systems, but as we will see in the next chapter, neo-Marxist theorists continue to adapt and adjust the basic Marxian model to the present.

NOTES

1. Barrington Moore, *Social Origins Of Dictatorship and Democracy: Lord and Peasant in the Making of the Modern World* (Boston: Beacon, 1966).

2. According to Moore, this was the case with China before the Communist Revolution—many landlords moved to the city and lived on rent collected from peasants and interests of the debt peasants borrowed from them (see pp. 218–221).

3. The Japanese army served as an external enemy that pulled the peasants together in China during the Japanese occupation in the Second World War (see p. 223).

4. Jeffrey Paige, *Agrarian Revolution: Social Movements and Export Agriculture in the Underdeveloped World* (New York: Free Press, 1975).

5. Ibid., pp. 33–34.

6. Charles Tilly, *From Mobilization to Revolution* (Reading, MA: Addison-Wesley, 1978). At the same time, John McCarthy and Meyer Zald were formulating another, more formal resource mobilization model; see their "Resource Mobilization in Social Movements: A Partial Theory," *American Journal of Sociology* 82 (1977), pp. 1212–1239.

7. Tilly, *From Mobilization to Revolution,* pp. 200–211.

8. Charles Tilly, *European Revolutions, 1492-1992* (Oxford, UK, and Cambridge, MA: Blackwell, 1993).

9. Tilly points out that military seizure was a common kind of revolutionary situation in Iberia (see pp. 86, 101).

10. Theda Skocpol, *States and Social Revolutions: A Comparative Analysis of France, Russia, and China* (New York: Cambridge University Press, 1979).

11. Ibid., p. 17.

12. Skocpol illustrates this point in her analysis of anti-seigneurial movement among the French peasants who welded together in the resistance against the seigneurs (see pp. 123–125).

13. According to Skocpol's analysis, this was the case with French and Russian peasants.

14. Skocpol argues that the military disintegration in Russia during the First World War was the main reason for state breakdown in Russia and the success of the 1917 Revolution that followed.

15. Jack Goldstone, *Revolution and Rebellion in the Early Modern World* (Berkeley: University of California Press, 1991).

17

The Continuing Tradition III: Neo-Marxian Theories

In the review of Karl Marx's theory of conflict in Chapter 11, much of the substance of Marx's ideas about the dynamics of capitalist societies was left out. Instead, we abstracted Marx's substantive categories to produce a general theory of conflict in systems of inequality. As Marx's ideas were used during the middle decades of the twentieth century, this more abstract approach to building on Marx was typical, as we saw in the work of Ralf Dahrendorf in Chapter 12 and J.H. Turner in Chapter 14. Yet, Marx considered himself to be a revolutionary[1], and his life's work was to develop a theoretical system that could explain the self-destructive dynamics of capitalism and the emergence of communism.[2] This more substantive and emancipatory thrust of Marx's work was never lost during the mid-century, but the theory had to be revised because the revolution by the proletariat did not occur and because the communist systems of the twentieth century were hardly emancipatory. As we will see in Chapter 42, this failure of Marxian predictions led to the emergence of critical theory in a rather pessimistic guise. In this chapter, however, we will review some theories that sustain both the substantive thrust of Marx's ideas and the view that the contradictions of capitalism will, perhaps, open the doors for new, more emancipatory forms of social organization.

A BRIEF REVIEW OF MARX'S SUBSTANTIVE ARGUMENT

To appreciate the creative directions that Marx's ideas have taken over the last two decades, we need to review, if only briefly, some substantive points of emphasis in Marx's analytical scheme. Marx believed the analytical goal was to explain the contradictions in capitalist modes of economic production and how these would lead to the conflict processes that would usher in communism. His predictions were wrong, perhaps because of some fatal errors in his logic, but his analysis is still useful, as we will see when we turn to more contemporary theories.

For Marx, capitalism is an economic system where those who own and control the means of production seek to make profits from the goods that they sell in competitive markets. In such a system, labor must sell itself as a commodity to capitalists and must do so under unfavorable circumstances where the supply of workers is high relative to demand, thereby driving the wages of labor down (and, hence, reducing the costs of labor for capitalists). The key to profits for capitalists is to make labor work beyond the actual labor time needed to make the goods that pay their wages; the difference is *surplus value* or profit for capitalists. The greater this surplus value is, the more workers are *exploited* by capitalists who, in essence, appropriate the surplus value of the goods produced by workers for their own gain. So, in Marx's view, labor power is the ultimate source of value of products as well as the source of profits for capitalists. Capitalism is thus sustained by exploitation.

But capitalists face a dilemma: They must compete with each other in markets, and hence, the price that they can get for products is driven down to the point where it becomes difficult to make a profit. As businesses fail, laborers are thrown out of work, which in turn means that they do not have wages to buy goods in markets. As a result, demand for goods declines; with this decline in demand, profits of capitalists also decline. Capitalists seek to get around this *declining rate of profit* by using more productive technologies and machinery in place of labor, thereby putting more laborers out of work and lessening demand for the goods produced by capitalists. Although such efforts by capitalists can give them a short-term advantage over their competitors and, thereby, increase their profits for a time, technologies and machinery are soon copied by competitors. As a consequence, a new round of competition over price occurs. This new round of competition, however, must confront the lessened capacity of unemployed labor to purchase goods in markets where prices are coming down.

In this process of declining profits, many capitalists go under, and indeed, their competitors might buy them out, creating oligopolies and monopolies that can better manipulate prices when they do not have to compete. But they must do so in conditions of lessening demand in markets where labor does not have wages to buy goods. Moreover, the many capitalists who have lost out in the competition are without money to buy goods and services, and indeed, many are driven into the proletariat. Eventually, even as monopolies gain control of key productive sectors, they have trouble making a profit; goods can go unsold because there is not enough money in the hands of workers and displaced bourgeoisie to buy them. As goods remain unsold, capitalists reduce production, but as workers are laid off, cutting back production only decreases the buying power of workers in markets. Thus, Marx felt that capitalism would, in the end, move toward a situation where profits and production were declining while unemployment and monopoly control of production were increasing.

From this situation would come the mobilization by the proletariat for revolution. Such revolutionary efforts had to confront, however, the power of capitalists to control the state and ideological system. But, as crises of unemployment deepen, mobilization by the proletariat can overcome the power of the state to control coercively and the capacity of ideology to impart *false consciousness* (see the propositions in Table 11-1 on page 157 for an overview of the conditions producing mobilization for conflict). Such mobilization is facilitated by the very acts of capitalists in concentrating workers so that they can communicate their grievances, in making them appendages to machines and thereby increasing their sense of alienation, and in subjecting them to a decline in, or uncertainty about, their living standards. All these acts of capitalists lead workers to question the legitimacy of political authority and the appropriateness of dominant ideologies.

This brief summary does not capture the depth, texture, or subtlety of Marx's analysis of capitalism, but it gives us the tools to place into context the two basic directions of theorizing that Marx's work continues to inspire. One prominent direction is revising Marx's conception of exploitation and its effects on class processes in modern societies. The other direction has been to move beyond the nation-state as a unit of analysis to a world system level where, it is argued, the contradictions of capitalism will ultimately emerge. We will begin with neo-Marxian class analysis, and, then explore world systems theorizing.

NEO-MARXIAN CLASS ANALYSIS

In the twentieth century, Marxist theory has confronted several problems. First, the predicted collapse of capitalism has not occurred, despite the Great Depression and periodic recessions. Second, the projected polarization of capitalist societies into bourgeoisie and proletarians has been muted by the growth of a large and varied middle class of managers, experts, small business operators, skilled manual workers, and others who do not seem highly disadvantaged and who do not see themselves as exploited. Third, capitalism has appeared to emerge, at least in recent history, as the clear victor in the contest between capitalism and communism, although state-managed societies operating under the ideological banner of communism were hardly what Marx had in mind. Nonetheless, the historical predictions of Marx and the trajectory of capitalism into socialism and communism have not occurred.

Yet, despite these troubling issues, Marxism has remained a viable intellectual tradition, driven perhaps by an emancipatory zeal emphasizing the elimination of exploitation of the disadvantaged by the advantaged. Still, Marxist intellectual circles have been in crisis for more than a decade, as many former Marxists have moved to other forms of radical thinking or have become critical theorists and post-modernists—perspectives that are explored in the chapters of Part VII. Although some still cling to the orthodox picture of Marx, briefly summarized earlier, most Marxists have changed Marx's core ideas to fit current historical realities. Among these Marxists are several important scholars,[3] but in an effort to summarize the issues with which they have had to deal, we will focus on the work of Erik Olin Wright who has, in his words, sought to do more than merely draw from the Marxian tradition but, instead, to contribute to "the reconstruction of Marxism." Wright has termed his and fellow travelers' approach "analytical Marxism." But before discussing the basic concepts in Wright's scheme, let us set the stage with his more metatheoretical assertions.

The Analytical Marxism of Erik Olin Wright

The goal of analytical Marxism is to shed some of the baggage of orthodox Marxist analysis, while retaining the core ideas that make Marx's theory unique.[4] Analytical Marxism retains the emancipatory thrust of Marx, stressing that the goal is to reduce, if not eliminate, inequalities and exploitation. Indeed, emphasis is on constructing scientific theory about how socialism can emerge from the dynamics inhering in capitalist exploitation. But this emancipatory thrust does not abandon a commitment to the conventional norms of science in which theoretical ideas are assessed against empirical observations. Moreover, the goal is to produce abstract formulations that specify the "mechanisms" generating empirical regularities in the world, and for the analytical Marxist, particular concern is with the mechanisms flowing from social class structures.

Social Class, Emancipation, and History

Marxist theory posits an historical trajectory: feudalism to capitalism, then capitalism to communism. In this historical trajectory, class inequalities will be eliminated, and communism will usher in a classless society. In this trajectory and outcome, class is the pivotal dynamic, mobilizing individuals to seek alternative social relations in which exploitation is eliminated. Wright believes that these three basic orienting assumptions—that is, historical trajectory, class emancipation, and class as the driving force of history—need to be mitigated somewhat.[5] Emphasis should be on how the dynamics of capitalism present possibilities for new, less exploitive social arrangements rather than on the inevitability of the forces driving human society toward communism. In this vein, emphasis on class emancipation should not blindly pursue the goal of a classless society but, rather, present a critique of existing social relations to reduce class inequalities and exploitation. Moreover, emphasis on class as the driving force of history must be tempered by a recognition that class is one of many forces shaping the organization of a society, both in the present and future.

Wright's theoretical work stresses this last consideration: What mechanisms revolving around social class generate what outcomes? As a Marxist, Wright views the *class structure* of a society as limiting the nature of *class formation* (the organization of individuals) and *class struggle* (the use of organization to transform class structures). He posits a simple model, as delineated in Figure 17-1.[6] In this model, class struggle transforms the nature of class formations and class structure, whereas class formations select or channel class struggle in certain directions depending on the nature of organization of class members. The key dynamic in Wright's program, however, is class structure; as the model outlines, class structure limits the nature of class formation and class struggle. The goal of a reconstituted Marxian analysis must, therefore, examine the properties of class structure in capitalist societies if class formations and emancipatory class struggles are to be understood.

Micro-Level versus Macro-Level Class Analysis

In Wright's view, Marxian class analysis must confront two impulses. One is to retain Marx's vision of the class structure in society as ultimately polarizing into two conflictual classes—in the capitalist historical epoch, these are the bourgeoisie who own the means of production versus the proletariat who are exploited. The other impulse is to explore "the complexity of the class structural concept itself in the hope that such complexity will more powerfully capture the explanatory mechanisms embedded in class relations."[7] Wright argues that this complexity is not as evident when analysis stays at the macro-level, focusing on the global characteristics of capitalism as composed ultimately of owners and workers. However, when research moves to the micro-level of the individual and seeks to understand the locations of individuals in the class system, a much more varied, complex, and contradictory picture emerges. If the causal effects of class relations are to be understood, then, it is necessary to explore class processes at the micro level.

Wright's basic strategy has been to examine the jobs that people hold, because in their jobs individuals connect to the system of production and the class relations that inhere in this system. For most of his work in the 1970s and 1980s,[8] Wright sought to construct a "relational map of locations of individuals" in the class system, but by the 1990s, he had recognized that "the simple

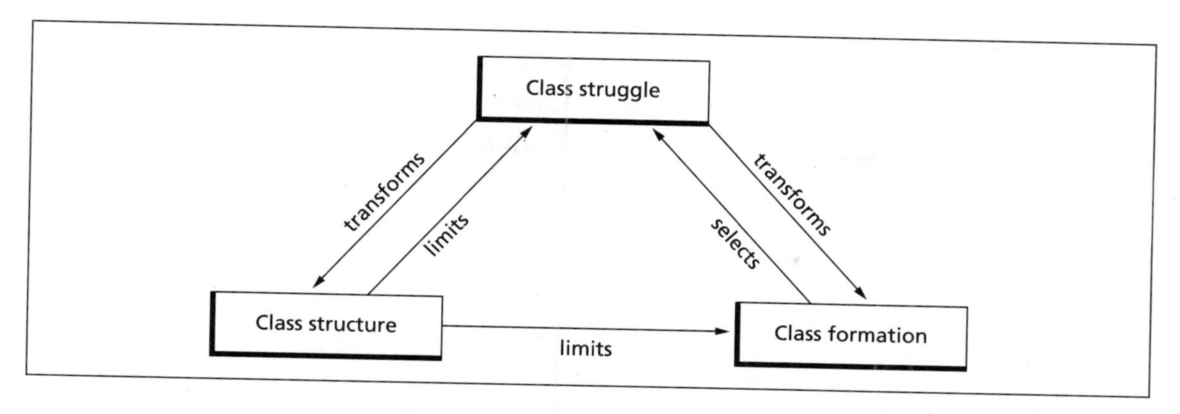

FIGURE 17-1 Wright's General Model of Class Analysis

linkage of individuals-in-jobs to classes" must be modified in several important ways: Individuals can occupy more than one job; they can have indirect and mediated relations to a job (as is the case with children and others who do not directly participate in the system of formal employment); they can move about jobs and cross class locations during the course of a career; and they can have contradictory locations in the class system.[9] In essence, as we will see shortly, Wright's conceptual scheme has tried to deal with these kinds of complexities that a micro-level analysis exposes; indeed, his conceptualizations have changed during the course of the last twenty years as he has confronted both data[10] and conceptual criticisms.[11]

When class analysis turns micro, the vagueness of much macro-level Marxian theory is revealed. Wright argues that it is important in a reconstructed Marxian analysis to be specific about the mechanisms involved in generating class formation and class struggle effects. He illustrates the differences between a macro-level and micro-level view of class mechanisms by summarizing the three ways in traditional Marxism that class structure is seen to exert effects on class formation and struggle:[12] (1) material interests, (2) lived experiences, and (3) collective capacities. We will examine each of these.

(1) *Material interests*. The first mechanism by which class structure exerts causal effects on class formation and struggle is through individuals' *material interests*. There are two basic types of material interests: (a) *economic welfare* or "the total package of toil-leisure-incomes available to a person,"[13] with people having an interest in reducing their toil and in increasing their leisure and consumption; and (b) *economic power* or the capacity to control how surplus products are distributed, with surplus product being defined as "that part of the total social product that is left over after all the inputs into production (both labor power and physical capital) have been reproduced."[14] Thus, the material interests of individuals are determined by their economic welfare and economic power.

The concept of *exploitation* ties these two types of material interests together. Those with economic power can use this power to appropriate productive surplus from those without power. In so doing, they increase their economic welfare at the expense of those whose surplus product they take. In Marxian class analysis, then, the material interests of classes are not just different; they are opposed in a kind of zero-sum game.

The material interests of individuals circumscribe their options. Depending on their respective material interests, people make different choices, employ diverse strategies, and make varying trade-offs. When individuals share material interests, their choices, strategies, and trade-offs should converge because they face similar dilemmas as they pursue economic welfare and economic power. In Marx's macro-level approach, social systems polarize around two classes with different material interests, fueled by exploitation. In a more micro-level approach to material interests, a more complex and, at times, contradictory picture emerges of individuals with diverse perceptions about what they should do in pursuing their economic welfare and in seeking economic power. Moreover, the lived experiences of individuals as they make choices and pursue strategies can also diverge, or at a minimum, individuals do not perceive that they have common material interests (a problem that Marx dismissed, perhaps too quickly, as "false consciousness").

(2) *Lived experience*. In Marxian theory, common material interests as dictated by class location lead to common experiences. Those who do not own capital are seen to have similar subjective understandings of the world because they are forced to sell their labor, because they are dominated and bossed around, and because they are unable to control the surplus products of their labor. When individuals are exploited this way, they are also *alienated,*[15] and together, these forces give alienated individuals a common experience that, Marx felt, would lead to their collective mobilization.[16] Yet, a more micro-level approach reveals that individuals might not have similar experiences, or at the very least, they do not see their lived experiences as similar because they occupy somewhat different jobs that locate them at different points in the class system.

(3) *Collective capacities*. The third class force in traditional Marxian analysis flows from the first two: People with common material interests and lived experiences possess a capacity for collective action. Moreover, as Marx emphasized, capitalists are forced by their own material interests to create many of the conditions that facilitate collective mobilization, such as concentrating workers in factories and urban areas, making them appendages to machine technologies, disrupting their life routines through lay-offs, providing literacy and media access, and encouraging other forces that shape class formation and struggle. This kind of scenario has a certain plausibility when examined at the macro level, but on the individual level, the contradictory locations of individuals in jobs and their differentiation across an array of middle-class locations make collective mobilization more problematic. Individuals do not see that they have common material interests or that they share common lived experiences; so, they do not naturally organize collectively. Even alliances and coalitions among individuals who see that they share some interests become problematic, although once formed these alliances have strength (because the costs of compromises have already been incurred).[17]

Thus, in Wright's micro approach, the three class forces in Marxian analysis do not reveal the same degree of coherence as when they are examined from a more macro level. Moreover, micro analysis shows that, in contrast with traditional Marxism, these forces are not necessarily correlated with each other. When analysis moves to the micro level, "there is no longer necessarily a simple coincidence of material interests lived experience, and collective capacity."[18] People who might seem to have common material interests do not perceive this to be the case, nor do they reveal a driving need to mobilize collectively. In a micro approach to understanding class mechanisms, then, the complexity of the class system makes traditional macro-level Marxian causal forces too global and crude to account for the specifics of class formation and struggle.

The Problem of the Middle Class

Wright's entry into Marxist analysis began with the question of how to account for the emergence and proliferation of jobs in the middle classes[19], and although he has sought to move in different directions, this question is still at the core of his theoretical formulations. In his effort to build a "map of the class structure," Wright recognizes that the array of middle-class positions posed the biggest challenge to traditional Marxian class analysis. If class structures are to be seen as relevant to the class formation and to the struggle by classes for social emancipation, then a more fine-grained analysis of this structure is necessary. Wright's analytical project has thus involved an effort to isolate the mechanisms being generated by this more complex class system. Through his empirical and theoretical work during the last two decades, he has proposed several ways to conceptualize class structures and the mechanism for class formation and struggle that inhere in these structures. Some of these have been rejected, others retained, and most modified and integrated in new ways by Wright. In contrast with orthodox Marxism, Wright has been willing to change his mind and to develop, elaborate, and qualify ideas in his effort to isolate the causal mechanisms of class structures in advanced capitalist societies. He has, in essence, proposed several models of class structures, the basic elements of which are briefly summarized here.

Contradictory Class Locations. Wright's first attempt at conceptualizing the middle classes in Marxian terms led him to develop the idea of *contradictory class locations*.[20] Individuals can occupy a class location that is contradictory because it puts people into different classes, presumably giving them contradictory material interests and diverse lived experiences and collective capacities. These locations are contradictory because, for example, many managers, semi-autonomous wage earners, professionals and experts, and small-scale employers can reveal varying amounts and combinations of (a) owning the means of their production, (b) purchasing the labor of others, (c) controlling and managing the labor of others, and (d) selling their labor. To illustrate, a manager sells labor to an owner of a business, but at the same time this manager will be involved in hiring and controlling the labor of others. Similarly a skilled consultant sells labor, but might own the facilities by which this labor is organized. Such variations can put individuals in contradictory class locations in the sense that they are neither owners of the means of production nor helpless sellers of their labor; they have an element of both, and, hence, they have contradictory material interests and, no doubt, lived experiences and collective capacities. They might not see themselves as exploiters or as exploited; indeed, they can be both in varying proportions.

This approach made Wright a prominent Marxist theorist, but it also brought criticism that, in turn, led Wright to pursue an alternative conceptualization. One problem was that domination (telling others what to do) and exploitation (extracting surplus product) were somewhat decoupled. For example, managers could give orders but they did not directly enjoy the economic welfare of appropriated surplus (this went to owners of the means of production). Another problem revolved around employment in the state or government. Are those employed by the state workers in the means of economic production? If not, then their status is unclear: They are paid labor, but their products are not directly appropriated by capitalists. They are managers and hence can direct not only each other but perhaps workers and even owners in the economy. These conceptual problems, coupled with the difficulties that Wright encountered in measuring contradictory locations of individuals, led him to propose a new conceptualization of the middle classes.[21]

Multiple Exploitation. Adopting ideas from John Roemer,[22] Wright's second solution to the problem of the middle classes was to posit an *exploitation nexus* that varies by the kinds of assets that individuals possess and their degree of ownership or control of these assets. The four assets are (a) *labor power assets*, (b) *capital assets*, (c) *organization assets*, and (d) *skill or credential assets*. Each of these leads to a particular type of exploitation: Individuals who only have labor assets are likely to be exploited because they depend on those who have the economic power to appropriate the surplus value of their labor; capital assets can be used to invest in equipment and labor as a means of extracting the surplus product generated by technology and labor; organization assets can be used to manage and control others in ways that extract surplus products; and skills or educational credentials can be employed to extract extra resources beyond the resources it took to acquire and maintain these skills and credentials.

This approach allowed Wright to address the issue of the middle class by recoupling the concepts of exploitation (of surplus) and domination (through control or order-giving). In essence what had been domination was translated into a new type of exploitation by those with organization assets, highly valued skills, and educational credentials. At the same time, the exploitation envisioned by Marx—extraction by capitalists of surplus value from workers—could remain close to Marx's original formulation. A particular society could then be typified by the combinations and configurations of these various types of exploitation.

Wright preferred this conceptualization of the middle classes to the contradictory class location formulation because it allowed him to put exploitation back as the central mechanism by which economic welfare and economic power are connected, thereby making his scheme more consistent with Marx's original formulation.[23] Moreover, it allowed him to conceptualize managers in the state, professionals, and other skilled workers with some degree of autonomy through Marxian-inspired class dynamics: they too are exploiters, just like capitalists, but they use different assets to extract surplus value and product from others.

This approach also came under heavy criticism, leading Wright to back down from some of the assertions in this model.[24] First, those with skills and credentials are not so much exploiters of the less skilled as advantaged workers who can prevent capitalists from exploiting them as much as their less-skilled counterparts are exploited (because capitalists value and need their skills and credentials). Second, managers inside government as well as in the private capitalist sector can move up the organizational hierarchy and use their higher salaries to buy into the capitalist sector (via purchase of stocks, bonds, and so on), thereby confusing just what their material interests would be. Third, the location of individuals in the state bureaucracy would determine the mix of assets; for example, those higher in the state might possess more capital assets or at least a mix of organization and capital assets, whereas those lower in the hierarchy would possess only organization assets, and those lower still would possess only labor assets to be exploited. Coupled with the problems of measuring these various types of assets and documenting empirically how they led to different forms of exploitation caused Wright to shift his scheme yet again.[25]

The Emerging Scheme. As Wright has dealt with the conceptual and empirical problems of measuring class locations at the micro level, he has responded to criticisms by elaborating new concepts or, perhaps more accurately, reformulating older ideas in a new guise. These new conceptualizations still revolve primarily around the problem of the middle classes, but they reveal a more eclectic character. Some of the key ideas are reviewed in the following discussion.[26]

One idea is the notion of *multiple locations*. Like most traditional Marxists, Wright had originally assumed that individuals had one class location, even if this location was contradictory and placed a person into classes with different material interests. But, people often have more than one job, and, hence, can actually have several class locations. For example, a person can have a salaried day job, then operate a small business at night or on weekends, thereby making this individual both a proletarian and capitalist.

Another idea deals with *mediated locations*. Individuals are often connected to a class via networks to others who hold a job or own capital. Children, wives, and husbands can all have a mediated relation to a class location of a parent or spouse, and these mediated relations can become complicated. For example, if a female manager is married to a carpenter, each has a mediated relation to the other's class location, and their children, if they have any, will bear mediated relations to both classes. Wright proposes the notion of an *overall class interest* in these situations, which is a "weighted combination of these direct and mediated locations," that may or may not be contradictory.

Still another idea in Wright's evolving scheme is a concern with *temporal locations*. Careers might not significantly change a person's class location, but often careers do involve movement across class locations, as when individuals move up government and corporate hierarchies, when a small business gets large, when workers begin to form companies, and when students move from school to job, then into a career track. People's class locations can thus change over time, giving them different material interests, lived experiences, and collective capacities.

Yet another idea is a conceptualization of distinct *strata* within classes. The argument is that within basic classes such as owners and workers, there are distinctive strata whose members might have somewhat different material interests and most likely have very different lived experiences and collective capacities. These strata can take a number of forms. For example, professionals and experts with skills and credentials can be seen as able to collect *rents* on their skills and credentials, which make them a distinct strata within the working class (they are still part of the working class because they must sell their labor power). Temporal mobility in a career can often increase these rents; if the rents are sufficiently high, they can be invested as capital, thereby giving a person a position within a stratum of the capitalist class as well as the working class. Similarly, managers or manual workers can translate their skills and career mobility into rents that can then become capital. All these workers are in contradictory locations because they are, on the one hand, workers who are paid a salary and, on the other hand, investors in businesses that hire workers. Thus, the basic classes envisioned by Marx—workers and owners of the means of production—can be reconceptualized by distinct strata within these broad classes, with the incumbents in various strata potentially having a contradictory class location as both workers and capitalists.

The issue of state employment can be translated into a *state mode of production*. Rather than view high-level state workers as organization exploiters, it is better to visualize the state as producing goods and services of a particular kind and, hence, as revealing distinct classes (and perhaps strata within these classes). The dominant class would be those who direct the appropriation and allocation of surplus productivity that the state acquires to support itself (in forms such as taxes, fees, tariffs, and the like). The subordinate class would be those who actually perform the services and produce the goods provided by the state. Within the state mode of production could be various combinations of contradictory positions. For example, a state manager can control the actions of other workers but be controlled by elite decision makers in the dominant class of the state; a state manager who can command a high salary can invest this rent in the private sector, thereby placing this individual in both the economic and state modes of production. Relations in the state mode of production can also be mediated, as when high-level members of the dominant class have relations with corporations doing business with government or being regulated by government. Moreover, the career path of many incumbents in both the economic and government modes of production can involve movement back and forth between these modes, thereby shifting the class location of individuals.

A further set of ideas developed in Wright's evolving project moves beyond the problem of the middle classes. The existence of an unemployable and often welfare-dependent segment of a population has led Wright to introduce a distinction between "nonexploitive economic oppression" and "exploitive economic oppression." In *exploitive economic oppression*, one group's economic welfare is increased by virtue of the exploitation of another whose welfare declines because the latter's surplus products are appropriated by the former. This situation is oppressive not just because of exploitation, per se, but because the exploiting group often uses morally sanctioned and legitimated coercion to get its way. Yet, even under these oppressive conditions of exploitation, the exploited have some power because their exploiters depend on them, so, the exploitation often involves implicit negotiation and consent. In *nonexploitive economic oppression*, there is no transfer of surplus productivity to an exploiter. Instead, the economic welfare of exploiters depends on the exclusion of the oppressed from access to valued resources being consumed by oppressors. Under these conditions, Wright argues, genocide can occur because the goal of exploiters is to get rid of those who might seek access to resources that they have, or covet. Yet, even here, the nonexploited oppressed have a resource: the capacity to disrupt efforts at consumption by exploiters. Thus, the nonexploited oppressed can often force exploiters to provide some resources (as happens with those who pay for the welfare of those who are kept out of the economy).

The Overall Marxian Project

At this writing, there are further elaborations of Wright's scheme that will be published soon. Thus, analytical Marxism is constantly changing in an effort to come to grips with the complexity of the class systems of advanced capitalist societies. The great asset of this theory program is the coherence that comes from efforts to conceptualize class mechanisms as exploitation and to sustain a common emancipatory vision. Thus, even as theorists and researchers argue and develop alternative schemes, the dialogue and debate stay within the extended boundaries of Marxian analysis.

This strength is also a weakness in at least one sense: It is hard to engage in the theoretical debate without buying into the key notion that at the core of class dynamics are not just differences in interests but material interests fueled by exploitation. Very different theorists share similar emancipatory goals—that is, to reduce inequalities—but if one follows the Marxian approach, emancipation can only come with reduction in exploitation. Yet, if another analyst does not see market systems distributing variously skilled labor to different jobs with varying degrees of remuneration as exploitative, then theoretical and emancipatory discussion will hit an impasse.

Wright's work has had wide appeal beyond Marxist circles not so much for the Marxism that, despite its flexibility and creativity, is just that: a Marxist analysis of stratification. Rather, the broader appeal has come from the carefully executed and comparative empirical work on class structures; this empirical work has proven useful to Marxists and non-Marxists alike. Moreover, Wright's insistence that Marxist analyses be subject to the same test of evidence as other scientific projects has added to the general appeal of Wright's and other analytical Marxists' work. Yet, such work will always be constrained by the desire to sustain, as much as is possible, basic Marxian categories of analysis—refined to be sure, but still squarely in the Marxian tradition—and by the need to orient theory to an ultimate emancipatory goal or, alternatively, to emancipatory possibilities. These concerns will bias inquiry, perhaps no more than other commitments by other theorists, but these biases will limit interchange with other theoretical perspectives.

NEO-MARXIAN WORLD SYSTEMS ANALYSIS

In the 1970s, Marx-inspired theory began to shift the unit of analysis from nation-states to relations among societies. Capitalism was viewed as a dynamic engine of transformation that would create a world-level economy. This world-capitalist economy, in turn, would reveal many of the same contradictions that Marx had predicted for capitalism within a particular society. The study of empires and imperialism had, of course, long been a major topic in a variety of disciplines—history, political science, economics, and sociology. Some of the flavor of what world system analysts would argue was captured early by the British economist, J.A. Hobson,[27] in his view that capitalist nations need to conquer and exploit other nations to stave off many of the problems predicted by Marx. But, Immanuel Wallerstein[28] codified Marxist ideas into a coherent conceptual scheme for the analysis of the world system. His work, in turn, has stimulated much further theoretical and empirical effort. We will concentrate on Wallerstein's scheme, but add conceptual points from others as we proceed.

Immanuel Wallerstein's Analysis of World System

World Empires and World Economy. Wallerstein begins his historical analysis on the emergence of a capitalist world system by distinguishing between two basic forms of interconnection among societies: (1) world empires and (2) world economy. A *world empire* is created by military conquest or threats of such conquest, then extraction of resources, usually in the form of tribute, from those populations that have been defeated or threatened. Often these conquered societies can retain considerable autonomy as long as they pay the tribute demanded by their conquerors. Whether through direct appropriation, taxation, or tribute, dominant nations can accumulate wealth and use their wealth to finance the privilege of elites and the military activities of the polity. Military empires are thus built around a strong state that administers the flow of taxes, franchises, and tributary wealth, while financing and coordinating the military for war-making and conquest. Wallerstein argues that this was, historically, the dominant form of societal interconnection among societies before the 1400s when the capitalist revolution had begun, although many debate this point and emphasize that trade-based and empire-based systems of domination existed long before modern European capitalism.[29] Moreover, many argue that imperial and trade forms of world system domination had come and gone in various cycles many times in history,[30] long before the modern world system began to develop in the 1400s. Nonetheless, the important point is that one form of connection among nations is through state-based imperialism. Indeed, this form has persisted for most of the twentieth century, as the Soviet Union before its breakup can attest.

Wallerstein, as well as other historical sociologists whose theories are examined in Chapter 16, emphasizes the structural dilemma for imperial forms of governance: sustaining the resource levels necessary to support privilege of elites and a large military and administrative bureaucracy despite resentful peasants and the need to control conquered populations. Corruption and graft only aggravate these problems, but in the end, the state leaders face fiscal crises and must confront a variety of enemies within and outside their homeland borders. Eventually, the empire collapses, often because two imperial empires come to a showdown conflict, as Collins emphasized in his analysis of geopolitics (see Table 15-6 on page 192).

In contrast with an empire, a *world economy* reveals a different structure composed of (a) multiple states at its core, some of which have approximately equal military power; (b) competition among these core states in both the military and economic arena, with the latter being dominated by markets; and (c) peripheral states whose cheap labor and raw resources are extracted through trade, but trade that is vastly unequal because of the military power and economic advantages of the core states in market transactions.

Core, Periphery, and Semiperiphery. Wallerstein's distinction among the core and periphery at the world-system level parallels in a rough way Marx's notion of capitalists and proletarians at the societal level. Core nations correspond to capitalist class at the societal level and, like the capitalist class, extract surplus through exploitation. The *core areas* of the world system are the great military powers of their time. Since military power ultimately rests on the economy's ability to support the use of coercive force, these military powers are also leading economic powers. There are *external areas* to the core, and these become the *periphery* when a core state decides to colonize them or engage in exploitive trade. The periphery consists of less developed countries whose resources are needed and, with threats of potential or actual military intervention, are taken in exploitive market transactions. Wallerstein also distinguishes what is termed the *semiperiphery,* which comprises (a) minor nations in the core area and (b) leading states in the periphery. These semiperipheral states have a higher degree of economic development and military strength than the periphery, but not as much as the core states; they are often used as intermediaries in trade between core and periphery.[31]

The semiperiphery can also be the origin of mobility among states, and, at times, areas of the periphery can become semiperipheral and, perhaps, even part of the core area (as with the history of the United States and Japan). Similarly, much of southeast Asia is moving today from either semiperipheral (China, for example) or even peripheral (India, for instance) to the core. Japan is clearly at the core in the current world system, whereas India and other parts of Asia are still somewhat semiperipheral but clearly capable of moving to a new core.

The basic connection that drives the world economy, however, is the relationship between the core and periphery. The core has a large consumer market for both basic and luxury goods, a well-paid labor force (at least relative to the labor force in the periphery), a comparatively low rate of taxation enabling the accumulation of private wealth, a high level of technology (both economic and military) coupled with market-driven needs to sustain technological innovation, and a set of large-scale firms that engage in trade with peripheral states. The periphery has resources that consumers in the core states desire, and because states in the periphery are at a trading disadvantage (because of their lack of military strength and their lack of technology and capital to develop their own resources), each exchange between core and periphery in a market transfers wealth to the core. This exploitation by the core perpetuates the problems of development in the periphery, because peripheral states do not receive sufficient money from the core to finance infrastructural development (roads, transportation, and communication) or to afford educational and other welfare-state needs.[32] Moreover, because of the lack of economic development and the high degree of economic uncertainty in peripheral nations, population growth places even greater burdens on the state as individual citizens view children as their only potential source of economic security in the future. As a consequence, peripheral states not only remain poor and underdeveloped, they are typically politically unstable, a situation that only sustains their problems of development and their dependence on core states for trade.

The Dynamics of the World Economy. A world economy reveals its own dynamics, some of which are much the same as in empires, but others of which are unique to capitalism. What the core states have all had in common with older forms of empire building is their constant wars with each other, especially over conquest and control of the periphery. Moreover, they encounter many of the same fiscal problems of empires in trying to sustain wealth, profits, and well-being at home along with a large military and administrative system to wage war and to control their own citizens and dissidents in their conquered or dominated territories. Indeed, core states are often just ahead of problems, as they colonize ever more territory to sustain the costs of their prosperity while financing the administrative and coercive basis of control at home and abroad. When these competing powers begin to fight closer to home, they often ruin their respective economies through the costs of financing war and then maintaining control. Such wars make core states vulnerable enough that new powers can move into the core and supplant them (as has occurred with, for example, Spain and Portugal).

Another dynamic, Wallerstein argues, is the cyclical tendencies of the world economy. Wallerstein emphasizes what are termed *Kondratieff waves,* which are long-term oscillations[33] in the world system, running approximately 150 years. At the beginning of a Kondratieff wave, the demand within core states for goods is high, which increases production and the need for ever more raw materials. This need for raw materials leads to the expansion of the core states into external areas, making the latter peripheral suppliers of resources to the core. The next step in the Kondratieff wave occurs when the supplies of raw materials and the production of goods exceeds demands for them, leading core states to reduce geographical expansion but, equally important, for businesses to reduce production and, thereby, set off the down cycle emphasized by Marx: lowered domestic demand, decreased production, intense competition for market share driving profits down, increased unemployment as production declines, and, consequently, even less demand for goods, further reductions in production, business failures, and growth of monopolies and oligopolies.

This concentration of capital, however, sets the stage for the next point in the wave: High unemployment generates class conflict as workers demand better working conditions and wages; such demands eventually lead to higher wages for workers as the state responds to political pressures and large corporations give in; concentrated capital meantime seeks new technologies and ever more efficient means of production to lower costs; with more wages, economic demand increases, and, with new technologies and capital investment, a new period of higher profits and relative prosperity ensues, leading to increased demand for raw materials from peripheral states. But eventually, this new round of prosperity falls victim to the forces predicted by Marx: Market saturation of goods, intense competition over price, increased unemployment, decreased demand, further decreases in production, increases in business failures, and crisis bring to a close this long 150 year wave.

Other Cyclical Dynamics in the World Economy. Within these long waves are shorter cycles that have been extensively studied by not only conventional economists but world-system analysts as well.[34] These all operate much as Marx had predicted, but without the great revolution at their end. The classic business cycle, sometimes termed *Juglar cycles*, appears to last from five to seven years. Production expands to increased market demand, unemployment declines, demand in markets increases further, production expands more, then oversupply of goods in relation to demand starts a recession. Some have argued that part of this cycle reflects the replacement costs of new machinery, which tends to wear out about every eight years, forcing new capital investments that can drive market demand, but once this demand is met, it can also set into motion the decline demand for capital goods. Such capital demand is especially important in high technology core nations where much employment revolves around making equipment and providing services for other businesses. Capital demand can become as significant, or more so, as household consumer demand for goods and services.[35]

Another cycle is what is termed the *Kuznet cycle*; these operate over a twenty-five-year period in core and semiperipheral states.[36] Just why these cycles occur is unknown, although there are several hypotheses. One is related to generational turnover, in which about every twenty years, demand for basic household purchases such as houses and other buildings declines—and, hence, decreases production and employment—until the next generation has sufficient money to drive up demand for these goods, thereby setting off another wave of prosperity.

Hegemonic Sequences. As Wallerstein argued,[37] but as others have developed further, there are oscillations in the degree of centralization among the core nations of the world system. Before capitalism,[38] these oscillations revolved around the rise and fall of empires through war, conquest, tribute, and collapse. With capitalism, however, the nature of the oscillation changes. Hegemonic core states seek to control trade, particularly trade across oceans, and thereby connect core and periphery in an exploitive trade arrangement. The dominant state or states can prevent military empires from encroaching on this trade and can force empires to act as capitalists in the system of world trade (as with the former Soviet Union and as is occurring in China today).

Thus, the cycle of centralization revolves around the rise and fall of hegemonic core states that have been able to dictate the terms of trade in the world system. Shifts in this domination by a core state can come with wars, but unlike pre-capitalist empire building, the domination that ensues is oriented toward dictating the terms of trade as much as toward outright conquest of territory or extraction of tribute in response to military threats. The rise of a new hegemonic state gives the state greater access to the resources of other peripheral and semiperipheral states, while enabling it to dominate other core states (as has been the case, for example, with the United States in the post-World War II period, at least to this point).

In addition to war, hegemonic states often rise because of new economic or military technology that gives them advantages. Under these conditions, states can charge "rents" for their innovations, or use them to control trade and, in the case of military technology, to make threats that improve the terms of trade. As these innovations are copied, however, the advantages can be lost or neutralized, setting the stage for another potential hegemonic state to emerge because of new technologies and other productive or military advantages.

The End of Capitalism?

Wallerstein and many other world-system analysts still accept Marx's vision that capitalism will collapse, but for world systems theory, capitalism must first penetrate the entire world for its contradictions to emerge. As long as peripheral states exist to be exploited by the core, capitalism can sustain itself by relying on the resources and the cheap labor of less developed countries. But once capitalism exists everywhere, there is no longer an escape from the processes outlined by Marx. The problems endemic to capitalism—saturation of markets, decreased demand, lowered production, and further decreases in demand—will lead to the collapse of capitalist modes of production, a period of conflict between old-line capitalists (along with their allies in the state) and the broader population that seeks a better way to distribute resources fairly. In the wake of these crises will come world-level socialism, and perhaps even world government. Although the details of this ultimate scenario vary among analysts, the emancipatory thrust of Marx's predictions remains. Whether these predictions are any more accurate than Marx's remains to be seen, but regardless of their accuracy, world-systems theory has provided important insights into basic dynamics of human organization.

CONCLUSION

Varieties of analytical class analysis along with diverse approaches to world-systems analysis attest to the continued viability of Marxian-inspired sociological theory. For unlike the more abstracted approaches of Dahrendorf and Turner, which took Marx's ideas and made them more abstract to fit changing empirical conditions, class and world-systems analyses retain much of the substantive flavor of Marx's theory and extend this substance in useful and creative ways. The key dynamic remains exploitation, whether by those who control capital within a society or by those nations who dominate other nations militarily and economically. Moreover, although a certain caution and moderation is evident, the emancipatory thrust of Marx is retained: In the long run, it is supposed, the contradictions of capitalism will indeed lead to a revolution in which there is less inequality in the distribution of resources among people within societies and between societies.

What makes these neo-Marxian approaches appealing is their willingness to confront data and to use empirical studies to modify, qualify, and change theoretical statements. This is not to say, however, that these theorists' emancipatory ideology does not make them reluctant to accept findings that might contradict the emphasis on exploitation as the central mechanism within and between societies, but at least there is a willingness to move outside the narrow confines of Marx's original formulation. In this way, they keep Marx alive as a viable source of theoretical development.

NOTES

1. Karl Marx and Friedrich Engels, *The Communist Manifesto* (New York: International, 1978; originally published in 1848).

2. Karl Marx, *Capital*, 3 volumes (New York: International, 1967; originally published in 1867, 1885, and 1894).

3. For example, Perry Anderson, *Considerations on Western Marxism* (London: New Left Review, 1976); Michael Buraway, *The Politics of Production* (London: Verso, 1985); Sam Bowles and Herbert Gintis, *Democracy and Capitalism* (New York: Basic Books, 1986); G.A. Cohen, *History of Labor and Freedom: Themes from Marx* (Oxford: Clarendon, 1988) and *Karl Marx's Theory of History: A Defense* (Princeton, NJ: Princeton University Press, 1978); John Elster, *Making Sense of Marx* (Cambridge: Cambridge University Press, 1978); Barry Hindess and Paul Q. Hirst, *Capital and Capitalism Today* (London: Routledge, 1977); Claus Offe, *Disorganized Capitalism: Contemporary Transformations of Work and Politics* (Cambridge: Cambridge University Press, 1985); Adam Przeworski, *Capitalism and Social Democracy* (Cambridge: Cambridge University Press, 1985); John A. Roemer, *A General Theory of Exploitation and Class* (Cambridge, MA: Harvard University Press, 1982) and *Analytical Foundations of Marxian Economic Theory* (Cambridge: Cambridge University Press, 1981).

4. Erik Olin Wright, "What Is Analytical Marxism?," *Socialist Review* 19 (1989), pp. 35–56.

5. Erik Olin Wright, "Class Analysis, History and Emancipation," *New Left Review* #202 (1993), pp. 15–35.

6. Ibid., p. 28.

7. Erik Olin Wright, "Rethinking, Once Again, the Concept of Class Structure," in *The Debate on Classes*, ed. E.O. Wright (London: Verso, 1989), p. 269.

8. See, in particular, Erik Olin Wright, *Class, Crisis and the State* (London: Verso, 1978), *Class Structure and Income Distribution* (New York: Academic, 1979), and *Classes* (London: Verso, 1985); Erik Olin Wright and Luca Perrone, "Marxist Class Categories and Income Inequality," *American Sociological Review* 42 (1977), pp. 32–55.

9. Wright, "Rethinking, Once Again, the Concept of Class Structure" (cited in note 7).

10. See, for examples, Wright's own comparative survey research on class structures, which can be found in "The Comparative Project on Class Structures and Class Consciousness: An Overview," *Acta Sociologica* 32 (1989), pp. 3–22; *Class Structure and Income Distribution* (cited in note 8); *Classes*; and Wright and Perrone, "Marxist Class Categories and Income Inequality" (cited in note 8).

11. See the chapters in Wright, *The Debate on Classes* (cited in note 7).

12. Wright, "Rethinking, Once Again, the Concept of Class Structure" (cited in note 7).

13. Ibid., p. 281.

14. Ibid., p. 282.

15. Marx conceived of *alienation* as the result of workers' inability to determine what they produce, how they produce, and to whom the products of their labor are sold.

16. See propositions in Table 11-2 on p. 158 for a list of these conditions.

17. Wright, "Class Analysis, History and Emancipation" (cited in note 5).

18. Wright, "Rethinking, Once Again, the Concept of Class Structure" (cited in note 7), p. 296.

19. Wright and Perrone, "Marxist Class Categories"; Wright, *Classes* and *Class, Crisis and the State* (cited in note 8).

20. Wright, *Class, Crisis and the State and Class Structure and Income Distribution* (cited in note 8).

21. Wright, *Classes* (cited in note 8).

22. Roemer, *A General Theory of Exploitation and Class* (cited in note 3).

23. Wright, "Rethinking, Once Again, the Concept of Class Structure" (cited in note 7).

24. Wright, "Class Analysis, History and Emancipation" (cited in note 5).

25. Unfortunately, his most recent effort, written in late 1996—Erik Olin Wright, *Class Counts* (Cambridge: Cambridge University Press, 1997)—was not available for review and inclusion in this chapter.

26. For brief reviews of the ideas presented here, see Wright, "Rethinking, Once Again, the Concept of Class" (cited in note 7) and "Class Analysis, History and Emancipation" (cited in note 5).

27. John Atkinson Hobson, *Capitalism and Imperialism in South Africa* (London: Contemporary Review, 1900); *The Conditions of Industrial Peace* (New York: Macmillan, 1927); *Confessions of an Economic Heretic* (London: G. Allen and Unwin, 1938); *The Economics of Distribution* (London: Macmillan, 1900).

28. Immanuel Wallerstein, *The Modern World System*, 3 volumes (New York: Academic, 1974, 1980, 1989). Earlier work by scholars such as Andre Gunder Frank on "dependency theory" anticipated much of what Wallerstein was to argue: Underdeveloped societies, especially those in Latin America, could not go through the stages to modernization because they were economically dependent on advanced economies, and this dependency and the corresponding exploitation by advanced industrial powers kept them

from becoming fully industrialized and modern. See for example, Frank's *Capitalism and Underdevelopment in Latin America* (New York: Monthly Review Press, 1967). See also his later work, *Dependent Accumulation* (New York: Monthly Review Press, 1979). Also, historians such as Fernand Braudel had conducted analyses of world-system processes (for his overview see *Civilization and Capitalism*, 3 volumes (New York: Harper & Row, 1964).

29. For a review, see Christopher Chase-Dunn and Peter Grimes, "World-Systems Analysis," *Annual Review of Sociology*, 21 (1995), pp. 387–417. See also Albert J. Bergesen, ed., *Studies of the Modern World System* (New York: Academic Press, 1980).

30. See Christopher Chase-Dunn and T.D. Hall's edited collection of essays on *Core/Periphery Relations in Precapitalist Worlds* (Boulder, CO: Westview, 1991). See also Andre Gunder Frank and B.K. Gills, eds., *The World System: Five Hundred Years or Five Thousand?* (London: Routledge, 1993).

31. For a somewhat different analyses, see Christopher Chase-Dunn, *Global Formation* (Cambridge: Blackwell, UK, 1989). See also Volker Bornschier and Christopher Chase-Dunn, *Transnational Corporations and Underdevelopment* (New York: Praeger, 1985).

32. This was the essential point of dependency theorists; cited in note 28.

33. See Chase-Dunn and Grimes, "World Systems Analysis" for a brief overview of wave analysis. For a short but very clear summary of Wallerstein's argument, see Randall Collins, *Theoretical Sociology* (New York: Harcourt Brace Jovanovich, 1988), pp. 96–97.

34. See Chase-Dunn and Grimes, "World Systems Analysis" (cited in note 29) for a brief review of empirical and conceptual work on these cycles.

35. Ibid., p. 404.

36. Ibid., pp. 404–405.

37. Wallerstein, *The Modern World System* (cited in note 28).

38. See Chase-Dunn and Grimes, "World Systems Analysis" (cited in note 29), pp. 411–414, for a useful review.

39. For a recent test of the world system model, see Ronan Van Rossem, "The World System Paradigm As General Theory of Development: A Cross-National Test," *American Sociological Review* 61 (1996), pp. 508–527.

18

The Continuing Tradition IV:

Theories of Gender Inequality and Stratification

In the 1960s and early 1970s, the obvious gender bias of sociological inquiry was exposed. Both theory and research in sociology had focused on males and, in a very real sense, contributed to patterns of gender inequality. Indeed, if one looks to the early masters of sociological theory, they were conspicuously quiet about gender issues, despite the fundamental facts that roughly one-half of the human population is female and that all patterns of social organization have historically revealed a gender-based division of labor. Only Herbert Spencer discussed gender questions with a surprisingly modern flair, whereas Karl Marx had relatively little to say about gendered inequality, although Frederick Engels' ideas are now often used by contemporary Marxist-oriented theorists who must rework the notion of reproduction and decide if reproductive and household labor are part of production. Max Weber wrote virtually nothing on gender questions, because his inner conflicts in relations with his own parents had left him confused about the sexuality of male-female relationships in his relations with his own wife. Similarly, others in sociology's early years did not theorize about what is obviously one of the universal properties of human organization. Indeed, sex and its elaboration into gender roles and associated patterns of inequality are universal features of human organization, and yet, these dynamics were not the subject of rigorous theoretical attention until the 1970s.

When sociological theorizing turned to gender questions, however, theorizing itself became confrontational and conflictual. As will be explored in Chapter 43, critical feminist thinking tended to reject science and "male modes" of thinking about the world in general, offering a variety of new ways to analyze social reality. Yet, many critical feminists stayed within the scientific fold, preferring to conceptualize gender processes in more neutral theoretical terms. In particular, those in this scientific camp worked within a conflict-theory approach, analyzing gender inequalities as yet another form of conflict-producing stratification. In this chapter, we will examine several of the most prominent theories of gender processes, then briefly summarize a creative effort by these various theorists to synthesize their respective approaches into one general model.

RANDALL COLLINS' EARLY GENDER STRATIFICATION THEORY

Part of Randall Collins general theory of social conflict, summarized in Chapter 15, is a theory of gender stratification.[1] In Collins' theory, use of physical force and bargaining over material and symbolic resources become critical variables. When one sex controls the means of coercion, that sex can use this power to dominate sexual encounters and to generate a system of gender inequality; the less powerful sex must then adopt strategies to mitigate this power advantage. Similarly, when one sex disproportionately controls material resources, as well as the economic processes by which such resources are generated, this sex has power to control sexual relations and to elaborate them into a system of gender stratification; the less powerful sex must then seek bargaining strategies to cope with disadvantage in economic power. These general dynamics of gender inequality have, historically, clearly favored males who have a slight but decisive coercive advantage over women and who have used this advantage to control economic and, for the most part, political as well as ideological resources. The general propositions of Collins' theory—reduced considerably from a much longer list—are presented in Table 18-1.[2]

These propositions emphasize that control over sexual activities as well as the associated talk and ritual between males and females is related to one sex's control over (1) the means of coercion and (2) the material resources in a society. We will examine these key elements of the theory:

(1) Generally, males are stronger than females, at least on average, but control of coercion involves more than person-to-person physical strength differences. A stronger sex's control over a weaker sex is mitigated by powers outside sexual partners, such as the legal system or the state, and by the presence or absence of the weaker sex's relatives. When powers external to the sexual partners exist and when relatives of the weaker sex are present in the household or community, then the coercive advantages of the dominant sexual partner are reduced. When these external allies are not available, however, the coercive advantages of the stronger sex are used and elaborated into high levels of gender inequality.

(2) Control over material resources is related to several forces. One is the amount of economic surplus generated by economic production. When there is little surplus because of low levels of technology, there is correspondingly little material wealth to be distributed unequally, whether by gender or some other criterion. Moreover, in simple economies, such as those among hunting and gathering populations, women contribute as much or more to the economic well being of the band as males; hence, males cannot easily claim control over these resources. As production expands with horticulture and agrarianism, however, an economic surplus is generated, with the result that males typically begin to use their coercive advantages to gain control of these material resources. Much of this control is achieved through male domination of key economic roles, and although females might still perform much of the economic labor, males control those positions that determine the distribution of economic outputs and surplus. Another critical force in control of material resources is the inheritance rules of the society; if only one sex can inherit wealth

TABLE 18-1 Abbreviated Summary of Collins' Propositions on Sex Stratification

I. Control over sexual activities between males and females as well as associated talk and ritual is a positive and additive function of
   A. The degree of one sex's control over the means of coercion, which is a negative function of
      1. The existence of coercive powers outside sexual partners and family groupings (such as the state).
      2. The presence of relatives of the subordinate sex.
   B. The degree to which one sex controls material resources, which is a positive and additive function of
      1. The level of economic surplus in a population.
      2. The degree to which key economic activities are performed by one sex.
      3. The degree to which resources are inherited rather than earned.

II. The greater the degree of control of sexual relations and related activities by one sex, the more likely are sexual relations to be defined as property relations, and the more likely are they to be normatively regulated by rules of incest, exogamy, and endogamy.

III. The greater the degree of control of sexual relations and related activities by one sex, the greater will be the efforts of the other sex to
   A. Reduce sexual encounters.
   B. Regulate them through ritual.

or the power used to generate wealth, then that sex has considerable control of material resources.

As one sex gains control over sexual and related activities, Collins argues that male-female relations often become translated into property relations, with the subordinate sex being defined as the property of the one that controls the means of coercion and material resources. As sexual relations are defined as property relations, they become increasingly regulated by normative rules, especially with respect to incest (those kindred with whom sex is prohibited), exogamy (those larger kin groups in which marriage and sex are not allowed), and endogamy (those kin groups within which marriage and sex are to occur). As this normative regulation occurs, the subordinate sex increasingly loses the capacity to alter the inequalities that are associated with one sex's ability to define normatively the other as property and to control coercive and material resources.

Yet, subordinates will develop strategies for mitigating against their position. Collins argues that the subordinate sex will seek to limit sexual encounters with the dominant sex. When encounters do occur, the subordinate sex will attempt to regulate them through rituals that provide prestige and honor as a means to reduce the coercive and material advantages of the dominant sex.

Many criticisms were leveled against Collins' theory, but it was one of the first efforts to develop a general theory of sexual stratification which was part of a larger neo-Weberian theory of inequality and conflict (see Chapter 15). Other theories were also being formulated at this time, and two of the more important were developed by Rae Lesser Blumberg and Janet Saltzman Chafetz—as we will now explore.

RAE LESSER BLUMBERG'S GENDER STRATIFICATION THEORY

In contrast with Collins' Weberian theory of gender stratification, Rae Lesser Blumberg developed a more Marxian-oriented theory, although only in the very muted sense of emphasizing women's degree of control of the means of production and the distribution of economic surplus.[3] Blumberg's theory is based on a broad empirical knowledge of diverse societal types, ranging from hunting and gathering through horticultural and agrarian systems to industrial societies. Her theory thus explains the position of women relative to men in all types of societies, from the earliest to the most complex societies of the late twentieth century.

Sexual stratification, Blumberg argues, is ultimately driven by the degree to which, relative to men, women control the means of production and the allocation of productive surplus or, in Marxian terms, "surplus value." Such control gives women *economic power* that, in turn, influences their level of political power, prestige, and other stratifying resources. In Blumberg's view, sexual inequalities are "nested" at diverse levels: Male-female relations are nested in households; households are nested in local communities; and if a society is sufficiently large to reveal a coercive state and a system of class stratification, household and community are nested inside of the class structure that, in turn, is lodged within a larger state-managed society. This nesting is important because women's control of economic resources can be located at different levels, and the level at which their economic power is strongest influences the power that women can command at the other levels of social organization.

Nesting of economic power is marked by what Blumberg calls a *discount rate* in which women's economic power will be reduced or enhanced depending on the level at which it is concentrated. If women's relative economic power is at the micro level of the household (for example, women work and contribute to family income), women will not have household authority proportionate to their economic contribution if males control more macro social spheres. Male control at these more macro levels "discounts" or reduces the power that women should have in the household. Conversely, if women possess power at more macro levels, then the discount rate turns positive and will enhance women's power at the more micro, household level. Thus, as a general proposition, the more women have economic power at macro levels of social organization, the more they will be able to gain access to other forms of power—political, coercive, ideological—and the more their economic contributions at the micro level will be appreciated and increase their authority within the household and their influence within a community.

Blumberg emphasizes that during times of transition, where women's economic power relative to men is growing, men are likely to perceive such changes as a threat and to repress physically and politically women's efforts to gain equal power. Yet, as women's relative economic power increases, this increase will translate into political influence: If women's economic power and political influence become consolidated, then political policies working against women will recede, male-supremacist ideologies will decline, and male violence against women will be punished.

Thus, gaining control of economic power—that is, the control of their means of production and the allocation of the fruits of their production—is the critical condition influencing women's position in the stratification system of a society. What, then, determines when and how women gain economic power? For Blumberg, the key variable is the *demand* for labor performed by women, especially the *strategic indispensability* of their labor. In simple societies without class stratification, women's participation in production is a function of (1) the demand for their labor relative to the supply of labor and (2) the degree of compatibility of this productive labor with the reproductive (especially breast feeding) labor that women must also perform. In class-stratified systems, the demand for labor becomes more important as other arrangements can be made for child care. In these complex systems, women can begin to move into male-dominated positions when there is a shortage of men to perform the activity (such as when men are at war, or the demographics of war have created a shortage of men in the population) or when the economic activity of women is sufficiently valuable for traditional norms about the sexual division of labor to change.

Thus, the *strategic indispensability* of women's labor becomes one of the most important determinants of women's access to economic power. When women's activities are defined as important, their indispensability increases, but other conditions influence indispensability: (a) the extent to which women control technical expertise, (b) the degree to which women can work autonomously from male supervision, (c) the size, scale, and level of organization in women's work groups, (d) the degree to which women can organize on their own behalf and pursue their interests, and (e) the degree to which women can avoid competition from workers (other groups of women, slaves, immigrants, and the like) who can be used as substitutes for women's labor.

The kinship system also influences women's capacity to gain economic power. Blumberg sees inheritance rules as the most critical: That is, can women inherit and control property? If so, then they can exercise economic power. Next most important is the residence rule of kinship: Does a bride continue to reside with or near her kin in their community? If so, then she can control domestic property and secure necessary support and coercive backup from her local family. Another influence from kinship comes from the descent rule: Is the system matrilineal (with property and power passing through the female's side of the family) or patrilineal (with property and power coming from the male's side)? If the system is matrilineal, women retain more power (even though much of the property and power goes through her male kin who will, nonetheless, provide support in dealings with her husband).

In addition to strategic indispensability and kinship rules, an important factor influencing women's economic power is the way surplus and other resources are distributed. When communal relations of production prevail, where men and women equally share work and its outputs, women will have more economic power than in systems where men control the means of production and distribution of its surplus. Such male control increases with class stratification; inequalities are high along every front, including the distribution of power, prestige, property, and opportunities for the respective sexes. Under these conditions, especially if women cannot inherit or control property, their ability to gain economic power is dramatically reduced. In these stratified systems, however, those at the bottom of the stratification ladder will have virtually no property, creating a situation where men and women will share their misery more equally. Conversely, as one moves up the class ladder in such systems, inequalities between the sexes increase, and females are denied access to economic power.

Without economic power, Blumberg argues, women are denied honor and prestige, and more important, they have less control over such basic matters as their fertility patterns (when and how many children to have), their marriages (when, if, and with whom), their rights to seek a divorce, their premarital sex, their access to extramarital sex, their household activities, their levels and types of education, and their freedom to move about and pursue diverse interests and opportunities. Thus, economic power has important consequences for what women can, or cannot, do in a society. And, Blumberg emphasizes, it is not just women's economic participation that matters; rather, does such economic participation translate into *control of one's own productive activities and the*

TABLE 18-2 Abbreviated Summary of Blumberg's Gender Stratification Theory

I. The degree of gender stratification is inversely related to the level of economic power women can mobilize, and conversely, the less economic power women can mobilize, the more likely are they to be oppressed physically, politically, and ideologically.

II. The level of economic power that women can mobilize is a positive and additive function of

   A. The ability to participate in economic production.

   B. The ability to control the distribution of their economic production.

   C. The ability to mitigate the "discount rate" for their labor which, in turn, is positively related to the degree to which they exert influence at macro levels of social organization.

   D. The strategic indispensability of their economic activities, which, in turn, is a positive and additive function of

      1. The demand for their labor relative to the supply.
      2. The comparability of their productive labor with reproductive obligations and options.
      3. The degree to which their labor is seen as important.
      4. The extent to which women control technical expertise.
      5. The degree to which women can work autonomously from male supervision.
      6. The size, scale, and level of organization in women's work groups.
      7. The degree to which women can organize on their own behalf and pursue their interests.
      8. The degree to which women can avoid competition from other sources of labor.

   E. The extent to which the kinship systems facilitate the acquisition of property by women that, in descending order of importance, depends on

      1. Rules of inheritance allowing women to inherit property.
      2. Rules of residence giving females kin allies.
      3. Rules of descent favoring the female side of the kinship system in passing property, power, and authority.

III. The greater women's economic power is relative to that enjoyed by men, the greater women's control of their own lives will be for

   A. Their fertility.

   B. Their marriages and divorces.

   C. Their premarital, marital, and extramarital sexual activities.

   D. Their household authority and duties.

   E. Their educational achievements.

   F. Their freedom to pursue diverse opportunities.

IV. The greater women's economic power is relative to men's and the more women control their own lives, the greater their access will be to other sources of value in stratified social systems, especially honor and prestige, political power, and ideological support for their rights.

distribution of the outputs from this productive activity? If women's work is strategically indispensable, if kinship facilitates their inheritance and acquisition of property, and if stratification at the macro level does not so blatantly favor men, then women can gain economic power. Without such economic power, sexual stratification will be high, but with this economic power, the degree gender inequality will decline. Table 18-2 summarizes Blumberg's main line of argument in propositional form somewhat shorter than her longer list of propositions or hypotheses.

JANET SALTZMAN CHAFETZ'S GENDER EQUITY THEORY

Janet Chafetz has been one of the most prominent feminist theorists committed to developing scientific explanations of gender stratification. Her most ambitious effort,

Gender Equity: An Integrated Theory of Stability and Change,[4] presents a set of models and propositions to explain both the forces maintaining a system of gender inequality as well as a theory of how such a system can be changed. As she emphasizes, the two are interrelated because "a theory of the maintenance and reproduction of gender systems is a theory of change targets, because it identifies the critical variables that sustain the status quo and, therefore, must be changed."[5]

The Maintenance of Gender Stratification

Chafetz argues that two types of forces sustain a system of gender inequality: (1) those that are coercive and (2) those that are voluntaristic acts by individuals. The two are interrelated, but Chafetz initially theorizes about them as separate forces.

Coercive Bases of Gender Inequality. Gender stratification, Chafetz contends, is ultimately related to the macro-level division of labor in a society. If this division of labor is gendered—that is, work is defined and distributed based on a person's sex—males will typically receive more resources than women, and this material resource advantage will translate into differences in power between men and women at the micro, interpersonal level. The more males have a material and power advantage over their wives stemming from the gendered division of labor, the more men will use this power in their relations with wives, and as a result, the more likely are wives to defer to, and comply with, the demands of their husbands.[6] Thus, once there is a macro-level division of labor that favors men, this system gives men power advantages for interpersonal demands and, as a result, makes them less likely to contribute to family and domestic work. Wives thus become burdened with domestic chores, even when they work, which makes it increasingly difficult for them to compete with men for resource-generating work outside the home—a situation that, in turn, sustains the macro-level gendered division of labor.

When men have advantages in the macro-level division of labor, they are also more likely to be incumbents in those elite societal positions to which power resources accrue. The more males control these elite power positions, the more likely the distribution of opportunities in both power and work roles outside the home will favor men over women.[7] Once this situation exists, the attitudes and behaviors in work roles will continue to give men advantages, because these roles will be viewed as attributes favoring men over women. Indeed, the attributes of women will often be negatively evaluated, thereby perpetuating the advantage of men in competition with women for those positions generating material and power resources.

These definitional processes at the macro level filter down not only to micro-level interactions between men and women, especially between husbands and wives, but to virtually all inter-gender interactions as well. As men control material and power resources, while using this resource advantage to define and, hence, ideologically control the work situation, men can use this same definitional power to regulate micro encounters between men and women. If women accept their male interaction partners' definitions of reality, they are more likely to defer to men and to play gender-traditional domestic roles that, in turn, support the macro-level bias in the division of labor.

This process is exacerbated because men control elite positions in the broader society and can, therefore, perpetuate definitions of worth that favor men; these definitions typically lead to the devaluation of the work that women perform, inside and outside of the domestic sphere. For example, domestic family work of a wife goes unpaid and, hence, undervalued, whereas traditional work roles for women, such as secretary, are underpaid because they are not valued as highly as roles performed by men.

Gender social definitions become, in Chafetz's model, a critical link between macro- and meso-level coercive processes and voluntaristic processes that operate more typically at the personal, decision-making level. Chafetz distinguishes among three types of gender definitions: (1) *gender ideology* or beliefs about the basic and, typically, presumably biological differences in the natures of men and women; (2) *gender norms* or expectations about the appropriate and proper ways for men and women to behave; and (3) *gender stereotypes* or accentuation of the differences between men and women in how they will generally respond in situations. For Chafetz, the greater the level of consensus is among members of a population on these gender definitions and the more the gendered differences are presumed by individuals to be the way the sexes are, the more power these definitions have to influence both macro- and micro-level social processes sustaining gender inequality. In general, Chafetz argues, the gender ideology sets the constraints for gender norms that then contribute to gender stereotyping about the differences between men and women.

All these definitional processes legitimate the distribution of opportunities for men and women in the division of labor, in positions of power and authority, and in micro encounters between men and women inside and outside of the home. These interrelated forces will, in turn, influence what Chafetz terms *engenderment* or the processes whereby men and women, but especially women, accept "voluntarily" their position in the system of gender stratification.

Voluntaristic Bases of Gender Inequality. The more the economic division of labor, the distribution of incumbents in elite positions, and the cultural definitions of a society reveal a gender bias, the more likely members of the adult generation are to evidence gender differentiation in their work and home activities. As a result, adults become both role models and active socializing agents for engendering the next generation in their behaviors, expectations for their future, and definitions of what is real and appropriate. To the extent that this engenderment occurs via family and other socialization forces, individuals "voluntarily" act in ways sustaining macro-level division of labor and cultural definitions about differences between men and women, while reproducing these gender differences in micro-level encounters between men and women.

These basic ideas, which Chafetz presents as a systematic list of propositions and as a series of causal models, constitute the core of her argument about the maintenance of gender inequality and stratification. Table 18-3 presents these propositions in altered and highly abbreviated form.

As Chafetz emphasizes, however, once the forces maintaining a system are understood, the critical targets for change are also identified. Much of her approach involves an effort to use understanding of the maintenance forces to construct what she terms "gender system change."[8] From the theory, the obvious targets for changing a system of gender inequality are (1) the gender division of labor, (2) the resulting superior resource power of men, (3) the social definitions comprising gender ideologies, norms, and stereotypes, and (4) the engenderment processes that differentiate the orientations, expectations, and behaviors of men and women.

Changing Systems of Gender Inequality and Stratification

Many of the processes generating changes in gender stratification, whether long-term change in the distribution of resources or shorter-term oscillations in the opportunities for women, are external to the gender system itself. Forces such as technological

TABLE 18-3: Abbreviated Summary of Chafetz's Theory on the Maintenance of Gender Stratification

I. The maintenance of a system of gender stratification is a joint function of the following conditions:
   A. The degree to which males have coercive power that, in turn, successively increases when
      1. The macro-level division of labor is gendered.
      2. The gendered division of labor gives males material resource advantages.
      3. The material resource advantages of men are used in micro encounters with women to extract compliance that, in turn, will
         a. Burden women in domestic and reproductive chores.
         b. Decrease their ability to compete with men in the macro-level division of labor.
      4. The macro-level and micro-level resource advantages of men translate into their control of elite positions in the broader society, which, in turn, leads to
         a. Distribution of opportunities in work, political, and other spheres favoring men over women.
         b. Definitions of positive and negative attributes in work, political, and other spheres favoring men over women.
      5. The definitions of men's and women's attributes are
         a. Accepted consensually by all members of a society.
         b. Organized into an integrated and hierarchical system of distinctions between men and women comprising
            1. Gender ideologies.
            2. Gender norms.
            3. Gender stereotypes.
         c. Used to legitimate the distribution of opportunities in
            1. The division of labor.
            2. The system of power.
            3. The micro encounters between men and women.
   B. The degree to which the coercive advantages of males, enumerated in I-A, cause engenderment or the voluntaristic acceptance of gender distinctions, which, in turn, is additively increased by
      1. Gendered division of domestic and nondomestic labor by adults in family settings.
      2. Gendered adult role models in family settings.
      3. Gender-biased socialization by adults in family settings.

changes, demographic shifts in the age and composition of the population, changes in the structure of the economy, and geopolitical forces like war and migration can all exert pressures on the system of gender inequality. Chafetz conceptualizes these as "unintentional change processes," and her argument runs as follows:

Demographic Variables. Expansion of the working-age population will tend to decrease the opportunities for women, holding steady the number of actual work roles available in a society, whereas a decline in the working-age population will increase opportunities for women in the economic division of labor. Sex ratios also have much the same effects: As the sex ratio of men to women drops below parity, opportunities for women increase; if the reverse is true and the proportion of men relative to women increases, opportunities for women will decline.

Technological Variables. Changes in technology have important effects on gender stratification. Technology can alter, Chafetz argues, the strength requirements, the mobility required, and the capacity to work outside the home and its domestic responsibilities. The greater is the amount of change along these variables, the greater will be the effects of technology on gender stratification. In general, when technologies reduce the strength requirements of work outside the home (which, as Chafetz emphasizes, are typically exaggerated by males anyway), when less physical mobility is required, and when the obligations of child rearing and household can be overcome by technology, then opportunities for women will increase, thereby reducing the effects of the gendered division of labor on gender inequality. Conversely, when strength requirements, mobility, and domestic burdens are all high, or perceived to be so by virtue of gender definitions, then opportunities for women decline. Moreover, technological change that renders obsolete the skills of any gender, whether the male or female, will work against women because men will enjoy a competitive advantage in seeking new resource-generating work roles.

Economic Variables. Structural changes in the economy, typically driven by both demographic and technological transformations, will also influence gender stratification. An expanding economy will help women gain access to resource-generating work roles, as long as men are fully employed. The reverse is also true: A retracting economy will decrease opportunities for women. De-skilling of jobs is an important form of economic change, as technologies (particularly information systems but mechanical ones as well) are used to perform tasks that once required skill. Because of men's favored position of power in the division of labor, their jobs will be less likely to be de-skilled than those occupied by women, or at the very least, men's jobs will be de-skilled less rapidly.

Political Variables. Conflicts in the political arena can be external or internal. Chafetz argues that, regardless of the case, conflict will tend to strengthen traditional gender definitions. Internal conflict is especially likely to do so, and the more prolonged the conflict, the more likely this is to be the case. Even when women can assume traditional male roles during a conflict, as when men are gone to war, Chafetz argues that traditional gender definitions will still be strengthened and that the opportunities presented to women because the sex ratio has dropped below parity will be only temporary. When men return from war or, if a large number of men have been killed, women will be displaced as the pool of male labor is replaced and engendered.

These demographic, technological, economic, and political forces are, as Chafetz stresses, often unintended; they simply occur as the world economy and technologies on which it runs change, as the political and the international politics among nations are altered, and as the international migration of populations unfolds over time. Yet, much change in gender stratification is intentional, involving deliberate acts to alter the distribution of resources among men and women.

One source of intended change is, at first glance, not easily predicted: Elite males who control key positions actively seeking to change gender stratification. Chafetz argues that two conditions cause this kind of elite-initiated change to occur: (1) when elites perceive that gender inequality threatens their incumbency as elites or thwarts their plans for the society; and (2) when competing factions of elites need to recruit women to their side to prevail in a conflict. Under these conditions, elites will attempt to mobilize women's support in exchange for promises to ameliorate women's disadvantages in the division of labor and in the system of gender definitions.

Whether or not initiated by such top-down efforts by elites, women's mobilization to pursue their interests is influenced by other forces, especially industrialization, urbanization, and expansion of the middle classes. When these events occur, middle class women might begin to seek expanded opportunities outside their domestic responsibilities, and they are more likely to have the resources (material, educational, and symbolic-ideological) to pursue opportunities. As they do so, however, they confront the existing system of gender stratification, which creates role dilemmas. As women overcome these dilemmas, they acquire a sense of efficacy and begin to change their frames of reference in ways counter to dominant gender definitions. As women pursue interests outside the domestic sphere with this changed frame of reference, especially if they do so in a context where other women are also in proximity pursuing the same goals, they will experience an escalated sense of deprivation as they encounter the gendered division of labor and the distribution of power. For now, they interpret these obstacles in light of a sense that they deserve more than old gender definitions allowed, and their sense of deprivation increases relative to their dramatically escalated expectations for what is possible, or what should be possible.

When women experience these sentiments collectively and in proximity, they begin to form women's movement organizations to pursue the interests of women in eliminating or at least mitigating gender inequality. Even if these movements split into diverse factions, the ideological and political ferment created will begin to erode old gender definitions and to instill those of the more moderate organizations of the women's movement. As public support increases for these moderate definitions, more expansive efforts to change the gender-biased division of labor, the system of gender ideology, norms, and stereotypes, and the distribution of power can proceed. And, if elites begin to support these efforts, the women's movement will proceed more rapidly and successfully.

Yet, Chafetz emphasizes, successful mobilization of women inevitably generates a sense of threat among powerful interest groups in a society. The greater this sense of threat is, the more likely those threatened are to lobby against efforts of the women's movement. Such threat need not always come from threatened males, but can also come from women who are committed to traditional gender roles and definitions. Such efforts to thwart the goals of women's movements can be influenced by the internal conflicts within the movement and its ideologies. Internal conflict, per se, consumes resources that could be mobilized to pursue the goals of the movement. Equally important, conflict creates alienation and disaffection from some part of the movement, giving anti-feminist men and women potential allies.

The basic ideas from Chafetz's propositions on change are summarized in abbreviated form in Table 18-4.

These dynamics of conflict mobilization appear within a broader cultural, political, technological, economic, and demographic context. Thus, political conflict, whether internal and external to the nation, produces cultural conservatism that, in turn, strengthens traditional gender definitions in ways that can work against efforts by women to mobilize. If gender definitions are highly conservative and biased against women, mobilization will encounter resistance, regardless of whether or not the movement occurs within the context of political conflict. Economic and demographic forces, described earlier, can also work against a movement; if the economy contracts, if technologies deskill, or if migration alters the supply of male workers or sex ratios, these forces can work against women's efforts to change gender definitions and the division of labor, as well as the distribution of power.

CONCLUSIONS: EFFORTS AT THEORETICAL SYNTHESIS

In what is unfortunately a rare exercise, the theorists examined in this chapter assembled—along with Scott Coltrane,[9] who has also worked in the area of gender social

TABLE 18-4: Abbreviated Summary of Chafetz's Theory on Change in Gender Stratification

I. Change in systems of gender stratification is determined by the following conditions:
   A. Unintended change processes stemming from
      1. Demographic changes in
         a. Work-age population relative to number of available jobs.
         b. Sex ratio.
      2. Technological innovations that cause changes in
         a. Strength requirements (or perceptions about such requirements) for resource-generating roles.
         b. Physical mobility requirements for resource generating roles.
         c. Domestic requirements of women so that they can pursue nondomestic resource-generating roles.
      3. Structural economic changes in
         a. Rate of economic growth and number of resource-generating work roles.
         b. De-skilling of work roles performed by men and women.
      4. Political conflicts causing a change in
         a. Salience of traditional gender definitions.
         b. Demographic shifts in sex ratio.
   B. Intentional change processes stemming from
      1. Efforts by elites to support changes in gender inequality when
         a. They perceive that gender stratification threatens their position as elites.
         b. They believe that gender stratification thwarts their societal plans and goals.
         c. They find themselves in competition with other factions of elites and require women as allies.
      2. Efforts by women to mobilize and pursue their interests in breaking down gender stratification, with such mobilization being a positive and additive function of
         a. Industrialization as it increases nondomestic roles.
         b. Urbanization as it congregates and concentrates women.
         c. Expansion of middle classes as it expands the pool of less domestically burdened women.
         d. Sense of efficacy and escalated expectations experienced by women with nondomestic options.
         e. Sense of relative deprivation as escalated expectations confront barriers in existing system of gender stratification.
         f. Experience of deprivation collectively by women.
         g. Ideological ferment that change public gender definitions toward moderate positions.
         h. Support from societal elites.
   C. Unintentional and intentional change processes in a system of gender stratification threaten interest groups that mobilize counter-change movements, with such movements gaining power when
      1. A large proportion of women are engendered within traditional gender definitions.
      2. A large proportion of men-dominated roles are threatened.
      3. A high degree of internal conflict within the women's movement has alienated former supporters.

processes, and Jonathan Turner,[10] who has developed more general theories of stratification—to reconcile differences among their respective theories and develop a synthetic and general theory of gender stratification. The resulting models are too complex to summarize here, but the general intent of the approach is delineated in Figure 18-1, where the general classes or "blocks" of variables operating on gender stratification are labeled. Within each block in the figure, causal models were drawn for key internal dynamics, especially as these are influenced by variables in the other blocks of variables. The resulting set of models is highly complex, but fully comprehensive. Let us not delve into the details, which can be better mastered by reading the published version of this effort at

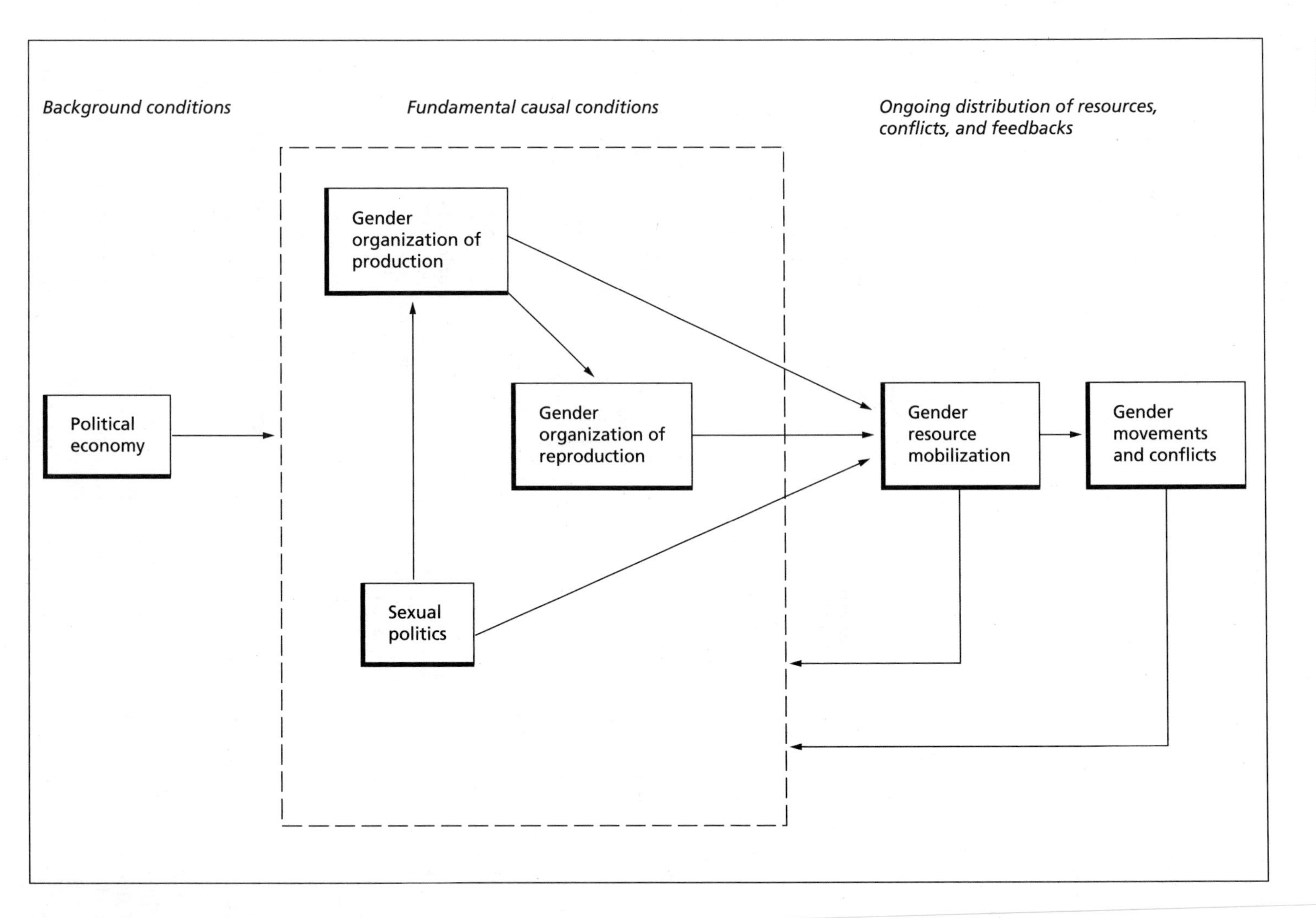

FIGURE 18-1 General Synthetic Model of Gender Stratification

theoretical synthesis,[11] but, rather, let us review the general nature of the variables in each block and the broad causal connections among blocks of variables.

The Effects of the Political Economy

At the far left of the figure is a block of variables labeled "political economy." This block includes such forces as the type and level of technology, the level of production, the concentration of power, the level of warfare with other societies, the degree of differentiation of kinship and household from production and politics, the level and degree of rigidity in stratification, and the degree and type of state formation.

Fundamental Causal Conditions

These political and economic variables operate as background forces on the fundamental causal conditions generating gender stratification, as outlined by the dotted lines containing the three blocks of variables in the middle portion of Figure 18-1: (1) the gender organization of production, (2) the gender organization of reproduction, and (3) sexual politics. Here is a brief summary of each:

(1) The gender organization block of variables includes the forces delineated in each of the three theories summarized in this chapter. One important variable is the degree of compatibility of productive with reproductive labor: Is economic production organized in ways that enable those involved in reproductive labor to participate in the economy? If so, to what degree? Another variable is the degree of segregation in work activities by gender, or the prominence of a gendered division of labor: Are economic roles played by men and women different? If so, to what degree? A third variable is the supply and demand for gender-linked labor: Is there high demand for labor associated with a particular gender, and is this demand high relative to the supply of individuals who are capable and willing to perform this labor? If so, what is the ratio of demand to supply and does this ratio provide opportunities for one gender to enter another's jobs? Another variable is demographic and concerns the rates of male or female migration in to and out of the society: Is the rate sufficiently great to affect the supply or demand for labor in general and gendered labor in particular? If so, by how much? A final variable in the gender organization of production block is the bargaining power of labor in general but gendered labor in particular: Does labor have bargaining rights and organization, and are those for men greater than those for women, or do these rights converge?

(2) The next fundamental block of variables relates to gender organization of reproduction. The division of labor involved in the production of goods and services is very much intertwined with the organization of reproductive labor (domestic chores, child-care, and child-socialization). There are several key variables in this block, including the level of technology for child birth and child-care that is available, the amount of time that must be spent in child-care by each sex, the birth rate as it increases or lowers the child-care burdens in family units, the death rate as it militates against high birth rates, the degree to which organizations or collective kinship patterns provide child-care options, the degree of class stratification of reproductive labor (as when lower-class individuals are used to care for the children in higher social classes). The values for these variables determine the way in which reproduction in a society is organized; in turn, the pattern of reproduction influences, and is influenced by, the gendered organization of production.

(3) The sexual politics block of variables denotes the way in which men and women are organized to exert influence and power over each other. One variable is the degree to

which all-male conflict groups exist in a society, giving males control of the coercive apparatus. Another is the degree to which males are able to define women as property. A third variable in this block is the extent to which sex becomes involved in alliance group politics, as when the sexes are exchanged in arranged marriages to form political alliances or when male elites solicit the support of women in factional conflicts with other male elites. A fourth variable is the extensiveness of erotic marketplaces where, for example, women can control the way in which erotic behavior is to proceed and the terms of sexual encounters between men and women.

Gender Resource Mobilization and Conflict

The previous three blocks of fundamental causal conditions are, of course, related to the nature of the political economy. Equally important, along with the effects of the political economy, these fundamental conditions have causal effects on the two blocks of variables on the right side of the model in Figure 18-1: (1) gender resource mobilization and (2) gender movements and conflict. Let us briefly discuss each of these:

(1) The capacity of women to mobilize resources to reduce the degree of gender stratification is related to a large number of variables whose values are influenced by key variables in the political economy block, the gender organization of production block, the gender organization of reproduction block, and the sexual politics block. One variable is the degree to which women can command measurable income that can give them leverage and resources to organize. A related variable is the degree to which women can control property and the ratio of their property relative to men's. Connected to this variable is the inheritance structure of the society: Does it allow women to inherit property, or does property generally pass through the male line? A fourth variable is the degree to which household organization or political forces allow women to have proximate allies to mitigate against the sexual violence that their husbands can vent on them. A fifth variable is the distinctiveness of the cultures (ideologies, norms, stereotypes, and other symbol systems) associated with men and women: How different are these cultures, and does the difference give men or women a consistent advantage in manipulating symbols in what spheres of activity? A related variable is the ranking of genders by worth, value, and other criteria: Are women given prestige and honor in cultural symbols that mark distinctions between men and women? If so, how much? A final variable is the ratio of men to women among the ranks of political, economic, and symbolic elites: Are women excluded or included, and, what is the ratio of their participation in what spheres of elite activity? These interconnected variables determine the level and types of resources that women can mobilize as they attempt to deal with a system of gender stratification. They also influence the nature and level of conflict that will emerge, as is explored below.

(2) Basically, the more resources women can mobilize, the more they can pursue conflict that will decrease the advantages males generally have in the political economy as these influence gendered production and reproduction, as well as sexual politics. This capacity to mobilize resources and pursue conflict is, however, influenced by some additional variables. One is the degree to which a counter-movement can be mobilized and the amount of success this movement can have in limiting gender mobilization and conflict. Another is the number of allies from other social movements that can be brought into gender conflict (for example, women can perhaps find allies in movements concerned with liberties for ethnic groups or certain classes of workers). The more the resources and organizations of these other social movements can be pulled into a

gender-oriented movement, the greater the level of mobilization will be and the more likely conflict is to be successful in changing background forces in the political economy and fundamental forces in the arenas of production, reproduction, and sexual politics.

This summary only describes the key variables. The actual theory will, of course, connect them in systematic ways. The propositions presented in Tables 18-1, 18-2, 18-3, and 18-4 provide some clues about the causal connections of those forces in gender inequality, stratification, and conflict. The summarized synthesis can provide even more causal detail—for those who are interested in the complex interconnections among the many forces involved in gender stratification. In closing, perhaps the key point is that the scientific theory of gender processes is now well developed, extending the conflict theoretical tradition in creative ways.

NOTES

1. Randall Collins, *Conflict Sociology: Toward an Explanatory Science* (New York: Academic, 1975), pp. 225–284 ; see also his "A Theory of Sexual Stratification," *Social Problems* 19 (1971), pp. 3–21.

2. For the longer lists of propositions from which these were extracted and reformulated, see both works cited in note 1.

3. Rae Lesser Blumberg, "A General Theory of Gender Stratification," *Sociological Theory* 2 (1984), pp. 23–101.

4. Janet Saltzman Chafetz, Gender Equity: An Integrated Theory of Stability and Change (Newbury Park, CA: Sage, 1990). See also her *Sex and Advantage: A Comparative Macro-Structural Theory of Sexual Stratification* (Totowa, NJ: Rowman and Allanheld, 1984).

5. Chafetz, *Gender Equity* (cited in note 4), p. 17.

6. Ibid., p. 48.

7. Ibid., p. 54.

8. Ibid., pp. 99–220.

9. See, for example, Scott Coltrane, "Father-Child Relationships and the Status of Women," *American Journal of Sociology* 93 (1988), pp. 1060–1095; "The Micropolitics of Gender in Nonindustrial Societies," *Gender and Society* 6 (1992), pp. 86–107; and *Family Man* (New York: Oxford University Press, 1996).

10. See, for example, Jonathan H. Turner, *Societal Stratification: A Theoretical Analysis* (New York: Columbia University Press, 1984).

11. For the fully developed theory, see Randall Collins, Janet Saltzman Chafetz, Rae Lesser Blumberg, Scott Coltrane, and Jonathan H. Turner, "Toward an Integrated Theory of Gender Stratification," *Sociological Perspectives* 36 (3, 1993), pp. 185–216.

PART IV

Exchange Theorizing

19

The Emerging Tradition: The Rise of Exchange Theorizing

From its very beginnings, sociological theory focused on the market forces that were transforming the modern social world. Adam Smith[1] was, of course, the first to formulate the "laws of supply and demand," but what is often forgotten is that Smith also presented to sociology one of the key questions that guided all nineteenth century sociological theory: What force or forces are to hold modern societies together as they differentiate and as actors pursue their narrow and specialized interests? His answer was a combination of moral and symbolic forces along with the "invisible hand of order" that comes when rational actors pursue their self interests in open and free markets.

Much of early sociology represented an effort to elaborate on this answer, or to formulate an alternative. In either case, social scientists of the nineteenth and twentieth centuries soon found themselves developing exchange-theoretic ideas.

EXCHANGE THEORY IN CLASSICAL ECONOMICS

Given Adam Smith's influence on sociology, and the impact of other British Isle thinkers, most of whom were Scottish, on nineteenth century social thought, it is perhaps best to begin with the ideas that come from this tradition. All these early classical economists considered themselves "moralists" and, hence, were concerned with broad ethical issues like justice, freedom, and fairness.[2] The label, utilitarianism, was meant to capture the broader moral concerns of these early moralists, but the term now tends to be associated with a narrow vision of their work, as is explored in Chapter 23 on how economic models of the marketplace have filtered into social theory. Today, these classical economic theorists portray humans as rational persons who seek to maximize their material benefits, or utility, from transactions or exchanges with others in a free and competitive marketplace. Free in the marketplace, people have access to necessary information; they can consider all available alternatives, and, on the basis of this consideration, rationally select the course of activity that will maximize material benefits. Entering into these rational considerations are calculations of the costs involved in pursuing various alternatives, with such costs being weighed against material benefits in an effort to determine which alternative will yield the maximum payoff or profit (benefits less costs). This view of the early utilitarians is narrow and ignores their concerns with morality, but this portion of their ideas endures today and inspires theory in not only sociology but economics and political science as well.

With the emergence of sociology as a self-conscious discipline, there was considerable borrowing, revision, and reaction to this conception of humans. In the end, many sociologists muted extreme utilitarian assumptions in the ways enumerated below:

1. Humans do not seek to maximize profits, as utilitarians argued, but they nonetheless attempt to make some profit in their social transactions with others.
2. Humans are not perfectly rational, but they do engage in calculations of costs and benefits in social transactions.
3. Humans do not have perfect information on all available alternatives, but they are usually aware of at least some alternatives, which form the basis for assessments of costs and benefits.
4. Humans always act under constraints, but they still compete with one another in seeking to make a profit in their transactions.
5. Humans always seek to make a profit in their transactions, but they are limited by the resources that they have when entering an exchange relation.

In addition to these alterations of utilitarian assumptions, exchange theory removes human interaction from the limitations of material transactions in an economic marketplace, requiring two more additions to the previous list:

6. Humans do engage in economic transactions in clearly defined marketplaces in all societies, but these transactions are only special cases of more general exchange relations occurring among individuals in virtually all social contexts.
7. Humans do pursue material goals in exchanges, but they also mobilize and exchange nonmaterial resources, such as sentiments, services, and symbols.

Aside from this revised substantive legacy, some forms of modern exchange theory have also adopted the strategy of the utilitarians for constructing social theory. In assuming humans to be rational, utilitarians argued that exchanges among people could also be studied by a rational science, one in which the "laws of human nature" would stand at the top of a deductive system of explanation. Thus utilitarians borrowed the early physical-science conception of theory as a logico-deductive system of axioms or laws and various layers of lower-order propositions that could be rationally deduced from the laws of "economic man." Most exchange theories are thus presented in a propositional format, as we will appreciate when more contemporary approaches are reviewed.

EXCHANGE THEORY IN ANTHROPOLOGY

Sir James Frazer

In 1919, Sir James Frazer's (1854–1941) second volume of *Folklore in the Old Testament* conducted what was probably the first explicit exchange-theoretic analysis of social institutions.[3] In examining a wide variety of kinship and marriage practices among primitive societies, Frazer was struck by the clear preference of Australian aborigines for cross-cousin over parallel-cousin marriages: "Why is the marriage of cross-cousins so often favored? Why is the marriage of ortho-cousins [that is, parallel cousins] so uniformly prohibited?"[4]

Although the substantive details of Frazer's descriptions of the aborigines' practices are fascinating in themselves (if only for their inaccuracy), the form of explanation marks his theoretical contribution. In a manner clearly indebted to utilitarian economics, Frazer launched an economic interpretation of the predominance of cross-cousin marriage patterns. In this explanation Frazer invoked the "law" of "economic motives": By having "no equivalent in property to give for a wife, an Australian aborigine is generally obliged to get her in exchange for a female relative, usually a sister or daughter."[5] Thus, the material or economic motives of individuals in society (lack of property and desire for a wife) explain various social patterns (cross-cousin marriages). Frazer went on to postulate that, once a particular pattern emanating from economic motives becomes established in a culture, it constrains other social patterns that can potentially emerge.

Frazer believed that the social and structural patterns that typify a particular culture reflect economic motives in humans, who, in exchanging commodities, attempt to satisfy their basic economic needs. Although Frazer's specific explanation was found to be sadly wanting by subsequent generations of anthropologists, especially Bronislaw Malinowski and Claude Lévi-Strauss, modern exchange theory in sociology invokes a similar conception of social organization:

1. Exchange processes are the result of efforts by people to realize basic needs.
2. When yielding payoffs for those involved, exchange processes lead to the patterning of interaction.
3. Such patterns of interaction not only serve the needs of individuals but also constrain the kinds of social structures that can subsequently emerge.

In addition to anticipating the general profile of modern explanations about how elementary exchange processes create more complex patterns in a society, Frazer's analysis also foreshadowed another concern of contemporary exchange theory: social systems' differentiation of privilege and power. Much as Karl Marx had done a generation earlier, Frazer noted that those who possess resources of high economic value can exploit those who have few such resources, thereby enabling the former to possess high privilege and presumably power. Hence the exchange of women among the aborigines was observed by Frazer to lead to the differentiation of power and privilege in at least two separate ways. First, "since among the Australian aboriginals women had a high economic and commercial value, a man who had many sisters or daughters was rich and a man who had none was poor and might be unable to procure a wife at all." Second, "the old men availed themselves of the system of exchange in order to procure a number of wives for themselves from among the young women, while the young men, having no women to give in exchange, were often obliged to remain single or to put up with the cast-off wives of their elders."[6] Thus, at least implicitly, Frazer supplemented the conflict theory contribution with a fourth exchange principle:

4. Exchange processes differentiate groups by their relative access to valued commodities,

resulting in differences in power, prestige, and privilege.

As provocative and seemingly seminal as Frazer's analysis appears, it had little direct impact on modern exchange theory. Rather, contemporary theory remains indebted to those in anthropology who reacted against Frazer's utilitarianism.

Bronislaw Malinowski and Nonmaterial Exchange

Despite Malinowski's close ties with Frazer, he developed an exchange perspective that radically altered the utilitarian slant of Frazer's analysis of cross-cousin marriage. Indeed, Frazer himself, in his preface to Malinowski's *Argonauts of the Western Pacific,* recognized the importance of Malinowski's contribution to the analysis of exchange relations.[7] In his now-famous ethnography of the Trobriand Islanders—a group of South Seas Island cultures—Malinowski observed an exchange system termed the *Kula Ring,* a closed circle of exchange relations among tribal peoples inhabiting a wide ring of islands. What was distinctive in this closed circle, Malinowski observed, was the predominance of exchange of two articles—armlets and necklaces—which the inhabitants constantly exchanged in opposite directions. Armlets traveling in one direction around the Kula ring were exchanged for necklaces moving in the opposite direction around the ring. In any particular exchange between individuals, then, an armlet would always be exchanged for a necklace.

In interpreting this unique exchange network, Malinowski distinguished material or economic from nonmaterial or symbolic exchanges. In contrast with the utilitarians and Frazer, who did not conceptualize nonmaterial exchange relations, Malinowski recognized that the Kula was not only an economic or material exchange network but also a symbolic exchange, cementing a web of social relationships: "One transaction does not finish the Kula relationship, the rule being 'once in the Kula, always in the Kula,' and a partnership between two men is a permanent and lifelong affair."[8] Although purely economic transactions did occur within the rules of the Kula, the ceremonial exchange of armlets and necklaces was observed by Malinowski to be the Kula's principal function.

The natives themselves, Malinowski emphasized, recognized the distinction between purely economic commodities and the symbolic significance of armlets and necklaces. However, to distinguish economic from symbolic commodities does not mean that the Trobriand Islanders failed to assign graded values to the symbolic commodities; indeed, they made gradations and used them to express and confirm the nature of the relationships among exchange partners as equals, superordinates, or subordinates. But, as Malinowski noted, "in all forms of [Kula] exchange in the Trobriands, there is not even a trace of gain, nor is there any reason for looking at it from the purely utilitarian and economic standpoint, since there is no enhancement of mutual utility through the exchange."[9] Rather, the motives behind the Kula were social psychological, for the exchanges in the ring were viewed by Malinowski to have implications for the needs of both individuals and society (recall from Chapter 2 that Malinowski was also a founder of functional theory). From his functionalist framework, he interpreted the Kula to mean "the fundamental impulse to display, to share, to bestow [and] the deep tendency to create social ties."[10] Malinowski, then, considered an enduring social pattern such as the Kula Ring to have positively functional consequences for satisfying individual psychological needs and societal needs for social integration and solidarity.

As Robert Merton and others were to emphasize (see Chapter 3), this form of functional analysis presents many logical difficulties. Nevertheless, Malinowski's analysis made several enduring contributions to modern exchange theory:

1. In Malinowski's words, "the meaning of the Kula will consist in being instrumental to dispel [the] conception of a rational being who wants nothing but to satisfy his simplest needs and does it according to the economic principle of least effort."[11]

2. Psychological rather than economic needs are the forces that initiate and sustain exchange relations and are therefore critical in the explanation of social behavior.

3. Exchange relations can also have implications beyond two parties, for, as the Kula demonstrates, complex patterns of indirect exchange can maintain extended and protracted social networks.

4. Symbolic exchange relations are the basic social process underlying both differentiation of ranks in a society and the integration of society into a cohesive and solidary whole.

With this emphasis, Malinowski helped free exchange theory from the limiting confines of utilitarianism. By stressing the importance of symbolic exchanges for both individual psychological processes and patterns of social integration, he anticipated the conceptual base for two basic types of exchange perspectives, one emphasizing the importance of psychological processes and the other stressing the significance of emergent cultural and structural forces on exchange relations.

Marcel Mauss and the Emergence of Exchange Structuralism

Reacting to what he perceived as Malinowski's tendency to overemphasize psychological instead of social needs, Marcel Mauss reinterpreted Malinowski's analysis of the Kula.[12] In this effort he formulated the broad outlines of a "collectivistic," or structural-exchange, perspective.[13] Mauss believed the critical question in examining an exchange network as complex as that of the Kula was, "In primitive or archaic types of societies, what is the principle whereby the gift received has to be repaid? What force is there in the thing which compels the recipient to make a return?"[14] The "force" compelling reciprocity was, Mauss believed, society or the group. As he noted, "It is groups, and not individuals, which carry on exchange, make contracts, and are bound by obligations."[15] The individuals actually engaged in an exchange represent the moral codes of the group. Exchange transactions among individuals are conducted in accordance with the rules of the group, thereby reinforcing these rules and codes. Thus, for Mauss, the over concern with individuals' self-interests by utilitarians and the overemphasis on psychological needs by Malinowski are replaced by a conception of individuals as representatives of social groups. In the end, exchange relations create, reinforce, and serve a group morality that is an entity *sui generis,* to borrow a famous phrase from Mauss's mentor, Émile Durkheim. Furthermore, in a vein similar to that of Frazer, once such a morality emerges and is reinforced by exchange activities, it regulates other activities in the social life of a group, beyond particular exchange transactions.

Mauss's work has received scant attention from sociologists, but he was the first to forge a reconciliation between the exchange principles of utilitarianism and the structural, or collectivistic, thought of Durkheim. In recognizing that exchange transactions give rise to and, at the same time, reinforce the normative structure of society, Mauss anticipated the structural position of some contemporary exchange theories. Mauss's influence on modern theory has been indirect, however. It is through Lévi-Strauss' structuralism that the French collectivist tradition of Durkheim and Mauss has influenced the exchange perspectives of contemporary sociological theory.

Claude Lévi-Strauss and Structuralism

In 1949, Lévi-Strauss launched an analysis of cross-cousin marriage in his classic work, *The Elementary Structures of Kinship*.[16] In restating Durkheim's objections to utilitarians, Lévi-Strauss took exception to Frazer's utilitarian interpretation of cross-cousin marriage patterns. And, similar to Mauss's opposition to Malinowski's emphasis on psychological needs, Lévi-Strauss developed a sophisticated structural-exchange perspective.

In rejecting Frazer's interpretation of cross-cousin marriage, Lévi-Strauss first questioned the substance of Frazer's utilitarian conceptualization. Frazer, he noted, "depicts the poor Australian aborigine wondering how he is going to obtain a wife since he has no material goods with which to purchase her, and discovering exchange as the solution to this apparently insoluble problem: 'men exchange their sisters in marriage because that was the cheapest way of getting a wife.'" In contrast, Lévi-Strauss emphasizes that "it is the exchange which counts and not the things exchanged." For Lévi-Strauss, exchange must be viewed by its functions for integrating the larger social structure. Lévi-Strauss then attacked Frazer's and the utilitarians' assumption that the first principles of social behavior are economic. Such an assumption contradicts the view that social structure is an emergent phenomenon that operates according to its own irreducible laws and principles.

Lévi-Strauss also rejected psychological interpretations of exchange processes, especially the position advocated by behaviorists (see later section). In contrast with psychological behaviorists, who see little real difference in the laws of behavior between animals and humans, Lévi-Strauss emphasized that humans possess a cultural heritage of norms and values that separates their behavior and societal organization from that of animal species. Human action is thus qualitatively different from animal behavior, especially in social exchange. Animals are not guided by values and rules that specify when, where, and how they are to carry out social transactions. Humans, however, carry with them into any exchange situation learned definitions of how they are to behave—thus assuring that the principles of human exchange will be distinctive.

Furthermore, exchange is more than the result of psychological needs, even those that have been acquired through socialization. Exchange cannot be understood solely through individual motives, because exchange relations are a reflection of patterns of social organization that exist as an entity, *sui generis*. Exchange behavior is thus regulated from without by norms and values, resulting in processes that can be analyzed only by their consequences, or functions, for these norms and values.

In arguing this view, Lévi-Strauss posited several fundamental exchange principles. First, all exchange relations involve costs for individuals, but, in contrast with economic or psychological explanations of exchange, such costs are attributed to society—to those customs, rules, laws, and values that require behaviors incurring costs. Yet individuals do not assign the costs to themselves, but to the "social order." Second, for all those scarce and valued resources in society—whether material objects, such as wives, or symbolic resources, like esteem and prestige—their distribution is regulated by norms and values. As long as resources are in abundant supply or are not highly valued in a society, their distribution goes unregulated, but, once they become scarce and highly valued, their distribution is soon regulated. Third, all exchange relations are governed by a norm of reciprocity, requiring those receiving valued resources to bestow on their benefactors other valued resources. In Lévi-Strauss's conception of reciprocity are various patterns of reciprocation specified by norms and values. In some situations, norms dictate "mutual" and direct rewarding of one's benefactor, whereas in other situations the reciprocity can be "univocal," involving diverse patterns of indirect exchange in which actors do not reciprocate directly but only through various third (fourth, fifth, and so forth) parties. Within these two general types of exchange reciprocity—mutual and univocal—numerous subtypes of exchange networks can be normatively regulated.

Lévi-Strauss believed that these three exchange principles offer a more useful set of concepts to describe cross-cousin marriage patterns, because these patterns can now be viewed by their functions for the larger social structure. Particular marriage patterns and other features of kinship organization no longer need be interpreted merely as direct exchanges among individuals but can be conceptualized as univocal exchanges between individuals and society. In freeing exchange from the analysis of only direct and mutual exchanges, Lévi-Strauss offered a tentative theory of societal integration and solidarity. His explanation extended Durkheim's provocative analysis and indicated how various subtypes of direct and univocal exchange both reflect and reinforce different patterns of societal integration and organization.

This theory of integration is, in itself, of theoretical importance, but it is more significant for our present purposes to stress Lévi-Strauss's impact on current sociological exchange perspectives. Two points of emphasis strongly influenced modern sociological theory.

1. Various forms of social structure, rather than individual motives, are the critical variables in the analysis of exchange relations.

2. Exchange relations in social systems are frequently not restricted to direct interaction among individuals but are protracted into complex networks of indirect exchange. On the one hand, these exchange processes are caused by patterns of social integration and organization; on the other hand, they promote diverse forms of such organization.

Lévi-Strauss's work represents the culmination of a reaction to economic utilitarianism as it was originally incorporated into anthropology by Frazer. Malinowski recognized the limitations of Frazer's analysis of only material or economic motives in direct exchange transactions. As the Kula Ring demonstrates, exchange can be generalized into protracted networks involving noneconomic motives that have implications for societal integration. Mauss drew explicit attention to the significance of social structure in regulating exchange processes and to the consequences of such processes for maintaining social structure. Finally, in this intellectual chain of events in anthropology, Lévi-Strauss began to indicate how different types of direct and indirect exchange are linked to different patterns of social organization. This intellectual heritage has influenced both the substance and the strategy of exchange theory in sociology, but it has done so only after considerable modification of assumptions and concepts by a particular strain of psychology: behaviorism.

PSYCHOLOGICAL BEHAVIORISM AND EXCHANGE THEORY

As a psychological perspective, behaviorism began from insights derived from observations of an accident. The Russian physiologist Ivan Petrovich Pavlov (1849–1936) discovered that experimental dogs associated food with the person bringing it.[17] He observed, for instance, that dogs on whom he was performing secretory experiments would salivate not only when presented with food but also when they heard their feeder's footsteps approaching. After considerable delay and personal agonizing, Pavlov undertook a series of experiments on animals to understand such "conditioned responses."[18] From these experiments he developed several principles that were later incorporated into behaviorism. These include the following: (1) A stimulus consistently associated with another stimulus producing a given physiological response will, by itself, elicit that response. (2) Such conditioned responses can be extinguished when gratifications associated with stimuli are no longer forthcoming. (3) Stimuli that are similar to those producing a conditioned response can also elicit the same response as the original stimulus. (4) Stimuli that increasingly differ from those used to condition a particular response will decreasingly be able to elicit this response. Thus, Pavlov's experiments exposed the principles of conditioned responses, extinction, response generalization, and response discrimination. Although Pavlov clearly recognized the significance of these findings for human behavior, his insights came to fruition in America under the tutelage of Edward Lee Thorndike and John B. Watson—the founders of behaviorism.

Thorndike conducted the first laboratory experiments on animals in America. During these experiments, he observed that animals would retain response patterns for which they were rewarded.[19] For example, in experiments on kittens placed in a puzzle box, Thorndike found that the kittens would engage in trial-and-error behavior until emitting the response that allowed them to escape. And, with each placement in the box, the kittens would engage in less trial-and-error behavior, indicating that the gratifications associated with a response allowing the kittens to escape caused them to learn and retain this response. From these and other studies, which were conducted at the same time as Pavlov's, Thorndike formulated three principles or laws: (1) the "law of effect," which holds that acts in a situation producing gratification will be more likely to occur in the future when that situation recurs; (2) the "law of use," which states that the situation-response connection is strengthened with repetitions and practice; and (3) the "law of disuse," which argues that the connection will weaken when practice is discontinued.[20]

These laws converge with those presented by Pavlov, but there is one important difference. Thorndike's experiments involved animals engaged in free trial-and-error behavior, whereas Pavlov's work was on the conditioning of physiological—typically glandular—responses in a tightly controlled laboratory situation. Thorndike's work could thus be seen as more directly relevant to human behavior in natural settings.

Watson was only one of several thinkers to recognize the significance of Pavlov's and Thorndike's work, but he soon became the dominant advocate of what was becoming explicitly known as behaviorism.[21] Watson's opening shot for the new science of behavior was fired in an article entitled "Psychology as the Behaviorist Views It":

> Psychology as the behaviorist views it is a purely objective experimental branch of natural science. Its theoretical goal is the prediction and control of behavior. Introspection forms no essential part of its methods, nor is the scientific value of its data dependent upon the readiness with which they lend themselves to interpretation in terms of consciousness. The behaviorist, in efforts to get a unitary scheme of animal response, recognizes no dividing line between man and brute.[22]

Watson thus became the advocate of the extreme behaviorism against which many vehemently reacted.[23] For Watson, psychology is the study of stimulus-response relations, and the only admissible evidence is overt behavior. Psychologists are to stay out of the "Pandora's box" of human consciousness and to study only observable behaviors as they are connected to observable stimuli.[24]

In many ways, behaviorism is similar to utilitarianism, because it operates on the principle that humans are reward-seeking organisms pursuing alternatives that will yield the most reward and the least punishment. Rewards are simply another way of phrasing the economist's concept of "utility," and "punishment" is somewhat equivalent to the notion of "cost." For the behaviorist, reward is any behavior that reinforces or meets the needs of the organism, whereas punishment denies rewards or forces the expenditure of energy to avoid pain (thereby incurring costs).

Modern exchange theories have borrowed the notion of reward from behaviorists and used it to reinterpret the utilitarian exchange heritage. In place of utility, the concept of reward has often been inserted, primarily because it allows exchange theorists to view behavior as motivated by psychological needs. However, the utilitarian concept of cost appears to have been retained in preference to the behaviorist's formulation of punishment, because the notion of cost allows exchange theorists to visualize more completely the alternative rewards that organisms forego in seeking to achieve a particular reward.

Despite these modifications of the basic concepts of behaviorism, its key theoretical generalizations have been incorporated with relatively little change into some forms of sociological exchange theory:

1. In any given situation, organisms will emit those behaviors that will yield the most reward and the least punishment.
2. Organisms will repeat those behaviors that have proved rewarding in the past.
3. Organisms will repeat behaviors in situations that are similar to those in the past in which behaviors were rewarded.
4. Present stimuli that on past occasions have been associated with rewards will evoke behaviors similar to those emitted in the past.
5. Repetition of behaviors will occur only as long as they continue to yield rewards.
6. An organism will display emotion if a behavior that has previously been rewarded in the same or similar situation suddenly goes unrewarded.
7. The more an organism receives rewards from a particular behavior, the less rewarding that behavior becomes (because of satiation) and the more likely the organism is to emit alternative behaviors in search of other rewards.

These principles were discovered in laboratory situations where experimenters typically manipulated the environment of the organism; so, it is difficult to visualize the experimental situation as interaction. The experimenter's tight control of the situation precludes the possibility that the animal will affect significantly the responses of the experimenter. This has forced modern exchange theories using behaviorist principles to incorporate the utilitarian's concern with transactions, or exchanges. In this way humans can be seen as mutually affecting one another's opportunities for rewards. In contrast with animals in a Skinner box or some similar laboratory situation, humans exchange rewards. Each person represents a potentially rewarding stimulus situation for the other.

As sociological exchange theorists have attempted to apply behaviorist principles to the study of human behavior, they have inevitably confronted the problem of the black box: Humans differ from laboratory animals in their greater ability to engage in a wide variety of complex cognitive processes. Indeed, as the utilitarians were the first to emphasize, what is distinctly human is the capacity to abstract, to calculate, to project outcomes, to weigh alternatives, and to perform a wide number of other cognitive manipulations. Furthermore, in borrowing behaviorists' concepts, contemporary exchange theorists have also had to introduce the concepts of an introspective psychology and structural sociology. Humans not only think in complex ways; their thinking is emotional and circumscribed by many social and cultural forces (first incorporated into the exchange theories of Mauss and Lévi-Strauss). Once it is recognized that behaviorist principles must incorporate concepts denoting both internal psychological processes and constraints of social structure and culture, it is also necessary to visualize exchange as frequently transcending the mutually rewarding activities of individuals in direct interaction. The organization of behavior by social structure and culture, coupled with humans' complex cognitive abilities, allows protracted and indirect exchange networks to exist.

When we review the impact of behaviorism on some forms of contemporary exchange theory, the vocabulary and general principles of behaviorism are clearly evident, but concepts have been redefined and the principles altered to incorporate the insights of the early utilitarians as well as the anthropological reaction to utilitarianism. The end result has been for proponents of an exchange perspective employing behaviorist concepts and principles to abandon much of what made behaviorism a unique perspective as they have dealt with the complexities introduced by human cognitive capacities and their organization into sociocultural groupings.

THE SOCIOLOGICAL TRADITION AND EXCHANGE THEORY

The vocabulary of exchange theory clearly comes from utilitarianism and behaviorism. Anthropological work forced the recognition that cultural and social dynamics need to be incorporated into exchange theory. When we look at early sociological work, however, the impact of early sociological theorists on modern exchange theory is difficult to assess for several reasons. First, much sociological theory represented a reaction against utilitarianism and extreme behaviorism and, therefore, has been reluctant to incorporate concepts from these fields. Second, the most developed of the early exchange theories—that provided by Georg Simmel in his *The Philosophy of Money*—was not translated into English until the 1970s.[25] (German-reading theorists, such as Peter Blau and Talcott Parsons, were to some degree influenced by Simmel's ideas.) Third, the topics of most interest to many sociological exchange theorists—differentiations of power and conflict in exchanges—have more typically been conceptualized as conflict theory than as exchange theory. But, as will become evident, sociological theories of exchange converge with those on conflict processes, and Marx's and Weber's ideas exerted considerable influence on sociologically oriented exchange theories.

Marx's Theory of Exchange and Conflict

Most contemporary theories of exchange examine situations where actors have unequal levels of resources with which to bargain. Those with valued resources are in a position to strike a better bargain, especially if others who value their resources do not possess equally valued resources to offer in exchange. This fact of social life is the situation described in Marx's conflict theory.[26] Capitalists have the power to control the distribution of material rewards, whereas all that workers have is their labor to offer in exchange. Although labor is valued by the capitalist, it is in plentiful supply, and thus no one worker is in a position to bargain effectively with an employer. As a consequence, capitalists can get labor at a low cost and can force workers to do what they want. As capitalists press their advantage, they create the very conditions that allow workers to develop resources—political, organizational, ideological—that workers can then use to strike a better bargain with capitalists and, in the end, to overthrow them.

Granted, this is simplifying Marx's implicit exchange theory, but the point is clear: Dialectical conflict theory is a variety of exchange theory. Let us list some of these exchange dynamics more explicitly:

1. Those who need scarce and valued resources that others possess but who do not have equally valued and scarce resources to offer in return will be dependent on those who control these resources.
2. Those who control valued resources have power over those who do not. That is, the power of one actor over another is directly related to (a) the capacity of one actor to monopolize the valued resources needed by other actors and (b) the inability of those actors who need these resources to offer equally valued and scarce resources in return.
3. Those with power will press their advantage and will try to extract more resources from those dependent on them in exchange for fewer (or the same level) of the resources that they control.
4. Those who press their advantage in this way will create conditions that encourage those who are dependent on them to (a) organize in ways that increase the value of their resources and, failing this, to (b) organize in ways that enable them to coerce those on whom they are dependent.

If the words *capitalist* and *proletarian* are inserted at the appropriate places in the previous list, Marx's exchange model becomes readily apparent. Dialectical conflict theory is thus a series of propositions about exchange dynamics in systems in which the distribution of resources is unequal. And, as will become evident in the next chapters, sociological exchange theories have emphasized these dynamics that inhere in the unequal distribution of resources. Such is Marx's major contribution to exchange theory.

Georg Simmel's Exchange Theory

In Simmel's *The Philosophy of Money*[27] is a critique of Marx's "value theory of labor"[28] and, in its place, a clear exposition of exchange theory. *The Philosophy of Money* is, as its title indicates, about the impact of money on social relations and social structure. For Simmel, social exchange involves the following elements:

1. The desire for a valued object that one does not have.
2. The possession of the valued object by an identifiable other.

TABLE19-1 Georg Simmel's Exchange Principles

I. *Attraction Principle:* The more actors perceive as valuable one another's respective resources, the more likely an exchange relationship is to develop among these actors.

II. *Value Principle:* The greater is the intensity of an actor's needs for a resource of a given type, and the less available is that resource, the greater is the value is of that resource to the actor.

II. *Power Principles:*

A. The more an actor perceives as valuable the resources of another actor, the greater is the power of the latter over the former.

B. The more liquid are an actor's resources, the greater will be the exchange options and alternatives and, hence, the greater will be the power of that actor in social exchanges.

IV. *Tension Principle:* The more actors in a social exchange manipulate the situation in an effort to misrepresent their needs for a resource or conceal the availability of resources, the greater is the level of tension in that exchange and the greater is the potential for conflict.

3. The offer of an object of value to secure the desired object from another.
4. The acceptance of this offer by the possessor of the valued object.[29]

Contained in this portrayal of social exchange are several additional points that Simmel emphasized. First, value is idiosyncratic and is, ultimately, tied to an individual's impulses and needs. Of course, what is defined as valuable is typically circumscribed by cultural and social patterns, but how valuable an object is will be a positive function of (a) the intensity of a person's needs and (b) the scarcity of the object. Second, much exchange involves efforts to manipulate situations so that the intensity of needs for an object is concealed and the availability of an object is made to seem less than what it actually is. Inherent in exchange, therefore, is a basic tension that can often erupt into other social forms, such as conflict. Third, to possess an object is to lessen its value and to increase the value of objects that one does not possess. Fourth, exchanges will occur only if both parties perceive that the object given is less valuable than the one received. Fifth, collective units as well as individuals participate in exchange relations and hence are subject to the four processes listed. Sixth, the more liquid the resources of an actor are in an exchange—that is, the more that resources can be used in many types of exchanges—the greater that actor's options and power will be. For if an actor is not bound to exchange with any other and can readily withdraw resources and exchange them with another, then that actor has considerable power to manipulate any exchange.

Economic exchange involving money is only one case of this more general social form, but it is a very special case. When money becomes the predominant means for establishing value in relationships, the properties and dynamics of social relations are transformed. This process of displacing other criteria of value, such as logic, ethics, and aesthetics, with a monetary criterion is precisely the long-term evolutionary trend in societies. This trend is both a cause and effect of money as the medium of exchange. Money emerged to facilitate exchanges and to realize even more completely humans' basic needs. But, once established, money has the power to transform the structure of social relations in society.

Thus, the key insight in *The Philosophy of Money* is that the use of different criteria for assessing value has an enormous impact on the form of social relations. As money replaces barter and other criteria for determining values, social relations are fundamentally changed. Yet they are transformed in accordance with some basic principles of social exchange, which are never codified by Simmel but are very clear. In Table 19-1, these ideas are summarized as abstract exchange principles.

CONCLUSION: EXCHANGE THEORY IN THE CONTEMPORARY ERA

Curiously, despite Adam Smith's influence on sociological theory in the nineteenth century, and behaviorists' impact on early social psychology, a clear sociological approach to exchange theory did not emerge until the 1960s. When it finally arrived, this approach has remained prominent within the sociological canon since the midcentury, and today, it is one of the most important perspectives within sociological theorizing. In the next three chapters on the maturing tradition during the 1960s and 1970s, we will explore how economic and behaviorist ideas were brought back into sociological theory and blended with the discipline's concern with social structure, power, and inequality. Then, we can explore the two surviving variants of this midcentury burst of creative activity, primarily rational choice theories (Chapter 24) and the exchange network approaches (Chapter 25).

NOTES

1. Adam Smith, *An Inquiry into the Nature and Causes of the Wealth of Nations* (London: Davis, 1805; originally published in 1776).
2. Charles Camic, "The Utilitarians Revisited," *American Journal of Sociology* 85 (1979), pp. 516–550.
3. Sir James George Frazer, *Folklore in the Old Testament*, vol. 2 (New York: Macmillan, 1919); see also his *Totemism and Exogamy: A Treatise on Certain Early Forms of Superstition and Society* (London: Dawsons of Pall Mall, 1968; originally published in 1910); and his Preface to Bronislaw Malinowski's *Argonauts of the Western Pacific* (London: Routledge & Kegan Paul, 1922), pp. vii–xiv.
4. Frazer, *Folklore* (cited in note 3), p. 199.
5. Ibid., p. 198.
6. Ibid., pp. 200–201 for this and immediately preceding quote.
7. Bronislaw Malinowski, *Argonauts of the Western Pacific* (London: Routledge & Kegan Paul, 1922), p. 81.
8. Ibid., pp. 82–83.
9. Ibid., p. 175.
10. Ibid.
11. Ibid., p. 516.
12. Marcel Mauss, *The Gift*, trans. I. Cunnison (New York: Free Press, 1954; originally published as *Essai sur le don en sociologie et anthropologie* [Paris: Presses universitaires de France, 1925]). It should be noted that Mauss rather consistently misinterpreted Malinowski's ethnography, but through such misinterpretation he came to visualize a "structural" alternative to "psychological" exchange theories.
13. In Peter Ekeh's excellent discussion of Mauss and Lévi-Strauss, *Social Exchange Theory and the Two Sociological Traditions* (Cambridge, MA: Harvard University Press, 1975), pp. 55-122, the term *collectivist* is used in preference to *structural* and is posited as the alternative to *individualistic* or psychological exchange perspectives. I prefer the terms *structural* and *psychological*; thus, although I am indebted to Ekeh's discussion, these terms will be used to make essentially the same distinction. My preference for these terms will become more evident in subsequent chapters, since, in contrast with Ekeh's analysis, I consider Peter M. Blau and George C. Homans to have developed, respectively, structural and psychological theories. Ekeh considers the theories of both Blau and Homans to be individualistic, or psychological.
14. Mauss, *The Gift* (cited in note 12), p. 1.
15. Ibid., p. 3.

16. Claude Lévi-Strauss, *The Elementary Structures of Kinship* (Boston: Beacon, 1969). This is a translation of Lévi-Strauss's 1967 revision of the original *Les structures élémentaires de la parenté* (Paris: Presses universitaires de France, 1949).

17. See, for relevant articles, lectures, and references, I.P. Pavlov, *Selected Works*, ed. K.S. Kostoyants, trans. S. Belsky (Moscow: Foreign Languages Publishing House, 1955); and *Lectures on Conditioned Reflexes*, 3rd ed., trans. W.H. Grant (New York: International, 1928).

18. I.P. Pavlov, "Autobiography," in *Selected Works* (cited in note 17), pp. 41–44.

19. Edward L. Thorndike, "Animal Intelligence: An Experimental Study of the Associative Processes in Animals," *Psychological Review Monograph*, Supplement 2 (1989).

20. See Edward L. Thorndike, *The Elements of Psychology* (New York: Seiler, 1905), *The Fundamentals of Learning* (New York: Teachers College Press, 1932), and *The Psychology of Wants, Interests, and Attitudes* (New York: D. Appleton, 1935).

21. Others who recognized their importance included Max F. Meyer, *Psychology of the Other-One* (Columbus, OH: Missouri Book, 1921); and Albert P. Weiss, *A Theoretical Basis of Human Behavior* (Columbus, OH: Adams, 1925).

22. J.B. Watson, "Psychology as the Behaviorist Views It," *Psychological Review* 20 (1913), pp. 158–177. For other basic works by Watson, see *Psychology from the Standpoint of a Behaviorist*, 3rd ed. (Philadelphia: Lippincott, 1929); *Behavior: An Introduction to Comparative Psychology* (New York: Henry Holt, 1914).

23. For example, in *Mind, Self, and Society* (Chicago: University of Chicago Press, 1934), Mead has eighteen references to Watson's work.

24. For a more detailed discussion of the emergence of behaviorism, see Jonathan H. Turner, Leonard Beeghley, and Charles Powers, *The Emergence of Sociological Theory* (Belmont, CA: Wadsworth, 1998).

25. Georg Simmel, *The Philosophy of Money*, trans. T. Bottomore and D. Frisby (Boston: Routledge & Kegan Paul, 1978; originally published in 1907).

26. Karl Marx and Frederick Engels, *The Communist Manifesto* (New York: International, 1971; originally published 1848); Karl Marx, *Capital: A Critical Analysis of Capitalist Production*, vol. 1 (New York: International, 1967; originally published in 1867).

27. See note 25.

28. Marx, *Capital* (cited in note 26).

29. Simmel, *The Philosophy of Money* (cited in note 25), pp. 85–88.

20

The Maturing Tradition I:

George C. Homans' Behavioristic Approach

In the early 1960s, exchange theory emerged as a distinctive perspective in sociology. Suddenly, some of America's most prominent theorists—George C. Homans, Peter M. Blau, James S. Coleman, and Richard Emerson, for example—were exploring the social universe using ideas borrowed from utilitarian economics and psychological behaviorism. For all these thinkers, a fundamental property of the social universe is the exchange of resources among actors, driven by needs to secure rewards or utilities in their relations with others. In this first chapter on the maturing exchange tradition, we will focus on the work of Homans who, perhaps more than any other theorist of his time, forced sociological theory to become attuned to the process of exchange. In subsequent chapters on the maturing exchange tradition, we will examine the important work of Blau, Emerson, and more generally, economic theorizing as it carried over to the contemporary period. The work of all these scholars in the 1960s and 1970s has made exchange theory one of the most prominent and productive branches of sociological theorizing.

THE BASIC APPROACH

Homans' exchange scheme first surfaced in a polemical reaction to Claude Lévi-Strauss' structural analysis of cross-cousin marriage patterns. In collaboration with David Schneider, Homans previewed what become prominent themes in his writings: (1) a skeptical view of any form of functional theorizing, (2) an emphasis on psychological principles as the axioms of social theory, and (3) a preoccupation with exchange-theoretic concepts.[1]

In their assessment of Lévi-Strauss's exchange functionalism, Homans and Schneider took exception to virtually all that made Lévi-Strauss's theory. First, they rejected the conceptualization of different forms of indirect, generalized exchange. In so conceptualizing exchange, Lévi-Strauss "thinned the meaning out of it." Second, Lévi-Strauss' position that different forms of exchange symbolically reaffirm and integrate different patterns of social organization was questioned, for an "institution is what it is because it results from the drives, or meets the immediate needs, of individuals or subgroups within a society."[2] The result of this rejection of Lévi-Strauss's thought was that Homans and Schneider argued that exchange theory must initially emphasize face-to-face interaction, focus primarily on limited and direct exchanges among individuals, and recognize that social structures are created and sustained by the behaviors of individuals.

With this critique of the anthropological tradition, Homans resurrected the utilitarian's concern with individual self-interest in the conceptual trappings of psychological behaviorism. Indeed, as Homans and Schneider emphasized. "We may call this an individual self-interest theory, if we remember that interests may be other than economic."[3] As became evident by the early 1960s, this self-interest theory is to be cast in the behaviorist language of B.F. Skinner. Given Homans' commitment to axiomatic theorizing and his concern with face-to-face interaction among individuals, it was perhaps inevitable that he would look toward Skinner and, indirectly, to the early founders of behaviorism—I.P. Pavlov, Edward Lee Thorndike, and J.B. Watson. But Homans borrowed directly from Skinner's reformulations of early behaviorist principles.[4] Stripped of its subtlety, Skinnerian behaviorism states as its basic principle that, if an animal has a need, it will perform activities that in the past have satisfied this need. A first corollary to this principle is that organisms will attempt to avoid unpleasant experiences but will endure limited amounts of such experiences as a cost in emitting the behaviors that satisfy an overriding need. A second corollary is that organisms will continue emitting certain behaviors only as long as they continue to produce desired and expected effects. A third corollary of Skinnerian psychology emphasizes that, as needs are satisfied by a particular behavior, animals are less likely to emit the behavior. A fourth corollary states that, if in the recent past a behavior has brought rewards and if these rewards suddenly stop, the organism will appear angry and gradually cease emitting the behavior that formerly satisfied its needs. A final corollary holds that, if an event has consistently occurred at the same time as a behavior that was rewarded or punished, the event becomes a stimulus and is likely to produce the behavior or its avoidance.

These principles were derived from behavioral psychologists' highly controlled observations of animals, whose needs could be inferred from deprivations imposed by the investigators. Although human needs are much more difficult to ascertain than those of laboratory pigeons and mice and, despite the fact that humans interact in groupings that defy experimental controls, Homans believed that the principles of operant psychology could be applied to the explanation of human behavior in both simple and complex groupings. One of the most important adjustments of Skinnerian principles to fit the facts of human social organization involved the recognition that needs are satisfied by other people and that people reward and punish one another. In contrast with Skinner's animals, which only indirectly interact with Skinner through the apparatus of the laboratory and which have little ability to reward Skinner (except perhaps to confirm his principles), humans constantly give and take, or exchange, rewards and punishments.

The conceptualization of human behavior as exchange of rewards (and punishments) among interacting individuals led Homans to incorporate, in altered form, the first principle of elementary economics: Humans rationally calculate the long-range consequences of their actions in a marketplace and attempt to maximize their material profits in their transactions. Homans qualified this simplistic notion, however:

> Indeed we are out to rehabilitate the economic man. The trouble with him was not that he was economic, that he used resources to some advantage, but that he was antisocial and materialistic, interested only in money and material goods and ready to sacrifice even his old mother to get them.[5]

Thus, to be an appropriate explanation of human behavior, this basic economic assumption must be altered in four ways: (1) People do not always attempt to maximize profits; they seek only to make some profit in exchange relations. (2) Humans do not usually make either long-run or rational calculations in exchanges, for, in everyday life, "the Theory of Games is good advice for human behavior but a poor description of it." (3) The things exchanged involve not only money but also other commodities, including approval, esteem, compliance, love, affection, and other less materialistic goods. (4) The marketplace is not a separate domain in human exchanges, for all interaction involves individuals exchanging rewards (and punishments) and seeking profits.

THE BASIC EXCHANGE PRINCIPLES

In Propositions I through III of Table 20-1, Homansian principles of Skinnerian psychology are restated. The more valuable an activity (III), the more often such activity is rewarded (I); also, the more a situation approximates one in which activity has been rewarded in the past (II), the more likely a particular activity will be emitted. Proposition IV indicates the condition under which the first three propositions fall into temporary abeyance. In accordance with the reinforcement principle of satiation or the economic law of marginal utility, humans eventually define rewarded activities as less valuable and begin to emit other activities in search of different rewards (again, however, in accordance with the principles enumerated in Propositions I through III). Proposition V introduces a more complicated set of conditions that qualify Propositions I through IV. From Skinner's observation that pigeons reveal anger and frustration when they do not receive an expected reward, Homans reasoned that humans will probably reveal the same behavior.

In addition to these principles, Homans introduces a "rationality proposition," which summarizes the stimulus, success, and value propositions. This proposition is also placed in Table 20-1 because it is so prominent in Homans' actual construction of deductive explanations. To translate the somewhat awkward vocabulary of Principle VI as Homans wrote it: People make calculations about various alternative lines of action. They perceive or calculate the value of the rewards that might be yielded by various actions. But they also temper this calculation through perceptions of how probable the receipt of rewards will be. Low probability of receiving highly valued rewards would lower their reward potential. Conversely, high probability of receiving a lower valued reward increases their overall reward potential. This relationship can be stated by the following formula:

$$\text{Action} = \text{Value} \times \text{Probability}$$

People are, Homans asserted, rational in the sense that they are likely to emit that behavior, or action, among alternatives in which value on the right side of the equation is largest. For example, if $Action_1$ is highly valued (say, 10) but the probability of getting it by emitting $Action_1$ is low (.20) and if $Action_2$ is less valued (say, 5) but the

TABLE 20-1 Homans' Exchange Propositions

I. *Success Proposition:* For all actions taken by persons, the more often a particular action of a person is rewarded, the more likely the person is to perform that action.

II. *Stimulus Proposition:* If in the past the occurrence of a particular stimulus or set of stimuli has been the occasion on which a person's action has been rewarded, then, the more similar the present stimuli are to the past ones, the more likely the person is to perform the action or some similar action now.

III. *Value Proposition:* The more valuable to a person the result of his or her action is, the more likely he or she is to perform the action.

IV. *Deprivation/Satiation Proposition:* The more often in the recent past a person has received a particular reward, the less valuable any further unit of that reward becomes for that person.

V. *Aggression/Approval Propositions:*

A. When a person's action does not receive the reward expected or receives punishment that was not expected, he or she will be angry and become more likely to perform aggressive behavior. The results of such behavior become more valuable to that person.

B. When a person's action receives the reward expected, especially greater reward than expected, or does not receive punishment expected, he or she will be pleased and become more likely to perform approving behavior. The results of such behavior become more valuable to that person.

VI. *Rationality Proposition:* In choosing between alternative actions, a person will choose that one for which, as perceived by him or her at the time, the value of the result, multiplied by the probability of getting that result, is greater.

probability of receiving it is greater (.50) than *Action*$_1$, then the actor will emit *Action*$_2$ (because $10 \times .20 = 2$ yields less reward than $5 \times .50 = 2.5$).

As summarized in Table 20-1, Homans believed these basic principles or laws explain, in the sense of deductive explanation, patterns of human organization. As is obvious, they are psychological in nature. What is more, these psychological axioms constitute from Homans' viewpoint the only general sociological propositions, because "there are no general sociological propositions that hold good of all societies or social groups as such." The fact that psychological propositions are the most general, however, does not make any less relevant or important the sociological propositions stating the relationships among group properties or between group properties and those of individuals. On the contrary, these are the very propositions that are to be deduced from the psychological axioms. Thus, sociological propositions will be conspicuous in the deductive system emanating from the psychological principles.

USING THE BASIC PRINCIPLES TO CONSTRUCT EXPLANATIONS

Homans himself never presented a well-developed deductive explanation, even though he championed their development.[6] He tended to simply invoke, in a rather ad hoc fashion, his axioms and to reconcile them in a very loose and imprecise manner with empirical regularities. Or he constructed brief deductive schemes to illustrate his strategy. One of Homans' explanations is reproduced later. Homans recognized that this is not a complete explanation, but he argued that it is as good as any other that exists. Moreover, although certain steps in the deductive scheme are omitted, it is the proper way to develop scientific explanations from his point of view. In this scheme, Homans sought to explain Golden's Law that industrialization and the level of literacy in the population are highly correlated. Such empirical generalizations are often considered to be theory, but, as a theorist, Homans correctly perceived that this "law" is only an empirical regularity that belongs at the bottom of the deductive scheme. Propositions move from the most abstract statement, or the axiom(s),[7] to the specific empirical regularity to be explained. Golden's Law is thus explained in the following manner:

1. Individuals are more likely to perform an activity the more valuable they perceive the reward of that activity to be.
2. Individuals are more likely to perform an activity the more successful they perceive the activity to be in getting that reward.[8]
3. Compared with agricultural societies, a higher proportion of individuals in industrial societies are prepared to reward activities that involve literacy. (Industrialists want to hire bookkeepers, clerks, and persons who can make and read blueprints, manuals, and so forth.)
4. Therefore a higher proportion of individuals in industrial societies will perceive the acquisition of literacy as rewarding.
5. And [by (1)] a higher proportion will attempt to acquire literacy.
6. The provision of schooling costs money, directly or indirectly.
7. Compared with agricultural societies, a higher proportion of individuals in industrial societies are, by some standard, wealthy.
8. Therefore a higher proportion are able to provide schooling (through government or private charity), and a higher proportion are able to pay for their own schooling without charity.
9. And a higher proportion will perceive the effort to acquire literacy as apt to be successful.
10. And [by (2) as well as by (1)] a higher proportion will attempt to acquire literacy.
11. Because their perceptions are generally accurate, a higher proportion of individuals in industrial societies will acquire literacy. That is, the literacy rate is apt to be higher in an industrial than in an agricultural society.[9]

Propositions 1 and 2 are an earlier statement of the rationality proposition that summarizes the success, stimulus, and value propositions. From these first two propositions, or axioms, others are derived in an effort to explain Golden's Law (Proposition 11 in the previous scheme). Examining some features of this explanation can provide insight into Homans' deductive strategy.

In this example, the transition between Propositions 2 and 3 ignores so many necessary variables as to simply describe in the words of behavioral psychology what Homans perceived to have occurred. Why are people in industrial societies prepared to reward literacy? This statement does not explain; it describes and thus opens a large gap in the logic of the deductive system. For Homans this statement is a given. It states a boundary condition, for the theory is not trying to explain why people are prepared to reward literacy. Another story is required to explain this event. Thus Homans argued that "no theory can explain everything" and that it is necessary to ignore some things and assume them to be givens for the purposes of explanation at hand. The issue remains, however: Has not Homans defined away the most interesting sociological issue—what makes people ready to reward literacy in a society's historical development? Homans believed people are just ready to do so.

Another problem in this scheme comes with the placement of the word *therefore*. This transitive is typically used immediately following a statement of the givens that define away important classes of sociological variables. For example, the "therefore" preceding key propositions (4 and 8) begs questions such as the following: Why do people perceive literacy as rewarding? What level of industrialization would make this so? What level of educational development? What feedback consequences does desire for literacy have for educational development? By ignoring the why and what of these questions, Homans could then in Propositions 5 and 10 reinsert the higher order axioms (1 and 2) of the explanation, thereby giving the scheme an appearance of deductive continuity. Answers to the critical sociological questions have been avoided, such as why people perceive as valuable and rewarding certain crucial activities.

Homans' reply to such criticisms is important to note. First, he emphasized that this deductive scheme was just an example or illustrative sketch, leaving out many details. Second, and more important, Homans contended that it is unreasonable for the critic to expect that a theory can and must explain every given condition (Propositions 3 and 7, for example). He offered the example of Newton's laws, which can help explain the movement of the tides (by virtue of gravitational forces of the moon relative to the axis of the earth), but these laws cannot explain why oceans are present and why the earth exists. This latter argument is reasonable, in principle, because deductive theories explain only a delimited range of phenomena. Still, even though Homans did not need to explain the givens of a deductive scheme, all the interesting sociological questions were contained in those givens. Moreover, these questions beg for the development of abstract *sociological* laws to explain them (Golden's Law is not an abstract principle, only an empirical generalization).

FROM BEHAVIOR TO MACROSTRUCTURE

Homans provided many illustrations of how behavioristic principles can explain research findings in social psychology. One of the most interesting chapters in *Social Behavior*,[10] comes at the end of the book, where he offered a view of how exchange processes can explain population-level and societal-level social phenomena. Phrased as a last orgy in his explication, Homans addressed the issue of how societies and civilizations are, ultimately, built from the face-to face exchanges of people in groups. His scenario went something like this:

At points in history, some people have the "capital" to reinforce or provide rewards for others, whether it comes from their possessing a surplus of food, money, a moral code, or valued leadership qualities. With such capital, "institutional elaboration" can occur, because some can invest their capital by trying to induce others (through rewards or threats of punishments) to engage in novel activities. These new activities can involve an "intermeshing of the behavior of a large number of persons in a more complicated or roundabout way than has hitherto been the custom." Whether this investment involves conquering territory and organizing a kingdom or creating a new form of business organization, those making the investment must have the resources—whether it be an army to threaten punishment, a charismatic personality to morally persuade followers, or the ability to provide for people's subsistence needs—to keep those so organized in a situation in which they derive some profit. At some point in this process, such organizations can become more efficient and hence rewarding to all when the rewards are clearly specified as generalized reinforcers, such as money, and when the activities expended to get their rewards are more clearly specified, such as when explicit norms and rules emerge. In turn, this increased efficiency allows greater organization of activities. This new efficiency increases the likelihood that generalized reinforcers and explicit norms will be used to regulate exchange relations and hence increase the profits to those involved. Eventually the exchange networks involving generalized reinforcers and an increasingly complex body of rules require differentiation of subunits—such as a legal and banking system—that can maintain the stability of the generalized reinforcers and the integrity of the norms.

From this kind of exchange process, then, social organization—whether at a societal, group, organizational, or institutional level—is constructed. The emergence of most patterns of organization is frequently buried in the recesses of history, but such emergence is typified by these accelerating processes: (1) People with capital (reward capacity) invest in creating more complex social relations that increase their rewards and allow those whose activities are organized to realize a profit. (2) With increased rewards, these people can invest in more complex patterns of organization. (3) Increasingly complex patterns of organization require, first of all, the use of generalized reinforcers and then the codification of norms to regulate activity. (4) With this organizational base, it then becomes possible to elaborate further the pattern of organization, creating the necessity for differentiation of subunits that assure the stability of the generalized reinforcers and the integrity of norms. (5) With this differentiation, it is possible to expand even further the networks of interaction, because there are standardized means for rewarding activities and codifying new norms as well as for enforcing old rules.

However, these complex patterns of social organization employing formal rules and secondary or generalized reinforcers can never cease to meet the more primary needs of individuals. Institutions first emerged to meet these needs, and, no matter how complex institutional arrangements become and how many norms and formal rules are elaborated, these extended interaction networks must ultimately reinforce humans' more primary needs. When these arrangements cease meeting the primary needs from which they ultimately sprang, an institution is vulnerable and apt to collapse if alternative actions, which can provide primary rewards, present themselves as a possibility. In this situation, low- or high-status persons—those who have little to lose by nonconformity to existing prescriptions—will break from established ways to expose to others a more rewarding alternative. Institutions might continue to extract conformity for a period, but they will cease to do so when they lose the capacity to provide primary rewards. Thus, complex institutional arrangements must ultimately be satisfying to individuals, not simply because of the weight of culture or norms but because they are constructed to serve people:

> Institutions do not keep going just because they are enshrined in norms, and it seems extraordinary that anyone should ever talk as if they did. They keep going because they have payoffs, ultimately payoffs for individuals. Nor is society a perpetual-motion machine, supplying its own fuel. It cannot keep itself going by planting in the young a desire for these goods and only those goods that it happens to be in shape to provide. It must provide goods that men find rewarding not simply because they are sharers in a particular culture but because they are men.[11]

That institutions of society must also meet primary needs sets the stage for a continual conflict between institutional elaboration and the primary needs of humans. As one form of institutional elaboration meets one set of needs, it can deprive people of other important rewards—opening the way for deviation and innovation by those presenting the alternative rewards that have been suppressed by dominant institutional arrangements. In turn, the new institutional elaborations that can ensue from innovators who have the capital to reward others will suppress other needs, which, through processes similar to its inception, will set off another process of institutional elaboration.

In sum, this sketch of how social organization is linked to elementary processes of exchange represents an interesting perspective for analyzing how patterns of social organization are built, maintained, altered, and broken down. Moreover, the broad contours of this kind of argument were repeated by other exchange theorists as they applied the principles of behaviorism or classical economics to explanations of larger-scale patterns of human organization. Yet, little of Homans' explicit imagery was retained, although the effort to move from micro principles about behavior of individuals to macro-level patterns of organization remained a critical concern in exchange theories.

NOTES

1. George C. Homans and David M. Schneider, *Marriage, Authority, and Final Causes: A Study of Unilateral Cross-Cousin Marriage* (New York: Free Press, 1955).

2. Ibid., p. 15.

3. Ibid.

4. George C. Homans, *Social Behavior: Its Elementary Forms* (New York: Harcourt Brace Jovanovich, 1961; second edition in 1972).

5. Ibid., p. 79.

6. Homans championed this conception of theory in a large number of works; see, for example, Homans, *Social Behavior* (cited in note 4); Homans, *The Nature of Social Science* (New York: Harcourt, Brace & World, 1967); "Fundamental Social Processes," in *Sociology*, ed. N.J. Smelser (New York: Wiley, 1967), pp. 27–78; "Contemporary Theory in Sociology," in *Handbook of Modern Sociology*, ed. R.E.L. Faris (Skokie, IL: Rand McNally, 1964), pp. 251–277; and "Bringing Men Back In," *American Sociological Review* 29 (December 1964), pp. 809–818. For an early statement of his position, see George C. Homans, "Social Behavior as Exchange," *American Journal of Sociology* 63 (August 1958), pp. 597–606; "Discovery and the Discovered in Social Theory," *Humboldt Journal of Social Relations* 7 (Fall-Winter 1979–1980), pp. 89–102.

7. The previous uses Homans' vocabulary, but, as emphasized in Chapter 1, "axiomatic" theory for sociology is unrealistic.

8. Note that "axioms" 1 and 2 here are simply an earlier version of Homans' rationality principle (VI) listed in Table 20-1.

9. George C. Homans, "Reply to Blain," *Sociological Inquiry* 41 (Winter 1971), pp. 19–24. This article was written in response to a challenge by Robert Blain for Homans to explain a sociological law: "On Homans' Psychological Reductionism," *Sociological Inquiry* 41 (Winter 1971), pp. 3–25.

10. Homans, *Social Behavior* (cited note 4), Chapter 16.

11. Ibid., p. 366

21

The Maturing Tradition II: Peter M. Blau's Dialectical Approach

A few years after George C. Homans' behavioristic approach appeared, another leading sociological theorist—Peter M. Blau—explored exchange theory.[1] Though he accepted the behavioristic underpinnings of exchange as a basic social process, Blau recognized that sociological theory had to move beyond simplistic behavioristic conceptualizations of human behavior. Similarly, crude views of humans as wholly rational had to be modified to fit the realities of human behavior. In the end, he developed a dialectical approach, emphasizing that within the strains toward integration arising from exchange are forces of opposition and potential conflict. Moreover, his analysis has a Simmelian thrust because, much like Georg Simmel, Blau tried to discover the form of exchange processes at both the micro and macro levels, and in so doing, he sought to highlight what was common to exchanges among individuals as well as among collective units of organization.

THE BASIC EXCHANGE PRINCIPLES

Although Blau never listed his principles in quite the same way as Homans did with his "axioms," the principles are nonetheless easy to extract from his discursive discussion. In Table 21-1, the basic principles are summarized. Proposition I, which can be termed the *rationality principle,* states that the frequency of rewards and the value of these rewards increase the likelihood that actions will be emitted. Propositions II-A and II-B on reciprocity borrow from Bronislaw Malinowski's and Claude Lévi-Strauss's initial discussion as reinterpreted by Alvin Gouldner.[2] Blau postulated that "the need to reciprocate for benefits received in order to continue receiving them serves as a 'starting mechanism' of social interaction."[3] Equally important, once exchanges have occurred, a "fundamental and ubiquitous norm of reciprocity" emerges to regulate subsequent exchanges. Thus,

TABLE 21-1 Blau's Implicit Exchange Principles

I. *Rationality Principle:* The more profit people expect from one another in emitting a particular activity, the more likely they are to emit that activity.

II. *Reciprocity Principles:*

A. The more people have exchanged rewards with one another, the more likely are reciprocal obligations to emerge and guide subsequent exchanges among these people.

B. The more the reciprocal obligations of an exchange relationship are violated, the more disposed deprived parties are to sanction negatively those violating the norm of reciprocity.

III. *Justice Principles:*

A. The more exchange relations have been established, the more likely they are to be governed by norms of "fair exchange."

B. The less norms of fairness are realized in an exchange, the more disposed deprived parties are to sanction negatively those violating the norms.

IV. *Marginal Utility Principle:* The more expected rewards have been forthcoming from the emission of a particular activity, the less valuable the activity is and the less likely its emission is.

V. *Imbalance Principle:* The more stabilized and balanced some exchange relations are among social units, the more likely are other exchange relations to become imbalanced and unstable.

inherent in the exchange process, per se, is a principle of reciprocity. Over time, and as the conditions of Principle I are met, a social "norm of reciprocity," whose violation brings about social disapproval and other negative sanctions, emerges in exchange relations.

Blau recognized that people establish expectations about what level of reward particular exchange relations should yield and that these expectations are normatively regulated. These norms are termed *norms of fair exchange* because they determine what the proportion of rewards to costs should be in a given exchange relation. Blau also asserted that aggression is forthcoming when these norms of fair exchange are violated. These ideas are incorporated into Principles III-A and III-B, and are termed the *justice principles*.[4] Following economists' analyses of transactions in the marketplace, Blau introduced a principle on "marginal utility" (Proposition IV). The more a person has received a reward, the more satiated he or she is with that reward and the less valuable are further increments of the reward.[5] Proposition V on imbalance completes the listing of Blau's abstract laws. For Blau, as for all exchange theorists, established exchange relations are seen to involve costs or alternative rewards foregone. Most actors must engage in more than one exchange relation, so the balance and stabilization of one exchange relation is likely to create imbalance and strain in other necessary exchange relations. Blau believes that social life is thus filled with dilemmas in which people must successively trade off stability and balance in one exchange relation for strain in others as they attempt to cope with the variety of relations that they must maintain.

ELEMENTARY SYSTEMS OF EXCHANGE

Blau initiated his discussion of elementary exchange processes with the assumption that people enter into social exchange because they perceive the possibility of deriving rewards (Principle I). Blau labeled this perception *social attraction* and postulated that, unless relationships involve such attraction, they are not relationships of exchange. In entering an exchange relationship, each actor assumes the perspective of another and thereby derives some perception of the other's needs. Actors then manipulate their presentation of self to convince one another that they have the valued qualities others appear to desire. In adjusting role behaviors in an effort to impress others with the resources that they have to offer, people operate under the principle of reciprocity, for, by indicating that one possesses valued qualities, each person is attempting to establish a claim on others for the receipt of rewards from them. All exchange operates under the presumption that people who bestow rewards will receive rewards in turn as payment for value received.

Actors attempt to impress one another through competition in which they reveal the rewards they have to offer in an effort to force others, in accordance with the norm of reciprocity, to reciprocate with an even more valuable reward. Social life is thus rife with people's competitive efforts to impress one another and thereby extract valuable rewards. But, as interaction proceeds, it inevitably becomes evident to the parties in an exchange that some people have more valued resources to offer than others, putting them in a unique position to extract rewards from all others who value the resources that they have to offer.

At this point in exchange relations, groups of individuals become differentiated by the resources they possess and the kinds of reciprocal demands they can make on others. Blau then asked an analytical question: What generic types or classes of rewards can those with resources extract in return for bestowing their valued resources on others? Blau conceptualized four general classes of such rewards: money, social approval, esteem or respect, and compliance. Although Blau did not make full use of his categorization of rewards, he offered some suggestive clues about how these categories can be incorporated into abstract theoretical statements.

Blau first ranked these generalized reinforcers by their value in exchange relations.

TABLE 21-2 Blau's Conditions for the Differentiation of Power in Social Exchange

I. The fewer services people can supply in return for the receipt of particularly valued services, the more those providing these particularly valued services can extract compliance.

II. The fewer alternative sources of rewards people have, the more those proving valuable services can extract compliance.

III. The less those receiving valuable services from particular individuals can employ physical force and coercion, the more those providing the services can extract compliance.

IV. The less those receiving the valuable services can do without them, the more those providing the services can extract compliance.

In most social relations, money is an inappropriate reward and hence is the least valuable reward. Social approval is an appropriate reward, but for most humans it is not very valuable, thus forcing those who derive valued services to offer with great frequency the more valuable reward of esteem or respect to those providing valued services. In many situations, the services offered can command no more than respect and esteem from those receiving the benefit of these services. At times, however, the services offered are sufficiently valuable to require those receiving them to offer, in accordance with the principles of reciprocity and justice, the most valuable class of rewards—compliance with one's requests.

When people can extract compliance in an exchange relationship, they have power. They have the capacity to withhold rewarding services and thereby punish or inflict heavy costs on those who might not comply. To conceptualize the degree of power possessed by individuals, Blau formulated four general propositions that determine the capacity of powerful individuals to extract compliance. These are listed and reformulated in Table 21-2.[6]

These four propositions list the conditions leading to differentiation of members in social groups by their power. To the extent that group members can supply some services in return, seek alternative rewards, potentially use physical force, or do without certain valuable services, individuals who can provide valuable services will be able to extract only esteem and approval from group members. Such groups will be differentiated by prestige rankings but not by power. Naturally, as Blau emphasized, most social groups reveal complex patterns of differentiation of power, prestige, and patterns of approval, but of particular interest to him are the dynamics involved in generating power, authority, and opposition.

Blau believed that, power differentials in groups create two contradictory forces: (1) strains toward integration and (2) strains toward opposition and conflict.

Strains Toward Integration

Differences in power inevitably create the potential for conflict. However, such potential is frequently suspended by a series of forces promoting the conversion of power into authority, in which subordinates accept as legitimate the leaders' demands for compliance. Principles II and III in Table 21-1 denote two processes fostering such group integration: Exchange relations always operate under the presumption of reciprocity and justice, forcing those deriving valued services to provide other rewards in payment. In providing these rewards, subordinates are guided by norms of fair exchange, in which the costs that they incur in offering compliance are to be proportional to the value of the services that they receive from leaders. Thus, to the extent that actors engage in exchanges with leaders and to the degree that the services provided by leaders are highly valued, subordination must be accepted as legitimate in accordance with the norms of reciprocity and fairness that emerge in all exchanges. Under these conditions, groups elaborate additional norms specifying just how exchanges with leaders are to be conducted to regularize the requirements for reciprocity and to maintain fair rates of exchange. Leaders who conform to these emergent norms can usually assure themselves that their leadership will be considered legitimate. Blau emphasized that, if leaders abide by the norms regulating exchange of their services for compliance, norms carrying negative sanctions typically emerge among subordinates stressing the need for compliance to leaders' requests. Through this process, subordinates exercise considerable social control over one another's actions and thereby promote the integration of super- and subordinate segments of groupings.

Authority, therefore, "rests on the common norms in a collectivity of subordinates that constrain its individual members to conform to the orders of a superior."[7] In many patterns of social organization, these norms simply emerge from the competitive exchanges among collective groups of actors. Frequently, however, for such "normative agreements" to be struck, participants in an exchange must be socialized into a common set of values that define not only what constitutes fair exchange in a given situation but also the way such exchange should be institutionalized into norms for both leaders and subordinates. Although it is quite possible for actors to arrive at normative consensus in the course of the exchange process itself, an initial set of common values facilitates the legitimation of power. Actors can now enter into exchanges with a common definition of the situation, which can provide a general framework for the normative regulation of emerging power differentials. Without common values, the competition for power is likely to be severe. In the absence of guidelines about reciprocity and fair exchange, considerable strain and tension will persist as definitions of these are worked out. For Blau, then, legitimation "entails not merely tolerant approval but active confirmation and promotion of social patterns by common values, either preexisting ones or those that emerge in a collectivity in the course of social interaction."[8]

With the legitimation of power through the normative regulation of interaction, as confirmed by common values, the structure of collective organization is altered. One of the most evident changes is the decline in interpersonal competition, for now actors' presentations of self shift from a concern about impressing others with their valuable qualities to an emphasis on confirming their status as loyal group members. Subordinates accept their status and manipulate their role behaviors to assure that they receive social approval from their peers as a reward for conformity to group norms. Leaders can typically assume a lower profile because they no longer must demonstrate their superior qualities in each and every encounter with subordinates—especially because norms now define when and how they should extract conformity and esteem for providing their valued services. Thus, with the legitimation of power as authority, the interactive processes (involving the way group members define the situation and present themselves to others) undergo a dramatic change, reducing the degree of competition and thereby fostering group integration.

With these events, the amount of direct interaction between leaders and subordinates usually declines, because power and ranking no longer must be constantly negotiated. This decline in direct interaction marks the formation of distinct subgroupings as members interact with those of their own social rank, avoiding the costs of interacting with either their inferiors or their superiors. In interacting primarily among themselves, subordinates avoid the high costs of interacting with leaders, and, although social approval from their peers is not a particularly valuable reward, it can be extracted with comparatively few costs—thus allowing a sufficient profit. Conversely, leaders can avoid the high costs (of time and energy) of constantly competing and negotiating with inferiors regarding when and how compliance and esteem are to be bestowed on them. Instead, by having relatively limited and well-defined contact with subordinates, they can derive the high rewards that come from compliance and esteem without incurring excessive costs in interacting with subordinates—thereby allowing for a profit.

Strains Toward Opposition

Thus far, Blau's exchange perspective is decidedly functional. Social exchange processes—attraction, competition, differentiation, and integration—are analyzed by how they contribute to creating a legitimated set of normatively regulated relations. Yet, Blau was keenly aware that social organization is always rife with conflict and opposition, creating an inevitable dialectic between integration and opposition in social structures.

Blau's exchange principles, summarized in Table 21-2, allow the conceptualization of these strains for opposition and conflict. As Principle II-B on reciprocity documents, the failure to receive expected rewards in return for various activities leads actors to attempt to apply negative sanctions that, when ineffective, can drive people to violent retaliation against those who have denied them an expected reward. Such retaliation is intensified by the dynamics summarized in Principle III-B on justice and fair exchange, because when those in power violate such norms, they inflict excessive costs on subordinates,

TABLE 21-3 Blau's Propositions on Exchange Conflict

I. The probability of opposition to those with power increases when exchange relations between super- and subordinates become imbalanced, with imbalance increasing when
   A. Norms of reciprocity are violated by the superordinates.
   B. Norms of fair exchange are violated by superordinates.

II. The probability of opposition increases as the sense of deprivation among subordinates escalates, with this sense of deprivation increasing when subordinates can experience collectively their sense of deprivation. Such collective experience increases when
   A. Subordinates are ecologically and spatially concentrated.
   B. Subordinates can communicate with one another.

III. The more subordinates can collectively experience deprivations in exchange relations with superordinates, the more likely are they to codify ideologically their deprivations and the more likely are they to oppose those with power.

IV. The more deprivations of subordinates are ideologically codified, the greater is their sense of solidarity and the more likely are they to oppose those with power.

V. The greater the sense of solidarity is among subordinates, the more they can define their opposition as a noble and worthy cause and the more likely they are to oppose those with power.

VI. The greater is the sense of ideological solidarity, the more likely are subordinates to view opposition as an end in itself, and the more likely are they to oppose those with power.

creating a situation that, at a minimum, leads to attempts to sanction negatively and, at most, to retaliation. Finally, Principle V on the inevitable imbalances emerging from multiple exchange relations emphasizes that to balance relations in one exchange context by meeting reciprocal obligations and conforming to norms of fairness is to put other relations into imbalance. Thus, the imbalances potentially encourage a cyclical process in which actors seek to balance previously unbalanced relations and thereby throw into imbalance currently balanced exchanges. In turn, exchange relations that are thrown into imbalance violate the norms of reciprocity and fair exchange, thus causing attempts at negative sanctioning and, under some conditions, retaliation. Blau posited, then, that sources of imbalance are built into all exchange relationships. When severely violating norms of reciprocity and fair exchange, these imbalances can lead to open conflict among individuals in group contexts.

In Table 21-3, Blau's ideas are summarized as propositions.[9] To appreciate the degree to which these propositions resemble those in conflict theory, we can compare these propositions with those in Tables 11-1 and 12-1. For, in the end, dialectical conflict theory and exchange theory are converging perspectives. It could be argued that conflict theory is a derivative of exchange theory, although many would argue just the reverse. In either case, the perspectives converge.

From the discursive context in which the propositions in Table 21-3 are imbedded comes a conceptualization of opposition. Blau hypothesized that, the more imbalanced exchange relations are experienced collectively, the greater is the sense of deprivation and the greater is the potential for opposition. Although he did not explicitly state the case, Blau appeared to argue that increasing ideological codification of deprivations, the formation of group solidarity, and the emergence of conflict as a way of life—that is, members' emotional involvement in and commitment to opposition to those with power—will increase the intensity of the opposition. These propositions offered a suggestive lead for conceptualizing inherent processes of opposition in exchange relations.[10]

EXCHANGE SYSTEMS AND MACROSTRUCTURE

Although the general processes of attraction, competition, differentiation, integration, and opposition are evident in the exchange among macrostructures, Blau saw several fundamental differences between these exchanges and those among microstructures.

1. In complex exchanges among macrostructures, the significance of "shared values" increases, for through such values indirect exchanges among macrostructures are mediated.
2. Exchange networks among macrostructures are typically institutionalized. Although spontaneous exchange is a ubiquitous feature of social life, there are usually well-established historical arrangements that circumscribe the basic exchange processes of attraction, competition, differentiation, integration, and even opposition among collective units.
3. Macrostructures are themselves the product of more elementary exchange processes, so the analysis of macrostructures requires the analysis of more than one level of social organization.[11]

Mediating Values

Blau believed that the "interpersonal attraction" of elementary exchange among individuals is replaced by shared values at the macro level. These values can be conceptualized as "media of social transactions" in that they provide a common set of standards for conducting the complex chains of indirect exchanges among social structures and their individual members. Such values are viewed by Blau as providing effective mediation of complex exchanges because the individual members of social structures have usually been socialized into a set of common values, leading them to accept these values as appropriate. Furthermore, when coupled with codification into laws and enforcement procedures by those groups and organizations with power, shared values provide a means for mediating the complex and indirect exchanges among the macrostructures of large-scale systems. In mediating indirect exchanges among groups and organizations, shared values provide standards for the calculation of (a) expected rewards, (b) reciprocity, and (c) fair exchange.

Thus, because individuals are not the units of complex exchanges, Blau emphasized that, for complex patterns of social organization to emerge and persist, a "functional equivalent" of direct interpersonal attraction must exist. Values assume this function and assure that exchange can proceed in accordance with the principles presented in Table 21-1. And even when complex exchanges do involve people, their interactions are frequently so protracted and indirect that one individual's rewards are contingent on others who are far removed, requiring that common values guide and regulate the exchanges.

Institutionalization

Whereas values facilitate processes of indirect exchange among diverse types of social units, institutionalization denotes those processes that regularize and stabilize complex exchange processes.[12] As people and various forms of collective organization become dependent on particular networks of indirect exchange for expected rewards, pressures for formalizing exchange networks through explicit norms increase. This formalization and regularization of complex exchange systems can be effective in three minimal conditions: (1) The formalized exchange networks must have profitable payoffs for most parties to the exchange. (2) Most individuals organized into collective units must have internalized through prior socialization the mediating values used to build exchange networks. And, (3) those units with power in the exchange system must receive a level of rewards that moves them to seek actively the formalization of rules governing exchange relations.

Institutions are historical products whose norms and underlying mediating values are handed down from one generation to another, thereby limiting and circumscribing the kinds of indirect exchange networks that can emerge. Institutions exert a kind of external constraint on individuals and various types of collective units, bending exchange processes to fit their prescriptions and proscriptions. Institutions thus represent sets of relatively stable and general norms regularizing different patterns of indirect and complex exchange relations among diverse social units.

Blau stressed that all institutionalized exchange systems reveal a counter-institutional component "consisting of those basic values and ideals that have not been realized and have not found expression in explicit institutional forms, and which are the ultimate source of social change."[13] To the extent that these values remain unrealized in institutionalized exchange relations, individuals who have internalized them will derive little payoff from existing institutional arrangements and will therefore feel deprived, seeking alternatives to dominant institutions. These unrealized values, even when codified into an opposition ideology advocating open revolution, usually contain at least some of the ideals and ultimate objectives legitimated by the prevailing culture. This indicates that institutional arrangements "contain the seeds of their potential destruction" by failing to meet all the expectations of reward raised by institutionalized values.

Blau did not enumerate the conditions for mobilization of individuals into conflict groups, but his scheme explicitly denoted the source of conflict and change: Counter-institutional values whose failure of realization by dominant institutional arrangements creates deprivations that can lead to conflict and change in social systems. Such tendencies for complex exchange systems to generate opposition can be explained by the basic principles of exchange. When certain mediating values are not institutionalized in a social system, exchange relations will not be viewed as reciprocated by those who have internalized these values. Thus, in accordance with Blau's principles on reciprocity (see Table 21-1), these segments of a collectivity are more likely to feel deprived and to seek ways of retaliating against the dominant institutional arrangements, which, from the perspective dictated by their values, have failed to reciprocate. For those who have internalized values that are not institutionalized, it is also likely that perceptions of fair exchange have been violated, leading them, in accordance with the principles of justice, to attempt to sanction negatively those arrangements that violate alternative norms of fair exchange. Finally, in institutionalized exchange networks, the balancing of exchange relations with some segments of a collectivity inevitably creates imbalances in relations with other segments (the imbalance principle in Table 21-1), thereby violating norms of reciprocity and fairness and setting into motion forces of opposition.

Unlike direct interpersonal exchanges, however, opposition in complex exchange systems is between large collective units of organization, which, in their internal dynamics, reveal their own propensities for integration and opposition. This requires that the analysis of integration and opposition in complex exchange networks be attuned to various levels of social organization. Such analysis needs to show, in particular, how exchange processes among macrostructures, whether for integration or for opposition, are partly influenced by the exchange processes occurring among their constituent substructures.

Levels of Social Organization

For Blau, the "dynamics of macrostructures rest on the manifold interdependences between the social forces within and among their substructures."[14] Blau simplifies the complex analytical tasks of examining the dynamics of substructures by positing that organized collectivities, especially formal organizations, are the most important substructures in the analysis of macrostructures. Thus, the theoretical analysis of complex exchange systems among macrostructures requires that primary attention be drawn to the relations of attraction, competition, differentiation, integration, and opposition among various types of complex organizations. In emphasizing the pivotal significance of complex organizations, Blau posited a particular image of society that should guide the ultimate construction of sociological theory.

Organizations in a society must typically derive rewards from one another, thus creating a situation in which they are both attracted to, and in competition with, one another. Hierarchical differentiation between successful and less successful organizations operating in the same sphere emerges from this competition. Such differentiation usually creates strains toward specialization in different fields among less successful organizations as they seek new sources of resources. To provide effective means for integration, separate political organizations must also emerge to regulate their exchanges. These political organizations possess power and are viewed as legitimate only as long as they are considered by individuals and organizations to follow the dictates of shared cultural values. Typically, political organizations are charged with several objectives: (1) regulating complex networks of indirect exchange by the enactment of laws; (2) controlling through law competition among dominant organizations, thereby assuring the latter of scarce resources; and (3) protecting existing exchange networks among organizations, especially those with power, from encroachment on these rewards by organizations opposing the current distribution of resources.

Blau, believed that differentiation and specialization occur among macrostructures because of the competition among organizations in a society. Although mediating values allow differentiation and specialization among organizations to occur, it is also necessary for separate political organizations to exist and regularize, through laws and the use of force, existent patterns of exchange among other organizations. Such political organizations will be viewed as legitimate as long as they normatively regulate exchanges that reflect the tenets of mediating values and protect the payoffs for most organizations, especially the most powerful. The existence of political authority inevitably encourages opposition movements, however, for now opposition groups have a clear target—the political organizations—against which to address their grievances. As long as political authority remains diffuse, opposition organizations can only compete unsuccessfully against various dominant organizations. With the legitimation of clear-cut political organizations charged with preserving current patterns of organization, opposition movements can concentrate their energies against one organization, the political system.

In addition to providing deprived groups with a target for their aggressions, political organizations inevitably must aggravate the deprivations of various segments of a population because political control involves exerting constraints and distributing resources unequally. Those segments of the population that must bear the brunt of such constraint

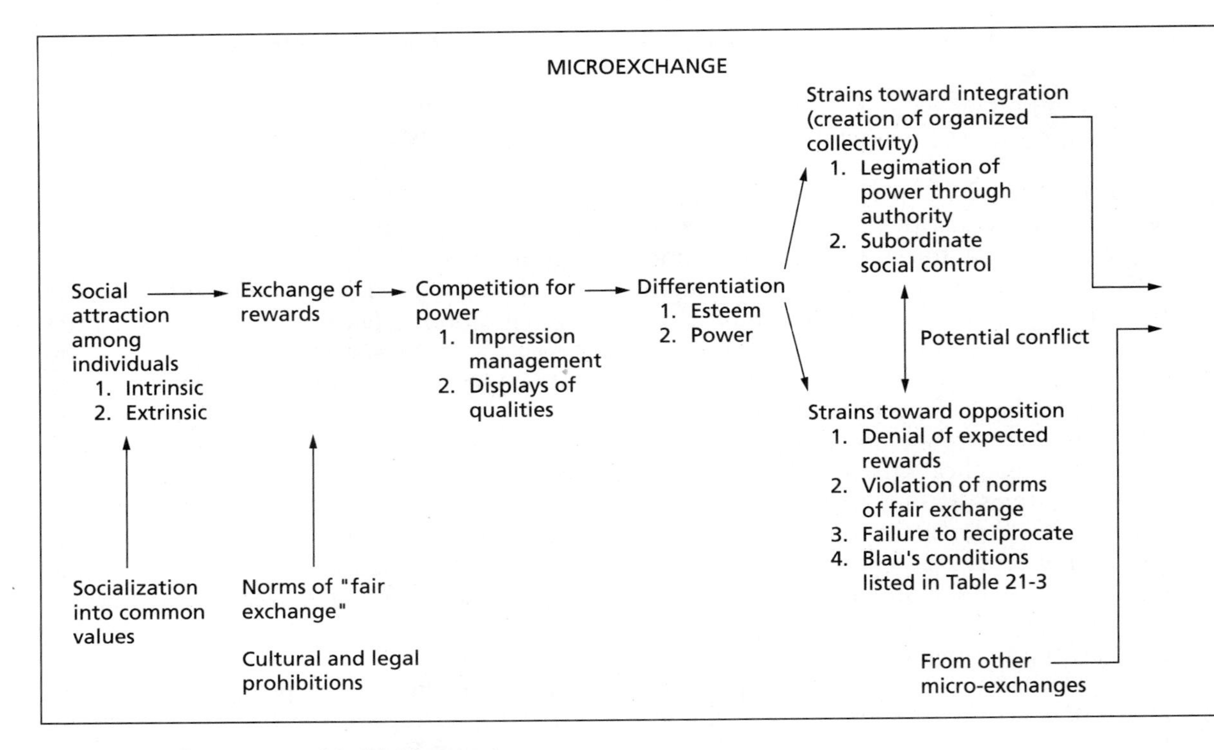

FIGURE 21-1 Blau's Image of Social Organization

and unequal distribution usually experience great deprivation of the principles of reciprocity and fair exchange, which, under various conditions, creates a movement against the existing political authorities. To the extent that this organized opposition forces redistribution of rewards, other segments of the population are likely to feel constrained and deprived, leading them to organize into an opposition movement. The organization of political authority ensures that, in accordance with the principle of imbalance, attempts to balance one set of exchange relations among organizations throw into imbalance other exchange relations, causing the formation of opposition organizations. Thus, built into the structure of political authority in a society are inherent forces of opposition that give society a dialectical and dynamic character.

BLAU'S IMAGE OF SOCIAL ORGANIZATION

Figure 21-1 summarizes Blau's view of social organization at the micro level and the macro organizational level. Clearly, the same processes operate at both levels of exchange: (1) social attraction, (2) exchange of rewards, (3) competition for power, (4) differentiation, (5) strains toward integration, and (6) strains toward opposition. Thus, the Simmelian thrust of Blau's effort is clear because he sees the basic form of exchange as much the same, regardless of whether the units involved in the exchange are individuals or collective units of organization. There are, of course, some differences between exchange among individuals and organizational units, and these are noted across the bottom of the figure.

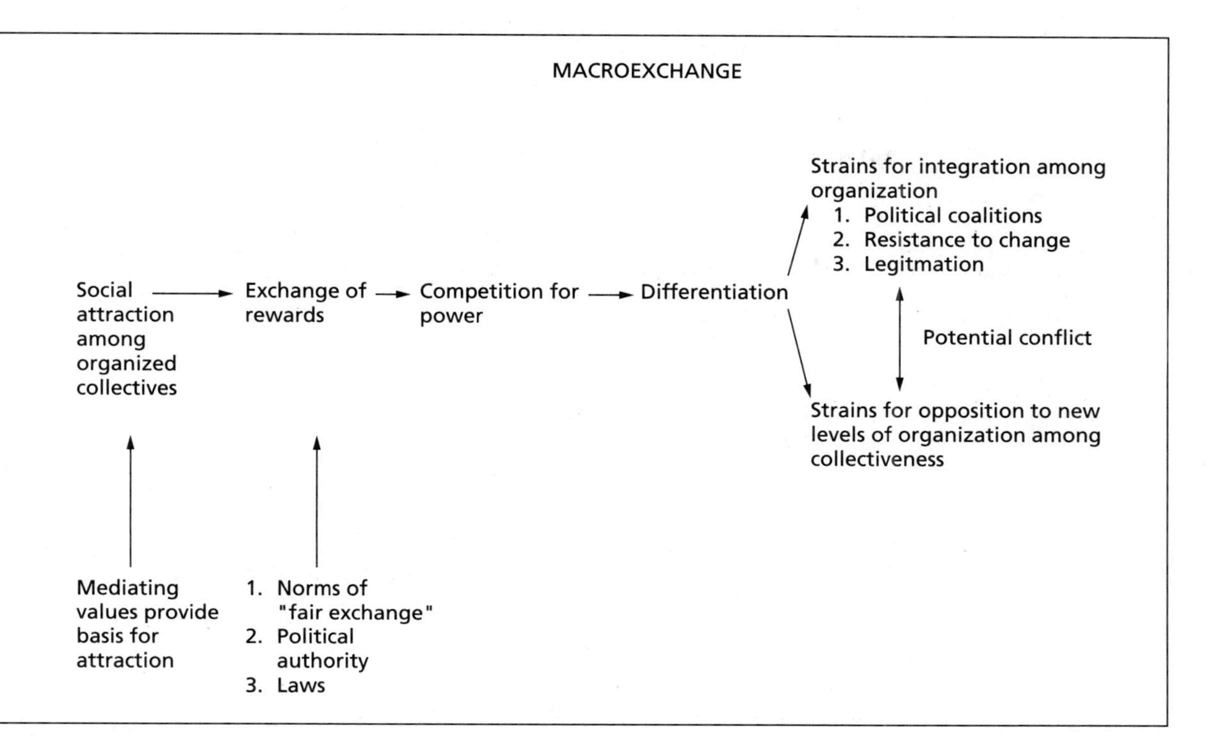

CONCLUSION

Despite Blau's effort to synthesize elements of functional and conflict theory, his approach did not endure. Indeed, even Blau himself increasingly believed that the isomorphism between micro-level and macro-level exchange processes was forced. And in the end, he abandoned the exchange approach at the macro-structural level, developing an alternative that we will examine in Chapter 39. Still, more than any theorist of the time, Blau recognized that exchange and conflict theory converge. For exchange processes produce both institutionalized patterns and forces of opposition to these patterns, primarily because exchange generates inequalities in power. But these leads were never followed by subsequent theorists who drew more from either network theory or more purely economic approaches. Let us now turn to the profile of these approaches as they stood during the early maturation of sociological exchange theory.

NOTES

1. Peter M. Blau's major exchange work is *Exchange and Power in Social Life* (New York: Wiley, 1964). This formal and expanded statement on his exchange perspective was anticipated in earlier works. For example, see Peter M. Blau, "A Theory of Social Integration," *American Journal of Sociology* 65 (May 1960), pp. 545–556; and Peter M. Blau, *The Dynamics of Bureaucracy*, 1st and 2nd eds. (Chicago: University of Chicago Press, 1955, 1963). It is of interest to note that George C. Homans in *Social Behavior: Its Elementary Forms* (New York: Harcourt, Brace & World, 1961) makes frequent reference to the data summarized in this latter work. For a more recent statement of Blau's position, see Peter M. Blau, "Interaction: Social Exchange," in *International Encyclopedia of the Social Sciences*, vol. 7 (New York: Macmillan, 1968), pp. 452–458.

2. Alvin W. Gouldner, "The Norm of Reciprocity," *American Sociological Review* 25 (April 1960), pp. 161–178.

3. Blau, *Exchange and Power* (cited in note 1), p. 92.

4. See Peter M. Blau, "Justice in Social Exchange," in *Institutions and Social Exchange: The Sociologies of Talcott Parsons and George C. Homans*, eds. H. Turk and R.L. Simpson (Indianapolis: Bobbs-Merrill, 1971), pp. 56–68. See also: Blau, *Exchange and Power* (cited in note 1), pp. 156–157.

5. Blau, *Exchange and Power* (cited in note 1), p. 90.

6. Ibid.

7. Ibid., p. 208.

8. Ibid., p. 221.

9. Ibid., pp. 224-252.

10. Peter M. Blau, "Dialectical Sociology: Comments," *Sociological Inquiry* 42 (Spring 1972), p. 185. This article was written in reply to an attempt to document Blau's shift from a functional to a dialectical perspective; see Michael A. Weinstein and Deena Weinstein, "Blau's Dialectical Sociology," *Sociological Inquiry* 42 (Spring 1972), pp. 173–182.

11. Blau, "Contrasting Theoretical Perspectives," in *The Micro-Macro Link*, eds. J.C. Alexander, B. Gisen, R. Münch, and N.J. Smelser. (Berkeley: University of California Press, 1987) pp. 253–311.

12. Blau, *Exchange and Power* (cited in note 1), pp. 273–280.

13. Ibid., p. 279.

14. Ibid., p. 284.

22

The Maturing Tradition III:

Richard M. Emerson's Exchange Network Approach

In the early 1960s, Richard M. Emerson followed Georg Simmel's lead in seeking a formal sociology of basic exchange processes. In essence, Emerson asked this: Could exchange among individual and collective actors be understood by the same basic principles? Emerson provided a creative answer to this question by synthesizing behaviorist psychology and sociological network analysis. The psychology gave him the driving force behind exchanges, whereas the network sociology allowed him to conceptualize the form of social relations among both individual and collective actors in the same terms. What emerged was exchange network analysis that, after Emerson's early death, was carried forward by colleagues and students, as we will see in Chapter 25.

THE BASIC STRATEGY

Emerson[1] borrowed the basic ideas of behaviorist psychology, but unlike many working in this tradition, he became more concerned with the form of relationships among the actors rather than the properties and characteristics of the actors themselves. This simple shift in emphasis profoundly affected how he built his exchange theory. The most significant departure from earlier exchange theories was that concern with why actors entered an exchange relationship in the first place given their values and preferences[2] was replaced by an emphasis on the existing exchange relationship and what is likely to transpire in this relationship in the future. Emerson believed that if an exchange relationship exists, this means that actors are willing to exchange valued resources, and the goal of theory is not so much to understand how this relationship originally came about but, instead, what will happen to it over time. Thus, the existing exchange relationship between actors becomes the unit of sociological analysis, not the actors themselves.

THE BASIC EXCHANGE PROCESSES

In Emerson's scheme, analysis thus began with an existing exchange relation between at least two actors. This relationship has been formed from (1) perceived opportunities by at least one actor, (2) the initiation of behaviors, and (3) the consummation of a transaction between actors mutually reinforcing each other. If initiations go unreinforced, then an exchange relation will not develop. And, unless the exchange transaction between actors endures for at least some period of time, it is theoretically uninteresting.

Emerson's approach thus started with an established exchange relation and then asked this question: To what basic processes is this relationship subject? His answer: (1) the *use of power* and (2) *balancing*. If exchange relations reveal high *dependency* of one actor, *B*, on another actor, *A*, for reinforcement, then *A* has what Emerson termed a *power advantage* over *B*. To have a power advantage is to use it, with the result that actor *A* forces increasing costs on actor *B* within the exchange relationship.

In Emerson's view, a power advantage represents an *imbalanced exchange* relation. A basic proposition in Emerson's scheme is that, over time, imbalanced exchange relations tend toward *balance*. For a situation where actor *A* enjoys a power advantage over *B* (that is, *B* is more dependent on the resources provided by *A* than vice versa), *balancing operations* are activated:

1. Actor *B* can decrease the value of reinforcers, or rewards, provided by actor *A*, thereby reducing *B*'s dependence on *A*.
2. Actor *B* can try to increase the number of its alternative sources for the reinforcers, or rewards, provided by *A*, thereby reducing *B*'s dependence on *A*.
3. Actor *B* can attempt to increase the value of the reinforcers that it provides for *A*, thereby increasing *A*'s dependence on *B*.
4. Actor *B* can seek to reduce the *A*'s alternative sources for the reinforcers provided by *B*, thereby making actor *A* more dependent on *B*.

Emerson stressed that, through at least one of these four balancing operations, the dependency of *B* and *A* on each other for rewards will reach an equilibrium. Thus exchange transactions reveal differences in power that, over time, tend toward balance. Naturally, in complex exchange relations involving many actors, *A*, *B*, *C*, *D*, . . . , *n*, the basic processes of dependence, power, and balance will ebb and flow as new actors and new reinforcers or resources enter the exchange relations.

THE BASIC EXCHANGE PROPOSITIONS

Emerson began with a relatively long list of behaviorist principles, which are not summarized here. The idea was to derive corollaries and theorems from these principles to explain the structural properties and dynamics of an existing exchange relationship. These dynamics revolve around (1) *dependency*, (2) *power use*, and (3) *balancing*. The initial set of derived theorems is summarized in Table 22-1. As is evident, these theorems describe the basic dynamics of exchange relations as power (P), which is related to the dependency (D) of actor *B* on *A* for valued resources; hence, the power of *A* over *B* is a positive function of *B*'s dependency on *A* for resources, or $P_{AB} = D_{BA}$. Balance is achieved in this A,B relation when, through the balancing operations listed, the dependency of *B* on *A* becomes equal to that of *A* on *B* for resources, or $D_{BA} = D_{AB}$.

With this understanding of the basic exchange process, Emerson then set out to introduce new corollaries and theorems to account for various structural forms of ongoing exchange relationships. In essence, Emerson added propositions, derived from the

TABLE 22-1 The Initial Theorems

Theorem 1: The greater the value of rewards to actor B in a situation, the more initiations by B reveal a curvilinear pattern, with initiations increasing over early transactions and then decreasing over time.

Theorem 2: The greater the dependency of *B* on a set of exchange relations, the more likely is *B* to initiate behaviors in this set of relations.

Theorem 3: The more the uncertainty of *B* increases in an exchange relation, the more the dependency of *B* on that situation increases, and vice versa.

Theorem 4: The greater the dependency of *B* on *A* for rewards in an *A*,*B* exchange relationship, the greater is the power of *A* over *B* and the more imbalanced is the relationship between *A* and *B*.

Theorem 5: The greater the imbalance of an *A*,*B* exchange relation at one point in time, the more likely it is to be balanced at a subsequent point in time.

basic theorems listed in Table 22-1 about dependency, power use, and balancing. At this point, Emerson began to use network analysis, summarized in Chapter 38, to describe various structural forms and their transformation in terms of the dynamics inhering in dependency, power, and balancing.

STRUCTURE, NETWORKS, AND EXCHANGE

Emerson's portrayal of social networks will be simplified; for our purposes here the full details of his network terminology need not be addressed. Although Emerson followed the conventions of graph theory and developed a number of definitions, only two definitions are critical:

> *Actors*: Points *A*, *B*, *C*, . . . , *n* in a network of relations. Different letters represent actors with different resources to exchange. The same letters—that is, A_1, A_2, A_3, and so forth—represent different actors exchanging similar resources.
>
> *Exchange relations*: *A*—*B*, *A*—*B*—*C*, A_1—A_2, and other patterns of ties that can connect different actors to each other, forming a network of relations.

The next conceptual task was to visualize the forms of networks that could be represented with these two definitions. For each basic form, new corollaries and theorems were added as Emerson documented the way in which the basic processes of dependence, power, and balance operate. His discussion was only preliminary, but it illustrated his perspective's potential.

Several basic social forms are given special treatment: (a) unilateral monopoly, (b) division of labor, (c) social circles, and (d) stratification.

Unilateral Monopoly

In the network illustrated in Figure 22-1, actor *A* is a source of valuable resources for actors B_1, B_2, and B_3. Actors B_1, B_2, and B_3 provide rewards for *A*, but, because *A* has multiple sources for rewards and the *B*s have only *A* as a source for their rewards, the situation is a unilateral monopoly.

Such a structure often typifies interpersonal as well as intercorporate units. For example, *A* could be a female date for three different men, B_1, B_2, and B_3. Or *A* could be a corporation that is the sole supplier of raw resources for three other manufacturing corporations, B_1, B_2, and B_3. Or *A* could be a governmental body and the *B*s could be dependent agencies. An important feature of the unilateral monopoly is that, by Emerson's definitions, it is imbalanced, and thus its structure is subject to change.

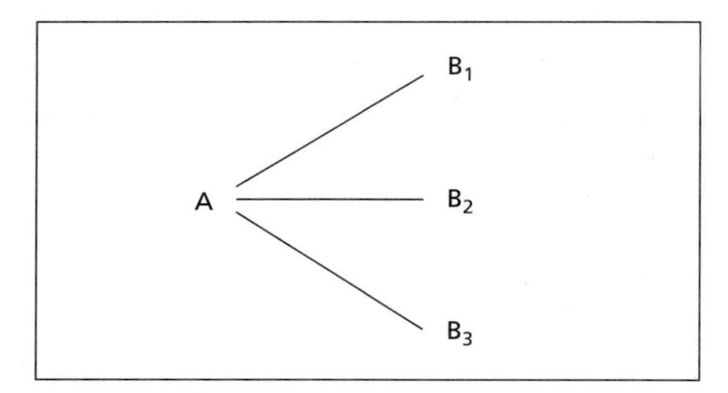

FIGURE 22-1 A Unilateral Monopoly

Emerson developed additional corollaries and theorems to account for the various ways this unilateral monopoly can change and become balanced. For instance, if no A_2, A_3, . . . , A_n exist and the *B*s cannot communicate with each other, the following proposition would apply (termed by Emerson *Exploitation Type I*):

> The more an exchange relation between *A* and multiple *B*s approximates a unilateral monopoly, the more additional resources each *B* will introduce into the exchange relation, with *A*'s resource utilization remaining constant or decreasing.

Emerson saw this adaptation as short-lived, because the network will become even more unbalanced. Assuming that the *B*s can survive as an entity without resources from *A*, then a new proposition applies (termed by Emerson *Exploitation Type II*):

> The more an exchange relation between *A* and multiple *B*s approximates a unilateral monopoly, the less valuable to *B*s are the resources provided by *A* across continuing transactions.

This proposition thus predicts that balancing operation 1—a decrease in the value of the reward for those at a power disadvantage—will balance a unilateral monopoly where no alternative sources of rewards exist and where *B*s cannot effectively communicate.

Other balancing operations are possible, if other conditions exist. If *B*s can communicate, they might form a coalition (balancing operation 4) and require *A* to balance exchanges with a united coalition of *B*s. If one *B* can provide a resource not possessed by the other *B*s, then a division of labor among *B*s (operations 3 and 4) would emerge. Or if another source of resources, A_2, can be found (operation 2), then the power advantage of A_1 is decreased. Each of these possible changes will occur under varying conditions, but these propositions provide a reason for the initiation of changes—a reason derived from basic principles of operant psychology (the details of these derivations are not discussed here).

Division of Labor

The emergence of a division of labor is one of many ways to balance exchange relations in a unilateral monopoly. If each of the *B*s can provide different resources for *A*, then they are likely to use these in the exchange with *A* and to specialize in providing *A* with these resources. This decreases the power of *A* and establishes a new type of network. For example, in Figure 22-2, the unilateral monopoly at the left is transformed to the division of labor form at the right, with B_1 becoming a new type of actor, *C*, with its own resources; with B_2 also specializing and becoming a new actor, *D*; and with B_3 doing the same and becoming actor *E*.

Emerson developed an additional proposition to describe this kind of change, in which each *B* has its own unique resources:

> The more resources are distributed *non*uniformly across *B*s in a unilateral monopoly with *A*, the more likely is each *B* to specialize and establish a separate exchange relation with *A*.

Several points should be emphasized. First, the units in this transformation can be

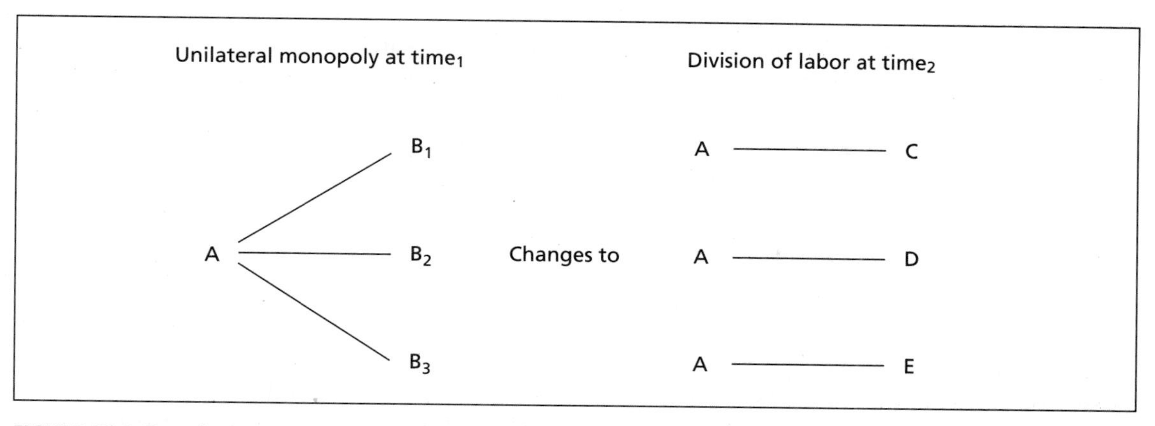

FIGURE 22-2 Transformation of Unilateral Monopoly to Division of Labor

individual or collective actors. Second, the change in the structure or form of the network is described as a proposition systematically derived from operant principles, corollaries, and other theorems. The proposition could thus apply to a wide variety of micro and macro contexts. For example, it could apply to workers in an office who specialize and provide *A* with resources not available from others. This proposition could also apply to a division in a corporation that seeks to balance its relations with the central authority by reorganizing itself in ways that distinguish it, and the services it can provide, from other divisions. Or this proposition could apply to relations between a colonial power (*A*) and its colonized nations (B_1, B_2, B_3), which specialize (become *C*, *D*, and *E*) in their predominant economic activities to establish a less dependent relationship with *A*.

Social Circles

Emerson emphasized that some exchanges are intercategory and others intracategory. An intercategory exchange is one in which one type of resource is exchanged for another type—money for goods, advice for esteem, tobacco for steel knives, and so on. The networks discussed thus far have involved *inter*-category exchanges between actors with different resources (*A*, *B*, *C*, *D*, *E*). An *intra*category is one in which the same resources are being exchanged—affection for affection, advice for advice, goods for goods, and so on. As indicated earlier, such exchanges are symbolized in Emerson's graph approach by using the same letter—A_1, A_2, A_3, and so forth—to represent actors with similar resources. Emerson then developed another proposition to describe what will occur in these intracategory exchanges:

> The more an exchange approximates an intracategory exchange, the more likely are exchange relations to become closed.

Emerson defined "closed" either as a circle of relations (diagrammed on the left in Figure 22-3) or as a balanced network in which all actors exchange with one another (diagrammed on the right in Figure 22-3). Emerson offered the example of tennis networks to illustrate this balancing process. If two tennis players of equal ability, A_1 and A_2, play together regularly, this is a balanced

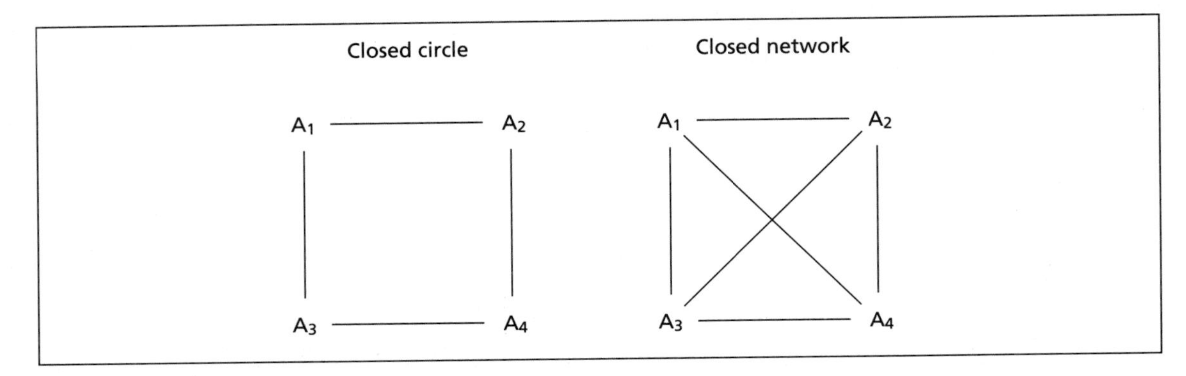

FIGURE 22-3 Closure of Intracategory Exchanges

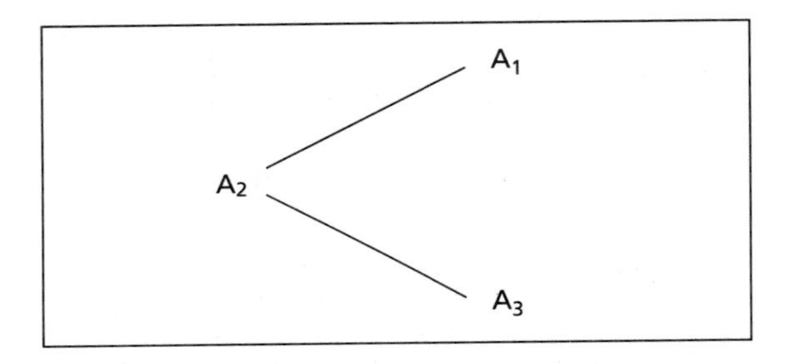

FIGURE 22-4 Imbalanced Intracategory Exchange

intracategory exchange—tennis for tennis. However, if A_3 enters and plays with A_2, then A_2 now enjoys a power advantage, as is diagrammed in Figure 22-4.

This is a unilateral monopoly, but, unlike those discussed earlier, it is an intracategory monopoly. A_1 and A_3 are dependent on A_2 for tennis. This relation is unbalanced and sets into motion processes of balance. A_4 might be recruited, creating either the circle or balanced network diagrammed in Figure 22-3. Once this kind of closed and balanced network is achieved, it resists entry by others, A_5, A_6, $A_7, \ldots, A_n$, because, as each additional actor enters, the network becomes unbalanced. Such a network, of course, is not confined to individuals; it can apply to nations forming a military alliance or common market, to cartels of corporations, and to other collective units.

Stratified Networks

The discussion about how intracategory exchanges often achieve balance through closure can help us understand processes of stratification. If, for example, tennis players A_1, A_2, A_3, and A_4 are unequal in ability, with A_1 and A_2 having more ability than A_3 and A_4, an initial circle might form among A_1, A_2, A_3, and A_4, but, over time, A_1 and A_2 will find more gratification in playing each other, and A_3 and A_4 might have to incur too many costs in initiating invitations to A_1 and A_2. An A_1 and A_3 tennis match is unbalanced; A_3 will have to provide additional resources—the tennis balls, praise, esteem, self-deprecation. The result will be for two classes to develop:

Upper social class A_1—A_2

Lower social class A_3—A_4

Moreover, A_1 and A_2 might enter into new exchanges with A_5 and A_6 at their ability level, forming a new social circle or network. Similarly, A_3 and A_4 might form new tennis relations with A_7 and A_8, creating social circles and networks with players at their ability level. The result is stratification that reveals the pattern in Figure 22-5.

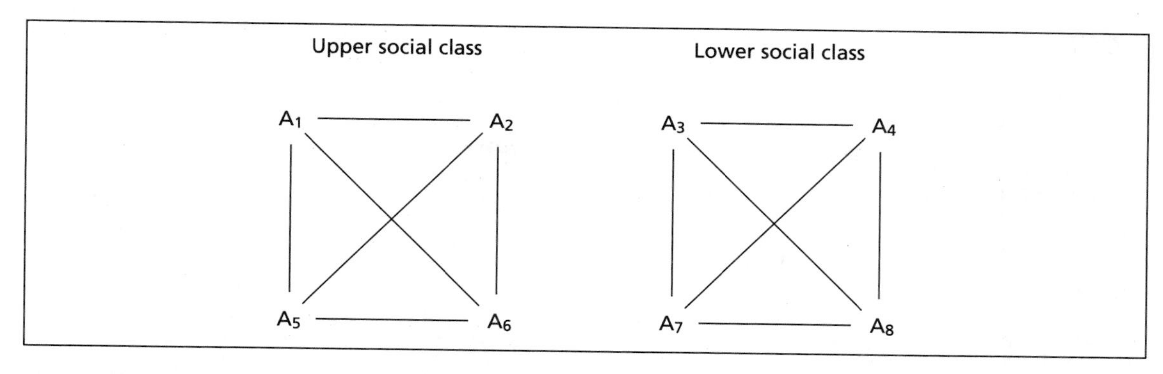

FIGURE 22-5 Stratification and Closure of Exchanges

Emerson's discussion of stratification processes was tentative, but he developed a proposition to describe these stratifying tendencies:

> The more resources are equally valued and the more resources are unequally distributed across a number of actors, the more likely is the network to stratify by resource magnitudes and the more likely are actors with a given level of resources to form closed exchange networks.

Again, this theorem can apply to corporate units as well as to individuals. Nations become stratified and form social circles, as is the case with the distinctions between the developed and underdeveloped nations and the alliances among countries within these two classes. Or this theorem can apply to traditional sociological definitions of class, because closed networks tend to form among members within, rather than across, social classes.

CONCLUSION

This blending of exchange theory and network analysis represented an important milestone in modern sociological theory. The loose deductive schemes of George Homans and others were replaced by considerably more rigor in making deductions from the principles of behaviorism, although we have not reviewed the actual deductions in this chapter. Moreover, network analysis gave exchange theory a way to talk about social structure as more than a metaphor. Structure could now be conceptualized in more precise terms and represented visually in graphs. Coupled with tests of these theoretical ideas in small group laboratories, a constant link between theory and research was maintained in Emerson's work as well as in the research of others who followed his lead.

Emerson's early death ended his contribution to the exchange theoretic program, but long before he died, he had taken on collaborators, most notably Karen Cook[3] who, in turn, has worked with colleagues, while training new generations of students. These students have continued the exchange theoretic tradition initiated by Emerson. Moreover, other established scholars not trained by Emerson have also ventured into exchange network theory. Some have drawn from Emerson; others have blended exchange theory and network analysis outside the Emerson tradition. The end result is that exchange network analysis is one of the most varied, vibrant research-theory traditions in contemporary sociology, as we will see in Chapter 25 on the current direction of this approach.

NOTES

1. Emerson's perspective is best stated in his "Exchange Theory, Part I: A Psychological Basis for Social Exchange" and "Exchange Theory, Part II: Exchange Relations and Network Structures," in *Sociological Theories in Progress*, eds. J. Berger, M. Zelditch, and B. Anderson (New York: Houghton Mifflin, 1972), pp. 38–87. Earlier empirical work that provided the initial impetus to, or the empirical support of, this theoretical perspective includes "Power-Dependence Relations," *American Sociological Review* 17 (February 1962), pp. 31–41; "Power-Dependence Relations: Two Experiments," *Sociometry* 27 (September 1964), pp. 282–298; John F. Stolte and Richard M. Emerson, "Structural Inequality: Position and Power in Network Structures," in *Behavioral Theory in Sociology*, ed. R. Hamblin (New Brunswick, NJ: Transaction Books, 1977). Other more conceptual works include "Operant Psychology and Exchange Theory," in *Behavioral Sociology*, eds. R. Burgess and D. Bushell (New York: Columbia University Press, 1969), and "Social Exchange Theory," in *Annual Review of Sociology*, eds. A. Inkeles and N. Smelser, vol. 2 (1976), pp. 335–362.

2. Curiously, Emerson returned to this question in his last article. See Richard M. Emerson, "Toward a Theory of Value in Social Exchange," in *Social Exchange Theory*, ed. Karen S. Cook (Newbury Park, CA: Sage, 1987), pp. 11–46. See, in the same volume, Jonathan H. Turner's critique of this shift in Emerson's thought: "Social Exchange Theory: Future Directions," pp. 223–239.

3. For example, see Karen S. Cook and Richard Emerson, "Power, Equity and Commitment in Exchange Networks," *American Sociological Review* 43 (1978), pp. 712–739; Karen S. Cook, Richard M. Emerson, Mary R. Gillmore, and Toshio Yamagishi, "The Distribution of Power in Exchange Networks," *American Journal of Sociology* 89 (1983), pp. 275–305; Karen S. Cook and Richard M. Emerson, "Exchange Networks and the Analysis of Complex Organizations," *Research in the Sociology of Organizations* 3 (1984), pp. 1–30. See also Karen S. Cook, "Exchange and Power in Networks of Interorganizational Relations," *Sociological Quarterly* 18 (Winter 1977), pp. 66–82; "Network Structures from an Exchange Perspective," in *Social Structure and Network Analysis*, eds. P. Marsden and N. Lin, pp. 23–46; and Karen S. Cook and Karen A. Hegtvedt, "Distributive Justice, Equity, and Equality," *American Sociological Review* 9 (1983), pp. 217–241. For recent overviews see Linda Molm and Karen S. Cook, "Social Exchange and Exchange Networks," in *Sociological Perspectives on Social Psychology*, eds. K.S. Cook, G.A. Fine, and J.S. House (Boston: Allyn & Bacon, 1995); Karen S. Cook, Linda D. Molm, and Toshio Yamagishi, "Exchange Relations and Exchange Networks: Recent Developments in Social Exchange Theory," in *Theoretical Research Programs: Studies in the Growth of Theory*, eds. J. Berger and M. Zelditch, Jr. pp. 296–323, (Stanford, CA: Stanford University Press, 1993).

23

The Maturing Tradition IV:

The Introduction of Economic and Game-Theoretic Models in Exchange Theory*

As emphasized in Chapter 19 on the emergence of exchange theory, the earliest theorists in this approach were what we today would call economists. In the late eighteenth and nineteenth centuries, however, these early theorists saw themselves as moralists or, perhaps, as political economists. And some like Karl Marx and Vilfredo Pareto became identified as sociologists. Thus, for more than two centuries there has been a persistent line of theorizing about how actors make choices in exchange relations with others in social situations. For what was to become *neoclassical economics*, these social situations were conceptualized as free markets in which rational and resource-maximizing actors exchanged resources in accordance with the laws of supply and demand. Sociologists have tended to view such conceptualizations of markets as overly simplified, but have not rejected economic theory altogether. Rather, social theorists across a broad range in political science, anthropology, and sociology have sought to reform economic theory, recognizing that perfect market conditions do not always prevail and that actors' rational calculations are often distorted by social and psychological forces. The most prominent of these efforts is what has become known as game theory, which is the main topic of this chapter. To appreciate why game theory emerged, however, we must first review the assumptions and pitfalls of neoclassical economic theory. And as we will see in Chapter 24 on rational choice theories, the core ideas of neoclassical economics and game theory undergird much exchange analysis in sociology.

* This chapter is coauthored with Charles H. Powers.

RATIONAL ACTORS IN MARKETS

Neoclassical Assumptions

Economists from the intellectual era of Adam Smith and Marx—that is, from the late eighteenth century through the middle of the nineteenth century—conceived of society as an integrated whole in which market activity, cultural values, and regulatory institutions were inexorably intertwined. This was the great period of "classical" economics. Disciplinary specialization was in its infancy and analytical frameworks tended to be sweeping in scope, so most social scientists of the time identified themselves as political economists.

By the end of the nineteenth century, economists began to narrow their intellectual focus. There were calls for a "pure" economics focusing exclusively on rational decision-making by people and organizations presumed to be maximizing the satisfaction of their own preferences.[1] This narrowing of disciplinary focus was hotly debated and has been justifiably described as an intellectual revolution. In its starkest form, contemporary neoclassical economists assume as a matter of axiom that market choices ultimately follow from conscious calculation aimed at maximizing preferences.[2] They also tend to assume that factors other than market choices are so difficult to measure and have such ambiguous meaning that they are better left out of economic analysis. Nevertheless, economists can conceive of all manner of behavior as by-products of rational decision-making aimed at satisfying preferences, thereby leading them to approach the subject matter of the social sciences within a market conceptual framework.[3] For example, Larry Iannaccone treats the choice of religious denominations as a trade-off between the demands associated with church membership ("costs" such as tithing and behavioral conformity) and the benefits they offer ("rewards" such as acceptance or help in locating a job).[4] Many noneconomists feel such analysis misconstrues people's true motives, but this line of criticism has done little to discourage economists who are convinced by the parsimony of their own characterizations. Extending neoclassical market analysis to what had previously been viewed as noneconomic behavior leaves, from an economist's point of view, economics as a "pure" science and, again from an economist's viewpoint, relegates the other social science disciplines to the realm of imprecise analysis and empty philosophizing.

Neoclassical economic theory is predicated on the assumption that people contrast expected costs and advantages before making each decision.[5] In doing so, it is assumed that people are searching for marginal advantages. The question is one of comparative advantage of one choice over another, without regard to previous decisions. This question might sound like a simple matter, but it is actually the fundamental point of departure between conventional economic perspectives, on the one side, and mainstream sociological perspectives, on the other. Sociologists believe that purely rational choices do not occur, for many reasons: Perceptions of likely rewards and probable costs are influenced by one's social structural position; values, habits, and traditions disturb clear hierarchies of preferences; emotions, cognitions, and imagination constantly alter perceptions of alternatives and preferences; and so on for all the sociological and psychological forces that constrain calculations of preferences, costs, and benefits.

Nonetheless, neoclassical economists and, as we will see in later chapters, some sociologists still use changes in preference, whether short- or long-term, to explain change in behavior. Moreover, the persistence in old patterns of behavior is taken by most economists to reflect an active preference for continuation in the same mode or a desire to avoid the costs and risks associated with searching out and experimenting with new alternatives.

That some people prefer repetition to change whereas others prefer change to repetition, and that some individuals are more averse to risk-taking than others, are generally treated as random "noise" falling outside neoclassical economic analysis. But, *opportunity costs*, which are the potential benefits people pass up by failing to select an alternative course of action, are also assumed to figure in human calculations. Altruistic behavior adds an additional element of complexity to any calculation of costs and benefits. Persons who forgo something for themselves to enable others to satisfy their needs can be thought of as gaining pleasure vicariously or reaping the intrinsic satisfaction that comes from knowing that they have been helpful to others. These less obvious types of intrinsic rewards are really the Achilles heel of neoclassical economics, because the perspective rests on an unproved assumption about people's motivations: *All behavioral choices reflect actors' current efforts to maximize preferred outcomes and minimize unwanted costs relative to possible behavioral alternatives.*

Neoclassical analysis begins with assumptions about individual decision-making, (microeconomics) but analysis quickly extrapolates to more the macro-level and explores systemic consequences (macroeconomics) following from assumptions about the conditions under which individual decisions are made. Armed with the assumption that each behavior represents a calculation of the best way to maximize *current* preferences under *current* conditions provides economics with a model of human behavior as ever flexible. Markets are thought to facilitate adaptive change; thus, markets in which people have choice are seen as preferable to bureaucratically structured hierarchies or other forms of organization that restrict choices.

Neoclassical economists have a particular notion of what a *market* is. To be considered a true market, actors must enjoy open choice among a wide range of alternative opportunities. This open *choice* among a *range* of meaningful alternatives makes it possible for people to *adjust* their actions to maximize their own preferences. In turn, the aggregate consequence of individuals adjusting in markets is assumed to lead to the maximization of collective welfare. An ideal market would position every actor to have nearly perfect current information, would allow unrestricted competition, and would minimize transaction costs and the frictional costs that come from making changes. This would allow people to adjust their behavior to produce the configuration of potential outcomes for which they have the strongest preferences. From this viewpoint, resources that could be more effectively used are redeployed, and inefficient practices are replaced.[6]

The Debate Over Neoclassical Assumptions

Few mainstream economists seem terribly troubled when forced to admit that real markets often fall short of the neoclassical "ideal-type." If a market is known to be oligopolistic, or controlled by a few actors, for example, most mainstream economists would argue that outcomes would be improved and collective welfare further enhanced if the market were to open more widely.[7] If the market does not produce the best possible results, neoclassical theory simply assumes that it is not truly free and open, and hence, the market is hampered from realizing the full promise and potential that a more pure version of an ideal-type market would produce.

Indeed, many neoclassical economists see markets as great engines for the true liberation of humans.[8] Discrimination, for instance, is antithetical to market behavior. It can even be argued that true markets promote equalization of outcomes because people and resources that flow into more rewarding endeavors increase competition and bring down prices, while reducing the supply of labor where demand is low.[9] Movement of people out of less valued jobs reduces downward pressure on wages by restricting labor supply, whereas movement of people toward attractive opportunities increases the supply of labor and talent, thereby creating downward job pressure by equalizing supply and demand across sectors. In this respect, the dynamics of a market can smooth out discrepancies and equalize outcomes.

Smoothly functioning markets also create new opportunities, allowing people to create new benefits for themselves through entrepreneurship. Indeed, economists often argue that aggregate welfare diminishes precisely when market freedoms are hampered by government or other restrictions that make it difficult for people to try new things.[10] In the hands of libertarians, market theory assumes the form of a kind of political theology with a constant generation of new opportunities, movement into those opportunities of less advantaged people, and the elimination of unequal returns unless inequality of returns are tied to differences in productivity and quality of output.[11]

Economists generally assume that people make and subsequently act on well-grounded cost/benefit comparisons among a full range of alternatives. Yet, not even the most committed neoclassical economists would argue that markets operate perfectly. Rather, their conception of the "market" is an ideal-type that, in its pure form, is an arena for exchange that attracts as many players as possible, that allows them easy entry and departure from the forum without significant transaction costs, that encourages players to make value-maximizing choices among potential exchange partners who are competing for the privilege of participating in exchange, that encourages players to consider the substitution of any pattern of exchange behavior with more advantageous courses of action should those become apparent, and, finally, that empowers players by providing them with enough information to make rational choices when selecting among possible exchange partners or possible substitute courses of action.[12]

What is most interesting to sociologists who follow literature in economics is the striking number of economists who are coming to recognize that the world is not as simple as a pristine neoclassical model suggests.[13] All kinds of cultural dynamics and institutional forces are at work in the social world.[14] These dynamics and forces fall outside the conventional neoclassical model of markets and, as Max Weber and Pareto understood, must be accounted for in order to provide an adequate grasp of economic phenomena. Economists outside the United States actually are beginning to sound more like sociologists. For example, *convention economists* in France recognize that economic behavior is constrained in a pervasive way by social values and beliefs.[15] Several Dutch economists have modified conventional neoclassical models with the inclusion of network variables.[16] And the *new institutional economists* centered in Germany place the historical evolution of banking systems, property rights, and other institutional forms in the vortex of economic analysis.[17] Yet, economists and sociologists alike have failed to appreciate the degree to which sociological principles could be used to explain deviations from the ideal-type market dynamics portrayed in the neoclassical perspective. Rather than being treated as "random noise," such deviations should be viewed as an indicator of one influence of sociocultural forces on decision-making.[18]

Sociology can thus offer economic theory realism about the ways in which market dynamics are seldom "pure."[19] After all, large bodies of sociological research have been generated about how people's choices are systematically restricted.[20] Before turning our attention to sociological exchange theory, it is important to consider one other theoretical development from economics: game theory.

GAME THEORY

The neoclassical economic model is built on an assumption that actors try to optimize preferential outcomes in markets that offer them full and timely information about a complete range of behavioral options and that allow actors to change any aspect of their performance at any time with few penalties. Moreover, *markets are assumed to be composed of such large numbers of potential exchange partners with free and open access to each other that outcomes for each individual are essentially independent of actions taken by any other single actor.* The market conditions of *supply and demand,* in neoclassical theory, are assumed to have pervasive impact on outcomes. The strategy of any single player is viewed as having little, if any, impact on market overall outcomes.

Game theory was developed by economists who wanted to retain certain elements from the neoclassical model, but who felt that neoclassical economists were inclined to make too many unrealistic assumptions about the perfection of markets.[21] At its most fundamental core, *game theory rests on the assumption that the outcomes that result from one actor's behavior are highly contingent on actions and reactions of other actors,* and the actions and reactions are uncertain and subject to change. For these reasons, *strategies* can have a significant impact on outcomes that are independent of prevailing market conditions. The strategies of other actors simply cannot be overlooked by any rational player and, consequently, should not be left out of socioeconomic theory.

With *strategy* as its focus, game theorists offered a "fundamental theorem" that, in fair two-person games of skill in which people can choose a range of strategies (games like chess), it is theoretically possible for either person to achieve at least "a draw." A draw occurs when evenly matched players follow the *minimax strategy* of minimizing an opponent's score while maximizing one's own score. But John von Neumann and Oskar Morgenstern caution that few situations are simple, nor do most involve a fair match of skills.[22] In the real world, variations in circumstances and uncertainty about the actions of others suggest that the wisest course of action might be to abandon the minimax strategy in favor of something less rational in its appearance: the *maximin strategy* of avoiding exposure to worst case outcomes.

Game theorists have developed a method for trying to predict behavior by using payoff matrices to assess the results that a player might expect from various strategies in a range of game plans. By way of illustration, John Mullen and Byron Roth discuss armed conflict between well armed government troops who have less than full popular support and poorly armed insurgents who enjoy significant pockets of popular support.[23] If government troops and insurgents follow the *maximin strategy* of avoiding exposure to worst case outcomes, they will fall into a civil war of attrition, with the insurgents avoiding all-out attacks of the government's military strongholds and the government minimizing the size and frequency of search and destroy missions unless total success is anticipated. Figure 23-1 shows the payoff matrix for this "game" (however deadly).

If each camp follows a minimax strategy, the game-plan is determined, although the final outcome is not. Insurgents will avoid massive confrontation against entrenched government positions in favor of waging a war of attrition using hit-and-run tactics

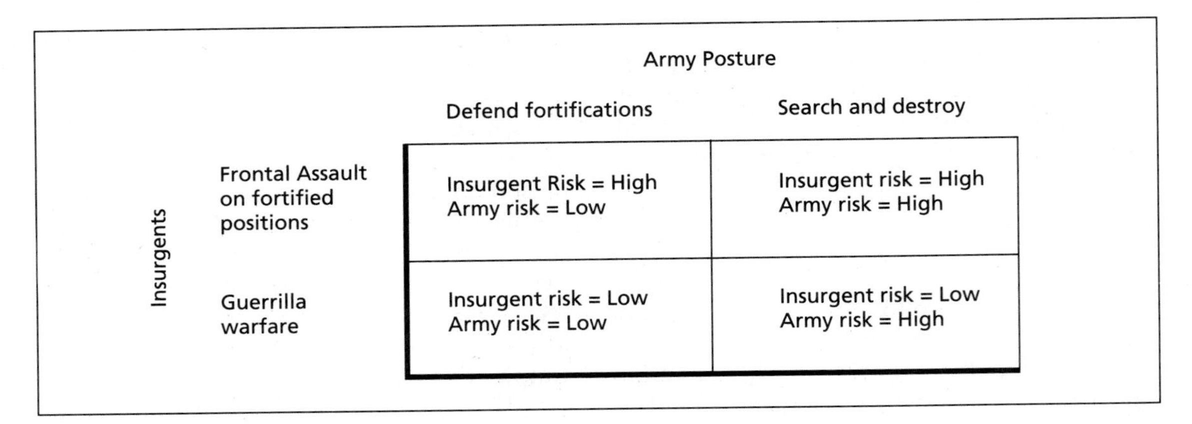

FIGURE 23-1 Payoff Matrix in a Game-Theoretic Analysis of Insurgency

and hoping to demoralize the army and further erode popular support for the government by exposing the government's inability to exercise effective control over the countryside or the cities. Meanwhile, the government will try to solidify control over fixed positions, starve the insurgents, and collect the intelligence needed to unleash military raids designed to kill key rebels or destroy key rebel units.

This is a useful example because it illustrates how game theorists approach the study of choice. Sometimes players with very different interests and objectives are led by *maximin strategic reasoning* to settle on one mode of engagement, or game plan. Although the final outcome is not predicted by this game-theoretical analysis, the choices of strategies of key players are. The strategic way in which such situations will begin to unfold has a "solution" as long as players follow a maximin strategy. This "solution" is what game theorists refer to as an *equilibrium saddlepoint* that forms the arena for continued engagement because no other terrain seems more desirable to either party or mutually acceptable to both the contesting parties. Von Neumann and Morgenstern did observe that mixing strategies can be a good idea, if simply to keep your opponents guessing, but the higher the stakes are, the harder this mixing becomes.[24] In contrast, where stakes are low, mixing strategies can be effective—as might be the case, for example, when taking an occasional shot from half court in a basketball game (a low-cost maneuver) opens up more opportunities for shots close to the rim by forcing the opposing side to spread its defense. However, risking an army in a dangerous confrontation increases the stakes considerably and decreases one's propensity to mix strategies.

Game theorists accept the neoclassical assumptions that people make calculations and try to maximize preferred outcomes. Indeed, game theorists regard these assumptions as fundamental axioms pivotal to understanding most social phenomena. But the "rational" course of action is far from obvious if, as game theorists believe, the defining characteristics of a pure market are almost always absent. The information on which people actually make decisions is typically quite limited, frequently misleading, and often untimely. Selection of exchange partners and behavioral responses to those partners are often limited and very tightly constrained. And most important, the strategies adopted by one actor can significantly affect the payoffs of other actors.

The discrepancies between the idealized market model of neoclassical economics and the real-life arenas in which exchange actually takes place are amplified because the rationality of an opponent cannot be assumed, and any opponent (rational or not) can engage in deceit or follow an irrational course of action for a time to create uncertainty and doubt. This means that the wisest response to a situation can only be determined after taking these deviations from the "pure" market model into account and after compensating for sources of uncertainty, such as alterations in strategy by other players.

Over several productive decades of work, game theorists have identified some strategic choices that seem to yield the best payoff under common circumstances. Game theorists have done this by, in essence, creating board games to match the conditions prevailing in different exchange environments. A critical aspect of these modeling exercises is the presumption that actors are usually able to alter their strategies.[25] Game theoretic models are further complicated when, as is so often the case, particular combinations of players are free to exchange information among themselves, to bargain, or even to collude. Far from dismissing such situations, game theorists have always found such cooperative situations intriguing subjects of analysis.[26]

Exercises in game theory are only useful if the normative and legal rules and the practical obstacles constraining behavior can be translated accurately into "rules of the game." In this sense, much of the analytical challenge of game theory is the preliminary work of defining the circumstances and conditions governing exchange. Once those circumstances and conditions have been amply defined, game theorists deductively search for the most "rational" playing strategies.[27]

Game theory has proven a useful approach in trying to define important exchange conditions that neoclassical economists might fail to highlight. For example, people have an assortment of interests that can either overlap or can conflict with the interests of others. Moreover, it is often in people's interests to form coalitions, and coalitions frequently invalidate the supply and demand model that defines neoclassical market analysis.[28] Game theorists are thus comfortable entertaining a rich stew of competing and complementary interests[29] because they make no assumptions about the natural order of interest in the real world. This means that some situations are zero-sum (one party gains only when another party loses), but others can be win/win (all parties gain) or lose/lose (all parties lose).[30]

Especially with the formation of coalitions, stable equilibria can be reached in which people with a variety of interests and objectives find themselves using interlocking strategies in which their preferences are satisfied at a level they can live with, and obvious behavioral alternatives would worsen their cost/benefit rations.[31] Yet, when coalitions do break up, the effectiveness of old strategies is undermined, whereas the payoffs for other strategies can change.[32]

Game theorists, like neoclassical economists, assume that people seek preferred outcomes. But the implicit neoclassical assumption of a giant free market conveys the impression that the outcomes that actors realize are almost entirely a function of how they use the resources at their disposal. In the real world, frequently the strategies that others follow significantly affect the outcomes produced by the strategies we choose, often mediated through complex and long-term feedback mechanisms.[33] By analyzing payoff matrices, like the simple one in Figure 23-1, game theorists have developed a method for assessing the impact on the payoffs for one party that come from the strategies adopted by other players. This approach can be further illustrated with some of the "classic" games that have been conceptualized by game theorists.

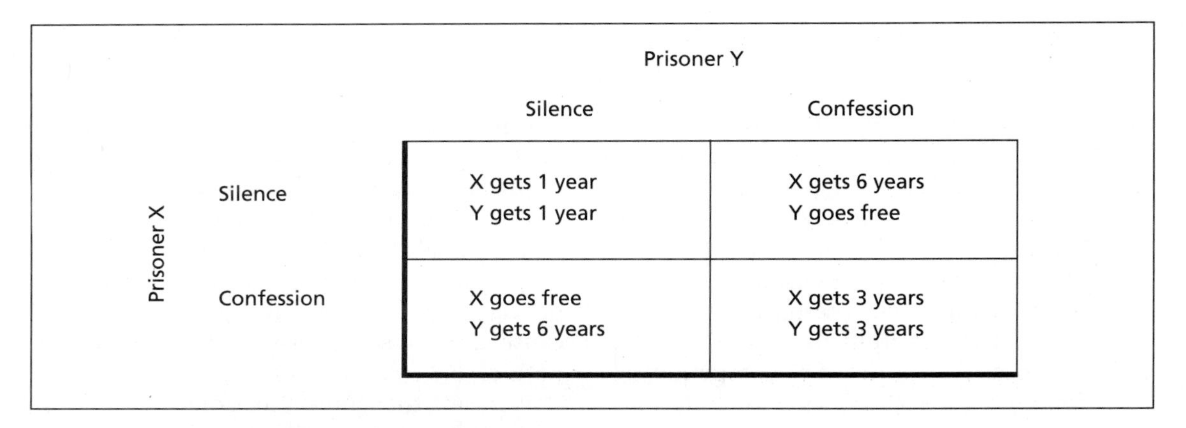

FIGURE 23-2 Payoff Matrix for a Game Analysis of the Prisoner's Dilemma

The Prisoner's Dilemma: Game Theorists' Favorite Illustration

Game theorists' favorite illustration is the "prisoner's dilemma." A *prisoner's dilemma* exists whenever both parties would be better off by cooperating, but each has an incentive to cheat. The prisoner's dilemma model is interesting because it illustrates how patterns of interest can create inherent instability in social relations.[34]

In its simplest form, the prisoner's dilemma presents itself when two suspected partners in crime have been arrested, but there is only evidence for a conviction of a lesser crime.[35] Each suspect is offered freedom in exchange for pleading guilty and testifying against the other suspect, but only on the condition that (s)he is the first to "cop a plea." If there are two confessions, both suspects will go to jail for the more serious crime. If there is only one confession, the judge (known throughout the criminal justice system as severe) is expected to make an example of the person who fails to cooperate. If both prisoners remain silent, they can expect to be convicted of the lesser crime, and receive a light sentence.

This scenario allows us to construct a payoff matrix in Figure 23-2 where 4 represents the promise of release from jail in exchange for coping a plea and testifying against one's partner, 3 represents the probability of being convicted of a lesser crime, 2 represents a standard sentence if both prisoners confess and receive the normal punishment for the crime, and 1 represents the maximum possible punishment for an uncooperative prisoner turned in by his or her accomplice in crime.

Prisoners are kept apart in situations like this, so it is hard for either to be confident about the other's actions. Mutual silence is the best overall solution (each gets a one-year sentence), whereas being the only silent partner is worst (leading to a six-year sentence). A prisoner trying to improve his or her own outcome cannot be sure whether confessing or silence will improve the outcome: If an actor confesses, and the partner does the same, each gets three years; if an actor remains silent and the partner confesses and cops the plea, then this actor gets six years; if an actor remains silent, and the partner does the same, each only gets one year. Remaining silent seems to be the preferred course of action, but still, if one does remain silent while one's partner confesses, a six-year sentence will ensue. Thus, there is a powerful incentive to consider confessing, but at some risk. Rationally speaking, what should one do? In the absence of other information, this is a genuine dilemma with no clear answer. The payoff matrix helps game theorists appreciate just how much of a dilemma this is, and therefore how unpredictable and unstable the outcomes will be even if the people involved are trying to be as rational as possible.

The prisoner's dilemma can actually be applied to a wide range of situations. For example, in capital-intensive industries with long lead times, one firm that acts alone to expand productive capacity and then reduces sales price will generally become more profitable by capturing a larger share of the market and reaping the benefits that come with economies of scale. But if all competing firms pursue aggressive expansion policies and reduce prices they will simultaneously increase costs and diminish revenue without capturing greater market share.[36] Of course, corporations often have informal ways of predicting each other's actions, but when that information is lacking, firms can become prisoners of their own uncertainty and face a strategic planning dilemma that can defy rational solution. The important point to grasp is that the relational situations modeled by game theorists, like a prisoner's dilemma, are evident in a wide variety of real-life situations.

Contrasting Prisoner's Dilemma with a Pareto Optimal Equilibrium

The prisoner's dilemma is, of course, only one of many types of situations making up the social universe. But game theorists have found payoff matrices to be a highly versatile tool for considering a wide range of situations encountered in the real world. In this section we illustrate the characteristics of a *Pareto optimal equilibrium* using the same method employed to understand the prisoner's dilemma, but illustrating the distinctive characteristics rather than similarities between these different situations.

A Pareto optimal equilibrium is the ideal-type economic solution in which no actor can change in a way that improves his or her situation. The market is viewed by economists as the best mechanism available for helping people adjust their situations until no further improvement is possible given their particular set of tastes, resources, and talents. Yet, the concept of optimal strategy applies whether interaction is organized in markets, in hierarchies, or in networks. Consider the example of Student I and Student II who each wish to do well on Professor X's final examination. *If* Professor X is known to expect exam takers to have prepared extensively by using study sheets passed out a week before the exam, *if* these study sheets stress a level of synthesis and integration not reached in class discussion and thus require new work rather than simple review, and *if* Professor X is known to grade substantially although not entirely on a curve, then a game theorist would predict that both students will study, based on the payoff matrix presented in Figure 23-3.

If each student follows a minimax strategy under the stipulated assumptions, each will pay attention to the study guide in preparing for the examination. A "Pareto optimal" stable equilibrium is reached because neither student can be advised to follow a different course of action, no matter what the other does. Payoff for student A's study strategy does indeed vary depending on what student B decides to do. But under the stipulated conditions, the study strategy is always best for A, regardless of B's course of action. It is true that A receives greater reward for studying if B does not study, but regardless of whether B studies or not, A is better off if A studies than if A does not study.

The choice of students to study offers a simple application of payoff matrices to what economists think of as conventional situations. During the last decades, game theorists have applied their analytical apparatus to more complex situations. For example, Francesco Forte and Charles Powers extrapolate the results that government regulators

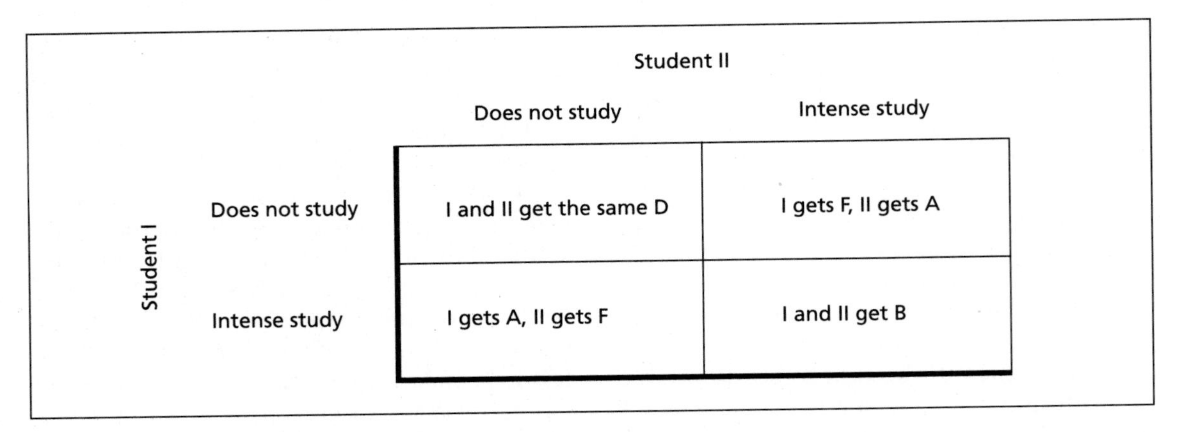

FIGURE 23-3 Payoff Matrix for a Game Analysis of Student Decision-Making

might expect from following different enforcement strategies.[37] They reason that concentrating resources on the most cost-effective enforcement activities will yield the highest level of compliance over time, even though that enforcement strategy means consciously overlooking some known violations. The use of game theory to expand discussion of such highly charged and practical issues suggests its vitality and utility as a theoretical approach.

CONCLUSION

In this chapter we have briefly examined two channels through which economic thought has influenced contemporary sociological theory. Neoclassical economics positing utilitarian calculation and behavioral adjustment provide the conceptual starting point for the rational choice theories examined in the next chapter. But rational choice sociologists generally eschew as unrealistic many of the simplifying assumptions made by economists and, instead, follow the lead of game theorists in economics. After trying to identify social structural and environmental constraints and actor orientations as carefully as possible, game theorists search for the behavioral strategies that seem to be optimal under real-world conditions rather than in the ideal typical world of neoclassical market imagery. As we will see in the next chapter, sociologists have used the assumptions of neoclassical economics about rationality and the techniques of game theorists to generate new insights into the traditional subject matter of sociology.

NOTES

1. Maffeo Pantaleoni, *Pure Economics* (New York: Macmillan, 1898).

2. Athanasios Asimakopulos, *Theories of Income Distribution* (Boston: Kluwer Academic, 1988).

3. Gary Becker, *A Treatise on the Family* (Cambridge, MA: Harvard University Press, 1981).

4. Larry Iannaccone, "Why Strict Churches are Strong," *American Journal of Sociology* 99 (1994), pp. 1006–1028.

5. Leon Walras, *Elements of Pure Economics or the Theory of Social Wealth,* trans. W. Jaffe (London: Allen and Unwin, 1954).

6. Charles Powers, "Bridging the Conceptual Gap Between Economics and Sociology," *Journal of Socio-Economics* (1996), pp. 225–243.

7. Robin Broadway and Neil Bruce, *Welfare Economics* (New York: Oxford University Press, 1984).

8. Wilhelm Roepke, *A Human Economy* (Chicago: H. Regnery, 1960).

9. Luigi Pasinetti, *Structural Change and Economic Growth* (Cambridge: Cambridge University Press, 1981).

10. Ludwig von Meises, *Human Action: A Treatise on Economics,* 3rd ed. (Chicago: H. Regnery, 1966).

11. Gerrit Meijer, "The History of Neo-Liberalism: A General View and Developments in Several Countries," *Rivista Internazionale de Scienze Economiche e Commerciali* 34 (1987), pp. 577–591.

12. Francis Bator, "The Simple Analytics of Welfare Maximization," *American Economic Review* 47 (1957) pp. 22–59.

13. Tjerk Huppes, ed., *Economics and Sociology: Towards an Integration,* 2nd ed. (The Hague: Nijhoff, 1982).

14. Mark Granovetter, "Economic Action and Social Structure: The Problem of Embeddedness," *American Journal of Sociology* 91 (1985), pp. 481–510.

15. Francois Eymard-Duvernay, "Comventions de qualite et formes de coordination," *Revue Economique* 40 (1989), pp. 329–359.

16. Chris de Neubourg and Maarten Vendrik, "An Extended Rationality Model of Social Norms in Labor Supply," *Journal of Economic Psychology* 15 (1994), pp. 93–126.

17. Erik Furubotn and Rudolf Richter, eds., *The New Institutional Economics* (Tubingen: J.C.B. Mohr, 1991).

18. See Powers, "Bridging the Conceptual Gap Between Economics and Sociology," pp. 225–243.

19. Joseph Stiglitz, "Another Century of Economic Science," *Economic Journal* 101 (1991), pp. 134–141.

20. Amitai Etzioni, *The Moral Dimension: Toward a New Economics* (New York: Free Press, 1988). Siegwart Lindenberg, "Homo Socio-economicus: The Emergence of a General Model of Man in the Social Sciences," *Journal of Institutional and Theoretical Economics* (Zeitschrift fur die Gesamte Staatswissenschaft) 146 (1990) pp. 727–748.

21. John von Neumann and Oskar Morgenstern, *Theory of Games and Economic Behavior* (Princeton, NJ: Princeton University Press, 1944).

22. Ibid.

23. John Mullen and Byron Roth, *Decision Making: Its Logic and Practice* (Savage, MD: Rowman and Littlefield, 1991), p. 250.

24. von Neumann and Morgenstern, *Theory of Games and Economic Behavior* (cited in note 21).

25. Ken Binmore, *Essays on the Foundations of Game Theory* (Cambridge, MA: Harvard University Press, 1981).

26. Ewald Burger and John Freund, *Introduction to the Theory of Games* (Englewood Cliffs, NJ: Prentice-Hall, 1959).

27. Thomas Schelling, *The Strategy of Conflict* (Cambridge, MA: Harvard University Press, 1962).

28. Werner Raub and Jeroen Weesie, "Symbiotic Arrangements: A Sociological Perspective," *Journal of Institutional and Theoretical Economics* 149 (1993) pp. 716–724.

29. John C. Harsanyi and Reinhard Selten, *A General Theory of Equilibrium in Games and Social Situations* (Cambridge: Cambridge University Press, 1977).

30. John Keegan, *The Face of Battle: A Study of Agincourt, Waterloo, and the Somme* (New York: Random House, 1977).

31. John Nash, "Two Person Cooperative Games," *Econometrica* 21 (1953) pp. 128–140.

32. Harsanyi and Selten, *A General Theory of Equilibrium Selection in Games* (cited in note 29).

33. Donella Meadows, Dennis Meadows, and Jorgen Randers, *Beyond the Limits: Confronting Global Collapse and Envisioning a Sustainable Future* (Post Mills, VT: Chelsea Green, 1992).

34. Richmond Campbell and Lanning Sowden, eds., *Paradoxes of Rationality and Cooperation* (Vancouver: University of British Columbia Press, 1985).

35. Anatol Rapoport and Albert Chammah, *The Prisoner's Dilemma: A Study of Conflict and Cooperation* (Ann Arbor: University of Michigan Press, 1969).

36. Mullen and Roth, *Decision Making: Its Logic and Practice* (cited in note 23), p. 257.

37. Francesco Forte and Charles Powers, "Applying Game Theory to the Protection of Public Funds; Some Introductory Notes," *European Journal of Law and Economics* 1:3 (1994) pp. 193–212.

24

The Continuing Tradition I: Rational Choice Theories

Adam Smith, who formulated the basic laws of supply and demand for free markets (see Chapter 23), is often viewed as the founder of utilitarian economics. In Smith's formulation, actors are conceptualized as rational and as seeking to maximize their utilities, or benefits.[1] Yet, his utilitarian ideas also led Smith to formulate the basic question guiding much sociological theory in the nineteenth century: What forces are to hold society together in a world where market-driven production and consumption lead individuals into ever more specialized social niches? What, then, was to keep specialization from splitting society apart? Adam Smith had two answers: (1) the laws of supply and demand in markets would operate as a kind of "invisible hand of order" in matching people's needs to production and in controlling fraud, abuse, and exploitation; and (2) differentiated societies would develop sentiments appropriate to the new social order being created by markets.[2]

Through the midpoint of this century, sociologists were suspicious of the first answer, wondering how social order could emerge from rational actors pursuing their own self interests in markets operating by profit and greed. The second answer—an emphasis on moral codes—was more appealing but sociologists soon forgot that Adam Smith had recognized the importance of cultural codes as an essential force of social organization. As a result, for more than half the twentieth century, sociologists were critical of the utilitarians' views. In essence, they asked this: How could selfish, resource-maximizing actors ever create the moral codes and modes of cooperation so necessary for an orderly society? Their answer was that such utilitarian theories could not explain the emergent forces holding societies together.

Yet, by midcentury, exchange theories were becoming a prominent part of the sociological canon, as we saw in Chapters 19, 20, and 21. Some were couched in behavioristic psychology, others in the utilitarian terminology of classical economics. In either case, the goal was to demonstrate how reward- or

TABLE 24-1 Assumptions of Rational Choice Theory

I. Humans are purposive and goal oriented.

II. Humans have sets of hierarchically ordered preferences or utilities.

III. In choosing lines of behavior, humans make rational calculations about
   A. The utility of alternative lines of conduct with reference to the preference hierarchy.
   B. The costs of each alternative in terms of utilities foregone.
   C. The best way to maximize utility.

IV. Emergent social phenomena—social structures, collective decisions, and collective behavior—are ultimately the result of rational choices made by utility-maximizing individuals.

V. Emergent social phenomena that arise from rational choices constitute a set of parameters for subsequent rational choices of individuals in the sense that they determine
   A. The distribution of resources among individuals.
   B. The distribution of opportunities for various lines of behavior.
   C. The distribution and nature of norms and obligations in a situation.

utility-seeking and rational actors could construct sociocultural systems. In effect, these new theorists took the challenge that had been presented by sociology's distrust of utilitarianism: Demonstrate how a model of an individualistic, rational, self-interested, and utility-maximizing actor can create the social and cultural forces that bind the members of society together. In this chapter, we will review two such efforts to meet this challenge through the theoretical approach that is now known as rational choice theorizing.[3]

MICHAEL HECHTER'S THEORY OF GROUP SOLIDARITY

Michael Hechter's theory of group solidarity seeks to explain how rational, resource-maximizing actors create and remain committed to the normative structure of groups.[4] Table 24-1 summarizes his utilitarian assumptions, but the basic ideas are straightforward: Individuals reveal preferences or hierarchies of utility (value); they seek to maximize these preferences; and under certain conditions, it is rational for them to construct cultural and social systems to maximize utilities. Most theories of culture and social structure, Hechter argues, simply assume the existence of emergent sociocultural phenomena but do not explain how and why they would ever emerge in the first place. Rational choice theory can, he believes, offer this explanation for why actors construct and then abide by the normative obligations of groups. And, if such a fundamental process as group solidarity can be explained by rational choice assumptions, virtually all emergent social phenomena can be similarly understood.

The Basic Problem of Order in Rational Choice Theorizing

In many contexts, especially those of interest to sociologists, individuals depend on others in a group context for those resources, or goods, that will maximize utilities. That is, these individuals cannot produce the good for themselves and, hence, must rely on others to produce it for them or join others in its production. For example, if companionship and affection are high preferences, this "good" can be attained only in interaction with others, usually in groups; if money is a preference, then this good can usually be attained in modern settings through work in an organizational context. Thus the "goods" that meet individual preferences can often be secured only in a group; groups are conceptualized in rational choice theory as existing to provide or produce "goods" for their members.

Those goods that are produced by the activities of group members can be viewed as *joint goods,* because they are produced jointly in the coordinated activities of group members. Such joint goods vary along a critical dimension: their degree of "publicness." A highly *public good* is available not only to the members of the group but not to others outside the group as well. Furthermore, once the good is produced, its use by one person does not diminish its supply for another. For instance, radio waves, navigational aids, and roads are public goods because they can be used by those who did not produce them and because their use by one person does not (at least to a point) preclude use by another. In contrast with public goods are *private goods,* which are produced for consumption by their producers. Moreover, consumption by one person decreases the capacity of others to consume the good. Private goods are thus kept out of the reach of others to ensure that only a person, or persons, can consume them.

The basic problem of order for rational choice theorists revolves around the question of public goods. This problem is described as the *free-rider* dilemma.[5] People are supposed to produce public goods jointly. It is rational to consume public goods without contributing to their production. To avoid the costs of contributing to production is free-riding. If everybody free-rides, then the joint good will never be produced. How, then, is this dilemma avoided?

An answer to this question has been controversial in the larger literature in economics, but the basic thrust of the argument is that, if a good is highly public, people can be coerced (through taxes, for example) to contribute to its production (say, national defense) or they can be induced to contribute by being rewarded (salaries, praise) for their contribution. Another way to prevent free-riding is to exclude those who do not contribute to production from consumption, thereby decreasing the degree of "publicness" of the good. This exclusion can result in a group that "throws out" noncontributing members or does not allow them to join in the first place. A final way to control free-riders is to impose user fees or prices for goods that are consumed.

Thus, for rational choice theory, the basis of social order revolves around creating group structures to produce goods that are consumed in ways that limit free-riding—that is, consumption without contributing in some way, directly or indirectly, to production. The sociologically central problem of social solidarity thus becomes one of understanding how rational egoists go about establishing groups that create normative obligations on their members to contribute and, then, enforcing their conformity—thereby diminishing the problem of free-riding. Solidarity is thus seen as a problem of social control.

The Basis of Social Control: Dependence, Monitoring, and Sanctioning

In rational choice theory, groups exist to provide joint goods. The more an individual depends on a group for resources or goods that rank high in his or her preference hierarchy, the greater is the potential power of the group over that individual. When people depend on a group for a valued good, it is rational for them to create rules and obligations that will ensure access to this joint good. Such is particularly likely to be the case when the valued joint good is not readily available elsewhere, when individuals lack information about alternatives, when the costs of exiting the group are high, when moving or transfer costs are high, and when personal ties, as unredeemable sunk investments, are strong.

Dependence is thus the incentive behind efforts to create normative obligations in order to ensure that actors will get a joint good. Groups thus have power over individuals who are dependent on the resources generated by the group; as a result of this power, *the extensiveness of normative obligations in a group is related to the degree of dependence.* Hence, dependence creates incentives not just for norms but for extensive norms that guide and regulate to a high degree.

Yet Hechter is quick to emphasize that "the extensiveness of a group alone ... has no necessary implications for group solidarity."[6] What is crucial is that group members will comply with these norms. Compliance is related to a group's *control capacity,* which in turn is a function of (1) *monitoring* and (2) *sanctioning.* Monitoring is the process of detecting nonconformity to group norms and obligations, whereas sanctioning is the use of rewards and punishments to induce conformity. When the monitoring capacity of a group is low, then it becomes difficult to ensure compliance to norms because conformity is a cost that rational individuals will avoid, if they can. And, without monitoring, sanctioning cannot effectively serve as an inducement to conformity.

Hechter believes, then, that solidarity is the product of dependence, monitoring, and sanctioning. But solidarity is also related to the nature of the group, which led Hechter to distinguish types of groups by the nature of joint goods produced.

Types of Groups

Hechter views control capacity—that is, monitoring and sanctioning—as operating differently in two basic types of groups. If a grouping produces a joint good for a market and does not itself consume the good, then control capacity can be potentially reduced because the profits from the sale of the good can be used to "buy" conformity. Conformity can be bought, for example, because members are compensated for their labor; if they are highly dependent on a group for this compensation, then it is rational to conform to norms. But, because the same compensation can be achieved in other groups, it is less likely that dependence on the group is high, which thereby reduces the extensiveness of norms. The result is that monitoring and sanctioning must be high in such groups, for it would be rational for the individual to free-ride and take compensation without a corresponding effort to produce the marketable good. Yet, if monitoring and sanctioning are too intrusive and impose costs on individuals, then it is rational to leave the group and seek compensation elsewhere. Moreover, as we will see, extensive monitoring and sanctioning are costly and cut into profits—hence limiting the social control capacity of the group. The control capacity of these *compensatory groups,* as Hechter calls them, is thus problematic and ensures that solidarity will be considerably lower than in *obligatory groups.*

Obligatory groups produce a joint good for their members' own consumption. Under these conditions, it is rational to create obligations for contributions from members; if dependence on the joint good is high, then there is considerable incentive for conformity because there is no easy alternative to the joint good (unlike the case in groups in which a generalized medium like money is employed as compensation). Moreover, monitoring and sanctioning can usually be more efficient because monitoring typically occurs as a by-product of joint production of a good that the members consume and because the ultimate sanction—expulsion from the group—is very costly to members who value this good. As Hechter notes[7]

> Due to greater dependence, obligatory groups have lower sanctioning and monitoring costs. Since every group has one relatively costless

> sanction at its disposal—the threat of expulsion—then the greater the dependence of group members, the more weight this sanctioning causes. [Moreover,] . . . monitoring and sanctioning are to some extent substitutable. If the value of the joint good is relatively large, the threat of expulsion can partly compensate for inadequate monitoring. The more one has to lose by noncompliance, the less likely one is to risk it. . . .

There is an implicit variable in Hechter's analysis: group size. In general terms, compensatory groups organize larger numbers of individuals to produce marketable goods, whereas obligatory groups are smaller and provide goods for their members that cannot be obtained (or, only at great cost) in a market. Thus, not only will dependence be higher in obligatory groups but monitoring and sanctioning also will be considerably easier, thereby increasing solidarity, which Hechter defines as the extent to which members' private resources are contributed to a collective end. High contributions of private resources can occur only with extensive norms and high conformity—two conditions unlikely to prevail in compensatory groups. In Hechter's terms, then, high solidarity can be achieved only in obligatory groups in which dependence on a jointly produced and consumed good is high, in which monitoring is comparatively easy because of small size and because members can observe one another's production and consumption of the good, and in which sanctioning is built into the very nature of the good (that is, receiving a good is a positive sanction, whereas expulsion or not receiving the good is a very costly negative sanction). Under these conditions, people will commit their private resources—time, energy, and self—to the production of the joint good and, in the process, promote high solidarity.

For Hechter, then, high degrees of solidarity are possible only in obligatory groups, in which dependence, monitoring, and sanctioning are high. Figure 24-1 represents his argument, modified in several respects. First, group size is added as a crucial variable. As obligatory groups get large, their monitoring and sanctioning capacity decreases. Second, on the far left is added a variable that is perhaps more typical of many human groupings: the ratio of consumption to compensation. That is, many groupings involve a mixture of extrinsic compensation for the production of goods consumed by others *and* goods consumed by members. For example, groups composed of members working for a salary in an organization often develop solidarity because they also produce joint goods—friendship, approval, assistance, and the like. Indeed, at times solidarity develops around obligations that run counter to the official work norms of the organization. We need to conceptualize groups not so much as two polar types, but as mixtures of (1) external compensation for goods jointly produced by and consumed by others and (2) internal consumption of goods jointly produced and consumed by members. The greater the proportion of compensation is to internal consumption, the less likely are the processes depicted in Figure 24-1 to operate; conversely, the less the ratio of compensation to internal consumption is, the more likely these processes are to be activated in ways that produce solidarity. This proposition does not violate the intent of Hechter's typology because he argues that control capacity of compensatory groups increases if such groups also produce a joint good for their members own consumption.

A larger-scale social system, such as a community or society, is a configuration of obligatory and compensatory groups. Solidarity will be confined mostly to obligatory groups, whereas the problems of free-riding will be most evident in compensatory

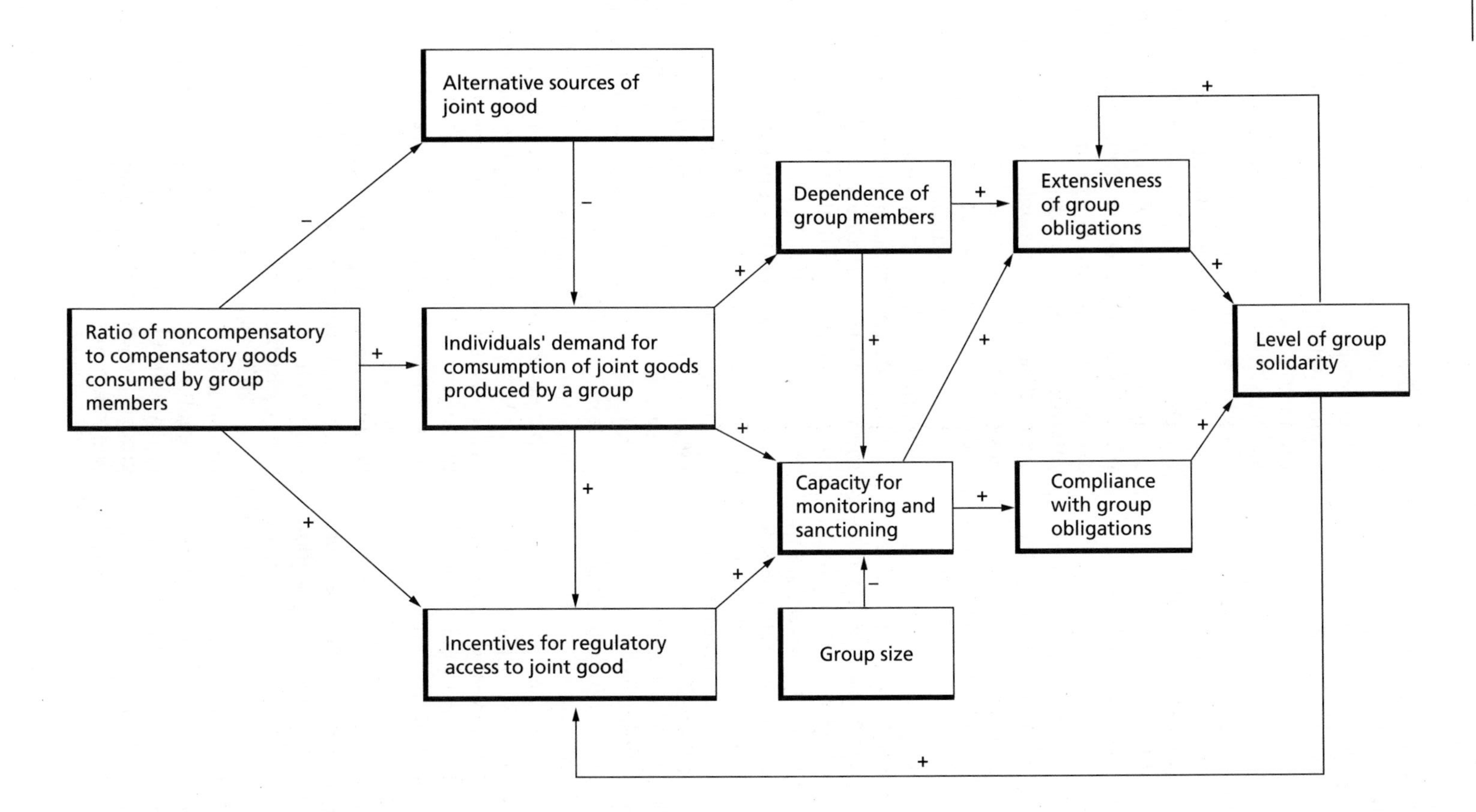

FIGURE 24-1 Hechter's Model of the Forces Producing Group Solidarity

groups, unless they also develop joint goods that are highly valued and consumed by group members. Hechter thus turns back to the basic distinctions that dominated early sociological theory—*gemeinschaft* versus *geselleschaft,* primary versus secondary groups, mechanical versus organic solidarity, traditional versus rational authority, folk versus urban—and has sought to explain these distinctions as the production of joint goods and the nature of the control process that stems from whether a joint good is consumed by members or produced for a market in exchange for extrinsic compensation. Hechter believes the nature of the joint good determines the level of dependence of individuals on the group and the control capacity of the group. High dependence and control are most likely when joint goods are consumed, and hence solidarity is high under these conditions. A society with only compensatory groups will, therefore, reveal low solidarity. What is distinctive about Hechter's conceptualization is that it is tied to a utilitarian theory in which both high and low levels of solidarity follow from rational choices of individuals.

Patterns of Control in Compensatory and Obligatory Groups

In a vein similar to classical sociological theory, Hechter examines the process of formalization.[8] As groups get larger, informal controls become inadequate, even in obligatory groups. Of course, if compensatory groups proliferate or grow in size, the process of creating formal controls escalates to an even greater degree. There is, however, a basic dilemma in this process: Formal monitoring and sanctioning are costly because they involve creating special agents, offices, procedures, and roles, thereby cutting into the production of goods. Obligatory groups can put off the process of formal controls because of high dependence and control that comes from joint consumption of a good that is valued, but when obligatory groups get too large, more formal controls become necessary. Compensatory groups can try to keep formal controls to a minimum, especially if they can create consumption of joint goods that reinforce the production norms for those goods that will be externally consumed; however, if they get too large, then they must also increase formal monitoring and sanctioning. In all these cases, it is rational for actors in groups to resist imposing formal controls because they are costly and cut into profits or joint consumption. But if free-riding becomes too widespread and cuts into production, then it is rational to begin to impose formal controls.

To some extent the implementation of formal controls can be delayed, or mitigated, by several forces. One is common socialization, and groups often seek members who share similar outlooks and commitments to reduce the risks of free-riding and cut the costs of monitoring. Another is selection for altruism, especially in obligatory groups, in which unselfish members are recruited. Yet there clearly are limits on how effective these forces can be in maintaining social control, especially in compensatory groups but also in obligatory groups as they get larger.

The result of this basic dilemma between the costs of free-riding, on the one side, and formalization of control, on the other, is for groups to seek "economizing" measures for monitoring and sanctioning.[9] These are particularly visible in compensatory groups in which formal social control is more essential to production, but elements of these "economizing tactics" can be seen in obligatory groups as they get larger or as they produce a joint good that creates problems of free-riding.

Hechter listed a number of monitoring economies. One way to decrease monitoring costs is to increase the visibility of individuals in the group through a variety of techniques: designing the architecture of the group so that people are physically visible to one another, requiring members to engage in public rituals to reaffirm their commitments to the group, encouraging group decision making so that individuals' preferences are exposed to others, and administering public sanctions for behavior that exemplifies group norms. Another set of techniques for reducing monitoring costs includes those that have members share the monitoring burden, as is the case when rewards are given to groups rather than individuals (under the assumption that, if your rewards depend on others, you will monitor their activity), when privacy is limited, when informants are rewarded, and when gossip is encouraged. A final economizing technique, which follows from socialization processes, is to minimize errors of interpretation of behavior through recruitment and training of members into a homogeneous culture.

One technique for economizing on sanctions is symbolic sanctioning through the creation of a prestige hierarchy and the differential rewarding of prestige to group members who personify group norms. Another technique is public sanctioning of deviance from group norms. A final sanctioning technique is to increase the exit costs of group members through geographical isolation from other groups, imposition of nonrefundable investments on entry to the group, and limitation of extragroup affiliations.

Yet there are limits to these monitoring and sanctioning economies, especially in compensatory groups. The result is that, at some point, a group must create formal agents and offices to monitor and control. Thus, a formal organization always reveals some of the economizing processes listed and also agents charged with monitoring and control—for example, comptrollers, supervisors, personnel offices, quality-control agents, and the like. Such monitoring and sanctioning are extremely costly, so it is not surprising that organizations try to economize here also.

In addition to the more general techniques previously listed, a variety of mechanisms for reducing agency costs, or increasing productive efficiency and hence profitability, can be employed. For example, inside and outside contractors are often used to perform work (and to incur their own monitoring and sanctioning costs) at a set price for an organization; standardization of tools, work flow, and other features of work is another way to reduce the need for monitoring; assessment of only outputs (and ignoring how these are generated) is yet another technique for economizing on at least some phases of monitoring; setting production goals for each stage in production is still another technique; and, perhaps most effective, the creation of an obligatory group within the larger compensatory organization is the most powerful economizing technique, as long as the norms of the obligatory group correspond to those of the more inclusive compensatory organization (sometimes, however, just the opposite is the case, which thereby increases monitoring costs to even higher levels).

The Theory Summarized

Such are the key elements in Hechter's theory. In Table 24-2, the theory is stated in more formal and abstract terms than in Hechter's work. When stated in this way, the theory can be applied to a wide range of empirical processes—social class formation, ethnic solidarity, complex organization, communities, and other social units. The propositions in Table 24-2 must be seen as building on the assumptions delineated in Table 24-1; when this is done, it is clear that Hechter has tried to explain, as he phrases the matter, "the micro foundations of the macro social order."

Macrostructural Implications

Hechter sees the basic ideas of rational choice theory as useful in understanding more

TABLE 24-2: Hechter's Implicit Principles of Social Structure

I. The more members of a group jointly produce goods for consumption outside the group, the more their productive efforts will depend on increases in the ratio of extrinsic compensation to intrinsic compensation.

II. The more members of a group jointly produce goods for their own consumption, the more their efforts will depend on the development of normative obligations.

III. The power of the group to constrain the decisions and behaviors of its members is positively related to the dependence of these members on the group for a good or compensation, with dependence increasing when

  A. More attractive alternative sources for a good or compensation are not available.

  B. Information about alternative sources for a good or compensation is not readily available.

  C. The costs of exiting the group are high.

  D. The moving or transfer costs to another group are high.

  E. The intensity of personal ties among group members is high.

IV. The more a group produces a joint good for its own consumption and develops normative obligations to regulate productive activity, the more likely are conditions III-A, III-B, III-C, III-D, and III-E to be met; conversely, the more a group produces a joint good for the consumption of others outside the group, the less likely these conditions are to be met.

V. The more a group produces a joint good for its own consumption and develops extensive normative obligations to regulate productive activity, the more likely is social control through monitoring and sanctioning to be informal and implicit and, hence, less costly.

VI. The more a group produces a good for external consumption and must rely on a high ratio of extrinsic over intrinsic compensation for members, the more likely is social control through monitoring and sanctioning to be formal and explicit and, hence, more costly.

VII. The larger is the size of a group, the more likely is social control to be formal and explicit and the greater will be its cost.

VIII. The greater is the cost of social control through monitoring and sanctioning, the more likely are economizing procedures to be employed in a group.

macrostructural processes among large populations of individuals. For example, the basic theory as outlined in Figure 24-1 and Tables 24-1 and 24-2 can be used to explain processes within nation-states. A state or government imposes relatively extensive obligations on its citizens—pay your taxes, be loyal, be willing to die in war, and so on. The reason for this capacity is that citizens are often highly dependent on the state for public goods and cannot leave the society easily (because they like where they live, cannot incur the exit and transfer costs, enjoy many benefits from the joint goods produced by the citizenry, and so on). Compliance with the demands of the state involves more than dependence and extensive obligations; compliance also hinges on the state's control capacity. But how is it possible for the state to monitor and sanction all its citizens who are organized into diverse configurations of obligatory and compensatory groups?

The answer to this question, Hechter argues, lies in economies of control. Such economies can be generated *within* and *between* groups. The key process is to get the citizens themselves monitoring and sanctioning one another within groups, then to link these groups together in ways that maximize dependence and control capacity.

JAMES S. COLEMAN'S THEORY OF GROUP SOLIDARITY

James S. Coleman was an early advocate of a rational choice perspective,[10] but not until the 1990s did he produce a general theory of social organization which was based on rational choice principles.[11] This more synthetic theory was published not long before Coleman's tragically early death. Coleman believed that actors have resources and interest in the resources of others; hence, interaction and ultimately social organization revolve around transactions between those who have and those who seek resources. These transactions can occur between individuals directly, they can also occur indirectly through intermediaries or chains of resource transfer, and they can occur in markets where resources are aggragated and bought and sold according to the laws of supply and demand.

Transferring Rights to Act

Coleman conceived of resources as *rights to act.* These rights can be given away in exchange for other rights to act. Thus, for example, authority relations consist of two types: *conjoint* authority where actors unilaterally give control of their rights to act to another because the vesting of others with authority is seen as in the best interests of all actors; and *disjoint* authority where actors give their rights away for extrinsic compensation, such as money. The same is true for norms that, in Coleman's eye, represent the transfer of rights of control to a system of rules that are sanctioned by others. Thus, social structures and cultural norms are ultimately built by virtue of individuals giving up their rights to control resources in exchange for expected benefits. For Coleman, then, the key theoretical questions in understanding social solidarity are (1) What conditions within a larger collectivity of individuals create a *demand* for rational actors to give their rights of control over resources to normative rules and the sanctions associated with these rules; and (2) what conditions make *realization* of effective control by norms and sanctions?

The Demand for Norms and Sanctions

These two questions—(1) the demand for norms and (2) their realization through effective sanctioning—are at the core of Coleman's theory. This theory goes in many directions, but we will confine our analysis to Coleman's discussion of the conditions producing group solidarity, so as to be in a position to compare Coleman's perspective with Hechter's theory of group solidarity. Thus, given the two questions guiding his approach, Coleman asked this: What basic conditions increase the demand for social norms and what makes them effective? In his answer, Coleman demonstrated how a view of actors as rational, self-interested, calculating, and resource-maximizing can explain emergent phenomena such as norms and group solidarity.

What conditions, then, increase the demand for norms? One is that actors are experiencing *negative externalities,* or harmful consequences in a particular context. Another is that actors cannot successfully bargain tit-for-tat or make offers or threats back and forth to reach agreements that reduce the negative externalities. Another condition is that there are too many actors involved for successful tit-for-tat bargaining, making bargaining cumbersome, difficult, and time consuming. And the most important condition is free-riding, where some actors do not contribute to the production of a joint good, and where, tit-for-tat bargaining cannot resolve the problem. Such free-riding, Coleman argued, is most likely in those groupings using disjoint authority or extrinsic compensation for productive activity (what Hechter termed a "compensatory group"). But, ultimately in Coleman's model, negative externalities give actors an *interest* in elaborating social structure and cultural systems. They begin to see that by giving up some of their rights of control over their resources and behaviors, they can reduce negative externalities and, thereby, increase their utilities. Free-riding can become a negative externality, to the degree that it imposes costs and harm on others. As we will see, free-riding became an important dynamic in Coleman's model of solidarity. But, there can be other sources of negative externalities—threats, conflict, abusive use of authority, or any source of punishment or costly action by others. Generally, actors will first engage in tit-for-tat bargaining to resolve a problem creating negative externalities. For example, a threat might be met by a counter-threat followed by an agreement to let the matter pass. But as the size of the group increases, this kind of bargaining becomes less viable: Pairwise bargaining among larger sets of actors to reduce negative externalities becomes difficult because of time and energy consumed; bargaining can create new negative externalities for actors who must constantly negotiate and for those who are left out of bargains.

Coleman argued that the creation of markets represents one solution to problems of pairwise, tit-for-tat bargaining among larger numbers of actors. As the volume of resources to be exchanged increases, and the number of buyers and sellers expands, markets determine the price or the resources that one must give up to get another valued resource, with price being determined by the relative supply and demand for resources, Yet, markets generate their own negative externalities associated with cheating, failure to meet obligations, unavailability of credit, and many other problems, and like all negative externalities, they can create demand for norms to regulate transactions.

Thus, wherever norms emerge (and other features of social systems like authority), they do so because actors have an interest in giving up certain rights to eliminate negative externalities and, thereby, to increase their utilities. Systems of norms, trust, and authority all represent ways to organize actors when tit-for-tat bargaining is difficult or unsuccessful, when the number of people involved grows beyond the capacity to bargain face-to-face, and when markets are no longer "frictionless" and begin to create their own negative externalities.

Under these conditions, actors will then bestow some of their rights to control resources to create *proscriptive norms*, or rules that prohibit certain types of behavior, and they will impose *negative sanctions* on those who violate these proscriptions or prohibitions. Such proscriptions and negative sanctions represent a solution by rational actors to the *first-order free-riding* problem, where actors are not contributing to the production of a jointly produced good. But, this solution to free-riding creates a *second-order free-rider* problem: Monitoring and sanctioning others are costly in time and energy, to say nothing of emotional stress and other negatives involved; hence, rational individuals are likely to engage in free-riding on the monitoring and administration of sanctions. The solution to first-order free-riding can, therefore, generate a new set of negative externalities associated with the costs and problems of sanctioning conformity to prohibitions.

The solution to this second-order free-rider problem is to create *prescriptive norms*, or norms that indicate what is supposed to be done (as opposed to what cannot be done, or is prohibited), coupled with positive sanctioning for conformity. Such positive sanctions can become, in themselves, a joint good and a source of positive externalities; indeed, receipt of positive sanctions (approval, support, congratulatory statements, esteem, and the like) increases the utilities that rational actors experience. And so, as the costs or negative externalities of the first-order free-rider problem create a demand for prescriptive norms and positive sanctions, actors can enjoy enhanced benefits and reduced costs when they give up their control of some resources and behavioral alternatives to normative prescriptions. Actors thus develop an interest in prescriptive normative control, and it becomes rational for them to do so.

The problem with systems of prescriptive norms and positive sanctions is that they are only viable in relatively small groupings and dense networks where monitoring and sanctioning can be part of the normal interaction among actors as they pursue a common goal

TABLE 24-3 Principles of Social Solidarity

I. The level of interest in creating norms among actors who are producing a joint good increases with
   A. The intensity of negative externalities that they collectively experience.
   B. The rate of free-riding in the production of the joint good.
   C. The level of actors' dependence on the production of the joint good.

II. The extensiveness of the norms created by actors with an interest in regulating the production of a joint good increases with
   A. The actors' level of dependence on the production of the joint good.
   B. The degree to which actors consume the joint good that they produce.
   C. The proportion of all actors receiving utilities for the production of the joint good.
   D. The rates of communication among members engaged in the production of a joint good which, in turn, is
      1. Negatively related to the size of the group.
      2. Positively related to the density of network ties among members of the group.

III. The ratio of prescriptive to proscriptive content of norms regulating the production of a joint good increases with
   A. The capacity to lower the costs of monitoring conformity to normative obligations, which, in turn, is positively related to
      1. Rates of communication among actors.
      2. Density of network ties.
      3. Ratio of informal to formal monitoring.
      4. Ratio of informal to formal sanctioning.
      5. Ratio of positive to negative sanctioning.
   B. The ratio of positive to negative sanctioning, which, in turn, is positively related to the ratio of informal to formal sanctioning.

IV. The level of solidarity among actors producing a joint good is, therefore, likely to increase when
   A. The actors' dependence on the production of the joint good is high.
   B. The extensiveness of normative obligations is great.
   C. The ratio of prescriptive to proscriptive content of norms is high.
   D. The ratio of positive to negative sanctions is high.
   E. The costs for monitoring and sanctioning are low.
   F. The proportion of actors receiving utilities from the production of the joint good is high.

or produce a joint good. Otherwise, the admission of positive sanctions becomes costly, as does the monitoring of conformity. High degrees of solidarity are thus only possible, Coleman argued, when relatively small numbers of actors give up control over their resources to prescriptive norms and rely heavily on positive sanctions that, themselves, become positive externalities that increases the utilities for actors.

Principles of Group Solidarity: Synthesizing Hechter's and Coleman's Theories

As is evident, Coleman's theory arrives at the same place as Hechter's rational choice approach to solidarity does. This convergence of perspectives can perhaps best be appreciated by bringing the two theories together in composite form, as is done in the principles enumerated in Table 24-3. In this table,[12] the respective vocabularies of Hechter's and Coleman's schemes are mixed together in ways that emphasize the original contributions of each. What is true for solidarity, both would argue, is true for other emergent social and cultural phenomena. But let us concentrate on the four principles of solidarity.

Proposition I simply states Coleman's and Hechter's views on what causes actors to have an interest in creating norms. The key conditions are commonly experienced negative externalities, high rates of free-riding, and dependence of actors on each other for the production of a joint good that gives them utility. Under these conditions, actors will give up some of their rights of control. Proposition II summarizes the conditions that, in Hechter's terms, create extensive regulatory norms where actors relinquish a wide range of rights of control to group norms. The basic conditions under which such extensive norms emerge include interests in creating norms (which are activated by the conditions listed in Proposition I), the ratio of "consumption of" to "extrinsic compensation for" the production of a joint good (that is, the more members themselves consume the joint good, the greater their interest is in creating extensive norms), the dependence of actors on the utilities offered by the joint good is high, the proportion of actors who receive utilities from the production of a joint good is high, and rates of communication among actors are high (with such rates being negatively related to the number of actors involved and positively related to the density of their networks).

Proposition III then addresses the issue of the amount of proscriptive versus prescriptive content of the norms that actors create under the conditions specified in Propositions I and II. The fundamental forces increasing *pre*scriptive content are the ability of actors to monitor each others' activities during the normal course of producing a joint good (which, in turn, is related to group size and density of networks), and the capacity to use positive instead of negative sanctions (which is related to the ability to use informal means of sanctioning).

And finally, Proposition IV summarizes how the conditions stated in the first three propositions all come together that increases social solidarity. The basic conclusion is when the actors' dependence on the production of the joint good is high, when they develop extensive norms, when the ratio of prescriptive to proscriptive content of such norms is high, when the ratio of positive to negative sanctions is also high, and when the proportion of actors receiving utilities is high, then social solidarity will increase.

CONCLUSION

Rational choice theories connect sociology to economic theory and seek to do what early sociologists felt was impossible: to conceptualize emergent social forms and structures through the behaviors of rational, self-interest actors. The two theories examined in this chapter are among the best in meeting the challenge that sociology has posed for theories that begin with assumptions of individual rationality. Just whether or not their answer is satisfactory depends on whether one is receptive to the view that humans always behave in terms of utilities and rewards and that social organization can only be fully understood in these terms.

The two theorists examined in this chapter have been part of a broad, worldwide effort to use economic theory to explain cultural and social processes. Indeed, these rational choice approaches influence not just sociology but political science as well, and there is little doubt that they are here to stay. As sociology closes the twentieth century, it has reembraced what it had rejected at the beginning of the century.

NOTES

1. Adam Smith, *An Inquiry into the Nature and Causes of the Wealth of Nations* (Indianapolis: Liberty Fund, 1981; originally published in 1775–1776).

2. Adam Smith, *The Theory of Moral Sentiments* (Indianapolis: Liberty Fund, 1974; originally published in 1759, and later revised in light of the questions raised in *The Wealth of Nations*).

3. This approach has different names in other disciplines but the core ideas come from the models presented in Chapter 23, where the incorporation of economic and game-theoretical models was introduced.

4. Michael Hechter, *Principles of Group Solidarity* (Berkeley: University of California Press, 1987); "Rational Choice Foundations of Social Order," in *Theory Building in Sociology*, ed. J.H. Turner (Newbury Park, CA: Sage, 1988).

5. Mancur Olson, *The Logic of Collective Action* (Cambridge, MA: Harvard University Press, 1965).

6. Hechter, *Principles of Group Solidarity* (cited in note 4), p. 49.

7. Ibid., p. 126.

8. Ibid., pp. 59–77, 104–124.

9. Ibid., pp. 126–146.

10. See, for examples of his earlier essay, James S. Coleman, *Individual Interests and Collective Action: Selected Essays* (Cambridge: Cambridge University Press, 1986).

11. James S. Coleman, *Foundations of Social Theory* (Cambridge, MA: Belknap, 1990).

12. See also Jonathan H. Turner, "The Production and Reproduction of Social Solidarity: A Synthesis of Two Rational Choice Theories," *Journal for the Theory of Social Behavior* 22 (1993), pp. 311–328.

25

The Continuing Tradition II: Exchange Network Theorizing

Richard M. Emerson was the first to connect exchange theory with network analysis,[1] as was explored in Chapter 22. This blending of exchange theory and network ideas gave Emerson a way to convert the nodes or points in graphically represented networks into actors who had certain behavioral propensities. Conversely, the structure of the network could be seen as determining the resources that actors use in exchange with each other. Thus, in Emerson's eye, social structure is composed of exchanges among actors seeking to enhance the value of their resources. And so behaviorism, which posited a dynamic but atomized actor, was blended with network sociology, which conceptualized structure without dynamic actors.

THE CORE IDEAS IN EMERSON'S APPROACH

The key dynamics in Emerson's theory are (1) power, (2) power use, and (3) balancing.[2] Actors have *power* to the extent that others depend on them for resources; hence, the power of Actor A over actor B is determined by the dependence of B on A for a resource that B values, and vice versa. *Dependence*, which is the ultimate source of power in Emerson's scheme, is determined by the degree to which (a) resources sought from other actors are highly valued and (b) alternatives for these resources are few or too costly to pursue. Under these conditions, where B values A's resources and where no attractive alternatives are available, the B's dependence on A is high; hence, the power of A over B is high. Conversely, where B has resources that A values and where alternatives for A are limited, B has power over A. Thus, both actors can reveal a high degree of mutual dependence, giving each *absolute power* over the other and, thereby, increasing *structural cohesion* because of the high amounts of *total* or *average power* in the exchange relationship.[3]

When one actor has more power than an exchange partner, however, this actor will engage in *power use* and exploit its exchange partner's dependence to secure additional resources or to reduce the costs it must incur in getting resources from this dependent partner. If A has power over B because of B's dependency on A, then A has the *power advantage* and will use it.

Such relations are *power imbalanced,* and Emerson felt that imbalance and power use would activate what he termed *balancing* operations. In a situation where A has a power advantage over B, B has four options: (1) Actor B can value less the resources provided by A; (2) B can find alternative sources for the resources provided by A; (3) B can increase the value of the resources it provides A; and (4) B can find ways to reduce A's alternatives for the resources that B provides. All these balancing mechanisms are designed to reduce dependency on A, or alternatively, to increase A's dependency on B in ways that balance the exchange relationship and give it a certain equilibrium.

Exchange in networks can be of two general types (Emerson, 1981):[4] (1) those where actors negotiate and bargain over the distribution of resources; and (2) those where actors do not negotiate but, instead, sequentially provide resources with the expectation that these rewards will be reciprocated. This distinction between what can be termed *negotiated exchanges* and *reciprocal exchanges* is important because it reflects different types of exchanges in the real world. When actors negotiate, they try to influence each other before the resources are divided, as when labor and management negotiate a contract or when individuals argue about whether to go to the movies or to the beach. The dynamics of negotiated exchange are distinctive because they typically take longer to execute, because they generally involve considerably more explicit awareness and calculation of costs and benefits, and because they are often part of conflicts among parties who seek a compromise acceptable to all. In contrast, reciprocal exchanges involve the giving of resources unilaterally by one party to another with, of course, some expectation that valued resources will be given back, as occurs when a person initiates affection with the intent that the other will respond with the same emotion. Reciprocal exchanges are thus constructed in sequences of contingent rewarding, whereas as negotiated exchanges unfold in a series of offers and counter-offers before resources are distributed.

These seminal ideas form the core of Emerson's theoretical scheme. Before his untimely death, Emerson had been collaborating with Karen Cook and their mutual students to test the implications of these ideas for different types of networks. The basic goal was to determine how the structure of the network—that is, the pattern of connections among actors—influences, and is influenced by, the distribution of power, power use, and balancing. This network-oriented program has continued, and indeed, rival programs have emerged outside the Emerson-Cook lineage.[5] And, although there has been a certain amount of cross-fertilization among these rival programs, they remain somewhat distinct, despite using the same basic conceptualization of social structure as networks and the same general view of actors as oriented to receiving value or utilities while avoiding costs and punishments.[6]

In this chapter, however, we will remain within the Emerson-Cook branch, examining first extensions by Emerson and Cook as well as some collaborative work by Cook with her students. Then, we will review the work of Edward Lawler and various collaborators who have de-emphasized somewhat Emerson's and Cook's emphasis on forms of social structure in favor of the bargaining strategies of actors and the emotion-arousing effects of the exchange process itself. Finally, we will examine Linda Molm's use of Emerson's and Cook's theory to explore punishment in exchange processes. Each of these programs employs Emerson's core ideas, but they also represent creative extensions and elaborations of the early theory summarized briefly earlier and in more detail in Chapter 22.

KAREN S. COOK'S THEORETICAL PROGRAM

Many of the problems of concern among network-exchange theorists were formulated by Karen Cook in collaboration with Richard Emerson and, later, with students working with either or both of them. The topics pursued by Cook and her collaborators vary, but several stand out: (1) the conditions under which actors in exchange networks develop commitments to partners; (2) the effects of centrality in networks on the distribution of power; (3) the relations among power use, equity, and justice considerations; and (4) the dynamics of restricted and generalized exchanges. Each of these areas of research is briefly examined in the following pages, where the important generalizations emerging from experimental studies will be emphasized.

Commitment Processes in Networks

In Emerson's original formulation, imbalanced power would lead the advantaged partner in an exchange to use this power to extract more resources from another actor, to the point where balancing operations push the exchange toward equilibrium. In early work with Emerson, Cook examined the process of commitment for its effects on power use, and vice versa. The reason for studying this topic is that from a network perspective, actors do not operate in a perfectly free and competitive market; rather they have connections to each other and, by virtue of these connections, they engage in exchange. *Commitment* to an exchange occurs when actors choose their current partners over potential alternatives, and in the case of extreme commitment, when they remain with partners who can give them less beneficial payoffs than alternative partners.[7]

From the perspective of a rational and calculating actor, it would be irrational for a more powerful actor to develop commitments to others because this commitment would decrease exploration of alternatives and the resulting ability to increase their payoffs from dependent actors. Conversely, on the surface, it would seem irrational for disadvantaged actors to develop attachments to those who are using their power to extract ever more resources from them. And yet, this is exactly what appears to happen. The early finding was that power use was inversely related to commitment: actors who develop commitments were less likely to use their power advantages against their less advantaged partners.[8] One early explanation for this tendency was that commitments reduce uncertainty; hence, as actors engage in frequent exchanges, they become more committed to the ratio of payoffs and, thereby, reduce the uncertainty inherent in exchange. Moreover, commitment tempers the use of power by the advantaged, because power-advantaged actors become less likely to seek alternatives, thereby enabling payoffs for power-disadvantaged actors to be regularized at predictable costs. A subsequent interpretation by Cook[9] emphasizes that efforts to entice commitment from power-advantaged actors is a balancing strategy to decrease power use by these advantaged actors.

Later, Peter Kollock[10] argued that under conditions of risk and uncertainty in receiving payoffs, commitment can be seen as a "profit-maximizing" strategy by both partners to the exchange. By establishing commitments, each actor lowers the costs associated with risk and uncertainty. Moreover, commitment itself becomes a positively valued resource that increases each actor's resource payoffs.[11] In particular, commitment can be seen as a power-balancing strategy

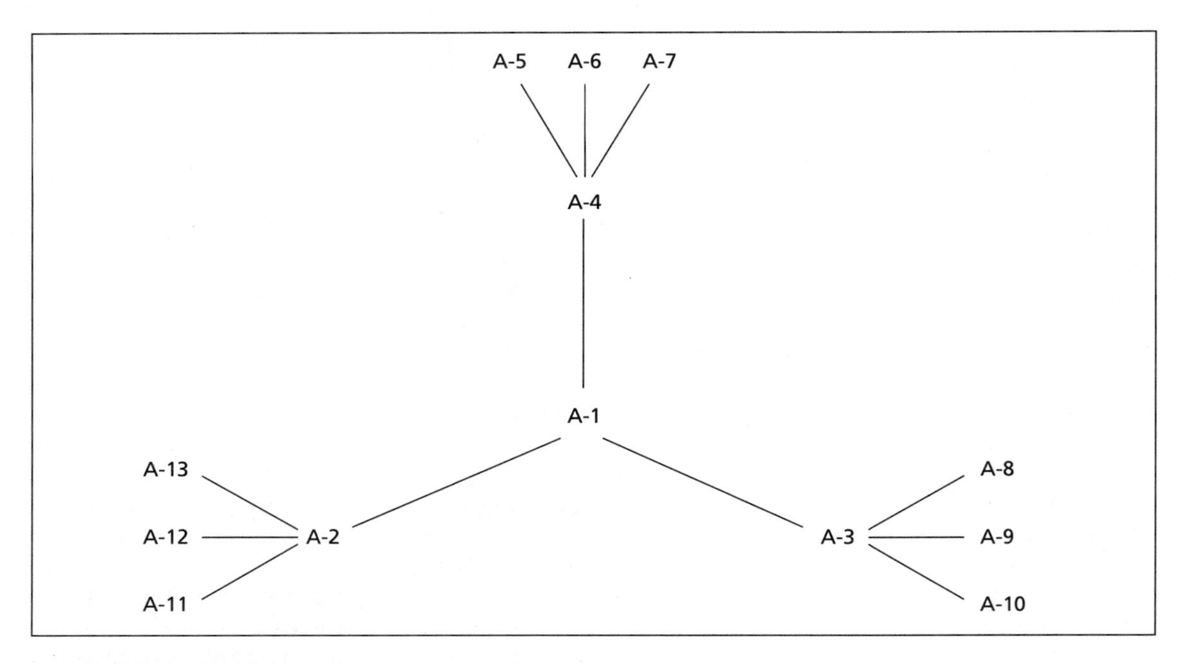

FIGURE 25-1 Power and Centrality in Exchange Networks

in which lower-power actors offer a new resource—commitment—to reduce power use by advantaged actors. This point, as we will explore shortly, was extended by Edward Lawler and Jeongkoo Yoon who began to see commitments as markers of emotions emerging from frequent exchanges among actors, creating an attachment to the relationship, per se, and thereby adding new kinds of emotional utilities among the exchange partners.

Centrality in Networks and the Distribution of Power

In Figure 25-1, Actor A-1 is in a position of centrality in relation to actors A-2, A-3, and A-4. Centrality can be measured and conceptualized in several ways (see Chapter 38 on network theory), but the basic theoretical idea is relatively straightforward: Some positions in a network mediate the flow of resources by virtue of being in the middle of ties to other points. Thus, in Figure 25-1, Actor A-1 mediates the flow of resources among Actors A-2, A-3, and A-4; hence, in this network, A-1 is in a position of high centrality. Similarly, further out on the network, actors A-2, A-3, and A-4 are also central between the peripheral actors at the ends of the network (A-5 through A-13) and the most central actor, A-1. For example, A-2 is central with respect to A-1, on the one side, and A-5, A-6, and A-7, on the other. The same is true for the other peripheral A's connected to A-3 and A-4. Cook, Emerson, and their coauthors then argued that in such a network, power would decentralize towards those actors who possess the highest degree of access to resources.[12] In the network in Figure 25-1, resources are flowing from actors A-5 through A-13 to A-4, A-3, and A-2, and then, on to A-1. Power-dependence theory predicts that this network will have a tendency to collapse around those actors who can control direct access to resources—in this case, Actors A-2, A-3, and A-4. Why should this be so?

The answer resides in the power advantage A-2, A-3, and A-4 have over the respective A's from whom they are getting resources. A-5 through A-13 have no other source for resources than A-2, A-3, and A-4, whereas each of these more central A's has alternative sources for resources. Thus, those actors at the edges of the network depend on A-2, A-3, and A-4; indeed, each of these three A's enjoys what Emerson termed *unilateral monopoly* (see pages 286–287 of Chapter 22). They have, therefore, the most direct access to resources, and they enjoy a power advantage because they can play the dependent A's (that is, A-5 through A-13) off against each other. A-1 ultimately depends on A-2, A-3, and A-4; hence, if the latter are willing to trade with A-1, then A-1 must have resources that they value. If not, the network will collapse into its three unilateral monopolies revolving around A-2, A-3, and A-4. For this network to remain stable, therefore, A-1 actually has to possess another resource that A-2, A-3, and A-4 value highly and that they cannot get readily elsewhere (hence, in the conventions of network diagrams, A-1 actually becomes B or some other letter indicating that it is providing different resources to A's).

Let us put some empirical content into these network dynamics. Let B represent the king in a feudal system, and A-2, A-3, and A-4 the lords of the king's realm. These lords provide resources to the king through their unilateral monopoly over peasants on their estates, and so, they pass on to the king some portion of the resources ultimately generated by peasants. What, then, does the king give back in exchange for this flow of resources? The answer is almost always problematic for kings, and this is why feudal systems tend to collapse, because the lords of the realm are closer to the material resources that sustain the king. Typically, the king provides the coordination of armies and other necessary activities among the lords (who are often feuding amongst themselves) for defense of the realm. This capacity to organize a kingdom is the resource that the king gives back to lords.

Many network structures approximate this one. The important point is that they are inherently unstable because those who enjoy the unilateral monopolies eventually become resentful that they have to pass on resources to a more central actor, and if they begin to perceive that this actor does not provide enough in return, they break off the exchange, and thereby change the distribution of power in the network. More recent work in networks, even in networks where actors are less likely to break off exchange, also found that the locus of power shifted toward the sources of supply of resources that were most highly valued.[13]

Equity and Justice in Exchange Networks

In their early collaborative work, Emerson and Cook had discovered that concerns about *equity*—or the distribution of outcomes from bargaining in accordance with each actor's respective contribution to the outcome—became interwoven with power dynamics.[14] If actors had knowledge of each other's payoffs and if equity considerations became salient, then power use by advantaged actors was curtailed somewhat. This finding was consistent with other research on the issue of justice in exchange—a consideration that was at the center of exchange theory from the beginning.[15]

In subsequent work with others, primarily Karen A. Hegtvedt, Cook has examined the role of justice in social exchange.[16] One of the most important conceptual distinctions highlighted by Cook and Hegtvedt is between distributive justice and procedural justice. *Distributive justice* denotes the norms or rules by which the allocation of resources among actors occurs. There are several types of such rules, including *equality* or the allocation of the same shares to all, *equity* or the distribution of resources relative to the respective inputs and contributions of actors to an outcome, or *need* or the distribution of resources to those who require them most. *Procedural justice* refers to the perceived fairness of the process of bargaining itself, rather than to the distributive outcomes that emerge from negotiations. Actors' sense of fairness in the procedures influences how they will respond to each other, thereby shaping the kinds of tactics and strategies that they will employ in bargaining.

How, then, does power influence considerations of justice? The findings are somewhat mixed, but several generalizations emerge from Cook's and Hegtvedt's experimental work as well as from their efforts to summarize the large literature on justice. One generalization is that more power-advantaged actors are, in general, likely to view the distribution of resources as fair, and these perceptions of fairness are related to their ratio of inputs to outcomes (that is, they see themselves as getting back an appropriate return on their investments and costs). Power-advantaged actors have, however, a tendency to constrain their power use when they know that their advantage rests with their structural position rather than with their abilities and talents as negotiators. In contrast, power-disadvantaged actors often tend to see situations as less fair and to bargain harder to overcome their low relative power.

Perceptions of equity, or the sense that payoffs correspond to one's contributions, have a somewhat greater effect on attributions of why one receives a payoff than does actual power advantage.[17] Individuals who see themselves as treated fairly are likely to have positive sentiments, and they are likely to attribute their fair outcomes to their personal characteristics (to maintain their identity in the situation) as well as to their partner's characteristics. Yet, this process of attribution is altered somewhat because those who have a power advantage (and correctly perceive that they have this advantage) are likely to attribute their success more to their personal characteristics and to the situation that gives them a power advantage than to the characteristics of their exchange partners. When weaker partners are not fully aware of their structurally based power disadvantage, they will bargain harder to overcome their lower outcomes. If they enjoy success in their efforts, they are likely to attribute this success to their personal characteristics.

The relationship between distributive justice or the allocation of benefits, on the one hand, and procedural justice, on the other hand, is far from clear. But Cook and Hegtvedt extract several generalizations from the literature.[18] One is that perceptions of unfair outcomes are almost always accompanied by perceptions that the procedures leading to these outcomes were also unfair. Yet, unfair procedures do not produce a corresponding sense of unjust outcomes as long as actors receive what they believe as a fair outcome. Thus, people tend to judge procedures by outcomes: When outcomes are high or perceived as fair, people are less likely to be critical of procedures, even when they are clearly unfair; whereas, if outcomes are low or perceived as unfair, procedures are almost always seen as unjust, even if they are fair.

Considerable work is yet to be done on this line of research, but it is a central dynamic in power. Perceptions of justice and fairness influence how power processes operate: If power use generates a sense of justice, or if it is used in accordance with accepted norms of distributive and procedural justice, then power use will produce more balanced exchanges. If power imbalance leads to actions by the advantaged actor that violate rules of justice, or if behaviors generate a perception of injustice, then power use will perpetuate imbalance, and disadvantaged actors will seek new strategies to rebalance the exchange, if they can.

Generalized Exchange Networks

In Chapter 24 on rational choice theories, the basic problem of order was defined as the problem or dilemma of *free-riding*. That is, it is rational for actors who are trying to maximize benefits and minimize costs to avoid contributing to the production of a joint good. For example, if a set of actors is to produce a product, such as a group-written term paper that will receive a grade, it is rational for each actor to avoid the costs of contributing to the production of the paper but to enjoy the benefits that a good grade bestows, or in other words, it is rational to *free-ride* on the labor of others. The dilemma, of course, is that if everyone free-rides, the joint good will not be produced, and no one will receive the benefits of the jointly produced good. As we saw in Chapter 24, rational choice theorists argue that under conditions of free-riding, actors create systems of norms, monitoring procedures, and sanctioning systems for controlling the negative results of free-riding, thereby assuring the production of the joint good.

In her more recent work, Cook[19] along with Toshio Yamagishi,[20] who has long worked "social dilemma" problems (like free-riding), have approached the same topic as rational choice theorists but from a different angle. Emerson made a distinction between (1) *elementary exchange* where actors directly give and take resources to enhance each actor's individual calculation of personal benefits, and (2) *productive exchange* where actors exchange resources to produce more benefits from the combination of their respective resources. But Emerson's portrayal did not capture a more fundamental distinction between (1) direct exchange and (2) generalized exchange. Most studies of exchange in networks focus on *direct exchanges* in which actors pass resources back and forth in a network that, ultimately, can be seen as a series of dyadic exchanges. Indeed, most studies producing the theoretical ideas summarized in this chapter come from experiments using direct, dyadic exchanges among partners. But, as we saw in Chapter 19 on the history of exchange theory, anthropologists have long been interested in *generalized exchange* systems in which actors give resources to others but do not receive resources directly back; rather, they acquire resources indirectly via actors who are a part of a pool of actors who feel obligated to pass resources on. For example, when a person offers to help someone whose car has broken down, this helpful person might receive positive rewards like thanks and gratitude directly, but also a more generalized exchange is operating: Helpful persons expect that others will help them when they have car trouble. The early anthropologists, such as Marcel Mauss and Claude Lévi-Strauss, had assumed that such generalized exchange systems created chains of social solidarity. Peter Ekeh,[21] in reviewing these early studies reached a similar conclusion, but more important, he made an important distinction between types of generalized exchanges that Cook and Yamagishi were to explore further. One basic type of generalized exchange is *group-generalized exchange,* in which group members pool their resources to receive the beneficial outcomes from this pooling. Another type is what Cook and Yamagishi prefer to label *network-generalized exchange,* in which each actor provides resources to another who, in turn, provides benefits for yet another actor and so on for however many actors are in the network; the original actor eventually receives resources back from one of these other actors in the network.

The group-generalized exchange comes closest to approximating what rational choice theorists postulate as the kind of situation that generates free-riding, which, in turn, leads to the emergence of norms and other systems of social control. It is rational for actors to avoid contributions of their share to the pooled resources, while enjoying the benefits of group production. If all members free-ride, however, no one receives benefits, and under these conditions, systems of norms, monitoring, sanctioning, and trust and other mechanisms emerge to limit free-riding. Yet, as Cook and Yamagishi point out, network-generalized exchange systems also present potential dilemmas for free-riding. If one person in the chain of resource flows takes resources from one party but does not pass the appropriate resources on to the next person in the chain, then the original actor and perhaps all those in between the defector and the first actor in the chain receive no benefits. Hence, the network collapses. Yet, because there are direct connections among actors in network exchange, Cook and Yamagishi argue that participants are more likely to cooperate in a network-generalized exchange system than in a group-generalized exchange.

The reason for this difference is that in a group-generalized exchange, where actors pool resources, the temptation to free-ride is high because each actor does not have a direct responsibility to pass resources on to a designated actor, but equally important, as the group gets larger, any one actor's decision not to participate is less consequential and perhaps less noticeable to the total group product, thereby diffusing the responsibility for each actor's noncontribution. This same tendency occurs in generalized-network exchanges as they become large, especially if the networks overlap and intersect in ways that make the chains complex and somewhat redundant, but control is still exercised because at least one actor will know if another did not make the appropriate contribution.

As a consequence, trust is more likely to evolve in the network-generalized system. Built into the nature of the network is a certain monitoring, and because of this, actors come to expect that each will exchange in the appropriate manner. This expectation becomes translated into trust; such trust creates a normative climate for participation and cooperation in the generalized exchange of resources. This tendency for trust and cooperation is greater in network-generalized exchanges than group-generalized exchanges, but, as we saw in the review of rational choice theories, free-riding can generate additional mechanisms beyond trust to assure cooperation in group-generalized exchanges. However, these mechanisms increase the costs of producing a joint good.

The significance of this recent work in Cook's evolving program is that it connects two branches of exchange theory—rational choice approaches and network approaches—which have during the last two decades gone their somewhat separate ways. Once problems of free-riding and other social dilemmas for actors are introduced into a network analysis of generalized-exchange systems, however, the opportunities for cross-fertilization increase. As we saw for James Coleman's[22] analysis of social solidarity in Chapter 24, small and dense networks represent a basic condition for generating social solidarity. But the significance for networks goes beyond small-group solidarities; the nature of the paths in a network-generalized system of exchange are important in all types of social structures. For example, the number of paths, the length of chains, the redundancy and density ties, their centrality, and other characteristics of the network will influence the level of free-riding. In turn, the level of free-riding will shape the kinds of mechanisms—from simple trust to more direct forms of social control such as monitoring and sanctioning—that will be used to sustain the production of joint benefits or goods. Much of the social order is constructed from such generalized-exchange systems, so the new direction in Cook's and her collaborators' program is perhaps the most promising.

	Distributive	Integrative
Tacit	Implicit perception of conflict of interest over a fixed amount of resources; tactics in negotiation about the allocation of resources occurs without open communication or even full acknowledgment by the parties.	Implicit perception of a conflict of interest, but despite the lack of clear and open communication, negotiation seeks to increase benefits for all parties.
Explicit	Conflict of interest over a fixed amount of resources is acknowledged; tactics in negotiation about the allocation of resources revolve around open communication with the full acknowledgment that they are bargaining.	Conflict of interest over a fixed amount of resources is perceived, but some agreements produce more total benefits than others; tactics in negotiation revolve around open communication designed to reach a solution where all parties gain.

FIGURE 25-2 Types of Bargaining

EDWARD J. LAWLER'S THEORETICAL PROGRAM

Among the more creative efforts to extend the original power-dependence theory developed by Emerson is Lawler's theoretically informed research program. With a variety of coauthors, he has sought to examine how the structure of power influences (1) the nature of bargaining in exchange relationships and (2) the emergence of commitments among exchange partners. These two bodies of work—bargaining and commitment—will be briefly reviewed here.

Power and Bargaining

In the early 1980s, Samuel Bacharach and Lawler[23] began to examine the process of bargaining in exchange relations; in more recent years, Lawler[24] and Rebecca Ford[25] have sought to consolidate theorizing on bargaining. The body of theory and research is complex, so, we will only outline the approach in broad contours. In this theory program, bargaining is conceptualized as the result of actors perceiving that they have a potential conflict of interest, with the result that goal-directed moves by one actor are intended to influence the goal-directed efforts of other actors. Bacharach and Lawler and, later, Lawler and Ford, developed a topology on bargaining, summarized in Figure 25-2. As the table underscores, bargaining can be either tacit or explicit, on the one side, and distributive and integrative, on the other. *Tacit bargaining* occurs when actors implicitly perceive that they have a conflict and anticipate each other's moves without open communication among them, whereas *explicit bargaining* is where actors acknowledge their conflict and agree to negotiate with lines of communication remaining open and with the expectation that they can reach a compromise. *Distributive bargaining* occurs when there is a fixed amount of a resource to split up, and where one party's gain is another's loss (a *zero-sum* situation), and *integrative bargaining* exists when parties seek common ground that can potentially increase or at least provide acceptable levels of benefits for all (a nonzero-sum situation). Of the four types summarized in Figure 25-2, tacit-distributive is the most difficult to resolve because parties do not fully acknowledge their conflict and yet must divide resources in a zero-sum fashion (that is, one party's gain will be another's loss). Tacit-integrative bargaining is somewhat less difficult because actors usually have common goals, and one party's loss is not necessarily another's gain (indeed, with common goals, agreements can increase the benefits to both parties, although they are limited in their negotiations by their lack of open communication). Explicit-distributive bargaining is typically the most straightforward because the conflict of interest is clear, fixed sums of a particular resource are being negotiated, and cycles of offers and counteroffers can proceed until a solution is reached. Explicit-integrative bargaining combines mutually acknowledged conflicts with an underlying common interest in solving the conflict and reaching a solution, and hence, such bargaining is likely to search for outcomes in which all parties gain.

For Lawler and his collaborators, the conflict that sets bargaining into motion is structurally based, with structure conceptualized as a network of interrelated positions that contain interests (in seeking certain kinds of resources) and as revealing cleavages in which different positions reveal varying interests. Actors occupying positions interpret their structurally generated interests, and on the basis of these interpretations, they make decisions in bargaining with incumbents in other positions. When actors perceive that their interests and those in other positions are in conflict, power processes are activated and bargaining begins. Such activation of power is *tactical* in the sense that actors use the power at their disposal to make a move, or set of moves, designed to influence another actor's definitions of its interests and behaviors. Tactics revolve around (1) perceptions by actors of resistance from other actors, (2) impression management in bargaining with them, and (3) most important, estimates of power.

Following Emerson, emphasis is on power as a function of the dependence of one actor on another for valued resources, with dependence being a reflection of (a) the value of the resources involved and (b) the number and quality of alternative sources for these resources. The more valued the resources are and the fewer the alternatives are, the more dependent an actor is on another who provides these resources, and, hence, the greater the power is of the resource provider over the one who is dependent. In making estimations of power in bargaining, several considerations are important, and these also follow from Emerson's original conceptual framework. First, actors can assess their *absolute power* or the dependence of the other party on them for valued resources (but without fully assessing the reverse: their dependence on the other for valued resources). Second, actors also develop a sense for the *total power* in the situation—that is, the degree to which they mutually depend on each other for valued resources (or combined sum of P_{ab} and P_{ba}). Third, actors can assess their *relative power* or absolute power of each actor as a proportion of total power (or P_{ab} as a proportion of [P_{ab} and P_{ba}]). As relative power increases, so must the inequality in power because one party is more dependent than another.

This research program emphasizes, however, that the power-dependence approach as originally conceptualized by Emerson does not fully account for the kinds of tactics that actors can use in bargaining. Emerson's approach can help explain more conciliatory bargaining tactics, where actors negotiate a solution to their conflict by making concessions, but the power-dependence approach does less well in explaining what are termed *punitive* or *coercive* tactics where threats of harm or actual behaviors inflicting damage on others occur. In a punitive situation, additional costs on other actors are being imposed, above those costs that would come by simply leaving the relationship and thereby depriving another actor of the resources that were being given in the exchange.

Most of the bargaining processes examined in this theory program are explicit because of the ways that the experiments testing the theories are constructed. Emphasis is generally on *nonzero-sum* power where the total power (mutual dependence among actors) and relative power (power of actors as a proportion of total power) can increase, or decrease. The findings from the many experiments conducted during the last two decades provide many generalizations, but we need only summarize the most important.

More *conciliatory tactics* in bargaining are most likely to be employed in situations where (1) total power among actors is high (that is, they mutually depend on each other for valued resources and have less attractive alternatives for these resources); (2) relative power is low or equality in their respective resources prevails, but equality only moves tactics toward conciliation if each party's respective coercive power (to engage in punitive tactics) is also equal; (3) the value of the resources being exchanged is high for both parties, especially when their relative power is equal; and (4) frequency of exchange among partners increases their commitments to the relationship, thereby providing new, emotional resources to the exchange that increase total power (or mutual dependence on the new resource).

Power relations also shape appeals to norms used as tactics by actors in the bargaining process. As the level of relative power among actors increases (that is, the level of inequality), higher-power actors are likely to make appeals to norms emphasizing equity or the distribution of resources through proportionate investments and contributions by the respective parties, whereas lower-power parties are likely to make appeals to norms stressing equality in the distribution of resources and to norms arguing for the responsibility of higher-power actors to provide benefits to lower-power actors. If a lower-power actor still has considerable power, despite its power disadvantage, then both parties become more likely to make appeals to equality norms. And, once bargains in the name of equality have been struck, they tend to be more enduring, even as relative power (inequality) increases.

Punitive use of power will occur with high relative power or inequality and, in general, will increase hostility and produce less conciliation in bargaining than a relationship with equal power. But the use of punishment is complicated by two contrasting dynamics, which are labeled, respectively, (1) bilateral deterrence and (2) conflict-spiral.[26] In the *bilateral deterrence* model, each party fears a hostile and punitive retaliatory attack by the other, with this mutual fear reducing the likelihood that either party will initiate an attack. In the *conflict-spiral* model, the expectation of one actor using its power leads the other to make a preemptive attack, which then leads to a counter-attack in a potentially escalating cycle. This distinction is significant because each model makes somewhat different predictions about what occurs when power shifts from relative equality (low relative power) to inequality (high relative power). In the bilateral deterrence model, the higher-power actor would exploit its advantage and attack because there is little to stop this actor from doing so, whereas the lower-power actor will also use its power because it expects exploitation by the higher-power actor and, hence, engages in punitive action to inform the higher-power actor that exploitive or punitive responses will not go unpunished. Thus, the prediction in the bilateral deterrence model is that as long as relative power is low or equal, regardless of the amount of total power, each party will have lower expectations for an attack from the other, while each party will be deterred from punitive tactics for fear of retaliation. But as inequality increases, both parties should use punitive tactics but for different reasons. In the conflict-spiral model, relative equality can create temptations that exceed the costs of retaliation and lead one actor to make a preemptive strike, setting off retaliation and a spiral of conflict. A shift to greater inequality, however, leads the lower-power party to perceive that there is little to be gained but much to lose if it attacks the higher-power party, whereas the higher-power party recognizes this reluctance, lowers its expectation of an attack, and thereby uses fewer exploitive and punitive tactics. Hence, the prediction in the conflict spiral model is that when inequality is low, each party expects an attack and might make a preemptive strike, whereas as inequality increases, the growing power differences decrease the likelihood of a punitive attack—the exact opposite prediction of the bilateral deterrence model. Which one is correct?

In general, the bilateral deterrence prediction appears to hold up better than the conflict-spiral prognosis, but each operates under specifiable conditions. Any condition that increases actors' perception of retaliation costs for its hostile acts will produce bilateral deterrence effects, whereas any condition that pushes the temptation to attack beyond fears of retaliation will tend to produce conflict-spiral effects. Most of the time, conditions favor considerations of the costs (in retaliation) of punitive behavior, and hence, increase the bilateral deterrence effect. The reasons for this bias toward the bilateral deterrence reside in the nature of power-dependence. High total power (mutual dependence) makes punitive tactics costly because the other party who provides valuable resources can counter-attack or, if it can, withdraw from the relationship and thereby deprive the attacking actor of valued resources. Only if the attacking party perceives the possibility for a dramatic gain in resources will its temptation to use punitive tactics exceed the costs of such use, thereby setting off the conflict-spiral. Even in situations of high relative power, where parties possess unequal power, the higher-power party might be reluctant to use punitive power if the lower-power party has some capacity to retaliate and thereby impose costs on the more powerful party.

The use of tactics in bargaining, whether punitive or conciliatory, can still be conceptualized within Emerson's analysis of balancing operations. Tactics in bargaining are, in the end, efforts to use power in ways that increase rewards, but the analysis of bargaining tactics can add to Emerson's original conception of balancing operations. When power is balanced—that is, the dependence of actors on each other for resources is equal—then conciliatory tactics will be used. Power is used when relations are imbalanced—that is, one actor has more power than the other—but this power use is mitigated when the less powerful actor resorts to one of Emerson's balancing strategies: (1) the dependent actor can value less the resources provided by the more powerful actor, and especially so if the more powerful actor employs punishment; (2) the more dependent actor can offer new resources to the more powerful actor, or as Lawler's research would add with the notion of bilateral deterrence, the more dependent actor can impose or threaten additional costs on the more powerful party if it uses punitive tactics; (3) the more dependent actor can seek new alternatives to resources provided by the more powerful, "extending the network" in Cook's terms or employing a "power change" strategy in Lawler's vocabulary, especially if punitive tactics by more powerful actors create incentives for the dependent actor to do so; and (4) the more

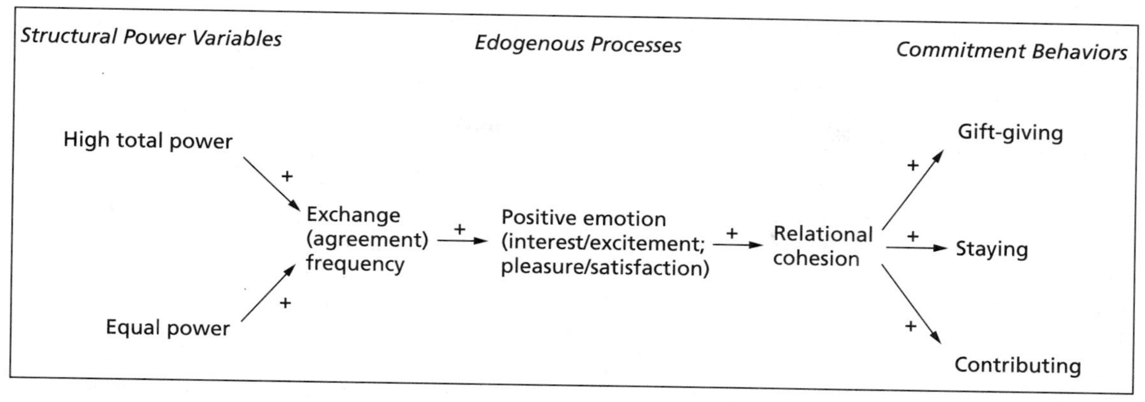

FIGURE 25-3 Lawler and Yoon's Model of Commitment

dependent actor can form a coalition among dependent actors providing resources to the more powerful actor and, in the process, reduce the alternatives for the more powerful actor, especially when the more powerful actor employs exploitive and punitive tactics on dependent actors and reduces their perception of costs in forming a coalition.

Commitment in Exchange Networks

In conjunction with Jeongkoo Yoon, Edward J. Lawler has more recently sought to understand how affective commitments develop in exchange relations.[27] As was noted for bargaining, frequency of bargaining and exchange lead to more conciliatory tactics of power use, leading Lawler to hypothesize that emotional commitments were involved. In contrast, Emerson and Cook's[28] formulation emphasized that commitment to exchange relations is related to the reduction of uncertainty over payoffs. In their model, structural power increases exchange frequency, which then reduces uncertainty about payoffs. In turn, the reduction of uncertainty increases actors' commitments to the exchange relationship because predictability assured expected levels of payoffs from the exchange. Lawler and Yoon, however, theorize that this presumed relationship among frequency, uncertainty-reduction, and commitment is really a proxy for more emotional dynamics. In their view, once these emotional forces could be isolated in experiments, the uncertainty-reduction argument would be obviated by a theory of emotional attachment. Although the uncertainty-reduction processes remains an independent force in producing commitments, their experiments confirm most of their hypotheses. The model tested by Lawler and Yoon is presented in Figure 25-3, and then we tease out the details of the argument.

On the left side of the model are the structural conditions that follow from Emerson's original formulation of power-dependence. *Total power* is the degree of actor's mutual dependence on each other for resources. The greater this mutual dependence is, then the higher is the total power in the relationship. Another concept that follows from Emerson's formulation is *relative power,* which is the level of inequality in the dependence of actors on each other for resources (that is, the relative power of Actor A is the ratio of A's power over B as a proportion of both A's and B's power; the greater A's proportion, the more is A's relative power and the greater is the inequality in the relationship). As shown in Figure 25-3, Lawler and Yoon only express relative power as "equal power," but the argument is really about structural situations where there is high total power and low relative power or equality. The greater is the mutual dependence of actors on resources that they provide for each other (high total power) and the more equal are their respective dependencies for these resources (equal power or low relative power), the more *structural cohesion* is built into their power relations. As a consequence, the more frequent are exchanges and agreements in these exchanges. Thus, the effects of structural cohesion (or high total power and low relative power) on emotions and commitments operate through *exchange frequency*.

Emotions for Lawler and Yoon are relatively short-term positive or negative evaluative states; Lawler and Yoon emphasize mild positive emotions and their effects on commitment. Two particular types of mild emotions are emphasized: (1) *interest and excitement* that revolve around anticipation of payoffs that give value; and (2) *pleasure and satisfaction* that orient actors to past and present payoffs yielding value. As the model outlines, exchange frequency under conditions of high total power and equality of power increases agreements in these exchanges, which, then, activate these two types of mild emotions, interest/excitement and pleasure/satisfaction.

As these emotions are activated, they increase relational cohesion. *Relational cohesion* is simply a combined function of the frequency of agreements in bargaining over resources and the positive emotions that are thereby aroused. Hence, the more frequent interactions and the more positive emotions generated from such interactions, the greater the level of relational cohesion is. Relational cohesion, in turn, produces *commitment behaviors* that, in the experimental settings developed by Lawler and Yoon, were divided into three types: (1) staying in the exchange relation even when attractive alternatives for resources were available; (2) giving token gifts to exchange partners unilaterally and without expectation for reciprocation; and (3) contributing to joint ventures with exchange partners, even under conditions of risk and uncertainty.

The results of the series of experiments supported the argument presented in Figure 25-3, but they did not obviate the effects of exchange frequency on uncertainty-reduction in generating commitment behaviors. Thus, commitment is related to the effects of exchange frequency on uncertainty reduction and arousal of mild positive emotions, although there is good reason to believe that the two are related: uncertainty reduction probably generates positive emotions, and vice versa. In explaining their model and research findings, Lawler and Yoon introduce concepts and causal relationships not specified in the model in Figure 25-3. One key concept is *objectification* of the relationship, or the perception by individuals that the exchange relationship, per se, is an object and source of the positive emotions experienced. Such objectification becomes a "force" furthering commitment behaviors, as people stay in the relationship because the relationship, per se, gives value, as individuals give token gifts to symbolize the relationship, and as they undertake joint ventures in the name of the relationship.

Once the discursive discussion around the model is finished, the model in Figure 25-3 becomes more complicated. These additional complications are presented in Figure 25-4, where (a) the direct, indirect, and reverse causal forces operating in Lawler and Yoon's theorizing are delineated, (b) the uncertainty-reduction argument from Emerson and Cook's work is retained, and (c) the discussion of objectification is added. The basic arguments remain the same, even in this more robust form. High total power increases exchange

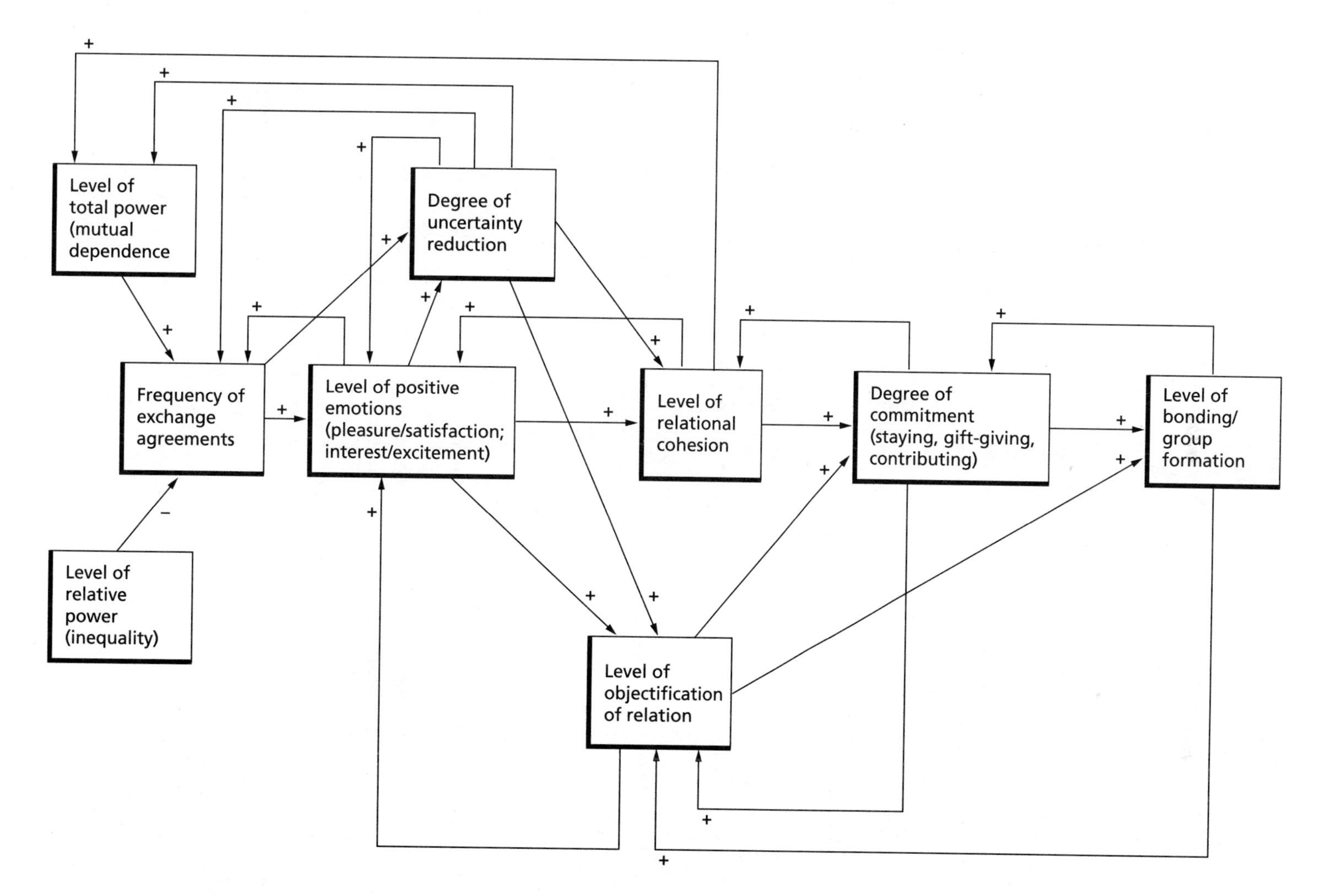

FIGURE 25-4 An Extension of Lawler's Model of Relational Cohesion and Commitment

frequency, whereas high relative power or inequality works against high rates of exchange agreement (as would be expected by virtue of the earlier discussion of Lawler's and various collaborators' analysis of bargaining). Thus, high mutual dependence and equality will increase exchange frequency, which then reduces uncertainty about payoffs and arouses mild positive emotions that, by themselves but also in combination, increase the level of objectification of the relationship as a gratifying entity, per se, beyond the specific bargains and payoffs in any particular round of exchange. Together, objectification and relational cohesion increase attachments and commitments to the exchange relationship that, in turn, increase the level of bonding and group formation. Some of these causal paths are inferred from, and some represent extensions of, Lawler and Yoon's theory; what is true of the direct causal paths is even more true of the reverse causal chains that flow from right to left in the model. These paths represent elaborations and extensions of the theory, but they do so in ways consistent with the theory.

As group bonds are formed from commitment behaviors, these feed back and increase the level of objectification, which, in turn, adds an extra dose of positive emotion. Similarly, bonding, commitments, and relational cohesion all feed back, increasing positive emotions that lower uncertainty. Together, these feed back to increase frequency of exchange agreements. As uncertainty is reduced and relational cohesion increased, this escalated cohesion increases mutual dependence for valued resources that sets the commitment process in motion once again. Obviously, these positive direct, indirect, and reverse causal effects cannot go on forever, making people committed, "emotional junkies," but the model emphasizes how structural conditions of high total and low relative power initiate a series of recursive processes that feed off each other and generate commitments to relationships.

LINDA D. MOLM'S THEORETICAL PROGRAM

Most experimental studies employing Emerson's power-dependence ideas have focused on *negotiated exchanges*, where the actors seek to influence each other's contributions before distributing their resources. Lawler's analysis of bargaining is a good example of this emphasis on negotiation, where rounds of offers and counter-offers are made before resources are exchanged. Another point of emphasis in the work of those following Emerson and Cook is *reward power*, where power use revolves around negotiating the ratio of rewards that actors provide for one another under power-dependence conditions. The program developed by Lawler and his collaborators actually has been one of the few to examine more punitive forms of power in exchange networks, but, like most work in the Emerson-Cook lineage, this program has remained focused on bargaining in negotiated exchanges.

In contrast, in a series of theoretically informed experiments, Linda D. Molm examined *reciprocal exchanges*, in which direct bargaining before the distribution of resources does not occur. Rather, in reciprocal exchanges, an actor provides benefits for the other without knowing in advance if, or how much, the other party will provide in return. Many exchanges are of this nature, as when a person does something beneficial for another and only afterward knows whether or not this favor will be reciprocated. Indeed, a great portion of exchange relations in the real world involves this kind of reciprocity, where benefits or punishments are offered and received sequentially without prior negotiation over the distribution of outcomes. Thus, Molm's program examined a somewhat different kind of exchange—reciprocal rather than negotiated.[29] As we will see, reciprocal exchanges reveal dynamics that differ from those revolving around bargaining. This difference is particularly evident in Molm's analysis of punishment in exchange relations, yielding findings that qualify those developed in the Lawler, Bacharach, and Ford program where coercion was examined in negotiated exchanges.

The Basic Question

The empirical findings in most research have been clear: Punitive tactics are rarely used in exchange processes, even when one actor has a clear resource advantage and could impose punishment without fear of costly retaliation.[30] For Molm, these findings have posed the two basic questions of her research program: (1) Why would punishment tactics not be used frequently in exchange relations, even when they could be used to an actor's advantage, and (2) what conditions would increase the use of punishment strategies and make them effective? Molm argues the reason that punishment is so infrequently used in exchange is not related to its ineffectiveness; rather, the ineffectiveness of punishment is because it is so infrequently used.

In drawing this conclusion, Molm's research challenges an assumption explicitly argued by early exchange theorists—namely, that coercion is counter-productive because it leads to retaliation and a conflict spiral. Alternatively, as Bacharach, Lawler, and Ford argued, the use of punishments reduces conciliatory bargaining tactics and obstructs conflict resolution. As a result, coercive power is rarely used because of the fear of retaliation that inheres in both the conflict-spiral and bilateral-deterrence models that they explored.

The Basic Concepts

Like all power-dependence theorists, Molm begins with the assumptions and basic propositions developed by Emerson and, later, by Emerson and Cook together. Actors initiate exchanges by giving rewards to others who control resources that they value; they increase rates of exchange with those partners who provide valued resources and decrease exchanges with those who do not; they explore alternatives when rewards decline or when punishments increase; the outcomes of an exchange are the rewards or punishments that actors have received from an exchange, with such outcomes reflecting the respective power of the actors; power is related to dependence of an actor on another, with dependence being a joint function of the value of the resources provided by an actor to another and the number of available alternatives for that resource; *average power* is the degree of mutual dependence of actors on each other for a valued resource (a notion that parallels Lawler's concept of *total power*); power imbalance denotes the state where one actor's dependence on another for resources is greater than the reverse; such imbalance is a reflection of *relative power*, or the difference between actors' respective dependencies on each other.

Molm's goal has been to extend these basic concepts to the analysis of punishment processes, but to do so, she expands the conceptualization of outcomes in an exchange. Drawing on Daniel Kahneman and Amos Tversky's theory,[31] she introduced the view that the *value of outcomes* from an exchange is influenced by three dynamics. One is what is termed *referential dependence*, or the tendency of actors to use the current status quo in exchange outcomes as the basic reference point for assessing gains or losses. When outcomes exceed the status quo, as established by past exchanges, they are seen as gains, but when they fall below this reference point, they are seen as losses or punishments. Another key dynamic in assessing outcomes is *diminishing sensitivity*, or the impact of gains or losses as they exceed or fall below the reference point created by the status quo. The first outcomes below or above this point have far more impact on actors' assessment of whether they have gained or incurred losses than do later ones; each increment of gain or loss above or below the reference point has less effect on the actors' sense of their outcomes. A third central dynamic is *loss aversion*, or the tendency of actors to assign greater subjective value and weight to losses than gains. A loss has more impact than an equivalent gain, and hence, actors tend to become more concerned with avoiding losses (because they count more subjectively) than with achieving an equivalent increment of gain. As we will see, these forces influencing assessments of outcomes in exchanges are central to Molm's analysis of punishment.

Punishment is the actual act of imposing costs on others, whereas *punishment power* is the capacity to impose costs; this capacity is related to the extent that others depend on this actor for avoiding punishments. Conceptually, punishment power would seem to be the same as reward power, which is a function of the level of one actor's dependency on another for valued resources. Thus, both reward and punishment power are defined by one actor's dependence on another to receive rewards or to avoid punishments. And an actor's dependence is related to the value of the resources provided by one actor to another (whether receiving rewards or avoiding punishments) and the alternative sources for a reward or for avoiding punishments. Hence, in accordance with Emerson's original formulation, a power imbalance on punishment power should lead to the use of this power advantage in the same way as actors possessing a reward power advantage. That is, rates of punishment by the resource-advantaged actor should increase as a means to extract more resources from others who depend on this actor for avoiding punishment and who do not have attractive alternatives for avoiding this punishment. But empirically, power use by punishment-advantaged actors does not occur as frequently as it does for a reward-power advantage. The question is, why?

The Theoretical Answer

Molm's answer begins with a basic insight into the nature of power-dependence exchanges revolving around reward power. In such exchanges, power use is *structurally induced*: Actor A's power resides in the dependence of Actor B on it for rewards; this dependence reflects that A has alternative sources for the rewards provided by B, whereas B's alternatives for the resources provided by A are not as attractive; hence, when A looks to its alternative sources for resources, A inevitably demonstrates its power advantage over B, even if A does not want to do so. It is thus in the nature of reward-power imbalance for actors to use this advantage, whether they do so intentionally or inadvertently. The very act of exchanging with alternative actors emphasizes the advantage of one actor over another, thereby forcing the latter to increase its resource offers to the advantaged actor.

In contrast, punishment power is not structurally induced in this way. Actors must decide to punish, or impose additional costs on another who has not provided benefits that were expected. Thus, in reward-power situations, A uses power by simply shopping alternatives, forcing B to provide more rewards if B wants to continue to exchange with A, whereas in punishment-power conditions, A must strategically use punishment to impose costs on B who is not providing enough of what A wants from B. Thus, use of punitive power is not so much structurally induced as strategically imposed when an actor feels that it is not getting what it should.

But, the question still remains: Why is the strategic use of punishment power so infrequently used compared with structurally induced use of reward power? The answer is that the strategic use of punishment power poses *risks.* As Lawler would emphasize, one risk is punitive retaliation by an actor on whom punitive tactics are tried, therefore increasing the costs on the actor who initiates punitive tactics. Another risk is that punishment tactics will cause the target of punishment to withdraw and stop exchanging, thereby denying the punitive actor resources that it values. Indeed, it appears that the risks associated with losing valued rewards are far more salient than the risks associated with retaliation in deterring use of punishment.

Risk is accelerated by *loss aversion,* which is the tendency of actors to be more concerned about not losing what they have been receiving in the status quo rather than about increasing, at risk, their resource shares beyond the status quo. If actors are relatively satisfied, they will be reluctant to use punitive strategies to get more, especially if there are risks involved.

Still another force inhibiting the use of punitive strategies is justice. In examining justice issues, Molm adds to findings reported earlier in Cook's and Hegtvedt's work. In Molm's analysis, actors' sense of injustice increases more when punitive tactics are used against them. As a consequence, negative emotions are aroused that, in turn, make them more disposed to use punitive tactics in retaliation. The recognition by an actor that others experiencing injustice will be more likely to engage in retaliatory tactics keeps this actor from using punishment in the first place.

What forces, then, overcome the inhibitory effects of risks, loss aversion, and retaliation by those experiencing injustice? Molm's answer is that an actor will initiate a punitive-power strategy against another when its dependence on resources provided by its target is high but its receipt of resources has been low—indeed, so low that the actor has little to lose by employing punitive tactics. Thus, in Molm's studies of reciprocal exchanges, the power disadvantaged will be most likely to use coercive tactics to get what they value from a power-advantaged exchange partner. Under these conditions, where an actor has not been receiving expected rewards, it has little to lose—because it is not getting much—in initiating more coercive tactics.

In Molm's experiments, such punitive strategies will be effective if used *contingently* when a partner fails to provide resources and if these tactics are used *consistently* each time a partner fails to provide needed and expected resources. Under these conditions of (1) contingency and (2) consistency, a power-disadvantaged actor can use punishment effectively. This effectiveness increases when norms or lack of alternatives constrain the power-advantaged partner's ability to withdraw readily from a relationship.

The use of punitive tactics by the disadvantaged actor will be less effective, however, when not used (1) contingently or (2) consistently. That is, punishment will be most effective if an actor employs coercive tactics with some regularity (consistency) when expected rewards are not forthcoming (contingency). If punitive tactics are used as a ploy to get a higher rate of return than justice norms or status quo considerations would dictate, they can be less effective in inducing an advantaged partner to provide resources. Indeed, these punitive tactics might activate negative justice sentiments and lead to withdrawal by the advantaged partner or retaliatory punishment. If the coercive power of the advantaged partner is greater than its reward power—that is, it is less costly to use coercion than to give rewards in response to punitive tactics by another—then the power-advantaged actor is more likely to retaliate with punitive tactics of its own. And the more costly it becomes for a power-advantaged actor to give rewards in response to the contingent and consistent punitive tactics by a power-disadvantaged actor, then the power-advantaged actor might resort to punitive tactics. Yet, Molm emphasizes that, even when retaliation occurs, consistent use of punishment can still be highly effective, especially if this use remains highly contingent on the targeted actor's failure to provide expected rewards.

From Molm's data, the level of positive sentiments will actually rise slightly, even from exchange relations that have been pushed to the point where the disadvantaged actor has resorted to punitive tactics to increase exchange frequency with a reluctant partner. This finding supports Lawler and Yoon's argument on effects of exchange frequency on commitment behaviors in bargaining situations, but in Molm's studies, the exchanges involved are reciprocal rather than negotiated through bargaining. More significantly, the effects of exchange frequency on emotions were generated in Molm's experiments under conditions of high relative power or inequality—thereby indicating that, in reciprocal exchanges the emotional effects are not destroyed by inequality as they appeared to be in Lawler and Yoon's analysis of negotiated exchanges. Still, although strong and consistent use of coercion will generate more positive sentiments than will weak and sporadic use of punishment, these positive sentiments will not be as strong as those for a partner who uses no coercion at all. Thus, punishment does not increase positive sentiments to the same degree as no punishment does.

New Theoretical Directions

Molm's analysis of punitive tactics in reciprocal exchanges supplements work by Lawler and others on bargaining in negotiated exchanges. Equally interesting, Molm's analysis converges with Lawler's more recent efforts to analyze how commitment emerges from frequently negotiated exchanges, as is modeled in Figure 25-3 on page 329. Molm has also approached the question of how exchange structures change over time, but she has sought to do so with a somewhat different model than Lawler and Yoon. She sees the emergence of commitments, trust, and normative expectations as being created by the transformation of exchange transactions from a structure of dependence to *inter*dependence.[32]

Structures of dependence in exchange generate greater risk, because outcomes cannot be guaranteed and, indeed, are contingent of what other actors do.[33] Among risk-aversive actors for whom the initial gains or losses are highly salient, there is comfort in creating and sustaining a status quo where benefits are predictable—an argument that converges with Cook and Kollock's analysis of commitment as a means to reduce uncertainty. One way to reduce risk and generate this comfortable status quo is to transform exchanges into *structures of interdependence*. This transformation involves *serial dependencies* in which actors' receipt of rewards depends on what they have given in the previous round of exchange. In reciprocal exchanges, actors must therefore set up mutual contingencies where each of their future payoffs depends on their partner's willingness to continue their past patterns of reward. When such serial dependencies become stabilized, they are transformed, in essence, into interdependencies. Once interdependencies are formed, *the relation itself* becomes a source of gratification, and, in Lawler and Yoon's notion of *objectification*,[34] the relation becomes an object outside of the exchange, per se, as well as an object that has value to the actors and thus keeps the exchanging partners in line.

Molm argues that this process of transformation is more readily achieved in the kinds of exchanges studied by Lawler and Yoon—namely, negotiated exchanges. In first bargaining over outcomes, a kind of serial dependence is created in the very structure of offer, counter-offer, and division of outcomes. As the relationship unfolds during bargaining, the actors become dependent on the relation itself to receive their benefits. If negotiated agreements become frequent, then the relation becomes even more "objectified" and, hence, more likely to be seen as a source of value in itself. In contrast, the reciprocal exchanges studied by Molm do not have the serial dependencies built into the nature of negotiation; instead, serial dependency leading to interdependence must occur as actors sequentially give and receive benefits. Such reciprocal exchanges do not have the dependency that is inherent in the process of negotiation and bargaining. So building trust, commitment, and social credit in ways that reduce risks for loss aversive actors, for whom the initial losses have high salience, takes longer and, perhaps, is more fragile.

Molm argues that because of the somewhat different ways that interdependencies are constructed in negotiated versus reciprocal exchanges, generating a ratio of exchange outcomes among actors will be more difficult in reciprocal than in negotiated exchanges. Moreover, actors will pursue less optimal and maximizing strategies in reciprocal than in negotiated exchanges. The reasons for these differences reside in the nature of bargaining as opposed to reciprocation. In bargaining, actors are constantly thinking about and calculating rewards, and the bargaining process itself keeps them engaged in exchange process. Thus, they are more likely to seek optimal solutions and worry less about the risks of failure because of the power of the bargaining situation to hold the participants together. In reciprocal exchanges, however, actors are far more concerned with risk aversion because they do not know if another will reciprocate until after they have offered benefits; as a result, they will be less concerned with precise calculations of utilities (which cannot be known until after the fact) and, hence, with maximization of outcomes. Rather, because of the uncertainty over reciprocation, actors will more typically try to keep outcomes at or above the status quo, a strategy that is less risky than efforts to maximize outcomes.

The Paradoxes of Reciprocal Exchanges

There are many interesting paradoxes in Molm's theory. One is that those with less reward power are more likely to use punitive tactics against more powerful partners. The double irony is that if they do so contingently and consistently, they will increase their resource gains from a more powerful partner, at least to a point.

Still another paradox is that punishment and, in the real world, its manifestation as violence are most likely to emerge in relations that have ceased being rewarding for the less dependent actor, where in essence this more powerful partner is decreasing exchange with a more dependent partner. Thus, violence comes when dependencies have declined, at least on one side, forcing the more dependent and less powerful actor to try coercion to keep a more powerful partner in the exchange.

A final paradox is that the most fragile relations, where actors are both experiencing a high sense of risk and are worried about losses, are often the most likely to develop solidarities as reciprocal exchange partners seek to reduce risk by transforming serial exchanges into interdependencies in which norms, trust, and commitments can emerge to reduce risk and mitigate against actors' aversion to loss.

CONCLUSION

In this chapter, we have explored three independent research programs, all drawing from the initial insights of Richard Emerson whose basic ideas were summarized in Chapter 22. There is, as might be expected, considerable overlap in the programs; yet, each adds something new to this lineage of exchange-network theory. Combined with rival exchange-network programs, a considerable amount of understanding of how power operates in exchange systems now exists—understandings that have been backed by clever experimental designs.

Unfortunately, there is surprisingly little integration of approaches within the Emerson-Cook lineage and even less so with other network approaches employing similar concepts but different vocabularies. Thus, the future of this entire approach to theoretical understanding of the dynamics of power and social structures rests not just with continuing the various lineages that have been spawned but with consolidating them.

Clearly, there are points of overlap that should encourage synthesis among at least three topics: What conditions generate use of reward and punitive power in both negotiated and reciprocal exchange networks? What conditions generate the formation of not only justice norms but also other normative systems that guide the actions of actors in both direct and generalized exchange networks? What conditions produce more affective states, revolving around emotions or, at the very least, trust and behavioral commitments in direct and generalized networks? Partial answers to these important questions can be found in the three theories examined here, as well as in rival programs outside the Emerson-Cook lineage. The goal for the next decade, as this approach enters the twenty-first century, should be theoretical consolidation as much as further experimentation.

NOTES

1. For a review of elementary concepts in network analysis, see Chapter 38.

2. See Chapter 22 for details. The core of Emerson's approach is found in a few seminal articles: "Power-Dependence Relations," *American Sociological Review* 17 (February 1962), pp. 31–41; "Power-Dependence Relations: Two Experiments," *Sociometry* 27 (September 1964), pp. 282–298; "Exchange Theory, Part I: A Psychological Basis for Social Exchange" and "Exchange Theory, Part II: Exchange Relations and Network Structures," in *Sociological Theories in Progress*, eds. J. Berger, M. Zelditch, and B. Anderson (New York: Houghton Mifflin, 1972), pp. 38–87; "Operant Psychology and Exchange Theory," in *Behavior Sociology*, eds. R. Burgess and D. Bushell (New York: Columbia University Press, 1969); "Social Exchange Theory," *Annual Review of Sociology* 2 (1976), pp. 535–562. See also John F. Stolte and Richard M. Emerson, "Structural Inequality: Position and Power in Network Structures," in *Behavioral Theory in Sociology*, ed. R. Hamblin (New Brunswick, NJ: Transaction, 1977).

3. As we will see, the notions of "absolute," "structural cohesion," "total," and "average" power are terms that others adopt for their respective schemes, although the meanings of the terms vary somewhat from Emerson's original usage.

4. Richard M. Emerson, "Social Exchange Theory," in *Social Psychology: Sociological Perspectives*, eds. M. Rosenberg and R.H. Turner (New York: Basic Books).

5. For examples see David Willer, "The Basic Concepts of the Elementary Theory," in *Networks, Exchange and Coercion*, eds. D. Willer and B. Anderson (New York: Elsevier, 1981); "Property and Social Exchange," *Advances in Group Processes* 2 (1985), pp. 123–142; and *Theory and the Experimental Investigation of Social Structures* (New York: Gordon and Breach, 1986); David Willer, Barry Markovsky, and Travis Patton, "Power Structures: Derivations and Applications of Elementary Theory," in *Sociological Theories in Progress: New Formulations*, eds. J. Berger, M. Zelditch, and B. Anderson (Newbury Park: Sage, 1989); David Willer and Barry Markovsky, "Elementary Theory: Its Development and Research Program," in *Theoretical Research Programs: Studies in the Growth of Theory*, eds. J. Berger and M. Zelditch, Jr. (Stanford, CA: Stanford University Press, 1993); Philip Bonacich, "Power and Centrality: A Family of Measures," *American Journal of Sociology* 92 (1987), pp. 1170–1082; Elisa Jayne Bienenstock and Philip Bonacich, "The Core as a Solution to Exclusionary Networks," *Social Networks* 14 (1992), pp. 231–243.

6. For overviews, see Linda D. Molm and Karen S. Cook, "Social Exchange and Exchange Networks,"

in *Sociological Perspectives on Social Psychology*, eds. K.S. Cook, G.A. Fine, and J.S. House (Boston: Allyn & Bacon, 1995); see also Karen S. Cook, Linda D. Molm, and Toshio Yamagishi, "Exchange Relations and Exchange Networks: Recent Developments in Social Exchange Theory" along with David Willer and Barry Markovsky, "Elementary Theory: Its Developments and Research Program," in *Theoretical Research Programs: Studies in the Growth of Theory*, eds. J. Berger and M. Zelditch (Stanford, CA: Stanford University Press, 1993).

7. It should be emphasized that most of this work, and until recently, virtually all work by Cook and most exchange theorists has been on *negatively connected networks* in which exchange with one actor precludes exchanging with another.

8. Karen S. Cook and Richard M. Emerson, "Power, Equity, and Commitment in Exchange Networks," *American Sociological Review* 43 (1978), pp. 721–739; Karen S. Cook and Richard M. Emerson, "Exchange Networks and the Analysis of Complex Organizations," *Research on the Sociology of Organizations* 3 (1984), pp. 1–30.

9. Ibid., (refer to both works in previous note).

10. Peter Kollock, "The Emergence of Exchange Structures: An Experimental Study of Uncertainty, Commitment and Trust," *American Journal of Sociology* 100 (1994), pp. 315–345.

11. See James Coleman's discussion in Chapter 24 on similar processes in his analysis of solidarity.

12. Karen S. Cook, Richard M. Emerson, Mary R. Gillmore, and Toshio Yamagishi, "The Distribution of Power in Exchange Networks: Theory and Experimental Results," *American Journal of Sociology* 87 (1983), pp. 275–305. More recently, Yamagishi, Gillmore, and Cook have sought to extend this line of work to positively connected networks where beneficial exchange with one partner does not preclude exchanges with others. Many of the same dynamics are revealed but with some modification: The value of resources becomes the only consideration in positively connected networks, and the locus of power shifts toward the source of supply of valued resources. See their "Network Connections and the Distribution of Power in Exchange Networks," *American Journal of Sociology* 93 (1988), pp. 833–851. For a more recent effort to theorize on these processes, see also Kazuo Yamaguchi, "Power in Networks of Substitutable and Complementary Exchange Relations: A Rational-Choice Model and An Analysis of Power Centralization," *American Sociological Review* 61 (1996), pp. 308–332. For rivals to the Cook program, also see John Skvoretz and David Willer, "Exclusion and Power: A Test of Four Theories of Power in Exchange Networks," *American Sociological Review* 58 (1993), pp. 801–818.

13. See text in previous footnote.

14. Cook and Emerson, "Power, Equity and Commitment in Exchange Networks" (cited in note 8).

15. See Chapters 20 and 21, respectively, on the work of George Homans and Peter Blau in the 1960s.

16. Karen A. Hegtvedt and Karen S. Cook, "The Role of Justice in Conflict Situations," *Advances in Group Processes* 4 (1987), pp. 109–136; Karen S. Cook and Karen A. Hegtvedt, "Distributive Justice, Equity, and Equality," *Annual Review of Sociology* 9 (1983), pp. 217–241; Karen Hegtvedt, Elaine Thompson, and Karen S. Cook, "Power and Equity: What Counts in Attributions for Exchange Outcomes," *Social Psychology Quarterly* 56 (1993), pp. 100–119.

17. Hegtvedt, Thompson, and Cook, "Power and Equity" (cited in note 16).

18. Hegtvedt and Cook, "The Role of Justice in Conflict Situations"; and Cook and Hegtvedt, "Distributive Justice, Equity, and Equality" (cited in note 16); for a review of the justice and injustice literature, see also Karen Hegtvedt and Barry Markovsky, "Justice and Injustice," in *Sociological Perspectives on Social Psychology*, eds. K.S. Cook, G.A. Fine, and J.S. House (Boston: Allyn & Bacon, 1995).

19. Toshio Yamagishi and Karen S. Cook, "Generalized Exchange and Social Dilemmas," *Social Psychology Quarterly* 56 (1993), pp. 235–249.

20. For relevant works by Yamagishi, see "The Provision of a Sanctioning System in the United States and Japan," *Social Psychology Quarterly* 51 (1988), pp. 267–270; "Seriousness of Social Dilemmas and the Provision of a Sanctioning System," *Social Psychology Quarterly* 51 (1988), pp. 32–42; "Unintended Consequences of Some Solutions to the Social Dilemmas Problem," *Sociological Theory and Method* 4 (1989), pp. 21–47.

21. Peter P. Ekeh, *Social Exchange Theory: The Two Traditions* (Cambridge, MA: Harvard University Press, 1974).

22. See pages 311–314 of Chapter 24.

23. Samuel B. Bacharach and Edward J. Lawler, *Bargaining: Power, Tactics, and Outcomes* (San Francisco: Jossey-Bass, 1981) and "Power and Tactics in Bargaining," *Industrial and Labor Relations Review* 34 (1981), pp. 219–233; see also Edward J. Lawler and Samuel B. Bacharach, "Power-Dependence in Individual Bargaining: The Expected Utility of Influence," *Industrial and Labor Relations Review* 32 (1979), pp. 196–204.

24. Edward J. Lawler, "Comparison of Dependence and Punitive Forms of Power," *Social Forces* 66 (1987), pp. 446–462; "Power in Bargaining Processes," *Sociological Quarterly* 33 (1992), pp. 17–34.

25. Edward J. Lawler and Rebecca Ford, "Metatheory and Friendly Competition in Theory Growth: The Case of Power Processes in Bargaining," in *Theoretical Research Programs: Studies in the Growth of Theory*, eds. J. Berger and M. Zelditch (Stanford, CA: Stanford University Press, 1993); "Bargaining and Influence in Conflict Situations," in *Sociological Perspectives on Social Psychology*, eds. K.S. Cook, G.A. Fine, and J.S. House, pp. 236–256 (Boston: Allyn & Bacon, 1995).

26. Lawler, "Comparison of Dependence and Punitive Forms of Power" and "Power in Bargaining Processes" (cited in note 24).

27. Edward J. Lawler and Jeongkoo Yoon, "Commitment in Exchange Relations: A Test of a Theory of Relational Cohesion," *American Sociological Review* 61 (1996), pp. 89–108; Edward Lawler, Jeongkoo Yoon, Mouraine R. Baker, and Michael D. Large, "Mutual Dependence and Gift Giving in Exchange Relations," *Advances in Group Processes* 12 (1995), pp. 271–298; Edward J. Lawler and Jeongkoo Yoon, "Power and the Emergence of Commitment Behavior in Negotiated Exchange," *American Sociological Review* 58 (1993), pp. 465–481.

28. Karen S. Cook and Richard M. Emerson, "Exchange Networks and the Analysis of Complex Organizations" (cited in note 8); see also Peter Kollock, "The Emergence of Exchange Structures" (cited in note 10).

29. For recent overviews of her program, see Linda D. Molm, *Coercive Power in Social Exchange* (Cambridge: University of Cambridge Press, 1997); "Risk and Power Use: Constraints on the Use of Coercion in Exchange," *American Sociological Review* 62 (1997), pp. 113–133; "Punishment and Coercion in Social Exchange," *Advances in Group Processes* 13 (1996), pp. 151–190; and "Is Punishment Effective?: Coercive Strategies in Social Exchange," *Social Psychology Quarterly* 57 (1994), pp. 79–94. For more general reviews of her work within the general context of exchange-network theory, see Linda D. Molm and Karen S. Cook, "Social Exchange and Exchange Networks" (cited in note 6).

30. Linda D. Molm, "The Structure and Use of Power: A Comparison of Reward and Punishment Power," *Social Psychology Quarterly* 51 (1988), pp. 108–122; "An Experimental Analysis of Imbalance in Punishment Power," *Social Forces* 68 (1989), pp. 178–203; "Punishment Power: A Balancing Process in Power-dependence Relations," *American Journal of Sociology* 94 (1989), pp. 1392–1428; and "Structure, Action, and Outcomes: The Dynamics of Power in Social Exchange," *American Sociological Review* 55 (1990), pp. 427–447.

31. Daniel Kahneman and Amos Tversky, "Choices, Values and Frames," *American Psychologist* 39 (1984), pp. 341–350. Their assumptions are integrated with assumptions from Emerson's original formulation. See Molm's "Risk and Power Use: Constraints on the Use of Coercion in Exchange" (cited in note 29).

32. Linda D. Molm, "Dependence and Risk: Transforming the Structure of Social Exchange," *Social Psychology Quarterly* 57 (1994), pp. 163–176.

33. This argument is consistent with other work in power-dependence. See notes 8 and 10 for references on other studies by Cook, Emerson, and Kollock.

34. An idea they borrow from Peter Berger and Thomas Luckman, *The Social Construction of Reality* (New York: Doubleday, 1967) who, in turn, had taken the idea from Émile Durkheim, *The Elementary Forms of the Religious Life* (New York: Free Press, 1954; originally published in 1912).

PART V

Interactionist Theorizing

26

The Emerging Tradition:

The Rise of Interactionist and Phenomenological Theorizing

The first sociological theorists in Europe were concerned primarily with macro-level phenomena, but with the beginning of the twentieth century, theorists in Europe and America turned to the analysis of micro-level processes. They began to understand that the structure of society is, in some ultimate sense, created and maintained by the actions and interaction of individuals, so increasingly, they sought to discover the fundamental processes of interaction among people. This burst of creative activity generated a wide range of micro-level theories that will, for the sake of simplicity, be labeled "interactionism." The rise of interactionism also marks the beginning of American theory as an active contributor to the theoretical canon of sociology. Hence, it is appropriate that we begin with the American contribution to interactionism, turning later to the European micro-oriented tradition that emerged in the early decades of this century.

EARLY AMERICAN INSIGHTS INTO INTERACTION

A philosopher at the University of Chicago, George Herbert Mead, made the great breakthrough in understanding the basic properties of human social interaction. His was not a blazing new insight but, rather, a synthesis of ideas that had been developed by others. Yet, without his synthesis, the study of interaction would have been greatly retarded. To appreciate Mead's genius, let us first review those from whom he drew inspiration, then explore how he pieced their ideas into a model of interaction that still serves as the basic framework for most interactionist theories.

William James' Analysis of "Self"

The Harvard psychologist William James (1842-1910) was perhaps the first social scientist to develop a clear concept of self. James recognized that humans have the capacity to view themselves as objects and to develop self-feelings and attitudes toward themselves. Just as humans can (a) denote symbolically other people and aspects of the world around them, (b) develop attitudes and feelings toward these objects, and (c) construct typical responses toward objects, so they can denote themselves, develop self-feelings and attitudes, and construct responses toward themselves. James called these capacities *self* and recognized their importance in shaping the way people respond in the world.

James developed a typology of selves: the "material self," which includes those physical objects that humans view as part of their being and as crucial to their identity; the "social self," which involves the self-feelings that individuals derive from associations with other people; and the "spiritual self," which embraces the general cognitive style and capacities typifying an individual.[1] This typology was never adopted by subsequent interactionists, but James' notion of the social self became a part of all interactionists' formulations.

James' concept of the social self recognized that people's feelings about themselves arise from interaction with others. As he noted, "a man has as many social selves as there are individuals who recognize him."[2] Yet James did not carry this initial insight very far. He was, after all, a psychologist who was more concerned with internal psychological functioning of individuals than with the social processes from which the capacities of individuals arise.

Charles Horton Cooley's Analysis of Self

Charles Horton Cooley offered two significant extensions in the analysis of self.[3] First, he refined the concept, viewing self as the process in which individuals see themselves as objects, along with other objects, in their social environment. Second, he recognized that self emerges from communication with others. As individuals interact with each other, they interpret each other's gestures and thereby see themselves from the viewpoint of others. They imagine how others evaluate them, and they derive images of themselves or self-feelings and attitudes. Cooley termed this process *the looking glass self:* the gestures of others serve as mirrors in which people see and evaluate themselves, just as they see and evaluate other objects in their social environment.

Cooley also recognized that self arises from interaction in group contexts. He developed the concept of "primary group" to emphasize that participation in front of the looking glass in some groups is more important in the genesis and maintenance of self than participation in other groups. Those small groups in which personal and intimate ties exist are the most important in shaping people's self-feelings and attitudes.

Cooley thus refined and narrowed James' notion of self and forced the recognition that it arises from symbolic communication with others in group contexts. These insights profoundly influenced George Herbert Mead.

John Dewey's Pragmatism

John Dewey (1859–1952) was, for a brief period, a colleague of Cooley's at the University of Michigan. But more important was Dewey's enduring association with George Herbert Mead, whom he brought to the University of Chicago. As the chief exponent of a school of thought known as *pragmatism,* Dewey stressed the process of human adjustment to the world, in which humans constantly seek to master the conditions of their environment. Thus, the unique characteristics of humans arise from the *process* of adjusting to their life conditions.

What is unique to humans, Dewey argued, is their capacity for thinking. Mind is not a structure but a process that emerges from humans' efforts to adjust to their environment. Mind for Dewey is the process of denoting objects in the environment, ascertaining potential lines of conduct, imagining the consequences of pursuing each line, inhibiting inappropriate responses, and, then, selecting a line of conduct that will facilitate adjustment. Mind is thus the process of thinking, which involves deliberation:

> Deliberation is a dramatic rehearsal (in imagination) of various competing possible lines of action. . . . Deliberation is an experiment in finding out what the various lines of possible action are really like. It is an experiment in making various combinations of selected elements . . . to see what the resultant action would be like if it were entered upon.[4]

Dewey's conception of mind as a process of adjustment, rather than as a thing or entity, was critical in shaping Mead's thoughts. Much as Cooley had done for the concept of self, Dewey demonstrated that mind emerges and is sustained through interactions in the social world.

Pragmatism, Darwinism, and Behaviorism in Mead's Thought

At the time that Mead began to formulate his synthesis, the convergence of several intellectual traditions was crucial because it appears to have influenced the direction of his thought. Mead considered himself a behaviorist, but not of the mechanical stimulus/response type. Many of his ideas actually were intended as a refutation of such prominent behaviorists as John B. Watson. Mead accepted the basic premise of behaviorism—that is, the view that reinforcement guides and directs action. However, he used this principle in a novel way. Moreover, he rejected as untenable the methodological presumption of early behaviorism that it was inappropriate to study the internal dynamics of the human mind. James', Cooley's, and Dewey's influence ensured that Mead would rework the principle of reinforcement in ways that allowed for the consideration of mind and self.

Another strain of thought that shaped Mead's synthesis is pragmatism, as it was acquired through exposure with Dewey. Pragmatism sees organisms as practical creatures that come to terms with the actual conditions of the world. Coupled with behaviorism, pragmatism offered a new way of viewing human life: Human beings seek to cope with their actual conditions, and they learn those behavioral patterns that provide gratification. The most important type of gratification is adjustment to social contexts.

This argument was buttressed in Mead's synthesis by yet another related intellectual tradition, Darwinism. Mead recognized that humans are organisms seeking a niche in which they can adapt. Historically, this was true of humans as an evolving species; more important, it is true of humans as they discover a niche in the social world. Mead's commitment to behaviorism and pragmatism thus allowed him to apply the basic principle of Darwinian theory to each human: That which facilitates survival or adaptation of the organism will be retained.

In this way, behaviorist, pragmatist, and Darwinian principles blended into an image of humans as attempting to adjust to the world around them and as retaining those characteristics—particularly mind and self—that enable them to adapt to their surroundings. Mind, self, and other unique features of humans evolve from efforts to survive in the social environment. They are thus capacities that arise from the processes of coping, adjusting, adapting, and achieving the ultimate gratification or reinforcement: survival. For this reason, Mead's analysis emphasizes the processes by which the infant organism acquires mind and self as an adaptation to society. But Mead did much more; he showed how society is viable only from the capacities for mind and self among individuals. From Mead's perspective, then, the capacities for mind, self, and society are intimately connected.

George Herbert Mead's Synthesis

The names of James, Cooley, and Dewey figure prominently in the development of interactionism, but Mead brought their related concepts together into a coherent theoretical perspective that linked the emergence of the human mind, the social self, and the structure of society to the process of social interaction.[5] Mead appears to have begun his synthesis with two basic assumptions: (1) the biological frailty of human organisms forces their cooperation with one another in group contexts to survive; and (2) those actions within and among human organisms that facilitate their cooperation, and hence their survival or adjustment, will be retained. Starting from these assumptions, Mead reorganized the concepts of others so that they denoted how mind, the social self, and society arise and are sustained through interaction.

Mind. Following Dewey's lead, Mead recognized that the unique feature of the human mind is its capacity to (1) use symbols to designate objects in the environment, (2) rehearse covertly alternative lines of action toward these objects, and (3) inhibit inappropriate lines of action and select a proper course of overt action. Mead termed this process of using symbols or language covertly *imaginative rehearsal,* revealing his conception of mind as a *process* rather than as a structure. Further, the existence and persistence of society, or cooperation in organized groups, were viewed by Mead as dependent on this capacity of humans to imaginatively rehearse lines of action toward one another and thereby select those behaviors that facilitate cooperation.

Much of Mead's analysis focused not so much on the mind of mature organisms as on how this capacity first develops in individuals. Unless mind emerges in infants, neither society nor self can exist. In accordance with principles of behaviorism, Darwinism, and pragmatism, Mead stressed that mind arises from a selective process in which an infant's initially wide repertoire of random gestures is narrowed as some gestures bring favorable reactions from those on whom the infant depends for survival. Such selection of gestures facilitating adjustment can occur either through trial and error or through conscious coaching by those with whom the infant must cooperate. Eventually, through either of these processes, gestures come to have common meanings for both the infant and those in its environment. With this development, gestures now denote the same objects and carry similar disposition for all the parties to an interaction. Gestures that have such common meanings are termed *conventional gestures* by Mead. These conventional gestures have increased efficiency for interaction among individuals because they allow more precise communication of desires and wants as well as intended courses of action—thereby increasing the capacity of organisms to adjust to one another.

The ability to use and to interpret conventional gestures with common meanings represents a significant step in the development of mind, self, and society. By perceiving and interpreting gestures, humans can now assume the perspective (dispositions, needs, wants, and propensities to act) of those with whom they must cooperate for survival. By reading and then interpreting covertly conventional gestures, individuals can imaginatively rehearse alternative lines of action that will facilitate adjustment to others. Thus, by being able to put oneself in another's place, or to "take the role of the other," to use Mead's concept, the covert rehearsal of action reaches a new level of efficiency, because actors can better gauge the consequences of their actions for others and thereby increase the probability of cooperative interaction.

Thus, when an organism develops the capacity (1) to understand conventional gestures, (2) to employ these gestures to take the role of others, and (3) to imaginatively rehearse alternative lines of action, Mead believed that such an organism possesses "mind."

Self. Drawing from James and Cooley, Mead stressed that, just as humans can designate symbolically other actors in the environment, so they can symbolically represent themselves as objects. The interpretation of gestures, then, facilitates human cooperation and also serves as the basis for self-assessment and evaluation. This capacity to derive images of oneself as an object of evaluation in interaction depends on the processes of mind. What Mead saw as significant about this process is that, as organisms mature, the transitory "self-images" derived from specific others in each interactive situation eventually become crystallized into a more or less stabilized "self-conception" of oneself as a certain type of object. With these self-conceptions, individuals' actions take on consistency, because they are now mediated through a coherent and stable set of attitudes, dispositions, or meanings about oneself as a certain type of person.

Mead chose to highlight three stages in the development of self, each stage marking not only a change in the kinds of transitory self-images an individual can derive from role taking but also an increasing crystallization of a more stabilized self-conception. The initial stage of role taking in which self-images can be derived is termed *play*. In play, infant organisms are capable of assuming the perspective of only a limited number of others, at first only one or two. Later, by virtue of biological maturation and practice at role taking, the maturing organism becomes capable of taking the role of several others engaged in organized activity. Mead termed this stage the *game* because it designates the capacity to derive multiple self-images from, and to cooperate with, a group of individuals engaged in some coordinated activity. (Mead typically illustrated this stage by giving the example of a baseball game in which all individuals must symbolically assume the role of all others on the team to participate effectively.) The final stage in the development of self occurs when an individual can take the role of the "generalized other" or "community of attitudes" evident in a society. At this stage, individuals are seen as capable of assuming the overall perspective of a community, or general beliefs, values, and norms. This means that humans can both (1) increase the appropriateness of their responses to other with whom they must interact and (2) expand their evaluative self-images from the expectations of specific others to the standards and perspective of the broader community. Thus, it is this ever-increasing capacity to take roles with an ever-expanding body of others that marks the stages in the development of self.

Society. Mead believed society or institutions represent the organized and patterned interactions among diverse individuals.[6] Such organization of interactions depends on mind. Without the capacities of mind to take roles and imaginatively rehearse alternative lines of activity, individuals could not coordinate their activities. Mead emphasized,

> The immediate effect of such role-taking lies in the control which the individual is able to exercise over his own response. The control of the action of the individual in a co-operative process can take place in the conduct

> of the individual himself if he can take the role of the other. It is this control of the response of the individual himself through taking the role of the other that leads to the value of this type of communication from the point of view of the organization of the conduct in the group.[7]

Society also depends on the capacities of self, especially the process of evaluating oneself from the perspective of the generalized other. Without the ability to see and evaluate oneself as an object from this community of attitudes, social control would rest solely on self-evaluations derived from role taking with specific and immediately present others—thus making coordination of diverse activities among larger groups extremely difficult.[8]

Although Mead was vitally concerned with how society and its institutions are maintained and perpetuated by the capacities of mind and self, these concepts also allowed him to view society as constantly in flux and rife with potential change. That role taking and imaginative rehearsal are ongoing processes among the participants in any interaction situation reveals the potential these processes give individuals for adjusting and readjusting their responses. Furthermore, the insertion of self as an object into the interactive process underscores that the outcome of interaction will be affected by the ways in which self-conceptions alter the initial reading of gestures and the subsequent rehearsal of alternative lines of behavior. Such a perspective thus emphasizes that social organization is both perpetuated by and altered through the adjustive capacities of mind and the mediating impact of self:

> Thus the institutions of society are organized forms of group or social activity—forms so organized that the individual members of society can act adequately and socially by taking the attitudes of others toward these activities. . . . [But] there is no necessary or inevitable reason why social institutions should be oppressive or rigidly conservative, or why they should not rather be, as many are, flexible and progressive, fostering individuality rather than discouraging it.[9]

This passage contains a clue to Mead's abiding distaste for rigid and oppressive patterns of social organization. He viewed society as a *constructed* phenomenon that arises from the adjustive interactions among individuals. As such, society can be altered or reconstructed through the processes denoted by the concepts of mind and self. However, Mead went one step further and stressed that change is frequently unpredictable, even by those emitting the change-inducing behavior. To account for this indeterminacy of action, Mead used two concepts first developed by James, the "I" and the "me."[10] For Mead, the "I" points to the impulsive tendencies of individuals, and the "me" represents the self-image of behavior after it has been emitted. With these concepts Mead emphasized that the "I," or impulsive behavior, cannot be predicted because the individual can only "know in experience" (the "me") what has actually transpired and what the consequences of the "I" have been.

In sum, then, Mead believes society represents those constructed patterns of coordinated activity that are maintained by, and changed through, symbolic interaction among and within actors. Both the maintenance and the change of society, therefore, occur through the processes of mind and self. Although many of the interactions causing both stability and change in groups are viewed by Mead as predictable, the possibility for spontaneous and unpredictable actions that alter existing patterns of interaction is also likely.

This conceptual legacy had a profound impact on a generation of American sociologists before the posthumous publication of Mead's lectures in 1934. Yet, despite the suggestiveness of Mead's concepts, they failed to address some important theoretical issues.

The most important of these issues concerns the vagueness of his concepts in denoting the nature of social organization or society and the precise points of articulation between society and the individual. Mead viewed society as organized activity, regulated by the generalized other, in which individuals make adjustments and cooperate with one another. Such adjustments and the cooperation are seen as possible by virtue of the capacities of mind and self. Whereas mind and self emerged from existent patterns of social organization, the maintenance or change of such organization was viewed by Mead as a reflection of the processes of mind and self. Although these and related concepts of the Meadian scheme pointed to the mutual interaction of society and the individual and although the concepts of mind and self denote crucial processes through which this dependency is maintained, they do not allow the analysis of variations in social organization patterns and in the various ways individuals are implicated in these patterns.

Conceptualizing Structure and Role

Though Mead's synthesis provided the initial conceptual breakthrough, it did not satisfactorily resolve the problem of how participation in the *structure* of society shaped individual conduct, and vice versa. In an effort to resolve this vagueness, sociological inquiry began to focus on the concept of role. Individuals were seen as playing roles associated with positions in larger networks of positions. With this vision, efforts to understand more about social structures and how individuals are implicated in them intensified during the 1920s and 1930s. This line of inquiry became known as *role theory.*

Robert Park's Role Theory. Robert Park, who came to the University of Chicago near the end of Mead's career, was one of the first to extend Mead's ideas through an emphasis on roles. As Park observed, "everybody is always and everywhere, more or less consciously, playing a role."[11] But Park stressed that roles are linked to structural positions in society and that self is intimately linked to playing roles within the confines of the positions of social structure:

> The conceptions which men form of themselves seem to depend upon their vocations, and in general upon the role they seek to play in communities and social groups in which they live, as well as upon the recognition and status which society accords them in these roles. It is status, i.e., recognition by the community, that confers upon the individual the character of a person, since a person is an individual who has status, not necessarily legal, but social.[12]

Park's analysis stressed that self emerges from the multiple roles that people play.[13] In turn, roles are connected to positions in social structures. This kind of analysis shifts attention to the nature of society and how its structure influences the processes outlined in Mead's synthesis.

Jacob Moreno's Role Theory. Inspired in part by Mead's concept of role taking and by his own earlier studies in Europe, Jacob Moreno was one of the first to develop the concept of role playing. In *Who Shall Survive* and in many publications in the journals that he founded in America, Moreno began to view social organization as a network of roles that constrain and channel behavior.[14] In his early works, Moreno distinguished different types of roles: (a) "psychosomatic roles," in which behavior is related to basic biological needs, as conditioned by culture, and in which role enactment is typically unconscious; (b) "psychodramatic roles," in which individuals behave in accordance with the specific expectations of a particular social context; and (c) "social roles," in which individuals conform to the more general expectations of various conventional social categories (for example, worker, Christian, mother, and father).

Despite the suggestiveness of these distinctions, their importance comes not so much from their substantive content as from their intent: to conceptualize social structures as organized networks of expectations that require varying types of role enactments by individuals. In this way, analysis can move beyond the vague Meadian conceptualization of society as coordinated activity regulated by the generalized other to a conceptualization of social organization as various *types* of interrelated role enactments regulated by varying *types* of expectations.

Ralph Linton's Role Theory. Shortly after Moreno's publication of *Who Shall Survive,* the anthropologist Ralph Linton further conceptualized the nature of social organization, and the individual's embeddedness in it, by distinguishing among the concepts of role, status, and individuals:

> A status, as distinct from the individual who may occupy it, is simply a collection of rights and duties. . . . A *role* represents the dynamic aspect of status. The individual is socially assigned to a status and occupies it with relation to other statuses. When he puts the rights and duties which constitute the status into effect, he is performing a role.[15]

This passage contains several important conceptual distinctions. Social structure reveals several distinct elements: (a) a network of positions, (b) a corresponding system of expectations, and (c) patterns of behavior that are enacted for the expectations of particular networks of interrelated positions. In retrospect, these distinctions might appear self-evident and trivial, but they made possible the subsequent elaboration of many interactionist concepts:

1. Linton's distinctions allow us to conceptualize society as clear-cut variables: the nature and kinds of interrelations among positions and the types of expectations attending these positions.
2. The variables Mead denoted by the concepts of mind and self can be analytically distinguished from both social structure (positions and expectations) and behavior (role enactment).
3. By conceptually separating the processes of role taking and imaginative rehearsal from both social structure and behavior, the points of articulation between society and the individual can be more clearly marked, because role taking pertains to covert interpretations of the expectations attending networks of statuses and role denotes the enactment of these expectations as mediated by self.

Thus, by offering more conceptual insight into the nature of social organization, Park, Moreno, and Linton provided a needed supplement to Mead's suggestive concepts. Now, it would be possible to understand more precisely the interrelations among mind, self, and society.

EARLY EUROPEAN INSIGHTS

Georg Simmel Analysis of Interaction

Georg Simmel was perhaps the first European sociologist to begin a serious exploration of interaction, or "sociability" as he called it. In so doing, he elevated the study of interaction from the taken-for-granted.[16] For Simmel, as for the first generation of American sociologists in Chicago, the macrostructures and processes studied by functional and some conflict theories—class, the state, family, religion, evolution—are ultimately reflections of the specific interactions among people. These interactions result in emergent social phenomena, but considerable insight into the latter can be attained by understanding the basic interactive processes that first give and then sustain their existence.

In Chapter 11 Simmel's analysis of the forms of conflict, and in Chapter 19, his exchange theory were summarized. But Simmel's study of interaction extends beyond just the analysis of conflict and exchange—he was concerned with understanding the forms and consequences of many diverse types of interactions. Some of his most important insights, which influenced American interactionists, concern the relationship between the individual and society. In his famous essay on "the web of group affiliations," for example, Simmel emphasized that human personality emerges from, and is shaped by, the particular configuration of a person's group affiliations.[17] What people are—that is, how they think of themselves and are prepared to act—is circumscribed by their group memberships. As he emphasized, "the genesis of the personality [is] the point of intersection for innumerable social influences, as the end-product of heritages derived from the most diverse groups and periods of adjustment."[18]

Although Simmel did not analyze the emergence of human personality in great detail, his "formal sociology" did break away from the macro concerns of early German, French, and British sociologists. He began in Europe a mode of analysis that became the prime concern of the first generation of American sociologists. Simmel thus could be considered one of the first European interactionists.

Émile Durkheim's Metamorphosis

In *Division of Labor in Society*, Émile Durkheim portrayed social reality as an emergent phenomenon, *sui generis*, and as not reducible to the psychic states of individuals. Yet, in his later works, such as *The Elementary Forms of the Religious Life*, Durkheim began to ask, How does society rule the individual? How is it that society "gets inside" individuals and guides them from within? Why do people share common orientations and perspectives?[19] Durkheim never answered these questions effectively because his earlier emphasis on social structures prevented him from seeing the micro reality of interactions among individuals implicated in macro social structures. But it is significant that the most forceful advocate of the sociologistic position became intrigued with the relationship between the individual and society. Two critical lines of interactionist thought emerged from *Elementary Forms*: (1) the analysis of ritual and (2) the concern with categories of thought.

1. As Durkheim became interested in the ultimate basis of social solidarity, he turned to the analysis of religion in simple societies. From his reading of secondary accounts on aborigines in Australia, he concluded in *Elementary Forms* that religious worship is, actually, the worship of society, a worship that is sustained by imputing a sacredness to the force of society as it constrained members and that is emotionally charged by the enactment of rituals. Thus, the basic behavioral mechanism by which solidarity is created and sustained, Durkheim argued, is the enactment of rituals that focus people's attention, arouse emotions, and create a common sense of solidarity. Durkheim never pursued the implications of this insight for the study of interaction, but later theorists began to recognize that interpersonal rituals are a key mechanism

for creating and sustaining patterns of interaction. And some even asserted that the ultimate bases of macrostructures are chains of interaction rituals.[20]

2. The concern with thought in *Elementary Forms* also influenced social theory. Durkheim emphasized that "the collective conscience" is not "entirely outside us" and that people's definitions of, and orientations to, situations are related to the organization of subjective consciousness. The categories of this consciousness, however, reflect the structural arrangements of society. Hence, varying macro structures generate different forms of thought and perception of the world. Such forms feed back and reinforce social structures.

The first line of thought exerted considerable influence in interactionist thinking, whereas the second formed the core idea behind much "structuralist" social theory (see Chapter 37 on structuralism). Although Durkheim's ideas on ritual and categories of thought were never as rigorously or systematically developed as his earlier work on macro processes (see Chapter 2 on the emergence of functionalism), these were perhaps his most original ideas.

Max Weber's Analysis of "Social Action"

Max Weber was also becoming increasingly concerned with the micro social world, although his most important insights were in macro and historical sociology. Yet Weber's definition of sociology was highly compatible with the flourishing American school of interactionism. For Weber, sociology is "that science which aims at the interpretative understanding of social behavior in order to gain an explanation of its causes, its course, and its effects."[21] Moreover, the behavior to be studied by sociology is seen by Weber as social action that includes

> All human behavior when and insofar as the acting individual attaches a subjective meaning to it. Action in this sense may be overt, purely inward, or subjective; it may consist of positive intervention in a situation, of deliberately refraining from such intervention, or passively acquiescing in the situation. Action is social insofar as by virtue of the subjective meaning attached to it by the acting individual (or individuals), it takes account of the behavior of others and is thereby oriented in its course.[22]

Thus, Weber recognized that the reality behind the macrostructures of society—classes, the state, institutions, and nations—are the meaningful and symbolic interactions among people. Moreover, Weber's methodology stresses the need for understanding macrostructures and processes "at the level of meaning." In the real world, actors interpret and give meaning to the reality around them and act on the basis of these meanings. Yet, despite this key insight, Weber's actual analysis of social structures—class, status, party, change, religion, bureaucracy, and the like—rarely follows his own methodological prescriptions. As with other European thinkers, he tended to focus on social and cultural *structures* and the impact of these structures on one another. The interacting and interpreting person is often lost amid Weber's elaborate taxonomies of structures and analyses of historical events. This failing attracted the attention of Alfred Schutz, who, more than any other European thinker, translated phenomenology into a perspective that could be incorporated into interactionist theory.

European Phenomenology

Phenomenology began as the project of the German philosopher Edmund Husserl (1859–1938).[23] In his hands, this project showed few signs of being anything more than an orgy of subjectivism.[24] The German social thinker Alfred Schutz, however, took Husserl's concepts and transformed them into an interactionist analysis that has exerted considerable influence on modern-day interactionism. Schutz's migration to the United States in 1939 facilitated this translation, especially as he came into contact with American interactionism, but his most important ideas were formulated before he came. His subsequent work in America involved an elaboration of basic ideas originally developed in Europe.

Edmund Husserl's Project. Husserl's ideas have been selectively borrowed and used in ways that he would not have condoned to develop modern phenomenology and various forms of interactionist thought. In reviewing Husserl's contribution, therefore, it is best to focus more on what was borrowed than on the details of his complete philosophical scheme. With this goal in mind, several features of his work can be highlighted: (1) the basic philosophical dilemma, (2) the properties of consciousness, (3) the critique of naturalistic empiricism, and (4) the philosophical alternative to social science.[25]

1. Basic questions confronting all inquiry are: What is real? What actually exists in the world? How is it possible to know what exists? For the philosopher Husserl, these were central questions that required attention. Husserl reasoned that humans know about the world only through experience. All notions of an external world, "out there," are mediated through the senses and can be known only through mental consciousness. The existence of other people, values, norms, and physical objects is always mediated by experiences as these register on people's conscious awareness. One does not directly have contact with reality; contact is always indirect and mediated through the processes of the human mind.

   Because the process of consciousness is so important and central to knowledge, philosophic inquiry must first attempt to understand how this process operates and how it influences human affairs. This concern with the process of consciousness—of how experience creates a sense of an external reality—became the central concern of phenomenology.

2. Husserl initially made reference to the "world of the natural attitude." Later he used the phrase *lifeworld*. In either case, with these concepts he emphasized that humans operate in a taken-for-granted world that permeates their mental life. It is the world that humans sense to exist. It is composed of the objects, people, places, ideas, and other things that people see and perceive as setting the parameters for their existence, for their activities, and for their pursuits.

   This lifeworld or world of the natural attitude *is* reality for humans. Two features of Husserl's conception of natural attitude influenced modern interactionist thought: (a) The lifeworld is taken for granted. It is rarely the topic of reflective thought, and yet it structures and shapes the way people act and think. (b) Humans operate on the presumption that they experience the same world. Because people experience only their own consciousness, they have little capacity to directly determine if this presumption is correct. Yet people act *as if* they experienced a common world.

   Human activity, then, is conducted in a lifeworld that is taken for granted and that is presumed to be experienced collectively. This brought Husserl back to his original problem: How do humans break out of their lifeworld and ascertain what is real? If people's lifeworld structures their consciousness and their actions, how is an objective science of human behavior and organization possible? These questions led Husserl to criticize what he termed *naturalistic science.*

3. Science assumes that a factual world exists, independent of, and external to, human senses and consciousness. Through the scientific method, this factual world can be directly known. With successive efforts at its measurement, increasing understanding of its properties can be ascertained. But Husserl challenged this vision of science: if one can know only through consciousness and if consciousness is structured by an implicit lifeworld, then how can objective measurement of some external and real world be possible? How can science measure objectively an external world when the only world that individuals experience is the lifeworld of their consciousness?

4. Husserl's solution to this problem is a philosophical one. He advocated what he termed the search for the *essence of consciousness.* To understand social events, the basic process through which these events are mediated—that is, consciousness—must be comprehended. The substantive *content* of consciousness, or the lifeworld, is not what is important; the abstract processes of consciousness, per se, are to be the topic of philosophic inquiry.

   Husserl advocated what he termed the *radical abstraction of the individual* from interpersonal experience. Investigators must suspend their natural attitude and seek to understand the fundamental processes of consciousness, per se. One must discover, in Husserl's words, "Pure Mind." To do this, it is necessary to perform "epoch"—that is, to see if the substance of one's lifeworld can be suspended. Only when divorced from the substance of the lifeworld can the fundamental and abstract properties of consciousness be exposed and understood. With understanding of these properties, real insight into the nature of reality would be possible. If all that humans know is presented through consciousness, it is necessary to understand the nature of consciousness in abstraction from the specific substance or content of the lifeworld.

   Husserl was not advocating Weber's method of *verstehen*, or sympathetic introspection into an investigator's own mind. Nor was Husserl suggesting the unstructured and intuitive search for people's definitions of situations. These methods would, he argued, only produce data on the substance of the lifeworld and would be no different than the structured measuring instruments of positivism. Rather, Husserl's goal was to create an abstract theory of consciousness that bracketed out, or suspended, any presumption of "an external social world out there."

   Not surprisingly, Husserl's philosophical doctrine failed. He never succeeded in developing an abstract theory of consciousness, radically abstracted from the lifeworld. But his ideas set into motion a new line of thought that became the basis for modern phenomenology and for its elaboration into ethnomethodology and other forms of theory.

Alfred Schutz's Phenomenological Interactionism. Schutz migrated to the United States in 1939 from Austria, after spending a year in Paris. With his interaction in American intellectual circles and the translation of his early works into English, Schutz's contribution to sociological theorizing has become increasingly recognized.[26]

This contribution resided in his ability to blend Husserl's radical phenomenology with Weber's action theory and American interactionism. This blend, in turn, stimulated the further development of phenomenology, the emergence of ethnomethodology, and the refinement of other interactionist theoretical perspectives.

Schutz's work began with a critique of his compatriot Weber, who employed the concept of social action in his many and varied inquiries.[27] Social action occurs when actors are consciously aware of each other and attribute meanings to their common situation. For Weber, then, a science of society must seek to understand social reality "at the level of meaning." Sociological inquiry must penetrate people's consciousness and discover how they view, define, and see the world. Weber advocated the method of *verstehen*, or sympathetic introspection. Investigators must become sufficiently involved in situations to be able to get inside the subjective world of actors. Causal and statistical analysis of complex social structures would be incomplete and inaccurate without such *verstehen* analysis.

Schutz's first major work addressed Weber's conception of action. Schutz's analysis is critical and detailed, and need not be summarized here, except to note that the basic critique turns on Weber's failure to use his *verstehen* method and to explore *why*, and through what processes, actors come to share common meanings. In Schutz's eye, Weber simply assumed that actors share subjective meanings, leading Schutz to ask, Why and how do actors come to acquire common subjective states in a situation? How do they create a common view of the world? This is the problem of "intersubjectivity," and it is central to Schutz's intellectual scheme.

Schutz departed immediately from Husserl's strategy of holding the individual in radical abstraction and of searching for Pure Mind or the abstract laws of consciousness. He accepted Husserl's notion that humans hold a natural attitude and lifeworld that is taken for granted and that shapes who they are and what they will do. He also accepted Husserl's notion that people perceive that they share the same lifeworld and act *as if* they lived in a common world of experiences and sensations. Moreover, Schutz acknowledged the power of Husserl's argument that social scientists cannot know about an external social world out there independently of their own lifeworld.[28]

Having accepted these lines of thought from Husserl, however, Schutz advocated Weber's strategy of sympathetic introspection into people's consciousness. Only by observing people in interaction, rather than in radical abstraction, can the processes whereby actors come to share the same world be discovered. Social science cannot understand how and why actors create a common subjective world independently of watching them do so. This abandonment of Husserl's phenomenological project liberated phenomenology from philosophy and allowed sociologists to study empirically what Schutz considered the most important social reality: the creation and maintenance of intersubjectivity—that is, a common subjective world among pluralities of interacting individuals.[29]

Unfortunately, Schutz died just as he was beginning a systematic synthesis of his ideas; as a result, only a somewhat fragmented but suggestive framework is evident in his collective work. But his early analysis of Weber, Husserl, and interactionism led to a concern with some key issues: (1) How do actors create a common subjective world? (2) What implications does this creation have for how social order is maintained?

All humans, Schutz asserted, carry in their minds rules, social recipes, conceptions of appropriate conduct, and other information that allows them to act in their social world. Extending Husserl's concept of lifeworld, Schutz views the sum of these rules, recipes, conceptions, and information as the individual's "stock knowledge at hand." Such stock knowledge gives people a frame of reference or orientation with which they can interpret events as they pragmatically act on the world around them.

Several features of this stock knowledge at hand are given particular emphasis by Schutz:

1. People's reality *is* their stock knowledge. For the members of a society, stock knowledge constitutes a "paramount reality"—a sense of an absolute reality that shapes and guides all social events. Actors use this stock knowledge and sense of reality as they pragmatically seek to deal with others in their environment.
2. The existence of stock knowledge that bestows a sense of reality on events gives the social world, as Schutz agreed with Husserl, a taken-for-granted character. The stock knowledge is rarely the object of conscious reflection but, rather, an implicit set of assumptions and procedures that are silently used by individuals as they interact.
3. Stock knowledge is learned. It is acquired through socialization within a common social and cultural world, but it becomes *the* reality for actors in this world.
4. People operate under a number of assumptions that allow them to create a sense of a "reciprocity of perspectives." That is, others with whom an actor must deal are considered to share an actor's stock knowledge at hand. And, although these others might have unique components in their stock knowledge because of their particular biographies, these can be ignored by actors.
5. The existence of stock knowledge, its acquisition through socialization, and its capacity to promote reciprocity of perspectives all give actors in a situation a *sense* or *presumption* that the world is the same for all and that it reveals identical properties for all. What often holds society together is this presumption of a common world.
6. The presumption of a common world allows actors to engage in the process of typification. Action in most situations, except the most personal and intimate, can proceed through mutual typification as actors use their stock knowledge to categorize one another and to adjust their responses to these typifications.[30] With typification, actors can effectively deal with their world; every nuance and characteristic of their situations do not have to be examined. Moreover, typification facilitates entrance into the social world; it simplifies adjustment because humans can treat each other as categories, or as "typical" objects of a particular kind.

These points of emphasis in Schutz's thought represented a blending of ideas from European phenomenology and American interactionism. The emphasis on stock knowledge is clearly borrowed from Husserl, but it is highly compatible with Mead's notion of the generalized other. The concern with the taken-for-granted character of the world as it is shaped by stock knowledge is also borrowed from Husserl but is similar to early interactionists' discussions of habit and routine behaviors. The emphasis on the acquired nature of stock knowledge coincides with early interactionists' discussions of the socialization process. The concern with the reciprocity of perspectives and with the process of typification owes much to Husserl and Weber but is compatible with Mead's notion of role taking, by which actors read one another's role and perspective.

But the major departure from much interactionist theory should also be emphasized: Actors operate on an unverified *presumption* that they share a common world, and this *sense* of a common world and the practices that produce this sense are crucial in maintaining social order. In other words, social organization might be possible not so much by the substance and content of stock knowledge, by the reciprocity of perspectives, or by successful typification as by the presumption actors share intersubjective states. Schutz did not carry this line of inquiry far, but he inspired new avenues of phenomenological inquiry.

BUILDING ON EARLY INTERACTIONIST INSIGHTS

Given its diverse sources, interactionist theorizing, by the twentieth century's end, diverged in many directions. At first, however, those who extended Mead's synthesis and those who expanded on the idea of role stayed within the broad parameters of Mead's views, but by the 1960s, new lines of interactionist thinking began to emerge, inspired more by the European masters than by Mead and those who had followed Mead in America. Thus, as interactionism matured at the midcentury, it might have exhausted the Meadian legacy, and, to a degree, purely Meadian micro sociology began to stagnate. Indeed, actual theorizing began to take a back seat to debates on the methodology of inquiry, particularly about whether or not a science of interaction was appropriate. At the same time, however, interesting empirical work was conducted by those following the Meadian tradition. Nonetheless, new theory within the strictly Meadian framework became increasingly hard to find in the literature.

As Mead-inspired interactionism stagnated, however, new alternatives emerged and reinvigorated the study of interaction. Today, it would be hard to visualize a coherent interactionist orientation. New micro-level ideas and interesting extensions and elaborations of old ones are still emerging, as we will see, but the midcentury coherence in the study of interaction has been lost. Yet, by the century's end, it would be reasonable to conclude that the process of interaction is probably the best understood dimension of the social universe. So what might seem like stagnation could, actually, be the natural slowdown that occurs whenever more complete knowledge of a phenomenon is achieved.

NOTES

1. William James, *The Principles of Psychology* (New York: Henry Holt, 1890), vol. 1, pp. 292–299.

2. Ibid., p. 294.

3. Charles Horton Cooley, *Human Nature and the Social Order* (New York: Scribner's, 1902) and *Social Organization: A Study of the Larger Mind* (New York: Scribner's, 1916).

4. John Dewey, *Human Nature and Human Conduct* (New York: Henry Holt, 1922), p. 190. For an earlier statement of these ideas, see John Dewey, *Psychology* (New York: Harper & Row, 1886).

5. Mead's most important sociological ideas can be found in the published lecture notes of his students. His most important exposition of interactionism is found in his *Mind, Self, and Society*, ed. C.W. Morris (Chicago: University of Chicago Press, 1934). Other useful sources include George Herbert Mead, *Selected Writings* (Indianapolis: Bobbs-Merrill, 1964); and Anselm Strauss, ed., *George Herbert Mead on Social Psychology* (Chicago: University of Chicago Press, 1964). For excellent secondary sources on the thought of Mead, see Tamotsu Shibutani, *Society and Personality: An Interactionist Approach* (Englewood Cliffs, NJ: Prentice-Hall, 1962); Anselm Strauss, *Mirrors and Masks: The Search for Identity* (Glencoe, IL: Free Press, 1959); Bernard N. Meltzer, "Mead's Social Psychology," in *The Social Psychology of George Herbert Mead* (Ann Arbor, MI: Center for Sociological Research, 1964), pp. 10–31; Jonathan H. Turner, "Returning to Social Physics: Illustrations from George Herbert Mead," *Perspectives in Social Theory* 2 (1981), and *A Theory of Social Interaction* (Stanford, CA: Stanford University Press, 1988); also see Jonathan H. Turner, Leonard Beeghley, and Charles Powers, *The Emergence of Sociological Theory* (Belmont, CA: Wadsworth, 1998). For a more global overview of Mead's ideas, see John D. Baldwin, *George Herbert Mead: A Unifying Theory for Sociology* (Beverly Hills, CA: Sage, 1986).

6. For a more detailed analysis, see Jonathan H. Turner, "A Note on G.H. Mead's Behavioristic Theory of Social Structure," *Journal for the Theory of Social Behavior* 12 (July 1982), pp. 213–222.

7. Mead, *Mind, Self, and Society* (cited in note 5), p. 254.

8. Ibid., pp. 256–257.

9. Ibid., pp. 261–262.

10. See James, *Principles of Psychology* (cited in note 1), pp. 135–176.

11. Robert E. Park, "Behind Our Masks," *Survey Graphic* 56 (May 1926), p. 135. For a convenient summary of the thrust of early research efforts in role theory, see Ralph H. Turner, "Social Roles: Sociological Aspects," *International Encyclopedia of the Social Sciences* (New York: Macmillan, 1968).

12. Robert E. Park, *Society* (New York: Free Press, 1955), pp. 285–286.

13. Indeed, Park studied briefly with Simmel in Berlin and apparently acquired insight into Simmel's study of the individual and the web of group affiliations (see later discussion). Coupled with his exposure to William James at Harvard, who also stressed the multiple sources of self, it is clear that Mead's legacy was supplemented by Simmel and James through the work of Robert Park.

14. Jacob Moreno, *Who Shall Survive* (Washington, DC, 1934); rev. ed. (New York: Beacon House, 1953).

15. Ralph Linton, *The Study of Man* (New York: Appleton-Century-Crofts, 1936), p. 28.

16. Georg Simmel, "Sociability," in *The Sociology of Georg Simmel*, ed. K.H. Wolff (New York: Free Press, 1950), pp. 40–57. For an excellent secondary account of Simmel's significance for interactionism, see Randall Collins and Michael Makowsky, *The Discovery of Society* (New York: Random House, 1972), pp. 138–142.

17. Georg Simmel, *Conflict and the Web of Group Affiliations*, trans. R. Bendix (Glencoe, IL: Free Press, 1955; originally published in 1922).

18. Ibid., p. 141.

19. Émile Durkheim, *The Elementary Forms of the Religious Life* (New York: Free Press, 1954; originally published in 1912).

20. Randall Collins develops this idea in *Conflict Sociology* (New York: Academic, 1975) and other works. See Chapter 32. But perhaps the most significant adoption of Durkheim's emphasis on ritual was by the late Erving Goffman, the subject of Chapter 30.

21. Max Weber, *Basic Concepts in Sociology* (New York: Citadel, 1964), p. 29.

22. Max Weber, *The Theory of Social and Economic Organization* (New York: Free Press, 1947; originally published after Weber's death), p. 88.

23. For some readable, general references on phenomenology, see George Psathas, ed., *Phenomenological Sociology* (New York: Wiley, 1973); Richard M. Zaner, *The Way of Phenomenology: Criticism as a Philosophical Discipline* (New York: Pegasus, 1970); Peter L. Berger and Thomas Luckman, *The Social Construction of Reality* (Garden City, NY: Doubleday, 1966); Herbert Spiegelberg, *The Phenomenological Movement*, vols. 1 and 2, 2nd ed. (The Hague: Martinus Nijhoff, 1969); Hans P. Neisser, "The Phenomenological Approach in Social Science," *Philosophy and Phenomenological Research* 20 (1959), pp. 198–212; Stephen Strasser, *Phenomenology and the Human Sciences* (Pittsburgh: Duquesne University Press, 1963); Maurice Natanson, ed., *Phenomenology and the Social Sciences* (Evanston, IL: Northwestern University Press, 1973); and Quentin Lauer, *Phenomenology: Its Genesis and Prospect* (New York: Harper Torchbooks, 1965).

24. Zygmunt Bauman, "On the Philosophical Status of Ethnomethodology," *Sociological Review* 21 (February 1973), p. 6.

25. Husserl's basic ideas are contained in the following: *Phenomenology and the Crisis of Western Philosophy* (New York: Harper & Row, 1965; originally published in 1936); *Ideas: General Introduction to Pure Phenomenology* (London: Collier-Macmillan, 1969; originally published in 1913); and "Phenomenology," in *The Encyclopedia Britannica*, 14th ed., vol. 17, col. 699–702, 1929. For excellent secondary analyses, see Helmut R. Wagner, "The Scope of Phenomenological Sociology," in *Phenomenological Sociology*, ed. G. Psathas, pp. 61–86, and "Husserl and Historicism," *Social Research* 39 (Winter 1972), pp. 696–719; Aron Gurwitsch, "The Common-Sense World as Social Reality," *Social Research* 29 (Spring 1962), pp. 50–72; Robert J. Antonio, "Phenomenological Sociology," in *Sociology: A Multiple Paradigm Science*, ed. G. Ritzer (Boston: Allyn & Bacon, 1975), pp. 109–112; Robert Welsh Jordan, "Husserl's Phenomenology as an 'Historical Science,'" *Social Research* 35 (Summer 1968), pp. 245–259.

26. For the basic ideas of Alfred Schutz, see his *The Phenomenology of the Social World* (Evanston, IL: Northwestern University Press, 1967; originally published in 1932); *Collected Papers*, vols. 1, 2, 3 (The Hague: Martinus Nijhoff, 1964, 1970, and 1971, respectively). For excellent secondary analyses, see Maurice Natanson, "Alfred Schutz on Social Reality and Social Science," *Social Research* 35 (Summer 1968), pp. 217–244.

27. Schutz, The Phenomenology of the Social World (cited in note 26).

28. Richard M. Zaner, "Theory of Intersubjectivity: Alfred Schutz," *Social Research* 28 (Spring 1961), p. 76.

29. For references to interactionists, see Schutz, *Collected Papers* (cited in note 26).

30. Ralph H. Turner's emphasis on role differentiation and accretion is an example of how these ideas have been extended by role theorists. See Chapter 29.

27

The Maturing Tradition I:

The Codification of Symbolic Interactionism

The first well-developed theoretical perspective to emerge from George Herbert Mead's synthesis was "symbolic interactionism," a term coined by Herbert Blumer who, on Mead's death, took over his social psychology course and who for some fifty years championed a particular interpretation of Mead's ideas. Blumer's advocacy did not go unchallenged, however, and an alternative school of thought emerged at the State University of Iowa, a challenge led by several scholars but principally by Manford Kuhn. The poles around which these two and their allies debated have been labeled the Iowa and Chicago Schools of symbolic interactionism.[1] These labels are misleading because Blumer had left Chicago by midcentury, and Kuhn died shortly thereafter. Moreover, much of the Iowa School tradition had shifted to symbolic interactionists at Indiana University by the 1960s, and so the Iowa-Chicago School dichotomy is more a label of convenience than a real indication of *where* the debate occurred. Still, in the 1950s and 1960s, the debate about how to develop theoretical explanations among those who identified with Mead's legacy was intense, if somewhat dispersed among universities. Indeed, today this very same debate has been replicated by new protagonists.[2]

POINTS OF CONVERGENCE IN SYMBOLIC INTERACTIONISM

Before turning to the points of divergence between Iowa and Chicago Schools, we should review the common legacy of Mead's assumptions that all symbolic interactionists employ. These points of convergence are what make symbolic interactionism a distinctive theoretical perspective.[3]

Humans as Symbol Users

Symbolic interactionists, as their name implies, place enormous emphasis on the capacity of humans to create and use symbols. In contrast with other animals, whose symbolic capacities are limited or nonexistent, the very essence of humans and the world that they create flows from their ability to symbolically represent one another, objects, ideas, and virtually any phase of their experience. Without the capacity to create symbols and to use them in human affairs, patterns of social organization among humans could not be created, maintained, or changed. Humans have become, to a very great degree, liberated from instinctual and biological programming and thus must rely on their symbol-using powers to adapt and survive in the world.

Symbolic Communication

Humans use symbols to communicate with one another. By virtue of their capacity to agree upon the meaning of vocal and bodily gestures, humans can effectively communicate. Symbolic communication is, of course, extremely complex, because people use more than word or language symbols in communication. They also use facial gestures, voice tones, body countenance, and other symbolic gestures that have common meaning and understanding.

Interaction and Role Taking

By reading and interpreting the gestures of others, humans communicate and interact. They become able to mutually read each other, to anticipate each other's responses, and to adjust to each other. Mead termed this basic capacity "taking the role of the other," or role taking—the ability to see the other's attitudes and dispositions to act. Interactionists still emphasize the process of role taking as the basic mechanism by which interaction occurs. For example, the late Arnold Rose, who was one of the leaders of contemporary interactionism, indicated that role taking "means that the individual communicator imagines—evokes within himself—how the recipient understands that communication."[4] Or, as another modern interactionist, Sheldon Stryker, has emphasized, role taking is "anticipating the responses of others with one in some social act."[5] And, as Alfred Lindesmith and Anselm Strauss stressed, role taking is "imaginatively assuming the position or point of view of another person."[6]

Without the ability to read gestures and to use these gestures as a basis for putting oneself in the position of others, interaction could not occur. And, without interaction, social organization could not exist.

Interaction, Humans, and Society

Just as Mead emphasized that mind, self, and society are intimately connected, so symbolic interactionists analyzed the relation between the genesis of "humanness" and patterns of interaction. What makes humans unique as a species and enables each individual to possess distinctive characteristics is the result of interaction in society. Conversely, what makes society possible are the capacities that humans acquire as they grow and mature in society.

Symbolic interactionists tended to emphasize the same human capacities as Mead: the genesis of mind and self. Mind is the capacity to think—to symbolically denote, weigh, assess, anticipate, map, and construct courses of action. Although Mead's term, *mind,* is rarely used today, the processes that this term denotes are given great emphasis. As Rose indicated: "Thinking is the process by which possible symbolic solutions and other future courses of action are examined, assessed for their relative advantages and disadvantages in terms of the values of the individual, and one of them chosen for action."[7]

Moreover, the concept of mind has been reformulated to embrace what W. I. Thomas termed the *definition of the situation.*[8] With the capacities of mind, actors can name, categorize, and orient themselves to constellations of objects—including themselves as objects—in all situations. In this way they can assess, weigh, and sort out appropriate lines of conduct.[9]

As the concept of the definition of the situation underscores, self remains the key concept in the interactionist literature. Emphasis in the interactionist orientation is on (a) the emergence of self-conceptions—relatively stable and enduring conceptions that people have about themselves—and (b) the ability to derive self-images—pictures of oneself as an object in social situations. Self is thus a major object that people inject into their definitions of situations. It shapes much of what people see, feel, and do in the world around them.

Society, or relatively stable patterns of interaction, is seen by interactionists as possible only by virtue of people's capacities to define situations and, most particularly, to view themselves as objects in situations. Society can exist by virtue of human capacities for thinking and defining as well as for self-reflection and evaluation.

In sum, these points of emphasis constitute the core of the interactionist approach. Humans create and use symbols. They communicate with symbols. They interact through role taking, which involves the reading of symbols emitted by others. What makes them unique as a species—the existence of mind and self—arises from interaction. Conversely, the emergence of these capacities allows for the interactions that form the basis of society.

AREAS OF DISAGREEMENT AND CONTROVERSY IN SYMBOLIC INTERACTION

From this initial starting point, Blumer and Kuhn often diverge, as have advocates of their respective positions during the last four decades.[10] The major areas of disagreement have revolved around the following issues: (1) What is the nature of the individual? (2) What is the nature of interaction? (3) What is the nature of social organization? (4) What is the most appropriate method for studying humans and society? And (5) what is the best form of sociological theorizing? Each of these five controversial questions will be examined in more detail.

The Nature of the Individual

Both Blumer and Kuhn emphasized the ability of humans to use symbols and to develop capacities for thinking, defining, and self-reflecting. However, they disagreed about the degree of structure and stability in human personality. Blumer emphasized that humans have the capacity to view themselves as objects and to insert any object into an interaction situation. Therefore, human actors are not pushed and pulled around by social and psychological forces, but rather, they are *active creators* of the world to which they respond. Interaction and emergent patterns of social organization can be understood only by focusing on these capacities of individuals to create symbolically the world of objects to which they respond. There is always the potential for spontaneity and indeterminacy in human behavior. If humans can invoke any object into a situation, they can radically alter their definitions of that situation and, hence, their behaviors. Self is but one of many objects to be seen in a situation; other objects from the past, present, or anticipated future can also be evoked and can provide a basis for action.

In contrast, Kuhn emphasized the importance of people's "core self" as an object. Through socialization, humans acquire a relatively stable set of meanings and attitudes toward themselves. The core self will shape and constrain the way people will define situations by circumscribing the cues that will be seen and the objects that will be injected into social situations. Human personality is thus structured and comparatively stable, giving people's actions a continuity and predictability. And, if it is possible to know the expectations of those groups that have shaped a person's core self and that provide a basis for its validation, then human behavior could, in principle, be highly predictable.[11]

The Nature of Interaction

Both Blumer and Kuhn stressed the process of role taking, in which humans mutually emit and interpret each other's gestures. From the information gained through this interpretation of gestures, actors can rehearse covertly various lines of activity and then emit those behaviors that can allow cooperative and organized activity. As might be expected, however, Blumer and Kuhn disagreed about the degree to which the interactions are actively constructed. Blumer's scheme, and that of most of the Chicago School, emphasizes the following points:

1. In addition to viewing each other as objects in an interaction situation, actors select and designate symbolically additional objects in any interaction. (a) One of the most important of these objects is the self. On the one hand, self can represent the transitory images that an actor derives from interpreting the gestures of others; on the other, self can denote the more enduring conceptions of one as an object that an actor brings to and interjects into the interaction. (b) Another important class of objects are the varying types of expectation structures—for example, norms and values—that can guide interaction. (c) Finally, because of the human organism's capacity to manipulate symbols, almost any other object—whether another person, a set of standards, or a dimension of self—can be inserted into the interaction.
2. Actors have various dispositions to act because of the objects in interaction situations. Thus, to understand the potentials for action among groups of individuals, it is necessary to understand the world of objects that they have symbolically designated.
3. For the particular cluster of objects and of the dispositions to act that they imply, each actor arrives at a definition of the situation. Such a definition serves as a general frame of reference within which the consequences of specific lines of conduct are assessed.
4. The selection of a particular line of behavior involves complex symbolic processes. At a minimum, actors typically evaluate: (a) the demands of others immediately present; (b) the self-images they derive from role taking, not only with others in the situation but also with those not actually present; (c) the normative expectations they perceive to exist in the situation; and (d) the dispositions to act toward any additional objects they might inject symbolically into the interaction.
5. Once behavior is emitted, redefinition of the situation and perhaps remapping of action can occur as the reactions of others are interpreted and as new objects are injected into, and old ones discarded from, the interaction.

Thus, by emphasizing these points, Blumer stressed the creative, constructed, and changeable nature of interaction. Rather than constituting the mere vehicle through which preexisting psychological, social, and cultural structures inexorably shape behavior, the symbolic nature of interaction assures that social, cultural, and psychological structures will be altered and changed through shifts in the definitions and behaviors of humans.

In contrast with Blumer's scheme, Kuhn stressed the power of the core self and the group context to constrain interaction.[12] Much interaction is released rather than constructed, as interacting individuals follow the dictates of the self-attitudes and the expectations of their respective roles. Although Kuhn would certainly not have denied the potential for constructing and reconstructing interactions, he tended to view individuals as highly constrained in their behaviors by virtue of their core self and the requirements of their mutual situation.

The Nature of Social Organization

Symbolic interactionism concentrates on the interactive *processes* by which humans form social relationships rather than on the *end products* of interactions. Moreover, both Blumer and Kuhn, as well as other interactionists, have tended to emphasize the micro processes among individuals within small-group contexts. Blumer consistently advocated a view of social organization as temporary and constantly changing, whereas Kuhn typically focused on the more structured aspects of social situations. In addition, Blumer argued for a view of social structure as merely one of many objects that actors employ in their definition of a situation. As Blumer emphasized

1. Behavior is a reflection of the interpretive, evaluational, definitional, and mapping processes of individuals in various interaction contexts, so social organization represents an active fitting together of action by those in interaction. Social organization must therefore be viewed more as a process than as a structure.
2. Social structure is an emergent phenomenon that is not reducible to the constituent actions of individuals, but it is difficult to understand patterns of social organization without recognizing that they represent an interlacing of the separate behaviors among individuals.
3. Although much interaction is repetitive and structured by clear-cut expectations and common definitions of the situation, its symbolic nature reveals the potential for new objects to be inserted or old ones altered and abandoned in a situation. The result is that reinteraction, reevaluation, redefinition, and remapping of behaviors can always occur. Social structure must therefore be viewed as rife with potential for alteration and change.
4. Thus, patterns of social organization represent emergent phenomena that can serve as objects that define situations for actors. However, the very symbolic processes that give rise to and sustain these patterns can also change and alter them.

In contrast with this emphasis, Kuhn usually sought to isolate the more structured features of situations. Kuhn saw social situations as constituting relatively stable networks of positions with attendant expectations or norms. Interactions often create such networks, but, once created, people conform to the expectations of those positions in which they have anchored their self-attitudes.

From this review of different assumptions of Blumer and Kuhn about the nature of individuals, interaction, and social organization, it is clear that Chicago School interactionists viewed individuals as potentially spontaneous, interaction as constantly in the process of change, and social organization as fluid and tenuous.[13] Iowa School interactionists were more prone to see individual personality and social organization as structured, with interactions being constrained by these structures.[14] These differences in assumptions resulted in, or perhaps have been a reflection of, varying conceptions about how to investigate the social world and how to build theory.

The Nature of Methods

E.L. Quarantelli and Joseph Cooper have observed that Mead's ideas provide for contemporary interactionists with a "frame of reference within which an observer can look at behavior rather than a specific set of hypotheses to be tested."[15] There can be little doubt that this statement is true. Yet much literature attempts to test some implications of Mead's ideas, especially those about self, with standard research protocols. This diversity in methodological approaches is underscored by the contrasting methodologies of Blumer and Kuhn. Indeed, the many students and students-of-students of these two figures tended to use Mead's ideas either as a sensitizing framework or as an inspiration for narrow research hypotheses.

Methodological approaches to studying the social world follow from thinkers' assumptions about what they can, or will, discover. The divergence of Blumer's and Kuhn's methodologies thus reflects their varying assumptions about the operation of symbolic processes. Ultimately, their differences boiled down to the question of causality. Whether or not events are viewed as the result of deterministic causes will influence the methodologies that are employed.

Diverging Assumptions About Causality. For Blumer, the Meadian legacy challenged the theoretical perspectives that underemphasize the internal symbolic processes of actors attempting to fit together their respective behaviors into an organized pattern.[16] Rather than being the result of system forces, societal needs, and structural mechanisms, social organization is the result of the mutual interpretations, evaluations, definitions, and mappings of individual actors. Thus, the symbolic processes of individuals cannot be viewed as a neutral medium through which social forces operate, but instead these processes must be viewed as shaping the ways social patterns are formed, sustained, and changed.

Similarly, Blumer's approach challenged theoretical perspectives that view behavior as the mere releasing of propensities built into a structured personality. Just as patterns of social organization must be conceptualized as in a continual state of potential flux through the processes of interpretation, evaluation, definition, and mapping, so the human personality must also be viewed as a constantly unfolding process rather than as a rigid structure from which behavior is mechanically released. Because humans can make varying and changing symbolic indications to themselves, they are capable of altering and shifting behavior. Behavior is not so much released as it is constructed by actors making successive indications to themselves.

For Blumer, social structures and normative expectations are objects that must be interpreted and then used to define a situation and to map out the prospective behaviors that ultimately create and sustain social structures. From this perspective, overt behavior at one point generates self-images that serve as objects for individuals to symbolically map subsequent actions at another point in time. At the same time, existing personality traits, such as self-conceptions, self-esteem, and internalized needs, mediate each successive phase of interpretation of gestures, evaluation of self-images, definition of the situation, and mapping of diverse behaviors.

In such a scheme, causality is difficult to discern. Social structures do not cause behavior, because they are only one class of objects inserted into an actor's seemingly unpredictable symbolic thinking. Similarly, self is only another object inserted into the definitional process. Action is thus created from the potentially large number of objects that actors can insert into situations. In this vision of the social world, then, behavior does not reveal clear causes. Indeed, the variables influencing an individual's definition of the situation and action are of the actor's own choosing and apparently not subject to clear causal analysis.

In contrast with this seemingly *in*deterministic view of causality, Kuhn argued that the social world is deterministic. The apparent spontaneity and indeterminacy of human behavior are simply the result of insufficient knowledge about the variables influencing people's definitions and actions. If the social experiences of individuals can be discerned, then it is possible to know what caused the emergence of their core self. With knowledge of the core self, of the expectations that have become internalized as a result of people's experiences, and of the particular expectations of a given situation, it is possible to understand and predict people's definitions of situations and their conduct. Naturally, this level of knowledge is impossible with current methodological techniques, but insight into deterministic causes of behavior and emergent patterns of social organization is possible *in principle*. For Kuhn, then, methodological strategies should therefore be directed at seeking the causes of behaviors.

Diverging Methodological Protocols. These differing assumptions about causality have shaped divergent methodological approaches within symbolic interactionism. Blumer mounted a consistent and persistent line of attack on sociological theory and research.[17] Rather than letting the nature of the empirical world dictate the kinds of research strategies used in its study, Blumer and others contended that research procedures too often determine what is to be studied.[18] Blumer believed that the fads of research protocol can blind investigators and theorists to the real character of the social world. Such research and theoretical protocols force analysis away from the direct examination of the empirical world in favor of preconceived notions of what is true and how these truths should be studied. In contrast, the processes of symbolic interaction dictate that research methodologies should respect the character of empirical reality and adopt methodological procedures that encourage its direct and unbiased examination.[19]

To achieve this end, the research act itself must be viewed as a process of symbolic interaction in which researchers take the role of those individuals whom they are studying. To do such role taking effectively, researchers must study interaction with a set of concepts that, rather than prematurely structuring the social world for investigators, sensitize them to interactive processes. This approach would enable investigators to maintain the distinction between the concepts of science and those of the interacting individuals under study. In this way the interpretive and definitional processes of actors adjusting to one another in concrete situations could guide the refinement and eventual incorporation of scientific concepts into theoretical statements on the interactive processes that make up society.[20]

Blumer[21] reasoned that the concepts and propositions of symbolic interactionism should allow for the direct examination of the empirical world; therefore, "their value and their validity are to be determined in that examination and not in seeing how they fare when subjected to the alien criteria of an irrelevant methodology." According to Blumer, these "alien criteria" embrace a false set of assumptions about just how concepts should be attached to events in the empirical world. In general, these "false" assumptions posit that, for each abstract concept, a set of operational definitions should guide researchers, who then examine the empirical cases denoted by the operational definition.

Blumer consistently emphasized current deficiencies in the attachment of sociological concepts to actual events in the empirical world:

> This ambiguous nature of concepts is the basic deficiency in social theory. It hinders us in coming to close grips with our empirical world, for we are not sure of what to grip. Our uncertainty as to what we are referring obstructs us from

asking pertinent questions and setting relevant problems for research.[22]

Blumer argued that only through the methodological processes of exploration and inspection can concepts be attached to the empirical. Rather than seeking a false sense of scientific security through rigid operational definitions, sociological theory must accept that the attachment of abstract concepts to the empirical world must be an *ongoing process* of investigators exploring and inspecting events in the empirical world.

In sum, then, Blumer's presentation of the methodological position of symbolic interactionism questions the current research protocols. As an alternative, he advocated (a) the more frequent use of the exploration and inspection process, whereby researchers seek to understand the symbolic processes that shape interaction, and (b) the recognition that only through ongoing research activities can concepts remain attached to the fluid interaction processes of the empirical world. In turn, this methodological position has profound implications for the construction of theory in sociology, as we will see shortly.

In contrast with Blumer's position, Kuhn's vision of a deterministic world led him to emphasize the commonality of methods in all the sciences. The key task of methodology is to provide operational definitions of concepts so that their implications can be tested against the facts of social life. Most of Kuhn's career was thus devoted to taking the suggestive but vague concepts of Mead's framework and developing measures of them. He sought to find replicable measures of such concepts as self, social act, social object, and reference group.[23] His most famous measuring instrument—the Twenty Statements Test (TST)—can illustrate his strategy. With the TST he sought to measure core self—the more enduring and basic attitudes that people have about themselves—by assessing people's answers to the question "What kind of person are you?"[24] For example, the most common variant of the TST reads as follows:

> In the spaces below, please give 20 different answers to the question, "Who Am I?" Give these as if you were giving them to yourself, not to somebody else. Write fairly rapidly, for the time is limited.

Answers to such questions were coded and scaled so that variations in people's self-conceptions can be linked to either prior social experiences or behaviors. Thus, Kuhn sought empirical indicators of key concepts. These indicators would allow recorded variations in one concept, such as self, to be linked to variations in other measurable concepts. In this way, Mead's legacy could be tested and used to build a theory of symbolic interactionism.

The Nature and Possibilities of Sociological Theory

Blumer's and Kuhn's assumptions about the nature of the individual, interaction, social organization, causality, and methodology were reflected in their different visions of what theory is, should be, and can be. Again, Blumer and Kuhn stood at opposite poles, with most interactionists standing between these extremes, yet leaning toward one or the other.

Blumer's Theory-Building Strategy. Blumer's assumptions, image of causal processes, and methodological position all dictated a particular conception of sociological theory. The recognition that sociological concepts do not come to grips with the empirical world is seen by Blumer as the result not only of inattention to actual evens in the empirical world but also of the kind of world it is. The use of more definitive concepts referring to classes of precisely defined events is perhaps desirable in theory building, but it may be impossible, given the nature of the empirical world. Because this world is composed of constantly shifting processes of symbolic interaction among actors in various contextual situations, the use of concepts that rip only some of the actual ongoing events from this context will fail to capture the contextual nature of the social world. More important, that social reality is ultimately "constructed" from the symbolic processes among individuals ensures that the actual instances denoted by concepts will shift and vary, thereby defying easy classification through rigid operational definitions.

These facts, Blumer argued, require the use of "sensitizing concepts," which, although lacking the precise specification of attributes and events of definitive concepts, do provide clues and suggestions about where to look for certain classes of phenomena. As such, sensitizing concepts offer a general sense of what is relevant and thereby allow investigators to approach flexibly a shifting empirical world and feel out and pick one's way in an unknown terrain. The use of this kind of concept does not necessarily reflect a lack of rigor in sociological theory but, rather, a recognition that, if "our empirical world presents itself in the form of distinctive and unique happenings or situations and if we seek through the direct study of this world to establish classes of objects and relations between classes, we are . . . forced to work with sensitizing concepts."[25]

The nature of the empirical world might preclude the development of definitive concepts, but sensitizing concepts can be improved and refined by flexibly approaching empirical situations denoted by sensitizing concepts and then by assessing how actual events compare with the concepts. Although the lack of fixed benchmarks and definitions makes this task more difficult for sensitizing concepts than for definitive ones, the progressive refinement of sensitizing concepts is possible through "careful and imaginative study of the stubborn world to which such concepts are addressed."[26] Furthermore, sensitizing concepts formulated in this way can be communicated and used to build sociology theory, and, although formal definitions and rigid classifications are not appropriate, sensitizing concepts can be explicitly communicated through descriptions and illustrations of the events to which they pertain.

In sum, the ongoing refinement, formulation, and communication of sensitizing concepts must inevitably be the building blocks of sociological theory. With careful formulation they can be incorporated into provisional theoretical statements that specify the conditions under which various types of interaction are likely to occur. In this way the concepts of theory will recognize the shifting nature of the social world and thereby provide a more accurate sets of statements about social organization.

The nature of the social world and the type of theory it dictates have profound implications for just how such theoretical statements are to be constructed and organized into theoretical formats. Blumer's emphasis on the constructed nature of reality and on the types of concepts that this necessitates led him to emphasize inductive theory construction. In inductive theory, generic propositions are abstracted from observations of concrete interaction situations. This emphasis on induction is considered desirable, because attempts at deductive theorizing in sociology usually do not involve rigorous derivations of propositions from each other or a scrupulous search for the negative empirical cases that would refute propositions.[27] These failings, Blumer contended, ensure that deductive sociological theory will remain unconnected to the events of the empirical world and, hence, unable to correct for errors in its theoretical statements. Coupled with the tendency for fads of research protocol to dictate research problems and methods used to investigate them, it is unlikely that deductive theory and the research in inspires can unearth those processes that would confirm or refute its generic statements. In the wake of this theoretical impasse, then, it is crucial that sociological theorizing refamiliarize itself with the actual events of the empirical world. As Blumer argued, "No theorizing, however ingenious, and no observance of scientific protocol, however meticulous, are substitutes for developing familiarity with what is actually going on in the sphere of life under study."[28] Without such inductive familiarity, sociological theory will remain a self-fulfilling set of theoretical prophecies bearing little relationship to the phenomena it is supposed to explain.

Kuhn's Theory-Building Strategy. Kuhn advocated a more deductive format for sociological theorizing than Blumer. Although his own work did not reveal great deductive rigor, he held that subsumption of lower-order propositions under more general principles is the most appropriate way to build theory. He visualized his "self-theory" as one step in the building of theory. By developing general statements about how self-attitudes emerge and shape social action, it would be possible to understand and predict human behavior. Moreover, less general propositions about aspects of self could be subsumed under general propositions about processes of symbolic interaction.

The ultimate goal of theory is consolidation or "coalescing" of testable lower-order theories under a general set of symbolic interactionist principles. Self-theory, as developed by Kuhn,[29] would be but one set of derivations from a more general system of interactionist principles. In contrast with Blumer, then, Kuhn believed theory should ultimately form a unified system from which specific propositions about different aspects or phases of symbolic interactionism could be derived.

CONCLUSION

The points of convergence and divergence of the Chicago and Iowa Schools, or of Blumer and Kuhn, are summarized in Table 27-1. The two right columns explore several issues about which interactionists disagree. It is important, however, to place these disagreements within the context of the points where symbolic interactionists agree, as is done in the second column.

These distinctions between the Chicago and Iowa Schools are hazardous when examining the work of a particular symbolic interactionist. These distinctions denote only *tendencies* to view humans, interaction, social organization, methods, and theory in a particular way. Few symbolic interactionists follow in any strict sense either Blumer's or Kuhn's positions. Blumer's and Kuhn's respective positions define the polar boundaries within which symbolic theorizing occurs. As we will see in the remaining chapters, however, symbolic interactionism was supplemented by other theoretical traditions, and yet, as we also will see, the issues raised by Blumer and Kuhn have remained with the interactionist tradition even as it expanded beyond Mead's initial insights.

TABLE 27-1 Convergence and Divergence in the Chicago and Iowa Schools of Symbolic Interactionism

Theoretical Issues	Convergence of Schools
The nature of humans	Humans create and use symbols to denote aspects of the world around them. What makes human unique are their symbolic capacities. Humans are capable of symbolically denoting and invoking objects, which can then shape their definitions of social situations and, hence, their actions. Humans are capable of self-reflection and evaluation. They see themselves as objects in most social situations.
The nature of interaction	Interaction depends on people's capacities to emit and interpret gestures. Role taking is the key mechanism of interaction because it enables actors to view the other's perspective, as well as that of others and groups not physically present. Role taking and mind operate together by allowing actors to use the perspectives of others and groups as a basis for their deliberations, or definitions of situations, before acting. In this way, people can adjust their responses to each other and to social situations.
The nature of social organization	Social structure is created, maintained, and changed by processes of symbolic interaction. It is not possible to understand patterns of social organization—even the most elaborate—without knowledge of the symbolic processes among individuals who ultimately make up this pattern.
The nature of sociological methods	Sociological methods must focus on the processes by which people define situations and select courses of action. Methods must focus on individual persons.
The nature of sociological theory	Theory must be about processes of interaction and seek to isolate out the conditions under which general types of behaviors and interactions are likely to occur.

TABLE 27-1 Convergence and Divergence in the Chicago and Iowa Schools of Symbolic Interactionism *(continued)*

Chicago School	Iowa School
Humans with minds can introject any object into a situation.	Humans with minds can define situations, but there tends to be consistency in the objects that they introject into situations.
Although self is an important object, it is not the only object.	Self is the most important object in the definition of a situation.
Humans weigh, asses, and map courses of action before action, but humans can potentially alter their definitions and actions.	Humans weight, assess, and map courses of action, but they do so through the prism of their core self and the groups in which this self is anchored.
Interaction is a constant process of role taking with others and groups.	Interaction depends on the process of role taking.
Others and groups thus becomes objects that are involved in people's definitions of situations.	The expectations of others and norms of the situation are important considerations in arriving at definitions of situations.
Self is another important object that enters into people's definitions.	People's core self is the most important consideration and constraint on interaction.
People's definitions of situations involve weighing and assessing objects and then mapping courses of action.	
Interaction involves constantly shifting definitions and changing patterns of action and interaction.	Interaction most often involves actions that conform to situational expectations as mitigated by the requirements of the core self.
Social structure is constructed by actors adjusting their responses to each other.	Social structures are composed of networks of positions with attendant expectations or norms.
Social structure is one of many objects that actors introject into their definitions of situations.	Although symbolic interactions create and change structures, once these structures are created they constrain interaction.
Social structure is subject to constant realignments as actors' definitions and behaviors change, forcing new adjustments from others.	Social structures are thus relatively stable, especially when people's core self is invested in particular networks of positions.
Sociological methods must penetrate the actors' mental world and see how they construct courses of action.	Sociological methods must measure with reliable instruments actors' symbolic processes.
Researchers must be attuned to the multiple, varied, ever-shifting, and often indeterminate influences on definitions of situations and actions.	Research should be directed toward defining and measuring those variables that causally influence behaviors.
Research must therefore use observational, biographical, and unstructured interview techniques if it is to penetrate people's definitional processes and consider changes in these processes.	Research must therefore use structure measuring instruments, such as questionnaires, to get reliable and valid measures of key variables.
Only sensitizing concepts are possible in sociology.	Sociology can develop precisely defined concepts with clear empirical measures.
Deductive theory is thus not possible in sociology.	Theory can thus be deductive, with a limited number of general propositions subsuming lower-order propositions and empirical generalizations on specific phases of symbolic interaction.
At best, theory can offer general and tentative descriptions and interpretations of behaviors and patterns of interaction.	Theory can offer abstract explanations that can allow predictions of behavior and interaction.

NOTES

1. Bernard N. Meltzer and Jerome W. Petras, "The Chicago and Iowa Schools of Symbolic Interactionism," in *Human Nature and Collective Behavior*, ed. T. Shibutani (Englewood Cliffs, NJ: Prentice-Hall, 1970).

2. Jonathan H. Turner, "The Rise of Scientific Sociology," *Science* 227 (March 15, 1985); Jonathan H. Turner and Stephen Park Turner, *American Sociology* (Warsaw: Polish Academy of Sciences, 1992) and *The Impossible Science* (Newbury Park, CA: Sage, 1990). See also Martin Bulmer, *The Chicago School of Sociology* (Chicago: University of Chicago Press, 1984), and Lester R. Kurtz, *Evaluating Chicago Sociology* (Chicago: University of Chicago Press, 1984).

3. For examples, consult Larry T. Reynolds, *Interactionism: Exposition and Critique*, 2nd ed. (Dix Hills, NY: General Hall, 1990); Jerome G. Manis and Bernard N. Meltzer, eds., *Symbolic Interaction: A Reader in Social Psychology* (Boston: Allyn & Bacon, 1972); J. Cardwell, *Social Psychology: A Symbolic Interactionist Approach* (Philadelphia: F.A. Davis, 1971); Alfred Lindesmith and Anselm Strauss, *Social Psychology* (New York: Holt, Rinehart & Winston, 1968); Arnold Rose, ed., *Human Behavior and Social Process* (Boston: Houghton Mifflin, 1962); Tamotsu Shibutani, *Society and Personality* (Englewood Cliffs, NJ: Prentice-Hall, 1961); C.K. Warriner, *The Emergence of Society* (Homewood, IL: Dorsey, 1970); Gregory Stone and H. Farberman, eds., *Symbolic Interaction: A Reader in Social Psychology* (Waltham, MA: Xerox Learning Systems, 1970); John P. Hewitt, *Self and Society: A Symbolic Interactionist Social Psychology* (Boston: Allyn & Bacon, 1976); Robert H. Laver and Warren H. Handel, *Social Psychology: The Theory and Application of Symbolic Interactionism* (Boston: Houghton Mifflin, 1977); Sheldon Stryker, *Symbolic Interactionism* (Menlo Park, CA: Benjamin/Cummings, 1980); Clark McPhail, "The Problems and Prospects of Behavioral Perspectives," *American Sociologist* 16 (1981), pp. 172–174.

4. Rose, *Human Behavior* (cited in note 3), p. 8.

5. Sheldon Stryker, "Symbolic Interaction as an Approach to Family Research," *Marriage and Family Living* 2 (May 1959), pp. 111–119. See also his "Role-Taking Accuracy and Adjustment," *Sociometry* 20 (December 1957), pp. 286–296.

6. Lindesmith and Strauss, *Social Psychology* (cited in note 3), p. 282.

7. Rose, *Human Behavior* (cited in note 3), p. 12.

8. W.I. Thomas, "The Definition of the Situation," in *Symbolic Interaction*, eds. J. Manis and B. Meltzer, pp. 331–336.

9. For clear statements on the concept of "definition of the situation" as it is currently used in interactionist theory, see Lindesmith and Strauss, *Social Psychology* (cited in note 3), pp. 280–283.

10. The following comparison draws heavily from Meltzer's and Petras' "The Chicago and Iowa Schools" but extends their analysis by drawing from the following works of Blumer and Kuhn. For Blumer, *Symbolic Interactionism: Perspective and Method* (Englewood Cliffs, NJ: Prentice-Hall, 1969); "Comment on 'Parsons as a Symbolic Interactionist,'" *Sociological Inquiry* 45 (Winter 1975), pp. 59–62. For Kuhn, "Major Trends in Symbolic Interaction Theory in the Past Twenty-Five Years," *Sociological Quarterly* 5 (Winter 1964), pp. 61–84; "The Reference Group Reconsidered," *Sociological Quarterly* 5 (Winter 1964), pp. 6–21; "Factors in Personality: Socio-Cultural Determinants as Seen Through the Amish," in *Aspects of Culture and Personality*, ed. F.L. Kittsu (New York: Abelard-Schuman, 1954); "Self-attitudes by Age, Sex, and Professional Training," in *Symbolic Interaction*, eds. G. Stone and H. Farberman, pp. 424–436; with T.S. McPartland, "An Empirical Investigation of Self-Attitude," *American Sociological Review* 19 (February 1954), pp. 68–76; "Family Impact on Personality," in *Problems in Social Psychology*, eds. J.E. Hulett and R. Stagner (Urbana: University of Illinois Press, 1953); with C. Addison Hickman, *Individuals, Groups, and Economic Behavior* (New York: Dryden, 1956).

11. For an interesting methodological critique of Kuhn's self-theory, see Charles W. Tucker, "Some Methodological Problems of Kuhn's Self Theory," *Sociological Quarterly* 7 (Winter 1966), pp. 345–358.

12. Hickman and Kuhn, *Individuals, Groups, and Economic Behavior*, pp. 224–225.

13. Prominent thinkers who leaned toward Blumer's position included Anselm Strauss, Alfred Lindesmith, Tamotsu Shibutani, and Ralph Turner. For a critique of Blumer's use of Mead, see Clark McPhail and Cynthia Rexroat, "Mead vs. Blumer," *American Sociological Review* 44 (1979), pp. 449–467.

14. Prominent Iowa School interactionists included Frank Miyamoto, Sanford Dornbusch, Simon Dinitz, Harry Dick, Sheldon Stryker, and Theodore Sarbin. For a list of studies by Kuhn's students, see Harold A. Mulford and Winfield W. Salisbury II, "Self-Conceptions in a General Population," *Sociological Quarterly* 5 (Winter 1964), pp. 35–46. Again, these individuals and those in note 12 would resist this classification, because none advocates as extreme a position as Kuhn or Blumer. Yet the tendency to follow either Kuhn's or Blumer's assumptions is evident in their work and in that of many others.

15. E.L. Quarantelli and Joseph Cooper, "Self-conceptions and Others: A Further Test of Meadian Hypotheses," *Sociological Quarterly* 7 (Summer 1966), pp. 281–297.

16. See particularly Blumer's "Society as Symbolic Interaction," in *Human Behavior and Social Process*, ed. A. Rose, pp. 179–192 (reprinted in Blumer's *Symbolic Interactionism*, cited in note 10); "Sociological Implications of the Thought of George Herbert Mead," in *Symbolic Interactionism*; and "The Methodological Position of Symbolic Interactionism," in *Symbolic Interactionism*.

17. See, in particular, Blumer's "Methodological Position" (cited in note 16).

18. Ibid., p. 33.

19. For an interesting discussion of the contrasts between these research strategies, see Llewellyn Gross, "Theory Construction in Sociology: A Methodological Inquiry," in *Symposium on Sociological Theory*, ed. L. Gross (New York: Harper & Row, 1959), pp. 531–563.

20. For an eloquent and reasoned argument in support of Blumer's position, see Norman K. Denzin, *The Research Act: A Theoretical Introduction to Sociological Methods* (Chicago: Aldine, 1970), pp. 185–218; and Norman K. Denzin, "Symbolic Interactionism and Ethnomethodology," in *Understanding Everyday Life: Toward a Reconstruction of Sociological Knowledge*, ed. J.D. Douglas (Chicago: Aldine, 1970), pp. 259–284.

21. Blumer, "Methodological Position" (cited in note 16), p. 49.

22. Herbert Blumer, "What Is Wrong with Social Theory?" *American Sociological Review* 19 (August 1954), pp. 146–158.

23. Meltzer and Petras, "The Chicago and Iowa Schools" (cited in note 1).

24. For an important critique of this methodology, see Tucker, "Some Methodological Problems" (cited in note 11).

25. Blumer, "What Is Wrong with Social Theory?" (cited in note 22), p. 150.

26. Ibid.

27. Blumer, "Methodological Position" (cited in note 16).

28. Ibid., p. 39.

29. Kuhn, "Major Trends" (cited in note 10), p. 80.

28

The Maturing Tradition II: Self and Identity Theories

In virtually all interactionist theorizing, the central dynamic of face-to-face relations is self, or the conceptions that individuals have about themselves as objects. Some of the most creative theorizing within both the Iowa and Chicago School traditions during the decades around the midcentury revolved around conceptualizing self in more detailed terms. The notion of "identity" became one prominent way to reconceptualize self. In general terms, self was viewed as a set or series of identities that respond to specific situations. Self was thus constructed from a number of identities, and these were viewed by theorists as arranged into hierarchies of salience and importance. Those identities high in the hierarchy are the most influential in individuals' efforts to organize their behaviors and present themselves in a certain light. At the same time, identities can act as filters of selective perception and interpretation as individuals mutually role-take with one another.

Thus, the effort to develop a more refined theory of self was the major thrust of much interactionist theorizing. In this chapter, we will explore two such theories of identity, one by Sheldon Stryker, who can be considered to be squarely in the Iowa School tradition, and the other by George P. McCall and J.L. Simmons, who, though trained in the Iowa tradition, moved more to the Blumer side in developing their theory.[1] In a real sense, however, both theories converged toward the middle of this Iowa-Chicago School polarization; for this reason that they had considerable impact in the two decades between 1960 and 1980.

THE IDENTITY THEORY OF SHELDON STRYKER

Designations and Definitions

In Sheldon Stryker's view,[2] human social behavior is organized by symbolic designations of all aspects of the environment, both physical and social. Among the most important of these designations are the symbols, and associated meanings, of the positions that people occupy in social structures. These positions carry with them shared expectations about how people are to enact roles and, in general, to comport themselves in relation to others. As individuals designate their own positions, they call forth in themselves expectations about how they are to behave, and as they designate the positions of others, they become cognizant of the expectations guiding the role behaviors of these others. They also become aware of broader frames of reference and definitions of the situation as these positional designations are made. And most important, individuals designate themselves as objects in relation to their location in structural positions and their perceptions of broader definitions of the situation.

Behavior is, however, not wholly determined or dictated by these designations and definitions. True, people are almost always aware of expectations associated with positions but, as they present themselves to others, the form and content of the interaction can change. The amount of such change will vary with the type of larger social structure within which the interaction occurs; some are open and flexible, whereas others are more closed and rigid. Still, all structures impose limits and constraints on what individuals do when engaged in face-to-face interaction.

Identities and the Salience Hierarchy

Stryker reasoned that identities are "parts" of self, and as such, they are internalized self designations that are associated with positions that individuals occupy within various social contexts. Identity is thus a critical link between the individual and social structure, because identities are designations that people make about themselves in relation to their location in social structures and the roles that they play by virtue of this location. Identities are organized into a *salience hierarchy*, and those identities high in the hierarchy are more likely to be evoked than those lower in this hierarchy. Not all situations will invoke multiple identities, but many do. The salience hierarchy determines those identities that respond as people orchestrate their roles and interpret the role behaviors of others. As a general rule, Stryker proposes that when an interaction situation is isolated from structural constraints, or these structural constraints are ambiguous, individuals will have more options in their choice of an identity, and, hence, they will be more likely to evoke more than one identity. But as a situation becomes embedded within social structures, the salience hierarchy becomes a good predictor of what identities will be used in interaction with others.

Commitment and Self

Stryker introduced the idea of *commitment* as a means for conceptualizing the link between social structure and self. Commitment designates the degree to which a person's relationship to others depends on being a certain kind of individual with a particular identity. The greater this dependence is, the more a person will be committed to a particular identity and the higher this identity will be in the person's salience hierarchy. Having an identity that is based on the views of others, as well as on broader social definitions, will tend to produce behaviors that conform to these views and definitions.

When people reveal such commitment to an identity in a situation, their sense of self-esteem becomes dependent on the successful execution of their identity. Moreover, when an identity is established by reference to the norms, values, and other symbols of the broader society, esteem is even more dependent on successful implementation of an identity. In this way, cultural definitions and expectations, social structural location,

TABLE 28-1: A Revised Formulation of Stryker's Hypotheses on Identity

The Salience of Identity

1. The more individuals are committed to an identity, the higher will this identity be in their salience hierarchy.
2. The degree of commitment to an identity is a positive and additive function of
   - A. The extent to which this identity is positively valued by others and broader cultural definitions.
   - B. The more congruent the expectations of others on whom one depends for an identity.
   - C. The more extensive the network of individuals on whom one depends for an identity.
   - D. The larger the number of persons in a network on whom one depends for an identity.

The Consequences of High Salience

3. The higher in a person's salience hierarchy is an identity, the more likely will that individual
   - A. Emit role performances that are consistent with the role expectations associated with that identity.
   - B. Perceive a given situation as an opportunity to perform in that identity.
   - C. Seek out situations that provide opportunities to perform in that identity.

The Consequences of Commitment to Identity

4. The greater the commitment to an identity, the greater will be
   - A. The effect of role performances on self-esteem.
   - B. The likelihood that role performances will reflect institutionalized values and norms.

Changing Commitments to Identity

5. The more external events alter the structure of a situation, the more likely are individuals to adopt new or novel identities.
6. The more changes in identity reinforce and reflect the value-commitments of the individual, the less the individual resists change in adopting a new identity.

identity, and esteem associated with that identity all become interwoven. And in this process, social structure constrains behavior and people's perceptions of themselves and others.

The Key Propositions

In the early version of the theory, Stryker developed a series of "hypotheses" about the conditions producing the salience of an identity, the effects of identities high in the salience hierarchy on role behaviors, the influence of commitment on esteem, and the nature of changes in identity. These are rephrased somewhat and summarized in Table 28-1. To state Stryker's argument more discursively, here is what he proposed: The more individuals reveal commitment to an identity, the higher this identity will be in the salience hierarchy. If this identity is positively evaluated in terms of the reactions of others and broader value standards, then this identity will move up a person's hierarchy. When the expectations of others are congruent and consistent, revealing few conflicts and disagreements, individuals will be even more committed to the identity presented to these others because they "speak with the same voice." And finally, when the network of these others on whom one depends for identity is large and extended, encompassing many others rather than just a few, the higher in the salience hierarchy will this identity become.

Once an identity is high in the salience hierarchy of an individual, role performances will become ever more consistent with the expectations attached to this identity. Moreover, when identities are high in the salience hierarchy, individuals will tend to perceive situations as opportunities to play out this identity in roles; and they will actively seek out situations where they can use this identity. In this way, the congruence between those identities high in people's hierarchies and the expectations of situations increases.

This congruence increases commitment because individuals come to see their identities as depending on the continued willingness of others to confirm their identities. As commitment increases, and as individuals become dependent on confirmation of their identities from others, their role performances have ever more consequences for their level of self-esteem. Moreover, as people become committed to identities and these identities move up in their salience hierarchy, they come to evaluate their role performances through broader cultural definitions and normative expectations; as people make such evaluations, they become even more committed to their identities.

External events can, however, erode commitments to an identity. When this occurs, people are more likely to adopt new identities, even novel identities. As individuals begin to seek new identities, change is likely to move in the direction of those identities that reflect their values. In this way, cultural values pull the formation of new identities in directions that will increase the congruence between cultural definitions and role performance as individuals develop new identity commitments and as their self-esteem becomes dependent on successful role performance of these identity commitments.

Over the last decade and a half, Stryker has continued to work on this identity theory, as we will see in Chapter 32 on theories of emotion. Yet, the theory remains much the same as reported here. In a sense, theories of identity peaked in the 1960s and 1970s, as did symbolic interactionism as a distinctive theoretical approach. Subsequent developments within interactionist theorizing moved beyond the paradigm provided by Mead and carried forward by symbolic interactionists. Alternative micro-level theories, drawing not only from Alfred Schutz but from European founders as well, have become a part of the interactionist theoretical approach, broadly conceived as the dynamics of face-to-face interaction in social settings. Still, before leaving the maturing interactionist tradition, we might examine a somewhat more elaborate theory of identity than Stryker's.

GEORGE P. MCCALL AND J.L. SIMMONS' THEORY OF IDENTITY

Role Identity and Role Support

In contrast with Stryker's more structural theory, where culture and social structure designate many of the identities held by individuals, McCall and Simmons[3] emphasized that roles are typically improvised as individuals seek to realize their various plans and goals. A *role identity* is, therefore, "the character and the role that an individual devises for himself (herself as well) as an occupant of a particular social position."[4] Role identity constitutes an "imaginative view of oneself" in a position, often a rather idealized view of oneself. Each role identity thus has a conventional portion linked to positions in social structure as well as an idiosyncratic portion constructed in people's imaginations.

Role identities become part of individuals' plans and goals because legitimating one's identity in the eyes of others is always a driving force of human behavior. Moreover, people evaluate themselves through the role performances intended to confirm a role identity. But, as McCall and Simmons emphasized, the most important audiences for a role performance are individuals themselves who assess their performances with respect to their own idealized view of their role identity. Still, people must also seek *role support* from relevant audiences outside their own minds for their role identities. This support involves more than audiences granting a person the right to occupy a position, and it includes more than approval from others for conduct by those in a position. For an individual to feel legitimated in a role, audiences must also approve of the more expressive content—the style, emotion, manner, and tone—of role performances designed to legitimate a role identity.

Because much of a role identity is rather idealized in the individual's mind and because a person must seek legitimation along several fronts, there is always discrepancy and disjuncture between the role identity and the role support received for that identity. People idealize too much, and they must seek support for performances that can be misinterpreted. As a result, there is almost always some dissatisfaction by individuals about how much their role identity has been legitimated by audiences. These points of disjuncture between identity and legitimating support motivate and drive individual behavior. Indeed, for McCall and Simmons, the most distinctive emotion among humans is the "drive to acquire support for (their) idealized conceptions of (themselves)."[5]

The Mechanisms for Maintaining Role Support

To overcome the discrepancy between what people desire and get in role support for an identity, several mechanisms are employed. One is the accumulation of "short-term credit" from interactions where discrepancies have been minimal; these emotional credits can then carry individuals through episodes where the responses from others provide less than whole-hearted role support. A second mechanism is "selective perception of cues" from others where individuals only see those responses confirming an identity. A third mechanism is "selective interpretation" of cues whereby the individual sees the cues accurately but puts "a spin" or interpretation on them that supports a role identity. A fourth mechanism is withdrawing from interactions that do not support an identity and seeking alternative situations where more support can be garnered. A fifth mechanism is switching to a new role identity whose performance will bring more support from others. A sixth mechanism is "scapegoating" audiences, blaming them for causing the discrepancy between performance and support. A seventh mechanism is "disavowing" unsuccessful performances that individuals had hoped to legitimate. And a final defensive mechanism is deprecating and rejecting the audience that withholds support for a role identity. When these mechanisms fail, individuals experience misery and anguish, and through such experiences, people learn to be cautious in committing themselves so openly and fully to particular role performances in front of certain audiences.[6]

The Hierarchy of Prominence

The *cohesiveness* role identities of individuals vary, McCall and Simmons argued, in how the elements of an identity fit together and in the compatibility among various role identities. There is also a *hierarchy of prominence* among role identities; although this hierarchy can shift and change as circumstances dictate, it tends to exist at any given point in an interaction. This prominence reflects the idealized view of individuals, the extent to which these ideals have been supported by audiences, the degree to which individuals have committed themselves to these identities, the extrinsic and intrinsic rewards (to be discussed shortly) associated with an identity, and the amount of previous investment in time and energy that has been devoted to an identity.

From this perspective, interaction revolves around each individual asserting through role performances identities that are high in their prominence hierarchy and that they seek to legitimate, in their own eyes as well as in the eyes of others. At the same time, each individual is interpreting the gestures of others to determine just what identity is high in the prominence hierarchy of others and whether or not the role performances of others are worthy of role support and other rewards. To some degree, the external structure of the situation provides the necessary information about what positions people occupy and what expectations are placed on them by virtue of incumbency in these positions. Yet, for McCall and Simmons, most interactions are to some degree ambiguous and unstructured, allowing alternative role performances and varying interpretations of these performances.

Much of the ambiguity in interaction is eliminated through simple role-taking in a person's *inner forum* or cognitive repertoire of vocabularies, gestures, motives, and other information that marks various identities and role performances. Humans have, therefore, the capacity to construct interpretations in light of the vast amounts of information that they accumulate in their "inner forum" or what Alfred Schutz called "stocks of knowledge at hand." This information might have to be assembled in somewhat "different proportions and balances" but humans' capacity for mind and thought enables them to do so with amazing speed and accuracy.

Individuals will often improvise a role, adjusting their identities and role performances in light of how they interpret the roles of others. As such improvisation occurs, various expressive strategies are employed; these strategies revolve around orchestrating gestures to present a certain image of self and to claim a particular identity that is high in the prominence hierarchy. Conversely, individuals read the dramaturgical presentations of others to "altercast" and determine the self that is being claimed by these others. In essence, then, interaction is the negotiation of identities, whereby people make expressive and dramaturgical presentations over identities that are high in their respective prominence hierarchies and that can be supported, or that can go unsupported, on the basis of role performances.

The Underlying Exchange Dynamic

This process of negotiation among individuals is complex and subtle, involving an initial but very tentative agreement to accept each others' claims. In this way, people avoid interrupting the expressive strategies that are being used to impart their respective identities. As this process unfolds, however, it moves into a real exchange negotiation whereby individuals seek the rewards that come with legitimation of their role performances. At this point McCall and Simmons merged their interactionist theory with exchange theory.

They begin by classifying three basic types of rewards: First, there are extrinsic rewards, such as money or other reinforcers that are visible to all. Second, there are intrinsic rewards that provide less visible means of reinforcement for the individual—rewards such as satisfaction, pride, and comfort. And third, and most important, there is role support for an identity, which McCall and Simmons believe is the most valuable of all rewards. Individuals are motivated to seek a profit—rewards less the costs in securing them—in all their interactions. Moreover, there are separate types of calculi for each of these three categories of reward, and there are *rules of the marketplace*: Rewards received by each party to an exchange should be roughly comparable in their type (whether extrinsic, intrinsic, or identity support), and rewards should be received in proportion to the investments individuals incur in receiving them (a principle of "distributive justice").

These negotiations are affected by what McCall and Simmons term the *salience of identities,* which are those identities that, for the immediate interaction at hand, are the most relevant in an individual's hierarchy of prominence. This salience of identities constitutes, in McCall and Simmons' words, a *situated self* that is most pertinent to the present interaction. This situational self determines a person's preferences about which role identities he or she will enact in a given situation, but the preferences of the situational self are fluid and changeable. In contrast, the *ideal self* is more stable than the situated self, while being the highest-order identity in the prominence hierarchy. A person's ideal self will thus influence which identities should be salient in an interaction and how they will be invoked to constitute a situated self. Besides the prominence hierarchy, other factors also influence the formation of a situated identity. The needs that an individual feels for support of an identity, the extrinsic and intrinsic rewards to be received by claiming a situated self, and the opportunity for profitable enactment of a role in relation to a situated self all shape identity formation.

All these factors are, in McCall and Simmons' view, potential reinforcers or payoffs for roles emitted in claiming an identity. These payoffs vary in value, however. Support of the ideal self brings greater rewards than either extrinsic or intrinsic rewards. The patterns of payoffs for rewards can also vary. For extrinsic and intrinsic types of rewards, when payoffs match expectations and desires, needs for them decline somewhat (in accordance with satiation or the principle of marginal utility). If people receive either more or less than they expected or desired of these two types of rewards, then their immediate need for these rewards suddenly escalates. In contrast, the payoff schedule for role support for an identity reveals a more complicated pattern. Role support for what was desired or expected does not increase the desire for further role support of an identity; a moderate discrepancy between the support sought and received increases the desire for support of an identity; but extreme discrepancies operate differently, depending on the sign of the discrepancy: If people receive support that greatly exceeds their expectations, they immediately desire more role support, whereas if they dramatically receive less role support than expected, their desire for this role support drops rapidly.

Because payoffs will almost always, or at least eventually, be less than expected, discrepancies will be chronic even after individuals have employed all the defense mechanisms to reduce discrepancies that were discussed earlier. Hence, people are constantly driven to overcome this discrepancy, but this search to reduce discrepancy is complicated by the payoff schedule for role support. Moderate discrepancies drive people to seek more role support, whereas large ones reduce efforts to secure role support for an identity. And when people have received more support than they expected for an identity, they want even more of this reward, raising this identity in salience and, over time, increasing its prominence in the hierarchy.

CONCLUSION

At about the same time that these identity theories were being developed, other researchers were working on conceptualizing self processes.[7] These two approaches can, therefore, be seen as exemplars of theoretical work within the symbolic interactionist canon during the decades surrounding the midpoint of the century. Drawing from George Herbert Mead's views, all these theories sought to understand the connection among the person, role behaviors, social structure, and culture. The key dynamic, they all argued, was self and identity. As people plug themselves into social structures and culture, take cognizance of each other, and make roles for themselves, they seek to confirm and reaffirm their self conceptions, generally conceptualized as a series of identities that vary in salience and that are arranged into hierarchies. Some theorists such as Stryker viewed culture and social structure as circumscribing these dynamics, whereas others like McCall and Simmons viewed the individual as exerting more latitude and freedom in constructing identities. But all symbolic interactionists tended to view self as the crucial force behind people's perceptions and actions in interactive situations, organizing and determining the flow of role-playing and role-taking in social contexts composed of others, cultural definitions, and social structural positions.

Along with these interactionist theories of self and identity, a related theoretical orientation emphasizing roles was emerging in the middle decades of the century. Self remained a key dynamic in these theories, but was not necessarily viewed as the driving dynamic of all role behaviors. Before moving to the more recent directions of interactionist theorizing, then, we need to pause and examine an equally prominent theoretical tradition: role theory.

NOTES

1. Aside from these figures, others seeking a theory of self and identity include Eugene Weinstein, Mary Glenn Wiley, and William DeVaughn, "Role and Interpersonal Style as Components of Interaction." *Social Forces* 45 (1966), pp. 210–216; Peter J. Burke and Judy C. Tully, "The Measurement of Role/Identity," *Social Forces* 55 (1977), pp. 881–897; Nelson N. Foote, "Identification as the Basis for a Theory of Motivation." *American Sociological Review* 16 (1951), pp. 14–21; Tamotsu Shibutani, *Society and Personality* (Englewood Cliffs, NJ: Prentice-Hall, 1961); Anselm Strauss, *Mirrors and Masks* (Glencoe, IL: Free Press, 1959); Gregory P. Stone, "Appearance and the Self" in *Behavior and Social Processes*, ed. Arnold M. Rose (Boston: Houghton Mifflin, 1962). For a review of the history and current profile of identity and self theories, see Viktor Gecas and Peter J. Burke, "Self and Identity," in *Sociological Perspectives on Social Psychology*, eds. Karen S. Cook, Gary Alan Fine and James S. House (Boston: Allyn & Bacon, 1955), pp. 41–67.

2. Sheldon Stryker, *Symbolic Interactionism: A Structural Version* (Menlo Park, CA: Benjamin/Cummings, 1980); "Identity Salience and Role Performance: The Relevance of Symbolic Interaction Theory for Family Research," *Journal of Marriage and the Family* (1968), pp. 558–564; "Fundamental Principles of Social Interaction" in *Sociology*, 2nd ed., Neil J. Smelser, ed. (New York: Wiley, 1973), pp. 495–547. For a more recent version of the theory, see Sheldon Stryker and Richard T. Serpe, "Commitment, Identity Salience, and Role Behavior" in *Personality, Roles, and Social Behavior*, eds. William Ickes and Eric Knowles (New York: Springer-Verlag, 1982), pp. 199–218; Richard T. Serpe and Sheldon Stryker, "The Construction of Self and the Reconstruction of Social Relationships," *Advances in Group Processes*, 4 (1987), pp. 41–66; and Sheldon Stryker, "Exploring the Relevance of Social Cognition for the Relationship of Self and Society" in *The Self-Society Dynamic: Cognition, Emotion, and Action*, eds. Judith Howard and Peter L. Callero (Cambridge: Cambridge University Press, 1991), pp. 19–41.

3. George P. McCall and J.L. Simmons, *Identities and Interactions* (New York: Basic Books, 1960). A second edition of this book was published in 1978, although the theory remained virtually unchanged.

4. Ibid., p. 67.

5. Ibid., p. 75.

6. Ibid., p. 101.

7. Perhaps the most significant of these was Peter J. Burke's approach, which sought to extend these identity theories in view of the methodological problems in measuring self and identity. See note 1 for relevant references for work at the century's midpoint. For more recent efforts by Burke, see his "An Identity Model for Network Exchange." *American Sociological Review* 62 (1997), pp. 134–150; "Attitudes, Behavior, and the Self" in *The Self-Society Dynamic*, eds. Judith Howard and Peter L. Callero, pp. 189–208, "Identity Processes and Social Stress," *American Sociological Review* 56 (1991), pp. 836–849.

29

The Maturing Tradition III:

Role Theories: Ralph H. Turner's Synthetic Approach

By the midcentury, the idea that individuals are connected to larger social structures by virtue of incumbency in status positions and role behaviors pervaded sociological theory. Role was the key concept that linked individuals and social structure, and, as a result of this emphasis, role theorizing became prominent. Yet, much of this role theorizing was, in reality, descriptions of expectations and behaviors in various empirical settings, or at best, more theoretical conceptualizations of a narrow range of role processes.

The thrust of much early role analysis, as it flowed from a mixture of Robert Park's, Georg Simmel's, Jacob Moreno's, Ralph Linton's, and George Herbert Mead's insights, was often captured in the midcentury by quoting a famous passage from Shakespeare's *As You Like It*:

> All the world's a stage
> and all the men and women merely
> players:
> They have their exits and their entrances;
> And one man in his time plays many
> parts.
> (Act II, scene vii)

The analogy was often drawn between the players on the stage and the actors of society.[1] Just as players have a clearly defined part to play, so actors in society occupy clear positions; just as players must follow a written script, so actors in society must follow norms; just as players must obey the orders of a director, so actors in society must conform to the dictates of those with power or those of importance; just as players must react to each other's performance on the stage, so members of society must mutually adjust their responses to one another; just as players respond to the audience, so actors in society take the role of various audiences or "generalized others"; and just as players with varying abilities and capacities bring to each role their unique interpretations, so actors with varying self-conceptions and role-playing skills have their own styles of interaction.

A general theory of roles never emerged from such metaphors, and during the last decades, efforts to do so appear to have waned. Yet, a great deal was learned about roles in the decades between 1940 and 1980, and the movement of micro-level theory to new topics does not so much reflect the failure of role theory as its success in uncovering many basic dynamics of role and such related processes as status and expectations. In this chapter, we will examine Ralph H. Turner's approach, which, more than any other role theory, has overcome the problems in earlier theories and which, at the same time, has sought to generate systematic theory.

Over the course of several decades, Ralph H. Turner mounted a consistent line of criticism against what he characterized as structural role theory.[2] This criticism incorporated several lines of attack: (1) Earlier role theory had presented an overly structured vision of the social world, with its emphasis on norms, status positions, and the enactment of normative expectations. (2) Role theory had tended to concentrate an inordinate amount of research and theory-building effort on "abnormal" social processes, such as role conflict and role strain, thereby ignoring the normal processes of human interaction. (3) Role theory was not theory but, rather, a series of disjointed and unconnected propositions and empirical generalizations. (4) Role theory did not recognize Mead's concept of role taking as its central dynamic. As a corrective to these problems, Turner offered in the decades around the midcentury a conceptualization of roles that emphasized the process of interaction over the dictates of social structures and cultural scripts.[3]

THE ROLE-MAKING PROCESS

Turner used Mead's concept of role taking to describe the nature of social action. Turner assumed that "it is the tendency to shape the phenomenal world into roles which is the key to the role-taking as the core process in interaction."[4] Turner stressed that actors emit gestures or cues—words, bodily countenance, voice inflections, dress, facial expressions, and other gestures—as they interact to "put themselves in the other's role" and, thereby to adjust their lines of conduct in ways that can facilitate cooperation. In emphasizing this point, Turner was simply following Mead's definition of taking the role of the other, or role taking.

Turner then extended Mead's concept. He first argued that cultural definitions of roles are often vague and even contradictory. At best, they provide a general framework within which actors must construct a line of conduct. Thus actors *make* their roles and communicate to others *what* role they are playing. Turner then argued that humans act *as if* all others in their environment are playing *identifiable roles*.[5] Humans assume others to be playing a role, and this assumption is what gives interaction a common basis. Operating with this folk assumption, people then read gestures and cues in an effort to determine what role others are playing.[6] This effort is facilitated by others creating and asserting their roles, with the result that they actively emit cues about what roles they are attempting to play.

For Turner, then, role taking was also *role making*. Humans make roles in three senses: (1) They are often faced with only a loose cultural framework in which they must make a role to play. (2) They assume others are playing a role and thus make an effort to discover the underlying role behind a person's acts. (3) Humans seek to make a role for themselves in all social situations by emitting cues to others that give them claim on a particular role. Interaction is, therefore, a joint and reciprocal process of role taking and role making.

THE "FOLK NORM OF CONSISTENCY"

As people interact with one another, Turner argued, they assess behavior less for its conformity to imputed norms or positions in a social structure and more for the consistency of behavior. Humans seek to group one another's behavior into coherent wholes or gestalts, and, by doing so, they can make sense of one another's actions, anticipate one another's behavior, and adjust to one another's responses. If another's responses are inconsistent and not seen as part of an underlying role, then interaction will prove difficult. Thus, there is an implicit "norm of consistency" in people's interactions with one another. Humans attempt to assess the consistency of others' actions to discern the underlying role that is being played.

THE TENTATIVE NATURE OF INTERACTION

Turner echoed Herbert Blumer's position when he stated that "interaction is always a tentative process, a process of continuously testing the conception one has of the role of the other."[7] Humans are constantly interpreting additional cues emitted by others and using these new cues to see if they are consistent with those previously emitted and with the imputed roles of others. If they are consistent, then the actor will continue to adjust responses in accordance with the imputed role of the other. But as soon as inconsistent cues are emitted, the identification of the other's role will undergo revision. Thus, the imputation of a particular role to another person will persist only as long as it provides a stable framework for interaction. The tentative nature of the role-making process points to another facet of roles: the process of *role verification*.

THE PROCESS OF ROLE VERIFICATION

Actors seek to verify that behaviors and other cues emitted by people in a situation do indeed constitute a role. Turner argued that such efforts at verification or validation are achieved by applying external and internal criteria. The most often used *internal criterion* is the degree to which an actor perceives a role to facilitate interaction. *External criteria* can vary, but in general they involve assessment of a role by important others, relevant groups, or commonly agreed-on standards. When an imputed role is validated or verified in this way, then it can serve as a stable basis for continued interaction among actors.

SELF-CONCEPTIONS AND ROLE

All humans reveal self-conceptions of themselves as certain kinds of objects. Humans develop self-attitudes and feelings from their interactions with others, but, as Turner and all role theorists emphasized, actors attempt to present themselves in ways that will reinforce their self-conceptions.[8] Because others will always seek to determine an individual's role, it becomes necessary for the individual to inform others, through cues and gestures, about the degree to which self is anchored in a role. Thus, actors will signal one another about their self-identity and the extent to which their role is consistent with their self-conception. For example, roles not consistent with a person's self-conception will likely be played with considerable distance and disdain, whereas those that an individual considers central to self-definitions will be played much differently.[9]

In sum, Turner's approach maintains much of the dramaturgical metaphor of early role theory, but with an emphasis on the behavioral aspect of roles, because actors impute roles to one another through behavioral cues. The notion in early analysis that roles are conceptions of expected behaviors is preserved, for assigning of a role to a person invokes an expectation that a certain type and range of responses will ensue. The view that roles are the norms attendant on status positions is given less emphasis than in early theories but not ignored, because norms and positions can be the basis for assigning and verifying roles.[10] And the conception in the dramaturgical metaphor of roles as parts that people learn to play is preserved, for people can denote one another's roles by virtue of their prior socialization into a common role repertoire.

TURNER'S STRATEGY FOR BUILDING ROLE THEORY

Although Turner accepted a process orientation, he was committed to developing interactionism into "something akin to axiomatic theory."[11] He had recognized that role theory was segmented into a series of narrow propositions and hypotheses and that role theorists had been reluctant "to find unifying themes to link various role processes."

Turner's strategy was to use propositions from the numerous research studies to build more formal and abstract theoretical statements. His goal was to maintain a productive dialogue between specific empirical propositions and more abstract theoretical statements. Turner advocated the use of what he termed *main tendency* propositions to link concepts to empirical regularities and to consolidate the thrust of these regularities.[12] What Turner sought was a series of statements that highlight what tends to occur in the normal operation of systems of interaction. To this end, Turner provided a long list of main tendency propositions on (a) roles as they emerge, (b) roles as an interactive framework, (c) roles in relation to actors, (d) roles in organizational settings, (e) roles in societal settings, and (f) roles and the person. The most important of these propositions will be examined.[13]

Emergence and Character of Roles

1. In any interactive situation, behavior, sentiments, and motives tend to be differentiated into units that can be termed roles; once differentiated, elements of behavior, sentiment, and motives that appear in the same situation tend to be assigned to existing roles. (Tendencies for role differentiation and accretion.)

2. In any interactive situation, the meaning of individual actions for ego (the actor) and for any alter (others) is assigned on the basis of the imputed role. (Tendencies for meaningfulness.)

3. In connection with every role, there is a tendency for certain attributes of actors, aspects of behavior, and features of situations to become salient cues for the identification of roles. (Tendencies for role cues.)

4. Every role tends to acquire an evaluation for rank and social desirability. (Tendencies for evaluation.)
5. The character of a role—that is, its definition—will tend to change if there are persistent changes in either the behaviors of those presumed to be playing the role or the contexts in which the role is played. (Tendencies for behavioral correspondence.)

In these five propositions, individuals are seen as viewing the world in terms of roles, as employing a "folk norm" to discover the consistency of behaviors and to assign behavioral elements to an imputed role (role differentiation and accretion), as using roles to interpret and define situations (meaningfulness tendency), as searching roles for signals about the attributes of actors as well as the nature of the situation (role cues), and as evaluating roles for their power, prestige, and esteem, while assessing them for their degree of social desirability and worth (tendency for evaluation). When role behaviors or situations are permanently altered, the definition of role will also undergo change (behavioral correspondence).

Role as an Interactive Framework

6. The establishment and persistence of interaction tend to depend on the emergence and identification of ego and alter roles. (Tendency for interaction in roles.)
7. Each role tends to form as a comprehensive way of coping with one or more relevant alter roles. (Tendency for role complementarity.)
8. There is a tendency for stabilized roles to be assigned the character of legitimate expectations and to be seen as the appropriate way to behave in a situation. (Tendency for legitimate expectations.)

In these three additional propositions, interaction is seen as depending on the identification of roles. Moreover, roles tend to be complements of others—as is the case with wife/husband, parent/child, boss/employee roles—and thus operate to regularize interaction among complementary roles. Finally, roles that prove useful and that allow stable and fruitful interaction are translated into expectations that future transactions will and should occur as in the past.

Role in Relation to Actor

9. Once stabilized, the role structure tends to persist, regardless of changes in actors. (Tendency for role persistence.)
10. There is a tendency to identify a given individual with a given role and a complementary tendency for an individual to adopt a given role for the duration of the interaction. (Tendency in role allocation.)
11. To the extent that ego's role is an adaptation to alter's role, it incorporates some conception of alter's role. (Tendency for role taking.)
12. Role behavior tends to be judged as adequate or inadequate by comparison with a conception of the role in question. (Tendency to assess role adequacy.)
13. The degree of adequacy in role performance of an actor determines the extent to which others will respond and reciprocate an actor's role performance. (Tendency for role reciprocity.)

Thus, once actors identify and assign one another to roles, the roles persist, and new actors will tend to be assigned to those roles that already exist in a situation. Humans also tend to adopt roles for the duration of an interaction, while having knowledge of the roles that others are playing. In addition, individuals carry with them general conceptions of what a role entails and what constitutes adequate performance. Finally, the adequacy of a person's role performance greatly influences the extent to which the role, along with the rights, privileges, and complementary behaviors that it deserves, will be acknowledged.

Role in Organizational Settings

14. To the extent that roles are incorporated into an organizational setting, organizational goals tend to become crucial criteria for role differentiation, evaluation, complementarity, legitimacy or expectation, consensus, allocation, and judgments of adequacy. (Tendency for organization goal dominance.)
15. To the extent that roles are incorporated into an organizational setting, the right to define the legitimate character of roles, to set the evaluations on roles, to allocate roles, and to judge role adequacy tends to be lodged in particular roles. (Tendency for legitimate role definers.)
16. To the extent that roles are incorporated into an organizational setting, differentiation tends to link roles to statuses in the organization. (Tendency for status.)
17. To the extent that roles are incorporated into an organizational setting, each role tends to develop as a pattern of adaptation to multiple alter roles. (Tendency for role-sets.)
18. To the extent that roles are incorporated into an organizational setting, the persistence of roles is intensified through tradition and formalization. (Tendency for formalization.)

When roles are lodged in an organization, then, its goals and key personnel become important in the role-making process. Moreover, it is primarily within organizations that status and role become merged. In this way, Turner incorporated Linton's and other early theorists' insight that status and role *can* become highly related, but Turner never abandoned Mead's and Blumer's emphasis that much interaction occurs in contexts where roles are not immersed within networks of clearly defined status positions. Turner also recognized that roles in structured situations develop as ways of adapting to a number of other roles that are typically assigned by role definers or required by organizational goals. Finally, roles within organizations tend to become formalized in that written agreements and tradition develop the power to maintain a given role system and to shape normative expectations.

Role in Societal Setting

19. Similar roles in different contexts tend to become merged, so they are identified as a single role recurring in different relationships. (Tendency for economy of roles.)
20. To the extent that roles refer to more general social contexts and situations, differentiation tends to link roles to social values. (Tendency for value anchorage.)
21. The individual in society tends to be assigned and to assume roles consistent with one another. (Tendency for allocation consistency.)

Many roles are identified, assumed, and imputed in relation to a broader societal context. In the tendencies listed previously, Turner first argued that people tend to group behaviors in different social contexts into as few unifying roles as is possible or practical. Thus, people will identify a role as a way of making sense of disparate behaviors in different contexts. At the societal level, values are the equivalent of goals in organizational settings for identifying, differentiating, allocating, evaluating, and legitimating roles. Finally, all people tend to assume multiple roles in society, but they tend to assume roles that are consistent with one another.

Role and the Person

22. Actors tend to act to alleviate role strain arising from role contradiction, role conflict, and role inadequacy and to heighten the gratifications of high role adequacy. (Tendency to resolve role strain.)

23. Individuals in society tend to adopt a repertoire of role relationships as a framework for their own behavior and as a perspective for interpretation of the behavior of others. (Tendency to be socialized into common culture.)

24. Individuals tend to form self-conceptions by selectively identifying certain roles from their repertoires as more characteristically "themselves" than other roles. (Tendency to anchor self-conception.)

25. The self-conception tends to stress those roles that supply the basis for effective adaptation to relevant alters. (Adaptation of self-conception tendency.)

26. To the extent that roles must be played in situations that contradict the self-conception, those roles will be assigned role distance, and mechanisms for demonstrating lack of personal involvement will be employed. (Tendency for role distance.)

The "person" was a concept employed by Turner to denote "the distinctive repertoire of roles" that an individual enacts in relevant social settings. In these generalizations, Turner emphasized that individuals seek to resolve tensions among roles and to avoid contradictions between self-conceptions and roles. These propositions are, to a very great extent, elaborations of Turner's assumptions about the relationship between self-conceptions and role.

Turner recognized that these and other tendency propositions do not specify the conditions under which the tendency actually would occur. Turner believed that to be true theory, role analysis would have to specify some of these conditions; otherwise, the tendency propositions simply describe what occurs without indicating *when* and *why* these tendencies should be evident. As Turner sought to expand his theory, he began to add explanatory content—that is, to indicate when and why the tendencies exist.

CONCLUSION: GENERATING EXPLANATORY LAWS

Turner began more explanatory efforts by specifying conditions under which particular tendencies become evident. For example, the tendencies for individuals to identify with a role and for others around the individual to make such an identification are likely to emerge when[14] (a) allocation of roles is less flexible, (b) roles are highly differentiated, (c) roles are implicated in conflictual relationships, (d) performance of roles is judged as competent, (e) roles are considered difficult, (f) roles are either of high or low rank, (g) power is vested in a role, and so on.[15] These kinds of propositions were very suggestive, but Turner began to feel that they were somewhat ad hoc. Hence, he was led to ask, Are there some underlying processes that can explain the tendencies and the conditions under which they become manifest? This question inspired him, along with various collaborators, to develop what he felt were true explanatory propositions that would stand at the top of a deductive system explaining the tendencies and the conditions under which they are activated.

These explanatory propositions bring Turner's work from the midcentury into the more contemporary theoretical time frame, but for ease of exposition, it is best to deal with these explanatory "laws" here. In their most recent reworking, Turner and Paul Colomy[16] posit three underlying processes that can explain the other propositions in Turner's developing theory: (1) functionality, (2) representation, and (3) tenability. We will briefly examined each of these.

Functionality

When activities are organized to meet explicit goals in an efficient and effective manner, then considerations of *functionality* are dominant. Such considerations are most likely when the organization of individuals to achieve goals can potentially involve conflicts of interest among the participants and when these participants must be recruited from diverse pools of individuals who differ in their abilities. Under these conditions, the differentiation and accretion of roles are organized in highly instrumental ways so that goals can be achieved within a minimum of conflict and friction.

Tenability

When roles form and differentiate in ways that allow individuals to gain personal rewards and gratifications, considerations of *tenability* are evident. Thus, as individuals calculate their costs and rewards, and indeed are encouraged by the organization of roles to do so, tenability dominates functionality.

Representation

When the accretion and differentiation of roles involve the embodiment of cultural values, considerations of *representation* are evident. Moreover, the salience of representation of roles increases, however, when roles are implicated in group conflict and when incumbents in roles are recruited from homogeneous pools of individuals. When representation dominates the organization of roles, tenability will become more important than functionality.

The propositions developed for functionality, tenability, and representation are more complex than this brief discussion, but the general intent is clear. These propositions serve as the higher-order "laws" in deductive systems that subsume the tendency propositions and statements on the conditions under which they hold true. Although this strategy is suggestive, Turner has never fully implemented it by subsuming all the tendency propositions under these three laws in tight deductive systems. Yet, unlike much role theorizing from the midcentury, Turner's approach recognized that generalizations about roles from empirical observations need to be organized in ways facilitating their explanation by more abstract principles.[17] For Turner, the relative amounts of functionality, tenability, and representation would provide the needed explanatory push.

NOTES

1. Jonathan H. Turner, "Role," *Blackwell Encyclopedia of 20th Century Social Thought* (Oxford: Blackwell, 1996).

2. See, for example, Ralph H. Turner, "Role-Taking: Processes versus Conformity," in *Human Behavior and Social Processes*, ed. A. Rose (Boston: Houghton Mifflin, 1962), pp. 20–40.

3. Ralph H. Turner, "Unanswered Questions in the Convergence between Structuralist and Interactionist Role Theories," in *Perspectives on Sociological Theory*, eds. S.N. Eisenstadt and H.J. Helle (London: Sage, 1985). In particular, this article addresses Warren Handel, "Normative Expectations and the Emergence of Meaning as Solutions to Problems: Convergence of Structural and Interactionist Views," *American Journal of Sociology* 84 (1979), pp. 855–881.

4. Turner, "Role-Taking: Processes versus Conformity" (cited in note 2).

5. Ibid.; "Social Roles: Sociological Aspects," *International Encyclopedia of the Social Sciences* (New York: Macmillan, 1968).

6. Ralph H. Turner, "The Normative Coherence of Folk Concepts," *Research Studies of the State College of Washington* 25 (1957).

7. Turner, "Role-Taking: Processes versus Conformity" (cited in note 2), p. 23.

8. Ralph H. Turner, "The Role and the Person," cited in *American Journal of Sociology* 84 (1978), pp. 1–23.

9. Turner, "Social Roles: Sociological Aspects" (cited in note 5). Turner has extensively analyzed this process of self-anchorage in roles. See, for example, Ralph H. Turner, "The Real Self: From Institution to Impulse," *American Journal of Sociology* 81 (1970), pp. 989–1016; Ralph H. Turner and Victoria Billings, "Social Context of Self-Feeling," in *The Self-Society Interface: Cognition, Emotion, and Action*, eds. J. Howard and P. Callero (Cambridge: Cambridge University Press, 1990); Ralph H. Turner and Steven Gordon, "The Boundaries of the Self: The Relationship of Authenticity to Inauthenticity in the Self-Conception," in *Self-Concept: Advances in Theory and Research*, eds. M. D. Lynch, A.A. Norem-Hebeisen, and Kenneth Gergen (Cambridge, MA: Ballinger, 1981).

10. Ralph H. Turner, "Rule Learning as Role Learning," *International Journal of Critical Sociology* 1 (1974), pp. 10–28.

11. Ralph H. Turner, "Strategy for Developing an Integrated Role Theory," *Humboldt Journal of Social Relations* 7 (1980), pp. 123–139 and "Role Theory as Theory," unpublished manuscript.

12. Turner, "Strategy" (cited in note 11), pp. 123–124.

13. Turner, "Social Roles: Sociological Aspects" (cited in note 5).

14. Turner, "Role and the Person" (cited in note 8).

15. In his most recent effort, "Strategy for Developing an Integrated Role Theory," Turner focused on "role allocation" and "role differentiation" because he views these as the two most critical tendencies. The example of "the person and role" results in the same explanatory principles. Thus, this illustration provides additional examples of Turner's strategy.

16. Ralph H. Turner and Paul Colomy, "Role Differentiation: Orienting Principles," *Advances in Group Processes* 5 (1987), pp. 1–47.

17. For representative examples of the variety of empirical research conducted by Turner, see "The Navy Disbursing Officer as a Bureaucrat," *American Sociological Review* 12 (1947), pp. 342–348; "Moral Judgment: A Study in Roles," *American Sociological Review* 17 (1952), pp. 70–77; "Occupational Patterns of Inequality," *American Journal of Sociology* 50 (1954), pp. 437–447; "Zoot-Suiters and Mexicans: Symbols in Crowd Behavior" (with S.J. Surace), *American Journal of Sociology* 62 (1956), pp. 14–20; "The Changing Ideology of Success: A Study of Aspirations of High School Men in Los Angeles," *Transactions of the Third World Congress of Sociology* 5 (1956), pp. 35–44; "An Experiment in Modification of the Role Conceptions," *Yearbook of the American Philosophical Society* (1959), pp. 329–332; "Some Family Determinants of Ambition," *Sociology and Social Research* 46 (1962), pp. 397–411; *The Social Context of Ambition* (San Francisco: Chandler & Sharp, 1964); and "Ambiguity and Interchangeability in Role Attribution" (with Norma Shosid), *American Sociological Review* 41 (1976); and "The True Self Method for Studying Self-Conception," *Symbolic Interaction* 4 (1981), pp. 1–20.

30

The Maturing Tradition IV:

Erving Goffman's Dramaturgical Approach

Erving Goffman was, perhaps, the most creative theorist of interaction processes in the second half of the twentieth century. In one of his last statements before his death in 1982, he defined the analysis of face-to-face interaction as the *interaction order,*[1] but unlike many micro-level theorists and researchers, Goffman did not proclaim that the interaction order is all that is real. Rather, he simply argued that this interaction order constitutes a distinctive realm of reality that reveals its own unique dynamics. For, "to speak of the relative autonomous forms of life in the interaction order . . . is not to put forward these forms as somehow prior, fundamental, or constitutive of the shape of macroscopic phenomena."[2] Goffman recognized that, at best, there is a "loose coupling" of the micro and macro realms. Macro phenomena, such as commodities markets, urban land-use values, economic growth, and societywide stratification, cannot be explained by micro-level analysis.[3] Of course, one can supplement macro-level explanations by recording how individuals interact in various types of settings and encounters, but these analyses will not supplant macro-level explanations. Conversely, what transpires in the interaction order cannot be explained solely by macroprocesses. Rather, macro-level phenomena are always transformed in ways unique to the individuals involved in interaction.

To be sure, macro phenomena constrain and circumscribe interaction and, at times, guide the general form of the interaction, but the inherent dynamics of the interaction itself preclude a one-to-one relation to these structural parameters. Indeed, the form of interaction can often be at odds with macrostructures, operating smoothly in ways that contradict these structures without dramatically changing them. Thus, the crude notion that interaction is constrained by macrostructures in ways that "reproduce" social structure does not recognize the autonomy of the interaction order. For interaction "is not an expression *of* structural arrangements in any simple sense; at best it is an expression advanced *in regard to* these arrangements. Social structures don't 'determine' culturally standard displays (of interaction rituals), they merely help select from the available repertoire of them."[4] Thus, there is a "loose coupling" of "interactional practices and social structures, a collapsing of strata and structures into broader categories, the categories themselves not corresponding one-to-one to anything in the structural world...."[5] There is, then, "a set of transformation rules, or a membrane selecting how various externally relevant social distinctions will be managed within the interaction."[6]

These "transformations" are, however, far from insignificant phenomena. Much of what gives the social world a sense of "being real" arises from the practices of individuals as they deal with one another in various situations. "We owe our unshaking sense of realities,"[7] to the rules of interpersonal contact. Moreover, although a single gathering and episode of interaction might not have great social significance, "through these comings together much of our social life is organized."[8] The interaction order is thus a central topic of sociological theory.

Goffman's approach to this domain is unique.[9] Although we must "keep faith with the spirit of natural science, and lurch along, seriously kidding ourselves that our rut has a forward direction,"[10] we must not become over-enamored with the mature sciences. Social life is "ours to study naturalistically," but our study should not be rigid.[11] Instead, ad hoc observation, cultivation of anecdotes, creative thinking, illustrations from literature, examination of books of etiquette, personal experiences, and many other sources of unsystematic data should guide inquiry into the micro order. Indeed, "human life is only a small irregular scab on the face of nature, not particularly amenable to deep systematic analysis."[12] Yet human life can be studied in the spirit of scientific inquiry.

THE DRAMATURGICAL METAPHOR

In both his first and last major works—*The Presentation of Self in Everyday Life* and *Frame Analysis,* respectively—Goffman analogized to the stage and theater. Hence the designation of his work as "dramaturgical" has become commonplace.[13] This designation is, however, somewhat misleading because it creates the impression that there is a script, a stage, an audience, props, and actors playing roles. Such imagery is more in tune with structural role theory (discussed in Chapter 29). Dramaturgy is used to denote Goffman's work in another sense: Individuals are actors who "put on" a performance, often cynical and deceptive, for one another and who manipulate the script, stage, props, and roles for their own purposes. This more cynical view of Goffman's dramaturgy is perhaps closer to the mark,[14] but it too is somewhat misleading. Commentators have often portrayed Goffman as presenting a kind of "con man"[15] view of human social interaction—a metaphor that captures some of the examples and topics of his approach but obscures the more fundamental processes denoted by Goffman's theorizing.

We can use the metaphor of dramaturgy in a less extreme sense. In Goffman's work there is concern with a cultural script, or normative rules; there is a heavy emphasis on how individuals manage their impressions and play roles; there is a concern with stages and props (physical space and objects); there is an emphasis on stag*ing,* of the manipulation of gestures as well as spacing, props, and other physical aspects of a setting; there is a view of self as situational, determined more by the cultural script, stage, and audience than by enduring and transituational configurations of self-attitudes and self-feelings; and there is a particular emphasis on how performances create a theatrical ambiance—a mood, definition, and sense of reality.

This metaphor provides only an orientation of Goffman's approach. We need to "fill in" this orientation with more details. To do so, we will review Goffman's most important works and try to pull the diverse vocabularies and concepts together into a more unified theoretical perspective—dramaturgical in its general contours but more than just a clever metaphor.[16]

THE PRESENTATION OF SELF

The Presentation of Self in Everyday Life[17] was Goffman's first major work and was largely responsible for the designation of Goffman as a dramaturgical theorist. The basic argument is that individuals deliberately "give" and inadvertently "give off" signs that provide others with information about how to respond. From such mutual use of "sign-vehicles," individuals develop a "definition of the situation," which is a "plan for cooperative activity," but which, at the same time, is "not so much a real agreement as to what exists but rather a real agreement as to whose claims concerning what issues will be temporarily honored."[18] In constructing this overall definition of a situation, individuals engage in performances in which each orchestrates gestures to "present oneself" in a particular manner as a person having identifiable characteristics and deserving of treatment in a certain fashion. These performances revolve around several interrelated dynamics.

First, a performance involves the creation of a *front*. A front includes the physical "setting" and the use of the physical layout, its fixed equipment like furniture and other "stage props" to create a certain impression. A front also involves (a) "items of expressive equipment" (emotions, energy, and other capacities for expression); (b) "appearance," or those signs that tell others of an individual's social position and status as well as the "ritual state" of the individual with respect to social, work, or recreational activity;[19] and (c) "manner," or those signs that inform others about the role that an individual expects to play.[20] As a general rule, people expect consistency in these elements of their fronts—use of setting and its props, mobilization of expressive equipment, social status, expression of ritual readiness for various types of activity, and efforts to assume certain roles.[21] There is a relatively small number of fronts, and people know them all. Moreover, fronts tend to be established, institutionalized, and stereotypical for various kinds of settings, with the result that "when an actor takes an established role, usually he (she) finds that a particular front has already been established for it."[22]

Second, in addition to presenting a front, individuals use gestures in what Goffman termed *dramatic realization,* or the infusion into activity of signs that highlight commitment to a given definition of a situation. The more a situation creates problems in presenting a front, Goffman argued, the greater will be efforts at dramatic realization.[23]

Third, performances also involve *idealizations,* or efforts to present oneself in ways that "incorporate and exemplify the officially accredited values of society."[24] When individuals are mobile, moving into a new setting, efforts at idealization will be most pronounced. Idealization creates a problem for individuals, however: If the idealization is to be effective, individuals must suppress, conceal, and underplay those elements of themselves that might contradict more general values.

Fourth, such efforts at concealment are part of a more general process of *maintaining expressive control.* Because minor cues and signs are read by others and contribute to a definition of a situation, actors must regulate their muscular activity, their signals of involvement, their orchestration of front, and their ability to be fit for interaction. The most picayune discrepancy between behavior and the definition of a situation can unsettle the interaction, because "the impression of reality fostered by a performance is a delicate, fragile thing that can be shattered by very minor mishaps."[25]

Fifth, individuals can also engage in *misrepresentation.* The eagerness of one's audience to read gestures and determine one's front makes that audience vulnerable to manipulation and duping.[26]

Sixth, individuals often attempt to engage in *mystification,* or the maintenance of distance from others as a way to keep them in awe and in conformity to a definition of a situation. Such mystification is, however, limited primarily to those of higher rank and status.

Seventh, individuals seek to make their performances seem *real* and to avoid communicating a sense of contrivance. Thus individuals must communicate, or at least appear to others, as sincere, natural, and spontaneous.

These procedures for bringing off a successful performance and thereby creating an overall definition of a situation are the core of Goffmanian sociology. They become elaborated and extended in subsequent works, but Goffman never abandoned the idea that fundamental to the interaction order are the efforts of individuals to orchestrate their performances, even in deceptive and manipulative ways, so as to maintain a particular definition of the situation. These ideas are presented propositionally in Table 30-1.

TABLE 30-1 Goffman's Propositions on Interaction and Performance

I. As individuals make visual and verbal contact, the more likely they are to use gestures to orchestrate a performance, with the success of this performance depending on
   A. Presentation of a coherent front that, in turn, is a positive and additive function of
      1. Control of physical space, props, and equipment in a setting.
      2. Control of expressive equipment in a setting.
      3. Control of signals marking propensity for types of ritual activity.
      4. Control of signals marking status outside and inside the interaction.
      5. Control of signals pertaining to identifiable roles.
   B. Incorporation and exemplification of general cultural values.
   C. Imbuing a situation with a personal mystique.
   D. Signaling sincerity.

II As individuals in a setting orchestrate their performances and, at the same time, accept one another's performances, they are likely to develop a common definition of the situation.

III. As a common definition of the situation emerges, the ease of the interaction increases.

Although the propositions in Table 30-1 constitute only an opening chapter in *The Presentation of Self,* they are by far the most enduring portions of this first major work. The rest of the book is concerned with performances sustained by more than one individual. Goffman introduced the concept of *team* to denote performances that are presented by individuals who must cooperate to effect a particular definition of the situation. Often two teams must present performances to each other, but more typically one team constitutes a performer and the other an audience. Team performers generally move between a *front region,* or frontstage, where they coordinate their performances before an audience, and a *back region,* or backstage, where team members can relax. Goffman also introduced the notion of *outside,* or the residual region beyond the frontstage and backstage. Frontstage behavior is polite, maintaining a decorum appropriate to a team performance (for example, selling cars, serving food, meeting students, and so on), whereas backstage behavior is more informal and is geared toward maintaining the solidarity and morale of team performers. When outsiders or members of the audience intrude on performers in the backstage, a tension is created because team members are caught in their nonperforming roles.[27]

A basic problem of all team performances is maintaining a particular definition of the situation in front of the audience. This problem is accentuated when there are large rank or status differences among team members,[28] when the team has many members,[29] when the front- and backstages are not clearly partitioned, and when the team must hide information contrary to its image of itself. To counteract these kinds of problems, social control among the team's members is essential. When members are backstage, such control is achieved through morale-boosting activities, such as denigrating the audience, kidding one another, shifting to informal address, and engaging in stage talk (talk about performances on frontstage). When they are onstage, control is sustained by realigning actions revolving around subtle communications among team members that, hopefully, the audience will not understand.

Breaches of the performances occur when a team member acts in ways that challenge the definition of the situation created by the team's performances. Attempts to prevent such incidents involve further efforts at social control, especially a backstage emphasis on (a) playing one's part and not emitting unmeant gestures, (b) showing loyalty to the team and not the audience, and (c) exercising foresight and anticipating potential problems with the team or the audience. Team members are assisted in social control by members of the audience who (a) tend to stay away from the backstage, (b) act disinterested when exposed to backstage behavior, and (c) employ elaborate etiquette (exhibiting proper attention and interest, inhibiting their own potential performances, avoiding faux pas) to avoid a "scene" with the team.

What is true of teams and audiences is, Goffman implied, also true of individuals. Interaction involves a performance for others who constitute an audience. One seeks to sustain a performance when moving to the frontstage and relax a front when moving to a backstage region. People try very hard to avoid mistakes and faux pas that could breach the definition of the situation, and they are assisted in this effort by others in their audience who exercise tact and etiquette to avoid a scene. Such are the themes of *The Presentation of Self,* and most of Goffman's work represented a conceptual elaboration of them. The notion of "teams" recedes, but general model elaborating these themes into a theory of interaction among individuals emerges.

FOCUSED INTERACTION

Goffman generally employed the terms *unfocused* and *focused* to denote two basic types of interaction. *Unfocused interaction* "consists of interpersonal communications that result solely by virtue of persons being in one another's presence, as when two strangers across the room from each other check up on each other's clothing, posture, and general manner, while each modifies his (her) own demeanor because he himself is under observation."[30] As will be explored later, such unfocused interaction is, Goffman argued, an important part of the interaction order, for much of what people do is exchange glances and monitor each other in public places. *Focused interaction,* in contrast, "occurs when people effectively agree to sustain for a time a single focus of cognitive and visual attention, as in a conversation, a board game, or a joint task sustained by a close face-to-face circle of contributors."[31]

Encounters

Focused interaction occurs within what Goffman termed *encounters,* which constitute one of the core structural units of the interaction order. Goffman mentioned encounters in his first work, *The Presentation of Self in Everyday Life,* but their full dimensions are explored in his next book, *Encounters.*[32] There, an encounter is defined as focused interaction revealing the following characteristics:[33]

1. A single visual and cognitive focus of attention.
2. A mutual and preferential openness to verbal communication
3. A heightened mutual relevance of acts
4. An eye-to-eye ecological huddle, maximizing mutual perception and monitoring
5. An emergent "we" feeling of solidarity and flow of feeling
6. A ritual and ceremonial punctuation of openings, closings, entrances, and exits
7. A set of procedures for corrective compensation for deviant acts

To sustain itself, an encounter develops a "membrane," or penetrable barrier to the larger social world in which the interaction is located. Goffman typically conceptualized the immediate setting of an encounter as a *gathering,* or the assembling in space of co-present individuals; in turn, gatherings are lodged within a more inclusive unit, the *social occasion,* or the larger undertaking sustained by fixed equipment, distinctive ethos and emotional structure, program an agenda, rules of proper and improper conduct, and preestablished sequencing of activities (beginning, phases, high point, and ending). Thus encounters emerge from episodes of focused interaction within gatherings that are lodged in social occasions.[34]

The membrane of an encounter, as well as its distinctive characteristics previously listed, are sustained by a set of *rules.* In *Encounters,* Goffman lists several; later, in what is probably his most significant work, *Interaction Ritual,* he lists several more.[35] Let us combine both discussions by listing the rules that guide focused interaction in encounters:

1. *Rules of irrelevance,* which "frame" a situation as excluding certain materials (attributes of participants, psychological states, cultural values and norms, etc.).[36]
2. *Rules of transformation,* which specify how materials moving through the membrane created by rules of irrelevance are to be altered to fit into the interaction.
3. *Rules of "realized resources,"* which provide a general schemata and framework for expression and interpretation for activities among participants.
4. *Rules of talk,* which are the procedures, conventions, and practices guiding the flow of verbalizations with respect to[37]
    a. Maintaining a single focus of attention.
    b. Establishing "clearance cues" for determining when one speaker is done and another can begin.
    c. Determining how long and how frequently any one person can hold the floor.
    d. Regulating interruptions and lulls in the conversation.
    e. Sanctioning participants whose attention wanders to matters outside the conversation.
    f. Ensuring that nearby people do not interfere with the conversation.
    g. Guiding the use of politeness and tact, even in the face of disagreements.

5. *Rules of self respect,* which encourage participants to honor with tact and etiquette their respective efforts to present themselves in a certain light.

Interaction is thus guided by complex configurations of rules that individuals learn how to use and apply in different types of encounters, logged in varying types of gatherings and social occasions. The "reality" of the world is, to a very great extent, sustained by people's ability to invoke and use these rules,[38] when these rules are operating effectively, individuals develop a "state of euphoria," or what Randall Collins has termed enhanced "emotional energy." However, encounters are vulnerable to "dysphoria" or tension when these rules do not exclude troublesome external materials or fail to regulate the flow of interaction. Such failures are seen by Goffman as *incidents* or *breaches;* when they can be effectively handled by tact and corrective procedures, they are then viewed as *integrations* because they are blended into the ongoing encounter. The key mechanism for avoiding dysphoria and maintaining the integration of the encounter is the use of ritual.

Ritual

In *Interaction Ritual,*[39] Goffman's great contribution is the recognition that minor, seemingly trivial, and everyday rituals—such as "Hello, how are you?" "Good morning," "Please, after you," and so on—are crucial to the maintenance of social order. In his words, he "reformulated Émile Durkheim's social psychology in a modern dress"[40] by recognizing that, when individuals gather and begin to interact, their behaviors are highly ritualized. That is, actors punctuate each phase of interpersonal contact with stereotypical sequences of behavior that invoke the rules of the encounter and, at the same time, become the medium or vehicle by which the rules are followed. Rituals are thus essential for (a) mobilizing individuals to participate in interaction; (b) making them cognizant of the relevant rules of irrelevance, transformation, resource use, and talk; (c) guiding them during the course of the interaction; and (d) helping them correct for breaches and incidents.

Among the most significant are those rituals revolving around deference and demeanor. *Deference* pertains to interpersonal rituals that express individuals' respect for others, their willingness to interact, their affection and other emotions, and their engagement in the encounter. In Goffman's words, deference establishes "marks of devotion" by which an actor "celebrates and confirms his (her) relationship to a recipient."[41] As a result, deference contains a "kind of promise, expressing in truncated form the actor's avowal and pledge to treat the recipient in a particular way in the on-coming activity."[42] Thus, seemingly innocuous gestures—"It's nice to see you again," "How are things?" "What are you doing?" "Good bye," "See you later," and many other stereotypical phrases as well as bodily movements—are rituals that present a demeanor invoking relevant rules and guiding the opening, sequencing, and closing of the interaction.

Deference rituals, Goffman argued, can be of two types: (1) avoidance rituals and (2) presentational rituals. *Avoidance rituals* are those that an individual uses to keep distance from another and to avoid violating the "ideal sphere" that lies around the other. Such rituals are most typical among unequals. *Presentational rituals* "encompass acts through which the individual makes specific attestations to recipients concerning how he regards them and how he will treat them in the on-coming interaction."[43] Goffman saw interaction as constantly involving a dialectic between avoidance and presentational rituals as individuals respect each other and maintain distance while trying to make contact and get things done.[44]

In contrast, *demeanor* is "that element of the individual's ceremonial behavior conveyed through deportment, dress, and bearing which serves to (inform) those in his immediate presence that he is a person of certain desirable or undesirable qualities."[45] Through demeanor rituals individuals present images of themselves to others and, at the same time, communicate that they are reliable, trustworthy, and tactful.

Thus, through deference and demeanor rituals individuals plug themselves into an encounter by invoking relevant rules and demonstrating their capacity to follow them, while indicating their respect for others and presenting themselves as certain kinds of individuals. The enactment of such deference and demeanor rituals in concrete gatherings, especially encounters but also including unfocused situations, provides a basis for the integration of society. For "throughout . . . ceremonial obligations and expectations, a constant flow of indulgences is spread through society, with others who are present constantly reminding the individual that he must keep himself together as a well demeaned person and affirm the sacred quality of these others."[46]

Roles

In presenting themselves to others, individuals also seek to play a particular role. Thus, as people present a front, invoke relevant rules, and emit rituals, they also try to orchestrate a role for themselves. For Goffman, then, a *role* is "a bundle of activities visibly performed before a set of others and visibly meshed into the activities these others perform."[47] In the terms of Ralph Turner's analysis (see Chapter 29), individuals attempt to "make a role" for themselves; if successful, this effort contributes to the overall definition of the situation.

In trying to establish a role, the individual "must see to it that the impressions of him that are conveyed in the situation are compatible with role-appropriate personal qualities effectively imputed to him."[48] Thus, individuals in a situation are expected to try to make roles for themselves that are consistent with their demeanor, their self as performed before others, and their front (stage props, expressive equipment, appearance). And, if the inconsistency between the attempted role and these additional aspects of a performance becomes evident, then others in the situation are likely to sanction the individual through subtle cues and gestures. These others are driven to do so because discrepancy between another's role and other performance cues disrupts the definition of the situation and the underlying sense of reality that this definition promotes. Thus, role is contingent on the responses and reactions of others, and, because their sense of reality partially depends on successful and appropriate role assumption, an individual will have difficulty changing a role in a situation once it is establish.

Often, however, people perceive a role to be incompatible with their image of themselves in a situation. Under these conditions they will display what Goffman termed *role distance,* whereby a "separation" of the person from a role is communicated. Such distancing, Goffman argued,[49] allows the individual to (a) release the tension associated with a role considered to be "beneath his (her) dignity," (b) present additional aspects of self that extend beyond the role, and (c) remove the burden of "complete compliance to the role," thereby making minor transgressions less dramatic and troublesome for others.

Role distance is but an extreme response to the more general process of *role embracement.* For any role, individuals will reveal varying degrees of attachment and involvement in the role. One extreme is role distance, whereas the other extreme is what

Goffman termed *engrossment,* or complete involvement in a role. In general, Goffman argued,[50] those roles in which individuals can direct what is going on are likely to involve high degrees of embracement, whereas those roles in which the individual is subordinate will be played with considerable role distance.

As is evident, then, the assumption of a role is connected to the self-image that actors project in their performance. Although the self that one reveals in a situation depends on the responses of others who can confirm or disconfirm that person's self in a situation, the organization of a performance onstage before others is still greatly circumscribed by self.

Self

Goffman's view of self is highly situational and contingent on the responses of others. Although one of the main activities of actors in a situation is to present themselves in a certain way, Goffman was highly skeptical about a "core" or "transituational" self-conception that is part of an individual's "personality." In almost all his works he took care to emphasize that individuals do not have an underlying "personality" or "identity" that is carried from situation to situation. For example, in his last major book, *Frame Analysis,*[51] he argued that people in interaction often presume that the presented self provides a glimpse at a more coherent and core self, but in reality this is simply a folk presumption because there is "no reason to think that all these gleanings about himself that an individual makes available, all these pointings from his current situation to the way he is in his other occasions, have anything very much in common."[52]

Yet, even though there is no transituational or core self, people's efforts to present images of themselves in a particular situation and others' reactions to this presentation are central dynamics in all encounters. Individuals constantly emit demeanor cues that project images of themselves as certain kinds of persons, or, in the vocabulary of *The Presentation of Self,* they engage in "performances." In *Interaction Ritual,* Goffman rephrased this argument somewhat, and, in so doing, he refined his views on self. In encounters, an individual acts out *a line,* which is "a pattern of verbal and nonverbal acts by which he expresses his view of the situation and, through this, his evaluation of the participants, especially himself."[53] In developing a line, an individual presents *a face,* which is "the positive social value a person effectively claims for himself by the line others assume he has during a particular contact."[54] Individuals seek to stay *in face* or to *maintain face* by presenting an image of themselves through their line that is supported by the responses of others and, if possible, sustained by impersonal agencies in a situation. Conversely, a person is *in wrong face* or *out of face* when the line emitted is inappropriate and unaccepted by others. Thus, although a person's social face "can be his most personal possession and the center of his security and pleasure, it is only on loan to him from society; it will be withdrawn unless he conducts himself in a way that is worthy of it. . . ."[55]

As noted earlier, Goffman argued that a key norm in any encounter is "the rule of self respect," which requires individuals to maintain face and, through tact and etiquette, the face of others. Thus, by virtue of tact or the "language of hint ..., innuendo, ambiguities, well-placed phrases, carefully worked jokes, and so on,"[56] individuals sustain each other's face; in so doing, they confirm the definition of the situation and promote a sense of a common reality. For this reason, a given line and face in an encounter are difficult to change, once established, because to alter face (and the line by which it is presented) would require redefining the situation and recreating a sense of reality. And, because face is "on loan" to a person from the responses of others, the individual must incur high costs—such as embarrassment or a breach of the situation—to alter a line and face.

Face engagements are usually initiated with eye contact, and, once initiated, they involve ritual openings appropriate to the situation (as determined by length of last engagement, amount of time since previous engagement, level of inequality, and so forth). During the course of the face engagement, each individual uses tack to maintain, if possible, each other's face and to sanction, if necessary, each other into their appropriate line. In particular, participants seek to avoid "a scene" or breach in the situation, so they use tact and etiquette to save their own face and that of others. Moreover, as deemed appropriate for the type of encounter (as well as for the larger gathering and more inclusive social occasion), individuals will attempt to maintain what Goffman sometimes termed *the territories of self,* revolving around such matters as physical props, ecological space, personal preserve (territory around one's body), and conversational rights (to talk and be heard), which are necessary for people to execute their line and maintain face.[57] In general, the higher the rank of individuals, the greater their territories of self in an encounter.[58] To violate such territories disrupts or breaches the situation, forcing remedial action by participants to restore their respective lines, face, definitions of the situation, and sense of reality.

Talk

Throughout his work, but especially in later books such as *Frame Analysis*[59] and in numerous essays (see those collected in *Forms of Talk*[60]), Goffman emphasized the significance of verbalizations for focusing people's attention. When "talk" is viewed interactionally, "it is an example of that arrangement by which individuals come together and sustain matters having a ratified, joint, current, and running claim upon attention, a claim which lodges them together in some sort of intersubjective, mental world."[61] Thus, in Goffman's view, "no resource is more effective as a basis for joint involvement than speaking" because it fetches "speaker and hearer into the same interpretation schema that applies to what is thus attended."[62]

Talk is thus a crucial mechanism for drawing individuals together, focusing their attention, and adjudicating an overall definition of the situation. Because talk is so central to focusing interaction, it is normatively regulated and ritualized. One significant norm is the prohibition against *self-talk,* because, when people talk to themselves, it "warns others that they might be wrong in assuming a jointly maintained base of ready mutual intelligibility. . . ."[63] Moreover, other kinds of quasi talk are also regulated and ritualized. For example, *response cues* or "exclamatory interjections which are not full-fledged words"—"Oops," "Wow," "Oh," and "Yikes"—are regulated as to when they can be used and the way they are uttered.[64] Verbal fillers—"ah," "uh," "um," and the like—are also ritualized and are used to facilitate "conversational tracking." In essence, they indicate that "the speaker does not have, as of yet, the proper word but is working on the matter" and that he or she is still engaged in the conversation. Even seemingly emotional cues and tabooed expressions, such as all the "four-letter words," are not so much an expression of emotion as "self-other alignment" and assert that "our inner concerns should be theirs." Such outbursts are normative and ritualized because this "invitation into our interiors tends to be made only when it will be easy to other persons present to see where the voyage takes them."[65]

In creating a definition of the situation, Goffman argued, talk operates in extremely complex ways. When individuals talk, they create what Goffman termed a *footing,* or assumed foundation for the conversation and the interaction. Because verbal symbols are easily manipulated, people can readily change the footing or basic premises underlying the conversation.[66] Such shifts in footing are, however, highly ritualized and usually reveal clear markers. For example, when a person says something like "Let's not talk about that," the footing of the conversation is shifted, but in a ritualized way; similarly, when someone utters a phrase like "That's great, but what about ...?" this person is also changing the footing through ritual.

Shifts in footing raise a question that increasingly dominated Goffman's later works: the issue of *embedding*. Goffman came to recognize that conversations are *layered* and, hence, embedded in different footings. There are often multiple footings for talk, as when someone "says one thing but means another" or when a person "hints" or "implies" something else. These "layerings" of conversations, which embed them in different contexts, are possible because speech is capable of generating subtle and complex meanings. For example, irony, sarcasm, puns, wit, double-entendres, inflections, shadings, and other manipulations of speech demonstrate the capacity of individuals to shift footings and contextual embeddings of a conversation (for example, think of a conversation in a work setting involving romantic flirtations; it will involve constant movement in footing and context). Yet, for encounters to proceed smoothly, these alterations in footing are, to some extent, normatively regulated and ritualized, enabling individuals to sustain a sense of common reality.

Talk is thus a critical dimension of focused interaction. Without it the gestures and cues that people can emit are limited and lack the subtlety and complexity of language. And, as Goffman began to explore this complexity in later works, earlier notions about "definitions of situations" seemed too crude because people could construct multiple, as well as subtly layered, definitions of any situation—an issue that we will explore near the end of this chapter. For our purposes here, the critical point is that talk focuses attention and pulls actors together, forcing their interaction on a face-to-face basis. But, despite the complexity of how this focusing can be done, talk is still normatively and ritually regulated in ways that produce a sense of shared reality for individuals.[67]

Table 30–2 summarizes Goffman's analysis of focused encounters. Focused encounters occur when social occasions put people in face-to-face contact; and the viability of the encounter depends on rules, rituals, and the capacity of individuals to present acceptable performances. The statements under I and II reveal the basic argument in Goffman's approach.

Disruption and Repair in Focused Interaction

Goffman stressed that disruption in encounters is never a trivial matter:[68]

> Social encounters differ a great deal in the importance that participants give to them but, whether crucial or picayune, all encounters represent occasions when the individual can become spontaneously involved in the proceedings and derive from this a firm sense of reality. And this feeling is not a trivial thing, regardless of the package in which it comes. When an incident occurs . . . then the reality is threatened. Unless the disturbance is checked, unless the interactants regain their proper involvement, the illusion of reality will be shattered. . . .

When a person emits gestures that contradict normative roles, present a contradictory front, fail to enact appropriate rituals, seek an inappropriate role, attempt a normatively or ritually incorrect line, or present a wrong face, there is potential for *a scene*. From the person's point of view, there is a possibility of

TABLE 30-2 Goffman's General Proposition on Focused Interaction

I. Encounters are created when
  A. Social occasions put individuals in physical proximity.
  B. Gatherings allow face-to-face contact revolving around talk.
II. The viability of an encounter is a positive and multiplicative function of
  A. The availability of relevant normative rules to guide participants with respect to such issues as
    1. Irrelevance, or excluded matters.
    2. Transformation, or how external matters are to be incorporated.
    3. Resource use, or what local resources are to be drawn from.
    4. Talk, or how verbalizations are to be ordered.
    5. Self-respect, or the maintenance of lines and face.
  B. The availability of ritual practices that can be used to
    1. Regulate talk and conversation.
    2. Express appropriate deference and demeanor.
    3. Invoke and punctuate normative rules.
    4. Repair breaches to the interaction.
  C. The capacity of individuals to present acceptable performances with respect to
    1. Lines, or directions of conduct.
    2. Roles, or specific clusters of rights and duties.
    3. Faces, or specific presentations of personal characteristics.
    4. Self, or particular images of oneself.

embarrassment, to use Goffman's favorite phrase; once embarrassed, an individual's responses can further degenerate in an escalating cycle of ever greater levels of embarrassment. From the perspective of others, a scene disrupts the definition of the situation and threatens the sense of reality necessary for them to feel comfortable. Individuals implicitly assume that people are reliable and trustworthy, that they are what they appear to be, that they are competent, and that they can be relied on; thus, when a scene occurs, these implicit assumptions are challenged and threaten the organization of the encounter (and, potentially, the larger gathering and social occasion).

For this reason an individual will seek to repair a scene caused by the use of inappropriate gestures, and others will use tact to assist the individual in such repair efforts. The sense of order of a situation is thus sustained by a variety of corrective responses by individuals and by the willingness of others to use tact in ignoring minor mistakes and, if this is not possible, to employ tact to facilitate an offending individual's corrective efforts. People "disattend" much potentially discrepant behavior, and, when this is no longer an option, they are prepared to accept apologies, accounts, new information, excuses and other ritually and normatively appropriate efforts at repair. Of course, this willingness to accept people as they are, to assume their competence, and to overlook minor interpersonal mistakes makes them vulnerable to manipulation and deceit.

UNFOCUSED INTERACTION

Goffman was one of the few sociologists to recognize that behavior and interaction in public places, or in unfocused settings, are important features of the interaction order and, by extension, of social organization in general. Such simple acts as walking down the street, standing in line, sitting in a waiting room or on a park bench, standing in an elevator, going to and from a public restroom, and many other activities represent a significant realm of social organization. These unfocused situations in which people are co-present but not involved in prolonged talk and "face encounters" represent a crucial topic of sociological inquiry—a topic that is often seen as trivial but that embraces much of peoples time and attention. In two works, *Relations in Public* and *Behavior in Public Places,* Goffman explored the dynamics of unfocused gatherings.[69]

Unfocused gatherings are like focused interactions in their general contours: They are normatively regulated; they call for performances by individuals; they include the presentation of a self; they involve the use of rituals; they have normatively and ritually appropriate procedures for repair; and they depend on a considerable amount of etiquette, tact, and inattention. Let us explore each of these features in somewhat greater detail.

Much like a focused interaction, unfocused gatherings involve normative rules concerning spacing, movement, positioning, listening, talking, and self-presentation. But, unlike focused interaction, norms do not have to sustain a well-defined membrane. There is no closure, intense focus of attention, or face-to-face obligations in unfocused encounters. Rather, rules pertain to how individuals are to comport themselves *without* becoming the focus of attention and involved in a face encounter. Rules are thus about how to move, talk, sit, stand, present self, apologize, and perform other actions necessary to sustain public order without creating a situation requiring the additional interpersonal "work" of focused interaction.

When in public, individuals still engage in performances, but, because the audience is not involved in a face engagement or prolonged tracks of talk, the presentation can be more muted and less animated. Goffman used a variety of terms to describe these presentations, two of the most frequent being *body idiom*[70] and *body gloss.*[71] Both terms denote the overall configuration of gestures, or demeanor, that an individual makes available and gleanable to others. (Conversely, others are constantly *scanning* to determine the content of others' body idiom and body gloss.) Such demeanor denotes a person's direction, speed, resoluteness, purpose, and other aspects of a course of action. In *Relations in Public,* Goffman enumerated three types of body gloss:[72] (1) *orientation gloss,* or gestures giving evidence to others confirming that a person is engaged in a recognizable and appropriate activity in the present time and place; (2) *circumspection gloss,* or gestures indicating to others that a person is not going to encroach on or threaten the activity of others; and (3) *overplay gloss,* or gestures signaling that a person is not constrained or under duress and is, therefore, fully in charge and control of his or her other movements and actions. Thus the public performance of an individual in unfocused interaction revolves around providing information that one is of "sound character and reasonable competency."[73]

In public and during unfocused interactions, the *territories of self* become an important consideration. Goffman listed various kinds of territorial considerations that can become salient during unfocused interaction, including[74] (a) fixed geographical spaces attached to a particular person, (b) egocentric preserves of non-encroachment that surround individuals as they move in space, (c) personal spaces that others are not to violate under any circumstances, (d) stalls or bounded places that an individual can temporarily claim, (c) use spaces that can be claimed as an individual engages in some instrumental activity, (f) turns or the claimed order of doing or receiving something relative to others in a situation, (g) possessional territory or objects identified with self and arrayed around an individual's body, (h) informational preserve or the body of facts about a person that is controlled and regulated, and (i) conversational preserve or the right to control who can summon and talk to an individual. Depending on the type of unfocused interaction, as well as on the number, age, sex, rank, position, and other characteristics of the participants, the territories of self will vary, but in all societies there are clearly understood norms about which configuration of these territories is relevant, and to what degree it can be invoked.

These territories of self are made visible through what Goffman termed *markers*. Markers are signals and objects that denote the type of territorial claim, its extent and boundary, and its duration. Violation of these markers involves an encroachment on a person's self and invites sanctioning, perhaps creating a breach or scene in the public order. Indeed, seemingly innocent acts—such as inadvertently taking someone's place, butting in line, cutting someone off, and the like—can become a violation or befoulment of another's self and, as a result, invite an extreme reaction. Thus, social organization in general depends on the capacity of individuals to read those markers that establish their territories of self in public situations.

Violations of norms and territories create breaches and potential scenes, even when individuals are not engaged in focused interaction. These are usually repaired through ritual activity, such as (a) *accounts* explaining why a transgression has occurred (ignorance, unusual circumstances, temporary incompetence, unmindfulness, and so on), (b) *apologies* (some combination of expressed embarrassment or chagrin, clarification that the proper conduct is known and understood, disavowal and rejection of one's behavior, penance, volunteering of restitution, and so forth), and (c) *requests,* or a preemptive asking for license to do something that might otherwise be considered a violation of a norm or a person's self.[75] The use of these ritualized forms of repair sustains the positioning, movement, and smooth flow of activity among people in unfocused situations; without these repair rituals, tempers would flair and other disruptive acts would overwhelm the public order.

The significance of ritualized responses for repair only highlights the importance of ritual in general for unfocused interaction. As individuals move about, stand, sit, and engage in other acts in public, these activities are punctuated with rituals, especially as people come close to contact with each other. Nods, smiles, hand gestures, bodily movements, and, if necessary, brief episodes of talk (especially during repairs) are all highly ritualized, involving stereotyped sequences of behavior that reinforce norms and signal individuals' willingness to get along with and accommodate each other.

In addition to ritual, much unfocused interaction involves tact and inattention. By simply ignoring or quietly tolerating small breaches of norms, self, and ritual practices, people can gather and move about without undue tension and acrimony. In this way, unfocused interactions are made to seem uneventful, enabling individuals to cultivate a sense of obdurate reality in the subtle glances, nods, momentary eye contact, shifting of direction, and other acts of public life.

In Table 30-3, the key propositions implied in Goffman's discussion are enumerated. The relationships among the variables listed in the propositions are multiplicative in that each one accelerates the effects of the other in maintaining order in public situations of unfocused gatherings. This interactive effect allows public order to be sustained without reliance on focused talk and conversation.

FRAMES AND THE ORGANIZATION OF EXPERIENCE

Goffman's last major work, *Frame Analysis: An Essay on the Organization of Experience,*[76] is

TABLE 30-3 Goffman's General Propositions on Unfocused Interaction

I. Order in unfocused interaction is a positive and multiplicative function of
   A. The clarity of normative rules regulating behavior in ways that limit face encounters and talk.
   B. The capacity of individuals to provide demeanor cues with respect to
      1. Orientation, or the appropriateness of activities at the present time and place.
      2. Circumspection, or the willingness to avoid encroachment on, and threat to, others.
      3. Overplays, or the capacity to signal that one can control and regulate conduct without duress and constraint.
   C. The capacity of individuals to signal with clear markers those configurations of normatively appropriate territories of self with respect to
      1. Fixed geographical spaces that can be claimed.
      2. Egocentric preserves of nonencroachment that can be claimed during movement in space.
      3. Personal spaces that can be claimed.
      4. Stalls of territory that can be temporarily used.
      5. Use spaces that can be occupied for instrumental purposes.
      6. Turns of performing or receiving goods that can be claimed.
      7. Possessional territory and objects identified with, and arrayed around, self.
      8. Informational preserves that can be used to regulate facts about individuals.
      9. Conversational preserves that can be invoked to control talk.
   D. The availability of configurations of normatively appropriate repair rituals revolving around
      1. Accounts, or explanations for transgressions.
      2. Apologies, or expressions of embarrassment, regret, and penance for mistakes.
      3. Requests, or redemptive inquiries about making a potential transgression.
   E. The availability and clarity of rituals to reinforce norms and to order conduct by restricting face engagements among individuals.
   F. The availability of ritualized procedures for ignoring minor transgressions of norms and territories of self (tact and etiquette).

hardly an "essay" but, rather, an 800-page treatise on phenomenology, or the subjective organization of experience in social situations. It is a dense and rambling work, but it nonetheless returns to a feature of interaction that guided Goffman's work from the very beginning: the construction of "definitions of situations." That is, how is it that people define the reality of situations?

What Is a Frame?

The concept of *frame* appeared in Goffman's first major work, *The Presentation of Self,* and periodically thereafter. Surprisingly, he never offered a precise definition of this term, but the basic idea is that people "interpret" events or "strips of activity" in situations with a "schemata" that cognitively encircles or *frames* what is occurring.[77] The frame is much like a picture frame in that it marks off the boundary of the pictured events, encapsulating and distinguishing them from the surrounding environment. Goffman's early discussion of the "rules of irrelevance" in encounters—that is, considerations, characteristics, aspects, and events in the external world to be excluded during a focused interaction—represented an earlier way of communicating the dynamics of framing. Thus, as people look at the world, they impose a frame that defines what is to be pictured on the inside and what is to be excluded by what Goffman termed the *rim of the frame* on the outside. Human experience is organized by frames, which provide an interpretive "framework" or "frame of reference" for designating events or "strips of activity."

Primary Frames

Goffman argued that, ultimately, interpretations of events are anchored in a *primary framework,* which is a frame that does not depend on some prior interpretation of events.[78] Primary frameworks are thus anchored in the *real world,* at least from the point of view of the individual's organization of experience. People tend to distinguish, Goffman emphasized, between natural and social frameworks.[79] A *natural frame* is anchored in purely physical means for interpreting the world—body, ecology, terrain, objects, natural events, and the like. A *social frame* is lodged in the world created by acts of intelligence and social life. These two kinds of primary frameworks can vary enormously in their organization; some are clearly organized as "entities, postulates, and roles," whereas most others "provide" only a lore of understanding, an approach, a perspective."[80] All social frameworks involve rules about what is to be excluded beyond the rim of the frame, what is to be pictured inside the frame, and what is to be done when acting within the frame. Yet, as humans perceive and act, they are likely to apply several frameworks, giving the organization of experience a complexity that Goffman only alluded to in his earlier works.

Although Goffman stressed that he was not analyzing social structure with the concept of frames, he was clearly developing a neo-Durkheimian argument, recasting Durkheim's social psychology—especially the notion of how the "collective conscience rules people from within"—into more complex and dynamic terms.[81] For, as he argued,[82]

> Taken all together, the primary frameworks of a particular social group constitute a central element of its culture, especially insofar as understandings emerge concerning principal classes of schemata, the relations of these classes to one another, and the sum total of forces and agents that these interpretive designs acknowledge to be loose in the world.

Yet, for the most part, Goffman concentrated on the dynamics of framing within the realms of personal experiences and the interaction order, leaving Durkheim's concerns about macrocultural processes to others. Indeed, as was so often the case with Goffman, he became so intrigued with the interpersonal manipulation of frames for deceitful purposes and with the fluidity and complexity of framing in contingent situations that it is often difficult to discern whether or not the analysis is still sociological.

Keys and Keying

What makes framing a complex process, Goffman argued, is that frames can be *transformed.* One basic way to transform a primary frame is to engage in *keying,* which is "a set of conventions by which a given activity, one already meaningful in terms of some primary framework, is transformed into something patterned on this activity but seen by the participants to be something quite else."[83] For example, a theatrical production of a family setting is a keying of "real families"; or a hobby, such as woodworking, is a keying of a more primary set of occupational activities; or a daydream about a love affair is a keying of real love; or a sporting event is a keying of some more primary activity (running, fleeing, fighting, and so on); or practicing and rehearsing are keyings of real performances; or joking about someone's "love life" is a keying of love affairs; and so on. Primary frameworks are seen by people as "real," whereas keyings are seen as "less real"; and the more one rekeys a primary framework—say, performing a keying of a keying of a keying (and so on)—the "less real" is the frame. The *rim* of the frame is still ultimately a primary framework, anchored in some natural or social reality, but humans have the capacity to continually rekey and *layer* or *laminate* their experiences. Thus terms like "definition of the situation" do not adequately capture this layering of experience through keying, nor do such terms adequately denote the multiplicity of frameworks that people can invoke because of their capacities for shifting primary frameworks and rekeying existing ones.

Fabrications

The second type of frame transformation—in addition to keying—is *fabrication,* which is "the intentional effort of one or more individuals to manage activity so that a party of one or more others will be induced to have a false belief about what it is that is going on."[84] Unlike a key, a fabrication is not a copy (or a copy of a copy) of some primary framework, but an effort to make others think that something else is going on. Hoaxes, con games, and strategic manipulations all involve fabrications—getting others to frame a situation in one way while others manipulate them as another, hidden, framework.

The Complexity of Experience

Thus, as people interact, they frame situations as primary frameworks, but they can also key these primary frames and fabricate new ones for purposes of deception and manipulation. From Goffman's viewpoint, an interaction can involve many keyings and rekeyings (that is, layers and laminations of interpretation) as well as fabrications. Once keying or fabrication occurs, further keying and fabrication actually are facilitated because they initiate an escalating movement away from a primary framework. Goffman never stated how much fabrication (and fabrication on fabrications) and keying (or keying on keyings) can occur in interaction, but he did see novels, dramatic theater, and cinema as providing the vehicles for the deepest layering of interaction (because each is an initial keying of some primary frame in the real world, which then opens the almost infinite possibilities for rekeying and fabrication).

Yet there are procedures in interaction—typically ritualized and normatively regulated—for bringing participants' experiences back to the original primary frame—that is, for wiping away layers of keyings and fabrications. For example, when people have become caught up in benign ridicule and joking about something, a person saying "seriously now . . ." is trying to wipe the slate clean of keyings (and rekeyings) and come back to the primary frame on which the mocking and joking were based. Such rituals (with the normative obligation to "try to be serious for a moment") seek to reanchor the interaction in "the real world."

In addition to keyings and deliberate fabrications, individuals can *misframe* events—whether from ignorance, ambiguity, error, or initial disputes over framing among participants. Such misframing can persist for a time, but eventually individuals seek to *clear the frame* by getting a correct reading of information so they can reframe the situation correctly. Such efforts to clear the frame become particularly difficult and problematic, however, when fabrication has been at least part of the reason for the misframing.[85]

Thus, framing is a very complex process—one that, Goffman contended, sociologists have not been willing to address seriously. The world that people experience is not unitary, and it is subject to considerable manipulation, whether for benign or deceptive purposes. And, because of humans' capacity for symbol use (especially talk), the processes of reframing, keying, and fabrication can create highly layered and complex experiences. Yet, during interaction, people seek to maintain a common frame (a need that, Goffman was all too willing to assert, makes them vulnerable to manipulation through fabrication). For without a common frame—even one that has been keyed or fabricated—the interaction cannot proceed smoothly. Unfortunately, Goffman concentrated on framing as "organizing experience," per se, rather than on framing as it organizes experience during focused and unfocused interaction. Hence the analysis of framing is provocative and suggestive, but too often it wanders away from the sociologically relevant topic—"the interaction order."

CONCLUSION

Erving Goffman's works represent a truly seminal breakthrough in the analysis of social interaction. The emphasis on self-presentations, norms, rituals, and frames presented sociological theory at the midcentury with many important conceptual leads that have been adopted by diverse theoretical traditions.

There are, however, some questionable points of emphasis that should, in closing, be mentioned. First, Goffman's somewhat cynical and manipulative view of humans and interaction—which have been consistently downplayed in this chapter—often takes analysis in directions that are not as fundamental to human organization as he implied. Second, Goffman's rather extreme situational view of self as only a projected image and perhaps a mirage in each and every situation is probably overdrawn. The denial of a core self or permanent identity certainly runs against the mainstream of interactionist theorizing and, perhaps, reality itself. And, third, Goffman's work tended, at times, to wander into a rather extreme subjectivism and interpersonal nihilism where experience is too layered and fickle and where interaction is too fluid and changeable by the slightest shift in frame and ritual. Yet, even with these points of criticism, Goffman's sociology represents a monumental achievement—certainly equal to that of George Herbert Mead, Alfred Schutz, and Émile Durkheim. Indeed, as the century draws to a close, Erving Goffman can be considered the premier micro-level theorist of the last six decades.

NOTES

1. Erving Goffman, "The Interaction Order," *American Sociological Review* (February 1983), pp. 1–17. See also Anne W. Rawls, "The Interaction Order Sui Generis. Goffman's Contribution to Social Theory," *Sociological Theory* 5 (2, 1987), pp. 136–149; and Stephan Fuchs, "The Constitution of Emergent Interaction Orders, A Comment on Rawls," *Sociological Theory* 5 (1, 1988), pp. 122–124.

2. Goffman, "The Interaction Order" (cited in note 1), p. 9.

3. Ibid.

4. Ibid., p. 11. Italics in original.

5. Ibid.

6. Ibid.

7. Erving Goffman, *Encounters: Two Studies in the Sociology of Interaction* (Indianapolis: Bobbs-Merrill, 1961), p. 81.

8. Erving Goffman, *Behavior in Public Places: Notes on the Social Organization of Gatherings* (New York: Free Press, 1963), p. 234.

9. For a useful set of essays on Goffman's analysis, see: Paul Drew and Anthony Wootton, eds., *Erving Goffman: Exploring the Interaction Order* (Cambridge, England: Polity, 1988); see also a recent symposium on Goffman in *Sociological Perspectives*, 39 (3, 1996).

10. Goffman, "The Interaction Order" (cited in note 1), p. 2.

11. Ibid., p. 17.

12. Ibid.

13. Erving Goffman, *The Presentation of Self in Everyday Life* (Garden City, NY: Anchor, 1959) and *Frame Analysis: An Essay on the Organization of Experience* (Boston: Northeastern University Press, 1986; originally published in 1974 by Harper & Row).

14. See, for example, Randall Collins, *Theoretical Sociology* (San Diego: Harcourt Brace Jovanovich, 1988), pp. 203–207, 291–298, and *Three Sociological Traditions* (New York: Oxford University Press, 1985).

15. For example, see: R.P. Cuzzort and E.W. King, *Twentieth-Century Social Thought*, 4th ed. (Fort Worth, TX: Holt, Rinehart & Winston, 1989), Chapter 12.

16. See Stephan Fuchs, "Second Thoughts on Emergent Interaction Orders," *Sociological Theory* 7 (1, 1989), pp. 121–123.

17. Goffman, *The Presentation of Self in Everyday Life* (cited in note 13).

18. Ibid., pp. 9–10.

19. Note: Randall Collins (see Chapter 32 in this work) took his idea of situations as either work-practical, ceremonial, or social from this discussion.

20. Note the similarity of this idea to Ralph H. Turner's discussion (see Chapter 29) on "role making."

21. Goffman, *The Presentation of Self* (cited in note 13), pp. 24–25.

22. Ibid., p. 27.

23. Ibid., p. 32.

24. Ibid., p. 35. Note how this idea parallels the notion of "representational" aspect of roles developed by Ralph H. Turner and Paul Colomy (see Chapter 29).

25. Ibid., p. 56.

26. This is a consistent theme in Goffman's work that, perhaps, he overemphasized. See, for example, Erving Goffman, *Strategic Interaction* (Philadelphia: University of Pennsylvania Press, 1969).

27. Goffman, *The Presentation of Self* (cited in note 13), pp. 137–138.

28. Ibid., p. 92.

29. Ibid., p. 141.

30. Goffman, *Encounters*, p. 7 (cited in note 7).

31. Ibid.

32. Ibid.

33. Ibid., p. 18. Note the similarity of this definition to Randall Collins' portrayal of "interaction rituals" (see Chapter 32).

34. Goffman, *Behavior in Public Places* (cited in note 8), pp. 18–20.

35. Goffman, *Encounters* (cited in note 7), pp. 20–33; and Erving Goffman, *Interaction Ritual: Essays on Face-to-Face Behavior* (Garden City, NY: Anchor, 1967), p. 33.

36. The concept of frame, which Goffman said that he took from Gregory Bateson, became central in Goffman's later work. See the closing section in this chapter.

37. These rules come from Goffman, *Interaction Ritual* (cited in note 35), p. 33.

38. Here Goffman anticipates much of ethnomethodology, which is examined in Chapter 31.

39. Goffman, *Interaction Ritual* (cited in note 35).

40. Particularly Durkheim's later work, as it culminated in *Elementary Forms of the Religious Life* (New York: Free Press, 1947; originally published in 1912).

41. Goffman, *Interaction Ritual* (cited in note 35), pp. 56–67.

42. Ibid., p. 60.

43. Ibid.

44. Ibid., pp. 75–76.

45. Ibid., p. 77.

46. Ibid., p. 91.

47. Goffman, *Encounters* (cited in note 7), p. 96.

48. Ibid., p. 87.

49. Ibid., p. 113.

50. Ibid., p. 107.

51. Goffman, *Frame Analysis* (cited in note 13).

52. Ibid., p. 299.

53. Goffman, *Interaction Ritual* (cited in note 35), p. 5.

54. Ibid.

55. Ibid., p. 10.

56. Ibid., p. 30.

57. Erving Goffman, *Relations in Public: Micro Studies of the Public Order* (New York: Harper Colophon, 1972); originally published in 1971 by Basic Books), pp. 38–41.

58. Ibid., pp. 40–41.

59. Goffman, *Frame Analysis* (cited in note 13).

60. Erving Goffman, *Forms of Talk* (Philadelphia: University of Pennsylvania Press, 1981).

61. Ibid., pp. 70–71.

62. Ibid., p. 71.

63. Ibid., p. 85.

64. Ibid., p. 120.

65. Ibid., p. 121.

66. Goffman termed this *reframing* (see later section) when it involved a shift in frame. However, footing and refooting can occur *within* an existing frame, so a person can change the footing of a conversation without breaking or changing a frame.

67. Again, this line of emphasis is what gives Goffman's work an ethnomethodological flair. See next chapter.

68. Goffman, *Interaction Ritual* (cited in note 35), p. 135.

69. cited in notes 57 and 8, respectively: Goffman, *Behavior in Public Places* and *Relations in Public.*

70. Goffman, *Behavior in Public Places* (cited in note 57), p. 35.

71. Goffman, *Relations in Public* (cited in note 8), p. 8.

72. Ibid., pp. 129–138.

73. Ibid., p. 162.

74. Ibid., Chapter 2.

75. Ibid., pp. 109–120.

76. Goffman, *Frame Analysis.* (cited in note 13)

77. See Goffman, *Frame Analysis* (cited in note 13), p. 10, for the vagueness of his portrayal.

78. Ibid., p. 21.

79. Ibid., pp. 21–24.

80. Ibid., p. 21.

81. See Durkheim, *Elementary Forms of the Religious Life* (cited in note 40).

82. Goffman, *Frame Analysis* (cited in note 13), p. 27.

83. Ibid., pp. 43–44.

84. Ibid., p. 83.

85. See ibid., p. 449, for a list of the conditions that make individuals inadvertently vulnerable to misframing. And see p. 463 for how misframing can be the result of deception and manipulation.

31

The Maturing Tradition V:

The Challenge of Ethnomethodology

In the 1960s, a more phenomenological form of interactionist theorizing emerged. This theorizing employed a number of labels, but the term "ethnomethodology" is the most common designation for this approach. Paraded as a radical alternative to not only other forms of interactionism but all sociological theory as well, ethnomethodology enjoyed considerable influence in the 1970s. Over time, however, the extremes of the original critique of mainstream sociology wore thin, and the perspective no longer remained at the critical edge. Yet, it became institutionalized as a form of sociological research—primarily revolving around the analysis of conversations. Our goal in this chapter is to appreciate the nature of the challenge ethnomethodology posed as it brought a more phenomenological emphasis to theories of social interaction.

As its label underscores, *ethnomethodology* examines the "folk" (ethno) "methods" that people use in dealing with one another. It should be emphasized at the outset, however, that early ethnomethodologists rejected being discussed as interactionists.[1] Indeed, they saw themselves as proposing a radically new paradigm that challenged existing conceptions of social reality—from functionalism to symbolic interactionism.

Why did ethnomethodologists so often see themselves as paradigmatic messiahs? The answer resides in ethnomethodologists' contention that they had discovered an alternative reality that could not be conceptualized by existing theoretical perspectives. And they raised a question that owes its inspiration to Edmund Husserl and Alfred Schutz: How do sociologists and other groups of humans create and sustain for each other the *presumption* that the social world has a real character? A "more real" phenomenon for those who proposed this question revolves around the complex ways people (laypersons and sociologists alike) go about consciously and unconsciously constructing, maintaining, and altering their "sense" of an external social reality. The cement that holds society together might not be the values, norms, common definitions, and exchange payoffs of current social theory but, rather, people's explicit and implicit "methods" for creating the presumption of a social order.

THE ORIGINS OF ETHNOMETHODOLOGY

Ethnomethodology borrowed and extended ideas from phenomenology and, despite disclaimers to the contrary, from Meadian-inspired symbolic interactionism. In extending the ideas of these schools of thought, however, ethnomethodology claimed to posit a different view of the world. And so, to appreciate just how ethnomethodology differed from more traditional forms of sociological theory, we must assess it in relation to those perspectives from which it tried to dissociate itself, but from which it still drew considerable inspiration.

Blumer's Interactionism and Ethnomethodology

As reviewed in Chapter 26, Herbert Blumer's interactionism emphasized the constructed and fluid nature of interaction. Because actors possess extensive symbolic capacities, they are capable of (a) introjecting new objects into situations, (b) redefining situations, and (c) realigning their joint actions. As Blumer stressed, "it would be wise to recognize that any given [act] is mediated by acting units interpreting the situations with which they are confronted."[2] Whereas these situations consist of norms, values, roles, beliefs, and social structures, these are merely types of many "objects" that can be symbolically introjected and reshuffled to produce new definitions of situations.

These ideas expressed a concern with how meanings, or definitions, are created by actors interacting in situations. The emphasis is on the *process* of interaction and on how actors create common meanings in dealing with one another. This line of inquiry was also pursued by ethnomethodologists, who focused on interaction and on the creation of meanings in situations. But there was an important shift in emphasis that corresponds to Alfred Schutz's analysis of the lifeworld: In what ways do people create a *sense* that they share a common view of the world? And how do people arrive at the *presumption* that there is an objective, external world? Blumer's interactionism stresses the process of creating meaning, but it acknowledged the existence of an external social order. Early ethnomethodology suspended, or "bracketed," in Husserl's terms, the issue of whether or not there is an external world of norms, roles, values, and beliefs. Instead, ethnomethodologists concentrated on how interaction creates among actors a *sense* of a factual world "out there."

Goffman's Dramaturgical Analysis and Ethnomethodology

To the dismay of many ethnomethodologists, Paul Attewell argued in the 1970s that the work of Erving Goffman represented a significant source of inspiration for ethnomethodology.[3] As was emphasized in Chapter 30, Goffman's work has often been termed the *dramaturgical school* of interactionism because it focused on the ways that actors manipulate gestures to create an impression in a particular social scene. Goffman tended to emphasize the process of impression management, per se, and not the purposes or goals toward which action is directed. Much of Goffman's analysis thus concentrated on the form of interaction itself rather than on the structures it created, sustained, or changed.[4] For example,

Goffman insightfully analyzed how actors validate self-presentations, how they justify their actions through gestures, how they demonstrate their membership in groups, how they display social distance, how they adjust to physical stigmas, and how they interpersonally manipulate many other situations.

This concern with the management of social scenes also became prominent in ethnomethodological analyses. Ethnomethodologists shared Goffman's concern with the techniques by which actors create impressions in social situations, but their interest was not with individuals' impression management, per se, but with how actors create a *sense of a common reality*. This too was an important aspect of Goffman's work. For Goffman, much of what individuals do in interaction revolves around efforts to sustain a common view and sense of reality. Like Schutz, Goffman argued that humans not only actively try to manipulate impressions to create and sustain a view of reality but also try to ignore discrepant information in an effort to keep a situation intact. Moreover, humans take for granted much information and work to keep from questioning the situation, lest their common view of reality come apart. All these themes are a part of ethnomethodology; although Goffman appears to have been highly critical of this perspective, he nonetheless contributed to its development.

Alfred Schutz's Phenomenological Analysis and Ethnomethodology

Schutz's phenomenology, as noted in Chapter 26, liberated phenomenology from Husserl's philosophical project.[5] Schutz asserted the importance of studying how interaction creates and maintains a "paramount reality." He was also concerned with how actors achieve reciprocity of perspectives and how they construct a taken-for-granted world that gives order to social life.

This emphasis on the taken-for-granted nature of the world and the importance of this lifeworld for maintaining an actor's sense of reality became the prime concern of ethnomethodologists. Indeed, many ethnomethodological concepts were borrowed or adapted from Husserl's and Schutz's phenomenology. Yet ethnomethodologists adapted phenomenological analysis to the issue of how social order is maintained by the practices that actors use to create a sense that they share the same lifeworld.

To summarize the argument thus far, ethnomethodology drew from and extended the concerns of interactionists such as Blumer and Goffman and the phenomenological projects of Husserl and Schutz. Ethnomethodology emphasized the process of interaction, the use of interpersonal techniques to create situational impressions, and the importance of perceptions of consensus among actors. In extending interactionism and phenomenology, ethnomethodologists believed that they had posited a different vision of the social world and an alternative orientation for understanding how social organization is created, maintained, and changed. As Hugh Mehan and Houston Wood noted, ethnomethodologists "have chosen to ask not how order is possible, but rather to ask *how a sense of order is possible*."[6]

THE NATURE OF ETHNOMETHODOLOGY

Metaphysics or Methodology?

Ethnomethodology was often misunderstood in the 1960s and 1970s. Part of the reason for this misunderstanding stemmed from the vagueness of the prose of some ethnomethodologists.[7] One form of such misinterpretation asserted that ethnomethodology represented a "corrective" to current sociological theorizing because it pointed to sources of bias among scientific investigators. From this position it was assumed that ethnomethodology could check the reliability and validity of investigators' observations by exposing not only their biases but also those of the scientific community accepting their observations. Although ethnomethodology was used for this purpose, those who advocated this use have failed to grasp the main thrust of the ethnomethodological position. For the ethnomethodologist, emphasis was not on questions about the reliability and validity of investigators' observations, but on the methods used by scientific investigators and laypersons alike to construct, maintain, and perhaps alter what each considers and believes to be a valid and reliable set of statements about order in the world. The methodology in the ethnomethodological perspective did not address questions about the proper, unbiased, or truly scientific search for knowledge; rather, ethnomethodology was concerned with the common methods people employ—whether scientists, homemakers, insurance salespersons, or laborers—to create a sense of order about the situations in which they interact. The best clue to this conceptual emphasis can be found in the word *ethnomethodology* itself—*ology*, "study of"; *method*, "the methods [used by]"; and *ethno*, "folk or people."

Another related source of misunderstanding in early commentaries on ethnomethodology came from those who assumed that this perspective simply used "soft" research methods, such as participant observation, to uncover some of the taken-for-granted rules, assumptions, and rituals of members in groups.[8] This interpretation would appear to transform ethnomethodology into a research-oriented variant of the symbolic interactionist perspective.[9] Such a variant of ethnomethods represented a more conscientious effort to get at actors' interpretative processes and the resulting definitions of the situation. By employing various techniques for observation of, and participation in, the symbolic world of those interacting individuals under study, a more accurate reading of how situations are defined, how norms emerge, and how social action is controlled could be achieved. Although ethnomethodologists did employ observation and participant methods to study interacting individuals, their concerns were not the same as those of symbolic interactionists. Like all dominant forms of sociological theorizing, symbolic interactionists posited that common definitions, values, and norms emerge from interaction and regulate how people perceive the world and interact with one another. For the interactionist, concern is with the conditions under which various types of explicit and implicit definitions, norms, and values emerge and thereby resolve the problem of how social organization is possible. In contrast, ethnomethodologists were interested in *how* members come to agree on an *impression* that there are such things as rules, definitions, and values. Just what types of rules and definitions emerge was not a central concern of the ethnomethodologist, because there are more fundamental questions: Through *what types of methods* do people go about seeing, describing, and asserting that rules and definitions exist? How do people use their beliefs that definitions and rules exist to describe for one another the social order?

Again, the methods of ethnomethodology did not refer to a new and improved technique by scientific sociology for deriving a more accurate picture of people's definitions of the situation and of the norms of social structure (as had been the case with symbolic interactionists). For the ethnomethodologist, emphasis was on the *methods employed by those under study* in creating, maintaining, and altering their presumption that a social order actually exists in the real world.

Concepts and Principles of Ethnomethodology

Schutz postulated one basic reality—the paramount—in which people's conduct of their everyday affairs occurs. Most early ethnomethodologists, however, were less interested in whether or not there is one or multiple realities, lifeworlds, or natural attitudes. Far more important in ethnomethodological analysis was the development of concepts and principles that could help explain how people's sense of reality is constructed, maintained, and changed. Although ethnomethodology never developed a unified body of concepts or propositions, a conceptual core in the ethnomethodological perspective was evident by the 1970s.

Reflexive Action and Interaction.[10] Much interaction sustains a particular vision of reality. For example, ritual activity directed toward the gods sustains the belief that gods influence everyday affairs. Such ritual activity is an example of reflexive action; it maintains a certain vision of reality. Even when the facts would seem to contradict a belief, human interaction upholds the contradicted belief. When intense prayer and ritual activity do not bring forth the desired intervention from the gods, the devout, rather than reject beliefs, proclaim that they did not pray hard enough, that their cause was not just, or that the gods in their wisdom have a greater plan. Such behavior is *reflexive*: It upholds or reinforces a belief, even in the face of evidence that the belief might be incorrect.

Much human interaction is reflexive. Humans interpret cues, gestures, words, and other information from one another in a way that sustains a particular vision of reality. Even contradictory evidence is reflexively interpreted to maintain a body of belief and knowledge. The concept of reflexivity thus focuses attention on how people in interaction go about maintaining the presumption that they are guided by a particular reality. Much of ethnomethodological inquiry has addressed this question of how reflexive interaction occurs. That is, what concepts and principles can be developed to explain the conditions under which different reflexive actions among interacting parties are likely to occur?

The Indexicality of Meaning. The gestures, cues, words, and other information sent and received by interacting parties have meaning in a *particular context*. Without some knowledge of the context—the biographies of the interacting parties, their avowed purpose, their past interactive experiences, and so forth—it would easily be possible to misinterpret the symbolic communication among interacting individuals. This was denoted by the concept of *indexicality*.[11] To say that an expression is indexical is to emphasize that the meaning of that expression is tied to a particular context.

This notion of indexicality drew attention to the problem of how actors in a context construct a vision of reality in that context. They develop expressions that invoke their common vision about what is real in their situation. The concept of indexicality thus directs investigators to actual interactive contexts to see how actors go about creating indexical expressions—words, facial and body gestures, and other cues—to create and sustain the presumption that a particular reality governs their affairs.

With these two key concepts, reflexivity and indexicality, the interactionists' concern with the process of symbolic communication was retained by ethnomethodology, and much of the phenomenological legacy of Schutz was rejuvenated. Concern was with how actors use gestures to create and sustain a lifeworld, body of knowledge, or natural attitude about what is real. The emphasis was not on the content of the lifeworld but on the methods or techniques that actors use to create, maintain, or even alter a vision of reality. As Mehan and Wood noted, "the ethnomethodological theory of reality constructor is about the *procedures* that accomplish reality. It is not about any specific reality."[12] This emphasis led ethnomethodologists to isolate the general types of methods employed by interacting actors.

Some General Interactive Methods. When analytical attention focuses on the methods that people use to construct a sense of reality, the task of the theorist is to isolate the general types of interpersonal techniques that people employ in interaction. Aaron Cicourel, for example, summarized several such techniques or methods isolated by ethnomethodologists: (1) searching for the normal form, (2) doing reciprocity of perspectives, and (3) using the et cetera principle.[13]

Searching for the Normal Form. If interacting parties sense that ambiguity exists about what is real and that their interaction is strained, they will emit gestures to tell each other to return to what is "normal" in their contextual situation. Actors are presumed to hold a vision of a normal form for situations or to be motivated to create one; hence much of their action is designed to reach this form.

Doing a Reciprocity of Perspectives. Borrowing from Schutz's formulation, ethnomethodologists emphasized that actors operate under the presumption, and actively communicate, that they would have the same experiences were they to switch places. Furthermore, until they are so informed by specific gestures, actors can ignore differences in perspectives that might arise from their unique biographies. Thus, much interaction will be consumed with gestures that seek to assure others that a reciprocity of perspectives does indeed exist.

Using the Et Cetera Principle. In examining an actual interaction, much is left unsaid. Actors must constantly "fill in" or "wait for" information necessary to "make sense" of another's words or deeds. When actors do so, they are using the et cetera principle. They are agreeing not to disrupt the interaction by asking for the needed information; they are willing to wait or to fill in. For example, the common phrase "you know," which often appears after an utterance, is often an assertion by one actor to another invoking the et cetera principle. The other is thus informed not to disrupt the interaction or the sense of reality in the situation with a counter utterance, such as "No, I do not know."

These three general types of folk methods were examples of what ethnomethodologists sought to discover. For some ethnomethodologists, the ultimate goal of theory was to determine the conditions under which these and other interpersonal techniques would be used to construct, maintain, or change a sense of reality. Yet few such propositions were to be found in the ethnomethodological literature.

If such propositions had been developed, what would they have looked like? Ethnomethodological propositions would follow from several assumptions: (1) Social order is maintained by the use of techniques that give actors a sense that they share a common reality. (2) The substance of the common reality is less important in maintaining social order than the actors' acceptance of a common set of techniques. With these assumptions, two examples of what ethnomethodological propositions might have looked like if ethnomethodologists had ever got around to generating them are the following:

1. The more actors fail to agree on the use of interactive techniques, such as the et cetera principle, the search for the normal

form, and the reciprocity of perspectives, the more likely is interaction to be disrupted and, hence, the less likely is social order to be maintained.

2. The more interaction proceeds on the basis of different, taken-for-granted visions of reality, the more likely is interaction to be disrupted and, hence, the less likely is social order to be maintained.

These propositions could perhaps be visualized as general laws from which more specific propositions about how actors go about constructing, maintaining, or changing their sense of reality could be developed. What was needed in ethnomethodology were statements on the *specific conditions* under which particular folk techniques are likely to be used to create a sense of a common world among interacting individuals. Such statements have been rare, although they are implied in some forms of conversational analysis.

Varieties of Ethnomethodological Inquiry

Garfinkel's Pioneering Inquiries. Harold Garfinkel's *Studies in Ethnomethodology* firmly established ethnomethodology as a distinctive theoretical perspective.[14] Although the book was not a formal theoretical statement, the studies and the commentary in it established the domain of ethnomethodological inquiry. Subsequent ethnomethodological research and theory began with Garfinkel's insights and took them in a variety of directions.

Garfinkel's work saw ethnomethodology as a field of inquiry that sought to understand the methods people employ to make sense of their world. He placed considerable emphasis on language as the vehicle by which this reality construction is done. Indeed, for Garfinkel, interacting individuals' efforts to account for their actions—that is, to represent them verbally to others—are the primary method by which the world is constructed. In Garfinkel's terms, *to do* interaction is *to tell* interaction, or, in other words, the primary folk technique used by actors is verbal description. In this way people use their accounts to construct a sense of reality.

Garfinkel placed enormous emphasis on indexicality—that is, members' accounts being tied to particular contexts and situations. An utterance, Garfinkel noted, indexes much more than it actually says; it also evokes connotations that can be understood only in the context of a situation. Garfinkel's work was thus the first to stress the indexical nature of interpersonal cues and to emphasize that individuals seek accounts to create a sense of reality.

In addition to laying much of the groundwork for ethnomethodology, Garfinkel and his associates conducted several interesting empirical studies to validate their assumptions about what is real. One line of empirical inquiry became known as the *breaching experiment,* in which the normal course of interaction was deliberately interrupted. For example, Garfinkel reported a series of conversations in which student experimenters challenged every statement of selected subjects. The end result was a series of conversations revealing the following pattern:

Subject: I had a flat tire.

Experimenter: What do you mean, you had a flat tire?

Subject: (appears momentarily stunned and then replies in a hostile manner): What do you mean, "What do you mean?" A flat tire is a flat tire. That is what I meant. Nothing special. What a crazy question![15]

In this situation the experimenter was apparently violating an implicit rule for this type of interaction and thereby aroused not only the hostility of the subject but also a negative sanction, "What a crazy question!"

Seemingly, in any interaction there are certain background features that everyone should understand and that should not be questioned so that all parties can "conduct their common conversational affairs without interference."[16] Such implicit methods appear to guide a considerable number of everyday affairs and are critical for the construction of at least the perception among interacting humans that an external social order exists. In this conversation, for example, the "et cetera principle" and the "search for the normal form" are invoked by the subject. Through breaching, Garfinkel hoped to discover the implicit ethnomethods being used by forcing actors to engage *actively* in the process of reality reconstruction after the situation had been disrupted.

Other research strategies also yielded insights into the methods parties use in an interaction for constructing a sense of reality. For example, Garfinkel and his associates summarized the "decision rules" jurors employed in reaching a verdict.[17] By examining a group such as a jury, which must by the nature of its task develop an interpretation of what really happened, the ethnomethodologist sought to achieve some insight into the generic properties of the processes of constructing a *sense of social reality*. From the investigators' observations of jurors, it appeared that "a person is 95 percent juror before [coming] near the court," indicating that, through their participation in other social settings and through instructions from the court, they had accepted the "official" rules for reaching a verdict. However, these rules were altered somewhat as participants came together in an actual jury setting and began the "work of assembling the 'corpus' which serves as grounds for inferring the correctness of a verdict."[18] Because the inevitable ambiguities of the cases before them made it difficult for strict conformity to the official rules of jury deliberation, new decision rules were invoked to allow jurors to achieve a "correct" view of "what actually happened." But, in their retrospective reporting to interviewers of how they reached the verdicts, jurors typically invoked the "official line" to justify the correctness of their decisions. When interviewers drew attention to discrepancies between the jurors' ideal accounts and their actual practices, jurors became anxious, indicating that somewhat different rules had been used to construct the corpus of what really happened.

In sum, these two examples of Garfinkel's research strategy illustrate the general intent of much early ethnomethodological inquiry: to penetrate natural social settings or to create social settings in which the investigator could observe humans attempting to assert, create, maintain, or change the rules for constructing the appearance of consensus over the structure of the real world. By focusing on the process or methods for constructing a reality rather than on the substance or content of the reality itself, research from the ethnomethodological point of view could potentially provide a more interesting and relevant answer to the question of "how and why society is possible." Garfinkel's studies stimulated a variety of research and theoretical strategies. Several of the most prominent early strategies will be briefly discussed.

Harvey Sacks's Linguistic Analysis. Until his untimely death in 1976, Harvey Sacks exerted considerable influence within ethnomethodology. Although his work was not well known outside ethnomethodological circles, it represented an attempt to extend Garfinkel's concern with verbal accounts, while eliminating some of the problems posed by indexicality.

Sacks was one of the first ethnomethodologists to articulate the phenomenological critique of sociology and to use this critique to build what he thought was an alternative form of theorizing.[19] The basic thrust of Sacks's critique can be stated as follows: Sociologists assume that language is a resource used in generating concepts and theories of the social world. However, sociologists are confusing resource and topic. In using language, sociologists are creating a reality; their words are not a neutral vehicle but *the* topic of inquiry for true sociological analysis—a point of emphasis that is, today, the core topic of ethnomethodological inquiry.[20]

Sacks's solution to this problem in sociology was typical of phenomenologists. If the pure properties of language can be understood, then it would be possible to have an objective social science without confusing resource with subject matter. Sacks's research tended to concentrate on the formal properties of language-in-use. Typically, Sacks took verbatim transcripts of actors in interaction and sought to understand the formal properties of the conversation while ignoring its substance. Such a tactic resolved the problem of indexicality, since Sacks simply ignored the substance and context of conversation and focused on its form. For example, "sequences of talk" among actors might occupy his attention.[21]

Sacks thus began to take ethnomethodology into formal linguistics, a trend that has continued and now seems to dominate current ethnomethodology. More important, he sought to discover universal forms of interaction—that is, abstracted patterns of talk—that might apply to all conversations. In this way he began to search for the laws of reality construction among interacting individuals.

Aaron Cicourel's Cognitive Approach. In *Method and Measurement in Sociology*, Aaron Cicourel launched an attack on sociology similar to that of Sacks's.[22] The use of mathematics, Cicourel argued, will not remove the problems associated with language because mathematics is a language that does not necessarily correspond to the phenomena that it describes. Instead, it "distorts and obliterates, acts as a filter or grid for that which will pass as knowledge in a given era."[23] The use of statistics similarly distorts: Events cannot be counted, averaged, and otherwise manipulated. Such statistical manipulations soon make sociological descriptions inaccurate and mold sociological analysis to the dictates of statistical logic.

In a less severe tone, Cicourel also questioned Garfinkel's assertion that interaction and verbal accounts are the same process.[24] Cicourel noted that humans see, sense, and feel much that they cannot communicate with words. Humans use "multiple modalities" for communicating in situations. Verbal accounts represent crude and incomplete translations of what is actually communicated in interaction. This recognition led Cicourel to rename his brand of ethnomethodology: *cognitive sociology*.

The details of his analysis are less important than the general intent of his effort to transform sociological research and theory. Basically, he sought to uncover the universal "interpretive procedures" by which humans organize their cognitions and give meaning to situations.[25] Through these interpretive procedures people develop a sense of social structure and can organize their actions. These interpretive procedures are universal and invariant in humans, and their discovery would allow understanding of how humans create a sense of social structure in the world around them.

Zimmerman, Pollner, and Wieder's Situational Approach. Sacks and Cicourel focused on the universal properties, respectively, of language use and cognitive perception/representation. This concern with invariance, or universal folk methods, became increasingly prominent in ethnomethodological inquiry. In several essays, for example,

Don Zimmerman, D. Lawrence Wieder, and Melvin Pollner developed an approach that sought the universal procedures people employ to construct a sense of reality.[26] Their position was perhaps the most clearly stated of all ethnomethodologies, drawing inspiration from Garfinkel but extending his ideas. To summarize their argument,

1. In all interaction situations, humans attempt to construct the appearance of consensus over relevant features of the interaction setting.
2. These setting features can include attitudes, opinions, beliefs, and other cognitions about the nature of the social setting in which they interact.
3. Humans engage in a variety of explicit and implicit interpersonal practices and methods to construct, maintain, and perhaps alter *the appearance* of consensus over these setting features.
4. Such interpersonal practices and methods result in the assembling and disassembling of what can be termed an *occasional corpus*—that is, the *perception* by interacting humans that the current setting has an orderly and understandable structure.
5. This appearance of consensus is not only the result of agreement on the substance and content of the occasioned corpus but also a reflection of each participant's compliance with the rules and procedures for assemblage and disassemblage of this consensus. In communicating, in however subtle a manner, that parties accept the implicit rules for constructing an occasioned corpus, they go a long way in establishing consensus about what is in the interaction setting.
6. In each interaction situation, the rules for constructing the occasioned corpus will be unique in some respects and hence not completely generalizable to other settings—thus requiring that humans in each and every interaction situation use interpersonal methods in search for agreement on the implicit rules for the assemblage of an occasioned corpus.
7. Thus, by constructing, reaffirming, or altering the rules for constructing an occasioned corpus, members in a setting can offer one another the appearance of an orderly and connected world, which compels certain perceptions and actions on their part.

From these kinds of assumptions about human interaction, Zimmerman, Pollner, and Wieder's ethnomethodology took its subject matter. Rather than focusing on the actual content and substance of the occasioned corpus and on the ways members believe it to force certain perceptions and actions, attention was drawn primarily to the *methods humans use* to construct, maintain, and change the *appearance* of an orderly and connected social world. These methods are directly observable and constitute a major portion of people's actions in everyday life. In contrast, the actual substance and content of the occasioned corpus are not directly observable and can only be inferred. Furthermore, in concentrating on the *process* of creating, sustaining, and changing the occasioned corpus, we can ask: Is not the process of creating the appearance of a stable social order for one another more critical to understanding how society is possible than the actual substance and content of the occasioned corpus? Is there anything more to society than members' presumption that society is "out there" forcing them to do and see certain things? Hence, order is not the result of the particular structure of the corpus, rather, order resides in the human capacity *to continually assemble and disassemble the corpus* in each and every interaction situation. This perspective suggested to ethnomethodologists that theoretical attention should therefore be placed on the ongoing process of assembling and disassembling the appearance of social order and on the particular methods people employ in doing so.

CONCLUSION

Ethnomethodology uncovered a series of interpersonal processes that traditional symbolic interactionists had failed to conceptualize. The implicit methods that people use to communicate a sense of social order are a very crucial dimension of social interaction and organization, and the theoretical goal of ethnomethodology was to specify the generic conditions under which various folk methods are used by individuals.

The rather extreme polemics about the "new reality" exposed by ethnomethodologists were soon forgotten, however.[27] The few findings of ethnomethodologists did not sustain their metaphysical vision of the world any more than these findings supported interactionist formulations. Indeed, the findings of ethnomethodologists could be blended very nicely with those of symbolic interactionists, role theorists, and other theoretical traditions. Ethnomethodologists had, therefore, overstated their case. People's sense of sharing a common world is, no doubt, an important property of interaction and social organization, but it is not the only interactive dynamic. To the degree that ethnomethodologists asserted that their domain of inquiry was *the only* reality, they made themselves look foolish.[28]

In the end, ethnomethodology became a rather isolated theoretical research program.[29] Its practitioners increasingly focused on conversational analysis—a mode of inquiry initiated by Sacks and carried forward by a number of creative scholars.[30] But their work has not had a great impact on mainstream sociological theory, inside or outside the interactionist tradition. Still, there was something important and fundamental about the assertions of ethnomethodologists, and for this reason much interactionist theorizing incorporated some aspect of this approach, even as the approach as a "new paradigm" was being pushed to the margins of sociological theorizing.

NOTES

1. Thomas P. Wilson, "Normative and Interpretative Paradigms in Sociology," in *Understanding Everyday Life*, ed. J. Douglas (London: Routledge, 1970).

2. Herbert Blumer, "Society as Symbolic Interaction," in *Human Behavior and Social Process*, ed. A. Rose (Boston: Houghton Mifflin, 1962).

3. Paul Attewell, "Ethnomethodology Since Garfinkel," *Theory and Society* 1 (1974), pp. 179–210.

4. This is not always true, especially in some of his more institutional works, such as *Asylums* (Garden City, NY: Anchor Books, 1961). Other representative works by Goffman include *The Presentation of Self in Everyday Life* (Garden City, NY: Doubleday, 1959); *Interaction Ritual* (Garden City, NY: Anchor, 1967); *Encounters* (Indianapolis: Bobbs-Merrill, 1961); *Stigma* (Englewood Cliffs, NJ: Prentice-Hall, 1963).

5. See, in particular, Alfred Schutz, *Collected Papers I: The Problem of Social Reality*, ed. Maurice Natanson (The Hague: Martinus Nijhoff, 1962); *Collected Papers II: Studies in Social Theory*, ed. Arvid Broderson (The Hague: Martinus Nijhoff, 1964); and *Collected Papers III: Studies in Phenomenological Philosophy* (The Hague: Martinus Nijhoff, 1966). For adaptations of these ideas to interactionism and ethnomethodology, see Alfred Schutz and Thomas Luckmann, *The Structure of the Lifeworld* (Evanston, IL: Northwestern University Press, 1973), as well as Thomas Luckmann, ed., *Phenomenology and Sociology* (New York: Penguin, 1978). See also Robert C. Freeman, "Phenomenological Sociology and Ethnomethodology," in *Introduction to the Sociologies of Everyday Life*, eds. J. Douglas and Patricia Adler (Boston: Allyn & Bacon, 1980).

6. Hugh Mehan and Houston Wood, *The Reality of Ethnomethodology* (New York: Wiley, 1975), p. 190. This is an excellent statement of the ethnomethodological perspective.

7. See, for example, Harold Garfinkel, *Studies in Ethnomethodology* (Englewood Cliffs, NJ: Prentice-Hall, 1967). However, later portrayals of the ethnomethodological position did much to clarify this initial vagueness. See, for example, Mehan and Wood, *The Reality of Ethnomethodology* (cited in note 6); D. Lawrence Wieder, *Language and Social Reality* (The Hague: Mouton, 1973); Don H. Zimmerman and Melvin Pollner, "The Everyday World as Phenomenon," in *Understanding Everyday Life*, ed. J.D. Douglas (Chicago: Aldine, 1970), pp. 80–103; Don H. Zimmerman and D. Lawrence Wieder, "Ethnomethodology and the Problem of Order: Comment on Denzin," in *Understanding Everyday Life*, pp. 285–295; Randall Collins and Michael Makowsky, *The Discovery of Society* (New York: Random House, 1972), pp. 209–213; George Psathas, "Ethnomethods and Phenomenology," *Social Research* 35 (September 1968), pp. 500–520; Roy Turner, ed., *Ethnomethodology* (Baltimore: Penguin, 1974); Thomas P. Wilson and Don H. Zimmerman, "Ethnomethodology, Sociology and Theory," *Humboldt Journal of Social Relations* 7 (Fall-Winter 1979–1980), pp. 52–88; Warren Handel, *Ethnomethodology: How People Make Sense* (Englewood Cliffs, NJ: Prentice-Hall, 1982); and George Psathas, ed., *Everyday Language: Studies in Ethnomethodology* (New York: Irvington, 1979).

8. For example, see Norman K. Denzin, "Symbolic Interactionism and Ethnomethodology," *American Sociological Review* 34 (December 1969), pp. 922–934.

9. For another example of this interpretation, see Walter L. Wallace, *Sociological Theory* (Chicago: Aldine, 1969), pp. 34–36.

10. For an early discussion of this phenomenon, see Garfinkel, *Studies in Ethnomethodology* (note 7). A more readable discussion can be found in Mehan and Wood, *The Reality of Ethnomethodology* (cited in note 7), pp. 137–178.

11. Garfinkel, *Studies in Ethnomethodology* (cited in note 7). Also see Harold Garfinkel and Harvey Sacks, "The Formal Properties of Practical Actions," in *Theoretical Sociology*, eds. J.C. McKinney and E.A. Tiryakian, pp. 84–97 (New York: Appleton-Century-Crofts, 1970).

12. Mehan and Wood, *The Reality of Ethnomethodology* (cited in note 6), p. 114.

13. Aaron V. Cicourel, *Cognitive Sociology* (London: Macmillan, 1973), pp. 85–88. It should be noted that these principles were implicit in Garfinkel's *Studies in Ethnomethodology* (cited in note 7).

14. Garfinkel, *Studies in Ethnomethodology* (cited in note 7).

15. Ibid., p. 42.

16. Ibid.

17. Ibid., pp. 104–115.

18. Ibid., p. 110.

19. Harvey Sacks, "Sociological Description," *Berkeley Journal of Sociology* 8 (1963), pp. 1–17.

20. Harvey Sacks, "An Initial Investigation of the Usability of Conversational Data for Doing Sociology," in *Studies in Interaction*, ed. D. Sudnow (New York: Free Press, 1972); see also Harvey Sacks, *Lectures on Conversation*, 2 volumes (New York: Blackwell, 1992). For a review of the current techniques of "conversational analysis, see Alain Coulon, *Ethnomethodology* (London: Sage, 1995); Douglas W. Maynard and Marilyn R. Whalen, "Language, Action, and Social Interaction," in *Sociological Perspectives on Social Psychology*, eds. K.S. Cook, G.A. Fine, and J.S. House (Boston: Allyn & Bacon, 1995).

21. Sack's best-known study, for example, is the coauthored article with Emmanuel Schegloff and Gail Jefferson, "A Simplest Systematics for the Analysis of

Turn Taking in Conversation," *Language* 50 (1974), pp. 696–735.

22. Aaron V. Cicourel, *Method and Measurement in Sociology* (New York: Free Press, 1964).

23. Ibid., p. 35.

24. Aaron V. Cicourel, "Cross Modal Communication," in *Linguistics and Language Science, Monograph 25*, ed. R. Shuy (Washington, DC: Georgetown University Press, 1973).

25. See, for example, Cicourel's *Cognitive Sociology* (cited in note 13) and "Basic Normative Rules in the Negotiation of Status and Role," in *Recent Sociology No. 2*, ed. H.P. Dreitzel (New York: Macmillan, 1970).

26. See, for example, Zimmerman and Pollner, "The Everyday World as a Phenomenon" (cited in note 7); Zimmerman and Wieder, "Ethnomethodology and the Problem of Order"; Wieder, *Language and Social Reality*.

27. For an interesting discussion of this assumption that challenges the ethnomethodological perspective, see Bill Harrell, "Symbols, Perception, and Meaning," in *Sociological Theory: Inquiries and Paradigms*, ed. L. Gross (New York: Harper & Row, 1967), pp. 104–127.

28. See Lewis Coser on this issue: "Two Methods in Search of a Substance," *American Sociological Review* 40 (1975), pp. 691–700.

29. For a conciliatory overview of ethnomethodology by an ethnomethodologist, see Deidre Boden, "The World as It Happens: Ethnomethodology and Conversation Analysis" in *Frontiers of Social Theory*, ed. George Ritzer (New York: Columbia University Press, 1990), pp. 185–213. See also John Heritage, *Garfinkel and Ethnomethodology* (Cambridge, UK: Polity, 1984). See also the extensive bibliographies of these works for further references.

30. For more recent efforts to develop conversational analysis, see Deidre Boden and Don H. Zimmerman, eds., *Talk and Social Structure: Studies in Ethnography and Conversation Analysis* (Cambridge, UK: Polity, 1991); Charles Goodwin and John Heritage, "Conversation Analysis," *Annual Review of Anthropology* 19 (1990), pp. 283–307; Douglas W. Maynard and Steven E. Clayman, "The Diversity of Ethnomethodology," *Annual Review of Sociology* 17 (1991), pp. 385–418.

32

The Continuing Tradition I: Theories of Emotion in Social Interaction

In the late 1970s, theories of interaction processes began to shift direction, focusing on how emotions are involved in interpersonal relations and the structuring of social encounters. Not all the theories that emerged over the next two decades were oriented to the legacy of George Herbert Mead and the Iowa and Chicago Schools of symbolic interactionism. Many were, to be sure, but other traditions were also drawn into theoretical formulations. Thus, in presenting representative samples, or exemplars, of theories on emotions, we cannot easily group them into a few theoretical camps. Instead, a variety of theories will be reviewed to communicate the diversity and eclecticism evident in the study of emotions. This diversity makes the sociology of emotions one of the very exciting areas of interactionist theorizing in sociology; indeed, it could be argued that the cutting edge of interactionism is how emotions shape the flow of interaction processes.

ARLIE HOCHSCHILD'S DRAMATURGICAL THEORY OF EMOTIONS

Arlie Russell Hochschild[1] was one of the first sociologists to develop a view of emotions as managed performances by individuals within the constraints of situational norms and broader cultural ideas about what emotions can be felt and presented in front of others. For Hochschild, the *emotion culture*[2] consists of a series of ideas about how and what people are supposed to experience in various types of situations, and this culture is filled with *emotional ideologies* about the appropriate attitudes and feelings for specific spheres and activities. *Emotional markers* are events in the biographies of individuals that personify and symbolize more general emotional ideologies.

In any context, Hochschild emphasizes, there are norms of two basic types: (1) *feeling rules* that indicate (a) the amount of appropriate emotion that can be felt in a situation, (b) the direction, whether positive or negative, of the emotion, and (c) the duration of the emotion; and (2) *display rules* that indicate the nature, intensity, and style of expressive behavior to be emitted. Thus, for any interaction, feeling and display rules circumscribe what can be done. These rules reflect ideologies of the broader emotion culture, the goals and purposes of groups in which interactions are lodged, and the distribution of power and other organizational features of the situation.

The existence of cultural ideologies and normative constraints on the selection and emission of emotions forces individuals to manage the feelings that they experience and present to others. At this point, Hochschild's analysis becomes dramaturgic, for much like Goffman before her, she sees actors as having to manage a presentation of self in situations guided by a cultural script of norms and broader ideologies. There are various types of what Hochschild terms *emotion work* or mechanisms for managing emotions and making the appropriate self presentation: (1) *body work* whereby individuals actually seek to change their bodily sensations in an effort to evoke the appropriate emotion (for example, deep breathing to create calm); (2) *surface acting* where individuals alter their external expressive gestures in ways that they hope will make them actually feel the appropriate emotion (for instance, emitting gestures expressing joy and sociality at a party in the attempt to feel happy); (3) *deep acting* where individuals attempt to change their internal feelings, or at least some of these feelings in the hope that the rest of the appropriate emotions will be activated and fall into place (for example, evoking feelings of sadness in an effort to feel sad at a funeral); and (4) *cognitive work* where the thoughts and ideas associated with particular emotions are evoked in an attempt to activate the corresponding feelings.

As Hochschild stresses, individuals are often put in situations where a considerable amount of emotion work must be performed. For example, in her pioneering study of airline attendants,[3] the requirement that attendants always be friendly, pleasant, and helpful even as passengers were rude and unpleasant placed an enormous emotional burden on the attendants. They had to manage their emotions through emotion work and present themselves in ways consistent with highly restrictive feeling and display rules. Virtually all encounters require emotion work, although some, such as the one faced by airline attendants, are particularly taxing and require a considerable amount of emotion management in self presentations.

Hochschild's work is one of several efforts in the late 1970s to examine emotions sociologically. The emphasis on how culture defines what people are to feel in situations and how these feelings are to be presented became part of all subsequent theories, and her analysis of the mechanisms of "emotion work" added a dramaturgical emphasis to emotional dynamics. At this time, other theories squarely within the symbolic interactionist tradition were also being formulated; these emphasized the centrality of self and identity, thereby providing important conceptual supplements to Hochschild's pioneering work.

SYMBOLIC INTERACTIONIST THEORIES OF EMOTION

Susan Shott's Role-Taking and Social Control Theory

One early statement on emotion within the symbolic interactionist tradition was made by Susan Shott.[4] Squarely in the Chicago School of symbolic interaction, Shott emphasizes that emotions involve both physiological arousal and labeling of this arousal as affect or emotion. Following Hochschild, she recognizes that this labeling is constrained by cultural forces, especially "feeling rules" that prescribe or discourage certain emotions and "vocabularies of emotion" that give individuals the terms to describe their emotions to themselves and to each other. Individuals thus must perform "emotion work" to arouse those emotions that are normatively appropriate to a situation. And, because the physiological arousal generating emotions is so undifferentiated, diffuse, and nonspecific, emotions must be socially constructed as actors define situations.

Within this general framework, Shott then tries to develop a theory of emotions that draws on basic symbolic interactionist tenets: (1) Individuals possess the capacity to respond to themselves as objects, having internalized conversations with themselves as they go about defining situations and responding to others. (2) Individuals develop conceptions of themselves by virtue of role-taking with (a) specific others (whether real or imagined) in a situation and (b) generalized others or broader communities of attitudes. And (3), individuals engage in self-control by virtue of self criticism about how they see themselves in situations, especially from the perspective of the generalized other.

Much role-taking is reflexive in that the individual has an internal conversation with self as an object, seen and evaluated from the perspective of specific and generalized others. In this evaluation process emotions are aroused and labeled; and if these emotions are negative, they mobilize the individual to adjust behavior. Shott sees several emotions as particularly important in controlling the actions of individuals: (1) guilt, (2) shame, (3) embarrassment, (4) pride, (5) vanity, and (6) empathy. Thus, as individuals role-take with others and generalized others, these six classes of emotions toward self and others are aroused and labeled; and they constrain the individual's responses.

(1) *Guilt* is an emotion that accompanies a negative self-evaluation that emerges when an individual recognizes that behavior is at variance with how one is obligated to behave. Guilt is thus activated when the individual commits or contemplates committing acts that are immoral from the point of view of significant others or the generalized other. When individuals experience guilt, they are motivated to avoid those persons who activate this feeling, but they are also motivated to engage in altruistic behavior to compensate for their inappropriate behavior, or even their thoughts about engaging in such behavior. As such, guilt not only brings the individual back into line; it also prompts the person to act in ways that promote moral actions that reaffirm cultural values and promote social solidarity.

(2) *Shame* is aroused not so much because the person considers self to be morally deficient, but rather because the individual senses that others see self as deficient. Shame is a particularly powerful emotion because it attacks people's fundamental sense of identity, or what Shott terms *general identity*. People experiencing shame are typically motivated to avoid those situations and others where they must endure this powerful emotion, but shame also motivates individuals to engage in altruistic behavior to demonstrate that self is not deficient. In so doing, shame breeds conformity and solidarity-producing behaviors.

(3) *Embarrassment* is seen by Shott as distinct from shame, because embarrassment arises when an individual's presentation of a *situational identity* is seen by the person and others as inept. Thus, shame is generally accompanied by embarrassment, but embarrassment need not include shame. Embarrassment is most directly connected to the immediate presence and responses of others in a situation. From role-taking with others, the individual perceives that the presentation

of self is inept. Like guilt and shame, embarrassment prompts people to avoid those situations where they feel this emotion, but at the same time, embarrassment can encourage individuals to engage in compensatory actions to demonstrate that their ineptitude is only a momentary lapse. Often, this compensatory behavior is altruistic in nature.

(4) *Pride* is the self-approval that accompanies role-taking with others and generalized others. In Shott's view, pride is a more permanent emotion that is less tied to each and every role-taking situation. Individuals carry pride with them from encounter to encounter, providing them with a sense that they stand well in the eyes of others and generalized others. When individuals become accustomed to experiencing pride, they regulate their behaviors to secure approving responses of others and, thereby, sustain not only this sense of pride but also the social order.

(5) *Vanity* is a more transitory and unstable sense of self-approval in which the individual is unsure of the approval of others. Vanity prompts people to become dependent on the immediate responses of others, constantly seeking their favorable reactions. This need to seek approval, however, controls the responses of individuals because they must continually act in ways that will elicit vanity-confirming responses.

(6) *Empathy* operates at two levels: (a) arousing in self the very same emotions that others are experiencing; and (b) generating a cognitive understanding of what it would feel if self were in another's place. Empathy connects individuals to others at an emotional level, leading people to share each other's emotions. If they experience positive emotions, these are shared and promote actions to sustain this positive arousal. If, on the other hand, others experience more negative emotions, these are also shared and encourage individuals to act in altruistic ways to help alleviate the burden of negative emotions and to, thereby, promote social solidarity.

These six classes of emotions control individuals, and because it is impossible to monitor and sanction every act or thought of a person, the process of role-taking as it produces these emotions is the most important mechanism of social control in society. Such control is self-control because people see themselves and others in certain emotional tones and respond to these emotional shadings in ways that bring their behaviors into conformity with norms and that, in some cases, prompt altruistic behavior.

At about the time that Shott was developing this constructivist view of emotions and social control, David Heise was initiating a more Iowa School strategy for understanding how emotions regulate and control people's responses. This "affect control theory" overlaps, to a degree, with Shott's Chicago School theory on role-taking, but the differences are significant, and they demark the two poles of theorizing within symbolic interactionism.

David Heise's Affect Control Theory

David Heise[5] and various collaborators,[6] particularly Lynn Smith-Lovin,[7] have developed a theory of emotions that builds on symbolic interactionist theories emphasizing the dynamics of identity. Much of this theory is stated in highly technical terms, involving equations and the language of computer simulations, but fortunately several more discursive summaries are now available.[8] The basic idea in affect control theory is that individuals react affectively or emotionally to every social event and that they do so to the degree of correspondence between *fundamental sentiments* or culturally established affective meanings that transcend particular situations, on the one side, and *transient impressions* or feelings that are specific emotional reactions to the particulars of a situation, on the other. Correspondence between fundamental sentiments and transient impressions will reduce emotional arousal, whereas discordance between fundamental sentiments and situational impressions arouses emotions and motivates individuals to bring their fundamental and situational sentiments into correspondence. Thus, in affect control theory, the basic motivating force in human action and interaction is the reduction of discrepancies, or what is termed *deflection*, between fundamental and transient sentiments and, as deflection is reduced, control of affect or emotion is achieved.

What makes this theory a variation of symbolic interactionism is its emphasis on the dynamics of "defining the situation." Humans are viewed as developing cognitions about every situation, with language being the primary symbolic medium through which cognitions are formed. Thus, people see situations as linguistic designations, and they represent, process, and communicate their cognitions to themselves and others through language. This emphasis on language is typical of symbolic interactionist theories, and it explains why the measurement of the affect associated with cognitions has used the "semantic differential" test in which pairs of words, such as good-bad, big-little, nice-awful, fast-slow, young-old, noisy-quiet, powerful-powerless, sad-happy, strong-weak, and so on—are presented to subjects as they respond to objects, persons, and situations. These linguistic designations for fundamental sentiments were seen in early formulations of the theory to vary along three basic dimensions:[9] (1) *evaluation* or an assessment that determines the degree of goodness or badness; (2) *potency* or the level of power or powerlessness; and (3) *activity* or the extent of animation or passivity. As people define situations, therefore, they invoke fundamental sentiments revolving around evaluations, potency, and activity, then they compare these fundamental sentiments to their more situation-specific impressions or their transient sentiments, which are also constructed in terms of evaluations, potency, and activity. And, as emphasized earlier, individuals are motivated to sustain a correspondence—or alternatively, to avoid deflections—between these two levels of sentiment.

What aspects of a situation are assessed through evaluation, potency, and activity? All definitions of a situation involve cognitions and associated emotions about (1) self or identity, (2) others and their identities, (3) behaviors, and (4) situation or setting. Thus, cognitions invoke affective associations for evaluation, potency, and activity relative to people's identity, their perceptions of others and their identity, their role behaviors, the nature of the situation, and the relevant rituals, scripts, norms, and other situational information. The unit of analysis in affect control theory is thus the *social event*, revolving around an individual (A) performing an act (B) on some object or person (O) in a setting (S), or ABOS. In all situations where social events occur, there are relevant *grammars* that limit the kinds of behaviors toward others that are possible and appropriate, while indicating the form and structure that behaviors are to take. For example, particular grammars indicate how actions toward others occur for situations as various as funerals, graduations, workplaces, or within the family or for each of the fundamental types of situations that constitute social life.

As a brand of symbolic interactionism, the notion of *identity* is central. Identity is composed of the cognitions and associated affect about oneself as an object as well as the person's assessment of the self being presented by others. Just what identities can be presented by self and other in an encounter are, of course, limited by the grammar of the situation. People select identities for presentation that conform to the grammar; in this way, they can minimize deflection. According to Heise, people's cognitions initially sort the range of possible and appropriate behaviors in a situation; then, the emotional reactions to these cognitions are activated along the evaluative, potency, and activity dimensions; and finally, these cognitions and emotions combined influence the kind of identity and role behavior that will be presented.

Thus, people assert identities through their role behaviors in a situation, and they assess the role behaviors of others to impute the identity being asserted by these others. As they do so, individuals present the emotions relevant and appropriate for their identity, and they read the emotions of others to facilitate imputation of the identity of others. The reciprocal reading of associated emotions enables them to achieve intersubjectivity.

More than mutual role-taking and role-making is involved in presenting and imputing identities, however. Identities are presented to avoid deflection or discordance between fundamental sentiments and transient sentiments in a situation. Identity becomes a prime vehicle for asserting and marking fundamental sentiments, as well as for assessing discordance between fundamental and transient sentiments. People see role behaviors as indicators of identities, then they determine if these identities are being played in ways that confirm, or disconfirm, fundamental sentiments. For instance, if a female presents herself in a situation as a mother, a designation that carries a certain evaluation in culturally established sentiments about evaluation, potency, and activity, but then proceeds to hit her child, a deflection has occurred. The fundamental sentiments associated with the identity of mother and the role behavior that mothers should exhibit are being contradicted by the actual role behaviors and transient impressions of what is going on. This deflection motivates people to reduce this discordance. There are three possible options: First, the role behavior can be reconceptualized, leaving the individual's identity in tack (for example, the mother is only playfully striking her child). Second, the setting can be reconceptualized (for instance, the child has been very bad and, for its own good, must be disciplined), but there are limits on such reconceptualizations imposed by the grammar of the situation. And third, the imputed identity of the person can be changed (for example, the person is a very bad mother).

This last option is quite common because it is often difficult to ignore or redefine role behaviors, and the grammars of situations limit redefinition of situations. What is left is to impute a new identity to the individual in the face of large deflections and the associated emotional arousal. There are two paths to such change. One is to impute *dispositional traits* to the actors (for instance, the mother was momentarily angry) that modify the identity (a mother with a short fuse, for instance) but leave much of the identity intact. A second and less frequent path is to impute a whole new identity that replaces the older one (this woman was not a mother but rather mean-spirited caretaker, for example). Because changes in identity are so traumatic, invoking considerable emotion and forcing people to redefine themselves and the situation, individuals can often avoid such reordering of cognition and emotion by emitting and recording emotions that confess to the disconfirming behavior; in essence, an apology offered (for example, the mother expresses how terrible she feels about losing control). When such emotional displays are accepted by others, the person can avoid being shoved into a new identity, and the others in the situation can forgo the cognitive and emotional gymnastics of imputing new identities. Indeed, people will try as long as possible to interpret and reinterpret social events so that deflection can be minimized. Humans seek routines and emotional stability, so they are motivated to keep deflections low by ignoring them—as long as they can.

From the point of view of affect control theory, then, the experience and expression of emotion can be a sign that deflection is occurring. The type of emotion, whether positive or negative, will be related to the nature of the deflection. Positive emotions can arise from deflections where a person's identity is seen by others as better, more potent, or more active than the person himself or herself perceives. Moreover, a negative identity held by a person (as unworthy, impotent, or inactive) will produce negative emotions when others confirm these self-images, even though deflection has not occurred. Thus, emotion and deflection can be related to each other in complicated ways, but affect control theory emphasizes that deflection will arouse people to bring "things back into line" and that emotion is what energizes people to do so.

This emotion mobilizes the individual asserting an identity to do something about the deflection, and the emotion forces others to begin redefining the situation. People will perhaps protest with shining modesty when deflection enhances their identity, or, perhaps if deflection is consistently positive, a new and more highly evaluated, potent, or active identity will emerge. Conversely, people feel "awful" when their role behaviors disconfirm their asserted identity, and they try to bring the two back into line, if they can. Others feel awkward and unsure when such disconfirmation occurs, and they try to redefine situations to bring fundamental and transient sentiments back into correspondence.

Sheldon Stryker's Identity Theory of Emotions

Sheldon Stryker's general theory of identity was summarized in Chapter 27, so here we only need to focus on the theory as it was later expanded to explain the dynamics of emotion. Stryker initially argued that individuals possess a hierarchy of identities,[10] and those identities high in this salience hierarchy are likely to guide role behaviors. Indeed, role behaviors are orchestrated to reveal an identity to which an individual is committed. The more committed to an identity an individual is, the more the individual's self-esteem rests on others' positive evaluation of the role performances marking this identity. Moreover, identities tend to be presented to conform to broader cultural values, normative expectations of social structures, and definitions of situations.

Emotions are implicated in these processes in several ways.[11] First, those role enactments that generate positive affect and reinforcement from others in a situation strengthen a person's commitment to an identity, moving it higher in the salience hierarchy. As individuals receive this positive feedback, their self-esteem is enhanced, which further increases commitment to the identity, raising it in the salience hierarchy and increasing the chances that this identity will shape subsequent role performance.

Second, when role performances of a person and others are judged inadequate in light of normative expectations, cultural values, definitions of the situation, or identities being asserted, negative emotional reactions mark this inadequacy. Conversely, when role performances are adequate or even more than adequate and exemplary, positive emotions signal this. Thus, emotions are markers of adequacy in role performances, telling individuals that their performances are acceptable or unacceptable. This marking function of emotions works in several ways. The individual reads the gestures of others to see if a role performance has been accepted, and if it has, then the person experiences positive emotions and will become further committed to the identity that was asserted in the role performance. If, on the other hand, the reaction is less than positive, then the individual will experience negative emotions—such as anger at self, shame, and guilt—and mobilize to improve the role performance, or if this is not possible, to lower the commitment to this identity being asserted in the role, moving it lower in the salience hierarchy and selecting a different identity that can be more adequately played out in a role. Not only do individuals get emotional about their own performances as they role-take with others, but they also inform others about the latter's role performances. Because role performances must be coordinated and meshed together to be effective, inadequacy by others will disrupt one's own role performance, and if this occurs, a person will manifest some form of anger and negatively sanction others. Thus, emotions become ways for individuals to mutually signal and mark the adequacy of their respective role performances in ways that facilitate the coordination and integration of roles.

Finally, emotions are also a sign of which identities are high in a person's salience hierarchy. If emotional reactions are intense when a role performance fails or when it is successful, this intensity indicates that a person is committed to the identity being played in a role and that it is high in the salience hierarchy. Conversely, if the emotional reaction of the individual is of low intensity, then this might signal that the identity is lower in the salience hierarchy and relatively unimportant to the individual.

In identity theory, then, emotions motivate individuals to play roles in which they receive positive reinforcement, and emotions also inform individuals about the adequacy of their performances and their commitments to identities in the salience hierarchy. Emotions thus drive individuals to play roles in ways that are consistent with normative expectations, definitions of the situation, cultural values, and highly salient feelings about self.

STATUS AND POWER THEORIES OF EMOTION

Randall Collins' Theory of Interaction Rituals

We encountered Randall Collins' theory of interaction rituals earlier in Chapter 15 on Weberian conflict theory.[12] Attention is now drawn to his micro-level theory of emotions. Collins' theory is about the levels of emotional energy that interaction rituals generate rather than about specific types of emotions, although as we will see shortly, he has teamed with Theodore Kemper to address the issue of how power and status produce particular types of emotions (see next section). But the basic theory, as Collins has developed it on his own, borrows from Émile Durkheim's and Erving Goffman's insights about the emotion-generating effects of rituals.

Collins believed that the micro unit of analysis is the *encounter* of at least two people who confront each other and interact. What transpires in such encounters is mediated by the exchange of resources and rituals. Rituals contain the following elements: (1) a physical assembly of co-present individuals; (2) mutual awareness of each other; (3) a common focus of attention; (4) a common emotional mood among co-present individuals; (5) a rhythmic coordination and synchronization of conversation and nonverbal gestures; (6) a symbolic representation of this group focus and mood with objects, persons, gestures, words, and ideas among interacting individuals; and (7) a sense of moral righteousness about these symbols marking group membership. Figure 32-1 portrays the dynamics of such rituals that Collins sees as the "emotional energizer" of interaction because, like Durkheim before him, Collins argues that co-presence, mutual awareness, common focus of attention, rhythmic coordination and synchronization of gestures and talk, common mood and symbolization arouse emotions that, in turn, feed back and heighten people's sense of co-presence, mood, attention, and symbolization.

Intermingled with these emotion-arousing properties of rituals are exchange processes. Collins visualizes two basic types of resources as crucial to understanding exchanges and rituals: (1) cultural capital and

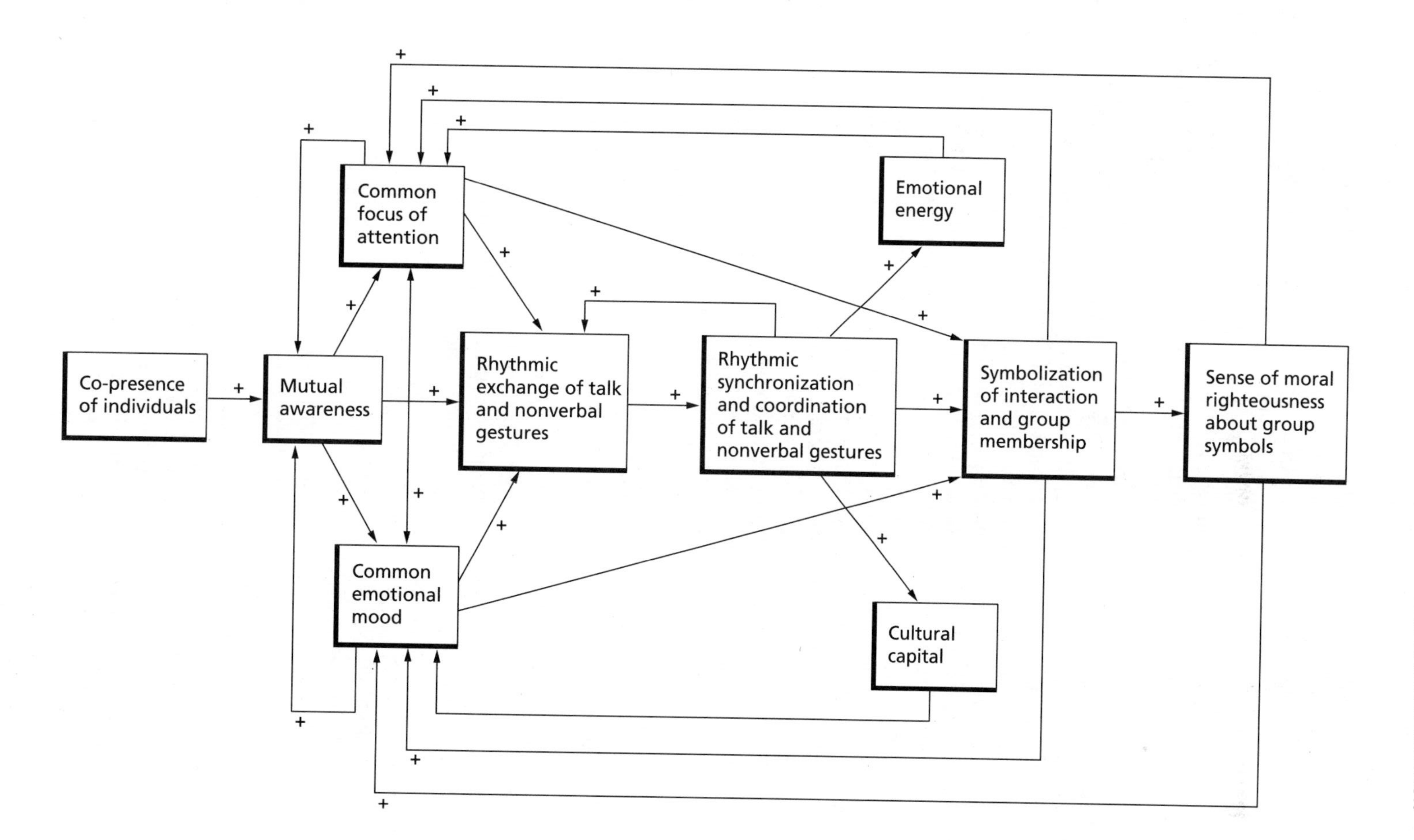

FIGURE 32-1 Collins' Conceptualization of Ritual

(2) emotional energy. *Cultural capital* consists of such resources as stored memories of previous conversations, vocal styles, special types of knowledge or expertise, the prerogatives to make decisions, and the right to receive honor. The concept of *generalized cultural capital* denotes those impersonal symbols that mark general classes of resources (for example, knowledge, positions, authority, and groupings), whereas the concept of *particularized cultural capital* refers to the memories that individuals have of the particular identities, reputations, and network or organizational positions of specific persons. *Emotional energy* is composed of the level and type of affect, feeling, and sentiment that individuals mobilize in a situation.

Interaction consists of individuals using their cultural capital and emotional energy to talk with one another. Such conversations involve an investment of capital and energy, with each individual attracted to situations that bring the best available payoff in cultural capital and emotional energy. Although individuals also seek a profit in the cultural capital that they spend and receive in interaction, Collins appears to emphasize emotional energy as the real driving force of interaction. Humans find positive emotional energy highly rewarding, and though other rewards are not irrelevant, positive emotional energy is the most valuable. Indeed, individuals seek to maximize their positive emotional energy by participating in those interaction rituals that generate a clear focus of attention, a common mood, emotional arousal, rhythmic synchronization of mood and arousal, and symbolization of these in terms of moral codes.

In Collins' view, there is a kind of market for interaction rituals: Individuals weigh the costs in time, energy, cultural capital, and other resources that they must spend to participate in the various rituals available to them, then they select those rituals that maximize emotional profits. In this sense Collins proclaimed emotional energy to be the common denominator of rational choice.[13] Thus, rather than representing an irrational force in human interaction, Collins sees the pursuit of emotions as highly rational: People seek out those interaction rituals in a marketplace of rituals that maximize profits (costs less the positive emotional energy produced by the ritual). The search for emotional energy is, therefore, the criterion by which various alternative encounters are assessed for how much emotional profit they can generate.

Humans are, in a sense, "emotional junkies," but they are implicitly rational about it. They must constantly balance those encounters where interaction rituals produce high levels of positive emotional energy (such as love making, family activities, religious participation, and gatherings of friends) with those more practical, work activities that give them the material resources to participate in more emotionally arousing encounters. Indeed, those who opt out of these work-practical activities and seek only high-emotion encounters (such as drop-outs in a drug culture) soon lose the material resources to enjoy emotion-arousing encounters. Moreover, within the context of work-practical activity, individuals typically seek out or create encounters that provide increases in emotional energy. For example, workers might create an informal subculture in which social encounters produce emotional energy that makes work more bearable, or as is often the case with professionals, they seek the rituals involved in acquiring power, authority, and status on the job as highly rewarding and as giving them an emotional charge (such is almost always the case, for instance, with "workaholics" who use the work setting as a place to charge up their levels of emotional energy).

Not only are there material costs as well as expenditures of cultural capital in interaction rituals, but emotional energy is, itself, a cost. People spend their emotional energy in interaction rituals, and they are willing to do so as long as they realize an emotional profit—that is, the emotional energy spent is repaid with even more positive emotions flowing from the common focus of attention, mood, arousal, rhythmic synchronization, and symbolization. When interaction rituals require too much emotional energy without sufficient emotional payoff, then individuals gravitate to other interaction rituals where their profits are higher.

What kinds of rituals provide the most positive emotional energy for the costs involved? For Collins, those encounters where individuals can have power (the capacity to tell others what to do) and status (the capacity to receive deference and honor) are the most likely to generate high emotional payoffs. Hence, those who possess the cultural capital to command respect and obedience are likely to receive the most positive emotional energy from interaction rituals. At this point Collins' theory begins to converge with that developed by Theodore Kemper, as we now explore.

Theodore Kemper's Status-Power Model

Theodore Kemper was one of the pioneers in the sociology of emotions. Originally, he termed his approach "a social interactional theory of emotions,"[14] and more recently, he has teamed with Randall Collins to develop the theory further.[15] The basic idea of the theory is quite simple: Individuals' relative power and status in social relationships, and changes in their power and status, have important effects on their emotional states. *Power* is the ability to compel others to follow one's wishes and dictates, whereas *status* is the giving and receiving of unforced deference, compliance, respect, and honor. All social relationships reveal power and status dimensions. According to Kemper and Collins, these dimensions are fundamental to understanding interaction and emotions. From Kemper's review of the literature on emotions,[16] he concluded that there are four basic and primary emotions: fear, anger, happiness/satisfaction, and sadness/depression. These primary emotions can, however, be elaborated into more complex combinations and variants, creating such complex and important emotions as guilt, shame, pride, jealousy, love, and other variants and combinations of the four primary emotions. Indeed, Kemper has argued that these primary emotions have a neurophysiological basis, having been selected in humans' evolutionary history because they promoted survival. Kemper thus attempts to explain how power and status generate the four basic emotions and important variants.

The theory distinguishes among (1) *structural emotions,* which are those affective states aroused by the relative power and status of individuals within social structures; (2) *situational emotions,* which arise by virtue of changes in power and status during the process of interaction; and (3) *anticipatory emotions,* which revolve around individuals' expectations for power and status in social relations. The dynamics of emotion thus inhere in the status and power that individuals actually hold, the status and power that is gained or lost during interaction, and the power and status that was expected in an interaction. The basic propositions of the theory are extracted and summarized in Table 32-1. What follows is a more discursive review of the argument is made.

When people have power and increase their power within social relationships, they experience a sense of satisfaction, security, and confidence. If individuals do not have power and, more important, lose power, they will experience anxiety and fear, leading them to lose confidence. Thus, to have power or to gain it gives people the confidence to secure and use more power, whereas to not possess power or to lose it does just the opposite and takes away the confidence so necessary to garnering and sustaining power.

TABLE 32-1: Kemper's Status-Power Theory of Emotions

Power and Emotions

I. The more a person experiences an increase in power in social relations, the greater will be this person's sense of security and confidence; conversely, the more a person experiences a loss of power in social relations, the greater will be this person's sense of anxiety and fear, leading to a loss of confidence.

II. The more a person anticipates a gain in power in social relations but does not receive this gain, the greater will be this person's sense of anxiety and, potentially, sense of fear; conversely, the more a person anticipates a loss of power in social relations but does not receive this loss, the greater will be this person's sense of satisfaction and, potentially, sense of confidence.

Status and Emotions

III. The more a person experiences an increase in status in social relations, the greater will be this person's sense of satisfaction and well being, and the greater will be this person's sense of liking for those who gave status and the more likely will the givers and receivers of status form bonds of solidarity.

IV. The more a person experiences a loss of status in social relations as a result of his or her own actions, the greater will be this person's sense of shame and embarrassment and, if the loss is sufficiently great, sense of depression as well.

V. The more a person experiences a loss of status in social relations as a result of another's actions, the greater will be the person's sense of anger and the more likely will this person be ready to fight, transforming this person's conduct to a power mode of behavior.

VI. A person's reaction to no change in status in social relations will

A. Lead to satisfaction, if this person anticipated no change.

B. Lead to shame and perhaps depression, if this person anticipated an increase in status and perceives that the failure to realize this anticipated increase was caused by his or her own actions.

C. Lead to anger and aggression, if this person anticipated an increase in status and perceives that the failure to realize this anticipated increase was caused by the actions of others.

D. Lead to satisfaction and well being, if this person receives more status than anticipated.

VII. The more a person gives status, the greater will be the sense of satisfaction, if this status is given freely, and the more the receiver of status will express appreciation and gratitude, which, in turn, increase the sense of satisfaction for the giver of status.

VIII. The more a person withholds status, when the giving of status is due, the greater will be the sense of guilt and shame.

Just what people expect before they interact, and what they actually get, are very important in the production of emotions. When individuals expect or anticipate a gain in power, but do not receive this gain, their anxiety and fear will escalate, and they begin to lose confidence in themselves. Conversely, when individuals have anticipated a loss of power, but do not lose power, they experience satisfaction, and they might begin to gain confidence.

For status, a more extensive set of arguments is presented. Much like power, when people experience a gain in status, they will also experience satisfaction. Moreover, they will come to like those who have given them status and, if the relationship persists, they will form stronger bonds with these supportive others. If individuals experience a loss of status in social relations, they will experience shame and embarrassment, especially if they believe that this loss of status is the result of their own actions and, hence, their own fault. If this loss is sufficiently great, they will experience not only shame but also depression. However, if they perceive that this loss of status is caused by the unfair and inappropriate actions of others, they will experience anger and become aggressive toward these others. As their anger rises, they move into a power mode of interaction—seeking to compel others to acknowledge their status.

The attribution of status loss to oneself or to others is, however, complicated by what people have anticipated before the interaction. If they have anticipated no gain or loss in status, and experience neither, then they feel satisfied (a mild version of happiness). If they have anticipated an increase in status but did not get it, and if they perceive that this failure to receive status has been the result of their own actions, then they will experience shame and, perhaps, depression. If the failure to receive expected status is seen as the result of others' inappropriate actions, then individuals will mobilize anger and become aggressive. And if individuals get more status than they anticipated, they will become satisfied.

The giving of status to others also generates emotions. When a person freely bestows status on another, the giver of status experiences satisfaction, and the recipient of the status will express appreciation and gratitude, which will feed back and increase the giver's sense of satisfaction. When a person withholds status that another deserves, this individual will feel guilt and shame for not doing what is appropriate, and persons who did not receive what they deserved (and expected) will experience anger if they correctly perceive that the other has wronged them.

In sum, then, for Kemper—and for Collins as well, in their collaborative work—emotion is ultimately tied to the relative power and status of individuals in social relationships, to shifts in status and power, and to both anticipation and attribution of who is responsible for these shifts. In the end, emotion is driven by those conditions (1) that give people the power to tell others what to do, or that take this power away, and (2) that bestow prestige, deference, and voluntary compliance, or that take this status away.

Cecilia Ridgeway's and Joseph Berger's Expectation States Theories of Affect

In Chapter 33, the core concepts of expectation states theory were reviewed. The central idea in these expectation states theories is that members of groups are assigned a status that, in turn, establishes expectations for their competence and for their role behaviors. Such assignment to a status and its corresponding expectations can be based on diffuse characteristics, such as age, ethnicity, gender, wealth, and other traits that individuals bring to the group from the outside, as well as be based on internal organizational forces establishing lines of authority, rank, and division of labor. In more recent years, one of the many creative branches in this general theoretical program has been the analysis of emotions; the work of Cecilia Ridgeway, Joseph Berger, and Cathryn Johnson can be used to illustrate the nature of this program.

Cecilia Ridgeway's Theory. Like most research and theory within the expectation states tradition, Ridgeway's[17] and Johnson's[18] work focuses mostly on task-oriented groups, where individuals are temporarily assembled to realize a clearly stated goal or purpose. As Ridgeway and Johnson argue, even the most transitory task-oriented group develops an affect structure as members agree and disagree on how to best accomplish the group's goals. Moreover, disagreements can also involve status challenges as lower-status members seek to raise their position and refuse to defer to the arguments of higher-status members.

In all groups, Ridgeway has argued,[19] the broader societal culture penetrates group structures as a set of norms or *blueprint rules* that indicate how people are supposed to behave in particular types of groups, whether task-oriented work groups or more informal and intimate groups. Following Hochschild's point of emphasis,[20] these blueprint rules include (a) "feeling rules" about what kinds of emotions are to be experienced in a situation and (b) "display rules" about what emotions can be expressed publicly and how they are to be expressed. An affect-arousing event will, therefore, activate these feeling and display rules; yet, status expectation processes are still important in organizing the way affect is mobilized and managed.

Disagreements represent one class of affect-arousing events, and when such disagreements among group members occur, status in the group determines whether individuals will blame themselves or others for the disagreement. If a person holds an equal or higher status compared with the other(s) with whom a disagreement exists, then this person will be likely to attribute the disagreement to other(s), and this person will feel and express annoyance (a mild form of anger). On the other hand, if an individual is of lower status than the person(s) with whom there is a disagreement, then this individual will tend to blame self and experience depression. For this reason, Ridgeway and Johnson note that superiors are far more likely to express anger and annoyance with subordinates than the reverse: The superior is expected to be more competent and, when authority is challenged, anger ensues because expectation states are being violated by inferiors, whereas when subordinates disagree with superordinates, expectation states force them to attribute the blame to themselves (because they are presumed to be less knowledgeable and competent). And, because no reward is to be gained from continuing to blame oneself, individuals tend not to pursue the disagreement. Instead, the lower-status person withdraws, often to the point of experiencing depression.

These disagreement-anger episodes can disrupt the solidarity of the group, however, even as the status order is reconfirmed. Anger and annoyance do not promote positive sentiments of liking and mutual trust. Depression by subordinates leads them to withdraw commitment from the group, thereby eroding solidarity further. In contrast, agreement among people of different status has the opposite effect: It promotes solidarity. If those who agree are of lower status, then the higher-status persons feel satisfaction (a moderate form of happiness) about the fact that subordinates agree with them. If those who agree are of equal or higher status, then individuals are likely to feel a more intense form of happiness, such as gratitude, that comes when superordinates agree with them. When people experience these positive emotions, Ridgeway and Johnson argue, they become motivated to reward those who agree with them; as they do so, all parties develop positive sentiments and become more likely to agree in the future. From this cycle of positive reinforcement comes enhanced group solidarity.

These status processes also help explain why groups typically reveal more positive than negative affect. As noted earlier, disagreements emanating from lower-status persons are sanctioned because they violate the expectations associated with lower- and higher-status group members, and hence, lower-status individuals soon stop punishing themselves and keep quiet. If disagreement can be avoided and agreement can be reached, then the positive emotions of satisfaction, happiness, and gratitude initiate cycles of positive reinforcement that evolve into solidarity that further orients group members to express agreement and to reward each other with positive sentiments.

An interesting dynamic, however, is initiated when a lower-status person is highly assertive and refuses to back down. Expectation states lead others to sanction the lower-status person negatively as being too pushy and self-interested—what is termed the "backlash effect" against those who violate the status order. Thus, lower-status individuals will almost always have difficulty in groups where they seek to raise their status by displaying competence; assertions of such competence will be defined as violating the status order and the expectations associated with this order. One strategy for overcoming this "backlash effect" is to couch assertions of competence in friendly and unthreatening ways, perhaps as suggestions rather than as challenges. In this manner, lower-status group members can, over time, slowly raise their status and change the expectations associated with them.

Another interesting facet of this theory is the argument that the degree of congruence between the status structure and the distribution of affect shapes rates of participation in making decisions. When high-status members are liked more than low-status members, their power to influence decisions is enhanced. Lower-status members not only see higher-status members as more competent, they also like them. Because of this congruence in status and sentiment, lower-status individuals are more likely to accept decisions of higher-status members. Congruence of affect and status thus amplifies the influence of status in task groups, and research by Janet R. Johnston demonstrates that such is also the case in more long-term and intimate groups as well.[21] Johnston's findings have, in turn, stimulated Joseph Berger's long-standing research program exploring expectation processes in more intimate and longer-term groups.

Joseph Berger's Affect-Expectations Theory. Joseph Berger argues that as emotional reactions emerge and are repeated over time within groups, these reactions generate *affect expectation states* that shape and guide the flow of interaction.[22] Such affect expectation states unfold in a series of stages: (a) Affect is aroused during the interaction; (b) an exchange of affect occurs and begins to set up expectation for what will happen in the current interaction; (c) these affect states also begin to generate an emotional orientation toward others in the situation that involves more stable sentiments—whether positive or negative—that will shape the course of subsequent interactions; and eventually, (d) assignment of personality characteristics consistent with the affect that is exchanged and the emotional orientations of individuals will occur, further circumscribing current and future interactions.

Through this approximate sequence, the affect structure of a group stabilizes the interaction, giving it consistency over time. Moreover, people develop situational identities as they exchange affect and impute personalities to each other. Once these identities are formed, they become expectations for what will occur in the situation, thereby further stabilizing the interaction. Individuals thus become driven and constrained to act in certain ways through the exchange of affect and the development of affect expectations. Yet, this kind of affective structure does not completely determine or circumscribe what occurs in the group. For, if expressive exchanges consistently do not confirm the affect expectations that have developed, then the affective orientations of individuals, the imputation of personality traits, and perhaps the situational identities of individuals will all change, and new affect expectations will begin to emerge.

NETWORK THEORIES OF EMOTION

Exchange theories all argue that individuals engage in transactions because they are rewarding, at least in relation to alternatives that they might otherwise pursue. Thus, from an exchange viewpoint, social solidarity is generated under certain conditions (see

Chapter 24 on rational choice theories): First, individuals find the resources produced and consumed by group members as rewarding. Second, they have incentives to develop norms to regulate their conduct to cut down on free-riding (that is, consumption of a "joint good" or the resources of the group without contributing to their production). Third, they develop procedures for informally monitoring each other and positively sanctioning conformity to norms. This approach implies that emotions are central to group solidarity, but the effects of emotions tend to be seen as just another jointly-produced good or resource that is exchanged. Such an argument seems reasonable, as far as it goes, but some exchange network theorists have sought to explain in more precise detail just how emotions generate social solidarity.

Barry Markovsky and Edward Lawler's Theory of Group Solidarity

Barry Markovsky and Edward J. Lawler have proposed one kind of network theory of group solidarity that makes the production of emotions a central dynamic. As summarized in Chapter 38, a network is a set of points or nodes that are, to varying degrees, connected to each other; the points can be individuals or collective actors; and the connection among actors involves the flow and exchange of resources. In Markovsky and Lawler's theory of group solidarity, the key feature of the network is "reachability" or the degree of connectiveness among the positions (sometimes denoted as the *density* of the network). If individuals can reach each other, make contact, and exchange resources, this "reachability" is the first structural condition necessary for producing high solidarity and is labeled by Markovsky and Lawler as *cohesion*. However, this structural condition of cohesion is not, by itself, sufficient to produce solidarity. Two additional conditions must be met: One is that the actors who can reach each other or who are in structural cohesion must form relatively strong bonds and attachments that endure over time. Another is that the individuals in the network have a *unity of structure* in which they remain reachable and in contact with each other, thereby avoiding a breakdown into subcliques within the larger network structure. What force, then, generates strong bonds and works against subclique formation? Markovsky and Lawler's answer is emotion.

Emotions are the "glue" that ties social structures together, establishing and maintaining bonds and attachments, and overcoming fracturing of the network into subcliques. The argument is presented as a series of formal propositions, which are rephrased and summarized here in a more discursive manner. The theory begins with the idea that expectations are an important force in an encounter among individuals: If people expect to experience positive emotions, they will be more likely to enter the encounter, whereas, they will avoid those encounters where they expect to receive negative emotions. If a person who expects to receive positive emotions enters the encounter and does indeed receive these emotions, then emotional attachment of the individual to the others in the encounter is strengthened. As the person-to-others attachment strengthens, these others experience positive emotions from the person and, as a consequence, they begin to develop attachments to the person.

As this person-encounter relationship develops, the person increasingly comes to expect positive emotions from others in the network, even if the person has yet to make contact with them. People, Markovsky and Lawler emphasize, tend to "overgeneralize" sentiments to the group or network as a whole, and as the strength of relations among all persons in the group increases, they all will "overgeneralize" positive sentiments to the larger social network as a whole.

Under these conditions, each actor begins to infer that relations among all others are positive and strong. If a person observes positive relations among others in the network, this person vicariously experiences the positive emotions, thereby increasing emotional attachments to the network as a whole.

There is, however, a counter-solidarity force in this constantly escalating attachment process: loss of identity. Markovsky and Lawler argue that individuals have a threshold level beyond which they begin to feel that they have lost the capacity to define who they are when confronted with the positive emotions that forge strong bonds of attachment and unity of social structure. When this sense of losing self increases, the person experiences negative emotions, and as these negative emotions surface, an individual will make efforts at detachment from the network—thereby reducing its solidarity.

Aside from this countervailing force of identity loss, however, structural cohesion (or reachability in a network) can set off solidarity-generating forces that can be difficult to arrest. Moreover, certain conditions accelerate solidarity-production. One is conflict with another group, which can break down cliques within a network and connect them in ways that increase reachability as they confront a common foe. Another is the use of positive sanctions over negative sanctions because positive sanctions (that is, rewarding individuals for abiding by norms) produce positive emotions and thereby strengthen social bonds and attachments.

Edward J. Lawler's Theory of Emotion and Commitment

Lawler in conjunction with others, particularly Jeongkoo Yoon, has developed a theory of affective commitment in exchange networks.[24] This theory builds on Richard Emerson's seminal analysis of power and social exchange, but it takes this approach in new directions, as we encountered earlier in Chapter 25. Here, we briefly summarize the theory to emphasize its importance in extending network theorizing. When actors mutually depend on each other for resources, valuing the resources and seeing these others as their best option among alternative sources of resources, these actors will begin to exchange more frequently; even if bargaining is involved, their frequency of agreements will increase. This effect is even more pronounced when actors are highly dependent on each other for resources and when their levels of resources are relatively equal (that is, one actor does not possess a larger power advantage over the other). Frequency of exchange under these conditions generates positive emotions—pleasure, satisfaction, and even excitement as well as anticipation. This frequency of interaction and the associated emotions increases what Emerson had termed *relational cohesion,* which, Lawler and Yoon argue, increases the degree of *commitment* of individuals to the interaction, even if attractive alternatives for the resources exchanged in the relation become available. As a consequence, individuals become more likely to stay in the relationship, to exchange token gifts, and to contribute to the relation even if risks and uncertainty about outcomes exist.

A related process is that the emotions generated in the exchanges lead to an *objectification* of the relationship in the sense that actors come to view the relationship, per se, as something beyond them and as providing for the positive emotions that increased commitment. This objectification is, of course, similar to interaction ritual theories where relations become "totemized" as something toward which moral sentiments and rituals are directed. And like interaction ritual theory, objectification is connected to power, but in a way that differs from power-status theories. The total power of the relationship (that is, the level mutual dependence of actors on each other for valued resources) and the equality power (that is, no actor is dramatically more dependent than another) do not directly affect emotions and commitment; rather, these structural features of the exchange network operate indirectly through *frequency of exchange* and agreements in bargaining to produce positive emotions, commitment, and objectification. Moreover, as Figure 25-3 on page 329 delineates, these forces feed back on each other, escalating their effects in generating emotions, objectification, and commitments as these increase attachments, bonds, and group formation.

EVOLUTIONARY-INTERACTION THEORIES OF EMOTION

Michael Hammond's Affect Maximization Theory

In recent years, Michael Hammond has proposed an evolutionary theory of human emotions that also evokes elements of exchange theory.[25] In Hammond's view, humans are biologically wired to seek positive emotional arousal and, conversely, to avoid negative emotions. Emotional arousal is what motivates individuals; in evolutionary terms, there was selective advantage to a species that could energize itself with emotional arousal. Unlike other animals, however, humans possess a relatively weak instinctual package to guide and direct behavior; instead, to energize their behavior, humans must create situations that arouse positive and decrease negative emotions. Human motivation is thus less directed than constructed as individuals seek social ties that generate positive affect. In Hammond's argument, individuals not only seek positive affect; they also attempt to maximize affect, leading them to construct various strategies for doing so.

One strategy is to build multiple affective ties so that the loss of one tie does not deny the actor all affect; instead there are alternative ties to fall back on. Another strategy is build ties of varying intensity, enabling individuals to modulate and select different levels of affect arousal as they want or need. A related strategy is to arrange affective ties into hierarchies of value so that some are more appealing and others less so, thereby allowing individuals to anticipate the level of affect they will receive as they enter situations. All these and other strategies are limited by two important considerations: (1) the constraints on the information processing capacities of actors; and (2) the constraints imposed by habituation. Each of these will be discussed.

(1) Humans have limited abilities to gather and process information about the social world; hence, they must economize in gathering information about possible sources of affect arousal. Moreover, some types of ties require a great deal more information than other ties to assess their affective potential. These constraints force actors to adopt strategies for limiting the costs of gathering information relative to the amount of affect that can be received. One strategy is to mix high and low information ties, with those requiring low information giving an assured level of moderate affect (as might be the case for a casual friendship with a colleague at work) and those requiring high information possessing the potential to offer high levels of affect (as might be for a relationship that could evolve into "love" or some other high affective state). In this way, the costs of time and energy in securing knowledge about moderate-affect ties are much less than those for a high-affect ties. Another strategy is to "embody" various relationships in objects, rituals, and other markers that inform a person that a certain level of affect can be achieved when these markers are present. For example, the objects associated with a "singles' bar" signal one level of potential affect; or, those with a birthday or wedding ritual indicate yet a different kind and level of

affect. Thus, association of various types of affective relationships with signs and signals is a short-hand way to reduce costs in securing awareness of the affect to be received in a situation. Still another strategy is to embed ties in collectively organized units, such as groups, communities, and social categories (like ethnicity), so that some ties always remain available even as others are severed. For instance, a person might identify with an ethnic category, viewing ties to fellow ethnics as providing a certain level of positive arousal, and even if one relationship is lost, fellow ethnics can be found to provide the positive arousal. Or, a person might visualize gender categories as providing a pool of ties offering a certain level of affect, thereby enabling this person to draw emotionally on all those in this category. Or, a club and organization can similarly offer multiple avenues for affective ties of a certain type and intensity. As various social units and categories are viewed by their capacity to offer affective arousal, individuals come to rely on them. If the emotions aroused are highly positive, then individuals see these units as highly salient because of the particular kind and level of affect to be secured.

(2) All situations that arouse affect confront *habituation* or the inevitable tendency for the emotion-arousing effects of a situation to dissipate over time.[26] Within exchange theories, habituation is phrased as marginal utility or satiation (the value of further increments of some reward begins to decrease). For actors seeking to maximize affect, habituation presents the same problem: the longer a high-affect interaction persists, the more accustomed to this type of affect a person becomes, and, hence, the less rewarding the affect will be over time. Thus, to maximize affect, individuals must find a way around habituation, or in Hammond's terms, develop *anti-habituation strategies*. The only ways to "cheat" on habituation and sustain high levels of affect are (a) to gain access to a greater variety of affect-arousing situations, and (b) to create novel situations that can provide new sources of affect that have not been diminished by habituation. These strategies are limited, however, by the amount of time and energy that people have and by the scale of the society in which they must seek sources of affect. There is only so much energy that a person can mobilize and only so much time within which to use this energy to create sources of affect, and as these time and energy requirements increase, the costs of affect rise and reduce the net level of affect that a person can get. The scale of a society can reduce these costs because large, highly differentiated societies offer many more options and opportunities for variety and novelty than smaller and less differentiated societies. Indeed, Hammond argues that the first human societies composed of small hunting and gathering bands greatly limited the amount of affect arousal people could experience because these societies could not offer much variety in sources of affect and because habituation inevitably lowered the affect received in the limited number of interpersonal spheres available in such small societies. Periodic religious rituals or perhaps a long-distance hunt might provide the most affect arousal, but much of the affect to be gained from repetitive, day-to-day activities and ties among people in simple societies had diminished because of habituation. Only when populations got larger and the complexity of social structure increased, Hammond argues, could humans consistently find ways around habituation.

When societies differentiate, they provide a wider variety of contexts—groups, social categories, communities, organizations—in which people can achieve affective arousal and mitigate against habituation. Inequality has been a particularly potent force of anti-habituation, at least for elites; high ranking individuals can avoid many of the costs (they can get others to incur them) involved in creating varieties of affect arousal. Thus, in Hammond's view, the creation of inequality can be viewed as an anti-habituation strategy. Thus, those who possess power and prestige can achieve much more affect arousal than those who do not have these resources. Power and prestige give actors the capacity to get others to do much of the dirty work in receiving rewards (hence lower the costs for receiving these rewards) and, at the same time, give them the resources to explore new avenues of affect.

Whether the evolutionary argument in Hammond's theory is correct might be less important than the general model proposed: Humans seek to maximize positive affect by (1) lowering the costs of gathering information and expending energy in finding reliable sources of affect and (2) cheating on habituation. When human behavior and social structure are viewed in these terms, theorizing on emotions takes a new direction: searching for the strategies employed by individuals to lower their costs and to get around habituation.

Jonathan Turner's Evolutionary Theory

As part of a larger effort to bring biological thinking back into sociology, Jonathan Turner has recently proposed an evolutionary theory about why humans evidence the capacity to read and use such a diverse array of emotions.[27] His theory begins with a controversial conclusion to be drawn from the data on primates: Those primates—the Great Apes such as orangutan, chimpanzee and gorilla—genetically closest to humans are not highly social. Hence, it is reasonable to conclude that at a basic genetic level, humans are not as social as is commonly assumed because, after all, humans and apes share a common ancestor in the not too distant past (and indeed, humans and chimpanzees still share 98 percent of their genes). Like their ape cousins, humans appear to desire autonomy, individuality, and mobility. Yet, as the distant ancestors of humans began to settle onto the African savanna, natural selection favored increased group organization to secure food and confront predators. How, then, were humans' ancestors to evolve increased sociality and solidarity given their ape legacy and propensity for low sociality?

Turner's answer is that natural selection reorganized the neurology of the brain in ways enabling greater cortical or rational control over emotional responses. Once this control existed (current apes do not have such control, it should be emphasized), emotions could be used to forge stronger social bonds and, thereby, increase fitness. In turn, once control and use of emotions increased chances of survival, selection worked to enhance this ability. Thus, in Turner's model, the dramatic changes in the size of the neocortex (the centers for rational thought) and the limbic system (the centers for emotion) in the human brain, and the rewiring of these centers to each other, occurred under intense selection pressures to produce an animal that could construct flexible and more tight-knit social bonds built on emotions.

At this point interactionist theory can inform biology because the selection pressures that guided the reorganization of the brain are sociological in nature. Thus, Turner asks, What kinds of sociological selection pressures were at work? His answer is that selection reorganized the brain to produce greater emotional capacities for (1) mobilization of energy, (2) attunement of responses, (3) sanctioning, (4) moral coding, (5) valuing-exchanging, and (6) decision-making. Each of these will be briefly discussed.

(1) Mobilization of Energy. All primates can mobilize emotional energy, so it is not so much the amount of emotion expressed that becomes important but, rather, the spectrum of emotions that can be evoked. To control and emit a wide array of emotions of varying levels of intensity gave humans the ability to develop diverse and flexible social structures. Yet, because of humans' genetic propensities for low sociality, the use of socially constructed rituals—as Émile Durkheim, Erving Goffman, and Randall Collins have emphasized—is still necessary to generate sufficient amounts and types of bond-producing emotions. For this reason, Turner concluded, all human encounters are ritualized: rituals ensure that emotions will mobilize sufficient energy to construct social bonds.

(2) Attunement of Responses. In the absence of genetic programming for high sociality, natural selection increased the capacity for emitting and reading a robust configuration of emotions. Humans can role-take and role-make in very fine-tuned ways because they can see and use many varied emotional cues to align and coordinate their actions. Such attunement is mostly visual and bypasses language centers. As a result, this reading of emotion-laden gestalts and configurations of gestures can occur very rapidly, enabling humans to process an incredible amount of information on emotions simultaneously. Such capacities were, no doubt, selected in humans' evolutionary history to facilitate flexible patterns of social bonding.

(3) Sanctioning. To sustain social bonds and group structures over time, sanctions become crucial. Without emotions, however, sanctions have no "teeth" or force behind them, so natural selection generated a wide array of emotions that could be used in diverse patterns of positive and negative sanctioning. Turner's theory explores the ways that specific emotions aroused during sanctioning sustain social structures, and create tensions in them as well, but the critical point is that natural selection organized humans' emotional capacities in ways enabling them to use and respond to sanctions.

(4) Moral Coding. Sanctions are, ultimately, a response to expectations about proper conduct. Such moral codes have no meaning unless they are imbued with emotional content. So, natural selection gave humans not only the cognitive capacity to formulate rules of conduct but also the neurological ability to attach emotions of varying types and intensities to these rules. Without the latter ability, rules have no moral force.

(5) Valuing-Exchanging. For exchanges of resources among humans to occur, they must possess value, or the capacity to bestow gratification or punishments. To the extent that exchanges of diverse resources are essential to social bonding and social organization, the human brain had to be rewired to associate emotions with resources. Without emotions, resources cannot possess value, and without this ability to attach new emotions to ever expanding resources, flexible and complex social structures cannot be produced or sustained. Indeed human social structure is not possible without the ability to attach emotions—and hence, value—to virtually all objects, symbols, and behaviors.

(6) Decision-Making. If the ability to think abstractly and make complex decisions is one of humans' unique abilities—at least in degree—the emotional tagging of options is essential. As recent neurological studies have demonstrated, decision-making is difficult without assigning emotions to various alternatives; and complex decision-making requires the use of many diverse emotions to sort among alternatives.

In sum, then, Turner argues that humans' emotional abilities are a compensatory mechanism for the low sociality inherited from the common ancestors of present-day apes and humans. In compensating for this propensity for low sociality, the brain was reorganized in ways enabling humans to use emotions to mobilize energy for building social bonds, to engage in fine-grained attunement of responses, to develop moral codes and use sanctions, to place value on exchanged resources, and to make rational decisions. Humans' incredible capacity for using and reading such a diverse array of emotions is thus the result of evolutionary history as it restructured the human brain. From Turner's viewpoint, the evolution of these emotional abilities is as important as is humans' capacity to use language or to think rationally. Indeed, the latter cannot occur effectively without emotions.

PSYCHOANALYTIC-INTERACTION THEORIES OF EMOTION

Thomas Scheff's Psychoanalytic Theory of Emotion

Thomas Scheff,[28] at times in collaboration with Suzanne Retzinger,[29] has blended interactionist ideas with psychoanalytic theory. Drawing from Charles Horton Cooley, Scheff emphasizes that people are in a constant state of self-feeling about pride or shame. As individuals read the gestures of others and see themselves in the "looking glass," they experience a sense of pride or, if the self images are more negative and critical, a sense of shame. Most of the time, individuals experience rather low levels of these emotions and, hence, easily sustain equanimity.

Scheff believes that pride is a positive emotion, promoting mutual attunement of interpersonal responses, mutual respect and deference, mutually agreed on conformity, strong social bonds, and high levels of social solidarity. In turn, these processes feed back and promote further pride among those in interaction. In contrast, shame has the potential to disrupt these solidarity-producing processes. Borrowing from Helen Lewis' psychoanalytic scheme, Scheff argues that when individuals experience a lack of deference from others and, indeed, experience criticism and other negative reactions from others, their self evaluations also become negative and cause shame. This experience of shame can be fully acknowledged and lead to a healthy reconstruction of bonds of trust, but if the shame is denied and repressed, it activates cycles of unacknowledged shame and outbursts of anger.

From Lewis' analysis, Scheff notes that there are two potential paths by which shame is denied. One is *overt, undifferentiated* shame where the individual has painful self feelings but hides from the underlying emotion of shame. The shame is, in essence, disguised as feeling "foolish," "stupid," "inadequate," "defective," "embarrassed," "awkward," "insecure," and so on. Under these conditions, individuals reveal disrupted and slowed speech, averted gazes, blushing, and oversoft speech; and though they have painful feelings, the source of these feelings—shame—remains unacknowledged. The second path for denial is termed *bypassed shame,* whereby the individual avoids the full experience of pain before it can be fully experienced. Speech might speed up but take on a repetitive quality that marks the effort to avoid the experience of shame. Later, Scheff termed these two paths of denying shame *under-* and *over-distancing* oneself from the shame. Overt, undifferentiated shame is "underdistanced" because intense feelings are experienced but not fully acknowledged for what they are, and bypassed shame is "overdistanced" because the pain is not perceived as the individual steps outside of self and acts as if the pain is not present.

Once shame is denied, individuals can find themselves locked into shame-anger cycles. Repressed or denied shame generates anger and hostility. Even as this anger is felt or expressed, it causes the individual to experience even more shame (though still denying the shame for what it is), which, in turn, can generate more anger and hostility that lead to more shame. This kind of cycle can last a lifetime, and it can be passed on to subsequent generations. Scheff even argues that it can be experienced collectively by large groups and even nation-states that vent their collective anger in conflict and war-making.

Structural conditions and cultural beliefs can thus inhibit people from acknowledging shame and pride, particularly shame. The more inhibitory these sociocultural forces, the more likely are individuals to use defense mechanisms and deny their shame, escalating the potential for anger and hostility. In turn, anger and hostility disrupt the attunement of individuals to each other, the development of mutual respect, the maintenance of strong and positive social bonds, and the production of social solidarity. For Scheff, then, the most important emotional substrate of micro-level interaction is the self-evaluations that individuals make relative to pride and shame.

Jonathan Turner's Psychoanalytic Theory

Drawing on his earlier theory of social interaction,[30] Jonathan Turner has developed in recent years a more elaborate version of Scheff's theory that adds elements from other theories of emotion.[31] Turner reasons that individuals have needs (1) to confirm their self-conceptions, (2) to secure some of the symbolic and material resources that a situation has to offer, (3) to experience a sense of being part of the flow of interaction, (4) to feel that others are experiencing the situation in much the same terms, and (5) to perceive that others can be trusted to behave in predictable ways. These needs influence each other. For example, to confirm one's self-conception, it is necessary to receive symbolic and material resources that mark a person as a certain type of individual; to feel a part of the flow of interaction requires that others are seen as experiencing the situation in the same way and that they can be relied on to behave in appropriate ways. Thus, if any one of these five motivating needs is not met, the other needs are activated. Thus, in Turner's theory, the mutual signaling and interpreting of gestures by individuals is motivated by needs to confirm self, to realize appropriate resources, to feel a part of the flow of interaction, to believe that others are experiencing the situation in much the same way, and to perceive that others can be trusted and counted on to behave in predictable ways.

For any situation, Turner argues, these five classes of needs create expectations about what is likely to occur. For instance, in entering a situation, people's past experiences or general knowledge of the situation will lead them to expect a certain type and level of self-confirmation, to expect certain resources in specified amounts, to expect feeling involved in the flow of interaction to a certain degree, to expect others to experience the situation in similar ways, at least to a degree, and to expect others to behave with a certain amount of predictability. When these expectations are exceeded, individuals will experience happiness and other variations of this positive emotion. When these expectations are not met, however, the more negative and destructive emotions of fear and anger are aroused. In turn, fear and anger become involved in complicated cycles of guilt, shame, repression, and depression.

People experience anger when their expectations are not met. When this anger becomes mixed with fears about the consequences of failing to meet expectations, shame emerges; if expressed anger makes one even more aware of the failure to meet expectations, or alienates others, then the level of experienced shame escalates. When anger is directed toward oneself and coupled with some fear about the consequences of failing to meet the expectations of others, individuals experience guilt. The greater their anger at self and fears about the consequences of their failure to meet the expectations of others, the greater is the sense of guilt.

Guilt and shame are thus intermixed with anger and fear. Guilt is mostly fear about failure to meet expectations of others, coupled with anger at oneself for such failure; shame is mostly anger that further disrupts a person's sense of meeting expectations, coupled with some fear about the consequences of having or expressing aggressive feelings. In contrast, pride is happiness at meeting or exceeding expectations, but coupled with a sense of relief that fears about something going wrong were unfounded. Like Scheff's theory, Turner sees happiness-pride, anger-shame, and fear-guilt as influencing an individual's capacity to role-take and role-make effectively, but he conceptualizes these cycles in a more complex way.

One complication is that the fear-guilt cycle and anger-shame cycle become interwoven. As guilt increases, it activates shame; conversely, as shame increases, it stimulates guilt. In so doing, the underlying emotions of fear and anger that lie below guilt and shame are also escalated, causing a further iteration of the anger-shame and fear-guilt cycles. But this time around, the two cycles form a more inclusive set of cycles in which anger, shame, guilt, and fear all begin to influence each other, escalating the intensity of negative emotions experienced by individuals.

Another complication comes from the effects of repression. Anger, fear, guilt, and shame are unpleasant emotions. Moreover, their emission is often socially unacceptable, and their expression can activate fear and anger in others, which, in turn, cause a person to experience guilt and shame. As a consequence, individuals often repress their anger and fear, as well as their shame and guilt. They simply do not acknowledge these emotions at a conscious level, but as they repress their feelings, these powerful negative emotions cumulate and begin to depress individuals as they seek to control highly charged negative feelings. These efforts to keep unpleasant emotions below the level of consciousness take emotional energy away from interactions. Although high status and power can enhance a person's emotional energy in a situation, the depressive effects of keeping a lid on anger, fear, guilt, and shame are never fully overcome by the possession of power and prestige.

Yet another complication comes from the pride, shame, and guilt that accumulate over a person's lifetime. When individuals have a biography filled with sensations of pride, these experiences become a kind of "emotional bank account" that can carry them through situations of low status and power or experiences of negative emotions like anger, fear, guilt, and shame without causing them to lose too much positive emotional energy. Conversely, when a person has a biography punctuated by anger, fear, shame, and guilt, these emotions will have accumulated, especially if they have gone unacknowledged and have been repressed, and these emotions will dampen emotional energy, even when a person has reason to experience pride or when the person can hold a position of power and status. Thus, people's emotional energy is more than purely situational; it is also influenced by their respective biographies as these have led them to experience pride, shame, and guilt (as well as the emotions of happiness, anger, and fear that underlie, respectively, pride, shame, and guilt).

CONCLUSION

As is evident, there are many different sociological theories on emotion. Some fall neatly within the symbolic interactionist tradition, others less so, but they all represent the most creative forefront of interactionist theorizing. George Herbert Mead never developed a theory of emotion, although he offered hints about how one could be developed. Some such as Cooley who saw pride as the basic emotion, and others like Durkheim who saw the connection between emotion and ritual as fundamental, provided valuable leads, but in the end, not until midcentury did sophisticated sociological theories of emotion began to surface. And it is really only during the last two decades that these theories have become highly prominent. Today, they mark the leading edge of interactionist theorizing, but as is clear from this selective but nonetheless representative review, these theories have not been pulled together into a more integrated theory. Still, the elements for such a synthetic theory are perhaps in place, and as the twenty-first century approaches, it will not be long until the emotional dynamics involved in human interaction become an integral part of the evolving interactionist theoretical tradition.

NOTES

1. Arlie R. Hochschild, "Emotion Work, Feeling Rules, and Social Structure," *American Journal of Sociology* 85 (1979), pp. 551–575; *The Managed Heart: The Commercialization of Human Feeling* (Berkeley: University of California Press, 1983).

2. Steven L. Gordon has provided a useful analysis of emotion culture, one that has been highly influential in not only Hochschild's work by other theorists and researchers as well. See "The Sociology of Emotion," in *Social Psychology: Sociological Perspectives*, eds. M. Rosenberg and R.H. Turner (New York: Basic Books, 1981), pp. 562–592; "Institutional and Impulsive Orientations in Selectively Appropriating Emotions to the Self," in *The Sociology of Emotions: Original Essays and Research Papers*, eds. D.D. Franks and E.D. McCarthy (Greenwich, CT: JAI, 1989), pp. 115–126; and "Social Structural Effects on Emotions," in *Research Agendas in the Sociology of Emotions*, ed. T.D. Kemper (Albany, NY: State University of New York Press, 1990), pp. 145–179.

3. Hochschild, *The Managed Heart* (cited in note 1).

4. Susan Shott, "Emotion and Social Life: A Symbolic Interactionist Analysis," *American Journal of Sociology* 84 (1979), pp. 1317–1334.

5. See David R. Heise, "Social Action as the Control of Affect," *Behavioral Science* 22 (1977), pp. 163–177; *Understanding Events: Affect and The Construction of Social Action* (New York: Cambridge University Press, 1979); "Effects of Emotion Displays on Social Identification," *Social Psychology Quarterly* 53 (1989), pp. 10–21.

6. David R. Heise and Richard Morgan, "The Structure of Emotions," *Social Psychology Quarterly* 51 (1988), pp. 19–31; Neil J. MacKinnon and David R. Heise, "Affect Control Theory: Delineation and History," in *Theoretical Research Programs: Studies in the Growth of Theory*, eds. J. Berger and M. Zelditch, Jr. (Stanford, CA: Stanford University Press, 1993), pp. 64–103.

7. Lynn Smith-Lovin and David R. Heise, eds., *Analyzing Social Interaction: Advances in Affect Control Theory* (New York: Gordon and Breach, 1988); David R. Heise and Lynn Smith-Lovin, "Impressions of Goodness, Powerfulness, and Liveliness From Discerned Social Events," *Social Psychology Quarterly* 44 (1981), pp. 93–106. For a comparison with expectation states theory, see Cecilia Ridgeway and Lynn Smith-Lovin, "Structure, Culture and Interaction: Comparing Two Theories," *Advances in Group Processes* 11 (1994), pp. 213–239.

8. Lynn Smith-Lovin, "The Sociology of Affect and Emotions." in *Sociological Perspectives on Social Psychology*, eds. K.S. Cook, G.A. Fine, and J.S. House (Boston: Allyn & Bacon, 1995), pp. 118–144. For a detailed but highly readable review of the affect control theory,

see Neil J. MacKinnon, *Symbolic Interactionism as Affect Control* (Albany: State University of New York Press, 1994).

9. See Heise and Smith-Lovin, "Impressions of Goodness, Powerfulness, and Liveliness" (cited in note 7).

10. Sheldon Stryker, *Symbolic Interactionism: A Social Structural Version* (Menlo Park, CA: Benjamin/Cummings, 1980).

11. Sheldon Stryker, "The Interplay of Affect and Identity: Exploring the Relationship of Social Structure, Social Interaction, Self and Emotions." Paper presented at the American Sociological Association meetings, Chicago, 1987; Sheldon Stryker and Richard Serpe, "Commitment, Identity Salience and Role Behavior: Theory and A Research Example," in *Personality, Roles and Social Behavior*, eds. W. Ickes and E. Knowles (New York: Springer-Verlag, 1982), pp. 199–218.

12. Randall Collins, *Conflict Sociology: Toward an Explanatory Social Science* (New York: Academic, 1975); "Stratification, Emotional Energy and the Transient Emotions," in *Research Agendas in the Sociology of Emotions*, ed. T.D. Kemper (Albany: State University of New York Press, 1990), pp. 27–57; "Interaction Ritual Chains, Power and Property: The Micro-Macro Connection as an Empirically-Based Theoretical Problem," in *The Micro-Macro Problem*, eds. J.C. Alexander, et al. (Berkeley: University of California Press); and *Theoretical Sociology* (New York: Academic, 1988), pp. 187–228; Randall Collins and Robert Hanneman, "Modeling Interaction Ritual Theory of Solidarity," *The Problem of Solidarity: Theories and Models*, eds P. Doreian and T.J. Fararo (New York: Gordon and Breach, 1998).

13. Randall Collins, "Emotional Energy as The Common Denominator of Rational Action," *Rationality and Society* 5 (1993), pp. 203–230.

14. Theodore D. Kemper, *A Social Interactional Theory of Emotions* (New York: Wiley, 1979); "Predicting Emotions from Social Relations," *Social Psychology Quarterly* 54 (1991), pp. 330–342.

15. Theodore D. Kemper and Randall Collins, "Dimensions of Microinteractionism," *American Journal of Sociology* 96 (1990), pp. 32–68.

16. Theodore D. Kemper, "How Many Emotions are There? Wedding the Social and the Autonomic Component," *American Journal of Sociology* 93 (1987), pp. 263–289.

17. Cecilia L. Ridgeway, "Status in Groups: The Importance of Emotion," *American Sociological Review* 47 (1982), pp. 76–88; "Affect," in *Group Processes: Sociological Analysis*, eds. M. Foschi and E. Lawler (Chicago: Nelson-Hall, 1994), pp. 205–220.

18. Cecilia L. Ridgeway and Cathryn Johnson, "What is the Relationship Between Socioemotional Behavior and Status in Task Groups?," *American Journal of Sociology* 95 (1990), pp. 1189–1212.

19. Ridgeway, "Status in Groups" (cited in note 17).

20. See both works in note 1.

21. Janet R. Johnston, "The Structure of Ex-Spousal Relations: An Exercise in Theoretical Integration and Application," in *Status Generalization: New Theory and Research*, eds. M. Webster, Jr. and M. Foschi (Stanford, CA: Stanford University Press, 1988), pp. 309–326.

22. Joseph Berger, "Directions in Expectation States Research," in *Status Generalization: New Theory and Research*, eds. M. Webster and M. Foschi (Stanford, CA: Stanford University Press, 1988), pp. 450–474.

23. Barry Markovsky and Edward J. Lawler, "A New Theory of Group Solidarity," *Advances in Group Processes*, 11 (1994), pp. 113–138.

24. Edward J. Lawler and Jeongkoo Yoon, "Commitment in Exchange Relations: A Test of a Theory of Relational Cohesion," *American Sociological Review* 61 (1996), pp. 89–108; Edward Lawler, Jeongkoo Yoon, Mouraine R. Baker, and Michael D. Large, "Mutual Dependence and Gift Giving in Exchange Relations," *Advances in Group Processes* 12 (1995), pp. 271–298; Edward J. Lawler and Jeongkoo Yoon, "Power and the Emergence of Commitment Behavior in Negotiated Exchange," *American Sociological Review* 58 (1993), pp. 465–481.

25. Michael Hammond, *The World is Not Enough: The Social Evolution of Emotion Rules*, forthcoming; "Affective Maximization: A New Macro-Theory in the Sociology of Emotions," in *Research Agendas in the Sociology of Emotions*, ed. T.D. Kemper (Albany: State University of New York Press, 1990), pp. 58–81; "The Footprints of Natural Selection in Human Emotion Rules: The Evolutionary Appeal of the Supernatural," paper presented at the American Sociological Association meetings, Pittsburgh, 1992; "Machines for the Production of Emotion," paper presented at the Human Behavior and Evolution Society meetings, Binghamton, NY, 1993.

26. Michael Hammond, "Emotions, Habitation, and Expanding Human Needs," paper presented at the American Sociological Association meetings, 1993; "Habituation and The Evolution of Emotion Rules," paper presented at the International Society for Research on Emotions, Pittsburgh, 1992; "Cheating on Evolution: Habituation and Emotion Rules," paper presented at the Society for Research on Emotions, Cambridge, England, 1994.

27. Jonathan H. Turner, "The Evolution of Emotions: A Darwinian-Durkheimian Analysis," *Journal for the Theory of Social Behavior*, 26 (1996), pp. 1–33; "The

Evolution of Emotions: The Nonverbal Basis of Human Organization," in *Biology with a Human Face: Nonverbal Communication and Social Interaction*, eds. U. Segerstrale and P. Molnar (New York: Lawrence Erlbaum, 1996), pp. 300–318; "The Nature and Dynamics of the Social among Humans," in *The Mark of the Social*, ed. J.D. Greenwood (New York: Rowman and Littlefield, 1996), pp. 105–133.

28. Thomas Scheff, "Shame and Conformity: The Deference-Emotion System," *American Sociological Review* 53 (1988), pp. 395–406; "Socialization of Emotion: Pride and Shame as Causal Agents," in *Research Agendas in the Sociology of Emotion*, ed. T.D. Kemper (Albany: State University of New York Press, 1990), pp. 281–304.

29. Thomas Scheff and Suzanne Retzinger, *Shame, Violence and Social Structure: Theory and Causes* (Lexington, MA: Lexington, 1992).

30. Jonathan H. Turner, *A Theory of Social Interaction* (Stanford, CA: Stanford University Press, 1988); "Toward a Sociological Theory of Motivation," *American Sociological Review* 57 (1986), pp. 15–27; "A Behavioral Theory of Social Structure," *Journal for the Theory of Social Behavior* 18 (1988), pp. 354–372; "A Theory of Microdynamics," *Advances in Group Processes* 7 (1989), pp. 1–26.

31. Jonathan H. Turner, "A General Theory of Motivation and Emotion in Human Interaction," *Österreichische Zeitschrift für Soziologie* 26 (1994), pp. 20–35; "Motivation, Emotional Energy, and Interaction Processes," working paper presented at the American Sociological Association meetings, New York, 1996.

33

The Continuing Tradition II: Expectation States Theory*

One of the most carefully and explicitly developed approaches to emerge in the second half of the twentieth century is a related set of theories known as *expectation states theory*. Although this program was originally the pursuit of Joseph Berger and his colleagues, students, and associates, expectation states theory has in recent years attracted the attention of many other investigators as well.[1]

Expectation states approaches are interactionist theories in several senses. First, behavior is assumed to be situationally dependent, with actors learning how to behave from social and cultural frames of reference available in the situation. Second, these frames of reference exert their influence through the perceptions of individuals, and, what people believe or expect to be true about the situation can override accurate perceptions of reality. Finally, these beliefs and expectations emerge and are maintained by the process of interaction itself.

There are, however, several respects in which expectation states theory deviates from traditional interactionist theorizing. One of the most important differences concerns the generally *episodic* nature of expectation states, where processes become *activated* under certain specifiable conditions, organize or re-organize behavior, and then are *de-activated,* ceasing to operate until the next time conditions arise to re-activate expectation state processes. As a result, individuals are not constantly "renegotiating the reality" of their situations, as is often asserted in some interactionist theories. To be sure, social realities are negotiated by individuals, but once they emerge they are likely to remain stable

*This chapter is coauthored with David G. Wagner.

until conditions sufficient to stimulate further negotiation appear once again.

THE CORE IDEAS

The most central concept to the theory is the notion of an *expectation state*. Expectation states represent stabilized anticipations for future behavior of one actor relative to that of another. Thus, a person is never simply expected to perform well or poorly, per se; rather, this person is expected to perform more capably or less capably relative to another individual, or other individuals. Furthermore, expectations are always both task-specific and person-specific in the sense that person A may be expected to perform more capably than B with respect to one task (say, repairing an automobile engine) and less capably than B on a second task (say, playing a Beethoven quartet). Similarly, A may be expected to perform more capably than B, but less capably than C, on the same task.

Because the self constitutes a social object, actors can also have expectations for *themselves* relative to specific others, although what actors are able to report about self-expectations is just as prone to error and misinterpretation as what they can report about expectations for others. Nevertheless, individuals behave *as if* they have adopted a specific set of expectations.

Expectation states are generated from a variety of different sources, such as evaluation of task performances during the course of an interaction, reflection on the appraisals of significant others, allocation of material or symbolic rewards, activation of differences in people's power and prestige, and assessment of justice and equity. These and other sources provide information from the broader social environment, and when individuals interact on a task, this information becomes salient in the immediate, local situation. Expectation states then emerge and organize information into a coherent picture, or definition, of the situation. This picture enables individuals to select behaviors that are appropriate to the situation and to avoid those that are not. Because these behaviors are generally consistent with definitions of the situation, they tend to reinforce established expectations, and typically, it requires the introduction of new information or a change in the local situation to break this self-perpetuating cycle. These core ideas can be summarized as follows:

1. Given certain conditions, individuals organize salient information from the social environment into expectations for behavior in the immediate situation of interaction. (*The salience of social information for expectation formation.*)
2. Individuals behave in accordance with their expectations regarding the immediate situation of interaction. (*The behavioral implications of expectations.*)
3. Individuals' behavior in the immediate situation of interaction tends to reinforce established expectations. (*The reinforcement of expectations.*)

Conditions sufficient to activate the information-organizing process are not always present, however. Consequently, behavior can be relatively stable over extended periods of time, changing only when the new conditions arise. Indeed, individuals do not continuously reorganize or renegotiate their definitions of the situation; rather they generally behave in accordance with their existing definitions, reorganizing or renegotiating only when necessary.

APPLICATIONS OF THE CORE IDEAS

Expectation states theories have been developed to account for at least fifteen distinct social processes. Some have been extensively investigated; others less so; and still others are too new to have had the opportunity to develop. We will discuss some of the most interesting of these processes.

Power and Prestige

The expectation states program began with Joseph Berger's interest in accounting for the emergence and maintenance of power and prestige differences in the behavior of two actors in initially undifferentiated groups. Berger's *behavior-expectation theory* was initially designed to explain why inequalities in power and prestige evolve so very quickly in groups even when members are initially similar in status. These inequalities include differences in opportunities to contribute to consideration of the group's task or problem (such as asking a question), actual attempts to provide solutions to these tasks and problems (such as answering a question), evaluations of the contributions (such as criticism of a proposed answer), and acceptance or rejection of influence (such as deferring to another actor with whom one has disagreed). Moreover, these inequalities were highly correlated, forming a single hierarchy of observable power and prestige differences among members of the group. Once such a hierarchy emerged, it tended to be stable—even over different group discussions on different days.

Behavior-expectation theories were developed initially to explain these phenomena as a consequence of an underlying structure of *expectations* for future task performance that seemed to emerge from the interactions among group members. Once these expectations exist, they determine the course of future interaction in the group, thereby reinforcing the existing structure of expectations. Thus, the inequalities are highly correlated because they are generated by the *same* underlying expectations. Further, unless other structural factors intervene, such as new group members and new information are introduced or the task focus of the group changes, the structure of expectations and the observable power and prestige hierarchy will remain stable. Adopting the principles described earlier, this process can be summarized as follows:

1. Actors behave toward others in a manner consistent with their expectations. An actor who is considered more capable than others is offered more opportunities to interact, makes more contributions to the interaction, more often has those contributions evaluated positively, and has more influence when disagreements occur. (*The behavioral consequences of expectations.*)
2. Expectations tend to remain the same as long as the actors involved in the situation and the tasks that they are performing remain the same. (*The persistence of expectations.*)

Although these expectation state processes occur under a wide variety of conditions, they are most evident in situations where actors who are initially similar in status work together collectively on a task that they value. Thus, we might observe expectation processes in jury deliberations, business conferences, family vacation planning, or a group of teenagers planning a trip to the movies. They might also be evident in the interaction among members of a basketball team, but might be less likely to occur in the interaction between members of opposing teams because they are not working toward a common goal.[2]

Other work in behavior-expectation theory has focused on the process by which expectations emerge from interaction.[3] This work suggests that resolving disagreements forces actors to evaluate their own and others' performances and to accept or reject influence from others on the basis of these evaluations. Any time a decision must be made, there is a some likelihood that the individuals will develop an expectation state for each other that is, in a very real sense, an anticipation for the quality of an actor's future contributions. This expectation state will be consistent with the preponderance of evaluations for the contributions of each individual in the group. More generally, we can summarize the process in these terms:

3. The more consistent the evaluations of an actor's past interaction are, the more likely the actor and others are to expect a level of capability from the actor in the future that realizes past evaluations. (*The emergence of expectations.*)

Later work has developed and extended these arguments. For example, inequalities with respect to *any* aspect of the group's interaction, not just the evaluation of an individual's performance, might generate expectations and spread to other aspects of the interaction.[4] Power and prestige hierarchies can emerge from interaction involving any number of actors as well as from different types of actors.[5] Finally, objective evaluations from external sources of information that contradict established expectations can help overcome existing expectations. Such changes depend on the number and extremity of the contradicting evaluations.[6]

Status Characteristics

Behavior-expectation theory primarily concerns task situations in which actors are similar with respect to concrete status distinctions such as "ethnicity," "age," "race," or "gender" that might be significant in society at large. But what if actors are *not* similar? *Status characteristics theory* considers how expectations can be generated by information and knowledge that actors bring with them from outside the group to the immediate interaction situation. More specifically, this theory describes how actors organize information about initial status differences that they use to generate expectations for performance. As with behavior-expectation theory, these expectations then govern the interaction, ensuring that power and prestige is distributed in accord with expectations.

Again, Joseph Berger (and his associates) found evidence of the status-organizing process in a wide variety of prior research. Basically, this research showed that external status distinctions become the basis for internal ones; status inequalities significant in the larger society become important in the task situation of a small group. These distinctions, present even as the group is being formed, govern the distribution of power and prestige, and, furthermore, these effects occur whether or not the status distinction is associated with the group's task.[7]

Status distinctions are characterized as *diffuse status characteristics*, with a characteristic being *diffuse* for a particular individual in the situation if (1) the characteristic has two or more states that the individual evaluates differently, (2) the individual associates a general expectation with each status state, and (3) the individual associates a distinct set of expectations for specific abilities or traits with each state.[8] Gender, race, ethnicity, educational attainment, occupation, and physical attractiveness are each examples of these kinds of diffuse status characteristic, because they generally meet the three features listed above. To illustrate, a person who believes that males are better than females and that men are generally more capable at tasks than women is treating gender as a diffuse status characteristic. Note that the properties that define a characteristic are based on attributions *made by the individual*, and obviously, these attributions need not have objective validity. More likely, they represent beliefs and values in cultural systems to which the individual is exposed.

How then does such diffuse status information become organized into expectations and translated into task behavior? An *activation*, or *salience*, proposition argues that any diffuse status information that distinguishes the actors in the situation (for example, gender in a mixed gender group) or that is believed to be task-relevant (for example, levels of education) will be activated. Thus,

1. If actors have differentiated diffuse status *or* if a status they share is culturally associated with a task that the actors perform, then the actors will attribute generalized expectations and specific abilities to themselves and each other consistent with their status. (*The salience of status information.*)

A second principle concerns establishing the relevance of salient status information to performance of the task. The *burden of proof* principle argues that status information that is salient will become relevant to the task, unless it is known or believed to be irrelevant; hence, the "burden of proof" principle requires a demonstration that the status information should not be considered as relevant. Diffuse status information will be assumed to be task-relevant unless there is specific information to the contrary.

2. Actors assume that salient status information applies to every new task and every new situation unless they have a specific knowledge or belief that demonstrates its inapplicability. (*The relevance of status information.*)

Individuals then assign expectations for task performance based on the relevant status information.

3. An actor with a status advantage is expected to perform more capably than the actor with a status disadvantage. (*The assignment of expectations.*)

Finally, in line with this basic expectation assumption, people behave in accordance with their expectations. Opportunities to initiate action, actual performance outputs, communicated evaluations of performance, and influence will all reflect the difference in self expectations and expectations for others. Thus, the presence of a single diffuse status characteristic that discriminates between actors is sufficient to generate differentiated power and prestige behavior, provided that the status differences are not dissociated from the task.

Multiple Characteristic Status Situations

Joseph Berger and his associates soon expanded the scope of this original theory to deal with situations where multiple status characteristics were present.[9] In essence, they asked this: What are the implications for expectations and behavior when individuals can be distinguished by *more than one* status characteristic, especially when the implications of these statuses are inconsistent?[10]

When multiple and inconsistent statuses become salient, the status information can be organized in either of two basic ways. Actors can *eliminate* some of the salient information and define their expectations and behavior on the basis of the remaining (usually consistent) subset of the information. Behavior and expectations should therefore be uniformly high or uniformly low, depending on which subset the individual uses to define the situation.[11] An equally plausible alternative, however, is that people *combine* all salient status information in arriving at expectations. That is, they use all the information that is available to them—provided that it is salient. Inconsistency in the implications of the salient status differences should then moderate expectations.

The combining process is governed by a *principle of organized subsets*. All status information implying successful task outcomes is combined to determine a value of positive performance expectations. In contrast, all information implying unsuccessful task outcomes is combined to determine a value of negative performance expectations. Further, within each of these subsets of status information, is an *attenuation* effect in which each additional increment of information has ever less effects on the expectation that will eventually emerge. The *aggregated expectations* for an individual are the sum of these positive and negative values, as moderated by the attenuation effect. The actor's *expectation advantage* (or disadvantage) relative to another is thus equal to the aggregated expectation for self minus that formed for the other. To state the argument more succinctly:

1. Actors combine information from multiple salient status differences to form aggregated expectations for self and others. (*The combining principle.*)
2. Actors combine positive status information into one (positively valued) set and negative information into a second (negatively valued) set. Expectations are aggregated by summing the values of these two sets. (*The principle of organized subsets.*)
3. Each additional piece of status information added to a set increases the value of that set at a decreasing rate. (*The attenuation principle.*)

Finally, the basic expectation assumption is modified to accommodate the multiple bases for the formation of expectations.

4. An actor's power and prestige behavior relative to another actor is a direct continuous function of the expectation (dis)advantage relative to this other.

Thus, the more extreme one's expectation (dis)advantage, the greater the differentiation in behavior is.

Distributive Justice

How do actors determine the justice or injustice of the rewards that they receive? How do they respond when they perceive an injustice in the distribution of rewards? The *status value theory* of distributive justice considers these questions.[12]

The status value view of justice was developed originally to challenge an earlier view known as *equity theory*.[13] Equity theory focuses on the exchange, or consummatory, value of rewards and assumes that evaluations of justice and injustice are based on comparisons of one actor's ratio of the values received to actual investments made to receive these rewards, compared with the ratio of at least one other actor in the immediate situation. If the ratios are equal, the situation is regarded as equitable and stable; if the ratios are unequal, however, the situation is seen as inequitable and individuals are motivated to reduce the inequity.

The *status value theory of distributive justice* argues that theories emphasizing comparisons of rewards to investments are inadequate. As an alternative explanation, status value theory considers the value of objects to be based on the status that they represent and signify rather than on consummatory value alone. For example, the value of a key to the executive washroom is viewed by the status, honor, esteem, and importance that it conveys to the person who possesses it (rather than by the reward value that comes with convenience and privacy). Justice issues, therefore, involve questions of status consistency and inconsistency between expectations and allocations of rewards; as long as actors receive the status value that they expect, the situation is seen as just. If actors receive a status value different from what they expect, however, then the situation is seen as unjust. Thus, one's sense of justice is not so much a matter of the ratio of rewards received to investments made, compared with the ratio for others in a situation as *an assessment of rewards in relation to what was expected.*

These expectations for reward depend on *referential structures*, and the activation of referential structures enables actors to relate their general cultural framework to their immediate situation. When a particular referential structure is activated, individuals expect to receive rewards commensurate with their relevant status. Basically, "what is" in general cultural definitions becomes "what is expected to be" in the immediate situation. Thus, a man who believes that men are generally better paid than women will expect a higher level of reward when he is interacting with a woman—provided, of course, that gender is a relevant and salient status characteristic to the actors and a relevant referential structure has been activated. To summarize more formally

1. If an actor activates referential structures, and thereby culturally associates different levels of reward with different status positions, then he or she is likely to develop expectations for the allocation of rewards in the immediate interaction situation that are consistent with this cultural association. (*The activation of referential structures.*)
2. If an actor receives the level of reward that is consistent with expectations created by the cultural association of referential structures to status characteristics, then the actor will regard the situation as just. (*The effect of expectations on assessments of justice.*)
3. If an actor receives a level of reward inconsistent with expectations, generated by the cultural association of referential structures with status characteristics, the actor will regard the situation as unjust. (*The effect of expectations on assessments of injustice.*)

A "reverse process" can also operate in the sense that an allocation of differentially evaluated rewards can generate performance expectations consistent with this allocation. This process is most likely to occur when a referential structure is activated and when the relevant status distinction is based primarily on performance.[14] To state this more formally,

4. If an actor is allocated a differentiated level of reward, differentiated performance expectations consistent with the allocated level of reward are likely to develop. (*The effect of differential rewards on expectations.*)

Sources of Self-Evaluation

Source theory considers how expectations can emerge through the reflected appraisals of significant others. Source theory uses ideas regarding unit evaluations and expectations developed in behavior-expectation theory, as well as ideas from status characteristics theory. As with status characteristics theory in general, the original version of source theory[15] dealt only with the simplest situation, involving a single evaluator. How do the appraisals of those with the *right to evaluate* affect the expectations and behavior among others in a situation? Under some circumstances, an evaluator can become a *source* of evaluations for the actor—that is, an evaluator whose assessments matter to the actor or, in other words, a "significant other." The likelihood of a particular evaluator becoming a source is directly related to an individual's expectations for this evaluator:

1. The higher is an actor's expectations for an evaluator, the greater is the likelihood that this evaluator will become a source of evaluations for the actor. (*The effect of external evaluations on expectations.*)

When this occurs, the actor's expectations and behavior are determined by the evaluations of the source:

2. Given a source as evaluator, an actor's evaluations, expectations, and behavior will be shaped and directed by this source's evaluations. (*The effects of sources of evaluation.*)

An actor's expectations for an evaluator are likely to be affected by any status characteristics that the evaluator possesses. An extension of the original source theory by Murray Webster showed that status is directly related to the likelihood of becoming a source:[16]

3. A high status evaluator is more likely to become a source for an actor than is a low status evaluator. (*The effect of evaluator status on evaluation importance.*)

If there are multiple evaluators and their evaluations conflict, actors apparently use source information in much the same way that they use status information. They process and combine all the salient cues to form composite expectations.[17]

4. Actors combine the unit evaluations of multiple conflicting sources to form self-expectations. (*Combining source evaluations.*)

Formation of Reward Expectations

Reward expectation theory[18] emphasizes that any or all referential structures can be activated in a particular task situation, and when activated, each provides a standard for the formation of expectations for reward.[19] The theory derives a set of theorems on the operation of such standards as determined by activation of referential structures. The first of these theorems concerns how information from multiple standards is treated in generating reward expectations.

1. If multiple referential structures establishing standards for allocation of rewards are activated in the immediate situation of interaction, actors combine the information from all activated structures in generating expectations for reward. (*Combining referential structures.*)

The second and third theorems consider how increases in the consistency or inconsistency of status characteristics affect the degree of inequality in reward expectations.

2. Increases in the amount of consistent status information salient for an actor in the immediate interaction situation increase the inequality in reward expectations. (*Status consistency and inequality in reward expectations.*)

3. Increases in the amount of inconsistent status information salient for an actor in the immediate interaction situation, however, decrease the inequality in reward expectations. (*Status inconsistency and equality in reward expectations.*)

The fourth theorem focuses on the interdependence of task and reward expectations.

4. Changes in an actor's task expectations (accomplished by adding or eliminating relevant status distinctions) produce correlated changes in the actor's reward expectations; in turn, changes in an actor's reward expectations (accomplished by adding or deleting standards of allocation) produce correlated changes in task expectations. (*The interdependence of expectations.*)

One consequence of the interdependence of task and reward expectations is that referential standards can produce differences in the *significance* of status characteristics. Differences in the relative importance of status characteristics often depend on whether or not these characteristics are the basis on which rewards and privileges are expected to be distributed. Statuses that differ in this respect should also differ in their significance or importance (or "weight") in the interaction. The fifth theorem establishes the differential significance of status characteristics.

5. Status characteristics have a greater effect on the actor's task expectations if they are the basis for referential standards of

allocation than if they are not. (*The social significance of status differences.*)

The Evolution of Status Expectations

Status characteristics theory generally describes status organizing processes in a single situation. Although individuals can join or leave the interaction, most of the status and performance information is available to them at the outset of the interaction. However, individuals often interact with the same or similar others across a sequence of tasks where new status, performance, and evaluational information is acquired at different stages. How do status expectations develop across such a sequence of tasks? A theory on the *evolution of status expectations* addresses this problem.[20] Specifically, this theory examines the conditions under which expectations and behaviors that emerge in one situation affect those of a subsequent situation. It also considers what causes major changes in the expectations as well as in the power and prestige differentials that evolve across situations. Finally, the theory discusses the conditions under which status expectations and behaviors stabilize. The theory presents a series of theorems that deal with these issues:

1. Expectations formed by an actor based on one situation will affect the actor's expectations and behavior in a current situation as long as the two situations are not dissociated. (*The effect of past expectations.*)
2. The effect of past expectations on the actor's expectations and behavior in a current situation will be greater if the status elements in the current situation are inconsistent with those in the past situation than if they are consistent. (*Inconsistent and consistent status transitions.*)
3. Given external evaluations of an actor's performance in a previous situation: (a) Differential evaluations that are consistent with the actors' power and prestige positions in the group increase the inequality; (b) differential evaluations that are inconsistent with the actors' power and prestige positions in the group decrease the inequality (and can even invert it); and (c) if the evaluations of both actors are similar, then any expectation advantage held by one of the actors will be reduced. (*Assignment of success or failure.*)
4. The expectations and behavior of actors will tend to become stable across consecutive interaction situations as long as no new information is introduced into the situation. (*Stability of status positions.*)
5. If actors have diffuse status characteristics from outside the group and specific status positions within the local situation that are inconsistent with each other, then their expectations and behavior will stabilize across consecutive interaction situations at a value between what would result from either the diffuse or specific statuses alone, again as long as no new information is introduced into the situation. (*The effects of interventions.*)

This last theorem has particular importance for overcoming the effects of diffuse status distinctions based on gender, race, and ethnic differences. It suggests that the assignment of specific competencies in a manner inconsistent with the implications of the diffuse characteristic can sometimes reduce the degree of inequality between the actors in a stable and relatively permanent manner.

Status Cues, Expectations, and Behavior

In many status situations, actors use a variety of social cues (such as patterns of speech, posture, direct references to background or experience, styles of dress) to help form expectations. In expectation states theory, *indicative* cues are distinguished from *expressive* cues, whereas *task* cues are separated from *categorical* cues. Indicative cues (for example, "I'm a doctor.")

directly label the actor's status state whereas expressive cues (a man's style of dress, for example) provide information from which status states can be inferred. The task/categorical distinction is independent of the indicative/expressive distinction. Task cues (for example, fluency of speech) provide information about the actor's capacities on an immediate task, whereas categorical cues (for example, language syntax) provide information about states of status characteristics that actors possess. Using these distinctions, status cues theory[21] yields the following propositions:

1. If no prior status differences exist in the situation, then differences in task cues will help generate expectation states, which in turn determine the distribution of power and prestige behavior in the situation. *(Task cues in the absence of status differences.)*

2. If status differences exist from the outset in a situation, then the differentiation in task cues will produce congruent differences in expectations, which in turn determine congruent differences in the rates of task cue behaviors. Consequently, rates of task cue behaviors will be consistent with the initial status differences. (*Status governance of task cues.*)

3. If for some reason, the differentiation in task cues is inconsistent with differentiation in categorical cues, then information from both sets combines to determine the actors' expectations and behavior. *(Combining inconsistent task and categorical cues.)*

Task cues provide information about capacities on an immediate task, whereas categorical cues provide information about status characteristics that might become relevant to the task. Hence, the strength of relevance of task cues is greater than that of categorical cues, signaling that, when inconsistent, the effect of task cues will be greater than the effect of categorical cues.[22]

Legitimation of Power and Prestige Hierarchies

Another theory in the expectation states program is concerned with the process of the legitimation of power and prestige orders. Legitimation processes are especially important for leaders with traditionally low statuses (for example, women or minorities) as they seek to engage successfully in the directive behaviors ordinarily expected of a leader. Legitimation is also likely to play a role in determining the effectiveness of controlling behaviors such as dominating and propitiating behaviors. For without being seen as having the right to engage in tasks, especially leadership activities, the power and prestige order will prove inviable.

Legitimation theory describes conditions under which a power and prestige order in a group becomes legitimated.[23] In accordance with the reward expectation theory, legitimacy theory assumes that part of any person's social framework is consensual beliefs, operating as referential structures to connect status positions with the diffuse statuses, task capacities, and task achievements. This information becomes activated in the local situation and helps generate expectations regarding status in the group. As actors behave in accordance with these expectations, they validate the expectations and establish their legitimacy.

1. When referential beliefs from the larger society are activated in the immediate interaction situation, actors create expectations about who will occupy high and low valued status positions in the situation. *(Referential beliefs and valued status positions.)*

2. Given expectations regarding the possession of high and low valued status positions in a situation, actors are likely to display differences in respect, esteem, and generalized deference behavior to others

that are consistent with these expectations. (*Valued status positions and behavior.*)

3. When behaviors consistent with expected status positions are validated by others and when they coincide with actual differences in power and prestige behaviors, the power and prestige hierarchy is likely to become legitimated. (*Behavioral validation and legitimacy.*)

An actor's behavior is *validated* by another if this other engages in supportive behavior or if this other's behavior does not contradict the behavior of the actor seeking validation and, hence, legitimacy.

Legitimation requires actors to make assumptions about "what ought to be" in the immediate situation. Expectations become normative, with the presumption that there will be collective support for these norms. A high status person has a *right* to expect a higher degree of esteem, respect, and generalized deference than does a low status individual. At the same time, others have the *right* to expect more valued contributions from this high status person than from low status actors. In addition, high status actors develop rights to exercise, if necessary, controlling behaviors—dominating and propitiating behaviors—over the actions of others.

CONCLUSION: STATE-ORGANIZING PROCESSES

In the course of developing the expectation states program, some have become involved in formulating *metatheoretical* ideas about the character and direction of the program as it has expanded.[24] For example, several have explored the importance of explicitly formulating *scope conditions* as part of the structure of the theory;[25] others have focused attention on *instantiation* of theory as part of the effort to distinguish between the theoretical elements of an abstract formulation and the factual elements that are involved in applying a theory to a concrete social setting. But perhaps the most basic of these metatheoretical ideas concerns the conception of a *state organizing process*.

Two levels of concern are identified in the state-organizing conception: (1) the *social framework* and (2) the *situation of action*. The elements in a social framework can be *cultural* (for example, norms, values, beliefs, and social categories), *formal* (institutionalized and formalized roles and authority positions, for example), or *interpersonal* (for example, enduring networks of sentiments, influence, and communication). A situation of action occurs when these elements of social framework are used in an interaction.

This distinction between social framework and situation of action suggests a concern with identifying what elements from the framework are accessed by the actors. The *salience* principles of status characteristics theory address this question by specifying how and under what conditions different types of status characteristics and referential structures become significant to the actors in their immediate situation of action.

Social processes occur in situations of immediate action that have specified properties and features. In the state-organizing process scheme, one of the most important components of such situations is the *activating events*. These include events or conditions that become the focus of the interaction process; they are the goal-states toward which actors are oriented. In status characteristics theory, activating events are the collective tasks that actors perform in their interaction, the outcomes of which can be evaluated by group success or failure.

Given appropriate activating conditions, a social process will evolve. *Behaviors* will occur that are addressed to the demands of the activating conditions, and *social information* will be used by the actors to define their situation. The outcome of this process is the formation of states—stable structures that define the relations of the actors to each other within their immediate situation. Once states are formed, these states determine the behaviors of the actors relative to each other and to the activating events. The theoretical concepts and principles that describe these relations in status characteristics theory are those of *expectation advantage*, the *observable power and prestige order*, and the *basic expectation assumption*. Given that an episode in a situation of action is completed—that, say, the task has been dealt with by the actors—the process is de-activated. In the case of status processes, this could mean that status distinctions that were operating in the situation become latent and that the power and prestige order becomes de-differentiated.

State-organizing processes can also have transsituational effects, involving outcomes from one interaction episode that become inputs to a succeeding episode (that is, *succession* effects), or they can involve outcomes from one or more interaction episodes that feed back to the level of the social framework (that is, *construction* effects). Some of the most recent work in the expectation states program deals with these transsituational effects. We have already described the work of Joseph Berger, M. Hamit Fisek, and Robert Z. Norman[26] concerned with how expectations formed in one task situation affect expectations formed in succeeding task situations. In addition, Barry Markovsky[27] has constructed diffusion models that describe how changes in expectations toward members of disadvantaged status classes created in a specific task situation can diffuse through a larger population and thus produce macro-level changes. Also, Cecilia Ridgeway[28] has formulated a theory that describes how the states of an initially nonvalued characteristic can acquire status value and generalized expectations and thus become a status characteristic (see Chapter 32 for a further discussion of Ridgeway's theory). These constructed status characteristics then become elements of the larger social framework.[29]

This conception of a state organizing process has also become a framework for subsequent work on existing expectation states theories and for constructing theories regarding new and different social processes. These ideas about state-organizing processes have thus become part of the ever-proliferating set of theoretical branches, each extending the theory and demonstrating its importance for ever more dimensions of social interaction.[30] Thus, as the twenty-first century approaches, expectation states theories remain one of the most active approaches in interactionist theorizing.

NOTES

1. Many others have been integral to the development of theories in the expectation states program, particularly Morris Zelditch, Jr., and Bernard P. Cohen. Nevertheless, Berger has been the intellectual force at the center of the program since its inception. Many of those who have since made separate contributions to the program first made contributions in collaboration with Berger. For examples of overviews of the progress, see Joseph Berger, Thomas L. Conner, and M. Hamit Fisek, eds., *Expectation States Theory: A Theoretical Research Program* (Cambridge, MA: Winthrop, 1974; reprinted by University Press of America in 1982); Joseph Berger, David G. Wagner, and Morris Zelditch, Jr., "Theory Growth, Social Processes, and Metatheory" in *Theory Building in Sociology,* ed. J.H. Turner (Newbury Park, CA: Sage, 1989), pp. 19–43; Joseph Berger and Morris Zelditch, Jr., eds., *Status Rewards and Influence* (San Francisco: Jossey-Bass, 1985); Joseph Berger, Bernard P. Cohen, and Morris Zelditch, Jr., "Status Characteristics and Expectation States" in *Sociological Theories in Progress*, eds. J. Berger, M. Zelditch, Jr., and B. Anderson (Boston: Houghton-Mifflin, 1966), pp. 29–46.

2. Berger first developed this argument in his doctoral dissertation: Joseph Berger, "Relations Between Performance, Rewards, and Action-Opportunities in Small Groups" (unpublished doctoral dissertation, Harvard University, 1958). See also Joseph Berger and Thomas L. Conner, "Performance Expectations and Behavior in Small Groups," *Acta Sociologica* 12 (1969), pp. 186–198.

3. See, for example, Joseph Berger, Thomas L. Conner, and W. McKeown, "Evaluations and the Formation and Maintenance of Performance Expectations," *Human Relations* 22 (1969), pp. 481–502; Thomas J. Fararo, "An Expectation States Process Model" in *Mathematical Sociology* (New York: Wiley, 1973), pp. 229–237.

4. Joseph Berger and Thomas L. Conner, "Performance Expectations and Behavior in Small Groups: A Revised Formulation" in *Expectation States Theory: A Theoretical Research Program*, eds. Berger, Conner, and Fisek, pp. 85–109; see also M. Hamit Fisek, "A Model for the Evolution of Status Structures," pp. 55–83, in the same volume.

5. Thomas J. Fararo and John Skvoretz, "E-State Structuralism: A Theoretical Method," *American Sociological Review* 51 (1986), pp. 591–602.

6. See, for example, Martha Foschi and R. Foschi, "A Bayesian Model for Performance Expectations: Extension and Simulation," *Social Psychology Quarterly* 42 (1979), pp. 232–241.

7. For a summary of some of this research see Bernard P. Cohen, Joseph Berger, and Morris Zelditch, Jr., "Status Conceptions and Interaction: A Case Study of the Problem of Developing Cumulative Knowledge" in *Experimental Social Psychology*, ed. C.G. McClintock (New York: Holt, Rinehart, and Winston, 1972), pp. 449–483.

8. See Joseph Berger, Bernard P. Cohen, and Morris Zelditch, Jr., "Status Characteristics and Expectation States" in *Sociological Theories in Progress*, pp. 29–46; the paper was updated, with a report on empirical results in *American Sociological Review* 37 (1972), pp. 241–255.

9. See Joseph Berger and M. Hamit Fisek, "A Generalization of the Theory of Status Characteristics and Expectation States" in *Expectation States Theory*, pp. 163–205; and Joseph Berger, M. Hamit Fisek, Robert Z. Norman, and Morris Zelditch, Jr., *Status Characteristics and Social Interaction* (New York: Elsevier, 1977).

10. The effects of another kind of status difference, referred to as a *specific status characteristic*, are also considered. Specific status differences apply only to a specific task or kind of task (for example, involving mathematical or artistic ability).

11. Elimination is the kind of assumption that has often been made by those who have investigated status inconsistency. Gerhard Lenski's status crystallization theory assumes that inconsistencies create tension and anxiety. Actors can reduce the tension by eliminating (for example, ignoring or hiding) inconsistent statuses. If they cannot reduce the inconsistency, then they might become isolated and be prone to various types of coping behaviors (for example, political radicalism). The kind of elimination most often assumed is based on a maximization principle where actors prefer status definitions that reflect positively on them. Thus, when confronted with inconsistent definitions, they are most likely to eliminate negative status information, maximizing the positive.

12. See Joseph Berger, Morris Zelditch, Jr., Bo Anderson, and Bernard P. Cohen, "Structural Aspects of Distributive Justice: A Status Value Formulation" in *Sociological Theories in Progress*, vol. II., eds. J. Berger, M. Zelditch, Jr., and B. Anderson (Boston: Houghton-Mifflin, 1972), pp. 119–146.

13. See J.S. Adams, "Inequity in Social Exchange" in *Advances in Experimental Social Psychology*, 2 (1965), pp. 267–299. See also E. Walster, E. Berschied, and G.W. Walster, "New Directions in Equity Research," *Journal of Personality and Social Psychology* 25 (1973), pp. 151–176.

14. See David G. Wagner, "Gender Differences in Reward Preference: A Status Based Account," *Small Group Research* 26 (1995), pp. 353–371 and "Reward Preferences in Mixed-Sex Interaction." Under review. See also the work under "Formation of Reward Expectations" cited in note 18.

15. Murray A. Webster, Jr., "Sources of Evaluations and Expectations for Performance," *Sociometry* 32 (1969), pp. 243–258.

16. Murray A. Webster, Jr., "Status Characteristics and Sources of Expectations." Report No. 82, Center for Social Organization of Schools, Johns Hopkins University, 1970.

17. See especially Murray A. Webster, Jr. and B. Sobieszek, *Sources of Self Evaluation* (New York: Wiley, 1974). For further extensions of this idea, see J.C. Moore, "Role Enactment and Self-Identity: An Expectation States Approach" in *Status, Rewards, and Influence: How Expectations Organize Interaction*, eds. J. Berger and M. Zelditch, Jr. (San Francisco: Jossey-Bass, 1985), pp. 262–316.

18. Joseph Berger, M. Hamit Fisek, Robert Z. Norman, and David G. Wagner, "The Formation of Reward Expectations in Status Situations" in *Status, Rewards and Influence*, pp. 215–261.

19. The current version of the reward expectation theory focuses on situations in which an ability standard is activated in the situation and in which one or more categorical structures can become activated.

20. Joseph Berger, M. Hamit Fisek, and Robert Z. Norman, "The Evolution of Status Expectations: A Theoretical Extension" in *Sociological Theories in Progress: New Formulations*, eds. J. Berger, M. Zelditch, Jr., and B. Anderson (Newbury Park, CA: Sage, 1989), pp. 100–130.

21. Joseph Berger, Murray A. Webster, Jr., Cecilia Ridgeway, and Susan J. Rosenholtz, "Status Cues, Expectations, and Behavior" in *Advances in Group Processes* 3 (1986), pp. 1–22.

22. This formulation responds to dominance theories that see dominating behaviors as the mechanism behind status differences in groups. See Alan Mazur, "A Biosocial Model of Status in Face-to-Face Primate Groups," *Social Forces* 64 (1985), pp. 377–402, and M. Lee and R. Ofshe, "The Impact of Behavioral Style and Status Characteristics on Social Influence: A Test of Two Competing Theories," *Social Psychology Quarterly* 44 (1981), pp. 73–82.

23. See Cecilia L. Ridgeway and Joseph Berger, "Expectations, Legitimacy, and Dominance in Task Groups," *American Sociological Review* 51 (1986), pp. 603–617; Cecilia L. Ridgeway and Joseph Berger, "The Legitimation of Power and Prestige Orders in Task Groups" in *Status Generalization: New Theory and Research*, eds. M.A. Webster and M. Foschi (Stanford, CA: Stanford University Press, 1988), pp. 207–231; Cecilia L. Ridgeway, "Gender Differences in Task Groups: A Status and Legitimacy Account," also in Webster and Foschi, 1988, pp. 188–206.

24. A much more complete listing of expectation states theories is available in David G. Wagner and Joseph Berger, "Status Characteristics Theory: The Growth of a Program" in *Theoretical Research Programs: Studies in the Growth of Theory*, eds. J. Berger and M. Zelditch (Stanford, CA: Stanford University Press, 1993), pp. 24–63. See also Joseph Berger, Morris Zelditch, Jr., and David G. Wagner, "Theory Growth, Social Processes, and Metatheory" in *Theory Building in Sociology*, pp. 19–42.

25. See, for example, Henry A. Walker and Bernard P. Cohen, "Scope Statements: Imperatives for Evaluating Theory," *American Sociological Review* 50 (1985), pp. 288–301.

26. See Berger, Fisek, and Norman, "The Evolution of Status Expectations" (cited in note 20).

27. Barry Markovsky, "From Expectation States to Macro Processes" in *Status Generalization: New Theory and Research*, eds. M.A. Webster and M. Foschi (Stanford, CA: Stanford University Press, 1988), pp. 351–365.

28. Cecilia L. Ridgeway, "The Social Construction of Status Value: Gender and Other Nominal Characteristics," *Social Forces* 70 (1991), pp. 367–386.

29. For a more comprehensive and detailed discussion of this conception of a state organizing process, see Joseph Berger, David G. Wagner, and Morris Zelditch, Jr., "A Working Strategy for Constructing Theories: State Organizing Processes" in *Metatheorizing*, ed. G. Ritzer (Newbury Park, CA: Sage, 1992), pp. 107–123.

30. Joseph Berger, "Directions in Expectation States Research" *Status Generalization: New Theory and Research* (cited in note 27), pp. 450–474, and David G. Wagner, "Status Violations: Toward an Expectation States Theory of the Social Control of Status Deviance," pp. 110–122.

PART VI

Structuralist Theorizing

34

The Emerging Tradition:

The Rise of Structuralist Theorizing

The notion of "social structure" or "structure" is central to sociology as an intellectual enterprise. Yet, despite this centrality, the concept of structure is only vaguely theorized; it tends to be used more as a metaphor than as a precisely defined theoretical term. In the chapters to follow, we will examine a variety of theoretical approaches where the topic of structure is central to the theory rather than being an implicit metaphor that sneaks in the back door. All these approaches draw from the early masters of sociology, extracting concepts or at least theoretical images of structure from sociology's first one hundred years. As this structural tradition matured in the middle decades of the twentieth century, these images of structure from the early masters began to diverge as various scholars sought to create their own "structural" or "structuralist" program. But even as these new theory programs emerged, they still drew on the key insights of sociology's first theorists.

STRUCTURAL ELEMENTS IN KARL MARX'S THEORIES

Karl Marx's ideas penetrate just about all macro-level theoretical traditions in sociology, so we would expect to find Marxian concepts in virtually all structural approaches. In particular, two key ideas appear to have been used by those developing a structural or structuralist theory program: (1) system reproduction and (2) system contradiction.[1]

1. Marx was concerned with how patterns of inequality "reproduce" themselves. That is, how is inequality in power and wealth sustained? And how are social relations structured to sustain these inequalities? This metaphor of social reproduction has been used by many structural theorists who like to view social structures as "reproduced" by the repetitive social encounters among individuals.

From this metaphorical use of Marx's ideas comes a vision of social structure as the distribution of resources among actors who use their respective resources in social encounters and, in the process, reproduce social structure and its attendant distribution of resources. Thus, structure *is* the symbolic, material, and political resources that actors have in their encounters. As actors employ these resources to their advantage, they reproduce the structure of their social relations because they sustain their respective shares of resources. For example, those who can control where others are physically located, who have access to coercion, who control communication channels, and who can manipulate the flow of information can "structure" successive encounters with others who have fewer resources in ways that reproduce inequalities. In so doing, those who have control reproduce structure, but structure is tied to the interactive encounters among individuals using their resources. It is not something "out there" beyond actors and their interactions in concrete situations.

2. Marx was, of course, interested in changing capitalist society, so he needed to introduce a concept that could break the vicious cycle of system reproduction. His analysis of Georg Wilhelm Friedrich Hegel's dialectics gave him the critical idea of "contradiction." Material social arrangements contain, Marx argued, patterns of social relations that are self-transforming. For example, capitalists must concentrate large pools of labor around machines in urban areas, which allows workers to communicate their grievances and to organize politically to change the very nature of capitalism. Thus, private expropriation of profits by capitalists contradicts the socially organized production process; over time this underlying contradiction will produce conflictual social relations—that is, a revolution by workers—that change and transform the nature of social relations and, hence, social structure.[2]

Again, like so many adaptations of early structural ideas, this notion of contradiction has been borrowed in a metaphorical way by contemporary theorists. It is used to introduce conflict and change into structural analysis. These theorists posit, in essence, that the distribution of resources that sustains or reproduces social relations is inherently subject to redistribution, under certain conditions. The main condition is inequality. That is, structures reproduced through unequal distribution of symbolic, material, and political resources will exhibit underlying contradictions that will, in turn, generate pressures for interactions that *do not* reproduce the structure but instead transform it in some way through redistribution of resources.

ÉMILE DURKHEIM'S FUNCTIONALISM AND THE EMERGENCE OF STRUCTURAL SOCIOLOGY

No one influenced the development of all forms of structural sociology more than Émile Durkheim. His key ideas have already been examined in several places, such as Chapter 2 on the emergence of functionalism and Chapter 26 on the emergence of interactionism. Here we will just concentrate on those ideas in Durkheim's work that had a significant impact on structural and structuralist sociology. Durkheim's functionalism was distinguished by its emphasis on the problem or requisite of social integration and on the mechanisms for meeting this one master requisite. In this regard, Durkheim stood directly in a long line of French thinkers, starting with Charles Montesquieu, proceeding through Jean Condorcet, Jacques Turgot, and Jean Jacques Rousseau, and then moving on to Claude-Henri de Saint-Simon and Auguste Comte. During his career, Durkheim posited four basic types of mechanisms for resolving integrative problems: (1) cultural (collective conscience, collective representations), (2) structural (structural interdependencies and subgroup formation), (3) interpersonal (ritual and the ensuing sense of effervescence and social solidarity), and (4) cognitive (classification, modes of symbolization). In essence, Durkheimian sociology examines how systems of cultural symbols, patterns of group formation and structural interdependence, ritual performances, and systems of cognitive classification integrate variously differentiated social structures. Cultural and structural mechanisms were emphasized in Durkheim's early works, before 1900,[3] whereas interpersonal and cognitive mechanisms became increasingly prominent in his later works in the early years of the twentieth century.[4]

Durkheim viewed structure in much the same way as Comte did—that is, as a form of "statical analysis"—but he used Montesquieu's term "social morphology."[5] For Durkheim, as for Montesquieu, morphological analysis should focus on the "number," the "nature," and the "interrelations" of parts or "elements." This view of structure was emphasized in *The Rules of the Sociological Method*,[6] where Durkheim viewed classification of social facts as involving attention to "the nature and number of the component elements and their mode of combination," whereas explanation "must seek separately the efficient cause (of a social fact) and the function it fulfills." Durkheim's positions on classification of social facts influenced those structuralist approaches that sought to map patterns of relations among the units of social systems. Eventually, this more material approach to structure evolved into network analysis via British anthropology, as we will see in Chapters 35 and 38.

Yet, later Durkheim shifted emphasis to the more mental aspects of structure. Indeed, Durkheimian sociology contained an ambiguity about whether structure when broken down to its morphological components is primarily mental, interpersonal, cultural, or material. This ambiguity became more pronounced in the first decade of the century as Durkheim studied how material structures become a part of the mental structure of individuals. For example, Durkheim's early essay on "Incest: The Nature and Origin of the Taboo"[7] along with his and Marcel Mauss' *Primitive Classification*[8] marked a clear movement to the social psychological aspects of structure. This shift became the foundation for much of Claude Lévi-Strauss' structuralism, as we will see in the next chapter, but unlike Lévi-Strauss who viewed material structures as reflection of mental categories of the mind, Durkheim always insisted that the structures of the mind reflect the material structure of actual social relations. For example, as Durkheim and Mauss noted in *Primitive Classification*:[9]

> Society was not simply a model which classificatory thought followed; it was its own division which served as the division the for the system of classification. The first logical categories were social categories; the first classes of things were classes of men, into which these things were integrated. It was because men were grouped, and thought of themselves in the form of groups, that in their ideas they grouped other things, and in the beginning the two modes of grouping were merged to the point of being indistinct. Moieties were the first genera; clans, the first species . . . And if the totality of things is conceived as a single system, this is because society itself is seen in the same way. It is a whole, or rather it is *the* unique whole to which everything else is related. Thus logical hierarchy is only another aspect of social hierarchy, and the unity of knowledge is nothing else

than the very unity of the collectivity, extended to the universe.

Moreover, Durkheim and Mauss provided Lévi-Strauss with yet another lead for his structuralism by emphasizing the importance of mythology, especially as derived from religion, as a reliable source for decoding the "logical hierarchy" and structure of thought. Finally, near the end of this long essay, Durkheim and Mauss introduced another element of structuralist thinking: mental structures are composed of logical connections that reflect how material and cultural "facts" are juxtaposed, merged, distinguished, and, most important, opposed. Although they do not pursue this thought, they clearly introduced Lévi-Strauss to what he was to conceptualize later as "binary oppositions." For example, Durkheim and Mauss noted,[10]

> There are sentimental affinities between things as between individuals, and they are classed according to these affinities. . . . All kinds of affective elements combine in the representation made of it. . . . Things are above all sacred or profane, pure or impure, friends or enemies, favourable or unfavourable; i.e., their most fundamental characteristics are only expressions of the way in which they affect social sensibility.

In sum, then, several key elements of structuralism were evident in Durkheim's work at the turn of the century: (1) Mental structures involve the logical ordering and generation of classificatory systems that, although modeled after society, become the basis for individuals' interpretation and action in society. (2) Such structures are designed to show the connectedness of phenomena as part of a coherent, systemic whole. (3) Finally, these structures are created by the logical relations of affinities and oppositions as they are encountered in the cultural and material structure of society.

On this latter point, others in the Durkheimian circle pursued the notion that mental structures are constructed from oppositions. Most notable was Robert Hertz, who was killed in World War I, like so many of Durkheim's younger colleagues. Hertz's best-known essays were published as *Death and the Right Hand*,[11] in which the notion of binary opposition is developed beyond Durkheim's and Mauss' conceptualization. According to Hertz, mental structures are built up from oppositions: strong-weak, night-day, left-right, natural-social, good-bad, and so on. Yet although the critical effort in *Death and the Right Hand* is to uncover the underlying principles beneath the surface structure of observed phenomena, an ambiguity exists: Are mental categories—such as "left" and "right"—reflections of social relations, or are they generated from some underlying cognitive capacity? On the surface, Hertz took a straight Durkheimian line—that mental categories reflect social structures—but his work gives consistent hints, supported by his examples, that mental processes per se produce their own structures.

Mauss, by himself, also might have provided Lévi-Strauss with implicit suggestions for reversing Durkheim's position. Although Mauss adhered very closely to Durkheimian principles, seeing himself as "the keeper of the Durkheimian tradition," his book (with Henri Beuchat) titled *Seasonal Variations of the Eskimo: A Study of Social Morphology*[12] emphasized the oppositional nature of thought—in this case the dualistic categories and behaviors created by the facts of winter and summer for the Eskimo. Moreover, in his most famous work, *The Gift*,[13] Mauss emphasized once again the search for underlying principles and practices—in this instance the principle of reciprocity—beneath surface structures and practices such as gift giving. Thus, because

Lévi-Strauss read Durkheim, Mauss, and others in the Durkheimian circle, such as Hertz, the basic elements of his structuralism were readily evident. Yet the question remains, What made Lévi-Strauss reverse Durkheim's and Mauss' position and argue that the material and the cultural structure of society (for example, kinship and mythology respectively) reflect innate capacities of the human mind for generating structures? We will try to answer this question in the next chapter because it is critical not only to sociological structuralism but also to the broad structuralist movement in the humanities and other social sciences.

Structuralist sociology was also influenced by one of the last ideas Durkheim developed in *The Elementary Forms of Religious Life*:[14] the view that conceptions of the sacred as well as the totems that symbolize the sacred and the rituals that are addressed to the sacred are, actually, the worship of society. As individuals feel and sense the power of society beyond them, they have a need to represent this power in the symbols of religion and to arouse the emotion surrounding these symbols through the enactment of rituals. Thus, in enacting religious rituals, individuals in society are not just confirming their faith in the supernatural, they are also legitimating the structure of society. In Durkheim's view, then, religion is the worship of society and of the structures that make up people's daily lives. Thus, social structure, cognitions and beliefs, and ritual practices were seen by Durkheim as intimately interconnected. This connection among these processes became a point of emphasis in many structuralist approaches, as is explored in Chapters 36 and 37.

GEORG SIMMEL'S FORMAL STRUCTURALISM

Although Durkheim was the most influential figure in the emergence of structuralism, the German sociologist Georg Simmel, was almost as influential. Simmel's emphasis on discovering the underlying forms of association among individuals and groups has exerted considerable influence on some modern schools of structural sociology. Simmel believed social structure consists of "permanent interactions," and formal sociology seeks the underlying pattern of these permanent interactions. The content or substantive nature of these interactions is far less significant, sociologically, than their basic form. Although interactions can reveal an enormous variety of contents, the underlying form of relations might be the same. As Simmel emphasized,[15]

> Social groups, which are the most diverse imaginable in purpose and general significance, may nevertheless show identical forms of behavior toward one another on the part of individual members. We find superiority and subordination, competition, division of labor . . . and innumerable similar features in the state, in the religious community, in a band of conspirators, in an economic association, in an art school, in the family. However diverse the interests are that give rise to these associations, the *forms* in which the interests are realized may yet be identical.

Social structure must thus be conceptualized as forms or configurations of interaction that undergird and make possible the wide variety of substantive activities of individuals. This point makes Simmel's eclectic approach to sociology more understandable, for although he studied widely diverse substantive matters, he was always seeking the underlying form of interaction. In all Simmel's most famous essays, this is the message: Whatever the surface substance and content of social relations, there is an underlying form or structure. For example, as Simmel examined conflict, he could see that conflicts between nation-states, individual people, and small groups all reveal certain basic elements, or forms; or, to illustrate further, as Simmel examined the effects of increasing numbers of people in an encounter, he could argue that the geometric increase in the number of possible relations (two people can have two ties, three can have six, four can have twelve, and so on) changes the form or structure of the encounter; or, as Simmel examined the effects of money (a neutral and nonspecific medium of exchange), he could conclude that the form of social relations is fundamentally altered.[16]

Simmel's essay, *The Web of Group Affiliations*,[17] has probably exerted the most influence on more contemporary structural thinking in sociology. Here, Simmel examined the effects of social differentiation on the affiliations of individuals. In less differentiated systems, the individual is absorbed and surrounded by one or just a few groups that are lodged inside each other, and as a result, the pattern of group affiliations pulls the individual in one direction. With social differentiation, however, the individual can now have multiple group affiliations, belonging to many different groups and being pulled in many different directions. As a result, no one group can absorb the individual, and the individual only gives a portion of self to any particular group. This process of multiple group affiliations sets the person free and increases individuality because each person can, to a degree, choose and select configurations of group affiliation. At the more macro level, as differentiation encourages multiple group affiliations, it generates many cross-cutting affiliations among individuals who might find themselves in conflict in one sphere and associates in another. As a result, the polarization of society is less likely because group affiliations place individuals in diverse social structures and, thereby, prevent them from being overly mobilized by any one group or small set of groups.

INTERACTIONISM AND MICROSTRUCTURALISM

As structural analyses have become more concerned with the interactive processes by which structure is produced and reproduced, they have turned to interactionist theory—primarily ethnomethodology, phenomenology, Chicago School symbolic interactionism, and the dramaturgical work of Erving Goffman. All these interactionist traditions are built from the conceptual base laid by George Herbert Mead and, to a lesser extent, Alfred Schutz. Let us first examine Mead's ideas that reappear in contemporary structural analysis and then turn to Schutz's approach.

George Herbert Mead's Behavioristic Structuralism

Mead was a social behaviorist,[18] by which he meant this: Those behavioral capacities facilitating adjustment to the social environment will be retained by a maturing individual. They are retained because they provide reinforcement or the rewards that come from adjustment to ongoing patterns of social organization. What, then, are these behavioral capacities that facilitate adjustment to social structures? Mead believed they are (1) the capacity to assume the perspective of a variety of others (*role taking*); (2) the capacity to see oneself as an object in situations and to adjust response in accordance with self-perceptions and evaluations (*self*); and (3) the capacity to engage in the process of "imaginative rehearsal" of alternative lines of conduct and their potential outcomes (*mind*). In a very real sense, Mead provided the mechanism for Durkheim's analysis of how the "collective conscience" is connected to individual behavior.

With these basic elements, Mead presented a general image of social organization as constructed and actively sustained through the mutual reading of gestures, the evaluation of oneself from the perspective of generalized others or "communities of attitudes," and the reflexive weighing of alternatives. These ideas are, today, very much a part of most microstructural approaches. Yet most micro approaches tend to underemphasize Mead's view that social structure constrains and circumscribes the options of individuals; in particular, Mead stressed that the ability to role-take with the generalized other and to use this community of attitudes as a basis of self-evaluation is crucial to the reproduction of structure. That is, in a very Durkheimian and structuralist fashion, Mead saw structure as being reproduced by people's use of general attitudes (values, beliefs, norms, and other cultural processes in modern terms). Thus, structuralists take only part of Mead's legacy, in which emphasis is on situational interpersonal practices. They tend to ignore Mead's more Durkheimian ideas on conformity to generalized others.

Alfred Schutz's Phenomenological Structuralism

Structuralists seem to prefer Schutz's conceptualization of stock knowledge at hand to Mead's notion of generalized others and to modern conceptions of roles, values, beliefs, and norms. This preference comes from the metaphorical idea of stocks that can be used selectively as the actor wants and desires. Schutz's conception would appear to de-emphasize explicit normative constraints and roles. But does it? Let us quote Schutz:[19]

> Now, to the natural man all his past experiences are present as *ordered* [italics in original], as knowledge or as awareness of what to expect, just as the whole external world is present to him as ordered. Ordinarily, and unless he is forced to solve a special kind of problem, he does not ask questions about how this ordered world was constituted. The particular patterns of order we are now considering are synthetic meaning—configurations of already lived experience.

Schutz's view emphasized constraint. Norms, values, beliefs, and roles are indeed highly salient parts of one's implicit and, if need be, explicit interpretation of a situation. They "order" experiences and lines of conduct. But for many contemporary theorists, the concept of *stocks of knowledge* is meant to connote something much less ordered and structured. Rather, these theorists see this concept as a series of implicit cognitions that are used to construct lines of conduct in strategic conduct.

THE EMERGING DIVERSITY OF STRUCTURAL AND STRUCTURALIST ANALYSIS

From these early roots, structural and structuralist sociology splintered in many different directions. This diversity began to emerge in the century's middle decades, as scholars pondered the fundamental nature of the structures organizing human life. At this time, French-inspired structuralism became a broad intellectual movement outside of sociology in areas as diverse as linguistics, literary criticism, and anthropology. Some of this movement penetrated sociology, primarily via the anthropology of Lévi-Strauss, and exerted considerable though waning[20] influence on what might be termed cultural structuralism. What remains of this emphasis is examined in Chapter 37 on cultural theories. Other approaches were more Simmelian, seeking to discover the forms of social relations in the actual connections among individuals or at least in their rates of interaction. Network sociology was one such Simmel-inspired tradition, although Durkheim's emphasis on structure as the number, nature, and manner of arrangement of parts was also an early source of inspiration (see Chapter 38). Similarly, the macrostructuralism of Peter Blau was also inspired by Simmel, as is explored in Chapter 39. Finally, several highly synthetic approaches have sought to blend ideas from all these early masters as well as from more contemporary theories. Anthony Giddens' "structuration theory" is the most prominent of these synthetic approaches, as we will see in Chapter 36.

NOTES

1. See Karl Marx, *A Contribution to the Critique of Political Economy* (New York: International, 1970); *The Economic and Philosophic Manuscripts of 1844* (New York: International, 1964); and *Capital: A Critical Analysis of Capitalist Production* (New York: International, 1967).

2. Karl Marx and Friedrich Engels, *The Communist Manifesto* (New York: International, 1971, originally published in 1847).

3. Émile Durkheim, *The Division of Labor in Society* (New York: Free Press, 1947, originally published in 1893) and *The Rules of the Sociological Method* (New York: Free Press, 1938, originally published in 1895).

4. Émile Durkheim, *The Elementary Forms of Religious Life* (New York: Free Press, 1947, originally published in 1912); Émile Durkheim and Marcel Mauss, *Primitive Classification* (London: Cohen and West, 1963, originally published in 1903).

5. Charles Montesquieu, *The Spirit of the Laws* (London: Colonial, 1900, originally published in 1748).

6. Durkheim, *The Rules of the Sociological Method* (cited in note 3), p. 96.

7. Émile Durkheim, "Incest: The Nature and Origin of the Taboo." *Anneé Sociologique* 1 (1898), pp. 1–70.

8. Durkheim and Mauss, *Primitive Classification* (cited in note 4).

9. Ibid., pp. 82–84.

10. Ibid., pp. 85–86.

11. Robert Hertz, *Death and the Right Hand* (London: Cohen and West, 1960, originally published in 1909).

12. Marcel Mauss and Henri Beuchat, *Seasonal Variations of the Eskimo: A Study of Morphology* (London: Routledge & Kegan Paul, 1979, originally published in 1904–1905).

13. Marcel Mauss, *The Gift: Forms and Functions of Exchange* (New York: Free Press, 1941, originally published in 1925).

14. Durkheim, *The Elementary Forms of Religious* Life (cited in note 4).

15. Georg Simmel, *Fundamental Problems of Sociology* (1918), portions of which are translated in Wolf, *The Sociology of Georg Simmel* (New York: Free Press, 1950). Quote is from p. 22 of original.

16. For a review of Simmel's work, see Jonathan H. Turner, Leonard Beeghley, and Charles Powers, *The Emergence of Sociological Theory* (Belmont, CA: Wadsworth, 1998).

17. Georg Simmel, "The Intersection of Social Spheres," first published in 1890, best communicates his idea. See *Georg Simmel: Sociologist and European*, trans. Peter Laurence (New York: Barnes and Noble,

1976), but the idea is most available in *Conflict and the Web of Group Affiliations*, trans. Reinhard Bendix (New York: Free Press, 1955).

18. George Herbert Mead, *Mind, Self, and Society* (Chicago: University of Chicago Press, 1934).

19. Alfred Schutz, *The Phenomenology of the Social World* (Evanston, IL: Northwestern University Press, 1967, originally published in 1932), p. 81.

20. See Anthony Giddens, "Structuralism, Post-structuralism and the Production of Culture" in *Social Theory Today,* eds., Anthony Giddens and Jonathan Turner, (Cambridge, England: Polity Press, 1987), pp. 195–223.

35

The Maturing Tradition:

French, British, and American Variations of Structural Theorizing

The ideas of Émile Durkheim, Karl Marx, and Georg Simmel inspired structuralist sociology, although Durkheim was the most influential figure in this trio of early sociologists. As Durkheim's functionalism shifted from a material to more mental view of structure, it shaped two different anthropological traditions, one that is embodied in the French anthropologist, Claude Lévi-Strauss, and the other the British anthropology as carried forward by A.R. Radcliffe-Brown and S.F. Nadel. Partly influenced by this British tradition, an American strain of structural analysis also began to emerge in the middle decades of the century. Later structural theories of the more contemporary era were often either extensions of, or reactions against, these three maturing structuralist traditions, as will become evident in the next block of chapters. For the present, let us outline in broad strokes how the ideas of early theorists, particularly those of Durkheim, were used to create a distinctly structuralist sociology.

THE FRENCH STRUCTURAL TRADITION OF CLAUDE LÉVI-STRAUSS

In Chapter 19 on the emergence of exchange theory, Lévi-Strauss' analysis of bridal exchanges was briefly examined. In *The Elementary Structures of Kinship*,[1] Lévi-Strauss only hinted at the more philosophical view of the world that his analysis of kinship implied.[2] Most of this book examined the varying levels of social solidarity that emerge from direct and indirect bridal exchanges among kin groups. Yet in many ways, *The Elementary Structures of Kinship* was a transitional work because it began to depart from the earlier foundations provided by Durkheim and Marcel Mauss.[3] Indeed, these departures from Durkheim and Mauss signaled that Lévi-Strauss was about to "turn Durkheim on his head" in much the same way that Marx was to revise Georg Wilhelm Friedrich Hegel. As

Durkheim and Mauss argued in *Primitive Classification*, human cognitive categories reflect the structure of society.[4] In contrast, Lévi-Strauss came to the opposite conclusion: the structure of society is but a surface manifestation of fundamental mental processes.

Lévi-Strauss came to this position under the influence of structural linguistics, as initially chartered by Ferdinand de Saussure[5] and Roman Jakobson. De Saussure is typically considered the father both of structural linguistics[6] and of Lévi-Strauss' structuralism as well. Commentators have often viewed Lévi-Strauss' interest in linguistics as decisive in his reversal of the Durkheimian tradition. Yet the Swiss linguist de Saussure saw himself as a Durkheimian. De Saussure's posthumously published lectures, *Course in General Linguistics* had a decidedly Durkheimian tone: He argued that the parts of language acquire their meaning only in relation to the structure of the whole; the units of language—whether sounds or morphemes—are only points in an overall structure that transcends the individual; language is "based entirely on the opposition of concrete units"[7]; the underlying structure of language (*langue*) can be known and understood only by reference to surface phenomena, such as speech (*parole*); and the structure of language is "no longer looked upon as an organism that developed independently but as a product of the collective mind of linguistic groups."[8] Thus, in reality, as a contemporary of Durkheim, de Saussure was far more committed to Durkheim's vision of reality than is typically acknowledged, but he made a critical breakthrough in linguistic analysis of the nineteenth century: Speech is but a surface manifestation of more fundamental mental processes. Language is not speech or the written word; rather, it is a particular way of thinking, which, in true Durkheimian fashion, de Saussure viewed as a product of the general patterns of social and cultural organization among people. This distinction of speech as a mere surface manifestation of underlying mental processes was increasingly used as a metaphor for Lévi-Strauss' structuralism. Of course, this metaphor is as old as Plato's view that reality is a mere reflection of universal essences and as recent as Marx's dictum that cultural values and beliefs, as well as institutional arrangements, are reflections of an underlying substructure of economic relations.

Lévi-Strauss also borrowed from the early-twentieth-century linguist Jakobson the notion that the mental thought underlying language occurs as *binary contrasts*, such as good/bad, male/female, yes/no, black/white, and human/nonhuman. Moreover, drawing from Jakobson and others, Lévi-Strauss viewed the underlying mental reality of binary opposites as organized, or mediated, by a series of "innate codes" or rules that could be used to generate many different social forms: language, art, music, social structure, myths, values, beliefs, and so on.[9]

In essence, as Lévi-Strauss received these very Durkheimian ideas, he appeared to focus primarily on the distinction between *langue* and *parole* and on the notion of language as constructed from oppositions, while ignoring de Saussure's emphasis on the social structural origins of *langue*.

Yet why did Lévi-Strauss find linguistic analysis so appealing, and why did he ignore the Durkheimian thrust of de Saussure's work? Lévi-Strauss' own self-reflective answers are not particularly revealing. For example, he claimed that he was probably born a structuralist, recalling that even when he was a two-year-old and still unable to read, he sought to decipher signs with similar groupings of letters. Another childhood influence, he claimed, was geology, in which the task was to discover the underlying geological operations for the tremendous diversity of landscapes.[10] He also constructed many genealogies, and for a time, Lévi-Strauss declared that he was an "anti-Durkheimian" and would embrace Anglo-American methods as an alternative to the Durkheimian approach.[11] Yet he always kept a foot in the French tradition. For example, he dedicated his essay "French Sociology" to Mauss and emphasized,[12]

> One could say that the entire purpose of the French school lies in an attempt to break up the categories of the layman, and to group the data into a deeper, sounder classification. As was emphasized by Durkheim, the true and only basis of sociology is social morphology, i.e., this part of sociology the task of which is to constitute and to classify social types.

Lévi-Strauss' goal increasingly became one of reworking the French tradition for analyzing morphology or structure. His earlier work seemed to lie squarely in the Durkheimian tradition. In *The Elementary Structures of Kinship*[13] (certainly a very Durkheimian-sounding title), he focused on how kinship rules regulate marriage, which owes a great deal to Mauss' *The Gift*.[14] Lévi-Strauss concluded that exchange is a "common denominator of a large number of apparently heterogeneous social activities," and like Mauss before him, he posited a universal structural "principle of reciprocity." Moreover, drawing from and criticizing Durkheim's early analysis of incest,[15] whereby incest was seen as the product of rules of exogamy, Lévi-Strauss viewed rules regarding incest as ordering principles in their own right. The details here are not as important as the recognition that his work was basically Durkheimian, but there were hints of significant additions. In particular, Lévi-Strauss postulated an unconscious mind involving a blueprint or model for coding operations. For example, "reciprocity" is perhaps a universal unconscious code, lodged in the neuroanatomy of the brain and existing *before* the material and cultural structure of society.

Why, then, did Lévi-Strauss make this change in Durkheimian sociology? The simple answer might be that he wanted to say something new. If he merely borrowed Durkheim's idea of morphology, Mauss' and Hertz's concern with underlying structural principles, Mauss' principle of reciprocity, Durkheim's and Mauss' concern with categories of thought, mythology, and ritual, and de Saussure's as well as Jakobson's basic ideas in linguistics, what would be original about his work? His strategy was simply to turn the Durkheimian school upside down and to view mental morphology as the underlying cause of cultural and material morphology. He decided essentially to convert what Durkheim saw as "real," "a thing," and a "social fact" into an unreality. In doing so, he changed what Durkheim saw as unreal into the ultimate reality. Thus structuralism was born as the result of Lévi-Strauss' search for something new to say in the long and distinguished French lineage. All elements of the French lineage remain, but they are reversed.

Thus, during the decades of the midcentury, Lévi-Strauss' structuralism became concerned with understanding cultural and social patterns as the universal mental processes that are rooted in the biochemistry of the human brain.[16] In this sense Lévi-Strauss' structuralism is mentalistic and reductionistic. To summarize what became the basic argument.[17]

1. The empirically observable must be viewed as a *system* of relationships among components—whether these components be elements of myths and folk tales or positions in a kinship system.
2. It is appropriate to construct "statistical models" of these observable systems to summarize the empirically observable relationships among components.

3. Such models, however, are only a surface manifestation of more fundamental forms of reality. These forms are the result of using various codes or rules to organize different binary opposites. Such forms can be visualized through the construction of "mechanical models," which articulate the logical results of using various rules to organize different binary oppositions.
4. The tendencies of statistical models will reflect, imperfectly, the properties of the mechanical model. But the latter is "more real."
5. The mechanical model is built from rules and binary oppositions that are innate to humans and rooted in the biochemistry and neurology of the brain.

Steps 1 and 2 are about as far as Lévi-Strauss had gone in the first publication of *The Elementary Structures of Kinship*. Subsequent work on kinship and on myths invoked at least the rhetoric of Steps 3, 4, and 5. What made structuralism distinctive, therefore, was the commitment to the assumptions and strategy implied in these last steps. The major problem with this strategy was that it is untestable. If mechanical models are never perfectly reflected in the empirical world, how is it possible to confirm or disconfirm the application of rules to binary opposites? As Marshall Sahlins sarcastically remarked, "What is apparent is false and what is hidden from perception and contradicts it is true."[18] Yet, despite such criticisms, the imagery communicated by Lévi-Strauss, especially in Steps 1 and 3, has influenced a great deal of structural theorizing. Although the extremes of Lévi-Strauss' approach have not been adopted by many, the idea of structure as involving "grammars" and "codes" that guide actors in their actions and in the production of social structures has remained appealing.

THE BRITISH STRUCTURAL TRADITION

Radcliffe-Brown's early works, such as his analysis of kinship among Australian tribes[19] and, more significantly, his analysis of ritual in *The Andaman Islanders*,[20] revealed many parallels to the late Durkheim's functional analysis of ritual.[21] Equally important, however, Radcliffe-Brown also developed Durkheim's views on social structure. This latter effort was particularly evident in Radcliffe-Brown's more theoretical work, especially on kinship. In his classic essay "On the Concept of Function in Social Science," Radcliffe-Brown asserted that "the concept of function . . . involves the notion of a structure consisting of a set of relations amongst unit entities, the continuity of the structure being maintained by a life-process made up of the activities of the constituent units."[22]

In Radcliffe-Brown's thinking, then, structural functionalism was to emphasize structure—that is, relations among entities—over function or the consequence of entities for system integration. Moreover, he emphasized his differences with Lévi-Strauss when he wrote to the latter[23]:

> I use the term "social structure" in a sense so different from yours as to make discussion so different as to be unlikely to be profitable. While for you, social structure has nothing to do with reality but with models that are built up, I regard the social structure as a reality.

In his last major theoretical statement, made in *A Natural Science of Society*,[24] Radcliffe-Brown echoed the sentiments in Durkheim's *The Rules of the Sociological Method*[25] and argued that social systems are an emergent natural system composed of the properties of relations among individuals. Therefore, they must be distinguished from psychological systems, which study relations *within* individuals. In the years between World Wars I and II, however, Radcliffe-Brown and other anthropologists were still welded to functional analysis, allowing notions of the functions of a structure to distort purely structural analysis. Whereas hints of a network approach can be found in a number of anthropological works,[26] S.F. Nadel's *The Theory of Social Structure*[27] was decisive for many anthropologists in separating "structure" and "function." In so doing, Nadel proposed a mode of analysis compatible with contemporary network analysis. Nadel began his argument with the assertion that conceptions of structure in the social sciences are too vague. Indeed, we should begin with a more precise, yet general, notion of all structure: "Structure indicates an ordered arrangement of parts which can be treated as transposable, being relations invariant, while the parts themselves are variable."[28] Thus, structure must concentrate on the properties of relations rather than actors, especially on those properties of relations that are invariant and always occur.

From this general conception of all structure, Nadel proposed that "we arrive at the structure of a society through abstracting from the concrete population and its behavior the pattern or network (or system) of relationships obtaining between actors in their capacity of playing roles relative to one another."[29] Within structures exist embedded "subgroups" characterized by certain types of relationships that hold people together. Thus, social structure is to be viewed as layers and clusters of networks—from the total network of a society to varying congeries of subnetworks. The key to discerning structure is to avoid what he termed "the distribution of relations on the grounds of their similarity and dissimilarity" and concentrate, instead, on the "interlocking of relationships whereby interactions implicit in one determine those occurring in others." That is, one should examine specific configurations of linkages among actors playing roles rather than the statistical distributions of actors in this or that type of role.

From these general ideas, several anthropologists, most notably J. Clyde Mitchell[30] and John A. Barnes,[31] welded the metaphorical imagery of work like Nadel's to the more specific techniques for conceptualizing the properties of networks. Coupled with pathbreaking empirical studies,[32] the anthropological tradition began to merge with work in sociology and social psychology. This merger came about, however, only after network analysis had developed in the United States within social psychology.

THE AMERICAN STRUCTURAL TRADITION IN SOCIAL PSYCHOLOGY

At about the same time that network analysis was emerging in England in anthropology, a parallel line of development was occurring in the United States, although some figures and ideas in this American program had roots in Europe. Most of this work came from social psychology, a discipline that at the time was unique to America, where the possibilities of experiments in group settings were beginning to create considerable excitement. As we saw in Chapter 25 on network exchange theory, this experimental tradition is very much alive today, but our concern here is with how the general network approach emerged in America within social psychological research.

Jacob Moreno and Sociometric Techniques

A transplanted European, Jacob Moreno was an eclectic thinker; we have already encountered his ideas on role and role playing in Chapter 28, but perhaps his more enduring

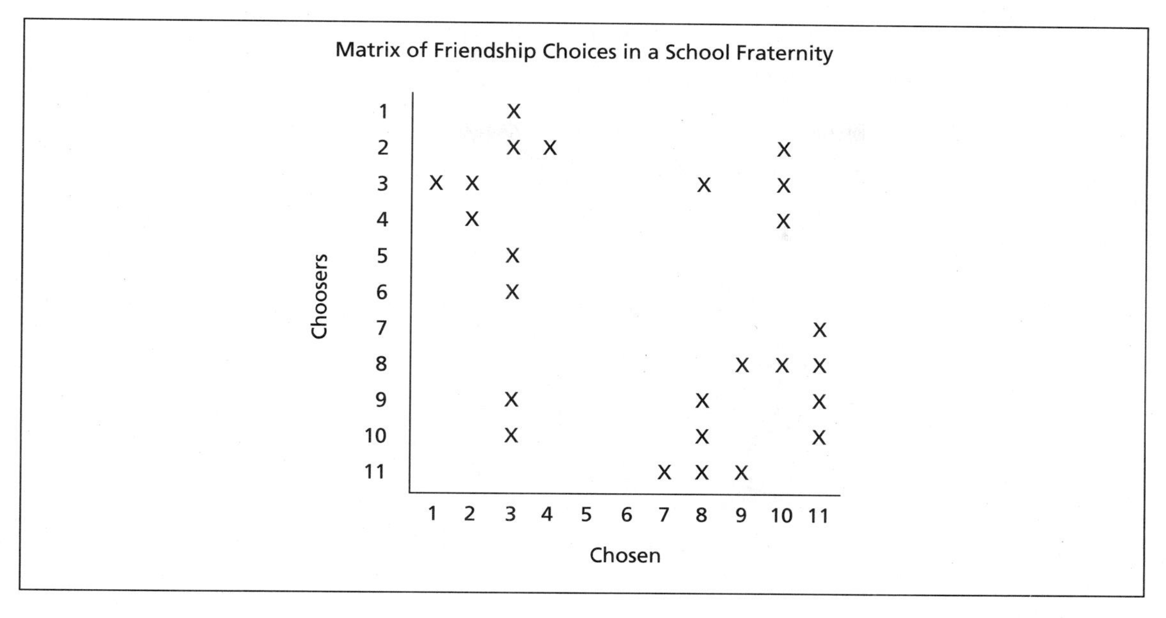

FIGURE 35-1 An Example of an Early Matrix

Source: Constructed from sociogram in J. Moreno, *Who Shall Survive?*, rev. ed., New York: Beacon House, 1953, p. 171.

contribution to sociology was the development of *sociograms*.[33] Moreno was interested in the processes of attraction and repulsion among individuals in groups, so he sought a way to conceptualize and measure these processes. What Moreno and subsequent researchers did was to ask group members about their preferences for associating with others in the group. Typically, group members would be asked questions about whom they liked and with whom they would want to spend time or engage in activity. Often subjects were asked to give their first, second, third, and so on, choices on these and related issues. The results could then be arrayed in a matrix (this was not always done) in which each person's rating of others in a group is recorded (see Figure 35-1 for a simplified example). The construction of such matrices became an important part of network analysis, but equally significant was the development of a sociogram in which group members were arrayed in a visual space, with their relative juxtaposition and connective lines representing the pattern of choices (those closest and connected being attracted in the direction of the arrows, and those distant and unconnected being less attracted to each other). Figure 35-2 illustrates the nature of Moreno's sociograms.

This visual representation of choices, as pulled from a matrix, captures the "structure" of preferences or, in Moreno's terms, the patterns of attraction and repulsion among individuals in groups. This visual array can be viewed as a network, because the "connections" among individuals are what is most significant. Moreover, in looking at the network, structural features emerge.

Moreno thus introduced some key conceptual ingredients of contemporary network analysis: the mapping of relations among actors in visual space to represent the structure of these relations. Yet, alongside Moreno's sociograms, other research and theoretical traditions were developing and pointing toward the same kind of structural analysis.

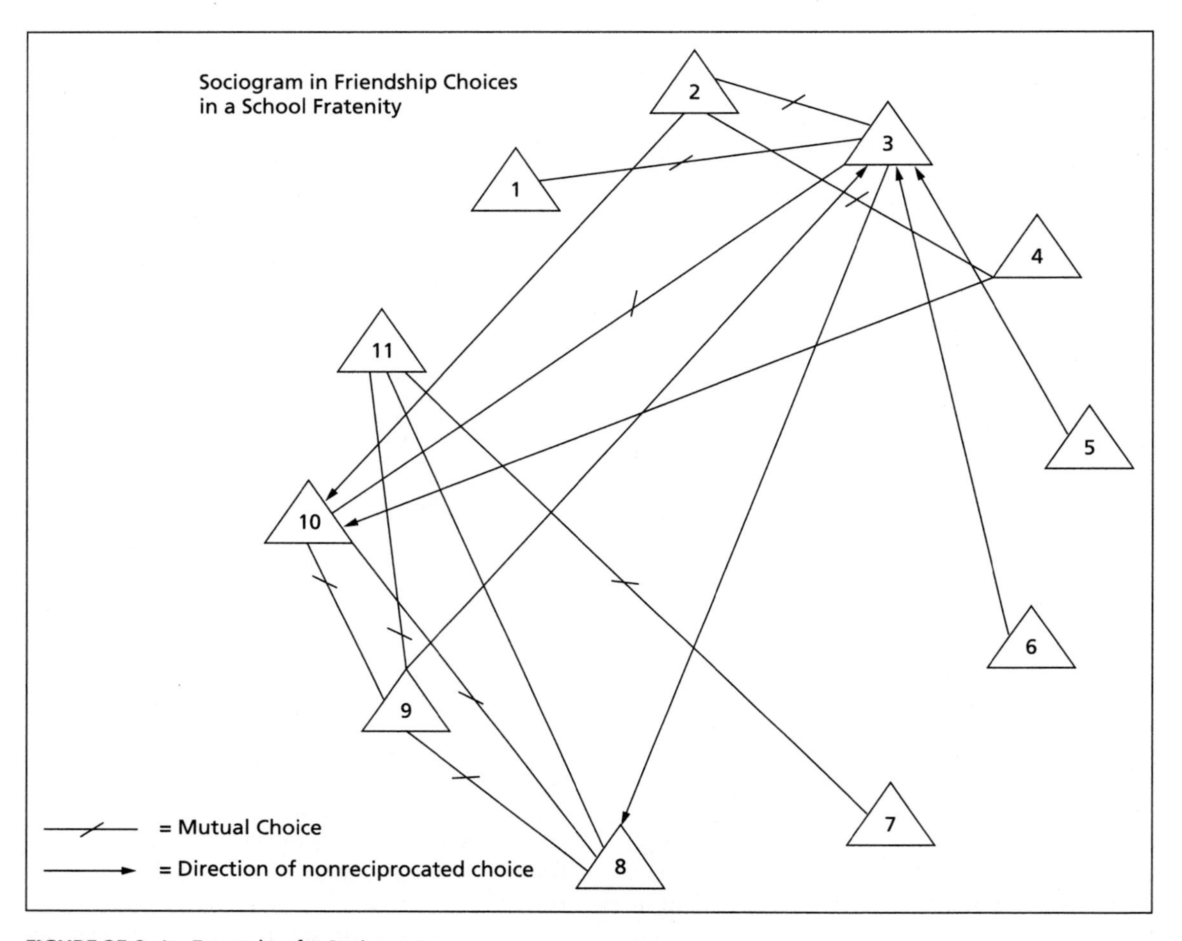

FIGURE 35-2 An Example of a Sociogram

Source: J. Moreno, , *Who Shall Survive?*, rev. ed., New York: Beacon House, 1953, p. 171.

Studies of Communications in Groups: Alex Bavelas and Harold Leavitt

Alex Bavelas[34] was one of the first to study how the structure of a network influences the flow of communication in experimental groups. Others such as Harold Leavitt[35] followed Bavelas' lead and also began to study how communication patterns influence the task performances of people in experimental groups. The network structure in these experiments usually involved artificially partitioning groups in such a way that messages could flow only in certain directions and through particular persons. Emerging from Bavelas' original study was the notion of *centrality*, which was evident when positions lie between other positions in a network. When communications had to flow through this central position, certain styles and levels of task performance prevailed, whereas other patterns of information flow produced different results. Figure 35-3 outlines some chains of communication flow that Bavelas originally isolated and that Leavitt later improved.

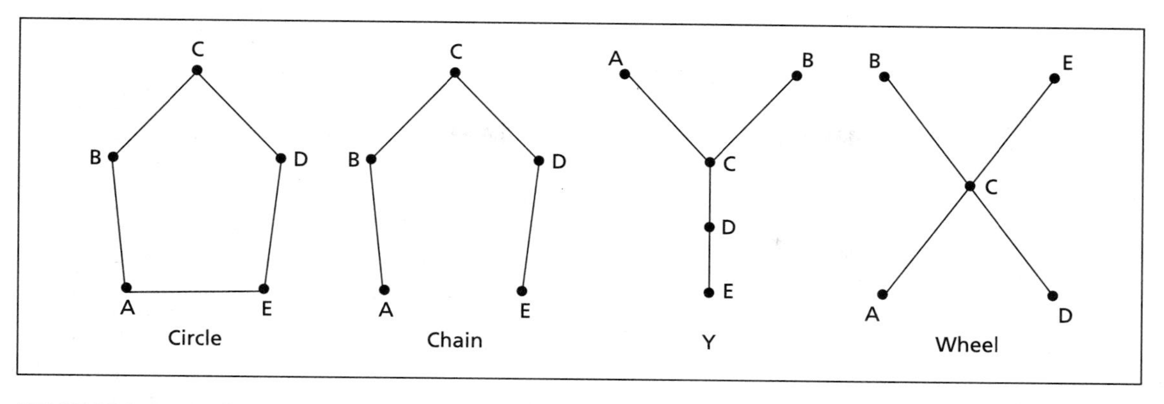

FIGURE 35-3 Types of Communication Structures in Experimental Groups.

Source: Harold J. Leavitt, "Some Effects of Certain Communication Patterns on Group Performance," *Journal of Abnormal and Social Psychology* 56,1951, p. 40.

The results of these experiments are perhaps less important than the image of structure that is offered, although we should note in passing that occupying central positions, such as *C* in Figure 35-3, exerted the most influence on the emergence of leadership, task performance, and effective communication. These diagrams in Figure 35-3 resemble the sociograms, but there are some important differences that became critical in modern network analysis. First, the network is conceptualized in the communication studies as consisting of positions rather than of persons, with the result that the pattern of relations among positions was viewed as a basic or generic type of structure. Indeed, different people could occupy the positions and the experimental results would be the same. Thus, there is a real sense that structure constitutes an emergent reality, beyond the individuals involved. Second, the idea that the links among positions involve flows of resources—in these studies, information and messages—anticipates the thrust of much network analysis. Of course, we could also see Moreno's sociograms as involving flows of affect and preferences among people, but the idea is less explicit and less embedded in a conception of networks as relations among positions.

Thus, these early experimental studies on communication created a new conceptualization of networks as (1) composed of positions, (2) connected by relations, and (3) involving the flows of resources.

Early Gestalt and Balance Approaches: Heider, Newcomb, Cartwright, and Harary

Fritz Heider,[36] who is often considered the founder of Gestalt psychology, developed some of the initial concepts in various theories of "balance" and "equilibrium" in cognitive perceptions. In Heider's view, individuals seek to balance[37] their cognitive conceptions; in his famous *P,O,X* model, Heider argued that a person (*P*) will attempt to balance cognitions toward an object or entity (*X*) with those of another person (*O*). If a person (*P*) has positive sentiments toward an object (*X*) and another person (*O*), but *O* has negative sentiments toward *X*, then a state of cognitive imbalance exists. A person has two options if the imbalance is to be resolved: (1) to change sentiments toward *X* or (2) to alter sentiments toward *O*. By altering sentiments to *X* toward the negative, cognitive balance is achieved, because *P* and *O* now reveal a negative orientation toward *X*, thereby affirming

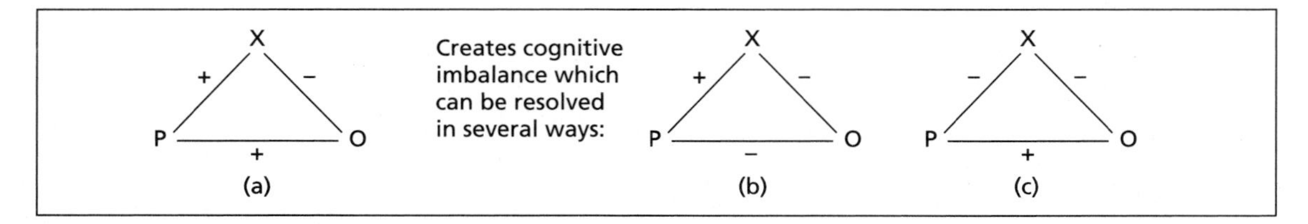

FIGURE 35-4 The Dynamics of Cognitive Balance

Source: Adapted from Fritz Heider (*The Psychology of Interpersonal Relations* [New York: Wiley, 1958]) with (+) and (–) used instead of Heider's notation.

their positive feelings toward each other. Or, by altering sentiments directed to *O* toward the negative, cognitive balance is achieved because *P* has a positive attitude toward *X* and negative feelings for *O*, who has a negative orientation to *X*.

Although Heider did not explicitly do so, this conception of balance can be expressed in algebraic terms, as is done in Figure 35-4 by multiplying the cognitive links in Figure 35-4(a): (+) x (–) x (+) = (–) or imbalance. This imbalance can be resolved by changing the sign of the links toward a (–) or a (+), as is done for Figures 35-4(b) and 35-4(c). By multiplying the signs for the lines in 35-4(b) or 35-4(c), a (+) product is achieved, indicating that the relation is now in balance.

Theodore Newcomb[38] extrapolated Heider's logic to the analysis of interpersonal communication. Newcomb argued that this tendency to seek balance applies equally to *inter*personal as well as to the *intra*personal situations represented by the *P,O,X* model, and he constructed an *A,B,X* model to emphasize this conclusion. A person (*A*) and another (*B*) who communicate and develop positive sentiments will, in an effort to maintain balance with each other, develop similar sentiments toward a third entity (*X*), which can be an object, an idea, or a third person. However, if *A*'s orientation to *X* is very strong in either a positive or a negative sense and *B*'s orientation is just the opposite, several options are available: (1) *A* can convince *B* to change its orientation toward *X*, and vice versa, or (2) *A* can change its orientation to *B*, and vice versa. Figure 35-5 represents this interpersonal situation for *A,B,X* in the same manner as Heider's *P,O,X* model in Figure 35-4. Situation 35-5(a) is in interpersonal imbalance, as can be determined by multiplying the signs (+) x (+) x (–) = (–) or imbalance. Figures 35-5(b), (c), and (d) represent three options that restore balance to the relations among *A*, *B*, and *X*. (In Figure 35-5(a), (b), and (c), the product of multiplying the signs now equals a (+), or balance.)

Heider's and Newcomb's approaches were to stimulate research that would more explicitly employ mathematics as a way to conceptualize the links in interpersonal networks. The key breakthrough had come earlier[39] in the use of the mathematical theory of linear graphs. Somewhat later, in the mid-1950s, Dorin Cartwright and Frank Harary[40] similarly employed the logic of signed-digraph theory to examine balance in larger groups consisting of more than three persons. Figure 35-6 presents a model developed by Cartwright and Harary for a larger set of actors.

The basic idea is much the same as in the *P,O,X* and *A,B,X* models, but now the nature of sentiments is specified by dotted (negative) and solid (positive) lines. By multiplying the signs—(+) = solid line; (–) = dotted line—across all the lines, points of imbalance and

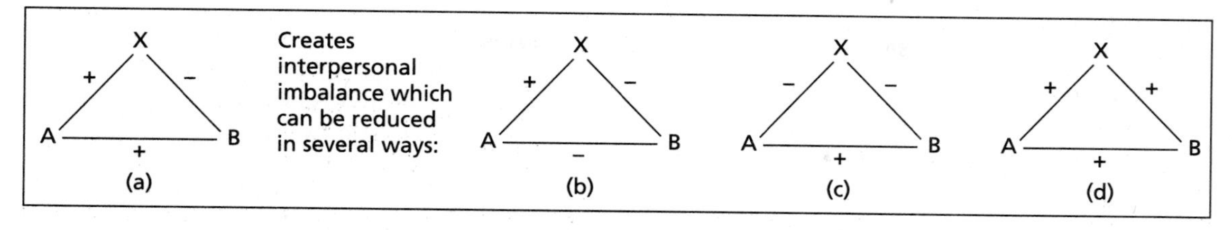

FIGURE 35-5 The Dynamics of Interpersonal Balance

Source: Adapted from Theodore Newcomb ("An Approach to the Study of Communicative Arts," *Psychological Review* 60 [1953], pp. 393–404) with alterations to Newcomb's system of notation.

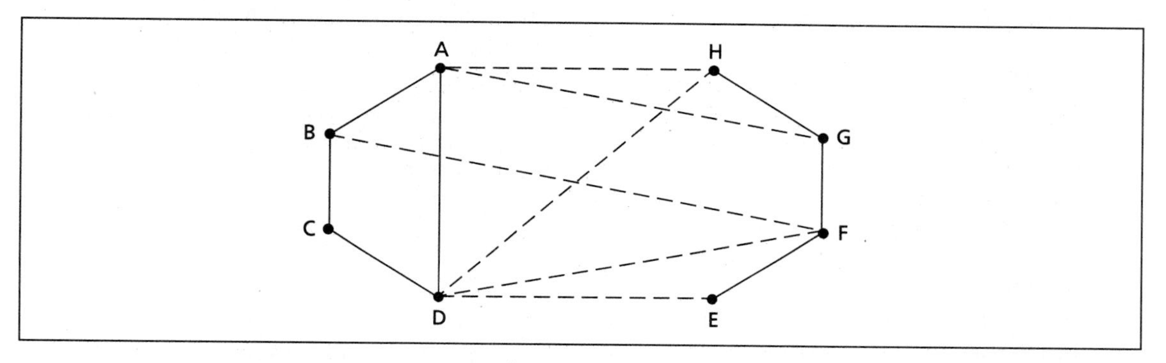

FIGURE 35-6 An *S*-Graph of Eight Points

Source: D. Cartwright and F. Harary, "Structural Balance: A Generalization of Heider's Theory," *Psychological Review* 63 (5), 1956, p. 286.

balance can be identified. For Cartwright and Harary, one way to assess balance is to multiply the various *cycles* on the graph—for example, *ABCD*; *ABCDEFGH*; *HDFG*; *DFE*, and so on. If multiplying the signs for each connection yields a positive outcome, then this structure is in balance. Another procedure is specified by a theorem: An *S*-graph is balanced if and only if all paths joining the same pair of points have the same sign.

The significance of introducing graph theory into balance models is that it facilitated the representation of social relations with mathematical conventions—something that Moreno, Heider, and Newcomb had failed to do. But the basic thrust of earlier analysis was retained: Graph theory could represent directions of links between actors (this is done by simply placing arrows on the lines as they intersect with a point); graph theory could represent two different types of relations between points to be specified by double lines and arrows; it could represent different positive or negative states (the sign being denoted by solid or dotted lines); it offered a better procedure for analyzing more complex social structures; and, unlike the matrices behind Moreno's and others' sociograms, graph theory would make them more amenable to mathematical and statistical manipulation. Thus, although the conventions of graph theory have not remained exactly the same, especially as adopted for network use, the logic of the analysis that graph theory facilitated was essential for the development of the network approach beyond crude matrices and sociograms or simple triadic relations to more complex networks involving the flows of multiple resources in varying directions.

CONCLUSION

In Chapter 37, we will examine the effort to continue the French structuralist tradition with a more scientific stance. But the emphasis on mental and cultural process remains, although with less mysticism than Lévi-Strauss' advocacy. In Chapter 38 on contemporary network analysis, the British and American structuralist traditions remain highly visible in contemporary theorizing, as we have already seen in Chapter 25 on exchange network analysis. In Chapter 36 on structuration theory, we can see a more synthetic blending of diverse traditions that draws from structuralism but interactionist and other theoretical traditions as well. And in Chapter 39 on macrostructural theory we see Simmel's formal approach carried forward as an alternative to both French structuralism and network analysis.

Thus, as structuralist theory matured, it went in many different directions. What is most evident in this divergence is the inability of sociological theorists to agree on their most fundamental idea—social structure. Depending on each theorist's adaptation of earlier traditions—or equally important, their rejection of traditions—very different theories of "structure" are proposed.

NOTES

1. Claude Lévi-Strauss, *The Elementary Structures of Kinship* (Paris: University of France, 1949).

2. Actually, an earlier work, "The Analysis of Structure in Linguistics and in Anthropology," *Word* 1 (1945), pp. 1–21, provided a better clue to the form of Lévi-Strauss' structuralism.

3. Marcel Mauss, *The Gift: Forms and Functions of Exchange* (New York: Free Press, 1954; originally published in 1924), is given particular credit. It must be remembered, of course, that Mauss was Durkheim's student and son-in-law.

4. Émile Durkheim and Marcel Mauss, *Primitive Classification*. (Chicago, IL: University of Chicago Press, 1963) Originally published in 1903, this is a rather extreme and unsuccessful effort to show how mental categories directly reflect the spatial and structural organization of a population. It is a horribly flawed work, but it is the most extreme statement of Durkheim's sociologistic position.

5. Ferdinand de Saussure, *Course in General Linguistics* (New York: McGraw-Hill, 1966); originally compiled posthumously by his students from their lecture notes in 1915.

6. Ferdinand de Saussure, *Course in General Linguistics* (New York: McGraw-Hill, 1966, originally published in 1915).

7. Ibid., p. 107.

8. Ibid., p. 108.

9. Actually, Jakobson simply argued that children's phonological development occurs as a system in which contrasts are critical—for example, "papa versus mama" or the contrasts that children learn between vowels and consonants. Lévi-Strauss appears to have added the jargon of information theory and computer technology. See Roman Jakobson, *Selected Writings 1: Phonological Studies* (The Hague: Mouton, 1962) and *Selected Writings 11: Word and Language* (The Hague: Mouton, 1971). For more detail, see A.R. Maryanski and Jonathan H. Turner, "The Offspring of Functionalism: French and British Structuralism," *Sociological Theory* 9 (1991), pp. 106–115.

10. Claude Lévi-Strauss, *Myth and Meaning* (New York: Schocken, 1979).

11. Claude Lévi-Strauss, *A World on the Wane* (London: Hutchinson, 1961).

12. Claude Lévi-Strauss, "French Sociology," in Georges Gurvitch and Wilbert E. Moore, eds., *Twentieth Century Sociology* (New York: Books for Libraries, 1945).

13. Lévi-Strauss, The Elementary Structures of Kinship (cited in note 1).

14. Mauss, *The Gift* (cited in note 3).

15. Émile Durkheim, "Incest: The Nature and Origin of the Taboo." *Anneé Sociologique* 1 (1898), pp. 1–70.

16. See, for example, Claude Lévi-Strauss, "Social Structure," in *Anthropology Today*, ed. A. Kroeber (Chicago: University of Chicago Press, 1953), pp. 524–553; *Structural Anthropology* (Paris: Plon, 1958; trans. 1963 by Basic Books); and *Mythologiques: le cru et le cuit* (Paris: Plon, 1964).

17. Mirian Glucksmann, *Structuralist Analysis in Contemporary Social Thought* (London: Routledge & Kegan Paul, 1974). A more sympathetic review of Lévi-Strauss, as well as a more general review of structuralist thought, can be found in Tom Bottomore and Robert Nisbet, "Structuralism," in their *A History of Sociological Analysis* (New York: Basic Books, 1978).

18. Marshall D. Sahlins, "On the Delphic Writings of Claude Lévi-Strauss," *Scientific American* 214 (1966), p. 134. For other relevant critiques, see Marvin Harris, *The Rise of Anthropological Theory* (New York: Crowell, 1968), pp. 464–513; and Eugene A. Hammel, "The Myth of Structural Analysis" (Addison-Wesley Module, no. 25, 1972).

19. A.R. Radcliffe-Brown, "Three Tribes of Western Australia," *Journal of Royal Anthropological Institute of Great Britain and Ireland* 43 (1913), pp. 8–88.

20. A.R. Radcliffe-Brown, *The Andaman Islanders* (Cambridge: Cambridge University Press, 1922, originally published in 1914).

21. Émile Durkheim, *The Elementary Forms of Religious Life* (New York: Free Press, 1947, originally published in 1912).

22. A.R. Radcliffe-Brown, "On the Concept of Function in Social Science." *American Anthropologist* 37 (1935), p. 396.

23. Quote cited in George P. Murdock, "Social Structure" in S. Tax, L. Eiseley, I. Rouse, and C. Voeglia, eds., *An Appraisal of Anthropology Today* (New York: Free Press, 1953).

24. A.R. Radcliffe-Brown, *A Natural Science of Society* (New York: Free Press, 1948).

25. Émile Durkheim, *The Rules of the Sociological Method* (New York: Free Press, 1938, originally published in 1895).

26. For examples, see Raymond Firth, *Elements of Social Organization* (London: Watts, 1952); E.E. Evans-Pritchard, *The Nuer* (London: Oxford University Press, 1940); and Meyer Fortes, *The Web of Kinship among the Tallensi* (London: Oxford University Press, 1949).

27. S.F. Nadel, *The Study of Social Structure* (London: Cohen and West, 1957).

28. Ibid., p. 8.

29. Ibid., p. 21.

30. J. Clyde Mitchell, "The Concept and Use of Social Networks," in Jeremy F. Boissevain and J. Clyde Mitchell, eds., *Network Analysis: Studies in Human Interaction* (The Hague: Mouton, 1973).

31. John A. Barnes, "Social Networks" (Addison-Wesley Module, no. 26, 1972). See also his "Network and Political Processes" in J.F. Boissevain and J.C. Mitchell, eds., *Network Analysis: Studies in Human Interaction* (The Hague: Mouton, 1973).

32. Perhaps the most significant was Elizabeth Bott, *Family and Social Network: Roles, Norms, and External Relationships in Ordinary Urban Families* (London: Tavistock, 1957, 1971).

33. Jacob L. Moreno, *Who Shall Survive?* (Washington, D.C.: Nervous and Mental Diseases Publishing, 1934; republished in revised form by Beacon House, New York, 1953).

34. Alex Bavelas, "A Mathematical Model for Group Structures," *Applied Anthropology* 7 (3) (1948), pp. 16–30.

35. Harold J. Leavitt, "Some Effects of Certain Communication Patterns on Group Performance," *Journal of Abnormal and Social Psychology* 56 (1951), pp. 38–50; Harold J. Leavitt and Kenneth E. Knight, "Most 'Efficient' Solution to Communication Networks: Empirical versus Analytical Search," *Sociometry* 26 (1963), pp. 260–267.

36. Fritz Heider, "Attitudes and Cognitive Organization," *Journal of Psychology* 2 (1946), pp. 107–112. For the best review of his thought as it accumulated over four decades, see his *The Psychology of Interpersonal Relations* (New York: Wiley, 1958).

37. The process of "attribution" was, along with the notion of "balance," the cornerstones of Heider's Gestalt approach.

38. Theodore M. Newcomb, "An Approach to the Study of Communicative Acts," *Psychological Review* 60 (1953), pp. 393–404. See his earlier work where these ideas took form: *Personality and Social Change* (New York: Dryden, 1943).

39. For example, D. König, *Theorie der Endlichen und Undlichen Graphen* (Leipzig, Teubner, 1936 but reissued, New York: Chelsea, 1950), is, as best we can tell, the first work on graph theory. It appears that the first important application of this theory to the social sciences came with R. Duncan Luce and A.D. Perry, "A Method of Matrix Analysis of Group Structure," *Psychometrika* 14 (1949), pp. 94–116, followed by R. Duncan Luce, "Connectivity and Generalized Cliques in Sociometric Group Structure," *Psychometrika* 15 (1950), pp. 169–190. Frank Harary's *Graph Theory* (Reading, MA: Addison-Wesley, 1969) later became a standard reference, which had been preceded by

Frank Harary and R.Z. Norman, *Graph Theory as a Mathematical Model in Social Science* (Ann Arbor: University of Michigan Institute for Social Research, 1953), and Frank Harary, R.Z. Norman, and Dorin Cartwright, *Structural Models: An Introduction to the Theory of Directed Graphs* (New York: Wiley, 1965).

40. Dorin Cartwright and Frank Harary, "Structural Balance: A Generalization of Heider's Theory," *Psychological Review* 63 (1956), pp. 277–293. For more recent work, see their "Balance and Clusterability: An Overview," in Holland and Leinhardt, eds., *Perspectives on Social Network Research* (New York: Academic, 1979).

36

The Continuing Tradition I:

Anthony Giddens' Structuration Theory

During the last thirty years, Anthony Giddens has been one of the most prominent critics of the scientific pretensions of sociology. Yet, at the same time, he has developed a relatively formal abstract conceptual scheme for analyzing the social world. In the mid-1980s, he brought elements of his advocacy together into an important theoretical synthesis of diverse theoretical traditions into what he had earlier titled "structuration theory."[1] This theory represents one of more creative theoretical efforts of the second half of the twentieth century. Although Giddens has developed theoretical interests in modernity and, indeed, has become an important contributor to the debate about modernity and post-modernity,[2] his theoretical contribution still resides primarily in the more formal statement of structuration theory. Thus, in this chapter we will briefly review Giddens' critique of positivists' natural science view of sociology and, then, devote most of our efforts to summarizing the key ideas in structuration theory.

THE CRITIQUE OF "SCIENTIFIC" SOCIAL THEORY

Giddens reasoned that there never can be any universal and timeless sociological laws, like those in physics or the biological sciences. Humans have the capacity for agency, and hence, they can change the very nature of social organization—thereby obviating any laws that are proposed to be universal. At best, "the concepts of theory should for many research purposes be regarded as sensitizing devices, nothing more."[3] Giddens buttresses this conclusion with two points of argument.

First, Giddens asserts that social theorizing involves a "double hermeneutic." Stripped of its jargon, this means that the concepts and generalizations used by social scientists to understand social processes can be employed by lay persons as agents who can alter these social processes. We must recognize, Giddens contends, that ordinary actors are also "social theorists who alter their theories in the light of their experience and are receptive to incoming information."[4] Thus, social science theories are not often "news" to individuals; when they are, such theories can be used to transform the very order they describe. Within the capacity of humans to be reflexive—that is, to think about their situation—is the ability to change it.[5]

Second, social theory is by its nature social criticism. Social theory often contradicts "the reasons that people give for doing things" and is, therefore, a critique of these reasons and the social arrangements that people construct in the name of these reasons. Sociology does not, therefore, need to develop a separate body of critical theory, as others have argued; it *is* critical theory by its very nature and by virtue of the effects it can have on social processes.

The implications, Giddens believes, are profound. We need to stop imitating the natural sciences. We must cease evaluating our success as intellectuals by whether or not we have discovered "timeless laws." We must recognize that social theory does not exist "outside" our universe. We should accept that what sociologists and lay actors do is, in a fundamental sense, very much the same. And, we must redirect our efforts to developing "sensitizing concepts" that allow us to understand the active processes of interaction among individuals as they produce and reproduce social structures while being guided by these structures.

THE "THEORY OF STRUCTURATION"

Because Giddens does not believe that abstract laws of social action, interaction, and organization exist, his "theory of structuration" is not a series of propositions. Instead, as Giddens' critique of science would suggest, his "theory" is a cluster of sensitizing concepts, linked together discursively. The key concept is *structuration*, which is intended to communicate the *duality of structure*.[6] That is, social structure is used by active agents; in so using the properties of structure, they transform or reproduce this structure. Thus the process of structuration requires a conceptualization of the nature of structure, of the agents who use structure, and of the ways that these are mutually implicated in each other to produce varying patterns of human organization.

Reconceptualizing Structure and Social System

Giddens believes structure can be conceptualized as *rules* and *resources* that actors use in "interaction contexts" that extend across "space" and over "time." In so using these rules and resources, actors sustain or reproduce structures in space and time.

Rules are "generalizable procedures" that actors understand and use in various circumstances. Giddens posits that a rule is a methodology or technique that actors know about, often only implicitly, and that provides a relevant formula for action.[7] From a sociological perspective, the most important rules are those that agents use in the reproduction of social relations over significant lengths of time and across space. These rules reveal certain characteristics: (1) they are frequently used in (a) conversations, (b) interaction rituals, and (c) the daily routines of individuals; (2) they are tacitly grasped and understood and are part of the "stock knowledge" of competent actors; (3) they are informal, remaining unwritten and unarticulated; and (4) they are weakly sanctioned through interpersonal techniques.[8]

The thrust of Giddens' argument is that rules are part of actors' "knowledgeability." Some can be normative in that actors can articulate and explicitly make reference to them, but many other rules are more implicitly understood and used to guide the flow of interaction in ways that are not easily expressed or verbalized. Moreover, actors can transform rules into new combinations as they confront and deal with one another and the contextual particulars of their interaction.

As the other critical property of structure, resources are facilities that actors use to get things done. For, even if there are well-understood methodologies and formulas—that is, rules—to guide action, there must also be the capacity to perform tasks. Such capacity requires resources, or the material equipment and the organizational ability to act in situations. Giddens visualizes resources as what generates power.[9] Power is not a resource, as much social theory argues. Rather, the mobilization of other resources is what gives actors power to get things done. Thus, power is integral to the very existence of structure: As actors interact, they use resources, and, as they use resources, they mobilize power to shape the actions of others.

Giddens visualizes rules and resources as "transformational" and as "mediating."[10] What he means by these terms is that rules and resources can be transformed into many different patterns and profiles. Resources can be mobilized in various ways to perform activities and achieve ends through the exercise of different forms and degrees of power; rules can generate many diverse combinations of methodologies and formulas to guide how people communicate, interact, and adjust to one another. Rules and resources are mediating in that they are what tie social relations together. They are what actors use to create, sustain, or transform relations across time and in space. And, because rules and resources are inherently transformational—that is, generative of diverse combinations—they can lace together many different patterns of social relations in time and space.

Giddens developed a typology of rules and resources that is rather vague and imprecise.[11] He sees the three concepts in this typology—domination, legitimation, and signification—as "theoretical primitives," which is, perhaps, an excuse for defining them imprecisely. The basic idea is that resources are the stuff of domination because they involve the mobilization of material and organizational facilities to do things. Some rules are transformed into instruments of legitimation because they make things seem correct and appropriate. Other rules are used to create signification, or meaningful symbolic systems, because they provide people with ways to see and interpret events. Actually, the scheme makes more sense if the concepts of domination, legitimation, and signification are given less emphasis and the elements of his discussion are selectively extracted to create the typology presented in Figure 36-1.

In the left column of Figure 36-1, structure is viewed by Giddens as composed of rules and resources. Rules are transformed into two basic types of mediating processes: (1) normative, or the creation of rights and obligations in a context; and (2) interpretative, or the generation of schemes and stocks of taken-for-granted knowledge in a context. Resources are transformed into two major types of facilities that can mediate social relations: (1) authoritative resources, or the organizational capacity to control and direct the patterns of interactions in a context; and (2) allocative resources, or the use of material features, artifacts, and goods to control and direct patterns of interaction in a context.

Giddens sees these types of rules and resources as mediating interaction via three modalities, as is portrayed in column 2 of Figure 36-1: rights and obligations, facilities, and interpretative schemes. The figure deviates somewhat from Giddens' discussion, but the idea is the same: that rules and resources are attached to interaction (or "social system" in Giddens' terms) via these three modalities. These modalities are then used to (a) generate the power that enables some actors to control others, (b) affirm the norms that, in turn, allow actors to be sanctioned

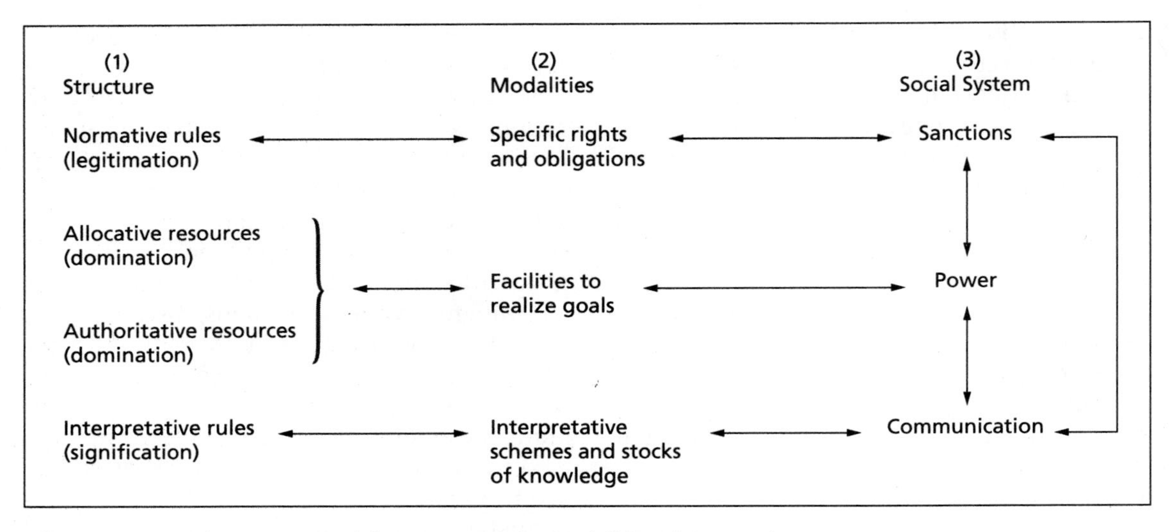

FIGURE 36-1 Social Structure, Social System, and the Modalities of Connection

for their conformity or nonconformity, and (c) create and use the interpretative schemes that make it possible for actors to communicate with one another.

Giddens also stresses that rules and resources are interrelated, emphasizing that the modalities and their use in interaction are separated only analytically. In the actual flow of interaction in the real empirical world, they exist simultaneously, thereby making their separation merely an exercise of analytical decomposition. Thus, power, sanctions, and media of communication are interconnected, as are the rules and resources of social structure. In social systems, where people are co-present and interact, power is used to secure a particular set of rights and obligations as well as a system of communication; conversely, power can be exercised only through communication and sanctioning.

Giddens, then, sees social structure as something used by actors, not as some external reality that pushes and shoves actors around. Social structure is defined as the rules and resources that can be transformed as actors use them in concrete settings. But, the question arises: How is structure to be connected to what people actually do in interaction settings, or what Giddens terms "social systems"? The answer is the notion of modalities, whereby rules and resources are transformed into power, sanctions, and communication. Thus, structure is not a mysterious system of codes, as Claude Lévi-Strauss and other structural idealists imply, nor is it a set of determinative parameters and external constraints on actors, as Peter Blau and other macrostructuralists contend. In Giddens' conceptualization, social structure is transformative and flexible, it is "part of" actors in concrete situations, and it is used by them to create patterns of social relations across space and through time.

Moreover, this typology allows Giddens to emphasize that, as agents interact in social systems, they can reproduce rules and resources (via the modalities) or they can transform them. Thus, social interaction and social structure are reciprocally implicated. Structuration is, therefore, the dual processes in which rules and resources are used to organize interaction across time and in space and, by virtue of this use, to reproduce or transform these rules and resources.

TABLE 36-1 The Typology of Institutions

Type of Institution		Rank Order of Emphasis on Rules and Resources
1. Symbolic orders, or modes of discourse, and patterns of communication	are produced and reproduced by	the use of interpretative rules (signification) in conjunction with normative rules (legitimation) and allocative as well as authoritative resources (domination).
2. Political institutions	are produced and reproduced by	the use of authoritative resources (domination) in conjunction with interpretative rules (signification) and normative rules (legitimation).
3. Economic institutions	are produced and reproduced by	the use of allocative resources (domination) in conjunction with interpretative rules (signification) and normative rules (legitimation).
4. Legal institutions	are produced and reproduced by	the use of normative rules (legitimation) in conjunction with authoritative and allocative resources (domination) and interpretative rules (signification).

Reconceptualizing Institutions

Giddens believes that institutions are systems of interaction in societies that endure over time and that distribute people in space. Giddens uses phrases like "deeply sedimented across time and in space in societies" to express the idea that, when rules and resources are reproduced over long periods of time and in explicit regions of space, then institutions can be said to exist in a society. Giddens offers a typology of institutions showing the weights and combinations of rules and resources that are implicated in interaction.[12] If signification (interpretative rules) is primary, followed, respectively by domination (allocative and authoritative resources) and then legitimation (normative rules), a "symbolic order" exists. If authoritative domination, signification, and legitimation are successively combined, political institutionalization occurs. If allocative dominance, signification, and legitimation are ordered, economic institutionalization prevails. And if legitimation, dominance, and signification are rank ordered, institutionalization of law occurs. Table 36-1 summarizes Giddens' argument.

In this conceptualization of institutions, Giddens seeks to avoid a mechanical view of institutionalization, in several senses. First, systems of interaction in empirical contexts are a mixture of institutional processes. Economic, political, legal, and symbolic orders are not easily separated; there is usually an element of each in any social system context. Second, institutions are tied to the rules and resources that agents employ and thereby reproduce; they are not external to individuals because they are formed by the use of varying rules and resources in actual social relations. Third, the most basic dimensions of all rules and resources—signification, domination, and legitimation—are all involved in institutionalization; it is only their relative salience for actors that gives the stabilization of relations across time and in space its distinctive institutional character.

Structural Principles, Sets, and Properties

The extent and form of institutionalization in societies are related to what Giddens terms *structural principles*.[13] These are the most general principles that guide the organization of societal totalities. These are what "stretch systems across time and space," and they allow for "system integration," or the maintenance of reciprocal relations among units in a society. For Giddens, "structural principles can thus be understood as the principles of organization which allow recognizably consistent forms of time-space distanciation on the basis of definite mechanisms of societal integration."[14] The basic idea seems to be that rules and resources are used by active agents in accordance with fundamental principles of organization. Such principles guide just how rules and resources are transformed and employed to mediate social relations.

On the basis of their underlying structural principles, three basic types of societies have existed: (1) "tribal societies," which are organized by structural principles that emphasize kinship and tradition as the mediating force behind social relations across time and in space; (2) "class-divided societies," which are organized by an urban/rural differentiation, with urban areas revealing distinctive political institutions that can be separated from economic institutions, formal codes of law or legal institutions, and modes of symbolic coordination or ordering through written texts and testaments; and (3) "class societies," which involve structural principles that separate and yet interconnect all four institutional spheres, especially the economic and political.[15]

Structural principles are implicated in the production and reproduction of "structures" or "structural sets." These structural sets are rule and resource bundles, or combinations and configurations of rules and resources, which are used to produce and reproduce certain types and forms of social relations across time and space. Giddens offers the example of how the structural principles of class societies (differentiation and clear separation of economy and polity) guide the use of the following structural set: *private property-money-capital-labor-contract-profit.* The details of his analysis are less important than the general idea that the general structural principles of class societies are transformed into more specific sets of rules and resources that agents use to mediate social relations. This structural set is used in capitalist societies and, as a consequence, is reproduced. In turn, such reproduction of the structural set reaffirms the more abstract structural principles of class societies.

As these and other structural sets are used by agents and as they are thereby reproduced, societies develop "structural properties," which are "institutionalized features of social systems, stretching across time and space."[16] That is, social relations become patterned in certain typical ways. Thus the structural set of private property-money-capital-labor-contract-profit can mediate only certain patterns of relations; that is, if this is the rule and resource bundle with which agents must work, then only certain forms of relations can be produced and reproduced in the economic sphere. Hence the institutionalization of relations in time and space reveals a particular form, or, in Giddens' terms, structural property.

Structural Contradiction

Giddens always emphasizes the inherent "transformative" potential of rules and resources. Structural principles, he argues, "operate in terms of one another but yet also contravene each other."[17] In other words, they reveal contradictions that can be either primary or secondary. A "primary contradiction" is one between structural principles that are formative and constitute a society, whereas a "secondary contradiction" is one that is "brought into being by primary contradictions."[18] For example, there is a contradiction between structural principles that mediate the institutionalization of private profits, on the one hand, and those that mediate socialized production, on the other. If workers pool their labor to produce goods and services, it is contradictory to allow only some to enjoy profits of such socialized labor.

Contradictions are not, Giddens emphasizes, the same as conflicts. Contradiction is a "disjunction of structural principles of system organization," whereas conflict is the actual struggle between actors in "definite social practices."[19] Thus, the contradiction between private profits and socialized labor is not, itself, a conflict. It can create situations of conflict, such as struggles between management and labor in a specific time and place, but such conflicts are not the same as contradiction.

For Giddens, then, the institutional patterns of a society represent the creation and use by agents of very generalized and abstract principles. These principles represent the development of particular rules and the mobilization of certain resources; such principles generate more concrete "bundles" or "sets" of rules and resources that agents actively use to produce and reproduce social relations in concrete settings; and many of these principles and sets contain contradictory elements that can encourage actual conflicts among actors. In this way, structure "constrains" but is not disembodied from agents. Rather, the "properties" of total societies are not external to individuals and collectivities but are persistently reproduced through the use of structural principles and sets by agents who act. Let us now turn to Giddens' discussion of these active agents.

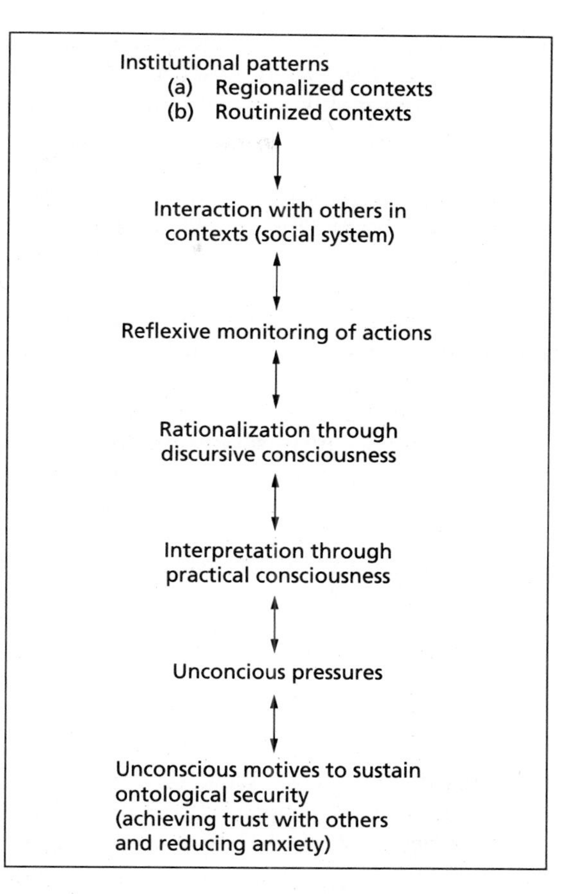

FIGURE 36-2 The Dynamics of Agency

Agents, Agency, and Action

As is evident, Giddens visualizes structure as a duality, as something that is part of the actions of agents. And so in Giddens' approach it is essential to understand the dynamics of human agency. He proposes a "stratification model," which is an effort to synthesize psychoanalytic theory, phenomenology, ethnomethodology, and elements of action theory. This model is depicted in the lower portions of Figure 36-2. For Giddens, "agency" denotes the events that an actor perpetrates rather than "intentions," "purposes," "ends," or other states. Agency is what an actor actually does in a situation that has visible consequences (not necessarily intended consequences). To understand the dynamics of agency requires analysis of each element on the model.

As drawn, the model in Figure 36-2 actually combines two overlapping models in Giddens' discussion, but his intent is reasonably clear: humans "reflexively monitor" their own conduct and that of others; in other words, they pay attention to, note, calculate, and assess the consequences of actions.[20] Monitoring is influenced by two levels of consciousness.[21] One is "discursive consciousness," which involves the capacity to give reasons for or rationalize what one does (and presumably to do the same for others' behavior). "Practical consciousness" is the stock of knowledge that one implicitly uses to act in situations and to interpret the actions of others. This knowledgeability is constantly used, but rarely articulated, to interpret events—one's own and those of others. Almost all acts are indexical in that they must be interpreted by their context, and this implicit stock of knowledge provides these contextual interpretations and frameworks.

There are also unconscious dimensions to human agency. There are many pressures to act in certain ways, which an actor does not perceive. Indeed, Giddens argues that much motivation is unconscious. Moreover, motivation is often much more diffuse than action theories portray. That is, there is no one-to-one relation between *an act* and *a motive*. Actors might be able to rationalize through their capacity for discursive consciousness in ways that make this one-to-one relationship seem to be what directs action. But much of what propels action lies below consciousness and, at best, provides very general and diffuse pressures to act. Moreover, much action might not be motivated at all; an actor simply monitors and responds to the environment.

In trying to reintroduce the *un*conscious into social theory, Giddens adopts Erik Erikson's psychoanalytic ideas.[22] The basic "force" behind much action is an unconscious set of processes to gain a "sense of trust" in interaction with others. Giddens terms this set of processes the *ontological security system* of an agent. That is, one of the driving but highly diffuse forces behind action is the desire to sustain ontological security or the sense of trust that comes from being able to reduce anxiety in social relations. Actors need to have this sense of trust. How they go about reducing anxiety to secure this sense is often unconscious because the mechanisms involved are developed before linguistic skills emerge in the young and because psychodynamics, such as repression, might also keep these fundamental feelings and their resolution from becoming conscious. In general, Giddens argues that ontological security is maintained through the routinization of encounters with others, through the successful interpretation of acts as practical or stock knowledge, and through the capacity for rationalization that comes with discursive consciousness.

As the top portions of Figure 36-2 emphasize, institutionalized patterns have an effect on, while being a consequence of, the dynamics of agency. As we will see shortly, unconscious motives for ontological security require routinized interactions (predictable, stable over time) that are regionalized (ordered in space). Such regionalization and routinization are the product of past interactions of agents and are sustained or reproduced through the present (and future) actions of agents. To sustain routines and regions, actors must monitor their actions while drawing on their stock knowledge and discursive capacities. In this way, Giddens visualizes institutionalized patterns implicated in the very nature of agency. Institutions and agents cannot exist without each other, for institutions are reproduced practices by agents, whereas the conscious and unconscious dynamics of agency depend on the routines and regions provided by institutionalized patterns.

Routinization and Regionalization of Interaction

Both the ontological security of agents and the institutionalization of structures in time and space depend on routinized and regionalized interaction among actors. Routinization of interaction patterns is what gives them continuity across time, thereby reproducing structure (rules and resources) and institutions. At the same time, routinization gives predictability to actions and, in so doing, provides a sense of ontological security. Thus, routines become critical for the most basic aspects of structure and human agency. Similarly, regionalization orders action in space by positioning actors in places relative to one another and by circumscribing how they are to present themselves and act. As with routines, the regionalization of interaction is essential to the sustenance of broader structural patterns and ontological security of actors, because it orders people's interactions in space and time, which in turn reproduces structures and meets an agent's need for ontological security.

Routines. Giddens sees routines as the key link between the episodic character of interactions (they start, proceed, and end), on the one hand, and basic trust and security, on the other hand.[23] Moreover, "the routinization of encounters is of major significance in binding the fleeting encounter to social reproduction and thus to the seeming 'fixity' of institutions."[24] In a very interesting discussion in which he borrows heavily from Erving Goffman (but with a phenomenological twist), Giddens proposed several procedures, or mechanisms, that humans use to sustain routines: (1) opening and closing rituals, (2) turn taking, (3) tact, (4) positioning, and (5) framing.[25]

1. Because interaction is serial—that is, it occurs sequentially—there must be symbolic markers of opening and closing. Such markers are essential to the maintenance of routines because they indicate when in the flow of time the elements of routine interaction are to begin and end. There are many such interpersonal markers—words, facial gestures, positions of bodies—and there are physical markers, such as rooms, buildings, roads, and equipment, that also signal when certain routinized interactions are to begin and end (note, for example, the interpersonal and physical markers for a lecture, which is a highly routinized interaction that sustains the ontological security of agents and perpetuates institutional patterns).

2. Turn taking in a conversation is another process that sustains a routine. All competent actors contain in their practical consciousness, or implicit stock of knowledge, a sense of how conversations are to proceed sequentially. People rely on "folk methods" to construct sequences of talk; in so doing, they sustain a routine and, hence, their psychological sense of security and the larger institutional context (think, for example, about a conversation that did not proceed smoothly in conversational turn taking; recall how disruptive this was for your sense of order and routine).

3. Tact is, in Giddens' view, "the main mechanism that sustains 'trust' or 'ontological security' over long time-space spans." By tact, Giddens means "a latent conceptual agreement among participants in interaction" about just how each party is to gesture and respond and about what is appropriate and inappropriate. People carry with them implicit stocks of knowledge that signal to them what would be "tactful" and what would be "rude" and "intrusive." And they use this sense of tact to regulate their emission of gestures, their talking, and their relative positioning in situations "to remain tactful," thereby sustaining their sense of trust and the larger social order. (Imagine interactions in which tact is not exercised—how they disrupt our routines, our sense of comfort, and our perceptions of an orderly situation).

4. Giddens rejects the idea of "role" as very useful and substitutes the notion of "position." People bring to situations a position or "social identity that carries with it a certain range of prerogatives and obligations," and they emit gestures in a process of mutual positioning, such as locating their bodies in certain points, asserting their prerogatives, and signaling their obligations. In this way interactions can be routinized, and people can sustain their sense of mutual trust as well as the larger social structures in which their interaction occurs. (For example, examine a student/student or professor/student interaction for positioning, and determine how it sustains a sense of trust and the institutional structure.)

5. Much of the coherence of positioning activities is made possible by "frames," which provide formulas for interpreting a context. Interactions tend to be framed in the sense that there are rules that apply to them, but these are not purely normative in the sense of precise instructions for participants. Equally important, frames are more implicitly held, and they operate as markers that assert when certain behaviors and demeanors should be activated. (For example, compare your sense of how to comport yourself at a funeral, at a cocktail party, in class, and in other contexts that are "framed.")

In sum, social structure is extended across time by these techniques that produce and reproduce routines. In so stretching interaction across time in an orderly and predictable manner, people realize their need for a sense of trust in others. In this way, then, Giddens connects the most basic properties of structure (rules and resources) to the most fundamental features of human agents (unconscious motives).

Regionalization. Structuration theory is concerned with the reproduction of relations not only across time but also in space. With the concept of regionalization of interaction, Giddens addresses the intersection of space and time.[26] For interaction is not just serial, moving in time; it is also located in space. Again borrowing from Goffman and also from time and space geography, Giddens introduces the concept of "locale" to account for the physical space in which interaction occurs as well as the contextual knowledge about what is to occur in this space. In a locale, actors are not only establishing their presence in relation to one another but they are also using their stocks of practical knowledge to interpret the context of the locale. Such interpretations provide them with the relevant frames, the appropriate procedures for tact, and the salient forms for sequencing gestures and talk.

Giddens classifies locales by their "modes." Locales vary in (1) their physical and symbolic boundaries, (2) their duration across time, (3) their span or extension in physical space, and (4) their character, or the ways they connect to other locales and to broader institutional patterns. Locales also vary in the degree to which they force people to sustain high public presence (what Goffman termed *frontstage*) or allow retreats to back regions where public presence is reduced (Goffman's *backstage*).[27] They also vary in how much disclosure of self (feelings, attitudes, and emotions) they require, some allowing "enclosure" or the withholding of self and other locales requiring "disclosure" of at least some aspects of self.

Regionalization of interaction through the creation of locales facilitates the maintenance of routines. In turn, the maintenance of routines across time and space sustains institutional structures. Thus, it is through routinized and regionalized systems of interaction that the reflexive capacities of agents reproduce institutional patterns.

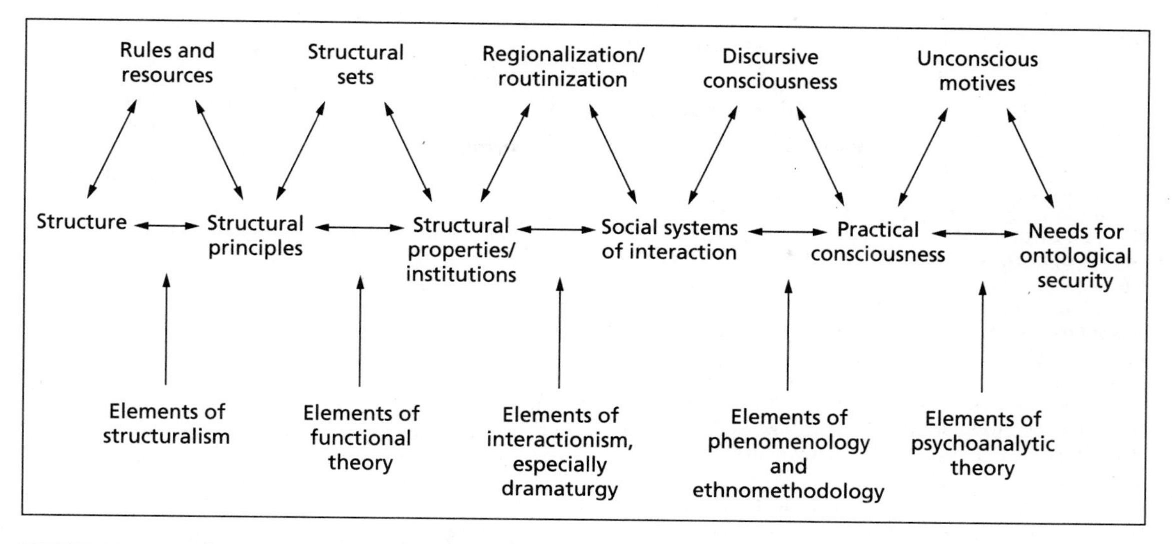

FIGURE 36-3 Key Elements of "Structuration Theory"

CONCLUSION

Figure 36-3 represents one way to visualize Giddens' conceptual scheme. In a rough sense, as one moves from left to right, the scheme gets increasingly micro, although Giddens would probably not visualize his theory in these macro versus micro terms. But the general message is clear: Rules and resources are used to construct structures; these rules and resources are also a part of structural principles that include structural sets; these structural properties are involved in institutionalization of systems of interaction; such interaction systems are organized by the processes of regionalization and routinization; and all these processes are influenced by practical and discursive consciousness that, in turn, are driven by unconscious motives, especially needs for ontological security.

Giddens would not consider his "theory" anything more than a conceptual scheme for describing, analyzing, and interpreting empirical events. Moreover, he would not see this scheme as representing timeless social processes, although the reason his works are read and respected is because these do seem like basic and fundamental processes that transcend time, context, and place.

NOTES

1. The basic theory is presented in numerous places, but the two most comprehensive statements are these: Anthony Giddens, *The Constitution of Society: Outline of the Theory of Structuration* (Oxford: Polity, 1984) and *Central Problems in Social Theory* (London: Macmillan, 1979). The University of California Press also has editions of these two books. For an excellent overview, both sociologically and philosophically, of Giddens' theoretical project, see Ira Cohen, *Structuration Theory: Anthony Giddens and the Constitution of Social Life* (London: Macmillan, 1989). For a commentary and debate on Giddens' work, see J. Clark, C. Modgil, and S. Modgil, eds., *Anthony Giddens: Consensus and Controversy* (London: Falmer, 1990). For a selection of readings, see *The Giddens Reader*, ed. Philip Cassell (Stanford, CA: Stanford University Press, 1993).

2. See, for examples, Anthony Giddens, *The Consequences of Modernity* (Stanford, CA: Stanford University Press, 1990); Ulrich Beck, Anthony Giddens, and Scott Lash, *Reflexive Modernization* (Stanford, CA: Stanford University Press, 1994); Anthony Giddens, *Modernity and Self-Identity* (Stanford, CA: Stanford University Press, 1991).

3. Giddens, *The Constitution of Society* (cited in note 1), p. 326.

4. Ibid., p. 335.

5. See, in particular, Anthony Giddens, *Profiles and Critiques in Social Theory* (London: Macmillan, 1982) and *New Rules of Sociological Method: A Positive Critique of Interpretative Sociologies*, 2nd ed. (Stanford, CA: Stanford University Press, 1993).

6. See *The Constitution of Society* (cited in note 1), pp. 207–213.

7. Ibid., pp. 20–21.

8. Ibid., p. 22.

9. Ibid., pp. 14–16.

10. Here Giddens seems to be taking what is useful from "structuralism" (see Chapter 35) and reworking these ideas into a more sociological approach. Giddens remains, however, extremely critical of structuralism; see his "Structuralism, Post-structuralism and the Production of Culture" in *Social Theory Today*, eds. A. Giddens and J. Turner (Cambridge, England: Polity, 1987).

11. *The Constitution of Society* (cited in note 1), p. 29 and *Central Problems in Social Theory*, pp. 97–107.

12. *Central Problems in Social Theory* (cited in note 1), p. 107 and *The Constitution of Society*, p. 31.

13. *The Constitution of Society* (cited in note 1), pp. 179–193.

14. Ibid., p. 181.

15. For an extensive discussion of this typology, see Giddens' *A Contemporary Critique of Historical Materialism: Power, Property and the State* (London: Macmillan, 1981).

16. *The Constitution of Society* (cited in note 1), p. 185.

17. Ibid., p. 193.

18. Ibid.

19. Ibid., p. 198.

20. Ibid., pp. 5–7; see also *Central Problems in Social Theory* (cited in note 1), pp. 56–59.

21. His debt to Schutz and phenomenology is evident here, but he has liberated it from its subjectivism. See Chapter 26 on the emergence of interactionism.

22. *The Constitution of Society* (cited in note 1), pp. 45–59.

23. Ibid., pp. 60–109.

24. Ibid., p. 72.

25. This list has been created from what is a much more discursive text.

26. Ibid., pp. 110–144.

27. See Erving Goffman, *The Presentation of Self in Everyday Life* (Garden City, NY: Doubleday, 1959).

37

The Continuing Tradition II: Cultural Theories*

French structuralism has exerted an enormous influence on modern social thought—from anthropology and sociology to literary criticism and many other fields of inquiry in between. These many types of structuralist analyses vary enormously, but they all have a common theme: There is a deep underlying structure to most surface phenomena, and this structure can be conceptualized as a series of generative rules that can create a wide variety of empirical phenomena. That is, empirically observable phenomena—from a literary text to a social structure—are constructed in conformity to an implicit logic. Some see this logic as lodged in the biology of the human brain, whereas others would view this underlying structure as a cultural product.

Purely structuralist analysis has had its greatest impact in linguistics and literary criticism, and for a time it enjoyed considerable popularity in anthropology and sociology.[1] But in recent years, rigid and orthodox structuralist approaches emphasizing searches for those deep and universal structures that order all phenomena have declined in sociology; in their place, several more eclectic perspectives have emerged. These more eclectic theories borrow elements of structuralist analysis and blend them with other conceptual traditions, such as conflict theory, interactionism, and phenomenology. There is still an emphasis on symbolic codes and the ways in which these are produced by underlying generative rules, but such codes are causally influenced by material conditions and are subject to interpretation by lay agents. Thus, just as Claude Lévi-Strauss "turned Durkheim on his head" in seeing social structure as a reflection of mental structures, many recent efforts have put Émile Durkheim back on his feet and augmented the emphasis on the structure of symbol systems with Marxian conflict analysis, interactionism and phenomenology, and

*This chapter is coauthored with Stephan Fuchs

other traditions in theoretical sociology. The result has been a revival of "cultural sociology" in a less ponderous guise than Parsonian functionalism. In these newer approaches, the structure of cultural codes is causally linked not only to the behavioral and interpersonal activities of individuals but also to the institutional parameters within which such activities are conducted.

Although several possible candidates can be seen as squarely in this more eclectic structuralist approach, two have been selected for review here—the French scholar Pierre Bourdieu and the American Robert Wuthnow. Their work can provide a sample of what will be termed here *cultural structuralism.* Each draws from Durkheim, incorporates some of the insights of Lévi-Strauss without viewing social structure as a mere surface manifestation of cultural codes, and connects the French tradition to theories emphasizing the causal priority of material social conditions.

CULTURAL ANALYSIS: ROBERT WUTHNOW

In recent years, there has been a revival of cultural analysis.[2] Unfortunately, the nature of "culture" is vaguely conceptualized in social theory, with the result that just about anything—physical objects, ideas, world views, subjective states, behaviors, rituals, thoughts, emotions, and so on—can be considered "cultural." One recent effort to narrow somewhat the domain of cultural analysis is the work of Robert Wuthnow. Although most of his research has focused on religion, he has sought a more general theoretical approach.[3]

This theoretical effort preaches against positivism, incorrectly portrayed as sheer empiricism, but it nonetheless synthesizes several theoretical traditions and, in so doing, develops some general propositions on cultural processes.[4] Wuthnow's approach is, therefore, one of the more creative approaches to structuralism, primarily because it blends structuralist concerns about relations among symbolic codes, per se, with other theoretical traditions. Among these other traditions are elements of dramaturgy, institutional analysis, and subjective approaches.

Cultural Structure, Ritual, and Institutional Context

In Wuthnow's view it is wise to avoid "radical subjectivity," for the "problem of meaning may well be more of a curse than a blessing in cultural analysis."[5] It is best, he argues, to move away from an overemphasis on attitudes, beliefs, and meanings of individuals, because these are difficult to measure. Instead, the structure of culture as revealed through *observable* communications and interactions is a more appropriate line of inquiry. In this way one does "not become embroiled in the ultimate phenomenological quest to probe and describe subjective meanings in all their rich detail."[6] Rather, the structure of cultural codes as produced, reproduced, or changed by interaction and communication is examined. Once emphasis shifts away from meaning, per se, to the structure of culture in social contexts and socially produced texts, other theoretical approaches become useful.

Dramaturgy is one essential supplement because of its emphasis on ritual as a mechanism for expressing and dramatizing symbols. In a sense, individual interpersonal rituals as well as collective rituals express deeply held meanings, but at the same time they affirm particular cultural structures. In so doing, ritual performs such diverse functions as reinforcing collective values, dramatizing certain relations, denoting key positions, embellishing certain messages, and highlighting particular activities.[7]

Another important theoretical supplement is institutional analysis. Culture does not exist as an abstract structure in its own right. Nor is it simply dramatic and ritualized performances; it is also embedded in organized social structures. Culture is produced by actors and organizations that require resources—material, organizational, and political—if they are to develop systems of cultural codes, ritualize them, and transmit them to others. Once the institutional basis of cultural activity is recognized, then the significance of inequalities in resources, the use of power, and the outbreak of conflict become essential parts of cultural analysis.

In sum, then, Wuthnow blends a muted subjective approach with structuralism, dramaturgy, and institutional analysis. He tries to view the subjective as manifested in cultural products, dramatic performances, and institutional processes. In attempting this synthesis, Wuthnow defines his topic as *the moral order.*

The Moral Order

Wuthnow views the moral order as involving the (1) construction of systems of cultural codes, (2) emission of rituals, and (3) mobilization of resources to produce and sustain these cultural codes and rituals. Let us examine each of these in turn.

The Structure of Moral Codes. A *moral code* is viewed by Wuthnow "as a set of cultural elements that define the nature of commitment to a particular course of behavior." Such sets of cultural elements have an "identifiable structure" involving not so much a "tightly organized or logically consistent system" as some basic "distinctions" that can be used "to make sense of areas in which problems in moral obligations may be likely to arise."[8] Wuthnow sees three such distinctions as crucial to structuring the moral order: (1) moral objects versus real programs, (2) core self versus enacted social roles, and (3) inevitable constraints versus intentional options. We will examine each of these.

1. The structure of a moral order distinguishes between (a) the *objects* of commitment and (b) the activities or *real programs* in which the committed engage. The objects of commitment can be varied—a person, a set of beliefs and values, a text, and so on—and the real programs can be almost any kind of activity. The critical point, Wuthnow argues, is that the objects of moral commitment and the behavior emitted to demonstrate this commitment are "connected" and, yet, "different." For example, one's object of commitment might be "making a better life for one's children," which is to be realized through "hard work" and other activities or real programs. For the structure of a moral order to be effective, it must implicitly distinguish and, at the same time, connect such objects and real programs.
2. The structure of moral codes must also, in Wuthnow's view, distinguish between (a) the person's "real self" or "true self" and (b) the various "roles" that he or she plays. Moral structures always link self-worth and behavior but, at the same time, allow them to be distinguished so that there is a "real me" who is morally worthy and who can be separated from the roles that can potentially compromise this sense of self-worth. For example, when someone reveals "role distance," an assertion is being made that a role is beneath one's dignity or self-worth.
3. Moral codes must also distinguish between (a) those forces that are out of people's control and (b) those that are within the realm of their will. That is, the *inevitable* must be distinguished from the *intentional.* In this way, cultural codes

> posit a moral evaluation of those behaviors that can be controlled through intent and will power, while forgiving or suspending evaluation for what is out of a person's control. Without this distinction it would be impossible to know what kinds of behaviors are to be subjected to moral evaluations.

Thus, the structure of a moral order revolves around three basic types of codes that denote and distinguish commitments with respect to moral objects/real programs, self/roles, and inevitable constraint/intentional options. Such codes indicate what is desirable by separating but also linking objects, behavior, self, roles, constraints, and intentions. Without this denotation of, and a distinction along, these three axes, a moral order and the institutional system in which it is lodged will reveal crises and will begin to break down. If objects and programs are not denoted, distinguished, and yet linked, then cynicism becomes rampant; if self and roles are confused, then loss of self-worth spreads; and if constraints and control are blurred, then apathy or frustration increases. Thus for Wuthnow,

> Morality . . . deals primarily with moral commitment—commitment to an object, ranging from an abstract value to a specific person, that involves behavior, that contributes to self-worth, and that takes place within broad definitions of what is inevitable or intentional. Moral commitment, although in some sense deeply personal and subjective, also involves symbolic constructions—codes—that define these various relations.[9]

The Nature of Ritual. Wuthnow believes ritual is "a symbolic-expressive aspect of behavior that communicates something about social relations, often in a relatively dramatic or formal manner."[10] A *moral ritual* "dramatizes collective values and demonstrates individuals' moral responsibility for such values."[11] In so doing, rituals operate to maintain the moral order—that is, the system of symbolic codes ordering moral objects/real programs, self/roles, and constraints/options. Such rituals can be embedded in normal interaction as well as in more elaborate collective ceremonies, and they can be privately or publicly performed.[12] But the key point is that ritual is a basic mechanism for sustaining the moral order.

However, as Wuthnow stresses, ritual is also used to cope with uncertainty in the social relations regulated by the codes of the moral order. Whether through increased options, uses of authority, ambiguity in expectations, lack of clarity in values, equivocality in key symbols, or unpredictability in key social relations, rituals are often invoked to deal with these varying bases of uncertainty. Uncertainty is thus one of the sources of escalated ritual activity. However, such uses of ritual are usually tied to efforts at mobilizing resources in institutional contexts to create a new moral order—a process that, as we will examine shortly, Wuthnow examines under the rubric of "ideology."

Institutional Context. For a moral order to exist, it must be produced and reproduced, and for new moral codes to emerge—that is, ideologies—they too must be actively produced by actors using resources. Thus systems of symbolic codes depend on material and organizational resources; if a moral order is to persist and if a new ideology is to become a part of the moral order, it must have a stable supply of resources for actors to use in sustaining the moral order, or in propagating a new ideology. That is, actors must have the material goods necessary to sustain themselves and the organizations in which they participate; they must have organizational bases that depend not only on material goods, such as money, but also on organizational "know-how," communication networks, and leadership; and at times they must also have power. Thus, the moral order is anchored in institutional structures revolving around material goods, money, leadership, communication networks, and organizational capacities.

Ideology. One of the central and yet ambiguous concepts in Wuthnow's analysis is his portrayal of ideology, which he defines as "symbols that express or dramatize something about the moral order."[13] This definition is very close to the one used for ritual, so it is somewhat unclear what Wuthnow has in mind.[14] The basic idea appears to be that an ideology is a subset of symbolic codes emphasizing a particular aspect of the more inclusive moral order. Ideologies are also the vehicles for change in the moral order because the moral order is altered through the development and subsequent institutionalization of new ideologies. The production and institutionalization of these subsets of symbolic codes depend on the mobilization of resources (leaders, communication networks, organizations, and material goods) and the creation and emission of rituals. New ideologies must often compete with one another, so, those ideologies with superior resource bases are more likely to survive and become a part of the moral order.

In sum, then, the moral order consists of a structure of codes, a system of rituals, and a configuration of resources that define the manner in which social relations should be constituted.[15] An important feature of the moral order is the production of ideologies, which are subsets of codes, ritual practices, and resource bases. With this conceptual baggage in hand, Wuthnow then turns to the analysis of dynamic processes in the moral order.

The Dynamics of the Moral Order

Wuthnow employs an ecological framework for the analysis of dynamics.[16] When a moral order does not specify the ordering of moral objects/real programs, self/roles, and inevitable constraints/intentional controls, when it cannot specify the appropriate communicative and ritual practices for its affirmation and dramatization, and when, as a result of these conditions, it cannot reduce the risks associated with various activities, the ambiguities of social situations, or the unpredictability of social relations, then the level of uncertainty among the members of a population increases. Under conditions of uncertainty, new ideologies are likely to be produced as a way of coping. Such ideological production is facilitated by (1) high degrees of heterogeneity in the types of social units—classes, groups, organizations, and so forth— in a social system and by high levels of diversity in resources and their distribution, (2) high rates of change (realignment of power, redistribution of resources, establishment of new structures, creation of new types of social relations), (3) inflexibility in cultural codes (created by tight connections among a few codes), and (4) reduced capacity of political authority to repress new cultural codes, rituals, and mobilizations of resources.

Wuthnow portrays these processes as an increase in "ideological variation" that results in "competition" among ideologies. Some ideologies are "more fit" to survive this competition and, as a consequence, are "selected." Such "fitness" and "selection" depends on an ideology's capacity to (1) define social relations in ways reducing uncertainty (over moral objects, programs, self, roles, constraints, options, risks, ambiguities, and unpredictability), (2) reveal a flexible structure consisting of many elements weakly connected, (3) secure a resource base (particularly

TABLE 37-1 Wuthnow's Principles of Cultural Dynamics

I. The degree of stability in the moral order of a social system is a positive function of its legitimacy, with the latter being a positive and additive function of
   - A The extent to which the symbolic codes of the moral order facilitate the ordering of
     1. Moral objects and real programs.
     2. Self and roles.
     3. Inevitable constraints and intentional control.
   - B. The extent to which the symbolic codes of the moral order are dramatized by ritual activities.
   - C. The extent to which the symbolic codes of the moral order are affirmed by communicative acts.

II. The rate and degree of change in the moral order of a social system are a positive function of the degree of ideological variation, with the latter being a positive and additive function of
   - A. The degree of uncertainty in the social relations of actors, which in turn is an additive function of
     1. The inability of cultural codes to order moral objects/real projects, self/roles, and constraints/options.
     2. The inability of rituals to dramatize key cultural codes.
     3. The inability of communicative acts to affirm key cultural codes.
     4. The inability of cultural codes to specify the risks associated with various activities and relations.
     5. The inability of cultural codes to reduce the ambiguity of various activities and relations.
     6. The inability of cultural codes to reduce the unpredictability of various acts and relations.
   - B. The level of ideological production and variation, which in turn is a positive and additive function of
     1. The degree of heterogeneity among social units.
     2. The diversity of resources and their distribution.
     3. The rate and degree of change in institutional structures.
     4. The degree of inflexibility in the sets of cultural codes, which is an inverse function of
        - a. The number of symbolic codes.
        - b. The weakness of connections among symbolic codes.
     5. The inability of political authority to repress ideological production.

III. The likelihood of survival and institutionization of new ideological variants is a positive and multiplicative function of
   - A. The capacity of an ideological variant to secure a resource base, which in turn is a positive and additive function of the capacity to generate
     1. Material resources.
     2. Communication networks.
     3. Rituals.
     4. Organizational footings.
     5. Leadership.
   - B. The capacity of an ideological variant to establish goals and pursue them.
   - C. The capacity of an ideological variant to maintain legitimacy with respect to
     1. Existing values and procedural rules.
     2. Existing political authority.
   - D. The capacity of an ideological variant to remain flexible, which in turn is a positive function of
     1. The number of symbolic codes.
     2. The weakness of connections among symbolic codes.

money, adherents, organizations, leadership, and communication channels), (4) specify ritual and communicative practices, (5) establish autonomous goals, and (6) achieve legitimacy in the eyes of political authority and in terms of existing values and procedural rules. The more that these conditions can be met, the more likely an ideology is to survive in competition with other ideologies, and the more likely it is to become institutionalized as part of the moral order. In particular, the institutionalization of an ideology depends on the establishment of rituals and modes of communication affirming the new moral codes within an organizational arrangement that allows for ritual dramatization of new codes reducing uncertainty, that secures a stable resource base, and that eventually receives acceptance by political authority.

Different types of ideological movements will emerge, Wuthnow appears to argue, under varying configurations of these conditions that produce variation, selection, and institutionalization.[17] Although he offers many illustrations of ideological movements, particularly of various kinds of religious movements as well as the emergence of science as an ideology, he does not systematically indicate how varying configurations of these general conditions produce basic types of ideological movements. Yet these variables all appear, in a rather ad hoc and discursive way, in his analysis of ideological movements. And so there is at least an implicit effort to test the theory.

In way of summary, then, Table 37-1 formalizes Wuthnow's theory. Wuthnow would probably reject this formalization as being too "positivistic," but, if his ideas are to be more explanatory and less discursive, formalization along these lines is desirable—hence, the provisional effort expressed in Table 37-1.

CONSTRUCTIVIST STRUCTURALISM: PIERRE BOURDIEU

Pierre Bourdieu's sociology defies each classification because it cuts across disciplinary boundaries—sociology, anthropology, education, cultural history, art, science, linguistics, and philosophy—and moves easily between empirical and conceptual inquiry.[18] Yet Bourdieu has characterized his work as *constructivist structuralism* or *structuralist constructivism;* in so doing, he distances himself somewhat from the Lévi-Straussian tradition:

> By structuralism or structuralist, I mean that there exists, within the social world itself and not only within symbolic systems (language, myth, etc.), objective structures independent of the consciousness and will of agents, which are capable of guiding and constraining their practices or their representations.[19]

Such structures constrain and circumscribe volition, but at the same time people use their capacities for thought, reflection, and action to *construct* social and cultural phenomena. They do so within the parameters of existing structures. These structures are not rigid constraints but, rather, materials for a wide variety of social and cultural constructions. Acknowledging his structuralist roots, Bourdieu analogizes to the relation of grammar and language in order to make this point: The grammar of a language only loosely constrains the production of actual speech; it can be seen as defining the possibilities for new kinds of speech acts. So it is with social and cultural structures: They exist independently of agents and guide their conduct, yet they also create options, possibilities, and paths for creative actions and for the construction of new and unique cultural and social phenomena. This perspective is best appreciated by highlighting Bourdieu's criticisms of those theoretical approaches from which he selectively borrows ideas.

Criticisms of Existing Theories

The Critique of Structuralism. Bourdieu's critique of structuralism is similar to symbolic interactionists' attacks on Parsonian functionalism and its emphasis on norms. According to Bourdieu, structuralists ignore the indeterminacy of situations and the practical ingenuity of agents who are not mechanical rule-following and role-playing robots in standard contexts. Rather, agents use their "practical sense" (*sens pratique*) to adapt to situational contingencies within certain "structural limits" that follow from "objective constraints." Social practice is more than the mere execution of an underlying structural "grammar" of action, just as "speech" (*parole*) is more than "language" (*langue*). What is missing, says Bourdieu, are the variable uses and contexts of speech and action.[20] Structuralism dismisses action as mere execution of underlying principles (lodged in the human brain or culture), just as normativism forgets that following rules and playing roles require skillful adjustment and flexible improvisations by creative agents.

Most important, Bourdieu argues that structuralism hypostatizes the "objectifying glance" of the outside academic observer. The "Homo academicus" transfers a particular *relation* to the world, the distant and objectifying gaze of the professional academic, onto the very properties of that world.[21] As a result, the outside observer constructs the world as a mere "spectacle," which is subject to neutral observation. The distant and uninvolved observer's relation to the world is not only systematic but also a passive *cognition,* so the world itself is viewed as consisting of cognition rather than active practices. According to Bourdieu, structuralism and other approaches that "objectify" the world do not simply research the empirical world "out there"; rather, they construct it *as* an "objective fact" through the distancing perspective of the outside observer.

Bourdieu does not, however, reject completely structuralism and other "objectifying" approaches that seek, in Durkheim's words, to discover external and constraining "social facts." As we will see, Bourdieu views social classes, and factions within such classes, as "social facts" whose structure can be objectively observed and viewed as external to, and constraining on, the thoughts and activities of individuals.[22] Moreover, Bourdieu at least borrows the metaphor, if not the essence, of structuralism in his efforts to discover the "generative principles" that people use to construct social and cultural phenomena—systems of classification, ideologies, forms of legitimating social practices, and other elements of "constructivist structuralism."[23]

The Critique of Interactionism and Phenomenology. Bourdieu is also critical of interactionism, phenomenology, and other subjective approaches.[24] Bourdieu believes there is more to social life than interaction, and there is more to interaction than the "definitions of situations" in symbolic interactionism or the "accounting practices" in ethnomethodology. The "actor" of symbolic interactionism and the "member" of ethnomethodology are abstractions that fail to realize that members are always incumbents in particular groups and classes. Interactions are always interactions-in-contexts, and the most important of these contexts is class location. Even such an elementary feature of interaction as the possibility that it might even occur among individuals varies with class background. Interaction is thus embedded in structure, and the structure constrains what is possible.

Moreover, in addition to this rather widespread critique of interactionism as "astructural," Bourdieu argues that interactionism is too cognitive in its overemphasis on the accounting and sense-making activities of agents. As a result, it forgets that actors have objective class-based interests. And, once again, the biases of "Homo academicus" are evident. It is in the nature of academics to define, assess, reflect, ponder, and interpret the social world; as a result of this propensity, a purely academic relation to the world is imposed on real people in social contexts. For interactionists, then, people are merely disinterested lay academics who define, reflect, interpret, and account for actions and situations. But lay interpretations, and academic portrayals of these lay interpretations, cannot accurately describe social reality, for two reasons. First, as noted, these interpretations are constrained by existing structures, especially class and class factions. Second, these interpretations are themselves part of objective class struggles as individuals construct legitimating definitions for their conduct.[25]

Bourdieu then borrows from Karl Marx the notion that people are located in a class position, that this position gives them certain interests, and that their interpretative actions are often ideologies designed to legitimate these interests. People's "definitions of situations" are neither neutral nor innocent, but are often ideological weapons that are very much a part of the objective class structures and the inherent conflicts of interests generated by such structures.[26]

The Critique of Utilitarianism. Rational economic theories also portray, and at the same time betray, Homo academicus' relation to the social world. Like academics in general, utilitarian economic theorists[27] see humans as rational, calculating, and maximizing (*sujets ravauts*); and rational exchange theories thus mistake a *model* of the human actor for real individuals, thereby reifying their theoretical abstractions.

Yet Bourdieu does not replace the economic model of rational action with an interpretative model of symbolic action. He does not argue that rational action theory is wrong because it is too rationalistic or because it ignores the interpretative side of action. To the contrary, he holds that rational action theory does not realize that even symbolic action is rational and based on class interests. Thus, according to Bourdieu, the error of the economic model is not that it presents all action as rational and interested; rather, the big mistake is to restrict interests and rationality to the immediate material payoffs collected by reflective and profit-seeking individuals.[28]

Bourdieu reasons that all social practices are "interested," even if individual agents are unaware of their interests and even if the stakes of these practices are not material profits. Social practices are attuned to the conditions of particular arenas in which actions might yield profits without deliberate intention. For example, in science it is the most "disinterested" and "pure" research that yields the highest cultural profits—that is, academic recognition and reputation. In social fields other than economic exchange it is the structural *denial* of any "interests" that often yields the highest gains. It is not that agents cynically deny being interested to increase their gains even more; rather, innocence assures that honest disinterestedness nevertheless is the most profitable practice.

For example, gift exchange economies, the subject of Bourdieu's early anthropological research,[29] might illustrate this complex idea. Gift exchange economies are typically embedded in larger social relations and solidarities so that exchange is not purely instrumental and material but has a strong moral quality to it. Economic exchanges are expected to follow the social logic of solidarity and group memberships at least as much as they follow the economic logic of material gain. From the narrowly economic perspective of rational action theory, the logic of solidarity would seem like an intrusion of "nonrational" forces, such as tradition or emotion, into an otherwise purely rational system of exchange; but, the logic of solidarity points to those processes by which symbolic and social capital is accumulated—a "social fact" that is missed by the narrow economic determinism of rational action theory. But, once the notion of "capital" is extended to include symbolic and social capital, apparently "irrational" practices can now be seen to follow their own interested logic, and, contrary to initial impressions, these practices are not irrational at all. The structural denial of narrowly economic interests in gift exchange economies conceals that social and economic capital can be increased the more the purely instrumental aspects of exchange move into the background. For instance, birthday and Christmas presents are socially more effective when they appear less material and economic; those who brag about the high costs of their presents do not understand the nature of gift exchanges and, as a consequence, are considered rude, thereby losing symbolic and social capital.

Thus, in broadening economic exchange to include social and symbolic resources, as all sociological exchange theories eventually do,[30] Bourdieu introduces a central concept in his approach: *capital*.[31] Those in different classes reveal not only varying levels or amounts of capital but also divergent types and configurations of capital. Bourdieu's view of capital recognizes that the resources individuals possess can be material, symbolic, social, and cultural; moreover, these resources reflect class location and are used to further the interests of those in a particular class position.

Bourdieu's Cultural Conflict Theory

Although Bourdieu has explored many topics, the conceptual core of his sociology is a vision of social classes and the cultural forms associated with these classes.[32] In essence, Bourdieu combines a Marxian theory of objective class position in relation to the means of production with a Weberian analysis of status groups (lifestyles, tastes, prestige) and politics (organized efforts to have one's class culture dominate). The key to this reconciliation of Karl Marx's and Max Weber's views of stratification is the expanded conceptualization of *capital* as more than economic and material resources, coupled with elements of French structuralism.

Classes and Capital. To understand Bourdieu's view of classes, it is first necessary to recognize a distinction among four types of capital.[33] (1) *economic* capital, or productive property (money and material objects that can be used to produce goods and services); (2) *social* capital, or positions and relations in groupings and social networks; (3) *cultural* capital, or informal interpersonal skills, habits, manners, linguistic styles, educational credentials, tastes, and lifestyles, and (4) *symbolic* capital, or the use of symbols to legitimate the possession of varying levels and configurations of the other three types of capital.

These forms of capital can be converted into one another, but only to a certain extent. The degree of convertibility of capital on various markets is itself at stake in social struggles. The overproduction of academic qualifications, for example, can decrease the convertibility of educational into economic capital ("credential inflation"). As a result, owners of credentials must struggle to get their cultural capital converted into economic gains, such as high-paying jobs. Likewise, the extent to which economic capital can be

TABLE 37-2 Representation of Classes and Class Factions in Industrial Societies*

***Dominant Class:* Richest in all forms of capital**	
Dominant faction:	Richest in economic capital, which can be used to buy other types of capital. This faction is composed primarily of those who own the means of production—that is, the classical bourgeoisie.
Intermediate faction:	Some economic capital, coupled with moderate levels of social, cultural, and symbolic capital. This faction is composed of high-credential professionals.
Dominated faction:	Little economic capital but high levels of cultural and symbolic capital. This faction is composed of intellectuals, artists, writers, and others who possess cultural resources valued in a society.
***Middle Class:* Moderate levels of all forms of capital**	
Dominant faction:	Highest in this class in economic capital but having considerably less economic capital than the dominant faction of the dominant class. This faction is composed of petite bourgeoisie (small business owners).
Intermediate faction:	Some economic, social, cultural, and symbolic capital but considerably less than the intermediate faction of the dominant class. This faction is composed of skilled clerical workers.
Dominated faction:	Little or no economic capital and comparatively high social, cultural, and symbolic capital. This class is composed of educational workers, such as schoolteachers, and other low-income and routinized professions that are involved in cultural production.
***Lower Class:* Low levels of all forms of capital**	
Dominant faction:	Comparatively high economic capital for this general class. Composed of skilled manual workers.
Intermediate faction:	Lower amounts of economic and other types of capital. Composed of semi-skilled workers without credentials.
Dominated faction:	Very low amounts of economic capital. Some symbolic capital in form of uneducated ideologues and intellectuals for the poor and working person.

*These portrayals are inferences from Bourdieu's more discursive and rambling text. The table captures the imagery of Bourdieu's analysis; however, because he is highly critical of stratification research in America, he would probably be critical of this "layered" portrayal of his argument.

converted into social capital is at stake in struggles over control of the political apparatus, and the efforts of those with economic capital to "buy" cultural capital can often be limited by their perceived lack of "taste" (a type of cultural capital).

The distribution of these four types of capital determines the objective class structure of a social system. The overall class structure reflects the total amount of capital possessed by various groupings. Hence the *dominant class* will possess the most economic, social, cultural, and symbolic capital; the *middle class* will possess less of these forms of capital; and the *lower classes* will have the least amount of these capital resources. The class structure is not, however, a simple lineal hierarchy. Within each class are *factions* that can be distinguished by (1) the composition or configuration of their capital and (2) the social origin and amount of time that individuals in families have possessed a particular profile or configuration of capital resources.

Table 37-2 represents schematically Bourdieu's portrayal of the factions in three classes. The top faction within a given class controls the greatest proportion of economic or productive capital typical of a class; the bottom faction possesses the greatest amount of cultural and symbolic capital for a class; and the middle faction possesses an intermediate amount of economic, cultural, and symbolic capital. The top faction is the dominant faction within a given class, and the bottom faction is the dominated faction for that class, with the middle faction being both superordinate over the dominated faction and subordinate to the top faction. As factions engage in struggles to control resources and legitimate themselves, they mobilize social capital to form groupings and networks of relations, but their capacity to form such networks is limited by their other forms of capital. Thus, the overall distribution of social capital (groups and organizational memberships, network ties, social relations, and so forth) for classes and their factions will correspond to the overall distribution of other forms of capital. However, the particular forms of groupings, networks, and social ties will reflect the particular configuration of economic, cultural, and symbolic capital typically possessed by a particular faction within a given class.

Bourdieu borrows Marx's distinction between a class "for itself" (organized to pursue its interests) and one "in itself" (unorganized but having common interests and objective location), then he argues that classes are not real groups but only "potentialities." As noted earlier, the objective distribution of resources for Bourdieu relates to actual groups as grammar relates to speech: It defines the possibilities for actors but needs actual people and concrete settings to become real. And, it is the transformation of class and class-faction interests into actual groupings that marks the dynamics of a society. Such transformation involves the use of productive material, cultural, and symbolic capital to mobilize social capital (groups and networks); even more important, class conflict tends to revolve around the mobilization of symbols into ideologies that legitimate a particular composition of resources.[34] Much conflict in human societies, therefore, revolves around efforts to manipulate symbols to make a particular pattern of social, cultural, and productive resources seem the most appropriate. For example, when intellectuals and artists decry the "crass commercialism," "acquisitiveness," and "greed" of big business, this activity involves the mobilization of symbols into an ideology that seeks to mitigate their domination by the owners of the means of production.

But class relations involve more than a simple pecking order. There are also homologies among similarly located factions within different classes. For example, the rich capitalists of the dominant class and the small business owners of the middle class are equivalent in their control of productive resources and their dominant position in their respective classes[35]; similarly, intellectuals, artists, and other cultural elites in the dominant class are equivalent to schoolteachers in the middle class because of their reliance on cultural capital and because of their subordinate position in relation to those who control the material resources of their respective classes. These homologies make class conflict complex, because those in similar objective positions in different classes—say, intellectuals and schoolteachers—will mobilize symbolic resources into somewhat similar ideologies—in this example, emphasizing learning, knowledge for its own sake, and life of the mind and, at the same time, decrying crass materialism. Such ideologies legitimate their own class position and attack those who dominate them (by emphasizing the importance of those cultural resources that they have more of). At the same time their homologous positions are separated by the different *amounts* of cultural capital owned: The intellectuals despise the strained efforts of schoolteachers to appear more sophisticated than they are, whereas the schoolteachers resent the decadent and irresponsible relativism of snobbish intellectuals. Thus, ideological conflict is complicated by the simultaneous convergence of factions within different classes and by the divergence of these factions by virtue of their position in different social classes.

Moreover, an additional complication stems from people sharing similar types and amounts of resources but having very different origins and social trajectories. Those who have recently moved to a class faction—say, the dominant productive elite or intermediate faction of the middle class—will have somewhat different styles and tastes than those who have been born into these classes, and these differences in social origin and mobility can create yet another source of ideological conflict. For example, the "old rich" will often comment on the "lack of class" and "ostentatiousness" of the "new rich"; or, the "solid middle class" will be somewhat snobbish toward the "poor boy who made good" but who "still has a lot to learn" or who "still is a bit crude."

All those points of convergence and divergence within and between classes and class factions make the dynamics of stratification complex. Although there is always an "objective class location," as determined by the amount and composition of capital and by the social origins of capital holders, the development of organizations and ideologies is not a simple process. Bourdieu often ventures into a more structuralist mode when trying to sort out how various classes, class factions, and splits of individuals with different social origins within class factions generate categories of thought, systems of speech, signs of distinction, forms of mythology, modes of appreciation, tastes, and lifestyle. The general argument is that objective location—class, faction within class, and social origin—creates interests and structural constraints that, in turn, allow different social constructions.[36] Such constructions might involve the use of "formal rules" (implicitly known by individuals with varying interests) to construct cultural codes that classify and organize "things," "signs," and "people" in the world. This kind of analysis by Bourdieu has not produced a fine-grained structuralist model of how individuals construct particular cultural codes, but it has provided an interesting analysis of "class cultures." Such "class cultures" are always the dependent variable for Bourdieu (with objective class location being the independent variable and rather poorly conceptualized structuralist processes of generative rules and cultural codes being the "intervening variables"). Yet the detailed description of these class cultures is perhaps Bourdieu's most unique contribution to sociology and is captured by his concept of *habitus*.

Class Cultures and Habitus. Those within a given class share certain modes of classification, appreciation, judgment, perception, and behavior. Bourdieu conceptualizes this mediating process between class and individual perceptions, choices, and behavior as *habitus*.[37] In a sense, habitus is the "collective unconscious" of those in similar positions because it provides cognitive and emotional guidelines that enable individuals to represent the world in common ways and to classify, choose, evaluate, and act in a particular manner.

The habitus creates syndromes of taste, speech, dress, manner, and other responses. For example, a preference for particular foods will tend to correspond to tastes in art, ways of dressing, styles of speech, manners of eating, and other cultural actions among those sharing a common class location. There is, then, a correlation between the class hierarchy and the cultural objects, preferences, and behaviors of those located at particular ranks in the hierarchy. For instance, Bourdieu devotes considerable attention to "taste," which is seen as one of the most visible manifestations of the habitus. Bourdieu views "taste" in a holistic and anthropological sense to include appreciation of art, ways of dressing, and preferences for foods.[38] Although taste appears as an innocent, natural, and personal phenomenon, it covaries with objective class location: The upper class is to the working class what an art museum is to television; the old upper class is to the new upper class what polite and distant elegance is to noisy and conspicuous consumption; and the dominant is to the dominated faction of the upper class what opera is to avant-garde theater. Because tastes are organized in a cultural hierarchy that mirrors the social hierarchy of objective class location, conflicts between tastes are class conflicts.

Bourdieu roughly distinguishes between two types of tastes, which correspond to high versus low overall capital, or high versus low objective class position.[39] The "taste of liberty and luxury" is the taste of the upper class; as such, it is removed from direct economic necessity and material need. The taste of liberty is the philosophy of art for its own sake. Following Immanuel Kant, Bourdieu calls this aesthetic the "pure gaze." The pure gaze looks at the sheer form of art and places this form above function and content. The upper-class taste of luxury is not concerned with art illustrating or representing some external reality; art is removed from life, just as upper-class life is removed from harsh material necessity. Consequently, the taste of luxury purifies and sublimates the ordinary and profane into the aesthetic and beautiful. The pure gaze confers aesthetic meaning to ordinary and profane objects because the taste of liberty is at leisure to relieve objects from their pragmatic functions. Thus, as the distance form basic material necessities increases, the pure gaze or the taste of luxury transforms the ordinary into the aesthetic, the material into the symbolic, the functional into the formal. And, because the taste of liberty is that of the dominant class, it is also the dominant and legitimate taste in society.

In contrast, the working class cultivates a "popular" aesthetic. Their taste is the taste of necessity, for working-class life is constrained by harsh economic imperatives. The popular taste wants art to represent reality and despises formal and self-sufficient art as decadent and degenerate. The popular taste favors the simple and honest rather than the complex and sophisticated. It is down-graded by the "legitimate" taste of luxury as naive and complacent, and these conflicts over tastes are class conflicts over cultural and symbolic capital.

Preferences for certain works and styles of art, however, are only part of "tastes" as ordered by habitus. Aesthetic choices are correlated with choices made in other cultural fields. The taste of liberty and luxury, for example, corresponds to the polite, distant, and disciplined style of upper-class conversation. Just as art is expected to be removed from life, so are the bodies of interlocutors expected to be removed from one another and so is the spirit expected to be removed from matter. Distance from economic necessity in the upper-class lifestyle not only corresponds to an aesthetic of pure form, but also entails that all natural and physical desires are to be sublimated and dematerialized. Hence, upper-class eating is highly regulated and disciplined, and foods that are less filling are preferred over fatty dishes. Similarly, items of clothing are chosen for fashion and aesthetic harmony, rather than for functional appropriateness. "Distance from necessity" is the motif underlying the upper-class lifestyle as a whole, not just aesthetic tastes as one area of practice.

Conversely, because they are immersed in physical reality and economic necessity, working-class people interact in more physical ways, touching one another's bodies, laughing heartily, and valuing straightforward outspokenness more than distant and "false" politeness. Similarly, the working-class taste favors foods that are more filling and less "refined" but more physically gratifying.

The popular taste chooses clothes and furniture that are functional, and this is so not only because of sheer economic constraints but also because of a true and profound dislike of that which is "formal" and "fancy."

In sum, then, Bourdieu has provided a conceptual model of class conflict that combines elements of Marxian, Weberian, and Durkheimian sociology. The structuralist aspects of Bourdieu's conceptualization of habitus as the mediating process between class position and individual behavior have been underemphasized in this review, but clearly Bourdieu places Durkheim "back on his feet" by emphasizing that class position determines habitus. But the useful elements of structuralism—systems of symbols as generative structures of codes—are retained and incorporated into a theory of class conflict as revolving around the mobilization of symbols into ideologies legitimating a class position and the associated lifestyle and habitus.

CONCLUSION

"Cultural structuralism" oscillates between the extreme positions of objectivism and subjectivism and, therefore, can avoid the conceptual problems of either position. The orthodox structuralism of Lévi-Strauss forgot that underlying generative codes need agents and contexts that transform a system of potentialities into actual historical action. Orthodox structuralism eliminates the "knowledgeable and capable agent" who can do more than simple execute latent schemes of practice. The panstructuralism of Lévi-Strauss opened up "culture" as a systematic area of research about which one can *theorize* but, at the same time, exaggerated the antihumanist notion of "subjectless" structures.

Cultural structuralism also avoids the indeterminate view of culture as a system of symbols that can be defined and negotiated at will by creative and spontaneous agents. Symbolic interactionism in particular exaggerates the capacity of agents to define situations and redefine self-identities in the spontaneous setting of unstructured "situations." Structuralism recovers the materialist notion that definitions of the situation and interpretations of symbols are constrained by external structures that cannot be totally controlled by the will of agents. Cultural practices and symbol use are not self-sufficient systems of unrestrained interpretive creativity; rather, cultural work has its own structures and constraints.

Cultural structuralism, then, does not fall into the traps of subjectless objectivism, and it also avoids the hermeneutic idealism of interpretive approaches to action. It has established the idea that culture has a structure that is itself a reality, *sui generis,* and that it can be analyzed like any other reality.

NOTES

1. For some general works reviewing structuralism, see Anthony Giddens, "Structuralism, Post-structuralism and the Production of Culture," in *Social Theory Today,* eds. A. Giddens and J.H. Turner (Stanford, CA: Stanford University Press, 1987); S. Clarke, *The Foundations of Structuralism* (Sussex, UK: Harvester, 1981); J. Sturrock, ed., *Structuralism and Science* (Oxford: Oxford University Press, 1979); W.G. Runciman, "What Is Structuralism?" in *Sociology in Its Place* (Cambridge: Cambridge University Press, 1970); Ino Rossi, *From the Sociology of Symbols to the Sociology of Signs* (New York: Columbia University Press, 1983), and Ino Rossi, ed., *Structural Sociology* (New York: Columbia University Press, 1982); Jacques Ehrmann, *Structuralism* (New York: Doubleday, 1970); Philip Pettit, *The Concept of Structuralism: A Critical Analysis* (Berkeley, CA: University of California Press, 1977); Charles C. Lemert, "The Uses of French Structuralism in Sociology," and Michelle Lamont and Robert Wuthnow, "Recent Cultural Sociology in Europe and the United States," in *Frontiers of Social Theory,* G. Ritzer, ed. (New York: Columbia University Press, 1990).

2. For a review, see Robert Wuthnow and Marsha Witten, "New Directions in the Study of Culture," *Annual Review of Sociology* 14 (1988), pp. 149–167. See also Robert Wuthnow, James Davidson Hunter, Albert Bergesen, and Edith Kurzweil, *Cultural Analysis: The World of Peter L. Berger, Mary Douglas, Michel Foucault, and Jurgen Habermas* (London: Routledge & Kegan Paul, 1984).

3. For examples of the work on religion, see Robert Wuthnow, *The Consciousness Reformation* (Berkeley: University of California Press, 1976), and *Experimentation in American Religion* (Berkeley: University of California Press, 1978); for a review of a more general theory, see Robert Wuthnow, *Meaning and Moral Order: Explorations in Cultural Analysis* (Berkeley, CA: University of California Press, 1987). For a review of this work, see Jonathan H. Turner, "Cultural Analysis and Social Theory," *American Journal of Sociology* 94 (July 1988), pp. 637–644.

4. Wuthnow would seemingly not view these as laws or principles, but his articulation of such laws (often implicitly) is what makes the work theoretically interesting.

5. Wuthnow, *Meaning and Moral Order* (cited in note 3), p. 64.

6. Ibid., p. 65.

7. Ibid., p. 132.

8. Ibid., p. 66.

9. Ibid., p. 70.

10. Ibid., p. 109.

11. Ibid., p. 140.

12. Here Wuthnow is drawing from the late Durkheimian tradition.

13. Wuthnow, *Meaning and Moral Order* (cited in note 3), p. 145.

14. See J.H. Turner, "Cultural Analysis and Social Theory" (cited in note 3), for a more detailed critique.

15. Wuthnow, *Meaning and Moral Order* (cited in note 3), p. 145.

16. See Chapters 8 and 9 for other theories using such a framework. See Wuthnow, *Meaning and Moral Order* (cited in note 3), for the outline of this ecological framework, especially Chapters 5 and 6.

17. Wuthnow offers examples in *Meaning and Moral Order,* Chapters 5–9, but the variables are woven into discursive text and rearranged in an ad hoc manner.

18. Indeed, Bourdieu has been enormously prolific, having authored some twenty five books and hundreds of articles in a variety of fields, including anthropology, education, cultural history, linguistics, philosophy, and sociology. His empirical work covers a wide spectrum of topics—art, academics, unemployment, peasants, classes, religion, sports, kinship, politics, law, and intellectuals. See Loic J.D. Wacquant, "Towards a Reflexive Sociology: A Workshop with Pierre Bourdieu," *Sociological Theory* 7 (1, Spring 1989), pp. 26–63. This article also contains a selected bibliography on Bourdieu's own works as well as secondary analyses and comments on Bourdieu.

19. Pierre Bourdieu, "Social Space and Symbolic Power," *Sociological Theory* 7 (1, Spring, 1989), p. 14.

20. Pierre Bourdieu, *Language and Symbolic Power* (Cambridge, MA: Harvard University Press, 1989).

21. Pierre Bourdieu, *Homo Academicus* (Stanford, CA: Stanford University Press, 1988).

22. Pierre Bourdieu, *Distinction: A Social Critique of the Judgement of Taste* (Cambridge, MA: Harvard University Press, 1984).

23. See Wacquant, "Towards a Reflexive Sociology" (cited in note 18); Bourdieu, "Social Space and Symbolic Power" (cited in note 19), pp. 14–25.

24. See Wacquant, "Towards a Reflexive Sociology" (cited in note 18).

25. Bourdieu, *Outline of a Theory of Practice* (Cambridge: Cambridge University Press, 1977), pp. 22 *ff.,* and *Distinction* (cited in note 20).

26. Ibid.

27. See chapters 23 and 24.

28. See Wacquant, "Towards a Reflexive Sociology" (cited in note 18), p. 43.

29. Bourdieu, *Outline of a Theory of Practice* (cited in note 25).

30. See chapters 23 and 24.

31. Pierre Bourdieu, "The Forms of Capital," in *Handbook of Theory and Research in the Sociology of Education,* ed. J.G. Richardson (New York: Greenwood, 1986). See also Michele Lamont and Annette P. Lareau, "Cultural Capital: Allusions, Gaps, and Glissandos in Recent Theoretical Developments," *Sociological Theory* 6 (2, Fall 1988), pp. 153–168.

32. Bourdieu, *Distinction* and *Outline of a Theory of Practice* (cited in notes 20 and 25).

33. Bourdieu, "The Forms of Capital" (cited in note 31).

34. Pierre Bourdieu, "Social Space and the Genesis of Groups," *Theory and Society* 14 (November 1985), pp. 723–744.

35. Bourdieu makes what in network analysis (see chapters 25 and 38) is termed *regular structural equivalence*. That is, those incumbents in positions that stand in an equivalent (similar) relation to other positions will act in a convergent way and evidence common attributes.

36. Bourdieu is not very clear about the issue of how the structural potentialities of a given objective class location become transformed into actual social groups capable of historical action. Like Lévi-Strauss, Bourdieu pursues the formal analogies between deep structures and actual practices, but he lacks a theory about how and when the transformations are going to be made, and made successfully.

37. Bourdieu, *Distinction* (cited in note 22).

38. Ibid.

39. Ibid. Actually, Bourdieu makes more fine-tuned distinctions, but we focus only on the main oppositions here.

38

The Continuing Tradition III: Network Analysis*

During the last thirty years, work within anthropology, social psychology, sociology, communications, psychology, geography, and political science has converged on the conceptualization of "structure" in "social networks." During this period, rather metaphorical and intuitive ideas about networks have been reconceptualized in various types of algebra, graph theory, and probability theory. This convergence has, in some ways, been a mixed blessing. On the one hand, grounding concepts in mathematics can give them greater precision and provide a common language for pulling together a common conceptual core from the overlapping metaphors of different disciplines. On the other hand, the extensive use of mathematics and computer algorithms far exceeds the technical skills of most social scientists. More important, the use and application of quantitative techniques, per se, have become a preoccupation among many who seem less and less interested in explaining how the actual social world operates.

Nonetheless, despite these drawbacks, the potential for network analysis as a theoretical approach is great because it captures an important property of social structure—patterns of relations among social units, whether people, collectivities, or positions. As Georg Simmel emphasized, at the core of any conceptualization of social structure is the notion that structure consists of relations and links among entities. Network analysis forces us to conceptualize carefully the nature of the entities and relations as well as the properties and dynamics that inhere in these relations.[1]

*This chapter is coauthored with Alexandra Maryanski.

BASIC THEORETICAL CONCEPTS IN NETWORK ANALYSIS

Points and Nodes

The units of a network can be persons, positions, corporate or collective actors, or virtually any entity that can be connected to another entity. In general, these units are conceptualized as *points or nodes*, and they are typically symbolized by letters or numbers. In Figure 38-1, a very simple network is drawn with each letter representing a point or node in the network. One goal of network analysis, then, is to array in visual space a pattern of connections among the units that are related to each other. In a mathematical sense, it makes little difference what the points and nodes are, and this has great virtue because it provides a common set of analytical tools for analyzing very diverse phenomena. Another goal of network analysis is to explain the dynamics of various patterns of ties among nodes, although this goal is often subordinated to developing computer algorithms for representing the connections among points and nodes in more complex networks than the one portrayed in Figure 38-1.

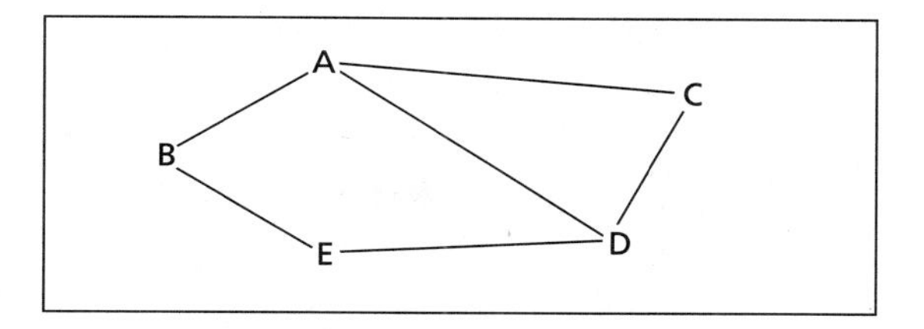

FIGURE 38-1 A Simple Network

Links, Ties, and Connections

The letters in Figure 38-1 represent the nodes or points of a structure. The lines connecting the letters indicate that these points are attached to each other in a particular pattern. The concept of *tie* is the most frequent way to denote this property of a network, and so in Figure 38-1, there are ties between *A-B*, *A-C*, *A-D*, *B-E*, *C-D*, and *D-E*. We not only need to know that points in a network are connected, but we also must have some idea of what it is that connects these points. That is, what is the nature of the tie? From the point of view of graph theory, it does not make much difference, but when the substantive concerns of sociologists are considered, it is important to know the nature of the ties. As we saw in Chapter 35 on the early sociograms constructed by Jacob Moreno, the ties involved emotional states such as liking and friendship, and the nodes themselves were individual people. But the nature of the tie can be diverse: the flow of information, money, goods, services, influence, emotions, deference, prestige, and virtually any force or resource that binds actors to each other.

Often, as we saw in Chapter 25 on exchange network theory, the ties are conceptualized as resources. When points or nodes are represented by different letters, this denotes that actors are exchanging different resources, such as prestige for advice, money for services, deference for information, and so on. Conversely, if they were exchanging similar resources, the nodes would be represented by the same letter and subscripted numbers, such as A_1, A_2, and A_3. But this is only one convention; the nature of the tie can also be represented by different kinds of lines, such as dotted, dashed, or colored lines. In graph theory, the lines can also reveal direction, as indicated by arrows. Moreover, if multiple resources are connecting positions in the graph, multiple lines (and, if necessary, arrows specifying direction) would be used. Thus, the graph represented in Figure 38-1 is obviously very simple, but it communicates the basic goal of network analysis: to represent in visual space the structure of connections among units.

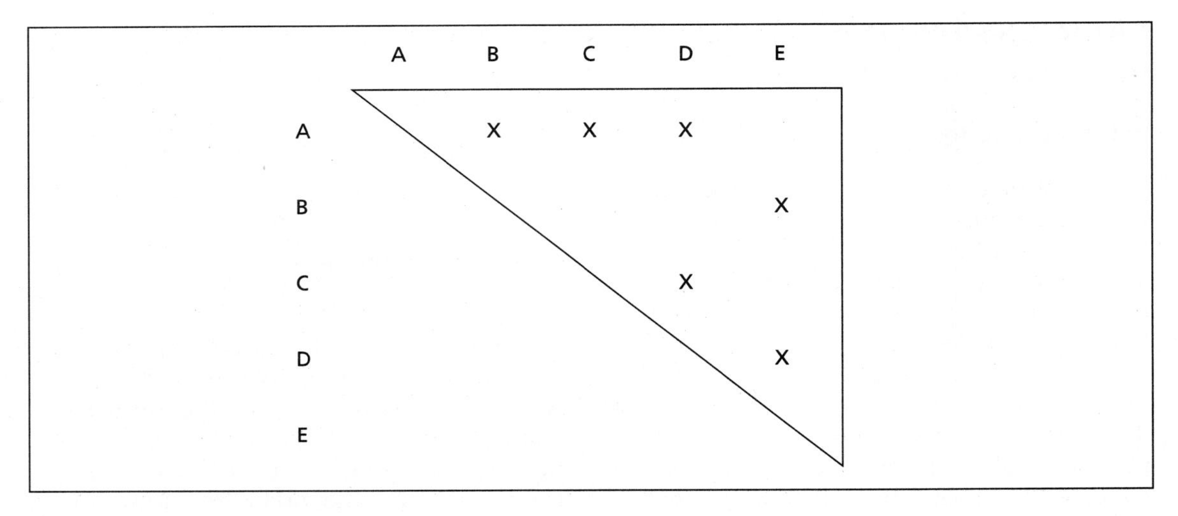

FIGURE 38-2 A Simple Matrix

One way to rise above the diversity of resources examined in network analysis is to visualize resource flows in networks for three generic types: materials, symbols, and emotions. That is, what connects persons, positions, and corporate actors in the social world is the flow of (1) symbols (information, ideas, values, norms, messages, etc.), (2) materials (physical things and perhaps symbols, such as money, that give access to physical things), and (3) emotions (approval, respect, liking, pleasure, and so forth). In nonsociological uses of networks the ties or links can be other types of phenomena, but, when the ties are social, they exist along material, symbolic, and emotional dimensions.

The configuration of ties can also be represented as a matrix, and in most network studies, the matrix is created before the actual network diagram. Moreover, when large numbers of nodes are involved, the matrix is often a better way to grasp the complexity of connections than a diagram, which would become too cumbersome to be useful. Figure 38-2 presents the logic of a matrix, using the very simple network represented in Figure 38-1. The mathematics of such matrices can become very complicated, but the general point is clear: to cross tabulate which nodes are connected to each other (as is done inside the triangular area of the matrix in Figure 38-2). If possible, once the matrix is constructed, it can be used to generate a graph, something like the one in Figure 38-1. With the use of sophisticated computer algorithms in network analysis, the matrix is the essential step for subsequent analysis; an actual diagram might not be drawn because the mathematical manipulations are too complex. Yet, most matrices will eventually be converted in network analysis into some form of visual representation in space—perhaps not a network digraph but some other technique, such as three dimensional bar graphs or clusters of points, will be used to express in visual space the relations among units.

PATTERNS AND CONFIGURATIONS OF TIES

From a network perspective, social structure is conceptualized as the form of ties among positions or nodes. That is, what is the pattern or configuration among what resources flowing among what sets of nodes or points in a graph? To answer questions like this, network sociology addresses several properties of networks. The most important of these are number of ties, directedness, reciprocity of ties, transitivity of ties, density of ties, strength of ties, bridges, brokerage, centrality, and equivalence.

Number of Ties

An important piece of information in performing network analysis is the total number of ties among all points and nodes. Naturally, the number of potential ties depends on the number of points in a graph and the number of resources involved in connecting the points. Yet, for any given number of points and resources, it is important to calculate both the actual and potential number of ties that are (and can be) generated. This information can then be used to calculate other dimensions of a network structure.

Directedness

It is important to know the direction in which resources flow through a network; so, as indicated earlier, arrows are often placed on the lines of a graph, making it a digraph. As a consequence, a better sense of the structure of the network emerges. For example, if the lines denote information, we would have a better understanding of how the ties in the network are constructed and maintained, because we could see the direction and sequence of the information flow.

Reciprocity of Ties

Another significant feature of networks is the reciprocity of ties among positions. That is, is the flow of resources one way, or is it reciprocated for any two positions? If the flow of resources is reciprocated, then it is conventional to have double lines with arrows pointing in the direction of the resource flow. Moreover, if different resources flow back and forth, this too can be represented. Surprisingly, conventions about how to represent this multiplicity of resource flows are not fully developed. One way to denote the flow of different resources is to use varying-colored lines or numbered lines; another is to label the points with the same letter subscripted (that is, A_1, A_2, A_3, and so forth) if similar resources flow and with varying letters (that is, A, B, C, D) if the resources connecting actors are different. But, whatever the notation, the extent and nature of reciprocity in ties become an important property of a social network.

Transitivity of Ties

A critical dimension of networks is the level of transitivity among sets of positions. *Transitivity* refers to the degree to which there is a "transfer" of a relation among subsets of positions. For example, if nodes A_1 and A_2 are connected with positive affect, and positions A_2 and A_3 are similarly connected, we can ask, will positions A_1 and A_3 also be tied together with positive affect? If the answer to this question is "yes," then the relations among A_1, A_2, and A_3 are transitive. Discovering patterns of transitivity in a network can be important because it helps explain other critical properties of a network, such as density and the formation of cliques.

Density of Ties

A significant property of a network is its degree of connectedness, or the extent to which nodes reveal the maximum possible number of ties. The more the actual number of ties among nodes approaches the total possible number among a set of nodes, the greater is

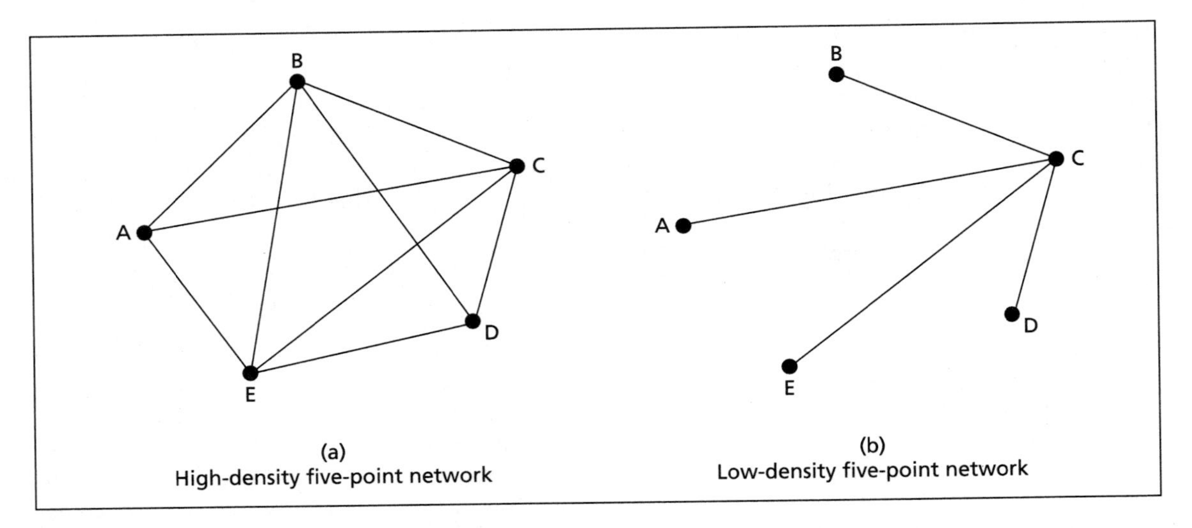

FIGURE 38-3 High- and Low-Density Networks

the overall *density* of a network.[2] Figure 38-3 compares the same five-node network under conditions of high and low density of ties.

Of even greater interest are subdensities of ties within a larger network structure. Such subdensities, which are sometimes referred to as *cliques*, reveal strong, reciprocated, and transitive ties among a particular subset of positions within the overall network.[3] For example, in Figure 38-4, there are three clusters of dense ties in the network, thus revealing three distinct cliques within the larger network.

Strength of Ties

Yet another crucial aspect of a network is the volume and level of resources that flow among positions. A weak tie is one where few or sporadic amounts of resources flow among positions, whereas a strong tie evidences a high level of resource flow. The overall structure of a network is significantly influenced by clusters and configurations of strong and weak ties. For example, if the ties in the cliques in Figure 38-4 are strong, the network is composed of cohesive subgroupings that have relatively sparse ties to one another. On the other hand, if the ties in these subdensities are weak, then the subgroupings will involve less intense linkages,[4] with the result that the structure of the whole network will be very different than would be the case if these ties were strong.

Bridges

When networks reveal subdensities, it is always interesting to know which positions connect the subdensities, or cliques, to one another. For example, in Figure 38-4, those ties connecting subdensities are bridges and are crucial in maintaining the overall connectedness of the network. Indeed, if one removed one of these positions or severed the tie, the structure of the network would be very different—it would become three separate networks. These bridging ties are typically weak,[5] because each position in the bridge is more embedded in the flow of resources of a particular subdensity or clique. But, nonetheless, such ties are often crucial to the maintenance of a larger social structure; it is

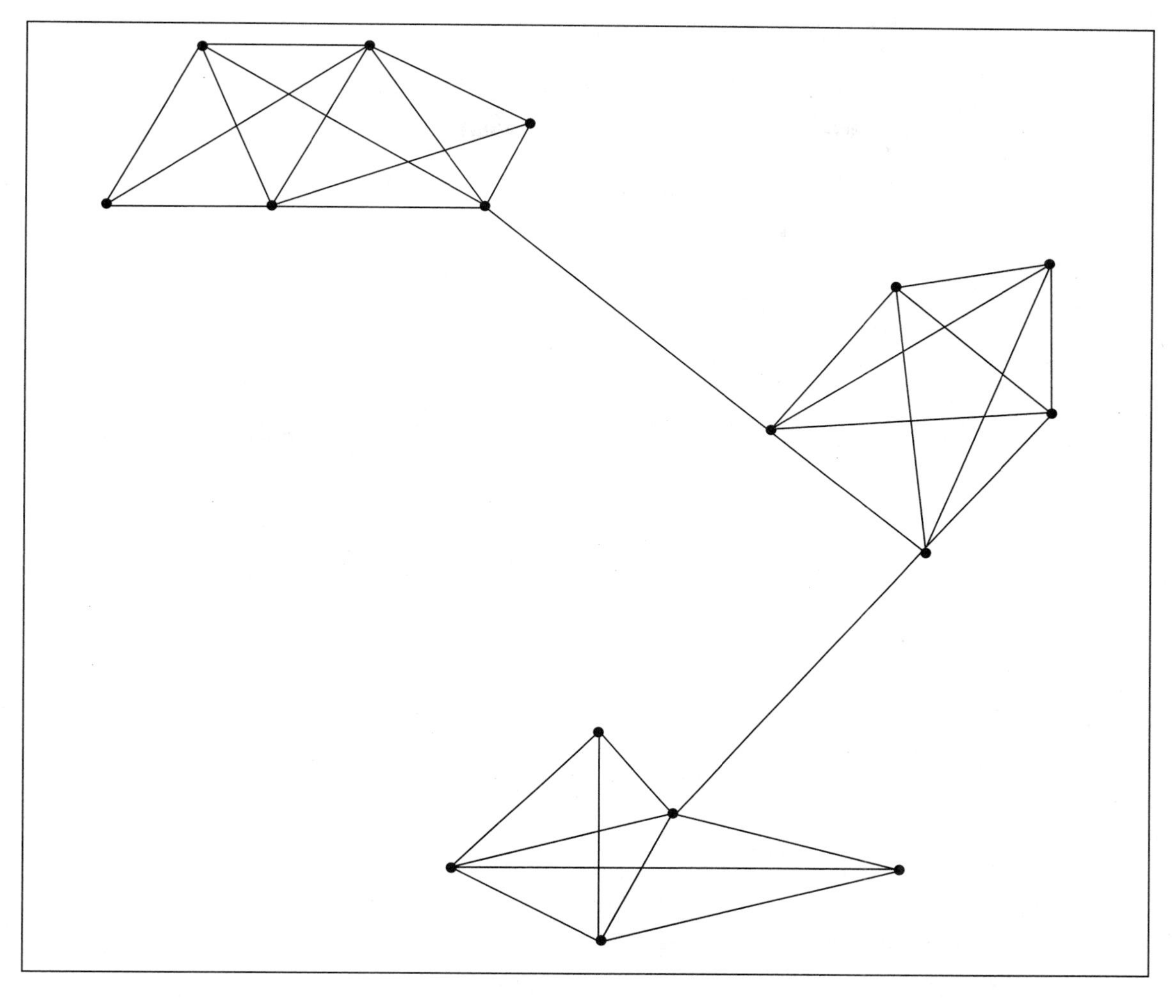

FIGURE 38-4 A Network with Three Distinct Cliques

not surprising that the number and nature of bridges within a network structure are highlighted in network analysis.

Brokerage

At times a particular position is outside subsets of positions but is crucial to the flow of resources to and from these subsets. This position is often in a brokerage situation because its activities determine the nature and level of resources that flow to and from subsets of positions.[6] In Figure 38-5, position A_7 is potentially a broker for the flow of resources from subsets consisting of positions A_1, A_2, A_3, A_4, and A_5 to B_1, B_2, B_3, B_4, B_5, and B_6. Position A_7 can become a broker if (1) the distinctive resources that pass to, and from, these two subsets are needed or valued by at least one of these subsets, and (2) direct ties, or bridges, between the two subsets do not exist. Indeed, a person or actor in a brokerage

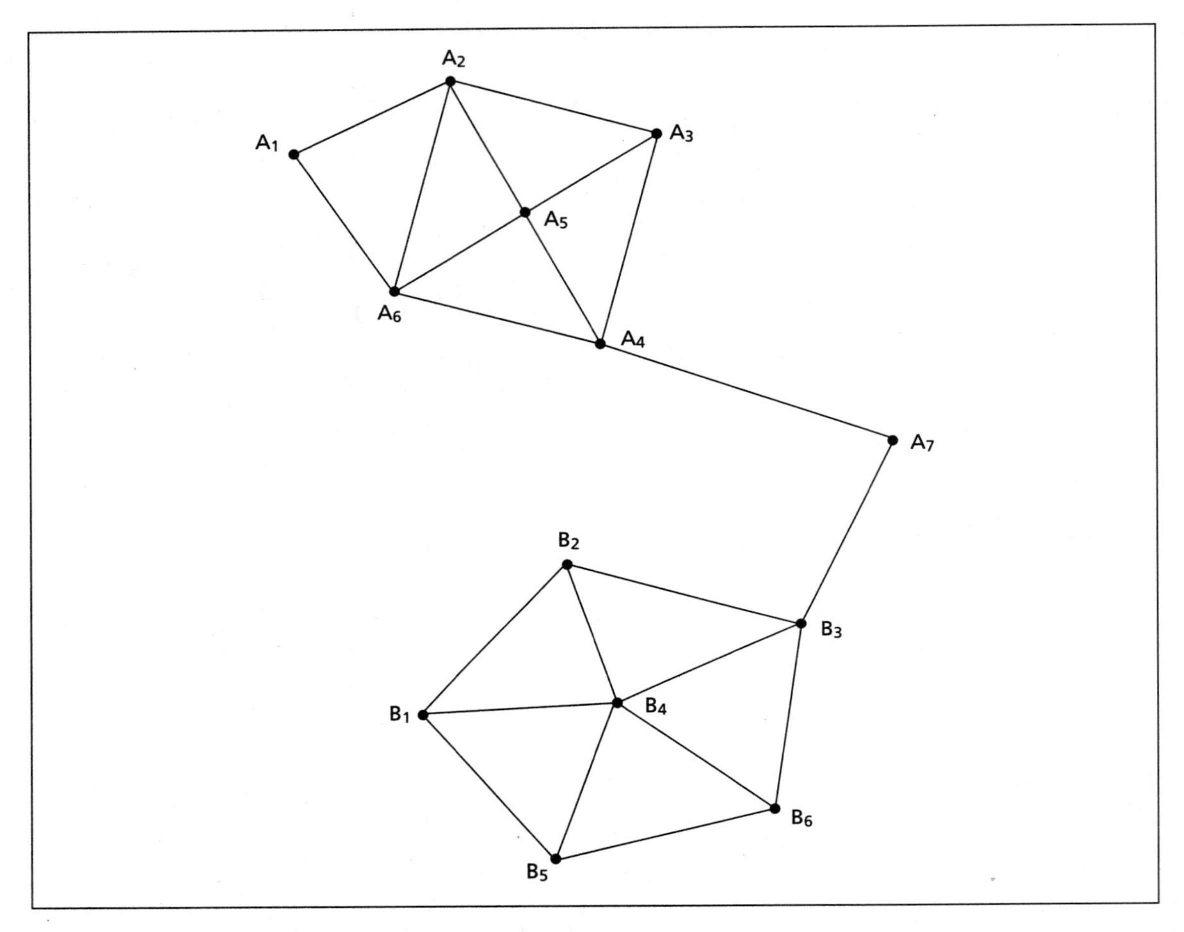

FIGURE 38-5 A Network with Brokerage Potential

position often seeks to prevent the development of bridges (like those in Figure 38-4) and to manipulate the flow of resources such that at least one, and if possible both, subsets are highly dependent on its activities.

Centrality

An extremely important property of a network is *centrality*, as was noted for Bavelas' and Leavitt's studies of communication in experimental groups. There are several ways to calculate centrality:[7] (1) the number of other positions with which a particular position is connected, (2) the number of points between which a position falls, and (3) the closeness of a position to others in a network. Although these three measures might denote somewhat different points as central, the theoretical idea is fairly straightforward: Some positions in a network mediate the flow of resources by virtue of their patterns of ties to other points. For example, in Figure 38-3 (b), point *C* is central in a network consisting of positions *A, B, C, D,* and *E*; or, to take another example, points A_5 and B_4 in Figure 38-5 are more central than other positions because they are directly connected to all, or

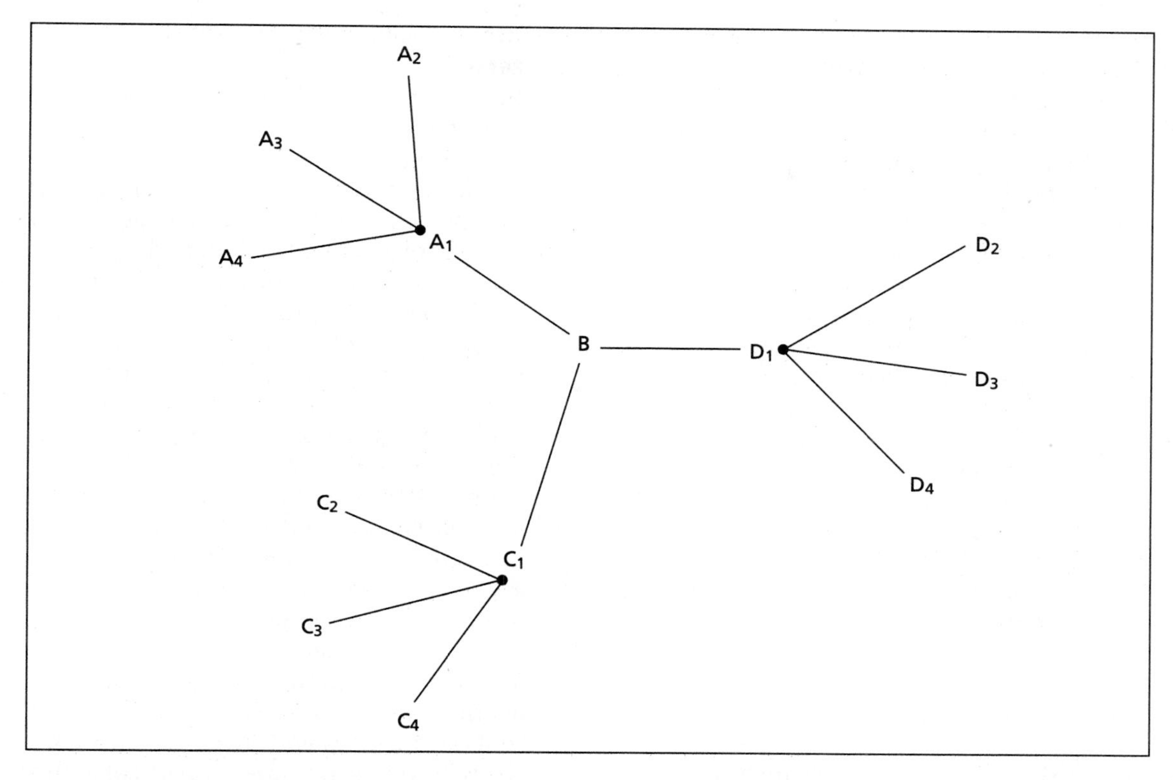

FIGURE 38-6 Equivalence in Social Networks

to the most, positions and because a higher proportion of resources will tend to pass through these positions. A network can also reveal several nodes of centrality, as is evident in Figure 38-6. Moreover, patterns of centrality can shift over time. Thus many of the dynamics of network structure revolve around the nature and pattern of centrality.

Equivalence

When positions stand in the same relation to another position, they are considered *equivalent*. When this idea was first introduced into network analysis, it was termed *structural equivalence* and restricted to situations in which a set of positions is connected to another position or set of positions in exactly the same way.[8] For example, positions C_2, C_3, and C_4 in Figure 38-6 are structurally equivalent because they reveal the same relation to position C_1. Figure 38-6 provides another illustration of structural equivalence, as well. A_2, A_3, and A_4 are structurally equivalent to A_1; similarly, D_2, D_3, and D_4 are equivalent to D_1; and A_1, C_1, and D_1 are structurally equivalent to B.

This original formulation of equivalence was limited, however, in that positions could be equivalent only when *actually connected to the same position*. We might also want to consider all positions as equivalent when they are connected to different positions but in the same form, pattern, or manner. For instance, in Figure 38-6, A_2, A_3, A_4, D_2, D_3, D_4, C_2, C_3, and C_4 can all be seen as equivalent because they bear the *same type* of relation to another position—that is to A_1, D_1, and C_1, respectively. This way of conceptualizing equivalence is termed *regular equivalence*[9] and, in a sense, subsumes the original notion of *structural equivalence.* That is, structural equivalence, wherein the equivalent *positions must actually be connected to the same position in the same way*, is a particular type of a more general equivalence phenomenon. These terms, "structural" and "regular," are awkward, but they have become conventional in network analysis, so we are stuck with them. The critical idea is that the number and nature of equivalent positions in a network have important influences on the dynamics of the network.[10] The general hypothesis is that actors in structurally equivalent or regularly equivalent positions will behave or act in similar ways.

CONCLUSION

The mathematics of network analysis can become quite complicated, as can the computer algorithms used to analyze data sets of the processes outlined above. This listing of concepts is somewhat metaphorical because it eliminates the formal and quantitative thrust of much network analysis. Indeed, much network analysis bypasses the conversion of matrices into graphs like those in the various figures presented and, instead, performs mathematical and statistical operations on just the matrices themselves. Yet, if network analysis is to realize its full theoretical (as opposed to methodological) potential, it might be wise to use concepts, at least initially, in a more verbal and intuitive sense.

Few would disagree with the notion that social structure is composed of relations among positions. But is this all that social structure is? Can the concepts denoting nodes, ties, and patterns of ties (number, strength, reciprocity, transitivity, bridges, brokerage, centrality, and equivalence) capture all the critical properties of social structure?

The answer to these questions is probably "no." Social structure probably involves other crucial processes that are not captured by these concepts. Yet a major property of social structure *is* its network characteristics, as Georg Simmel was perhaps the first to really appreciate. For, whatever other dimensions social structure might reveal—cultural, behavioral, ecological, temporal, psychological, and so forth—its backbone is a system of interconnections among actors who occupy positions relative to one another and who exchange resources. And, so, network analysis has great potential for theories of social structure. Has this potential been realized? Probably not, for several reasons.

First, as just noted, network analysis is overly methodological and concerned with generating quantitative techniques for arraying data in matrices and then converting the matrices into descriptions of particular networks (whether as graphs or as equations). As long as this is the case, network sociology will remain primarily a tool for empirical description.

Second, there has been little effort to develop principles of network dynamics, per se. Few[11] seem to ask theoretical questions within the network tradition itself. For example, how does the degree of density, centrality, equivalence, bridging, and brokerage influence the nature of the network and the flow of relations among positions in the network? There are many empirical descriptions of events that touch on this question but few actual theoretical laws or principles.[12]

Third, network sociology has yet to translate traditional theoretical concerns and concepts into network terminology in a way that highlights the superiority, or at least the viability, of using network theoretical constructs for mainstream theory in sociology. For example, power, hierarchy, differentiation, integration, stratification, conflict, and many other concerns of sociological theory have not been adequately reconceptualized in network terms, and hence it is unlikely that sociological theory will adopt or incorporate a network approach until this translation of traditional questions occurs.

All these points, however, need to be qualified because numerous sociologists have actually sought to develop laws of network processes and to address traditional theoretical concerns with network concepts. Although these efforts are far from constituting a coherent theory of network dynamics, they do illustrate the potential utility of network sociology, as we saw, for example, in our review of network exchange theory in Chapter 25.

NOTES

1. For some readable overviews on network analysis, see Barry Wellman, "Network Analysis: Some Basic Principles," *Sociological Theory* (1983), pp. 155–200; Jeremy F. Boissevain and J. Clyde Mitchell, eds., *Network Analysis* (The Hague: Mouton, 1973) and *Social Networks in Urban Situations* (Manchester: Manchester University Press, 1969); J.A. Barnes, "Social Networks" (Addison-Wesley Module, no. 26, 1972); Barry S. Wellman and S.D. Berkowitz, *Social Structures: A Network Approach* (Cambridge: Cambridge University Press, 1988). Somewhat more technical summaries of recent network research can be found in Samuel Leinhardt, ed., *Social Networks: A Developing Paradigm* (New York: Academic, 1977); Paul Holland and Samuel Leinhardt, eds., *Perspectives in Social Network Research* (New York: Academic, 1979); Ronald S. Burt, "Models of Network Structure," *Annual Review of Sociology* 6 (1980), pp. 79–141; Peter Marsden and Nan Lin, eds., *Social Structure and Network Analysis* (Newbury Park, CA: Sage, 1982). For advanced research on networks, consult recent issues of the journal *Social Networks*.

2. There are other ways to measure density; this definition is meant to be illustrative of the general idea.

3. The terminology on subdensities varies. "Clique" is still the most prominent term, but "alliances" has been offered as an alternative. Moreover, the old sociological standbys "group" and "subgroup" seem to have made a comeback in network analysis.

4. At one time, "intensity" appears to have been used in preference to "strength." See Mitchell, "The Concept and Use of Social Networks." It appears that Granovetter's classic article shifted usage in favor of "strength" and "weakness." See note 5.

5. See Mark Granovetter, "The Strength of Weak Ties," *American Journal of Sociology* 78 (1973), pp. 1360–1380; and "The Strength of Weak Ties: A Network Theory Revisited," *Sociological Theory* (1983), pp. 201–233. The basic network "law" from Granovetter's original study can be expressed as follows: *The degree of integration of a network composed of highly dense subcliques is a positive function of the extensiveness of bridges, involving weak ties, among these subcliques.*

6. Ronald S. Burt has, perhaps, done the most interesting work here. See, for example, his *Toward a Structural Theory of Action* (New York: Academic, 1982) and "A Structural Theory of Interlocking Corporate Directorships," *Social Networks* 1 (1978–1979), pp. 415–435.

7. The definitive works here are Linton C. Freeman, "Centrality in Social Networks: Conceptual Clarification," *Social Networks* 1 (1979), pp. 215–239; and Linton C. Freeman, Douglas Boeder, and Robert R. Mulholland, "Centrality in Social Networks: Experimental Results," *Social Networks* 2 (1979),

pp. 119–141. See also Linton C. Freeman, "Centered Graphs and the Structure of Ego Networks," *Mathematical Social Sciences* 3 (1982), pp. 291–304, and Philip Bonacich, "Power and Centrality: A Family of Measures," *American Journal of Sociology* 92 (1987), pp. 1170–1182.

8. François Lorrain and Harrison C. White, "Structural Equivalence of Individuals in Social Networks," *Journal of Mathematical Sociology* 1 (1971), pp. 49–80; Harrison C. White, Scott A. Boorman, and Ronald L. Breiger, "Social Structure from Multiple Networks: I. Block Models of Roles and Positions," *American Journal of Sociology* 8 (1976), pp. 730–780.

9. Lee Douglas Sailer, "Structural Equivalence," *Social Networks* 1 (1978), pp. 73–90.

10. In many ways Karl Marx's idea that those who stand in a common relationship to the means of production have common interests is an equivalence agreement. Thus, the idea of equivalence is not new to sociology—just the formalism used to express it is new.

11. There are, of course, some notable exceptions to this statement. For an example of what we see as the kinds of laws that need to be formulated, see our formal statement of Granovetter's hypothesis in note 5.

12. Mark Granovetter, "The Theory-Gap in Social Network Analysis" in *Perspectives on Social Network Research*, eds. P. Holland and S. Leinhardt (New York: Academic, 1979).

39

The Continuing Tradition IV: Peter M. Blau's Macrostructural Theory

Peter M. Blau's theory of exchange was examined in Chapter 21. In the late 1970s, however, Blau abandoned his exchange strategy and began to advocate a purely macrostructural form of analysis. Rather than view macro-level phenomena as built from interpersonal exchange processes, he argued that the structural properties of a population cannot be conceptualized as their micro-foundations. Indeed, it is inappropriate to even conceptualize the macro-level structural properties of populations as interpersonal social processes. In his new vision, interpersonal processes among individuals are constrained by the emergent properties of macrostructure, and theory should concentrate on how macrostructural forces increasingly affect ever more micro-level phenomena rather than the reverse. In 1977, Blau made this forceful argument in his book, *Inequality and Heterogeneity*,[1] and in 1994 he presented the argument again after further reflection and in response to his critics.[2] In this chapter, we will present the theory in its more recent formulation, although we will draw on ideas in the first statement of the theory.

BLAU'S CONCEPTION OF MACROSTRUCTURE

Blau's vision of macrostructure is deceptively simple: The members of a population are distributed in a number of positions that, in turn, circumscribe their chances and opportunities for social association. Thus, macrostructure offers both opportunities for social association and constraints on who is likely to associate with whom. The goal of theory is to conceptualize the properties of macrostructure and, in so doing, to develop some basic laws of macrostructural dynamics.[3]

Again, Blau offers very simple definitions of key concepts, which, as we will see, have far reaching implications. The first key concept is the notion of *parameter* or *structural parameter* by which he means those characteristics that members of a population use to make distinctions among themselves. There are two types of parameters: (1) *nominal parameters* that distinguish members of a population by discrete categories, such as people's gender, ethnicity, and religion; and (2) *graduated parameters* that place members along a continuous scale or rank order, such as their income, years of education, age, power, wealth, socioeconomic status, prestige, and other graduated scales that rank order people.

For either of these types of parameters, members of a population are distributed. For example, a certain number of individuals will be counted as members of various ethnic groups, or varying numbers of individuals will be located at different income levels. Thus, at the most general level, macrostructure can be conceptualized as the distributions of population members across all of the nominal and graduated parameters that distinguish people. The dynamics of macrostructure ultimately inhere, therefore, in the number of parameters, their nature as either nominal or graduated, and the distributions of population members among these parameters.

Nominal parameters are population distributions among discrete categories, or "groups," which is the term Blau uses to denote all those who are distinguished by a nominal parameter. The degree of differentiation among nominal parameters determines the *level of heterogeneity* of macrostructure, which is defined as "the chance expectation that two randomly chosen persons belong to different groups."[4] The greater is the number of categories into which people can be placed, and the more even their distribution among these categories is, the greater their heterogeneity is.

Graduated parameters are distributions of people on rank-ordered scales. A population's differentiation by such graduated parameters points to the *level of inequality* in the population, which is defined as "the chance expectation that the absolute difference in given resources between two randomly chosen persons relative to the mean resource difference in the population."[5] In other words, the more the resources of two randomly chosen individuals diverge from the population average on some graduated parameter, such as income, the greater the inequality is in the distribution of this resource. The theoretical minimum of inequality for a resource would be an even distribution of resources along a scale, whereas the maximum would be for all resources of a kind to be held by one person. And, so the more resources are held in the fewer hands, the more unequal their distribution is.

Thus far, macrostructure is viewed by respective distributions of population members across all the nominal and graduated parameters that distinguish people. In turn, the number of parameters and the respective distributions of people at various grades and in particular categories determine, respectively, the degree of inequality and heterogeneity. The most important emergent property of macrostructure, Blau argues, is the degree to which parameters are correlated. Does a position on one parameter predict the position on another? For example, if members of an ethnic group are disproportionately located on lower positions of graduated parameters, such as income, wealth, prestige, and power, then parameters are correlated with each other. The higher the correlation among parameters, the more they are, in Blau's terms, *consolidated* with each other, and the more consolidated parameters are, the greater barrier they pose to social interaction and social mobility among the diverse members of a population. Conversely, when the correlation among parameters is low or nonexistent, the greater the *intersection* of parameters is, and the more they promote social relations and mobility among differentiated members of a population.

Thus, macrostructure is conceptualized by Blau by the degree of differentiation and distribution of a population along nominal and graduated parameters that determine the degree of heterogeneity and inequality. The degree of consolidation (high correlation) or intersection (low correlation) among these parameters will influence the rates of contact, interaction, and mobility among the members of a population. Thus, the three key dynamics of macrostructure revolve around (1) heterogeneity, (2) inequality, and (3) intersection or consolidation as they influence rates of social association. These ideas are, as we noted in Chapter 34 and as Blau acknowledges, extensions of Georg Simmel's early insight that the intersection of various social affiliations is an important property of social structure. Like Simmel, Blau's goal is to extract the underlying form of relationship among the properties of social structure and to develop explicit propositions that explain these basic forms. Thus, the power of Blau's theory resides in the propositions that he develops from this view of macrostructure.

THE FORMAL THEORY OF MACROSTRUCTURE

The Theoretical Strategy

In all his work, Blau has been committed to developing laws of human organization. In his view, scientific theory is a "hierarchical system of propositions of increasing levels of generality. All lower order propositions follow in strict logic from higher order ones alone."[6] In developing such a theory, several steps are critical, and their successful implementation is what typifies good theory.

First, theory must begin with a conception of the basic subject matter—that is, what is to be explained? In deductive theory the phenomenon to be explained by a theory is termed the *explicandum*. Blau calls the explicandum the "pattern of relations between people in society, particularly those of their social relations that integrate society's diverse groups and strata into a distinct coherent social structure."[7]

In addressing this range of phenomena, Blau offers a narrow definition of macrostructure. Discussions of social structure, Blau notes, typically recognize "that there are differences in social positions, that there are social relations among these positions, and that people's positions and corresponding roles influence their social relations."[8] But most theoretical efforts extend the conceptual inventory, adding such notions as psychological needs and cultural values, beliefs, and norms. In contrast with these tendencies, Blau prefers to postulate a narrow view of social structure as consisting of the distributions of people among different social positions and their associations.

> To speak of social structure is to speak of differentiation among people. For social structures, as conceptualized, are rooted in the social distinctions people make in their role relations and social associations.[9]

Yet, when one examines an entire society, a community, or any macrostructure, analysis becomes more complicated. The number of people and positions to be analyzed is great. Moreover, people occupy many different positions simultaneously—religious, familial, neighborhood, work, recreational, political, educational, and the like. Thus, for each type of position, there is a separate distribution of the population, making a macrostructure a "multidimensional space of social positions among which people are distributed and which affect their social relations."[10] But how are all these relations among large numbers of people occupying simultaneously many different positions to be analyzed? In Blau's portrayal, network analysis would analyze each and every relation among all people in all positions, but, when the numbers of people, positions, and relations become large, macrostructural analysis must be concerned with the general "patterns of social relations among different social positions occupied by many persons, not with the networks of all relations between individuals."[11] This is done by defining social positions by common attributes of people—age, sex, ethnicity, occupation, income, religion—and then examining their overall *rates* of association. Note that not every person or association is examined—only general categories or strata of people and overall rates of association among people in these categories. In this way, macrostructural theory is distinguished from the microstructural approaches, sociometry, and network analysis.

A second critical element in building theory is "familiarity with relevant empirical knowledge." The often rigid line between deduction and induction is artificial, Blau argues, because deduction rarely occurs in an empirical vacuum. Indeed, the very questions to be answered by a deductive theory are typically the result of interesting empirical findings. Thus, to know what it is that a theory should explain requires familiarity with existing empirical generalizations.

A third element in a theory is the "creative insight" that is typically embodied in a "central theoretical term" or "operator." An operator "supplies the organizing principle for the major theoretical propositions."[12] One central operator is the *intersection of parameters,* which reduces divisions in a society by enabling people to associate with a diversity of others in at least some of their positions, and another is the *consolidation of parameters,* which tends to create divisions in a society because positions cluster together in ways that create barriers to associations with others outside a particular circle of associations. Thus, for Blau, "the degree to which parameters intersect, or alternatively consolidate differences in social positions through their strong correlations, reflects the most important structural conditions in a society, which have crucial consequences for conflict and for social integration."[13]

A fourth central element of theory building is conceptual clarification or key terms, especially of the operator. For Blau, such clarification involves distinguishing patterns of differentiation between two basic types of parameters. As we have seen, one basic type is a "nominal parameter," which divides people into different groups that are not rank ordered, and the other type is the "graduated parameter," which differentiates people by hierarchical status rankings. Thus, all macrostructures reveal varying degrees of differentiation for nominal and graduated parameters. When people are differentiated into many groups by numerous nominal parameters, then a high degree of "heterogeneity" exists, and when people are differentiated into widely different status positions on the basis of graduated parameters, then "inequality" is high.[14]

A fifth important element in theory construction is the development of a deductive system—that is, the incorporation of concepts into propositions and then the hierarchical ordering of them. Although Blau uses the vocabulary of axiomatic theory, his system of propositions does not evidence the rigor of true axiomatic theory, but instead, it is a system of formal statements that are only loosely "deducted" from higher order axioms. Thus, in Blau's view, the ordering of propositions involves (a) the articulation of several simple assumptions as axioms and (b) the development of theorems that apply these assumptions to generic types of social relations.

The Basic Assumptions and Theorems

In Blau's initial formulation of the theory,[15] he presented many assumptions (what he saw as "axioms") and theorems. Soon, the whole system became rather unwieldy as assumptions and theorems proliferated.[16] In the more recent version of the theory, Blau has dramatically reduced the assumptions to three, and the theorems that follow from these assumptions to fourteen.[17] The theory is now much more manageable. Let us begin with the assumptions, then move to the theorems that Blau derives from these assumptions.

TABLE 39-1: Blau's Basic Assumptions

A-1: The probability of social association among individuals depends on their opportunities for contact.

A-2: The proximity of individuals in multidimensional space increases the probability of social associations.

- A-2(1): The rates of associations of persons in the same nominal position are higher than their rates with outsiders.
- A-2(2): The average social distance in graduated positions between associates is lower than in the population at large.

A-3: Associates in other groups or strata facilitate mobility of associates to these groups or strata.

The Basic Assumptions. The first assumption in Blau's view of the social universe is that the probability of association among individuals in a population depends on their opportunities for social contact (A-1 in Table 39-1). This assumption seems rather obvious, of course, but it is fundamental to what Blau wants to do: explain how macrostructure influences the opportunities for contact among individuals. To realize this goal, he introduced a second basic assumption that the proximity of individuals in a multidimensional space (organized by nominal and graduated parameters) increases the probability of social associations among individuals (A-2 in Table 39-1). To this assumption, he added two subassumptions: (1) Rates of association among people in the same position as determined by a nominal parameter will be higher than their rates with outsiders, and (2) average social distance among individuals in a position established by a graduated parameter will be lower than with the population at large. Later, after introducing several theorems, Blau added a third assumption, which we will state here to keep all the assumptions together. This third assumption states that associates in other groups (as determined by nominal parameters) and strata (as defined by graduated parameters) will facilitate movement or mobility to these groups and strata (A-3 in Table 39-1).

What, then, is the image connoted by these assumptions which, for convenience are listed in Table 39-1? For people to associate and move, they must have opportunities to do so; these opportunities are determined by location in macrostructure as determined by nominal and graduated parameters; people will tend to associate with those close to them in multidimensional space because they have the most opportunity to do so; yet, when opportunities to know and associate with others in more distant groups and strata are present, these associations often become the bridges for mobility to new groups and strata. With these very simple assumptions, Blau then developed a series of theorems. These theorems explain how the dynamics inhering in the properties of macrostructure—that is, heterogeneity, inequality, and intersection—influence rates of social association and mobility among individuals.

The Basic Theorems. The theorems are listed in their numerical order in Table 39-2 on page 539, but we will discuss them somewhat out of order because Blau does not introduce them in numerical sequence. Theorem 1 (T-1 in Table 39-2) is not derived from any of the assumptions but, instead, is a mathematical truism: The rate of association of members in one group with those of another, as defined by some parameter, is inversely related with its size relative to the other group. Thus, members of a large group will have in all probability lower rates of association with members of a smaller group than the reverse. The reason for this truism is that if members of two groups of varying size have relations, the smaller group has fewer people who can have relations, whereas the large group has more people who can have relations; as a result, the same number of reciprocal relations between members of these two groups will constitute a greater proportion of the smaller group's total possible relations than the larger group's total relations. This basic truism is translated into the theorem that, probabilistically, throughout all the groups determined by parameters dividing and classifying population members, the rates of interaction and group size will be inversely related. Thus, for example, because blacks in the United States constitute a much smaller group than whites (as defined by the parameter of skin color), the rates of association of blacks with whites will, on average, be higher than those of whites with blacks. This truism is just that—a truism—but it has many implications for the dynamics of macrostructure, because it affects rates of association and mobility among groups in a population.

The second theorem follows from Assumption 1: The greater the level of heterogeneity is, the greater chances are for fortuitous encounters involving members of different groups. The logic behind this theorem is that, if social associations depend on opportunities for contact and if heterogeneity is defined as the chance that two randomly selected individuals will belong to different groups, then it is more likely that people from different groups will come into contact. Many of these contacts will be of low salience and soon forgotten, but heterogeneity creates opportunities for contact and, hence, the development of more lasting relations—which has important implications for other macrostructural processes.

The third theorem is the same as the second, but this time the underlying assumption is applied to inequality. The theorem is, however, less intuitively obvious. It states, in essence, that the probability of social contacts and associations among people from different positions will increase with inequality. Again, social associations depend on opportunities for contact, and because inequality is defined as the chance that two randomly selected individuals will diverge from the average level of resources held by members of a population, it is more likely, therefore, that people with very different resources on some graduated parameter will come into contact under conditions of inequality.

The fourth theorem is about mobility, which involves the movement of individuals from one position to another, as defined by either nominal or graduated parameters. This mobility can take many substantive forms, such as geographical movement, migration into a country or out of it, movement to new occupations, movement to new positions of power or prestige, increases in income, and in general, the ability to occupy new positions in groups defined by nominal parameters or places on rank-ordered scales as defined by graduated parameters. The theorem states that mobility increases the chances of intergroup and interstatus contact. The reason for this tendency is that as people move, they will tend to keep some of their old associations at the same time that they acquire new ones, and this increases the probability that both the old and new associates will, themselves, come into direct contact.

Blau then skipped to theorem 11 on the intersection of parameters. This theorem states that the more pronounced the intersection of parameters is—that is, the less consolidated they are and, hence, the less a position on one parameter is correlated with positions on other parameters—the more pronounced the intersection of social differences will be and, therefore, the more people's choices of ingroup associations will also involve them in intergroup associations. From this basic idea, Blau developed a set of important corollaries, although a little later in his exposition of the revised theory. These corollaries concern what he terms *penetrating differentiation*, or the extent to which differentiation at one level of macrostructure is retained at a lower level. For example, if differentiation of ethnicity at the national level is retained or replicated at the community level, then the differentiation has penetrated a lower level structure. But if differentiation of ethnicity at a national level is not retained at the community level—that is, different ethnic groups live in different communities—then the differentiation has not penetrated to the lower level structure. Actually, the parameter of ethnicity has become consolidated with another parameter, community of residence. In Blau's view, the more differences can penetrate into successive levels of structure—say, for example, ethnic differences at the national level are reproduced at state, community, neighborhood, household levels—the greater will be the associations among diverse members of the population. To phrase these corollaries more or less in Blau's terms: (1) The more heterogeneity penetrates into lower-level substructures, the more probable are intergroup relations; (2) the more inequality penetrates into lower-level substructures, the greater is the likelihood of status distance between associates; (3) the more intersecting differences penetrate into lower-level substructures, the more probable are intergroup relations; (4) the rates of inter-subunit associations increase with increasing penetration of heterogeneity into these subunits; (5) similarly, rates of inter-subunit associations increase with penetration of inequality into such subunits; and (6) rates of inter-subunit associations increase with penetration of intersections of other parameters into such subunits.

As they flow from the mathematical truism and the first two assumptions, this set of four theorems—T-1, T-2, T-3, and T-11—are the core of Blau's theory, although the six corollaries to Theorem 11 on penetrating differentiation should be included in this core. Blau added Assumption 3 (as shown in Table 39-1) as he developed other theorems on social mobility and conflict, but in a real sense these theorems follow from the core.

The first set of these more derivative theorems (T-6 through T-10 and T-12) concerns mobility processes, or the movement of individuals horizontally and vertically in the multidimensional space created by nominal and graduated parameters. Theorem 6 uses Assumption 3 and states that high rates of intergroup and status-distant contacts increase the chances of mobility from one group or status to another. The next two theorems, numbers 7 and 8, state what is implicit in Theorem 6, namely, that heterogeneity and inequality increase the probability of mobility. The reason for this conclusion is that heterogeneity and inequality increase rates of intergroup and status-distant relations, and because associates in other groups and strata facilitate mobility to these positions (Assumption 3), associations generated by heterogeneity and inequality will be likely to increase rates of mobility. Theorem 12 follows the same logic in asserting that intersection of parameters increases rates of mobility because such intersection increases intergroup and inter-strata associations.

The final theorem on mobility—number 5—covers how the varying sizes of groups and strata affect mobility. Because the rates of association of members in the smaller group or strata will be higher than those of members in the larger group or strata (the mathematical truism specified in Theorem 1), the large size of a group or strata distinguished by a nominal or graduated parameter will lower rates of mobility, and conversely, small size will increase the probability of mobility. Somewhat later in his presentation, Blau examined the effects of mobility on the respective sizes of groups or strata, arguing in Theorem 9 that an excess of moves from larger to smaller groups raises heterogeneity, and conversely, large-scale movements from smaller to larger groups increases homogeneity. For example, if whites flee the central city in large numbers to smaller suburbs, the heterogeneity of the city increases (because whites and other ethnic populations are now more equal in their respective numbers and, hence, more likely to be selected randomly), and the homogeneity of the smaller suburb escalates (because the larger proportion of whites increases the probability that only whites would be randomly selected). The same logic applies to inequality, although Theorem 10 is stated in a more complicated form: Mobility up or down strata specified by graduated parameters toward the boundary that separates the upper and middle strata will reduce inequality because this movement is toward the largest segment of the distribution. The reason for this conclusion resides in the shape of distributions revolving around inequality. Such distributions will be pyramids, with the very few at the top of the resource distribution, and ever more toward the middle and bottom. Hence, mobility of top members down increases the size of middle strata, as does mobility up from the bottom. As the middle stratum becomes more populous, the probability of selecting at random two individuals from this same stratum increases, thereby indicating that the level of inequality has declined.

Blau also embarked on a discussion of conflict, but he introduced only one new theorem in doing so. His argument is that as rates of association increase, so do potential rates of conflict because conflict is, indeed, a type of social relation. Hence, from Theorem 1, members of smaller groups or strata are more likely to be involved in conflict, whether as victims or malefactors. From Theorems 2 and 3, both heterogeneity and inequality are also likely to increase rates of conflict, and from Theorem 4, higher rates of mobility will likely increase the incidence of conflict associations. Theorem 13 is new and addresses the question of what consolidation and intersection of parameters do to conflict. This is an important theorem because it qualifies the effects on conflict specified in Theorems 1, 2, 3, and 4. The argument is that the consolidation of parameters intensifies conflict in frequency and animosity, whereas intersection decreases the intensity of conflict. The argument comes from Simmel who observed that cross-cutting cleavages keep conflicts in bounds because those in conflict along one parameter might be in a more positive association along another. For example, members of different ethnic groups might be colleagues at work, thereby taking some of the edge off ethnic conflict.

These, then, are the theorems in Blau's theory. We have not examined them in order because Blau does not present them in order, which, of course, raises the question of why they were numbered the way that they were. Nonetheless, Table 39-2 presents the theorems in numerical order, and by reading down the theorems, a sense of what Blau is trying to communicate emerges. Although the theorems are very abstract, they have empirical implications for understanding rates of association among members of groups and strata, for rates of mobility across groups and strata, and for the rates and intensity of conflict between groups and strata. Let us take a couple of hypothetical examples of how the theorems can be used to understand substantive processes.

Using the Theorems to Explain Empirical Events. If, for example, a society reveals high rates of mobility toward the middle

TABLE 39-2: Blau's Basic Theorems

T-1: The probability of intergroup relations declines with proportionate increases in group size (from a mathematical truism).

T-2: The greater the heterogeneity, the greater the chances are that fortuitous encounters involve members of different groups (from A-1, shown in Table 39-1).

T-3: The probability of status-distant social contacts and associations increases with increasing inequality (from A-1).

T-4: Mobility improves chances of intergroup contact because mobile persons are likely to bring their old and new contacts together (from A-1).

T-5: The smaller the group, the better the chances of mobility by its members have (from T-1 and A-3).

T-6: High rates of intergroup or status-distant relations increase the probability of social mobility (from A-3 and T-3).

T-7: Heterogeneity increases the probability of intergroup mobility (from A-1 and T-2).

T-8: Inequality increases the probability of status-distant mobility (from A-1 and T-3).

T-9: An excess of moves from larger to smaller groups raises, and an excess of moves from smaller to larger ones lowers, the level of heterogeneity (from definition of heterogeneity and T-1).

T-10: Mobility, up or down, toward the boundary between the upper and middle classes reduces inequality (from definition of inequality and T-1).

T-11: The more pronounced the intersection of social differences is, the greater the probability is that people's ingroup choices involve them in intergroup relations (from A-1, A-2, and T-2).

- T-11(1): The more heterogeneity penetrates into lower-level substructures, the more probable are intergroup relations.
- T-11(2): The more inequality penetrates into low-level substructures, the greater is the likely status distance between associates.
- T-11(3): The more intersecting differences penetrate into low-level substructures, the more probable are intergroup relations.
- T-11(4): The rates of inter-subunit associations increase with increasing penetration of heterogeneity.
- T-11(5): The rates of inter-subunit associations increase with increasing penetration of inequality.
- T-11(6): The rates of inter-subunit associations increase with increasing penetration of intersection.

T-12: Multiple intersection of parameters makes mobility more probable (from A-3 and T-2).

T-13: Intersecting social differences reduce the likelihood of intergroup conflict (from A-1, A-2, and T-2).

T-14: Mobility increases the consolidation of social differences (from A-2 and A-3).

strata, what would be our prediction about rates of contact and conflict among members of different strata? First, we would want to know the respective sizes of the various strata specified by the graduated parameter (say, income), and second, we would want to know how much movement up or down these strata was actually occurring. But even without this more fine-grained knowledge, we know in general terms that the distribution of people in various rank-ordered strata will look like a pyramid, so that mobility will reduce inequality and, at the same time, reduce rates of association and conflict among those in status-distant positions. As the size of the middle stratum increases, people will be more likely to interact with other middle stratum members, although upper-strata people (as members of an ever smaller group) will have higher rates of interaction with those below them than the reverse. Similarly, if the lower stratum is reduced in size as a result of mobility to the middle strata, their rates of association with members of the middle stratum will exceed those of the middle to lower. Thus, members in the middle stratum will be more likely to associate with each other, losing contact with the upper and lower strata. As this contact is lost, rates of association, conflict, and mobility to the upper and lower strata will decline, unless some exogenous economic or political force alters the mobility pattern. Such a pattern of mobility was typical of industrial nations in the first two-thirds of the twentieth century, and it helps explain why the middle classes have little understanding of the wealthy and rather low appreciation for the problems of the poor and disadvantaged at the bottom of the income distribution.

To take another example, suppose that there are high rates of immigration into large cities by a particular ethnic group. What could we predict for the city? Again, we would want to know the respective sizes of various ethnic groups already in the city as well as the rate of immigration for the mobile ethnic group. But even without this knowledge, we can make some predictions. As the size of the immigrant population increases, so does heterogeneity, and, hence, so will rates of association and conflict among ethnic groups. And if ethnicity is consolidated with another graduated parameter such as income or power, then the intensity of such conflicts should increase. Or if the city-wide heterogeneity is not replicated at the neighborhood level—that is, there is ethnic segregation by neighborhood—then rates of association will be less because the heterogeneity has not penetrated to the neighborhood substructures of the city, and rates of conflict and mobility will decline. Yet, because ethnicity is now consolidated with neighborhood as well as with graduated parameters like income and power, the intensity of conflicts when they do occur should increase.

It is not hard to plug recent experiences of American or any western society into these abstract scenarios, but the point should be clear: just knowing some very basic information about macrostructures—the parameters, the distribution of people across these parameters, the degree of intersection or consolidation of parameters, and the level of penetration of parameters—allows predictions to be made about important dynamics such as rates of association, conflict, and mobility. And these predications can be made without detailed knowledge of the culture, history, politics, and other unique features of the situation. This is the power of deductive theory, especially a theory that operates with a few key ideas.

Exogenous Forces and Blau's Theory of Macrostructure

Blau's theory cannot ignore history or the empirical particulars of situations because, as he recognizes, the number of parameters, their nature, the distribution of people in them, and the degree of consolidation, intersection, and penetration of parameters are influenced by historical forces.[18] Yet, in his most recent formulation of the theory, Blau sees two types of exogenous forces as most important: (1) demographic trends and (2) economic developments. Demographic trends influence rates of migration (both immigration and emigration), fertility, and mortality. These rates are important because they determine the respective sizes of groups and strata, and as Theorem 1 emphasizes, relative size of groups and strata has an enormous influence on rates of association and, as other theorems indicate, on mobility and conflict as well. Economic development determines the number of positions in the division of labor and rates of mobility to and from various positions. What these two exogenous forces do is load the variables in Blau's theory, telling us the number and nature of parameters, the respective sizes and rates of growth of groups and strata, and the movements of people across positions.

Once this information is known, the theory makes predictions about associations, conflict, and mobility trends. And it does so without delving into the micro-level interactions of individuals, without prying into people's psychological dispositions, without analyzing culture, and without addressing many other forces that are prominent in other structural theories, to say nothing of functional, interactionist, conflict, and critical theories. This is the great strength of the theory: its parsimony and emphasis on just a few critical macrostructural forces. This can also be considered a weak point because the theory ignores much that might be important in understanding macro-level processes. Yet, Blau has carried Simmel's structural program to a new level, refining and executing the formal sociology that Simmel advocated but only sporadically executed. Moreover, Blau has generated a testable theory—which is more than most structuralist theorists have been able to do. And as he summarizes in his recent formulation of the theory, tests of the approach have thus far been confirmatory.[19]

CONCLUSION

In his recent book, Blau has sought to demonstrate how his macrostructural theory can be used to shed light on his earlier work on exchange theory. Moreover, he has interpreted and criticized meso-level theories in such areas as organizations, network analysis, and human ecology in light of the insights generated by the assumptions and theorems of his theory. These arguments have not been discussed here, where concern has been on summarizing the theory itself, but the basic argument is the same for both micro-level and meso-level analysis: Interpersonal relations, process of exchange among individuals and groups, formation of networks, ecological distributions of organizational forms, structure and operation of organizations, processes of class formation, dynamics of ethnic relations, and all other social processes are ultimately constrained by heterogeneity, inequality, and intersection as Blau has conceptualized these in his theorems. Rather than view macrostructure as built from the micro-level and meso-level processes, it is more proper to view these processes as constrained by the nature and number of parameters, the distributions of population members across these parameters, and the patterns of intersection and penetration of the parameters into substructures. Blau does not argue, of course, that other processes are unimportant, only that they cannot be fully understood without knowledge of the macrostructural forces specified in his theory. Moreover, macrostructure constitutes an emergent reality that cannot be understood by the psychology of individuals, their exchanges of resources, or the organizational forms that they create; instead, to examine a population as a whole, it is necessary to focus theory on the properties of its macrostructure, ignoring for the moment the meso and micro processes that are also operating in the social universe. Blau has thus become the most forceful advocate of a purely macrostructural sociology.

NOTES

1. See Peter M. Blau, *Inequality and Heterogeneity: A Primitive Theory of Social Structure* (New York: Free Press, 1977); see also "A Macrosociological Theory of Social Structure," *American Journal of Sociology* 83 (July 1977), pp. 26–54; "Contrasting Theoretical Perspectives," in J.C. Alexander, B. Gisen, R. Münch, and N. J. Smelser, eds., *The Micro-Macro Link* (Berkeley: University of California Press, 1987); and "Structures of Social Positions and Structures of Social Relations," in J.H. Turner, ed., *Theory Building in Sociology* (Newbury Park, CA: Sage, 1988).

2. Peter M. Blau, *Structural Context of Opportunities* (Chicago: University of Chicago Press, 1994).

3. For a statement on Blau's theory-building strategy, see Peter M. Blau, "Elements of Sociological Theorizing," *Humboldt Journal of Social Relations* 7 (Fall-Winter, 1979–1980), p. 105. For a recent review of how Blau developed his theory, see Peter M. Blau, "A Circuitous Path to Macrostructural Theory," *Annual Review of Sociology* 21 (1995).

4. Blau, *Structural Context* (cited in note 2), pp. 13–14.

5. Ibid., p. 14.

6. Blau, "Elements of Sociological Theorizing" (cited in note 3).

7. Ibid., p. 112.

8. Blau, "A Macrosociological Theory of Social Structure" (cited in note 1), p. 27.

9. Ibid., p. 28.

10. Ibid.

11. Ibid.

12. Blau, "Elements of Sociological Theorizing" (cited in note 3), p. 107.

13. Blau, "A Macrosociological Theory of Social Structure" (cited in note 1), p. 32.

14. Blau, *Inequality and Heterogeneity* (cited in note 1), p. 281.

15. Ibid.

16. For a critique and effort to reduce the number of assumptions and theorems, see Jonathan H. Turner, "A Theory of Social Structure: An Assessment of Blau's Strategy," *Contemporary Sociology* 7 (1978), pp. 698–705.

17. Blau, *Structural Context of Opportunities* (cited in note 2), pp. 21–50.

18. Ibid., pp. 173–203.

19. Ibid., pp. 53–169.

PART VII

Critical Theorizing

40

The Emerging Tradition:

The Rise of the Critical Analysis of Modernity

Virtually all early sociologists were influenced by a broad intellectual movement, often termed "The Enlightenment," which grew out of both the Renaissance and, later, the Age of Science in the seventeenth century.[1] As we have seen for the emergence of most theoretical perspectives in sociology, The Enlightenment still inspires thinkers, in at least two respects. First, the social universe has often been seen as "progressing," moving from one stage of development to another. To be sure, theorists have disagreed about the stages, and many have had doubts about the notion of "progress," but it would be hard to deny that sociologists see directional movement of society or world systems as a central theme. A second legacy from The Enlightenment has been the belief that science can be used to further social progress. As with the idea of progress, this faith in science has not been universal, but even those who have doubted that science is the key to social progress still tend to believe that analysis of the human condition and its pathologies can be used for human betterment.

These two points of emphasis from The Enlightenment were part of a more general effort to come to terms with what is often termed "modernity" or the transformations associated with the rise of commerce and industrial capitalism from the debris of the old feudal order. Indeed, the central problem for all early sociologists was to understand the dramatic transformations of the social order being caused by the expansion of commerce and markets, the industrialization of production, the urbanization of labor, the decline of cohesive and local communities, the rise of the bureaucratic state, the decreasing salience of sacred symbols as a result of expanding secular law and science, the conflicts among new social classes, and many other disruptive transformations. These were changes that early theorists sought to comprehend. Some were pessimistic and worried about what was occurring; others were optimistic about the new modern age; still others believed that things would get better after the current turmoil subsided. But no one, who was considered a serious social thinker, could ignore "modernity."

Critical theorizing in all its forms enters this old debate about modernity from a number of different directions. As the name implies, most theorists in this "critical" tradition view industrial capitalism in negative terms, and some have even posited a new stage of history, "postmodernity," which is similarly viewed in a negative light. Almost all critical theorists disparage the optimism of The Enlightenment, seeing the use of science for constructing a better society as naive, as pursuit of an illusion, or even as harmful. For most, science is part of a broader culture of commerce and capitalism, which, to critical theorists, are the cause of the problems in the modern or postmodern era, not part of their solution. Yet, ironically, these very same critics often appear to be figures of The Enlightenment because they address the very same problems of the earlier Enlightenment-inspired theorists, because they use analysis and reason to pronounce the problems of the modern or postmodern era, and because they often propose solutions to the ills of the current era, even as they drown their pronouncements in pessimism. True, most critical theorists maintain a hearty disdain for science and the implicit Enlightenment projects of theories examined in earlier chapters, but they have not escaped the mood, tone, and problematic of The Enlightenment.

CRITICAL STRAINS IN MARX'S THOUGHT

In 1846 Karl Marx and Friedrich Engels completed *The German Ideology*, which was initially turned down by the publisher.[2] Much of this work is an attack on the "Young Hegelians," who were advocates of the German philosopher Georg Hegel, and is of little interest today. Yet this attack contained certain basic ideas that have served as the impetus behind "critical theory," or the view that social theory must be critical of oppressive arrangements and propose emancipatory alternatives. This theme exists, of course, in all of Marx's work,[3] but the key elements of contemporary critical theory are most evident in this first statement.

Marx criticized the Young Hegelians severely because he had once been one of them and was now making an irrevocable break. Marx saw the Hegelians as hopeless idealists, in the philosophical sense. That is, they saw the world as reflective of ideas, with the dynamics of social life revolving around consciousness and other cognitive processes by which "ideal essences" work their magic on humans. Marx saw this emphasis on the "reality of ideas" as nothing more than a conservative ideology that supports people's oppression by the material forces of their existence. His alternative was "to stand Hegel on his head," but in this early work there is still an emphasis on the relation between consciousness and self-reflection, on the one hand, and social reality, on the other. This dualism became central to contemporary critical theory.

Actually, Marx's "standing of Hegel on his head" has been reversed by some contemporary theorists who, in essence, have put Hegel back on his feet. Indeed, for many who commented on the condition of modernity or postmodernity near the close of the twentieth century, the world has been transformed into a sea of symbols that have lost anchorage in material conditions and that have, as a result, changed the very nature of society from one driven by control of the means of material production to one dominated by signs and texts symbolizing little but themselves. For critical theorists schooled in the Marxian tradition, even those who call themselves postmodernists, such arguments go too far, but there can be little doubt that Marx's dismissal of Hegel and the Young Hegelians was not the final word on the place of ideas, symbols, and signs in societal evolution.

Marx was a modernist, not a postmodernist, so he went a different direction. For Marx, humans are unique by virtue of their conscious awareness of themselves and their situation; they are capable of self-reflection and, hence, assessment of their positions in society. Such consciousness arises from people's daily existence and is not a realm of ideas that is somehow independent of the material world, as much German philosophy argued or as later versions of postmodernism implied. For Marx, people produce their ideas and conceptions of the world because of social structures in which they are born, raised, and live.

The essence of people's lives is the process of production, because, for Marx, human "life involves, before anything else, eating and drinking, a habitation, clothing, and many other material things."[4] To meet these contingencies of life, production is necessary, but, as production satisfies one set of needs, new needs arise and encourage alterations in the ways that productive activity is organized. The elaboration of productive activity creates a division of labor, which, in the end, is alienating because it increasingly deprives humans of their capacity to control their productive activities. Moreover, as people work, they are exploited in ways that generate private property and capital for those who enslave them. Thus, as people work as alienated cogs in the division of labor, they produce that which enslaves them: private property and profits for those who control the modes and means of production. Marx provided a more detailed discussion of the evolution of productive forces to this capitalist stage, and like any Enlightenment thinker, he argued that this capitalist stage would lead to a new era of human organization.

Marx believed that the capacity to use language, to think, and to analyze their conditions would enable humans to alter their environment. People do not merely have to react to their material conditions in some mechanical way; they can also use their capacities for thought and reflection to construct new material conditions and corresponding social relations. Indeed, the course of history involved such processes as people actively restructured the material conditions of their existence. The goal of social theory, Marx implicitly argued, is to use humans' unique facility to expose those oppressive social relations and to propose alternatives. Marx's entire career was devoted to this goal, and this emancipatory aspect of Marx's thought forms the foundation for critical theory, even in some of its postmodern manifestations.

Marx used the somewhat ambiguous term "praxis" to describe this blending of theory and action. The basic notion is that action to change social conditions generates increased knowledge that can then be used to mount more effective change-producing action. Thus, the interplay between action and theoretical understanding can eventually lead individuals to a better social life. Although those with power can impose their ideologies on subordinates and, thereby, distort the latter's perceptions of their true interests, Marx had typical Enlightenment-inspired faith that subordinates possessed the capacity for praxis and that they would eventually use their capacities for agency to change the nature of modernity.

Today, contemporary critical theorists appear somewhat divided on the question of whether analysis of modernity and postmodernity can be used to improve the human condition. As we will see shortly, many confronted Max Weber's pessimism about the ever tightening "cage" of rational-legal authority and state domination. Others sustained the emancipatory faith of Marx's belief in praxis.

Still others emphasized an inherent force articulated in Marx's analysis of capitalism—the capacity of money-driven markets to "commodify" all things, symbols, and ideals—as a basis for a renewed pessimism about the human condition. To *commodify* means that symbols, signs, objects, cultures, relationships, and virtually anything can be turned into a marketable thing, to be bought and sold for a price stated in terms of money. Hence, as capitalists seek profits, they buy and sell not just the material objects necessary for human survival, but they produce and sell symbols and signs that, as commodities, lose their power to provide meaning to human life. Coupled with information technologies that Marx could never have visualized, as well as markets for services and cultural symbols that Marx did not fully anticipate, the social world is now dominated by the production and distribution of signs, symbols, texts, and other cultural commodities. This transformation has changed the very nature of humans' capacities to understand and respond to their conditions.

WEBER'S PESSIMISM: THE BASIC DILEMMA FOR EARLY CRITICAL THEORISTS

Max Weber was concerned with the historical transition to modern capitalist societies, and his description and explanation of this transition represent a devastating critique of Marx's optimism about revolutionary movements toward a new utopian society. Weber's analysis is complex, and the historical detail that he presented to document his case is impressive, but his argument is captured by the concept of *rationalization*.[5] Weber argued that the rationality that defines modern societies is "means/ends rationality" and, hence, involves a search for the most efficient means to achieve a defined end. The process of rationalization, Weber felt, involves the ever increasing penetration of means/ends rationality into more spheres of life, thereby destroying older traditions. As bureaucracies expand in the economic and governmental sphere, and as markets allow individuals to pursue their personal ends rationally, the traditional moral fabric is broken. Weber agreed with Georg Simmel that this rationalization of life brings individuals a new freedom from domination by religious dogmatism, community, class and other traditional forces; but in their place it creates a new kind of domination by impersonal economic forces, such as markets and corporate bureaucracies, and by the vast administrative apparatus off the ever-expanding state. Human options were, in Weber's view, becoming ever more constrained by the "iron cage" of rational and legal authority. Unlike Marx, Weber did not see such a situation as rife with revolutionary potential; rather, he saw the social world as increasingly administered by impersonal bureaucratic forces.

This pessimistic view seemed, by the early 1930s, to be a far more reasonable assessment of modernity than was Marx's utopian dream. Indeed, the communist revolution in Russia had degenerated into Stalinism and bureaucratic totalitarianism by the Communist Party; in the West, particularly the United States, workers seemed ever more willing to sell themselves in markets and work in large-scale organizations; and political fascism in Germany and Italy was creating large authoritarian bureaucracies. How, then, was the first generation of critical theorists to reconcile Weber's more accurate assessment of empirical trends with Marx's optimistic and emancipatory vision? This became the central question of early critical theory.

SIMMEL'S IMPLICIT ATTACK ON MARX'S EMANCIPATORY PROJECT

Many of Georg Simmel's ideas represent an important qualification to Marx's reasoning and, to a lesser extent, to Weber's as well. Marx's more emancipatory side saw capitalism as producing the conditions that would lead to a revolution, ushering in a new form of human organization in which individuals are freed from the capitalists' domination. Thus, as capitalism expands, the division of labor makes workers appendages to machines, concentrates workers in urban areas, quantifies social relations through money and markets, and forces workers to be mere role players (rather than fully involved participants) in social relations. In so doing, capitalism generates the personal alienation and resentments as well as the social structural conditions that will lead subordinates to become aware of their domination and to organize in an effort to change their plight.

Simmel challenged much of Marx's analysis in his *The Philosophy of Money*.[6] This critique revolves around one of the themes in Marx's writing: Capitalism quantifies social life with money and, in so doing, makes exchanges in markets paramount; the result is that human social relations are increasingly commodified. Such commodification is personified in the labor market, where workers sell themselves as a thing, and coupled with the growing division of labor, workers become mere cogs in an impersonal organizational machine. Such processes, Marx believed, would be so oppressive as to initiate revolutionary pressures for their elimination. Simmel, however, looked at these forces much differently. Although a certain level of alienation from work and "commodification" of relations through the use of money are inevitable with increasing differentiation and expansion of productive forces and markets, Simmel saw these forces of modernity as liberating individuals from the constraints of tradition. In Simmel's Enlightenment-oriented view, people have more options about how they spend their money and what they do; they can move about with more freedom and form new and varied social relations; they can live lifestyles that reflect their tastes and values; and, in general, they are more liberated than their counterparts in less complex, traditional societies.

This critique of Marx was, however, rejected by the early critical theorists who did not want to visualize modern societies as liberating. And yet these theorists were confronted with the failure of Marx's predictions about the communist revolution and the coming emancipation of society. In an attempt to reconstruct Marx's vision of humans' capacity to make history, they were forced to accept Weber's highly pessimistic view of the constraints of modern society and to reject Simmel's more optimistic diagnosis. But, in so doing, they became trapped in a dilemma: If capitalism is not self-transforming as Marx's revolutionary model indicates, if modern life is not so liberating as Simmel felt, and if Weber's analysis of increasing constraint in societies must therefore be accepted as true, then how is liberation to occur? What force will drive people's emancipation from domination? Early critical theorists would not accept Simmel's judgment—that is, people are more *free* than in traditional societies—and so they conceptually retreated into a contemplative subjectivism. They viewed the liberating force as somehow springing from human nature and its capacity for conscious reflection—a kind of "watered down," even impotent, sense of praxis.

CRITICAL THEORY, LIBERATION, AND POSTMODERNISM

The somewhat indirect debate among Marx, Weber, and Simmel—or at least, Weber and Simmel with Marx—about the condition of modernity has been replicated throughout the twentieth century. The vocabularies have shifted, but the debate itself will not go away. Marx's emancipatory vision still dominates, even as Marx has been turned on his head and Hegel put back on his feet. Weber's gloom and doom pessimism has, surprisingly, become less significant in the more recent debate on postmodernity, despite state bureaucratic power having extended everywhere and totalitarian regimes existing in most societies of the world. Simmel has enjoyed some resurgence of interest, but most seem to get his message backwards: The rapid spread of social differentiation, the destruction of traditional and oppressive cultural symbols, the ever widening use of money in all exchanges, the sense of value created by free choice in markets, the new options and freedoms enjoyed by actors, and many other features of modernity that Simmel saw, on net balance, in a very positive light are now viewed by many postmodernists as somehow the sources of a new evil. These new pessimists are, ironically, still captured by The Enlightenment and its fears and concerns about modernity, but unlike many who first engaged these issues, there is a Weberian sense of hopeless despair about a world in which the forces of postmodernity erode meaning from people's lives. True, the Marxists-turned postmodernists usually see emancipation as just around the corner, but none has accepted what Simmel saw: Perhaps the changes of modernity are, themselves, the very liberating forces that so many have sought in the "next stage" of human evolution.

Emerging at the same time that critical theory was enjoying a rebirth in the United States during the 1960s and 1970s—essentially importing the Marx versus Weber disagreement from the Frankfurt School of the 1930s—was a series of "liberation" movements among obvious victims of societal discrimination. The Civil Rights movement in America was the first to capture attention inside and outside of academia; students also decided in the 1960s that they were oppressed or, at least, the society at large was oppressive; soon, women's liberation movements emerged to challenge traditional patterns of male domination in key institutional spheres; and eventually, sexual preference movements arose and forced Americans to examine prejudices and patterns of discrimination in these areas. These actual movements stimulated a great deal of intellectual debate, inside and outside academia, but as the close of the twentieth century nears, the academic side of these conflicts has become somewhat routinized within academic politics over departmental status, resources, faculty, budgets, and the normal internal fights that any bureaucratic organization reveals. More significantly for our purposes, these broader intellectual and political debates overlap with the concerns of sociological theory.

This overlap with theory is critical in one sense: virtually all the "voices"—to use what was a common cliché in the 1990s—are antiscience, seeing it as somehow contributing to sexism, patriarchy, racism, homophobia, postmodernity, class oppression, and virtually any pattern of oppression that humans have had to confront. In the chapters to follow, we will explore only a limited sample of what could be a whole book of critical theories, but we will limit ourselves to those "theories" that now appear to occupy a good deal of "discourse"—to use yet another cliché from the early 1990s—among those who define themselves as "theorists."

NOTES

1. Jonathan H. Turner, "Founders and Classics: A Canon in Motion" in *The Student Sociologist's Handbook*, ed. C. Middleton, J. Gubbay, and C. Ballard (Oxford: Blackwell, 1997).

2. Karl Marx and Friedrich Engels, *The German Ideology* (New York: International, 1947; written in 1846).

3. Karl Marx, *Capital: A Critical Analysis of Capitalist Production*, volume 1 (New York: International, 1967; originally published in 1867); Karl Marx and Friedrich Engels, *The Communist Manifesto* (New York: International, 1971; originally published in 1848).

4. Marx and Engels, *The German Ideology* (cited in note 2), p. 15.

5. Max Weber, *Economy and Society*, trans. G. Roth (Berkeley: University of California Press, 1978).

6. Georg Simmel, *The Philosophy of Money*, trans. T. Bottomore and D. Frisbie (Boston: Routledge & Kegan Paul, 1978; originally published in 1907).

41

The Maturing Tradition:

The Frankfurt School and the Cultural Turn

The spirit of Georg Wilhelm Friedrich Hegel (1770–1831) could not be exorcised by much twentieth century critical theory. Key elements of the Marxian scheme—such as the dialectical view of history, the concept of alienation, or the notion of praxis, for example—came from Hegel. Karl Marx converted these basic ideas to a materialism emphasizing that the alienation created by inequalities in material relations of production generated an inherent dialectic of history led by humans who possessed the capacities for agency and praxis and who, thereby, would move human society to its final state of communism. In contrast with Hegel, Marx believed that ideas, politics, and other institutional systems were "superstructures" reflecting and, indeed, being controlled by a "substructure" lodged in the patterns of organization and ownership of economic production.

More than any Enlightenment-inspired sociologist of his time, Marx saw that the critique of existing relations of domination, the emergence of class conflict, the emancipation of humans, and the progress of society were all interwoven. The critique was to begin with attacks on the inequalities generated by the economic system and, then, on the political and ideological superstructures that legitimated the means and modes of production. Yet, by the third decade of the twentieth century, even the Great Depression had not produced the predicted revolution by the proletariat, nor was it possible to see humans as progressing. Even when times were better after World War II, capitalism was not collapsing but, if anything, gaining converts. And China's "communist revolution" and subsequent "cultural revolution" had begun to look very much like the Stalin purges in Russia.

The collapse of communism in the 1990s forced further adjustments by critical theory.[1] Indeed, criticism moved from one revolving around the "immiseration" of the population to its tasteless overconsumption and manipulation by the symbols of advertising. In the end, critical theory and postmodernism began to blend together as concerns with symbols, signs, culture, and ideology seemed to hold sway over older Marxian views about the economic substructure. Or, at the very least, critical theorists were working very hard to find problems in the cultural products of mature capitalist systems and the developing capitalist world order. Marx was perhaps turning over in his grave, but Hegelian themes were nonetheless reemerging.

We will pause, therefore, and offer a few representative samples of how the critical theorists working in the decades around the twentieth century's midpoint were trying to keep their Marxian faith given a reality that was no longer on its Marxian trajectory. In so doing, Marxian materialism and Hegelian idealism were put back together in an uneasy accommodation.

THE FRANKFURT SCHOOL

Thus the first generation of critical theorists, who are frequently referred to as the Frankfurt School because of their location in Germany and their explicit interdisciplinary effort to interpret the oppressive events of the twentieth century, confronted a real dilemma: how to reconcile Marx's emancipatory dream with the stark reality of modern society as conceptualized by Max Weber.[2] Indeed, when the Frankfurt Institute for Social Research was founded in 1923, there seemed little reason to be optimistic about developing a theoretically informed program for freeing people from unnecessary domination. The defeat of the left-wing working-class movements, the rise of fascism in the aftermath of World War I, and the degeneration of the Russian Revolution into Stalinism had, by the 1930s, made it clear that Marx's analysis needed drastic revision. Moreover, the expansion of the state, the spread of bureaucracy, and the emphasis on means/ends rationality through the application of science and technology all signaled that Weber's analysis had to be confronted.

The members of the Frankfurt School wanted to maintain Marx's views on praxis—that is, a blending of theory and action or the use of theory to stimulate action, and vice versa. And they wanted theory to expose oppression in society and to propose less constrictive options. Yet they were confronted with the spread of political and economic domination of the masses. Thus modern critical theory in sociology was born in a time when there was little reason to be optimistic about realizing emancipatory goals.

Three members of the Frankfurt School are most central: György Lukács, Max Horkheimer, and Theodor Adorno.[3] Lukács' major work appeared in the 1920s,[4] whereas Horkheimer[5] and Adorno[6] were active well into the 1960s. In many ways, Lukács was the key link in the transition from Marx and Weber to modern critical theory, because Horkheimer and Adorno were reacting to much of Lukács analysis and approach.

All these scholars are important because they directly influenced the intellectual development and subsequent work of Jürgen Habermas, the most prolific contemporary critical theorist whose work is examined in the next chapter.[7]

György Lukács

Lukács blended Marx and Weber together by seeing a convergence of Marx's ideas about commodification of social relations through money and markets with Weber's thesis about the penetration of rationality into ever more spheres of modern life. Borrowing from Marx's analysis of the "fetishism of commodities," Lukács employed the concept of *reification* to denote the process by which social relationships become "objects" that can be manipulated, bought, and sold. Then, reinterpreting Weber's notion of "rationalization" to mean a growing emphasis on the process of "calculation" of exchange values, Lukács combined Weber's and Marx's ideas. As traditional societies change, he argued, there is less reliance on moral standards and processes of communication to achieve societal integration; instead, there is more use of money, markets, and rational calculations. As a result, relations are coordinated by exchange values and by people's perceptions of one another as "things."[8]

Lukács painted himself into a conceptual corner, however. If indeed such is the historical process, how is it to be stopped? Lukács' answer was to resurrect a contrite Hegel; that is, rather than look to contradictions in material conditions or economic and political forces, one must examine the dialectical forces inherent in human consciousness. There are limits, Lukács argued, to how much reification and rationalization people will endure. Human subjects have an inner quality that keeps rationalization from completely taking over.[9]

This emphasis on the process of consciousness is very much a part of critical theory that borrows much from the early Marx[10] and that, at the Frankfurt School, had a heavy dose of Freud and psychoanalytic theory. As a result, unlike its sources of inspiration, Marx and Weber, early critical theory was subjectivist and failed to analyze intersubjectivity, or the ways people interact through mutually shared conscious activity. Emphasizing the inherent resistance of subjects to their total reification, Lukács could only propose that the critical theorist's role is to expose reification at work by analyzing the historical processes that have dehumanized people. As a consequence, Lukács made critical theory highly contemplative, emphasizing that the solution to the problem of domination resides in making people more aware and conscious of their situation through a detailed, historical analysis of reification.

Max Horkheimer and Theodor Adorno

Both Horkheimer and Adorno were highly suspicious of Lukács' Hegelian solution to the dilemma of reification and rationalization. These processes do not imply their own critique, as Hegel would have suggested. Subjective consciousness and material reality cannot be separated. Consciousness does not automatically offer resistance to those material forces that commodify, reify, and rationalize. Critical theory must, therefore, actively (1) describe historical forces that dominate human freedom and (2) expose ideological justifications of these forces. Such is to be achieved through interdisciplinary research among variously trained researchers and theorists who confront one another's ideas and use this dialogue to analyze concrete social conditions and to propose courses of ameliorative action. This emphasis on praxis—the confrontation between theory and action in the world—involves developing ideas about what oppresses and what to do about it in the course of human struggles. As Horkheimer argued, "[The] value of theory is not decided alone by the formal criteria of truth . . . but by its connection with tasks, which in the particular historical moment are taken up by progressive social forces."[11] Such critical theory is, Horkheimer claimed, guided by a "particular practical interest" in the emancipation of people from class domination.[12] Thus, critical theory is tied, in a sense that Marx might have appreciated, to people's practical interests.

As Adorno and Horkheimer interacted and collaborated, their positions converged (although by the late 1950s and early 1960s Horkheimer had seemingly rejected much of his earlier work). Adorno was more philosophical and, yet, research oriented than Horkheimer; Adorno's empirical work on "the authoritarian personality" had a major impact on research in sociology and psychology, but his theoretical impact came from his collaboration with Horkheimer and, in many ways, through Horkheimer's single-authored work.[13] Adorno was very pessimistic about the chances of critical theory making great changes, although his essays were designed to expose patterns of recognized and unrecognized domination of individuals by social and psychological forces. At best, his "negative dialectics" could allow humans to "tread water" until historical circumstances were more favorable to emancipatory movements. The goal of negative dialectics was to sustain a constant critique of ideas, conceptions, and conditions. This critique could not by itself change anything, for it operates only on the plane of ideas and concepts. But it can keep ideological dogmatisms from obscuring conditions that might eventually allow emancipatory action.

Both Horkheimer and Adorno emphasized that humans' "subjective side" is restricted by the spread of rationalization. In conceptualizing this process, they created a kind of dualism between the subjective world and the realm of material objects, seeing the latter as oppressing the former. From their viewpoint, critical theory must expose this dualism, and it must analyze how this "instrumental reason" (means/ends rationality) has invaded the human spirit. In this way some resistance can be offered to these oppressive forces.

Within the Frankfurt School, then, the idealism of Lukács had been brought partially back into a more orthodox Marxian position, but not completely so. The damage had been done to pure Marxian materialism, and outside of the narrow confines of Frankfurt, critical theory once again turned to idealism, even among those critical Marxists who had emigrated to America from Frankfurt during the rise of Nazism.[14]

THE HEGELIAN TURN IN CRITICAL THEORY

Gramsci's Theory of Ideological Hegemony

Antonio Gramsci was an Italian Marxist who, obviously, cannot be considered part of the Frankfurt School. Yet, he is a key figure in continuing what the Frankfurt School emphasized: Criticism acknowledging that the capitalist systems of the twentieth century's midpoint were generating prosperity and that the working classes in these systems did not seem particularly disposed to revolution. Gramsci completed the turning of Marx's ideas back into a more Hegelian mode.[15] Marx believed that ideology and the "false consciousness" of workers were ideological obfuscations created and maintained by those who controlled the material (economic) "substructure." Marx had argued—see Chapter 16 for more details about how this idea was carried forward in the twentieth century—that those who control the means and modes of production also control the state which, in turn, generates ideologies justifying this control and power. In this way, the proletariat is kept, for a time until the full contradictions of capitalism are manifest, from becoming a class "for themselves" ready to pursue revolutionary conflict with their oppressors. Gramsci simply turned this argument around: The "superstructure" of state and ideology drives the organization of society and the consciousness of the population.

Gramsci believed the ruling social class is *hegemonic*, controlling not only property and power, but ideology as well. Indeed, the ruling class holds onto its power and wealth by virtue of its ability to use ideologies to manipulate workers and all others. The state is no longer a crude tool of coercion, nor an intrusive and insensitive bureaucratic authority; it has become the propagator of culture and the civic education of the population, creating and controlling key institutional systems in more indirect, unobtrusive and, seemingly, inoffensive ways. Thus, the views of capitalists become the dominant views of all, with workers believing in the appropriateness of the market-driven systems of competition; the commodification of objects, signs, and symbols; the buying and selling of their labor; the use of law to enforce contracts favoring the interests of the wealthy; the encouragement of private charities, the sponsorship of clubs and voluntary organizations; the state's conceptions of a "good citizen"; the civics curriculum of the schools; and virtually all spheres of institutional activity that are penetrated by the ideology of the state. Culture and ideology are, in Albert Bergesen's words,[16] "no longer the thing to be explained but . . . now a thing that does the explaining." A dominant material class rules, to be sure, but it does so by cultural symbols, and the real battle in capitalist societies is over whose symbols will prevail. Or, more accurately, can subordinates generate alternative ideologies to those controlled by the state?

This view of critical theory takes much of the mechanical menace out of Weber's "iron cage" metaphor, because the state's control is now "soft" and "internal." It has bars that bend flexibly around those whose perceptions of the world it seeks to control. The Marxian view of emancipation is still alive in Gramsci's theories, because the goal of "theory" is to expose the full extent to which ideology has been effectively used to manipulate subordinates. Moreover, the recognition that systems of symbols become the base of society is a theme that resonated well with later postmodernists (see Chapter 44) and structuralists (see Chapter 37) who began to conceptualize modernity as the production of signs and symbols.

Althusser's Structuralism

Initially, Louis Althusser seems more strictly orthodox in his Marxism than Gramsci,[17] yet, he was also a French scholar in a long line of structuralists whose emphasis is on the logic of the deeper, underlying structure of surface empirical reality.[18] Althusser remains close to Marx in this sense: The underlying structure and logic of the economy is ultimately determinative. But, having said this, he then developed a theory of "The Ideological State Apparatus"[19] which gave prominence to the state's use of ideology to sustain control within a society.

For Althusser, economic, political, and ideological systems reveal their own structures, hidden beneath the surface and operating by their own logics. The economic might be the dominant system, circumscribing the operation of political and ideological structures, but these latter have a certain autonomy. History is, in essence, a reshuffling of these deep structures, and the individual actor becomes merely a vessel through which the inherent properties of structures operate. Individual actions, perceptions, beliefs, emotions, convictions, and other states of consciousness are somehow "less real" than the underlying structure that cannot be directly observed. To analogize the structuralist theories from which Althusser drew inspiration, social control comes from individuals perceiving that they are but words in a grammatical system generated by an even more fundamental structure. Each actor is at a surface place in the economic and political structures of a society, and their perceptions of these places also put them within an ideological or cultural sphere. But these places and spheres are only one level of reality; people also see themselves as part of a deeper set of structures that, in essence, defines who and what they are. Under these conditions, ideology has even more power because it is doing much more than blinding the subjects to some other reality, such as their objective class interests. Ideology is also defining actors' places in a reality beyond their direct control and a reality operating by its own logic of structure.

Thus, unlike Marx or Gramsci who believe ideology is a tool—an invidious and insidious one—used by those in power, Althusser sees the Ideological State Apparatus as more controlling because it is perceived not just as conventions, rules, mores, traditions, and beliefs, but instead as the essence of order and persons' place in this order. The subject is thus trapped in the deeper logics of economic, political, and ideological systems that erode human capacities for praxis and agency.

CONCLUSION

By the middle of the twentieth century, Marx's emancipatory project had been turned into something very different than he had visualized. His and Engel's *The Communist Manifesto* was a call to arms, based on a view of the inherent contradictions in the nature of capitalist systems. Within one hundred years of this call, critical theory had become decidedly more philosophical. Indeed, Marx's dismissal of the "young Hegelians" in *The German Ideology* had apparently not worked; they were back in different forms and guises, but they increasingly dominated critical theorizing in the twentieth century. The "young Hegelians" so viciously criticized by Marx and Engels had considered themselves revolutionaries, but Marx saw them as more concerned with ideas about reality than with reality itself. They were accused of "blowing theoretical bubbles" about ideals and essences, and it could be imagined that he and Engels might make the very same criticisms of the critical theories that developed in the second half of the twentieth century, especially as these theories began to merge with postmodernism.

NOTES

1. See, for examples, David Hoy and Thomas McCarthy, *Critical Theory* (Oxford: Blackwell, 1994), or Stephen Regan, ed., *The Year's Work in Critical and Cultural Theory* (Oxford: Blackwell, 1995). Earlier reviews and analyses of critical theory and sociology include Paul Connerton, ed., *Critical Sociology* (New York: Penguin Books, 1970); Raymond Geuss, *The Idea of a Critical Theory* (New York: Cambridge University Press, 1981); David Held, *Introduction to Critical Theory* (Berkeley: University of California Press, 1980); Trent Schroyer, *The Critique of Domination: The Origins and Development of Critical Theory* (New York: Braziller, 1973); Albrecht Wellmer, *Critical Theory of Society* (New York: Seabury, 1974); Ellsworth R. Fuhrman and William E. Snizek, "Some Observations on the Nature and Content of Critical Theory," *Humboldt Journal of Social Relations* 7 (Fall–Winter 1979–1980), pp. 33–51; Zygmunt Bauman, *Towards a Critical Society* (Boston: Routledge & Kegan Paul, 1976); Robert J. Antonio, "The Origin, Development and Contemporary Status of Critical Theory," *Sociological Quarterly* 24 (Summer 1983), pp. 325–351; Jim Faught, "Objective Reason and the Justification of Norms," *California Sociologist* 4 (Winter 1981), pp. 33–53.

2. For descriptions of this activity, see Martin Jay, *The Dialectical Imagination* (Boston: Little, Brown, 1973), and "The Frankfurt School's Critique of Marxist Humanism," *Social Research* 39 (1972), pp. 285–305; David Held, *Introduction to Critical Theory* (Berkeley: University of California Press, 1980), pp. 29–110; Robert J. Antonio, "The Origin, Development, and Contemporary Status of Critical Theory," *Sociological Quarterly* 24 (Summer 1983), pp. 325–351; Phil Slater, *Origin and Significance of The Frankfurt School* (London: Routledge & Kegan Paul, 1977).

3. Other prominent members included Friedrich Pollock (economist), Erich Fromm (psychoanalyst, social psychologist), Franz Neumann (political scientist), Herbert Marcuse (philosopher), and Leo Loenthal (sociologist). During the Nazi years, the school relocated to the United States, and many of its members never returned to Germany.

4. György Lukács, *History and Class Consciousness* (Cambridge, MA: MIT Press, 1968; originally published in 1922).

5. Max Horkheimer, *Critical Theory: Selected Essays* (New York: Herder and Herder, 1972), is a translation of essays written in German in the 1930s and 1940s; *Eclipse of Reason* (New York: Oxford University Press, 1947; reprinted by Seabury in 1974) was the only book by Horkheimer originally published in English. It takes a slightly different turn than earlier works, but it does present ideas that emerged from his association with Theodor Adorno. See also Horkheimer, *Critique of Instrumental Reason* (New York: Seabury,

1974). See David Held, *Introduction to Critical Theory* (cited in note 1), pp. 489–491, for a more complete listing of Horkheimer's works in German.

6. Theodor W. Adorno, *Negative Dialectics* (New York: Seabury, 1973; originally published in 1966), and, with Max Horkheimer, *Dialectic of Enlightenment* (New York: Herder and Herder, 1972; originally published in 1947). See Held, *Introduction to Critical Theory* (cited in note 1), pp. 485–487, for a more complete listing of his works.

7. "From Lukács to Adorno: Rationalization as Reification," pp. 339–399 in Jürgen Habermas, *The Theory of Communicative Action*, vol. 1 (Boston: Beacon, 1984), contains Habermas' critique of Lukács, Horkheimer, and Adorno.

8. Lukács, *History and Class Consciousness* (cited in note 4).

9. Ibid., pp. 89–102. In a sense, Lukács becomes another "Young Hegelian" whom Marx would have criticized. Yet, in Marx's own analysis, he sees alienation, per se, as producing resistance by workers to further alienation by the forces of production. This is the image that Lukács seems to take from Marx.

10. Karl Marx and Friedrich Engels, *The German Ideology* (New York: International, 1947; originally written in 1846).

11. Max Horkheimer, "Zum Rationalismusstreit in der gegenwartigen Philosophie," originally published in 1935; reprinted in *Kritische Theorie*, vol. 1, ed. A. Schmidt (Frankfurt: Fischer Verlag, 1968), pp. 146–147. This and volume 2, by the way, represent a compilation of many of the essays Horkheimer wrote while at the Institute in Frankfurt.

12. Habermas used this idea, but he extended it in several ways.

13. Theodor W. Adorno, Else Frenkel-Brunswick, Daniel Levinson, and R. Nevitt Sanford, *The Authoritarian Personality* (New York: Harper & Row, 1950).

14. See note 3.

15. Antonio Gramsci, *Selections from the Prison Notebooks* (New York: International, 1971; originally published in 1928).

16. Albert Bergesen, "The Rise of Semiotic Marxism," *Sociological Perspectives* 36 (1993), p. 5.

17. Louis Althusser, *For Marx* (New York: Pantheon, 1965); *Lenin and Philosophy* (New York: Monthly Review Press, 1971); Louis Althusser and Etienne Balabar, *Reading Capital* (London: New Left, 1968).

18. See Chapters 34, 35, and 37 for examples of this French lineage of structuralism.

19. "Ideology and Ideological Status Apparatus" in Althusser, *Lenin and Philosophy* (cited in note 17).

42

The Continuing Tradition I:
Jürgen Habermas' Frankfurt-School Project

The German philosopher-sociologist, Jürgen Habermas, undoubtedly has been the most prolific descendant of the original Frankfurt School. As with the earlier generation of Frankfurt School social theorists, Habermas' work revolves around several important questions: (1) How can social theory develop ideas that keep Karl Marx's emancipatory project alive, yet, at the same time, recognize the empirical inadequacy of his prognosis for advanced capitalist societies? (2) How can social theory confront Max Weber's historical analysis of rationalization[1] in a way that avoids his pessimism and thereby keeps Marx's emancipatory goals at the center of theory? (3) How can social theory avoid the retreat into subjectivism of earlier critical theorists, such as György Lukács, Max Horkheimer, and Theodor Adorno, who increasingly focused on states of subjective consciousness *within individuals* and, as a consequence, lost Marx's insight that society is constructed from, and must therefore be emancipated by, the processes that sustain *social relations among* individuals? (4) How can social theory conceptualize and develop a theory that reconciles the forces of material production and political organization with the forces of intersubjectivity among reflective and conscious individuals in such a way that it avoids (*a*) Weber's pessimism about the domination of consciousness by rational economic and political forces, (*b*) Marx's naive optimism about inevitability of class consciousness and revolt, and (*c*) early critical theorists' retreat into the subjectivism of Georg Hegel's dialectic, where oppression mysteriously mobilizes its negation through increases in subjective consciousnesses and resistance?

At different points in his career, Habermas has focused on one or another of these questions, but all four have always guided his approach, at least implicitly. Habermas has been accused of abandoning the critical thrust of his earlier works, but this conclusion is too harsh. For, in trying to answer these questions, he has increasingly recognized that mere critique of oppression is not enough. Such critique becomes a "reified object itself." Although early critical theorists knew this, they never developed conceptual schemes that accounted for the underlying dynamics of societies. For critique to be useful in liberating people from domination, it is necessary, Habermas seems to say, for the critique to discuss the fundamental processes integrating social systems. In this way the critique has some possibility of suggesting ways to create new types of social relations. Without theoretical understanding about how society works, critique is only superficial debunking and becomes an exercise in futility. This willingness to theorize about the underlying dynamics of society, to avoid the retreat into subjectivism, to reject superficial criticism and instead to base critique on reasoned theoretical analysis, and to incorporate ideas from many diverse theoretical approaches make Habermas' work theoretically significant.[2]

HABERMAS' ANALYSIS OF "THE PUBLIC SPHERE"

In his first major publication, *Structural Transformation of the Public Sphere,* Habermas traced the evolution and dissolution of what he termed *the public sphere.*[3] This sphere is a realm of social life where people can discuss matters of general interest; where they can discuss and debate these issues without recourse to custom, dogma, and force; and where they can resolve differences of opinion by rational argument. To say the least, this conception of a public sphere is rather romanticized, but the imagery of free and open discussion that is resolved by rational argumentation became a central theme in Habermas' subsequent approach. Increasingly throughout his career, Habermas came to see emancipation from domination as possible through "communicative action," which is a reincarnation of the public sphere in more conceptual clothing.

In this early work, however, Habermas appeared more interested in history and viewed the emergence of the public sphere as occurring in the eighteenth century, when various forums for public debate—clubs, cafés, journals, newspapers—proliferated. He concluded that these forums helped erode the basic structure of feudalism, which is legitimated by religion and custom rather than by agreements that have been reached through public debate and discourse. The public sphere was greatly expanded, Habermas argued, by the extension of market economies and the resulting liberation of the individual from the constraints of feudalism. Free citizens, property holders, traders, merchants, and members of other new sectors in society could now be actively concerned about the governance of society and could openly discuss and debate issues. But, in a vein similar to Weber's analysis of rationalization, Habermas argued that the public sphere was eroded by some of the very forces that stimulated its expansion. As market economies experience instability, the powers of the state are extended in an effort to stabilize the economy; with the expansion of bureaucracy to ever more contexts of social life, the public sphere is constricted. And, increasingly, the state seeks to redefine problems as technical and soluble by technologies and administrative procedures rather than by public debate and argumentation.

The details of this argument are less important than the fact that this work established Habermas' credentials as a critical theorist. All the key elements of critical theory are there—the decline of freedom with the expansion of capitalism and the bureaucratized state as well as the seeming power of the state to construct and control social life. The solution to these problems is to resurrect the public sphere, but how is this to be done given the growing power of the state? Thus, in this early work, Habermas had painted himself into the same conceptual corner as his teachers in the Frankfurt School. The next phase of his work extended this critique of capitalist society, but he also tried to redirect critical theory so that it does not have to retreat into the contemplative subjectivism of Lukács, Horkheimer, and Adorno. Habermas began this project in the late 1960s with an analysis of knowledge systems and a critique of science.

THE CRITIQUE OF SCIENCE

In *The Logic of the Social Sciences*[4] and *Knowledge and Human Interest*,[5] Habermas analyzes systems of knowledge in an effort to elaborate a framework for critical theory. The ultimate goals of this analysis is to establish the fact that science is but one type of knowledge that exists to meet only one set of human interests. To realize this goal, Habermas posits three basic types of knowledge that encompass the full range of human reason: (1) There is *empirical/analytic* knowledge, which is concerned with understanding the lawful properties of the material world. (2) There is *hermeneutic/historical* knowledge, which is devoted to the understanding of meanings, especially through the interpretations of historical texts. (3) There is *critical* knowledge, which is devoted to uncovering conditions of constraint and domination. These three types of knowledge reflect three basic types of human interests: (1) a *technical* interest in the reproduction of existence through control of the environment, (2) a *practical* interest in understanding the meaning of situations, and (3) an *emancipatory* interest in freedom for growth and improvement. Such interests reside not in individuals but in more general imperatives for reproduction, meaning, and freedom that presumably are built into the species as it has become organized into societies. These three interests create, therefore, three types of knowledge. The interest in material reproduction has produced science or empirical/ analytic knowledge, the interest in understanding of meaning has led to the development of hermeneutic/historical knowledge, and the interest in freedom has required the development of critical theory.

These interests in technical control, practical understanding, and emancipation generate different types of knowledge through three types of media: (1) "work" for realizing interests in technical control through the development of empirical/analytic knowledge, (2) "language" for realizing practical interests in understanding through hermeneutic knowledge, and (3) "authority" for realizing interests in emancipation through the development of critical theory. There is a kind of functionalism in this analysis: needs for "material survival and social reproduction," for "continuity of society through interpretive understanding," and for "utopian fulfillment" create interests. Then, through the media of work, language, and authority, these needs produce three types of knowledge: the scientific, hermeneutical, and critical.

This kind of typologizing is, of course, reminiscent of Weber and is the vehicle through which Habermas makes the central point: Positivism and the search for natural laws constitute only one type of knowledge, although the historical trend has been for the empirical/analytic to dominate the other type types of knowledge. Interests in technical control through work and the development of science have dominated the interests in understanding and emancipation. And so, if social life seems meaningless and cold, it is because technical interests in producing science have dictated what kind of knowledge is permissible and legitimate. Thus Weber's "rationalization thesis" is restated with the

TABLE 42-1 Types of Knowledge, Interests, Media (and Functional Needs)

Functional Needs	Interests	Knowledge	Media
Material survival and social reproduction generate pressures for	technical control of the environment, which leads to the development of	empirical/analytic knowledge, which is achieved through	work
Continuity of social relations generates pressures for	practical understanding through interpretations of other's subjective states, which leads to the development of	hermeneutic and historical knowledge, which is achieved through	language
Desires for utopian fulfillment generate pressures for	emancipation from unnecessary domination, which leads to the development of	critical theory, which is achieved through	authority

typological distinction among interest, knowledge, and media. Table 42-1 summarizes Habermas' argument.

This typology allowed Habermas to achieve several goals. First, he attacked the assumption that science is value free because, like all knowledge, it is attached to a set of interests. Second, he revised the Weberian thesis of rationalization in such a way that it dictates a renewed emphasis on hermeneutics and criticism. These other two types of knowledge are being driven out by empirical/analytic knowledge, or science. Therefore it is necessary to reemphasize these neglected types of knowledge. Third, by viewing positivism in the social sciences as a type of empirical/analytic knowledge, Habermas associated it with human interests in technical control. He therefore visualized social science as a tool of economic and political interests. Science thus becomes an ideology; actually, Habermas sees it as the underlying cause of the legitimation crises of advanced capitalist societies (more on this shortly). In dismissing positivism in this way, he oriented his own project to hermeneutics with a critical twist. That is, he visualized the major task of critical theory as the analysis of those processes by which people achieve interpretative understanding of one another in ways that give social life a sense of continuity. Increasingly, Habermas came to focus on the communicative processes among actors as the theoretical core for critical theorizing. Goals of emancipation cannot be realized without knowledge about how people interact and communicate. Such an emphasis represents a restatement in a new guise of Habermas' early analysis of the public sphere, but now the process of public discourse and debate is viewed as the essence of human interaction in general. Moreover, to understand interaction, it is necessary to analyze language and linguistic processes among individuals. Knowledge of these processes can, in turn, give critical theory a firm conceptual basis from which to launch a critique of society and to suggest paths for the emancipation of individuals. Yet, to justify this emphasis on hermeneutics and criticism, Habermas must first analyze the crises of capitalist societies through the overextension of empirical/analytic systems of knowledge.

LEGITIMATION CRISES IN SOCIETY

As Habermas had argued in his earlier work, there are several historical trends in modern societies: (1) the decline of the public sphere, (2) the increasing intervention of the state into the economy, and (3) the growing dominance of science in the service of the state's interests in technical control. These ideas are woven together in *Legitimation Crisis*.[6]

The basic argument in *Legitimation Crisis* is that, as the state increasingly intervenes in the economy, it also seeks to translate political issues into "technical problems." Issues thus are not topics for public debate; rather, they represent technical problems that require the use of technologies by experts in bureaucratic organizations. As a result, there is a "depoliticization" of practical issues by redefining them as technical problems. To do this, the state propagates a "technocratic consciousness" that Habermas believed represents a new kind of ideology. Unlike previous ideologies, however, it does not promise a future utopia; but, like other ideologies, it is seductive in its ability to veil problems, to simplify perceived options, and to justify a particular way of organizing social life. At the core of this technocratic consciousness is an emphasis on "instrumental reason," or what Weber termed *means/ends rationality*. That is, criteria of the efficiency of means in realizing explicit goals increasingly guide evaluations of social action and people's approach to problems. This emphasis on instrumental reason displaces other types of action, such as behaviors oriented to mutual understanding. This displacement occurs in a series of stages: Science is first used by the state to realize specific goals; then, the criterion of efficiency is used by the state to reconcile competing goals of groupings; next, basic cultural values are themselves assessed and evaluated for their efficiency and rationality; finally, in Habermas' version of *Brave New World,* decisions are completely delegated to computers, which seek the most rational and efficient course of action.

This reliance on the ideology of technocratic consciousnesses creates, Habermas argues, new dilemmas of political legitimation. Habermas believes, capitalist societies can be divided into three basic subsystems: (1) the economic, (2) the politico-administrative, and (3) the cultural (what he later calls *lifeworld*). From this division of societies into these subsystems, Habermas then posits four points of crises: (1) an "economic crisis" occurs if the economic subsystem cannot generate sufficient productivity to meet people's needs; (2) a "rationality crisis" exists when the politico-administrative subsystem cannot generate a sufficient number of instrumental decisions; (3) a "motivation crisis" exists when actors cannot use cultural symbols to generate sufficient meaning to feel committed to participate fully in the society; and (4) a "legitimation crisis" arises when actors do not possess the "requisite number of generalized motivations" or diffuse commitments to the political subsystem's right to make decisions. Much of this analysis of crises is described in Marxian terms but emphasizes that economic and rationality crises are perhaps less important than either motivational or legitimation crises. For, as technocratic consciousness penetrates all spheres of social life and creates productive economies and an intrusive state, the crisis tendencies of late capitalism shift from the inability to produce sufficient economic goods or political decisions to the failure to generate (a) diffuse commitments to political processes and (b) adequate levels of meaning among individual actors.

In *Legitimation Crisis* is an early form of what becomes an important distinction: "Systemic" processes revolving around the economy and the politico-administrative apparatus of the state must be distinguished from "cultural" processes. This distinction will later be conceptualized as *system* and *lifeworld,* respectively, but the central point is this: In tune with his Frankfurt School roots, Habermas is shifting emphasis from Marx's analysis of the economic crisis of production to crises of meaning and commitment; if the problems or crises of capitalist societies are in these areas, then critical theory must focus on the communicative and interactive processes by which humans generate understandings and meanings among themselves. If instrumental reason, or means/ends rationality, is driving out action based on mutual understanding and commitment, then the goal of critical theory is to expose this trend and to suggest ways of overcoming it, especially because legitimation and motivational crises make people aware that something is missing from their lives and, therefore, receptive to more emancipatory alternatives. So the task of critical theory is to develop a theoretical perspective that allows the restructuring of meaning and commitment in social life. This goal will be realized, Habermas argues, by further understanding of how people communicate, interact, and develop symbolic meanings.

EARLY ANALYSES OF SPEECH AND INTERACTION

In 1970 Habermas wrote two articles that marked a return to the idea of the public sphere, but with a new, more theoretical thrust. They also signaled an increasing emphasis on the process of speech, communication, and interaction. In his "On Systematically Distorted Communication," Habermas outlined the nature of undistorted communication.[7] True to Habermas' Weberian origins, this outline is an ideal type. The goal is to determine the essentials and essence of undistorted communication so that those processes that distort communication, such as domination, can be better exposed. What, then, are the features of undistorted communication? Habermas lists five: (1) expressions, actions, and gestures are noncontradictory; (2) communication is public and conforms to cultural standards of what is appropriate; (3) actors can distinguish between the properties of language, per se, and the events and processes that are described by language; (4) communication leads to, and is the product of, intersubjectivity, or the capacity of actors to understand one another's subjective states and to develop a sense of shared collective meanings; and (5) conceptualizations of time and space are understood by actors to mean different things when externally observed and when subjectively experienced in the process of interaction. The details of his analysis on the distortion of communication are less essential than the assertions about what critical theory must conceptualize. For Habermas the conceptualization of undistorted communication is used as a foil for mounting a critique against those social forces that make such idealized communication difficult to realize. Moreover, as his subsequent work testifies, Habermas emphasizes condition 4, or communication and intersubjectivity among actors.

This emphasis became evident in his other 1970 article, "Toward a Theory of Communicative Competence."[8] The details of this argument are not as critical as the overall intent, especially because his ideas undergo subsequent modification. Habermas argues that, for actors to be competent, they must know more than the linguistic rules of how to construct sentences and to talk; they must also master "idealogue-constitutive universals," which are part of the "social linguistic structure of society." Behind this jargon is the idea that the meaning of language and speech is contextual and that actors use implicit stores or stocks of knowledge to interpret the meaning of utterances.

Habermas then proposes yet another ideal type, "the ideal speech situation," in which actors possess all the relevant background knowledge and linguistic skills to communicate without distortion.

Thus, in the early 1970s, Habermas begins to view the mission of critical theory as emphasizing the process of interaction as mediated by speech. But such speech acts draw on stores of knowledge—rules, norms, values, tacit understandings, memory traces, and the like—for their interpretation. These ideals of the speech process represent a restatement of the romanticized public sphere, where issues were openly debated, discussed, and rationally resolved. What Habermas has done, of course, is to restate this view of "what is good and desirable" in more theoretical and conceptual terms, although it could be argued that there is not much difference between the romanticized portrayal of the public sphere and the ideal-typical conceptualization of speech. But with this conceptualization, the goal of critical theory must be to expose those conditions that distort communication and that inhibit realization of the ideal speech situation. Habermas' utopia is thus a society where actors can communicate without distortion, achieve a sense of one another's subjective states, and openly reconcile their differences through argumentation that is free of external constraint and coercion. In other words, he wants to restore the public sphere but in a more encompassing way—that is, in people's day-to-day interactions.

Habermas moved in several different directions in trying to construct a rational approach for realizing this utopia. He borrows metaphorically from psychoanalytic theory as a way to uncover the distortions that inhibit open discourse,[9] but this psychoanalytic journey is far less important than his growing concentration on the process of communicative action and interaction as the basis for creating a society that reduces domination and constraint. Thus, by the mid-1970s, he labels his analysis *universal pragmatics,* whose centerpiece is the "theory of communicative action."[10] This theory will be discussed in more detail shortly, but let us briefly review its key elements. Communication involves more than words, grammar, and syntax; it also involves what Habermas terms *validity claims.* There are three types of claims: (1) those asserting that a course of action as indicated through speech is the most effective and efficient means for attaining ends; (2) those claiming that an action is correct and proper in accordance with relevant norms; and (3) those maintaining that the subjective experiences as expressed in a speech act are sincere and authentic. All speech acts implicitly make these three claims, although a speech act can emphasize one more than the other two. Those responding to communication can accept or challenge these validity claims; if challenged, then the actors contest, debate, criticize, and revise their communication. They use, of course, shared "stocks of knowledge" about norms, means/ends effectiveness, and sincerity to make their claims as well as to contest and revise them. This process (which restates the public sphere in yet one more guise) is often usurped when claims are settled by recourse to power and authority. But if claims are settled by the "giving of reasons for" and "reasons against" the claim in a mutual give-and-take among individuals, then Habermas sees it as "rational discourse." Thus, built into the very process of interaction is the potential for rational discourse that can be used to create a more just, open, and free society. Such discourse is not merely means/ends rationality, for it involves adjudication of two other validity claims: those concerned with normative appropriateness and those concerned with subjective sincerity. Actors thus implicitly assess and critique one another for effectiveness, normative appropriateness, and sincerity of their respective speech acts; so the goal of critical theory is to expose those societal conditions that keep such processes from occurring for *all three types* of validity claims.

In this way, Habermas moves critical theory from Lukács', Horkheimer's, and Adorno's emphasis on subjective consciousness to a concern with *inter*subjective consciousness and the interactive processes by which intersubjectivity is created, maintained, and changed through the validity claims in each speech act. Moreover, rather than viewing the potential for liberating alternatives as residing in subjective consciousness, Habermas could assert that emancipatory potential inheres in each and every communicative interaction. Because speech and communication are the basis of interaction and because society is ultimately sustained by interaction, the creation of less restrictive societies will come about by realizing the inherent dynamics of the communication process.

HABERMAS' RECONCEPTUALIZATION OF SOCIAL EVOLUTION

All critical theory is historical in the sense that it tries to analyze the long-term development of oppressive arrangements in society. Indeed, the central problem of critical theory is to reconcile Marx's and Weber's respective analyses of the development of advanced capitalism. It is not surprising, therefore, that Habermas produces a historical/evolutionary analysis, but, in contrast with Weber, he sees emancipatory potential in evolutionary trends. Yet at the same time he wants to avoid the incorrect prognosis in Marx's analysis and to retain the emancipatory thrust of Marx's approach. Habermas' first major effort to effect this reconciliation appeared in his *The Reconstruction of Historical Materialism,* parts of which have been translated and appear in *Communication and the Evolution of Society.*[12]

Habermas' approach to evolution pulls together many of the themes discussed earlier, so a brief review of his general argument can set the stage for an analysis of his most recent theoretical synthesis, *The Theory of Communicative Action.*[13] In many ways Habermas reintroduces traditional functionalism into Marx's and Weber's evolutionary descriptions, but with both a phenomenological and a structuralist emphasis.

As have all functional theorists, he views evolution as the process of structural differentiation and the emergence of integrative problems. He also borrows from Herbert Spencer, Talcott Parsons, and Niklas Luhmann when he argues that the integration of complex systems leads to an adaptive upgrading, increasing the capacity of the society to cope with the environment.[14] That is, complex systems that are integrated are better adapted to their environments than are less complex systems. The key issue, then, is: What conditions increase or decrease integration? For, without integration, differentiation produces severe problems.

Habermas' analysis of system integration argues that contained in the world views or stocks of knowledge of individual actors are learning capacities and stores of information that determine the overall learning level of a society. In turn, this learning level shapes the society's steering capacity to respond to environmental problems. At times Habermas refers to these learning levels as *organization principles*. Thus, as systems confront problems of internal integration and external contingencies, the stocks of knowledge and world views of individual actors are translated into organization principles and steering capacities, which in turn set limits on just how a system can respond. For example, a society with only religious mythology will be less complex and less able to respond to environmental challenges than a more complex society with large stores of technology and stocks of normative procedures determining its organization principles. But societies can "learn"[15] that, when confronted with problems beyond the capacity of their current organization principles and steering mechanisms, they can draw upon the "cognitive potential" in the world views and stocks of knowledge of individuals who reorganize their actions. The result of this learning creates new levels of information that allow the development of new organization principles for securing integration despite increased societal differentiation and complexity.

The basis for societal integration lies in the processes by which actors communicate and develop mutual understandings and stores of knowledge. To the extent that these interactive processes are arrested by the patterns of economic and political organization, the society's learning capacity is correspondingly diminished. One of the main integrative problems of capitalist societies is the integration of the material forces of production (economy as administered by the state), on the one side, and the cultural stores of knowledge that are produced by communicative interaction, on the other side. Societies that differentiate materially in the economic and political realms without achieving integration on a normative and cultural level (that is, shared understandings) will remain unintegrated and experience crises.

Built into these dynamics, however, is their resolution. The processes of "communicative interaction" that produce and reproduce unifying cultural symbols must be given equal weight with the "labor" processes that generate material production and reproduction. At this point Habermas developed his more synthetic approach in *The Theory of Communicative Action.*

THE THEORY OF COMMUNICATIVE ACTION

The two-volume *The Theory of Communicative Action* pulls together into a reasonably coherent framework various strands of Habermas' thought.[16] Yet, true to his general style of scholarship, Habermas wandered over a rather large intellectual landscape. In Thomas McCarthy's words, Habermas develops his ideas through "a somewhat unusual combination of theoretical constructions with historical reconstructions of the ideas of 'classical' social theorists."[17] Such thinkers as Marx, Weber, Durkheim, Mead, Lukács, Horkheimer, Adorno, and Parsons are, for Habermas, "still very much alive" and are treated as "virtual dialogue partners."[18] As a consequence, the two volumes meander through selected portions of various thinkers' work critiquing and yet using key ideas. After the dust settles, however, the end result is a very creative synthesis of ideas into a critical theory.

Habermas' basic premise is summarized near the end of volume 1:

> If we assume that the human species maintains itself through the socially coordinated activities of its members and that this coordination is established through communication—and in certain spheres of life, through communication aimed at reaching agreement—then the reproduction of the species also requires satisfying the conditions of a rationality inherent in communicative action.[19]

In other words, intrinsic to the process of communicative action, where actors implicitly make, challenge, and accept one another's validity claims, is a rationality that can potentially serve as the basis for reconstructing the social order in less oppressive ways. The first volume of *The Theory of Communicative Action* thus focuses on action and rationality in an effort to reconceptualize both processes in a manner that shifts emphasis from the subjectivity and consciousness of the individual to the process of symbolic interaction. In a sense, volume 1 is Habermas' microsociology, whereas volume 2 is his macrosociology. In the second volume, Habermas introduces the concept of system and tries to connect it to microprocesses of action and interaction through a reconceptualization of the phenomenological concept of lifeworld.

The Overall Project

Let us begin by briefly reviewing the overall argument, and then return to volumes 1 and 2 with a more detailed analysis. There are four types of action: (1) teleological, (2) normative, (3) dramaturgical, and (4) communicative. Only communicative action contains the elements whereby actors reach intersubjective understanding. Such communicative action—which is, actually, *inter*action—presupposes a set of background assumptions and stocks of knowledge, or, in Habermas' terms, a *lifeworld.* Also operating in any society are "system" processes, which revolve around the material maintenance of the species and its survival. The evolutionary trend is for system processes and lifeworld processes to become internally differentiated *and* differentiated from each other. The integration of a society depends on a balance between system and lifeworld processes. As modern societies have evolved, however, this balance has been upset as system processes revolving around the economy and the state (also law, family, and other reproductive structures) have "colonized" and dominated lifeworld processes concerned with mutually shared meanings, understandings, and intersubjectivity. As a result, modern society is poorly integrated.

These integrative problems in capitalist societies are manifested in crises concerning the "reproduction of the lifeworld"; that is, the acts of communicative interaction that reproduce this lifeworld are displaced by "delinguistified media," such as money and power, that are used in the reproduction of system processes (economy and government). The solution to these crises is a rebalancing of relations between lifeworld and system. This rebalancing is to come through the resurrection of the public sphere in the economic and political arenas and in the creation of more situations in which communicative action (interaction) can proceed uninhibited by the intrusion of system's media, such as power and money. The goal of critical theory, therefore, is to document those facets of society in which the lifeworld has been colonized and to suggest approaches whereby situations of communicative action (interaction) can be reestablished. Such is Habermas' general argument, and now we can fill in some of the details.

The Reconceptualization of Action and Rationality

In volume 1 of *The Theory of Communicative Action,* Habermas undertakes a long and detailed analysis of Weber's conceptualization of action and rationalization. He wants to reconceptualize rationality and action in ways that allow him to view rational action as a potentially liberating rather than imprisoning force.[20] In this way, he feels, he can avoid the pessimism of Weber and the retreat into subjectivity of Lukács, Adorno, and Horkheimer. There are, Habermas concludes, several basic types of action:[21]

1. *Teleological action* is behavior oriented to calculating various means and selecting the most appropriate ones to realize explicit goals. Such action becomes strategic when other acting agents are involved in one's calculations. Habermas also calls this action "instrumental" because it is concerned with means to achieve ends. Most important, he emphasizes that this

kind of action is too often considered to be "rational action" in previous conceptualizations of rationality. As he argues, this view of rationality is too narrow and forces critical theory into a conceptual trap: if teleological or means/ends rationality has taken over the modern world and has, as a consequence, oppressed people, then how can critical theory propose rational alternatives? Would not such a rational theory be yet one more oppressive application of means and ends rationality? The answers to these questions lie in recognizing that there are several types of action and that true rationality resides not in teleological action but in communicative action.

2. *Normatively regulated action* is behavior that is oriented to common values of a group. Thus, normative action is directed toward complying with normative expectations of collectively organized groupings of individuals.

3. *Dramaturgical action* is action that involves conscious manipulation of oneself before an audience or public. It is ego-centered in that it involves actors mutually manipulating their behaviors to present their own intentions, but it is also social in that such manipulation is done in the context of organized activity.

4. *Communicative action* is interaction among agents who use speech and nonverbal symbols as a way of understanding their mutual situation and their respective plans of action to agree on how to coordinate their behaviors.

These four types of action presuppose different kinds of "worlds." That is, each action is oriented to a somewhat different aspect of the universe, which can be divided into the (1) "objective or external world" of manipulable objects; (2) "social world" of norms, values, and other socially recognized expectations; and (3) "subjective world" of experiences. Teleological action is concerned primarily with the objective world; normatively regulated action with the social; and dramaturgical with the subjective and external. But only with communicative action do actors "refer simultaneously to things in the objective, social, and subjective worlds in order to negotiate common definitions of the situation."[22]

Such communicative action is therefore potentially more rational than all of the others because it deals with all three worlds and because it proceeds as speech acts that assert three types of validity claims. Such speech acts assert that (1) statements are true in "propositional content," or in reference to the external and objective world; (2) statements are correct with respect to the existing normative context, or social world; and (3) statements are sincere and manifest the subjective world of intention and experiences of the actor.[23] The process of communicative action in which these three types of validity claims are made, accepted, or challenged by others is inherently more rational than other types of action. If a validity claim is not accepted, then it is debated and discussed in an effort to reach understanding without recourse to force and authority.[24] The process of reaching understanding through validity claims, their acceptance, or their discussion takes place against

> The background of a culturally ingrained preunderstanding. This background remains unproblematic as a whole; only that part of the stock of knowledge that participants make use of and thematize at a given time is put to the test. To the extent that definitions of situations are negotiated by participants *themselves,* this thematic segment of the lifeworld is at their disposal with the negotiation of each new definition of the situation.[25]

Thus, in the process of making validity claims through speech acts, actors use existing definitions of situations or create new ones that establish order in their social relations. Such definitions become part of the stocks of knowledge in their lifeworlds, and they become the standards by which validity claims are made, accepted, and challenged. Thus, in reaching an understanding through communicative action, the lifeworld serves as a point of reference for the adjudication of validity claims, which encompass the full range of worlds—the objective, social, and subjective. And so, in Habermas' eyes, there is more rationality inherent in the very process of communicative interaction than in means/ends or teleological action.[26] As Habermas summarizes,

> We have . . . characterized the rational structure of the processes of reaching understanding in terms of (a) the three world-relations of actors and the corresponding concepts of the objective, social, and subjective worlds; (b) the validity claims of propositional truth, normative rightness, and sincerity or authenticity; (c) the concept of a rationally motivated agreement, that is, one based on the intersubjective recognition of criticizable validity claims; and (d) the concept of reaching understanding as the cooperative negotiation of common definitions of the situation.[27]

Thus, as people communicatively act (interact), they use and at the same time produce common definitions of the situation. Such definitions are part of the lifeworld of a society; if they have been produced and reproduced through the communicative action, then they are the basis for the rational and nonoppressive integration of a society. Let us now turn to Habermas' discussion of this lifeworld, which serves as the "court of appeals" in communicative action.

The Lifeworld and System Processes of Society

Habermas believes the lifeworld is a "culturally transmitted and linguistically organized stock of interpretative patterns." But what are these "interpretative patterns" about? What do they pertain to? His answer, as one expects from Habermas, is yet another typology. There are three different types of interpretative patterns in the lifeworld: There are interpretative patterns with respect to culture, or systems of symbols; there are those pertaining to society, or social institutions; and there are those oriented to personality, or aspects of self and being. That is, (1) actors possess implicit and shared stocks of knowledge about cultural traditions, values beliefs, linguistic structures and their use in interaction; (2) actors also know how to organize social relations and what kinds and patterns of coordinated interaction are proper and appropriate; and (3) actors understand what people are like, how they should act, and what is normal or aberrant.

These three types of interpretative patterns correspond, Habermas asserts, to the following functional needs for reproducing the lifeworld (and, by implication, for integrating society): (1) reaching understanding through communicative action transmits, preserves, and renews cultural knowledge; (2) communicative action that coordinates interaction meets the need for social integration and group solidarity; and (3) communicative action that socializes agents meets the need for the formation of personal identities.[28]

Thus, the three components of the lifeworld—culture, society, personality—meet corresponding needs of society—cultural reproduction, social integration, and personality formation—through three dimensions along with communicative action is conducted: reaching understanding, coordinating interaction, and effecting socialization. As Habermas summarizes in volume 2,

> In coming to an understanding with one another about their situation, participants in communication stand in a cultural tradition which they use and at the same time renew; in coordinating their actions via intersubjective recognition of criticizable validity claims, they rely upon their membership in groupings and at the same time reenforce their integration; through participating in interaction with competent persons, growing children internalize value orientations and acquire generalized capacities for action.[29]

These lifeworld processes are interrelated with system processes in a society. Action in economic, political, familial, and other institutional contexts draws on, and reproduces, the cultural, societal, and personality dimensions of the lifeworld. Yet evolutionary trends are for differentiation of the lifeworld into separate stocks of knowledge with respect to culture, society, and personality and for differentiation of system processes into distinctive and separate institutional clusters, such as economy, state, family, and law. Such differentiation creates problems of integration and balance between the lifeworld and system.[30] And therein reside the dilemmas and crises of modern societies.

Evolutionary Dynamics and Societal Crises

In a sense, Habermas blends traditional analysis by functionalists on societal and cultural differentiation with a Marxian dialectic whereby the seeds for emancipation are sown in the creation of an ever more rationalized and differentiated society. Borrowing from Durkheim's analysis of mechanical solidarity, Habermas argues that "the more cultural traditions predecide which validity claims, when, where, for what, from whom, and to whom must be accepted, the less the participants themselves have the possibility of making explicit and examining the potential groups in which their yes/no positions are based."[31] But "as mechanical solidarity gives way to organic solidarity based upon functional interdependence," then "the more the worldview that furnishes the cultural stock of knowledge is decentered" and "the less the need for understanding is covered *in advance* by an interpreted lifeworld immune from critique," and therefore "the more this need has to be met by the interpretative accomplishments of the participants themselves." That is, if the lifeworld is to be sustained and reproduced, it becomes ever more necessary with growing societal complexity for social actions to be based on communicative processes. The result is that there is greater potential for rational communicative action because less and less of the social order is preordained by a simple and undifferentiated lifeworld. But system processes have reduced this potential, and the task of critical theory is to document how system processes have colonized the lifeworld and thereby arrested this potentially superior rationality inherent in the speech acts of communicative action.

How have system processes restricted this potential contained in communicative action? As the sacred and traditional basis of the lifeworld organization has dissolved and been replaced by linguistic interaction around a lifeworld differentiated along cultural, social, and personality axes, there is a countertrend in the differentiation of system processes. System evolution involves the expansion of material production through the greater use of technologies, science, and "delinguistified steering mechanisms" such as money and power to carry out system processes.[32] These media do not rely on the validity claims of communicative action; when they become the media of interaction in ever more spheres of life—markets, bureaucracies, welfare state policies, legal systems, and even family relations—the processes of communicative action so essential for lifeworld reproduction are invaded and colonized. Thus, system processes use power and money as their media of integration, and in the process they "decouple the lifeworld" from its functions for societal integration.[33] There is an irony here because differentiation of the lifeworld facilitated the differentiation of system processes and the use of money and power,[34] so "the rationalized lifeworld makes possible the rise of growth of subsystems which strike back at it in a destructive fashion."[35]

Through this ironical process, capitalism creates market dynamics using money, which in turn spawn a welfare state employing power in ways that reduce political and economic crises but that increase those cries revolving around lifeworld reproduction. For the new crises and conflicts "arise in areas of cultural reproduction, of social integration and of socialization."[36]

THE GOAL OF CRITICAL THEORY

Habermas has now circled back to this initial concerns and those of early critical theorists. He has recast the Weberian thesis by asserting that true rationality inheres in communicative action, not teleological (and strategic or instrumental) action, as Weber claimed. And he has redefined the critical theorist's view on modern crises; they are not crises of rationalization, but crises of colonization of those truly rational processes that inhere in the speech acts of communicative action, which reproduce the lifeworld so essential to societal integration. Thus, built into the integrating processes of differentiated societies (note: not the subjective processes of individuals, as early critical theorists claimed) is the potential for a critical theory that seeks to restore communicative rationality despite impersonal steering mechanisms. If system differentiation occurs in delinguistified media, like money and power, and if these reduce the reliance on communicative action, then crises are inevitable. The resulting collective frustration over the lack of meaning in social life can be used by critical theorists to mobilize people to restore the proper balance between system and lifeworld processes. Thus, crises of material production will not be the impetus for change, as Marx contended. Rather, the crises of lifeworld reproduction will serve as the stimulus to societal reorganization. And returning to his first work, Habermas sees such reorganization as involving (1) the restoration of the public sphere in politics, where relinguistified debate and argumentation, rather than delinguistified power and authority, are used to make political decisions (thus reducing "legitimation crises"), and (2) the extension of communicative action back into those spheres—family, work, and social relations—that have become increasingly dominated by delinguistified steering media (thereby eliminating "motivational crises").

The potential for this reorganization inheres in the nature of societal integration through the rationality inherent in the communicative actions that reproduce the lifeworld. The purpose of critical theory is to release this rational potential.

NOTES

1. That is, the spread of means/end rationality into ever more spheres of life.

2. Jürgen Habermas, *The Theory of Communicative Action* (Boston: Beacon, 1981, 1984). Some useful reviews and critiques of Habermas' work include John B. Thompson and David Held, eds., *Habermas: Critical Debates* (London: Macmillan, 1982); David Held, *An Introduction to Critical Theory* (London: Hutchinson, 1980), chapters 9–12.

3. Jürgen Habermas, *Struckturwandel der Offentlichkeit* (Neuwied, Germany: Luchterhand, 1962).

4. Jürgen Habermas, *Zur Logik der Sozialwissenschaften* (Frankfurt: Suhrkamp, 1970).

5. Jürgen Habermas, *Knowledge and Human Interest,* trans. J. Shapiro (London: Heinemann, 1970; originally published in German in 1968). The basic ideas in *Zur Logik der Sozialwissenschaften* and *Knowledge and Human Interest* were stated in Habermas' inaugural lecture at the University of Frankfurt in 1965 and were first published in "Knowledge and Interest," *Inquiry* 9 (1966), pp. 285–300.

6. Jürgen Habermas, *Legitimation Crisis*, trans. T. McCarthy (London: Heinemann, 1976; originally published in German in 1973).

7. Jürgen Habermas, "On Systematically Distorted Communication," *Inquiry* 13 (1970); pp. 205–218.

8. Jürgen Habermas, "Toward a Theory of Communicative Competence," *Inquiry* 13 (1970), pp. 360–375.

9. Habermas sometimes calls this aspect of his program "depth hermeneutics." The idea is to create a methodology of inquiry for social systems that parallels the approach of psychoanalysis—that is, dialogue, removal of barriers to understanding, analysis of underlying causal processes, and efforts to use this understanding to dissolve distortions in interaction.

10. For an early statement, see "Some Distinctions in Universal Pragmatics: A Working Paper," *Theory and Society* 3 (1976), pp. 155–167.

11. Jürgen Habermas, *Zur Rekonstruktion des Historischen Materialismus* (Frankfurt: Suhrkamp, 1976).

12. Jürgen Habermas, *Communication and the Evolution of Society*, trans. T. McCarthy (London: Heinemann, 1979).

13. For an earlier statement, see Jürgen Habermas, "Towards a Reconstruction of Historical Materialism," *Theory and Society* 2 (3, 1975), pp. 84–98.

14. He borrows from Niklas Luhmann here (see chapter 5 in this work), although much of Habermas' approach is a reaction to Luhmann.

15. Habermas analogizes here to Jean Piaget's and Lawrence Kohlberg's analysis of the cognitive development of children, seeing societies as able to "learn" as they become more structurally complex.

16. Jürgen Habermas, *The Theory of Communicative Action*, 2 vols. (cited in note 2). The subtitle of volume 1, *Reason and the Rationalization of Society*, gives some indication of its thrust. The translator, Thomas McCarthy, has done an excellent service in translating very difficult prose. Also, his "Translator's Introduction" to volume 1, pp. v–xxxvii, is the best summary of Habermas' recent theory that I have come across.

17. Thomas McCarthy, "Translator's Introduction" (cited in note 16), p. vii.

18. Ibid.

19. Jürgen Habermas, *The Theory of Communicative Action* (cited in note 16), vol. 1, p. 397.

20. Recall that its subtitle is *Reason and the Rationalization of Society.*

21. Jürgen Habermas, *The Theory of Communicative Action* (cited in note 16), pp. 85–102.

22. Ibid., p. 95.

23. Ibid., p. 99.

24. Recall Habermas' earlier discussion of nondistorted communication and the ideal speech act. This is his most recent reconceptualization of these ideas.

25. Ibid., p. 100. Emphasis in original.

26. Ibid., p. 302.

27. Ibid., p. 137.

28. We are now into volume 2, ibid., pp. 205–240, entitled *System and Lifeworld: A Critique of Functionalist Reason*, which is a somewhat ironic title because of the heavily functional arguments in volume 2. But, as noted earlier, Habermas' earlier work has always had an implicit functionalism.

29. Ibid., p. 208

30. This is the old functionalist argument of "differentiation" producing "integrative problems," which is as old as Spencer and which is Parsons reincarnated with a phenomenological twist.

31. All quotes here are from p. 70 of volume 1.

32. Here Habermas is borrowing from Simmel's analysis in *The Philosophy of Money* (see Chapter 19 of this work) and from Parsons' conceptualization of generalized media (see Chapter 4).

33. Volume 2 of *The Theory of Communicative Action,* pp. 256–276.

34. Habermas appears in these arguments to borrow heavily from Parsons' analysis of evolution (see Chapter 4 of this work).

35. Volume 2, p. 227.

36. Ibid., p. 576.

43

The Continuing Tradition II:

The Feminist Critique of Sociological Theory: Gender, Politics, and Patriarchy*

Since the early 1970s, one of the most sustained challenges to "mainstream" sociological theory has come from critical feminist theorists. Appearing shortly after the beginning of the "second wave" of the women's movement in the mid-to-late 1960s, the first feminist critiques focused on the underrepresentation of women and women's experiences within sociology, both as the subjects of research and the producers of theory. Concurrently, feminist theorists examined the construction of gender and sex roles in modern society to demonstrate the existence of a "female world" that sociology had hitherto ignored.[1] Subsequent critiques went further, as feminists used the concepts of gender and patriarchy to reveal masculine (or androcentric) stances in social research methodologies and in sociological theory. These more radical critiques questioned the capacity of sociological research and theory, as a body of knowledge constructed from the experiences of men, to address the experiences of women. Critical feminist theorists proposed alternative methodological approaches, including the construction of a "feminist standpoint" or women's sociology that would begin with the social universe of women and reflect women's perspectives on society.[2] In the past decade, several radical feminist theorists have also used the epistemological issues inherent in feminist methodologies to critique the "positivistic" foundations of sociological understanding and to lay the groundwork for a "feminist epistemology."[3]

The feminist critique, like much of sociological theory at present, does not form a coherent paradigm. There is little consensus among critical feminist theorists about what constitutes "the feminist critique" or about

*This chapter is coauthored with P.R. Turner

how sociological understandings should be restructured to obviate the criticisms of feminists. Nevertheless, several common threads distinguish the feminist position from other forms of critical theory. The critical feminist theorists discussed in this chapter share with their more scientifically oriented colleagues examined in Chapter 18 the conviction that gender represents a fundamental form of social division within society. They also share a commitment to analyzing the sources of oppression and inequality for women, the most important being the patriarchal structure of society and its institutions. These two groups differ, however, in their focus, methods, and purpose. Instead of examining gender inequalities as a form of conflict-producing stratification, the feminist theorists examined in this chapter have focused on gender and forms of patriarchy to criticize social research practices and the production of sociological theory itself.[4] In their analyses of theory production as a form of patriarchy, many of these feminist theorists have not felt bound by "positivistic" methods of inquiry that guided the theorists examined in Chapter 18; indeed, they have questioned the legitimacy and objectivity of these methods by exploring the ways in which they embody androcentric (male-oriented) modes of thought. Finally, these critical feminist theorists are often self-consciously aware of the political and practical implications of their attacks for feminist politics and the status of women. As Mary Jo Neitz stated, "the critical questions for feminist scholarship came out of the women's movement, not out of the disciplines."[5] Consequently, most critical feminist theorists want to preserve the emancipatory dimension of feminist theory—that is, its ability to serve, like Marxism, as "both a mode of understanding and a call to action."[6] This self-conscious conflation of theory, method, politics, and praxis, combined with a focus on gender and patriarchy as the primary sources of oppression and inequality, constitute the common denominators distinguishing the feminist critique of sociological theory.

REPRESENTATION AND THE CONSTRUCTION OF GENDER

Early Challenges to Social Science

Early feminist critiques of sociological research and theory were concerned primarily with the issue of representation. Kathryn Ward and Linda Grant, in an analysis of more than 700 gender-related articles in sociology journals between 1974 and 1983, asserted that the feminist critique of gender inequities within sociology dealt with the following four issues:[7]

1. Omission and underrepresentation of women as research subjects.
2. Concentration on masculine-dominated sectors of social life.
3. Use of paradigms, concepts, methods and theories that more faithfully portray men's than women's experiences.
4. Use of men and male lifestyles as the norms against which social phenomena were interpreted.

One of the most influential early works that addressed all of these issues is Ann Oakley's survey of housewives and their opinions of housework. Oakley argued that discrimination against women in society is mirrored by sexism in sociology, and she used the academic neglect of housework as *work* to address the broader issue of sexual bias within sociological research and theory in general.[8] Oakley stated that women's "invisibility" in sociology can be seen in all the major subject areas of sociology. In the subject area of deviance, for example, Oakley argued that until the mid-1970s very little data had been collected on women and that "theories of deviance may include some passing reference to women, but interpretations of female behavior are uncomfortably subsumed under the umbrella of explanation geared to the model of masculine behavior."[9] She also questioned whether or not standard definitions of deviance consider patterns of behavior that are gender-related or associated only with women.[10] With regard to social stratification theory, to take an example from another prominent subject area of sociology, Oakley posited that the following untested assumptions about class membership effectively render women invisible and irrelevant: (1) The family is the unit of stratification; (2) the social position of the family is determined by the status of the man in it; and (3) only in rare circumstances is a woman's social position not determined by the men to whom they are attached by marriage or family of origin.[11] Oakley argued that these assumptions often do not reflect social reality, because many people do not live in families, many families are headed by women, and many husbands and wives do not have identical social status rankings. The problematics of these assumptions would be revealed, according to Oakley, if the significance of gender as a criterion for social differentiation and stratification is recognized by sociologists. Without such recognition, women's roles and position in the social stratification system would continue to be hidden and misrepresented.[12]

Oakley attributed the inherent sexism in sociological theory and research to the male-oriented attitudes of its "founding fathers,"[13] the paucity of women social researchers and theorists, and the pervasiveness of ideologies advocating gender roles within contemporary societies. The ideology of gender, she argued, contains stereotypical assumptions about women's social status and behavior that are uncritically reproduced within sociology, and these stereotypes will only be overcome if women's experiences are made the focus of analysis and viewed from their perspective.[14]

Most of these early critiques of sociological research and theory shares with Oakley two related assumptions about the primacy and construction of gender: First, gender is a fundamental determinant of social relations and behavior, and second, gender divisions in a society shape the experiences and perspectives of each sex, with the result that women's experiences are distinctly different from those of men. Some feminist scholars went further than Oakley and began to argue that society is gendered in ways that segregate men and women into distinct and often exclusionary homosocial worlds. Building on Georg Simmel's insight that "women possess a world of their own which is not comparable with the world of men,"[15] Jessie Bernard argued in *The Female World* that society is divided into "single-sex" worlds. Sociology and other disciplines in the humanities and human sciences have dealt heretofore almost exclusively with the male world. Bernard has sought to correct this imbalance by tracing the historical development of the female world and the uniquely female experiences and perspectives that have emerged from it. She has argued that the female world differs subjectively and objectively from that of men, and hence, the female world must be examined "as an entity in its own right, not as a byproduct of the male world."[16] The neglect of the female world and women's experiences by sociology and other disciplines, Bernard has asserted, deprives public debate of perspectives that might provide innovative solutions and approaches to contemporary problems.[17]

Other feminist theorists have used the concept of gendered social spheres to explore more specifically the stratification of sex roles and the social construction of gender. Jean

Lipman-Blumen proposed a homosocial theory of sex roles to account for the traditional barriers that restrict women's entry into male spheres—such as politics, military, and major league sports—and that confine women to the domestic sphere.[18] She hypothesized that men are socialized to be attracted to, and to be interested in, other men. This attraction is reinforced and perpetuated by patriarchal social institutions that traditionally value men over women and give to men nearly exclusive control of resources.[19] With the important exceptions of reproductive and sexual needs, men look to men for support, whereas women are forced to transform themselves into sex objects to acquire resources and support from men. Lipman-Blumen argued the preeminence of men in the exchange and protection of resources creates "dominance hierarchies" that persist even when technology eliminates the need for the differentiation and stratification of sex roles.[20]

In her highly influential book, *The Reproduction of Mothering: Psychoanalysis and the Sociology of Gender*, Nancy Chadorow theorized that women's responsibility for childrearing (and the collateral absence of male domestic roles) has profound consequences for the construction of gender identity and the sexual division of labor. Chadorow posits, "all societies are constituted around a structural split, growing out of women's mothering, between the private, domestic world of women and the public, social world of men."[21] This structural split is created by a sexual division of labor that is itself reproduced each generation by gender personality differences between men and women. These behavioral differences or "intrapsychic structures" are not biological; rather, they stem from the distinct social relationships girls and boys develop with their mothers. Girls learn to be women and to mother by identifying with their mothers, whereas boys must develop a masculine gender identification in opposition to their mothers and often in the absence of affective, on-going relationships to a father figure. This results in different "relational capacities" and "senses of self" in men and women that prepare them "to assume the adult gender roles which situate women primarily within the sphere of reproduction in a sexually unequal society."[22]

A Sociology for Women: Feminist Methodologies, Epistemologies, and "Standpoint" Theories

Efforts in the late 1970s and 1980s to devise a sociology for women arose from a growing conviction among more radical critical feminist theorists that to simply refocus the discipline's theoretical and methodological lenses on gender and women's domains would do little to correct the androcentric (male) biases and patterns of patriarchal thought inhering in sociological work. Because of these biases, traditional forms of explanation within sociology were increasingly seen as fundamentally incapable of representing accurately a social world in which, many feminist critics began to assert, all social relations are gendered.[23] Recognizing the link between feminist thought and politics, John Shotter and Josephine Logan argue that only by finding a "new voice" can feminist scholars and the women's movement as a whole escape the "pervasiveness of patriarchy":

> The women's movement must of necessity develop itself within a patriarchal culture of such a depth and pervasiveness that, even in reacting to or resisting its oppressive nature, the women's movement continually "reinfects" or "contaminates" itself with it. All of us, women and men alike, are "soaked" in it. . . . Patriarchy is enshrined in our social practices, in our ways of positioning and relating ourselves to one another, and in the re-

> sources we use in making sense of one another. . . . We must find a different voice, a new place currently unrecognized, from which to speak about the nature of our lives together.[24]

This kind of more radical attack on the sociology of knowledge was fueled by a growing frustration among feminist critics in general about what they saw as a continued resistance within sociology—and sociological theory in particular—to study gender and to draw out the conceptual and theoretical implications of this research. Grant and Ward, for example, found in their analysis of articles published in ten sociology journals between 1974 and 1983 that the number of theoretical papers, reviews, and critiques focusing on gender in mainstream journals was still relatively small compared with other traditional topics—an indication, they feel, that journal editors continue to view gender as a peripheral in contemporary sociology.[25]

Other feminist critics have echoed Grant and Ward's concerns. Acker's review of stratification literature asserts that, aside from Rae Lesser Blumberg's landmark book (see Chapter 18),[26] stratification texts still do not "successfully integrate women into the analysis and generally evade the problem by including brief descriptions of sex-based inequality generally ungrounded in a conceptualization of societal-wide stratification."[27] Judith Stacey and Barrie Thorne conclude that although feminist scholars have made valuable contributions to numerous traditional branches of sociological research (for example, organizations, occupations, criminology, deviance and stratification), and have pioneered work in many others (for example, sexual harassment, feminization of poverty), they had yet to effect significant conceptual transformations in the field.[28] Feminist scholars in sociology—unlike their counterparts in anthropology, history, and literary criticism—have not succeeded in influencing the discipline to the point where women are being put regularly at the center of analysis. Instead, Stacey and Thorne argue that "feminist sociology seems to have been both co-opted and ghettoized, while the discipline as a whole and its dominant paradigms have proceeded relatively unchanged."[29]

Feminist scholars propose a variety of explanations to account for the "ghettoization" of gender issues within sociological research and theory. In their study of women's involvement in theory production, Ward and Grant emphasize that the relative lack of theorizing on gender might be partly because of the scarcity and low profile of women theorists. Compared with their male colleagues, Ward and Grant found that women sociologists (1) affiliate with the ASA theory section less, (2) self-identify as theorists less, (3) write fewer textbooks and journal articles, (4) receive less visibility in textbooks and popular teaching materials, and (5) serve less frequently as editors or board members of theory journals.[30] As possible explanations for the "peculiar eclipsing" of women as theorists Ward and Grant cite (1) the comparatively high status of theory production which, in turn, can lead to higher barriers against the entry of women, and (2) the fragmentation of contemporary sociological theory into multiple and competing paradigms that can restrict the spread of feminist thought.

Ward and Grant also speculate that differences in how women sociologists approach their research subjects might decrease the likelihood of their work on gender being accepted for publication. They have found that there are "systematic links" between gender and methods within sociology, with women scholars publishing in major sociological journals employing qualitative methods more often than their male colleagues.[31] Many feminist theorists argue that qualitative methods (for example, intensive interviews and participant observation) are more appropriate for exploring gender and women's issues, which tend to be more private, context-bound, and hence, less easily quantifiable.[32] Yet Grant and Ward have failed to find a correlation between gender and methods in published research on gender among mainstream sociological journals. In gender articles, both sexes preferred quantitative methods, leading them to speculate that "qualitative papers on gender might have presented double nonconformity, reducing the likelihood of acceptance for publication."[33] Finally, Ward and Grant also refer to Judith A. Howard's contention that the influence of feminist perspectives is limited by "the inability of extant social theories (including Marxism) to conceptualize gender as a major organizing principle of society and culture and by the predominance of positivist epistemological traditions that encourage emphasis on gender as a variable rather than a concept."[34]

Acker takes a different tack, suggesting that sociological theory "continues in a prefeminist mode" because of both institutional resistance within the discipline and the underdevelopment of feminist alternatives to mainstream theoretical concepts and methodologies.[35] Sociology, Acker argues, shares with other academic disciplines

> A particular connection to power in society as a whole or to the relations of ruling: The almost exclusively male domain of academic thought is associated with abstract, intellectual, textually mediated processes through which organizing, managing and governing are carried out. . . . The perspectives that develop their concepts and problematics from within what is relevant to the relations of ruling are successful.[36]

Acker questions whether critical feminist theories can ever effect a paradigm shift within sociological theory as long as societal institutions remain patriarchally structured and power relations continue to be dominated by men.[37] Yet, Acker also holds feminist theory accountable for its relative lack of influence within mainstream sociological theory. Feminist scholars, she argues, "have not, as yet, been able to suggest new ways of looking at things that are obviously better than the old ways for comprehending a whole range of problems. . . ."[38] Acker outlines her vision of a feminist paradigm capable of competing with or supplanting established theoretical positions. Such a paradigm would[39]

1. Provide a better understanding of class structure, the state, social revolution, and militarism, as well as a better understanding of the sex segregation of labor, male dominance in the family, and sexual violence.
2. Place women and their lives in a central place in understanding social relations as a whole, while creating a more accurate and comprehensive account of industrial, capitalist society.
3. Contain a methodology that produces knowledge *for* rather than *of* women in all their diverse situations.

Implicit in Acker's proposal for a feminist paradigm is a radical critique of the very concepts, methodologies, and epistemological assumptions that form the foundation of sociological understandings of the social universe. As espoused by feminist theorists such as Dorothy Smith, Sandra Harding, and Evelyn Fox Keller, feminist theorists have failed thus far to transform prevailing mainstream theoretical paradigms because the underlying epistemologies and methodologies give privilege to the male experience.[40] Dorothy Smith, one of the first feminist critical theorists to call for a "woman-centered" sociology, asserts that sociological thought has been "based on and built up within the male social universe,"[41] and as a consequence, sociology contains unexamined androcentric (male) modes of thought that serve the interests of men and that are by definition gender biased and exclusionary of women's perspectives.[42] She argues that as long as feminist scholars work within forms of thought made or controlled by men, women will be constrained to view themselves not as subjects but as the "other" and their experience will be marginalized accordingly.[43] It is not enough, she claims,

> To supplement an established sociology by addressing ourselves to what has been left out, overlooked, or by making sociological issues of the relevances of the world of women. That merely extends the authority of the existing sociological procedures and makes of a women's sociology an addendum.[44]

This radical feminist critique has taken three main forms. First, it criticizes standard methodological practices and proposes to replace them with feminist methodologies. Second, it questions the epistemological assumptions of "positivistic" science because science, like the broader society, is gendered, and, therefore, new epistemologies need to be developed that eliminate gender bias while recognizing the primacy of gender and its implications for knowledge production. Finally, radical feminist theorists propose the construction of an independent feminist "standpoint" that can avoid both the androcentric biases inherent in social research practices and the "positivistic" epistemological foundations of mainstream sociological theory.[45]

One of the first detailed feminist critiques of contemporary social methodology was a volume of essays edited by sociologists Marcia Millman and Rosabeth Moss Kanter.[46] Viewing the various critiques collectively, Millman and Moss identify six standard methodological practices that can lead to gender bias in social research. During the past twenty years, these practices have remained central to the feminist critique of social methodology:

1. The use of conventional field-defining models that overlook important areas of social inquiry.
2. The focus by sociologists on public, official, visible role players to the neglect of unofficial, supportive, private, and invisible spheres of social life and organization that are equally important.
3. The assumption in sociology of a "single society," in which generalizations can be made that will apply equally to men and women.
4. The neglect in numerous fields of study of sex as an important explanatory variable.
5. The focus in sociology on explaining the status quo, which tends to provide rationalizations for existing power relations.
6. The use of certain methodological techniques, often quantitative, and research situations that might systematically prevent the collection of certain kinds of data.[47]

First, Millman and Kanter argue that sociologists' reliance on models of social structure and action has led to a "systematic blindness to crucial elements of social reality."[48] For example, they assert that the sociological focus on Weberian rationality as an explanation for human action and social organization effectively removes the equally important element of emotion from consideration. They also question the veracity of sociological models that do not focus on the individual and his or her subjective experience as the center of analysis and instead emphasize issues of agency.[49] Bernard had earlier made a distinction between "agency," which emphasizes variables, and "communion," which focuses on individuals[50]:

> Agency operates by way of mastery and control; communion with naturalistic observation, sensitivity to qualitative patterning, and greater personal participation by the investigator. . . . The specific processes involved in agentic (sic) research are typically male preoccupations. . . . The scientist using this approach creates his own controlled reality. He can manipulate it. He is master. He has power. The communal approach is much humbler. It disavows control, for control spoils the results. Its value rests precisely on the absence of controls.

Millman and Kanter make the point that research based exclusively on "agentic" quantitative methods fails to represent accurately crucial segments of the social world.[51]

Second, sociology overlooks important arenas of social life by focusing only on "official actors and actions," and ignoring private, unofficial and local social structures where women often predominate.[52] Millman, for example, posits that research on deviance and social control often emphasizes locations such as courtrooms and mental hospitals, but fails to recognize "the importance of studying everyday, interpersonal social control and the subtle, continuous series of maneuvers that individuals use to keep each other in line during ordinary mundane activities."[53]

Third, the assumption by sociologists that all humans inhabit a "single society" runs counter to evidence collected by feminist sociologists such as Bernard and Oakley that men and women often inhabit their own social worlds.[54] Focusing on a more narrow issue, Thelma McCormack posits that voting studies have erroneously assumed that men and women inhabit a single political culture, with the result that women tend to appear more conservative or apathetic.[55]

Fourth, Millman and Kanter cite several studies that demonstrate the failure of social researchers to consider sex as an explanatory variable. They include Sarah Lightfoot's analysis of the sociology of education, in which she asserts that researchers do not consider issues raised by the fact that most teachers are women.[56]

Fifth, Millman and Kanter argue that by seeking to explain the status quo, sociologists need to be more sensitive to the ways in which their research might also legitimate existing social relations and institutions. They argue that researchers should focus more attention on social transformation.[57] Arlene Daniels goes further to assert that research on women should not only concern itself with revealing the sources of women's oppression but should also engage in exploring the concrete ways in which their status and lives can be improved.[58]

Finally, Millman and Kanter point out that unquestioned methodological assumptions and techniques can adversely affect findings and conclusions. They cite David Tresemer's analysis of statistical studies of sex differences. Tresemer asserts that most of these studies are misleading because they improperly use bipolar, unidimensional continuous, normal distributions that have the effect of exaggerating differences.[59] Although this particular problem might best be corrected by adopting another less biased quantitative method, Millman and Kanter claim that in many cases qualitative techniques might be more suitable and yield more balanced results than the standard quantitative approaches.[60]

Critical feminist theorists have developed several methodological approaches designed to address the topic of gender asymmetry and to avoid the possible gender biases in standard sociological practices. Judith Cook and Mary Fonow assert that feminist methodologies frequently employ the following seven research strategies and techniques[61]:

1. Visual techniques, such as photography and videotaping to collect or elicit data.
2. Triangulation, or the use of more than one research technique simultaneously.
3. The use of linguistic techniques in conversational analysis.
4. Textual analysis as a means to identify gender bias.
5. Refined quantitative approaches to measure phenomena related to sexual asymmetry and women's worlds.
6. Collaborative strategies or collective research models to enhance feedback and promote cooperative, egalitarian relations among researchers.
7. Situation-at-hand research practices that use an already existing situation as a focus for sociological inquiry or as a means of collecting data.

Cook and Fonow stress that none of these techniques is explicitly or exclusively feminist; however, their innovative character is revealed in their application and in the degree to which they incorporate and are informed by five basic principles that govern the production of feminist knowledge.[62] Cook and Fonow identify these principles as follows[63]:

1. The necessity of continuously and reflexively attending to the significance of gender relations as a basic feature of social life, including the conduct of research.
2. The centrality of consciousness-raising as a specific methodological tool and as a "way of seeing."
3. The need to challenge the norm of "objectivity" that assumes a dichotomy between the subject and object of research.
4. The concern for the ethical implications of research.
5. An emphasis on the transformation of patriarchy and the empowerment of women.

Feminist standpoint theories, as espoused by Dorothy Smith, Hilary Rose, Nancy Hartsock, and others, build on these epistemological and methodological principles by explicitly focusing on women and their direct experience as the center of analysis. *Standpoint theorists* argue that they can use women's experience to analyze social relations in ways that overcome the androcentric dichotomies of Enlightenment "positivism"—such as culture versus nature, rational mind versus irrational emotions, objectivity versus subjectivity, public versus private—that have structured knowledge production in the social and natural sciences.[64] Smith asserts that "women's perspective discredits sociology's claim to constitute an objective knowledge independent of the sociologist's situation."[65]

Feminist standpoint theorists articulate a feminist methodology that at once privileges the feminist standpoint as more inherently objective, challenges "mainstream" sociological inquiry, and provides a paradigmatic alternative. Smith disputes that such a paradigm shift would entail a "radical transformation of the subject matter"[66]; rather, she argues that what is involved is the restructuring of the relationship between the sociologist and the object of her research:

> What I am suggesting is more in the nature of a re-organization which changes the relation of the sociologist to the object of her knowledge and changes also her problematic. This re-organization involves first placing the sociologist where she is actually situated, namely at the beginning of those acts by which she knows or will come to know; and second, making her direct experience of the everyday world the primary ground of her knowledge.[67]

Critical theorists also suggest that a feminist standpoint methodology rooted in women's experience would elucidate the epistemological connections among the production of knowledge, everyday experiences, and political praxis that "positivistic" epistemologies often deny. Rose asserts in "Women's Work: Women's Knowledge" that human knowledge and consciousness "are not abstract or divorced from experience or 'given' by some process separate from the unitary material reality of the world. Human knowledge . . . comes from practice, from working on and changing the world."[68]

Viewed collectively, feminist empiricism—in the form of feminist methodologies, epistemologies, and standpoints—challenges key tenets of traditional empiricism embodied in "positivistic" science. Sandra Harding argues that feminist empiricism specifically questions three central assumptions:

1. The assumption that the social identity of the observer is irrelevant to the "goodness" of the results of research, asserting that the androcentrism of science is both highly visible and damaging. It argues that women *as a social group* are more likely than men *as a social group* to select problems for inquiry that do not distort human social experience.

2. The assumption that science's methodological and sociological norms are sufficient to eliminate androcentric biases by suggesting that the norms themselves appear to be biased insofar as they have been incapable of detecting androcentrism.

3. The assumption that science must be protected from politics. Feminist empiricism argues that *some* politics—the politics of movement for emancipatory social change—can increase the objectivity of science.[69]

Critiquing the Critique: Challenges to Critical Feminist Theory

Feminist methodologies, epistemologies, and standpoint theories have received their own share of criticism. The most formidable of these have come from feminist critics themselves, who use feminist concepts and positions to, in effect, critique the "critique." What links many of these counter critiques is an emphasis on the role of ideology in the structuring of feminist theory, epistemology, and practice.

The foundation for all critical feminist theory is the belief in the primacy of gender as a fundamental division that structures social relations. Sarah Matthews questions the importance of gender dichotomy, arguing that "this dichotomy does not match social reality as closely as assumed."[70] The existence of gender identities coded masculine and feminine does not, she asserts, necessarily mean that gender is the critical variable determining social behavior[71]:

> All of these [feminist] critiques have in common as their beginning point the assumption that distinguishing between two genders is the appropriate foundation from which to build research questions and theory. To say that women have been excluded from sociological research; that research on women must be done to parallel research on men; that different methodologies must be utilized to understand women in society; that boys and girls are socialized differently; and that women as a group are oppressed, is to accept and to re-enforce the taken-for-granted assumption that there are in fact two gender categories into which it is important to sort all human beings.[72]

Matthews goes on to argue that gender or sex is not "an immutable fact" and that feminist research and theory can be interpreted as supportive of this position. That this support has not been acknowledged, she asserts, is due to the ideological basis of feminism that is committed to seeing two genders.[73] She concludes by positing that sociologists can overcome sexual bias in research by developing paradigms "that do not include gender as having *a priori* significance."[74]

Attempts to construct feminist methods that espouse a privileged position for an independent "women's standpoint" have also been criticized by feminist critics such as Elizabeth Spelman. Spelman sees the phrase "as a woman" as "the Trojan horse of feminist ethnocentrism" because it embodies the assumption that "gender identity exists in isolation from race and class identity."[75] She argues that the feminist perspective or standpoint "obscures the heterogeneity of women"[76] and hence serves as little more than a methodological means of privileging white, middle-class women's experience. Spelman, in particular, challenges the following five assumptions inherent in feminist discussions of gender[77]:

1. Women can be talked about "as women."
2. Women are oppressed "as women."
3. Gender can be isolated from other elements of identity that affect one's social, economic, and political position such as race, class, ethnicity; hence, sexism can be isolated from racism and classism.
4. Women's situation can be contrasted with men's.
5. Relations between men and women can be compared with relations between other oppressor/oppressed groups, and, hence, it is possible to compare the situation of women with the situation of blacks, Jews, the poor, and others.

Like Matthews, Spelman attributes the current feminist research agenda, with its emphasis on a unitary concept of "woman," as a result of an ideology within feminism asserting that "the possibility of a coherent feminist politics seems to require a singleness of voice and purpose."[78]

THE FEMINIST CRITIQUE AND THE RECONCEPTUALIZATION OF SOCIOLOGICAL THEORY

Potential problems within critical feminist theory have not prevented theorists from applying feminist concepts and perspectives in their efforts to critique and reevaluate other theories. Beginning in the early 1970s, feminist critical theorists have attacked such diverse theoretical approaches as functionalism, macrostructural theories of stratification, network theories, theories of emotions, and symbolic interactionism, sociobiology, and rational choice theory.[79] The aim of many of these revisionist critiques has been to point out androcentric and patriarchal biases, but others have sought to redeem aspects of these mainstream theoretical traditions by demonstrating their contributions to, as well as applications for, gender analysis. In recent years, feminist critical theorists have focused much of their attention on other varieties of critical theory, delineating the connections and tensions between feminist critical thought and classical Marxism, the Marxist-Weberian critical theory associated with the Frankfurt School, Jürgen Habermas' theories of communicative action, and theories of postmodernism.[80] What follows is an overview of some points and issues raised by critical feminists on four specific theoretical traditions within sociology: Parsonian functionalism (see Chapter 4), rational choice theory (Chapter 24), Marxist theory (Chapters 11, 17, 40), and theories of postmodernism (Chapter 44).

Parsonian Functionalism

One of the first theoretical traditions to be targeted by feminist critical theorists was functionalism and, more specifically, the theories of Talcott Parsons. Parsons' theory of gender socialization has been characterized by feminist theorists as oppressive to both sexes but especially to women.[81] In particular, Parsons is criticized for ignoring the issue of power as it relates to the control that husbands have over wives.[82] Other feminist critics attacked Parsons for "justifying male dominance by describing an instrumental (occupational) role for men and an expressive (domestic) role for women as a division of labor that was functional for family solidarity."[83]

Ruth Wallace criticized Parsons' emphasis on four functions—adaptation, goal attainment, integration, and latency (AGIL)—challenging the empirical basis of his AGIL model that evolved from his experiments on leadership in small groups with Robert F. Bales. Wallace asserts that the homogeneity of Bales' data base, which was composed entirely of small groups of all-male Harvard undergraduates, raises serious theoretical issues that Parsons overlooked when he made this data base the empirical foundation for the four functional prerequisites of his AGIL model for all action systems (see pages 34–40 in Chapter 4). Wallace questions whether the patterns arising from more heterogeneous groups—which might reflect differences in quality and sequence of activity—would have produced AGIL.[84]

In contrast, feminist sociologist Miriam Johnson argued against depictions of Parsonian theory as essentially conservative and gender biased, particularly in light of the ways in which left-oriented neo-functionalism builds on the multidimensional aspect of Parsons' model.[85] She asserts that three of the most common criticisms leveled against functionalism—namely, that it is circular, teleological, and supportive of the status quo—are inaccurate.[86] Evaluating Parsons from a feminist standpoint, Johnson posits that, in contrast with other theoretical traditions, Parsonian functionalism has considered gender more explicitly and provides a "broader and more multidimensional approach":

> Parsons had more to say about gender, the family, socialization, and evolutionary change in the family and gender relations than have other approaches, either macro or micro. . . . In contrast to both Marxist and interactionist theories, functionalism can at least potentially analyze patterns that are functional and dysfunctional for women in certain structural positions in a manner that treats women as neither a residual category of worker, nor as equally privileged "members" in an interaction.[87]

Finally, Johnson argues for the continued relevancy of Parsonian theory for feminist thought, asserting that Parsons' evolutionary schema for understanding social change can effectively be applied to the goals of the feminist movement. As it has evolved since the early 1960s, Johnson sees the feminist agenda as having "evidenced all four of the basic evolutionary processes of structural differentiation, inclusion, upgrading, and value generalization."[88]

Rational Choice Theory

Rational choice theory has been referred to as "the straw man of feminist theory"—that is, as the theory that has perhaps drawn the most consistent attacks from critical feminist theorists in their effort to distinguish themselves and their work from the "mainstream."[89] In particular, feminists have questioned the epistemological concepts implicit in the theory's attempt to explain social outcomes in terms of purposive agents subject to both external restraints and internal preferences (see Chapters 23 and 24). Basing her analysis on the work of Nancy Chodorow and Carol Gilligan,[90] Paula England posits that men and women are socialized to experience a different relationship between self and other, with women experiencing connectedness and men feeling separated from the other.[91] England argues that this distinction has consequences for one version of rational choice theory that draws on neoclassical economic theory and its assumption of a separative model of self.[92] Using a connective self model, England critiques the following four assumptions of neoclassical rational choice theory[93]:

1. Individuals act on the basis of self-interest.
2. Interpersonal utility comparisons are impossible.
3. Tastes are exogenous to economic models.
4. Individuals are rational.

According to England, the first three propositions are much more problematic if seen from a model of connective versus separative selves because (1) separative selves have a greater tendency to be selfish; (2) connective selves can be more empathic, which would facilitate utility comparisons; and (3) tastes are exogenous only if selves are viewed as separate.[94] Finally, she argues that rationality is problematic and must be revised to the extent that it is based on the assumption that cognition is radically separate from emotion[95]:

> Desires or "emotions" appear in economic theory only in the realm of tastes that are seen as exogenous. The theory thus creates a radical separation between two spheres of subjective events. In one sphere are the "tastes" (preferences, emotions, desires, values) that determine one's ends. In the other sphere are the cognitions, the calculations about what means will achieve the ends satisfying the demands of the first sphere. The rationality principle resides in this second sphere. In this way, economists have reproduced the reason/emotion dichotomy so common to Western thought. . . . In reality, there is probably much more commingling of the realms of emotion and cognition, so this conceptual separation is artificial. In this sense, the general feminist critique of false separations requires a revision of concept of rationality.[96]

Like functionalism, rational choice theory does have its defenders among feminist critical theorists. Debra Friedman and Carol Diem, for example, attempt to rebut the notion that rational choice theories are inherently sexist by contesting England's assertion that they are based on a separative model of self (and, hence, biased toward men) and by specifically reevaluating England's critique of the four assumptions of neoclassical economics.[97] With regard to England's critique of rationality, for example, Friedmand and Diem argue that

> To the extent that a separation of tastes and rationality exists then, it is a function of the assumptions regarding tastes. Since we have argued . . . that the assumptions that tastes are stable and exogenous cannot be said to stem from a separative bias, it follows that a separation of rationality and tastes does not result from a separative bias.[98]

Friedman and Diem then assert that rational choice theory can be helpful to feminist theorists by providing explanations for institutional-level gender inequality, as illustrated by empirical studies conducted by Mary Brinton, Kathleen Gerson, and Kristin Luker.[99]

Friedman and Diem conclude that although rational choice theory has biases, its biases are not essentially sexist.[100] Friedman and Diem concur with other feminist critics that to correct these biases selfishness and altruism should be incorporated as a variable and that the role of emotions and interpersonal preferences should be reassessed and revised.[101] However, they argue against the "wholesale rejection" of rational choice theory, asserting that "like most good theories, rational choice elucidates issues that others fail to see."[102]

Marxist Theory

Of all the approaches within sociological theory, Marxist analysis has until quite recently received the most attention (positive and negative) from feminist theorists. Only in the 1990s have theories of postmodernism gradually replaced Marxism as the most influential critical theoretical approach among the more radical feminist theorists. There exists even today a very distinct and divisive fault line between, on the one side, feminist critical theorists who seek in various ways to incorporate or build on Marxist theories or concepts (often called "socialist-feminists," "left feminists" or "capitalist-patriarchy feminists") and, on the other side, "radical feminists" who reject Marxism in favor of postmodernist epistemologies (see Chapter 44).[103] Between these two extremes is a whole range of feminist positions that are influenced to some degree by one or both of these critical traditions.

The "marriage," as it has often been called,[104] between feminist and Marxist theory could be characterized as having always been one of love and hate. On one hand, these theories share common emphases on the division of labor, power relations and inequality, the sources of oppression, and the importance of political struggle. They each seek to embody "a mode of understanding and a call to action."[105] On the other hand, many critical feminists have attacked Marxist analysis for being (1) insensitive to gender as a concept and explanatory variable, (2) phallocentric (that is, identifying male interests with human interests),[106] (3) dismissive of unwaged labor such as housework,[107] and (4) overly dependent on global transformation or "totalizing" explanations.[108] Ironically, the ideological determination of both Marxism and feminism to unite theory, praxis, and politics accounts in part for the conflict between them. Catherine MacKinnon summarizes this fundamental tension between Marxists and feminists as follows:

> Marxists have criticized feminism as bourgeois in theory and in practice, meaning that it works in the interest of the ruling class. They argue that to analyze society in terms of sex ignores class divisions among women, dividing the proletariat. Feminist demands, it is claimed, could be fully satisfied within capitalism. . . . Feminists charge that Marxism is male defined in theory and in practice, meaning that it moves within the world view and in the interest of men. Feminists argue that analyzing society exclusively in class terms ignores the distinctive social experiences of the sexes, obscuring women's unity. Marxist demands, it is claimed, could be (and in part have been) satisfied without altering women's inequality to men.[109]

MacKinnon goes further, however, to argue that attempts to form a synthesis between Marxism and feminism are destined for failure. Each theory, according to her, is grounded in radically different methods. The Marxist method is dialectical materialism, whereas the quintessential feminist method is "consciousness raising," which she defines as "the collective critical reconstitution of the meaning of women's social experience, as women live through it."[110] These two methods are, she asserts, epistemologically opposed:

> Marxism and feminism . . . posit a different relation between thought and thing, both in terms of the relationship of the analysis itself to the social life it captures and in terms of the participation of thought in the social life it analyzes. To the extent that materialism is scientific it posits and refers to a reality outside thought which it considers to have an objective . . . content. Consciousness raising, by contrast, inquires into an intrinsically social situation, into that mixture of thought and materiality which is women's sexuality in the most generic sense. . . . This method stands inside its own determinations in order to uncover them. . . .[111]

MacKinnon concludes by arguing that consciousness raising can be equally applied to class as well as gender relations, and the failure of Marxism to realize this might account for the failure of workers in advanced capitalist economies to embrace socialism.[112] For this reason, she sees the relationship of feminism with Marxism as analogous to that of Marxism and classical political economy: Feminism turns Marxism "inside out and on its head" and hence serves as "its final conclusion and ultimate critique."[113]

MacKinnon's critique represents an early, seminal version of what has in recent years emerged as a thoroughgoing epistemological attack on the assumptions maintained by Marxist theoretical approaches. Influenced by French structuralism and theories of postmodernism, radical feminist critics have charged that Marxism shares the following philosophical positions with the bourgeois system it seeks to challenge:

1. An ideal of society as a unitary totality.
2. A concept of universal (equal) subjectivity.
3. The idea that global explanations are superior to particularized or local explanations.
4. A unitary view of truth, reason, reality, and causality as ultimate neutral and objective principles in knowledge.[114]

Mia Campioni and Elizabeth Grosz argue that Marxism's adherence to these propositions renders it fundamentally phallocentric and, hence, incompatible with a "women-centered" epistemology that "entails 'seeing women first' [and] conceptualizing a space which allows women to be considered autonomous shapers and creators of meaning."[115]

Other critical feminist theorists, however, have taken a much more synergetic stance toward Marxism, building connections between feminist thought and Marxist analysis in a variety of ways. Some, such as Rae Blumberg (see Chapter 18) have done so in a muted sense by focusing on issues such as the control of the means of production and the distribution of economic surplus in their analyses of gender inequality and stratification.[116] Similarly, Nancy Hartsock has built on the concepts of historical materialism and the division of labor to argue for a privileged feminist standpoint from which feminist theorists can critique "the phallocratic institutions and ideology that constitute the capitalism form of patriarchy"[117]:

> Like the lives of proletarians according to Marxian theory, women's lives make available a particular and privileged vantage point on male supremacy. . . . Just as Marx's understanding of the world from the standpoint of the proletariat enabled him to go beneath bourgeois ideology, so a feminist standpoint can allow us to descend further into materiality to an epistemological level at which we can better understand both why patriarchal institutions and ideologies take such perverse and deadly forms and how both theory and practice can be redirected in more liberatory directions.[118]

Some critical feminist theorists, often identified as "socialist-feminists" or "capitalist-patriarchy" feminists, have argued for the uniting of feminist and Marxist theory and politics, positing that they are functionally interrelated.[119] Socialist-feminists tend not to debate the primacy of gender or class but instead make a place for both in their analysis. They see capitalism and patriarchy as separate but connected systems of oppression and seek to build on their complementarity by expanding Marx's labor theory of value to include unwaged labor and by repudiating the concept of separate public and private spheres.[120]

Finally, at the other extreme, some theorists advocate dropping the hyphen separating Marxism and feminism altogether, arguing that capitalism and patriarchy are not only complementary, but conceptually have so much in common that it makes little sense to treat them as separate systems, structures, or practices.[121] Beth Shelton and Ben Agger reject (1) orthodox-Marxist perspective with its dismissal of unwaged labor, (2) "dualistic" forms of Marxist feminism that argue that women's oppression and class-based oppression are analytically distinct and pertain to different spheres, and (3) capitalist-patriarchy feminism that posits capitalism and patriarchy as interrelated yet clearly differentiable.[122] Instead, Shelton and Agger aim to reconceptualize Marxism and feminism as "the same discourses, the same theoretical, empirical, and political practices in terms of how they deal with gender."[123] They assert that Marxism is not merely a theory of class, but "a theory of everything, including women" when it is conceived as a critique of the hierarchy of "value over valuelessness," of "productive over reproductive activity."[124] Shelton and Agger counter criticism that they want to subordinate feminism to Marxism by arguing that such a position assumes that they are different:

> While most Marxists and feminists resolve the domestic-labor debate differently, theirs are not fundamentally different perspectives. . . . We are convinced that Marxism in its best sense is feminism (and that feminism can become Marxism once it learns not to hyphenate itself). The hyphenation of Marxism and feminism is not simply a rhetorical move; it separates men and women who should be united in the conquest of their common enemy.[125]

Theories of Postmodernism

Theories of postmodernism (see Chapter 44) have had a strong influence on radical feminist theorists who seek to construct independent feminist methodologies and epistemologies. As important political-cultural currents, feminism and postmodernism attempt to develop new paradigms of social criticism that do not depend on traditional philosophical positions regarding the meaning and production of knowledge.[126] But, as Nancy Fraser and Linda Nicholson have argued, feminists and postmodernists tend to approach these questions from opposite directions: Postmodernists begin by elaborating "antifoundational metaphilosophical perspectives" that have consequences for the shape and character of social criticism, while feminists develop critical perspectives about social criticism that lead them to reconsider the epistemological basis of philosophy.[127] Where these two positions meet—and wherein lies the points of consensus and disagreement between them—are the subjects of lively debate among adherents of both theoretical approaches.

At one end, political scientist Jane Flax argues that feminist theory should be viewed as a "type of postmodern philosophy."[128] It has a "special affinity" with postmodern philosophy, sharing with it an uncertainty about the appropriate methods and epistemological grounding for explaining human experience.[129] Postmodernist thought, Flax asserts, challenges the following tenets of The Enlightenment:

1. The existence of a stable, coherent self.
2. The conviction that reason and "science" can provide an objective, reliable, and universal foundation for knowledge.
3. The belief that the knowledge acquired from the use of reason will be "True" and will represent something real and unchanging (universal) about human minds or the structure of the natural world.
4. The assertion that reason itself has transcendental and universal qualities.
5. The sense that there are connections among reason, autonomy, and freedom, with all claims to truth and rightful authority being submitted to the tribunal of reason.
6. The belief that by grounding claims to authority in reason conflicts among truth, knowledge, and power can be overcome. Truth can serve power without distortion; in turn, by using knowledge in the service of power, both freedom and progress will be assured.
7. The use of science as the exemplar of the right use of reason and as the paradigm for all true knowledge.
8. The use of language as merely the neutral medium through which representation of things in the world occurs.[130]

Flax argues that postmodern philosophy deconstructs these beliefs in ways that feminist theorists can profitably use to evaluate the fundamental goals of feminist theory—that is, the analysis of gender relations, the construction and experience of gender, and the structure of gender as a social category. She posits that feminist theorists can

> Enter into and echo postmodernist discourses as we have begun to deconstruct notions of reason, knowledge, or the self and to reveal the effects of the gender arrangements that lay beneath their "neutral" and universalizing facades. . . . As a practical social relation, gender can be understood only by close examination of the meanings of "male" and "female" and the consequences of being assigned to one or the other gender within concrete social practices.[131]

Fraser and Nicholson prefer not to subsume feminism to postmodernism; instead, they argue that each has respective strengths and weaknesses that can be reinforced or obviated by synthesizing the two into a "postmodernist feminism." A postmodern feminism would bring together the best of both approaches by combining "a postmodernist incredulity toward metanarratives with the social-critical power of feminism."[132] Such a synthesis, they argue, would involve adopting the following theoretical positions and strategies:

1. Postmodern critiques need not oppose large historical narratives nor analyses of societal macrostructures. Postmodern feminists need not abandon the large theoretical tools needed to address large political problems.
2. Postmodern-feminist theory must, however, be explicitly historical, attuned to the cultural specificity of different societies and periods and to that of different groups within societies and periods. The categories of postmodern-feminist theory must eschew where ever possible ahistorical, functionalist categories like "reproduction" and "mothering" in favor of historically specific institutional categories.
3. Postmodern-feminist theory would be nonuniversalist. Its mode of attention would be comparativist rather than

universalizing, attuned to changes and contrasts instead of to "covering laws."

4. Postmodern-feminist theory would dispense with the idea of a subject of history. It would replace unitary notions of "woman" and "feminine gender identity" with plural and constructed conceptions of social identity, treating gender as one relevant strain among others such as class, race, ethnicity, age, and sexual orientation.
5. Postmodern-feminist theory would be pragmatic. It would tailor its methods and categories to the specific task at hand, using multiple categories and forswearing the metaphysical comfort of a single "feminist method" or "feminist epistemology."[133]

Finally, at the other end of the debate, critical feminists have expressed concerns about what they see as the potential for nihilism and extreme epistemological relativism inherent in postmodernist attacks on "positivistic" philosophy. Scientist and feminist Evelyn Fox Keller argues in favor of a feminist critique that would expose androcentric bias within the social and natural sciences, but warns of the dangers of adopting a relativistic epistemological position that would reduce science to a mere social construct and undermine feminism's emancipatory agenda:

> Joining feminist thought to other social studies of science brings the promise of radically new insights, but it also adds to the existing intellectual dangers a political threat. The intellectual danger resides in viewing science as pure social product; science then dissolves into ideology, and objectivity loses all intrinsic meaning. In the resulting culture relativism, any emancipatory function of modern science is negated, and the arbitration of truth recedes into the political domain. . . . Feminist relativism is just the kind of radical move that transforms the political spectrum into a circle. By rejecting objectivity as a masculine ideal, it simultaneously lends its voice to an enemy chorus and dooms women to residing outside of the real politik modern culture; it exacerbates the very problem it wishes to solve.[134]

Concerns about the potential of postmodernism to reduce knowledge and meaning to the status of rhetoric have been raised even by radical feminist theorists who have otherwise taken strong constructivist positions about the sources and meanings of truth in the social and natural sciences. In her book, *Simians, Cyborgs and Women*, historian of science Donna Haraway expresses the difficult dilemma radical feminist theorists face in their efforts to espouse a feminist epistemology that can empower women, challenge androcentric bias in all knowledge claims, and create a feminist version of objectivity, while avoiding extreme relativism. Indeed, her words represent a pointed way to conclude this chapter:

> We wanted a way to go beyond showing bias in science . . . and beyond separating the good scientific sheep from the bad goats of bias and misuse. It seemed promising to do this by the strongest possible constructionist argument that left no cracks for reducing the issues to bias versus objectivity, use versus misuse, science versus pseudo-science. We unmasked the doctrines of objectivity . . . and we ended up with one more excuse for not learning any post-Newtonian physics and one more reason to drop the old feminist self-help practices of repairing our own cars. They're just texts anyway, so let the boys have them back. . . . Our problem is how to have *simultaneously* an account of radical historical contingency for all knowledge claims . . ., a critical practice for recognizing our own "semiotic technologies" for making meanings, *and* a no-nonsense commitment to faithful accounts of a "real world. . . ."[135]

NOTES

1. Jessie Bernard, *The Female World* (New York: Free Press, 1981); Ann Oakley, *The Sociology of Housework* (New York: Pantheon, 1974).

2. Dorothy Smith, "Women's Perspective as a Radical Critique of Sociology," *Sociological Inquiry* 44 (1974), pp. 7–15; "Sociological Theory: Methods of Writing Patriarchy" in *Feminism and Sociological Theory* (Newbury Park, CA: Sage, 1989), pp. 34–64; *The Everyday World as Problematic: A Feminist Sociology* (Boston: Northeastern University Press, 1987). Also Nancy Hartsock, *Money, Sex, and Power* (New York: Longman, 1983).

3. For example, see Chris Weedon, *Feminist Practice & Poststructuralist Theory* (New York: Basil Blackwell, 1987); Judith Butler, "Contingent Foundations: Feminism and the Question of Post-Modernism" in *The Postmodern Turn: New Perspectives on Social Theory* (Cambridge: Cambridge University Press, 1994), pp. 152–170; Susan Hekman, *Gender and Knowledge: Elements of a Postmodern Feminism* (Boston: Northeastern University Press, 1990); Sondra Farganis, "Postmodernism and Feminism" in *Postmodernism and Social Inquiry* (New York: Guilford, 1994); and Thomas Meisenhelder, "Habermas and Feminism: The Future of Critical Theory" in *Feminism and Sociological Theory*, ed. Ruth A. Wallace (Newbury Park, CA: Sage, 1989), pp. 119–134.

4. This has led some to question whether many feminist critiques should properly be called "theory." See Mary Jo Neitz, "Introduction to the Special Issue Sociology and Feminist Scholarship," *The American Sociologist* 20 (1989), p. 5.

5. Ibid., p. 4.

6. Sondra Farganis, "Social Theory and Feminist Theory: The Need for Dialogue," *Social Inquiry* 56 (1986), p. 56.

7. Kathryn Ward and Linda Grant, "The Feminist Critique and a Decade of Published Research in Sociology Journals," *The Sociological Quarterly* 26 (1985), p. 140. On the issue of inadequate representation of women as the subjects of research, see Arlie Russell Hochschild, "A Review of Sex Role Research," *American Journal of Sociology* 78 (1973), pp. 1011–1029 and Cynthia Fuchs Epstein, "A Different Angle of Vision: Notes on the Selective Eye of Sociology," *Social Science Quarterly* 55 (1974), pp. 645–656. Critiques on the neglect of female-dominated social sectors include Ann Oakley, *Sociology of Housework* (New York: Pantheon, 1974), and Jessie Bernard, "My Four Revolutions: An Autobiographical History of the ASA," *American Journal of Sociology* 78 (1973), pp. 773–791. For early critiques of sociological methods and their failure to reflect women's experiences and perspectives see Dorothy Smith, "Women's Perspective as a Radical Critique of Sociology," *Sociological Inquiry* 44 (1974), pp. 7–15, and Arlie Russell Hochschild, "The Sociology of Feeling and Emotion: Selected Possibilities" in *Another Voice: Feminist Perspectives on Social Life and Social Science* (New York: Octagon, 1976), pp. 280–307. Finally, among the early feminist theorists to criticize the use of men and their lifestyles as "normative" were Joan Acker, "Women and Social Stratification: A Case of Intellectual Sexism," *American Journal of Sociology* 78 (1973), pp. 936–945 and Jessie Bernard, "Research on Sex Differences: An Overview of the State of the Art" in *Women, Wives, Mothers* (Chicago: Aldine Publishing Company, 1975), pp. 7–29.

8. Oakley, *Sociology of Housework* (cited in note 1), p. 2.

9. Ibid., p. 5.

10. Ibid., p. 8.

11. Ibid., pp. 8–9. In a similar critique of social stratification literature, Joan Acker adds two additional assumptions: "(a) Women determine their own social status only when they are not attached to a man; and (b) women are unequal to men in many ways, are differentially evaluated on the basis of sex, but this is irrelevant to the structure of stratification systems." See Acker, "Women and Social Stratification: A Case of Intellectual Sexism," *American Journal of Sociology* 78 (1973), p. 937.

12. Oakley, *Sociology of Housework* (cited in note 1), pp. 12–13.

13. Ibid., p. 21. Oakley argues that of the five "founding fathers" of sociology—Marx, Comte, Spencer, Durkheim and Weber— only two, Marx and Weber, had "emancipated views" about women. Terry Kandal, in a more recent analysis of the "woman question" in classical sociological theory, offers a less condemnatory accounting, asserting that there were "complex and contradictory variations in the writings about women by different classical theorists." *The Woman Question in Classical Sociological Theory* (Miami: Florida International University Press, 1988), p. 245.

14. Ibid., pp. 21–28.

15. As quoted in opening epigraphs in Jessie Bernard's *The Female World* (New York: Free Press, 1981).

16. Ibid., p. 3.

17. Ibid., pp. 12–15.

18. See, for example, Jean Lipman-Blumen, "Toward a Homosocial Theory of Sex Roles: An Explanation of the Sex Segregation of Social Institutions," *Signs* (1976), pp. 15–31.

19. Ibid., p. 16.

20. Ibid., p. 17.

21. Nancy Chadorow, *The Reproduction of Mothering: Psychoanalysis and the Sociology of Gender* (Berkeley: University of California Press, 1978), pp. 173–174.

22. Ibid., p. 173.

23. Joan Acker, "Making Gender Visible" in *Feminism and Sociological Theory,* ed. R.A. Wallace (Newbury Park, CA: Sage, 1989), p. 73.

24. John Shotter and Josephine Logan, "The Pervasiveness of Patriarchy: On Finding a Different Voice" in *Feminist Thought and the Structure of Knowledge* (New York: New York University Press, 1988), pp. 69–70.

25. Linda Grant and Kathryn Ward, "Is There an Association Between Gender and Methods in Sociological Research?" *American Sociological Review* 52 (1987), p. 861.

26. Rae Lesser Blumberg, *Stratification: Social, Economic and Sexual Inequality* (Dubuque, Iowa: William C. Brown, 1978).

27. Joan Acker, "Women and Stratification: A Review of Recent Literature," *Contemporary Sociology* 9 (1980), p. 26.

28. Judith Stacey and Barrie Thorne, "The Missing Feminist Revolution in Sociology," *Social Problems* 32 (1985), pp. 301–316.

29. Ibid., p. 302.

30. Kathryn Ward and Linda Grant, "On a Wavelength of Their Own? Women and Sociological Theory," *Current Perspectives in Social Theory* 11 (1991), p. 134.

31. Grant and Ward, "Is There an Association Between Gender and Methods in Sociological Research?" (cited in note 25), pp. 856–862.

32. See Marlene Mackie, "Female Sociologists' Productivity, Collegial Relations, and Research Style Examined through Journal Publications," *Social and Social Research* 69 (1985), pp. 189–209; Ellen Carol Dubois, ed., *Feminist Scholarship: Kindling in the Groves of Academe* (Urbana: University of Illinois Press, 1985); and Rhoda Unger, "Through the Looking Glass: No Wonderland Yet! The Reciprocal Relationship between Methodology and Models of Reality," *Psychology of Women Quarterly* 8 (1983), pp. 9–32.

33. Grant and Ward, "Is There an Association Between Gender and Methods in Sociological Research?" (cited in note 25), p. 861.

34. See J.A. Howard, "Dilemmas in Feminist Theorizing: Politics and the Academy," *Current Perspectives in Social Theory* 8 (1987), pp. 279–312.

35. Joan Acker, "Making Gender Visible" (cited in note 23), pp. 65–81.

36. Ibid., pp. 68-69. As Acker readily acknowledges, her analysis of sociology's relationship to ruling power structures owes much to Dorothy Smith. See note 2: Smith, *The Everyday World as Problematic.*

37. Ibid., Acker, p. 78.

38. Ibid., Acker, p. 72.

39. Ibid., Acker, p. 67.

40. Dorothy Smith, *The Everyday World as Problematic* (cited in note 2). Also see Sandra Harding, *The Science Question in Feminism* (Ithaca, NY: Cornell University Press, 1986); Evelyn Fox Keller, "Feminism and Science," *Signs* 7 (1982), pp. 589–602.

41. Smith, "Women's Perspective" (cited in note 2), p. 7.

42. Stacey and Thorne, "The Missing Revolution" (cited in note 28), p. 309.

43. Smith, *The Everyday World as Problematic* (cited in note 2), p. 52.

44. Smith, "Women's Perspective" (cited in note 2), p. 7. A similar criticism is made by Liz Stanley and Sue Wise in their book *Breaking Out: Feminist Consciousness and Feminist Research* (London: Routledge, 1983), p. 28.

45. Shotter and Logan, "The Pervasiveness of Patriarchy" (cited in note 24), pp. 69–86. Also see Dorothy Smith, "A Sociology for Women" in *The Prism of Sex* (Madison: University of Wisconsin Press, 1979), pp. 135–188.

46. Marcia Millman and Rosabeth Moss Kanter, eds., *Another Voice: Feminist Perspectives on Social Life and Social Science* (New York: Octagon, 1976).

47. Ibid., pp. ix–xvii.

48. Ibid., p. ix.

49. See Arlie Russell Hochschild's essay "The Sociology of Feeling and Emotion: Selected Possibilities" in *Another Voice: Feminist Perspectives on Social Life and Social Science,* eds. Marcia Millman and Rosabeth Moss Kanter (New York: Octagon, 1976), pp. 280–307.

50. Jessie Bernard, "My Four Revolutions" (cited in note 7), p. 785. For more discussion of agency and communion approaches to research see David Bakan, "Psychology Can Now Kick the Science Habit," *Psychology Today* 5 (1972), pp. 26, 28, 86–88; and Rae Carlson, "Sex Differences in Ego Functioning: Exploratory Studies of Agency and Communion," *Journal of Consulting and Clinical Psychology* 37 (1971), pp. 267–277.

51. Millman and Kanter, eds., *Another Voice* (cited in note 46), p. x.

52. Ibid., p. xi.

53. Ibid.

54. Ibid. See also note 1: Bernard, *The Female World,* and Oakley, *The Sociology of Housework.*

55. Millman and Kanter, eds., *Another Voice,* p. xiii. Also Thelma McCormack, "Toward a Nonsexist Perspective on Social and Political Change," *Another Voice* (cited in note 49), pp. 1–33.

56. Millman and Kanter, eds., *Another Voice,* p. xiv.

57. Ibid., p. xv.

58. Ibid. Also see Arlene Daniels, "Feminist Perspectives in Sociological Research," *Another Voice* (cited in note 49), pp. 340–380.

59. Millman and Kanter, eds., *Another Voice,* p. xv. See also David Tresemer, "Assumptions Made About Gender Roles," *Another Voice* (cited in note 49), pp. 308-39.

60. Millman and Kanter, eds., *Another Voice,* p. xvi.

61. Judith Cook and Mary Fonow, "Knowledge and Women's Interests: Issues of Epistemology and Methodology in Feminist Sociological Research," *Sociological Inquiry* 56 (1986), pp. 2–29.

62. Ibid., p. 14.

63. Ibid., p. 2.

64. Sandra Harding, *The Science Question in Feminism* (Ithaca: Cornell University Press, 1986), pp. 136–162.

65. Smith, "Women's Perspective" (cited in note 2), p. 11.

66. Ibid.

67. Ibid.

68. Hilary Rose, "Women's Work: Women's Knowledge" in *What is Feminism,* eds. Juliet Mitchell and Ann Oakley (New York: Pantheon, 1986), p. 161.

69. Harding (cited in note 64), p. 162.

70. Sarah Matthews, "Rethinking Sociology through a Feminist Perspective," *American Sociologist* 17 (1982), p. 29.

71. Ibid.

72. Ibid., p. 30.

73. Ibid., p. 29.

74. Ibid.

75. Elizabeth Spelman, *Inessential Woman: Problems of Exclusion in Feminist Thought* (Boston: Beacon, 1988), p. x.

76. Ibid., p. ix.

77. Ibid., p. 165.

78. Ibid., p. 161.

79. See, for example, the following essays in Paula England, ed., *Theory on Gender/Feminism on Theory* (New York: Aldine de Gruyter, 1993): Lynn Smith-Lovin and J. Miller McPherson, "You are Who You Know: A Network Approach to Gender," pp. 223–251; Dana Dunn et al., "Macrostructural Perspectives on Gender Inequality," pp. 69–90; Miriam Johnson, "Functionalism and Feminism: Is Estrangement Necessary?," pp. 115–130; Debra Friedman and Carol Diem, "Feminism and the Pro-(Rational-) Choice Movement: Rational-Choice Theory, Feminist Critiques and Gender Inequality," pp. 91–114. Other works include Miriam Johnson, "Feminism and the Theories of Talcott Parsons" in *Feminism and Sociological Theory,* ed. R.C. Wallace (Newbury Park, CA: Sage, 1989), pp. 101–118; Marian Lowe and Ruth Hubbard, "Sociobiology and Biosociology: Can Science Prove the Biological Bias of Sex Differences in Behavior" in *Genesis and Gender* (New York: Gordian, 1979), pp. 91–112; Arlie Russell Hochschild, "The Sociology of Feeling and Emotion: Selected Possibilities" in *Another Voice: Feminist Perspectives on Social Life and Social Science,* eds. Marcia Millman and Rosabeth Moss Kanter (New York: Octagon, 1976), pp. 280–307; Paula England, "A Feminist Critique of Rational-Choice Theories: Implications for Sociology," *American Sociologist* 20 (1989), pp. 14–28; and Paula England and Barbara Kilbourne, "Feminist Critique of the Separative Model of the Self: Implications for Rational Choice Theory," *Rationality and Society* 2 (2), pp. 156–171.

80. In addition to the work cited at the beginning of this chapter, for Foucault and post-structuralism, see Caroline Ramazanoglu, ed., *Up Against Foucault: Explorations of Some Tensions between Foucault and Feminism* (New York: Routledge, 1993); and Linda Alcoff, "Cultural Feminism versus Post-Structuralism: The Identity Crisis in Feminist Theory" in *Feminist Theory in Practice and Process,* eds. Micheline Malson, Jean O'Barr, Sarah Westphal-Wihl, Mary Wyer (Chicago: University of Chicago Press, 1986), pp. 295–326. For critical feminist analyses of Marxist theory and the connections between it and feminist theory, see Catherine MacKinnon, "Feminism, Marxism, Method, and the State: An Agenda For Theory," *Signs* 7 (1982), pp. 515–544; Mia Campioni and Elizabeth Grosz, "Love's Labours Lost: Marxism and Feminism" in *A Reader in Feminist Knowledge* (New York: Routledge, 1991), pp. 366–397; Beth Anne Shelton and Ben Agger, "Shotgun Wedding, Unhappy Marriage, No-Fault Divorce? Rethinking Feminism-Marxism Relationship" in *Theory on Gender/Feminism on Theory,* pp. 25–42; Zillah Eisenstein, ed. *Capitalist Patriarchy and the Case for Socialist Feminism* (New York: Monthly Review Press, 1979); Heidi Hartmann, "The Unhappy Marriage and Marxism and Feminism: Towards a More Progressive Union," *Capital and Class* 8, pp. 1–33; Annette Kuhn and Ann Marie Wolpe, eds., *Feminism and Materialism: Women and Modes of Production* (London: Routledge, 1978); Ben Agger, *Fast Capitalism: A Critical Theory of Significance* (Urbana, IL: University of Illinois Press,

1989); and Alison Jagger, *Feminist Politics and Human Nature* (Sussex: Harvester, 1983). Finally, additional works on feminist and postmodernism include Jane Flax, "Postmodernism and Gender relations in Feminist Theory" in *Feminist Theory in Practice and Process*, pp. 51–74; and Imelda Whelehan, *Modern Feminist Thought* (New York: New York University Press, 1995), pp. 194–215.

81. Ruth Wallace, "Introduction" in *Feminism and Sociological Theory* (Newbury Park, CA: Sage, 1989), p. 12.

82. Miriam Johnson, "Functionalism and Feminism" (cited in note 79). p. 124.

83. Ibid., p. 115.

84. Wallace, "Introduction" (cited in note 81), p. 12.

85. Johnson, "Feminism and the Theories of Talcott Parsons" (cited in note 79), p. 115.

86. Ibid., pp. 117–119.

87. Ibid., pp. 119–120.

88. Ibid., p. 123.

89. Friedman and Diem, "Feminism and the Pro-(Rational-)Choice Movement" (cited in note 79), p. 91.

90. See Chodorow, *The Reproduction of Mothering* (1978) in note 21; and Carol Gilligan, *In a Different Voice: Psychological Theory and Women's Development* (Cambridge, MA: Harvard University Press, 1982).

91. Neitz, "Introduction to Special Issue: Sociology and Feminist Scholarship" (cited in note 4), p. 5.

92. England, "A Feminist Critique of Rational-Choice Theory" (cited in note 79), p. 14.

93. Ibid., p. 17.

94. Neitz, "Introduction to Special Issue: Sociology and Feminist Scholarship" (cited in note 4), p. 8.

95. England, "A Feminist Critique of Rational-Choice Theory" (cited in note 79), p. 21.

96. Ibid.

97. Friedman and Diem (cited in note 79), pp. 92–100.

98. Ibid., p. 100.

99. Ibid., p. 92. See also Mary Brinton, "The Social-Institutional Bases of Gender Stratification: Japan as an Illustrative Case," *American Journal of Sociology* 94 (1988), pp. 300–334; Kathleen Gerson, *Hard Choices: How Women Decide about Work, Career and Motherhood* (Berkeley: University of California Press, 1985); and Kristen Luker, *Abortion and the Politics of Motherhood* (Berkeley: University of California Press, 1984).

100. Friedman and Diem (cited in note 79), p. 109.

101. Ibid., p. 110.

102. Ibid.

103. It should be noted that these labels are extremely fluid and there exists as yet no consensus regarding their precise meanings or which feminist theorists fall into each (or neither) camp.

104. Campioni and Grosz, "Love's Labours Lost" (both cited in note 80), p. 366; and Shelton and Agger, "Shotgun Wedding."

105. Farganis, "Social Theory and Feminist Theory: The Need for Dialogue" (cited in note 6), p. 56.

106. Campioni and Grosz (cited in note 80), p. 367.

107. Shelton and Agger (cited in note 80), p. 25.

108. Ibid., p. 35.

109. MacKinnon, "Feminism, Marxism, Method, and the State: An Agenda for Theory" (cited in note 80), pp. 517–518.

110. Ibid., p. 543.

111. Ibid.

112. Ibid., pp. 543–544.

113. Ibid., p. 544.

114. Campioni and Grosz, "Love's Labours Lost" (cited in note 80), p. 368.

115. Ibid., p. 367.

116. Blumberg, *Stratification: Social, Economic and Sexual Inequality* (cited in note 26).

117. Hartsock, *Money, Sex and Power* (cited in note 2), p. 231.

118. Ibid.

119. Shelton and Agger (cited in note 80), p. 33.

120. Ibid. See also Jagger, *Feminist Politics and Human Nature* (cited in note 80), and Eisenstein, ed., *Capitalist Patriarchy and the Case for Socialist Feminism.*

121. Shelton and Agger (cited in note 80), p. 34.

122. Ibid., pp. 25–33.

123. Ibid., pp. 25–27.

124. Ibid., pp. 36–37.

125. Ibid., p.36.

126. Nancy Fraser and Linda Nicholson, "Social Criticism without Philosophy: An Encounter between Feminism and Postmodernism" in *The Post-Modern Turn: New Perspectives on Social Theory* (Cambridge: Cambridge University Press, 1994), p. 242.

127. Ibid., pp. 242–243.

128. Flax, "Postmodernism and Gender Relations in Feminist Theory" (cited in note 80), p. 54.

129. Ibid.

130. Ibid., pp. 54–55.

131. Ibid., pp. 56, 60. Flax cites as examples of pertinent work in this area Alice Jardine, *Gynesis: Configurations of Woman and Modernity* (Ithaca, NY: Cornell University Press, 1985) and Donna Haraway, "A Manifesto for Cyborgs: Science, Technology, and Socialist Feminism in the 1980s," *Socialist Review* 80 (1983), pp. 65–107.

132. Ibid., p. 258.

133. Ibid.

134. Evelyn Fox Keller, "Feminism and Science," *Signs* 7 (1982), p. 593.

135. Donna Haraway, *Simians, Cyborgs and Women* (New York: Routledge, 1991), pp. 186–187.

44

The Continuing Tradition III: Theories of Postmodernism*

Sociology was created as a self-conscious discipline to explain the transformations associated with "modernity," especially the rise of industrial capitalism and the corresponding decline of the agrarian feudal system in Europe and, eventually, elsewhere in the world. Some theorists—indeed, most examined in this book—have taken a scientific stance and have searched for fundamental and basic social forces that operate in all times and places and that, as a consequence, explain social processes in not only the modern era but also the past and future as well. Others have preferred a more restricted time framework and more modest goals, but even these scholars have sought to develop analytical frameworks for, at the very least, interpreting events in the modern world.

We are now at the end of this section of chapters that challenges the scientific and, in some cases, even the analytical pretensions of sociology. The label "postmodern" encompasses many divergent points of view, but the term contains two common themes: (1) a critique of sociology as a science and (2) a decisive break with modernity in which cultural symbols, media-driven images, and other forces of symbolic signification have changed the nature of social organization and the relation of individuals to the social world.

THE POSTMODERN CRITIQUE OF SCIENCE

The Age of Science, as it emerged from The Enlightenment, posited that it would be possible to use language to denote key properties of the universe and to communicate among scientists the nature and dynamics of these

*This chapter is coauthored with Kenneth Allan.

properties. Indeed, it was believed that, as knowledge accumulated by testing general theories against empirical cases, ever more formal languages, such as mathematics, could be used by theorists and ever more precise measuring instruments could be developed by experimental researchers. In this manner, the accumulation of knowledge about properties and dynamics of the universe would accelerate. For such accumulation to occur, the degree of correspondence between theories stated in languages and the actual nature of the universe would have to increase. No scientist assumed, of course, that there could ever be a perfect correspondence, but there was a faith that the use of more precise languages and measuring instruments calibrated for these languages could make representations of the universe increasingly accurate.

This faith in science was one of the cornerstones of "modernism," in at least this sense: Scientific knowledge could be used to forge a better society. As knowledge about the world accumulated, it could be used to increase productivity, democracy, and fairness in patterns of social organization. As with all critical approaches, postmodernism attacks modernity's faith in science. Postmodernism poses three interrelated problems associated with human knowledge.

First, the problem of representation. Postmodernists often question the view that science could be used to demystify the world by discovering and using the law-like principles governing its operation. Undergirding this modern belief was the notion that there is a single best mode—scientific theory and research findings—for expressing "truth" about the world. Postmodernists typically challenge this assumed correspondence between the signs of scientific language and obdurate reality. Does the language of science, or any language, provide a direct window through which we can view reality; that is, does language simply represent reality? Or, is language a social construction that by its very existence distorts the picture of reality? To the degree that language is a social construction, and, thus related to social groups and their interests, the assumption of the direct representation of language is rendered problematic.

Second, the problem of power and vested interests. Though some postmodernists may concede that the physical world might operate by laws, the very process of discovering these laws creates culture that, in turn, is subject to interests, politics, and forms of domination. For example, law-like knowledge in subatomic physics has reflected political interests in war-making or the laws of genetics can be seen to serve the interests of biotechnology firms. What is true of the laws of the physical and organic worlds is even more true of social laws whose very articulation reflects moral, political, economic, and other interests within the social sphere. From a postmodernist's view, "truth" in science, especially social science, is not a correspondence between theoretical statements and the actual social universe, but a cultural production like any other sign system. Science cannot, therefore, enjoy a "privileged voice" because it is like all cultural texts.

Third, the problem of continuity. Postmodernists question the view that knowledge accumulates in ways that increase continuity among understandings about the world and that can be used to advance society. This faith in knowledge was a hallmark of "modernity," but to postmodernists, who see discontinuities in knowledge as tied to shifts in the interests of dominant factions in society, such faith in the progressing continuity of knowledge and culture is not only misguided but empirically wrong. Postmodernists argue that because there is not a truth that exists apart from the ideological interests of humans, discontinuity of knowledge is the norm, and a permanent pluralism of cultures is the only real truth that humans must continually face.

In the end, the postmodern philosophical attack on science denies privileged status to any knowledge system, including its own "because any place of arrival is but a temporary station. No place is privileged, no place better than another, as from no place the horizon is nearer than from any other."[1] Postmodernism emphasizes that, as a human creation, knowledge is relative to, and contingent on, the circumstances in which it was generated. Because knowledge is ultimately a system of signs, or a language of human expression, it is about itself as much as an external world "out there."

This critique has many of the same elements as expressed by the first critical theorists examined in Chapter 41, or more recent incarnations of critical theory examined in Chapters 42 and 43, but it adds new twists and takes new turns. The postmodern feminists examined in Chapter 43 can be seen as examples of this shift in emphasis within the postmodern critique, but we should pursue the more general argument further by examining the key founding thinkers and contemporary figures.

Jean-François Lyotard

Jean-François Lyotard built much of his critique of science[2] on a notion borrowed from Ludwig Wittgenstein: language as a game.[3] Wittgenstein had posited an analogous relationship between the manner in which language functions and the way games are played. He argued that language, like a game, is an autonomous creation that requires no justification for its existence other than itself and is subject only to its own rules. With the term language, Wittgenstein was denoting not simply words and their syntactical arrangements but, rather, all that is wrapped up in the presentation, reception, and enactment of human expression. And this expression exists for no other reason than itself and is subject to no other rules than its own.

Lyotard, using Wittgenstein's analogy, proposes a comparison between the narrative form of knowledge, on the one side, and the denotative, scientific form of knowledge, on the other. Narrative is a form of expression that is close to the social world of real people; it is expressed within a social circle for the purpose of creating and sustaining this social circle. In this view of narratives, Lyotard means something akin to the oral histories of small familial-based groups, in which there are internal rules concerning who has the right to speak and who has the responsibility to listen. Verification of the knowledge created in such a narrative is reflexive because such verification of the narrative refers to its own rules of discourse.

In contrast, the denotative, scientific form of knowledge does not originate from social bonds, but rather, science proposes to simply represent what is in the physical universe, thereby subjugating the narrative form of knowledge. Lyotard argues that narrative and tradition are no longer needed in science because theory and research will reveal the true nature of the universe. Yet, there is a problem with science: To be heard, science must appeal to a form of narrative knowledge, in this case a *grand narrative.* This grand narrative is based on The Enlightenment's promise of human emancipation and encompasses the vision of progress most clearly expressed by Georg Hegel. Lyotard maintains that postmodernism is defined by a diffuse sense of doubt concerning any such grand narrative and that this need to appeal to a narrative reveals science to be a language game like any other. Thus, science has no special authority or power to supervise other language games. According to Lyotard's vision, dissension must now be emphasized rather than consensus; heterogeneous claims to knowledge, in which one voice is not privileged over another, are the only true basis of knowledge.

Richard Rorty

The philosopher Richard Rorty[4] pushes the philosophical critique of science further to the extreme—indeed, to the point of appearing to assert that there is no external, obdurate reality "out there." Science argues, of course, that there is a reality "out there," that languages can be used in increasingly precise ways to denote and understand its properties and dynamics, that language can be used to communicate the discoveries of science to fellow scientists and others as well, that truth is the degree of correspondence between theory and data, and that efforts to increase this correspondence make science self-correcting. Hence, science does use language in ways that make it a more "objective" representation of the external world, thereby ensuring that certain languages are more objective than others. But, Rorty asserts, the issue can never be which language is more objective or scientific because the use of language is always directed toward pragmatic ends. Underlying every claim that one language brings a better understanding to a phenomenon than another is the assumption that it is more useful for a particular purpose. As Rorty claims, "vocabularies are useful or useless, good or bad, helpful or misleading, sensitive or coarse, and so on; but they are not 'more objective' or 'less objective' nor more or less 'scientific.'"[5]

In addition, even if there is a true reality and a true language with which to represent that reality, the moment *humans use a language,* it becomes evaluative. Thus, discovering the "true nature" of reality is not the goal of language use, scientific or otherwise, but rather, this use is directed at practical concerns, and such pragmatic concerns are always value based. Social scientists can tell stories that converge and reflect their common value concerns, thereby increasing their sense of solidarity and community as social scientists. Social scientists can also tell stories of power or of forces that divide them, thereby revealing the discordant features of any collective. But these stories are about this community of individuals more than about any "real world out there."

What, then, is left for social analysts? Rorty's answer is typical of many postmodernists: Deconstruct texts produced by communities of individuals using language for some value-laden, practical purpose. *Deconstruction* refers to the process of taking apart the elements of a text, and in Rorty's view, such deconstruction can be (a) critical, exposing the interests and ideology contained in the text, or (b) affirming, revealing elements of a text to enlighten and inform.

As becomes evident, then, Lyotard and Rorty would take sociological theory in a very different direction. Emphasis would be on language and texts,[6] and scientific explanations of an obdurate social reality would be seen as yet another form of text. This new direction—which represents a philosophical assault on scientific sociological theory—can be further appraised with illustrative examples of sociologists who have used the thrust of this philosophical critique in their more explicitly sociological works.

Illustrative Elaborations and Extensions within Sociology

Richard Harvey Brown's "Society as Text." Richard Harvey Brown[7] specifically applies the philosophical critique of postmodernism to social science, positing that social and cultural realities, and social science itself, are linguistic constructions. Brown advocates an approach termed *symbolic realism*[8] in which the universe is seen as existing for humans through communicative action. Moreover, each act of communication is based on previously constructed communicative actions—modes and forms of discourse, ideologies, world views, and other linguistic forms. Hence the search for the first or ultimate reality is a fruitless endeavor. Indeed, the "worlds" that are accepted as "true" at any one point in human history are constituted through normative, practical, epistemological, political, aesthetic, and moral practices which, in turn, are themselves symbolic constructions. Symbolic realism is thus a critical approach that seeks to uncover the ideologies surrounding the premises on which knowledge, senses of self, and portrayals of realities are constructed.

In this critical vein, Brown concludes that sociological theories themselves are "practices through which things take on meaning and value, and not merely as representations of a reality that is wholly exterior to them."[9] Theory in sociology ought to be critical and reflexive, with theorists recognizing that their own theories are themselves rhetorical constructions and, thus, texts constructed in time, place, and context as well as through linguistic conventions. This kind of approach to social theory as rhetoric sees its own discourse as not being about society—that is, as an effort to construct "true" representations on the nature of society—but, rather, as being simply part of what constitutes society, particularly its forms of textual discourse and representation.[10]

Charles C. Lemert's Emphasis on Rhetoric. For Charles Lemert,[11] all social theory is inherently discursive, stated in loose languages and constructed within particular social circles. What discursive theoretical texts explain—that is, "empirical reality"—is itself a discursive text, constructed like any text and subject to all the distorting properties of any text production. That theoretical texts depend on empirical texts, and vice versa, for their scientific value increases science's discursiveness. Each scientific portrayal of reality—whether from the "theory" or "research findings" side—is a text and, actually, a kind of "text on top of another text." This compounding of texts has implications for social scientific explanations: Social science explanations are often less adequate to understanding reality than are ordinary discursive texts that are less convoluted and compounded.

This understanding of the primacy of language leads to the use of irony in social theory. But the social theorist's use of irony is more than a literary device; it is a position from which the theorist views reality and its relationship to language. The physical universe is known to humans only through representation, culture, or language. Thus, reality shifts in response to different modes of representation. For example, in the eyes of humanity, the earth has shifted from being the center of the universe to being simply one of many planets circling one of a multitude of stars. But this position of the theorist also views language as humanity's certainty: Language is certain, and in that sense real, because the fundamental qualities and meanings of a linguistic system are known, valid, and replicable *within itself.* Because language is certain, theorists can make general statements concerning the properties of the physical universe, but the certitude of those statements rests in language and not in the physical reality or even the assumed relationship between culture and reality. The only position then for a social theorist is an ironic one: General theoretical statements can be made but with a "tongue-in-cheek" attitude because the author of such statements is reflexively aware that the only certitude of the statements rests in the linguistic system and not in the physical or social reality.

Mark Gottdiener's and Steven Seidman's Critique. Steven Seidman[12] adopts Rorty's distinction between enriching and critical forms of text deconstruction. And, along with Mark Gottdiener,[13] and many others as well, Seidman questions any social theory that posits a *foundationalism,* or the view that knowledge accumulates such that one level of knowledge can serve as the base on which ever more knowledge is built. Foundationalism is, of course, at the very core of science, so, this critique is truly fundamental for the activities that intellectuals can pursue. For Seidman and Gottdiener, however, foundationalism is just another effort to impose a grand narrative as a privileged voice.

Gottdiener sees foundationalism in sociological theory as an ideology, based on a *logocentrism* where "the classics" in theory are seen as the base on which a theoretical position is built. Such logocentrism is nothing more than a political ploy by established theorists to maintain their privileged position. Thus, sociological theory is about language and power-games among theorists, seeking to construct a grand narrative that also sustains their privilege and authority within an intellectual community.[14]

If such is the case, Seidman argues, the hope of human emancipation through sociological theory must be replaced by "the more modest aspiration of a relentless defense of immediate, local pleasures and struggles for justice."[15] But Gottdiener sees this position as akin to Rorty's concern with the use of vocabularies to sustain communities of scholars. Because of the inherent and ultimate privileging of one morality over another in such communities—not all moralities, even local ones, are commensurable one with another—theory does not become sufficiently critical. Instead, Gottdiener advocates a continuous, critical, and reflexive cycle of evaluation relative to the relationship between power and knowledge for evaluating all theories—postmodern or otherwise—that seek some grand narrative or self-legitimating tradition.

In sum, these representative commentaries offer a sense for the postmodern critique of science. Yet, postmodernism is much more than a philosophical critique. If this were all that postmodernists had to offer, there would be little point in examining these criticisms. Moreover, not all postmodernist theorists accept the philosophical critique just outlined, although most are highly suspicious of the "hard-science" view of sociological theory.[16] Indeed, most are committed to analysis of contemporary societies, especially the effects that dramatic changes in distribution, transportation, and information systems have had on the individual self and patterns of social organization. We can begin with "economic postmodernists" who generally employ extensions of Karl Marx's ideas and who, to varying degrees, still retain some of Marx's emancipatory zeal or, if not zeal, guarded hope that a better future that may lie ahead.

ECONOMIC POSTMODERNISM

Economic postmodernists could all be seen as part of the Hegelian turn of critical theory examined in Chapter 42. There is a concern with capital, especially its overaccumulation (that is, overabundance) as well as its level of dispersion and rapid movement in the new world system of markets driven and connected by information technologies. Moreover, culture or systems of symbols are seen to emerge from economic processes, but they exert independent effects on not only the economy but also every other facet of human endeavor. Indeed, for some economic postmodernists, advanced capitalism has evolved into a new stage of human history[17] that, like earlier modernity, is typified by a series of "problems," including the loss of a core or essential sense of self, the use of symbolic as much as material means to control individuals, the increased salience of cultural resources as both tools of repression and potential resistance, the emotional disengagement of individuals from culture, and the loss of national identities and a corresponding shift to local and personal identities. This list, and other "pathologies" of the postmodern era, sound much like those that concerned early sociologists when they worried about such matters as anomie and egoism (Émile Durkheim), alienation (Marx), marginal and fractured self (Georg Simmel), ideological control and manipulation by the powerful (Marx and, later, Antonio Gramsci and Louis Althusser), political-ideological mobilization as resistance (Marx), over-differentiation and fragmentation of social structure (Adam Smith, Herbert Spencer, and Durkheim), rationalization and domination by overconcerns with efficiency (Max Weber), and so on. Thus, economically oriented postmodernists evidence many of the same analytical tendencies of those who first sought to theorize about modernity.

Fredric Jameson

Among the central figures in economic postmodernism, Fredric Jameson is the most explicitly Marxist.[18] Although his theory is about the complex interplay among multinational capitalism, technological advance, and the mass-media, "the truth of postmodernism . . . [is] the world space of multinational capital."[19] He posits that capitalism has gone through three distinct phases, with each phase linked to a particular kind of technology. Early-market capitalism was linked to steam-driven machinery; mid-monopoly capitalism was characterized by steam and combustion engines; and, late-multinational capitalism is associated with nuclear power and electronic machines.

Late-multinational capitalism is the subject of postmodern theory. In particular, the nature of *praxis,* or the use of thought to organize action to change conditions and the use of experiences in action to reexamine thought, is transformed and confounded by the changed nature of signification that comes when the machines of symbolic reproduction—cameras, computers, videos, movies, tape recorders, fax machines—remove the direct connection between human production and its symbolic representation. These machines generate sequences of signs on top of signs that alter the nature of praxis—that is, how can thought guide action on the world, and vice versa, when concepts in such thought are so detached from material conditions?

Drawing from Marx's philosophy of knowledge, Jameson still attempts to use the method of praxis to critique the social construction of reality in postmodernity. Marx argued that reality did not exist in concepts, ideas, or reflexive thought but in the material world of production. Indeed, he broke with the young Hegelians over this issue, seeing them as "blowing theoretical bubbles" about the reality of ideas (see Chapter 41). But, like the earlier generation of critical theorists in the first decades of the twentieth century (see Chapter 42), those of the late twentieth century begin to sound even more Hegelian. According to Jameson, the creation of consciousness through production was unproblematically represented by the aesthetic of the machine in earlier phases of capitalism, but in multinational capitalism, electronic machines like movie cameras, videos, tape recorders, and computers do not have the same capacity for signification because they are machines of *reproduction* rather than of production. Thus, the foundation of thought and knowledge in postmodernity is not simply false, as Marx's view of "false consciousness" emphasized, it is nonexistent. Because the machines of late capitalism reproduce knowledge rather than produce it, and because the reproduction itself is focused more on the medium than on the message, the signification chain from object to sign has broken down. Jameson characterizes this breakdown as the schizophrenia of culture. Based on de Saussure's notion that meaning is a function of the relationship between signifiers (see Chapter 35), the concept of a break in the signification chain indicates that each sign stands alone, or in a relatively loose association with fragmented groups of other signs, and that meaning is free-floating and untied to any clear material reality.

Moreover, in a postmodern world dominated by machines of reproduction, language loses the capacity to ground concepts to place, to moments of time, or to objects in addition to losing its ability to organize symbols into coherent systems of concepts about place, time, and object. As language loses these capacities, time and space become disassociated. If a sign system becomes detached and free-floating and if it is fragmented and without order, the meaning of concepts in relation to time and space cannot be guaranteed. Indeed, meaning in any sense becomes problematic. The conceptual connection between the "here-and-now" and its relation to the previous "there-and-then" has broken down, and the individual experiences "a series of pure and unrelated presents in time."[20]

Jameson goes on to argue that culture in the postmodern condition has created a fragmented rather than alienated subject. Self is not so much alienated from the failure to control its own productive activities; rather, self is now a series of images in a material world dominated by the instruments of reproduction rather than production. In addition, the decentering of the postmodern self produces a kind of emotional flatness or depthlessness "since there is no longer any self to do the feeling . . . [emotions] are now free-floating and impersonal."[21] Subjects are thus fragmented and dissolved, having no material basis for consciousness or narratives about their situation; under these conditions, individuals' capacity for praxis—using thought to act, and action to generate thought—is diminished. Of course, this capacity for praxis is not so diminished that Jameson cannot develop a critical theory of the postmodern condition, although the action side of Marx's notion of praxis is as notably absent, if not impotent, as it was for the first generation of Frankfurt critical theorists.

David Harvey

In a manner like that of Jameson, David Harvey[22] posits that capitalism has brought about significant problems associated with humans' capacity to conceptualize time and space. Yet, for Harvey the cultural and perceptual problems associated with postmodernism are not new. Some of the same tendencies toward fragmentation and confusion in political, cultural, and philosophical movements occurred around the turn of the century. And, in Harvey's view, the cultural features of the postmodern world are no more permanent—as many postmodernists imply—than were those of the modernity that emerged in the nineteenth and early twentieth centuries.

Unlike Jameson, Harvey does not see the critical condition of postmodernity as the problem of praxis—of anchoring signs and symbols in a material reality that can be changed through thought and action—but, rather, as a condition of *overaccumulation,* or the modes by which too much capital is assembled and disseminated. All capitalist systems—as Marx recognized—have evidenced this problem of overaccumulation, because capitalism is a system designed to grow through exploitation of labor, technological innovation, and organizational retrenchment. At some point there is overabundance: too many products to sell to nonexistent buyers, too much productive capacity that goes unused, or too much money to invest with insufficient prospects for profits.

This overaccumulation is met in a variety of ways, the most common being the business cycle where workers are laid off, plants close, bankruptcies increase, and money is devalued. Such cycles generally restore macro-level economic controls (usually by government) over money supply, interest rates, unemployment compensation, bankruptcy laws, tax policies, and the like. But Harvey emphasizes another response to overaccumulation: absorption of surplus capital through "temporal" and "spatial" displacement. *Temporal displacement* occurs when investors buy "futures" on commodities yet to be produced, when they purchase stock option in hopes of stock prices rising, when they invested in other financial instruments (mortgages, long-term bonds, government securities), or when they pursue any strategy for using time and the swings of all markets to displace capital and reduce overaccumulation. *Spatial displacement* involves moving capital away from areas of overaccumulation to new locations in need of investment capital. Harvey argues that *displacement* is most effective when both its temporal and spatial aspects are combined, as when money raised in London is sent to Latin America to buy bonds (which will probably be resold again in the future) to finance infrastructural development.

The use of both spatial and temporal displacement to meet the issue of overaccumulation can contribute to the more general problem of time and space displacement. Time and space displacement occurs because of four factors: (1) advanced communication and transportation technologies, (2) increased rationalization of distribution processes, (3) meta- and world-level money markets that accelerate the circulation of money, and (4) decreased spatial concentration of capital in geographical locations (cities, nations, regions). These changes create a perceived sense of time and space compression that must be matched by changes in beliefs, ideologies, perceptions, and other systems of symbols. As technologies combine to allow us to move people and objects more quickly through space—as with the advent of travel by rail, automobile, jet, rocket—space becomes compressed; that is, distance is reduced and space is not as forbidding or meaningful as it was at one time. Ironically, as the speed of transportation, communication, market exchanges, commodity distribution, and capital circulation increases, the amount of available time decreases, because there are more things to do and more ways to do them. Thus, our sense of time and space compresses in response to increases in specific technologies and structural capacities. If these technological and structural changes occur gradually, then the culture that renders the resulting alterations in time and space understandable and meaningful will evolve along with the changes. But, if the changes in structure and technology occur rapidly, as in postmodernity, then the modifications in symbolic categories will not keep pace, and people will be left with a sense of disorientation concerning two primary categories of human existence, time and space. The present response to overaccumulation, "flexible capitalism," helps create a sense of time and space compression as capital is rapidly moved and manipulated on a global scale in response to portfolio management techniques.

In addition, because the new mode of accumulation is designed to move capital spatially and temporally in a flexible and thus ever-changing manner, disorientation ensues as the mode of regulation struggles to keep up with the mode of accumulation. For example, if capital sustaining jobs in one country can be immediately exported to another with lower-priced labor, beliefs among workers about loyalty to the company, conceptions about how to develop a career, commitments of companies to local communities, ideologies of government, import policies, beliefs about training and retraining, conceptions of labor markets, ideologies of corporate responsibility, laws about foreign investment, and many other cultural modes for regulating the flow of capital will all begin to change. Thus, in postmodernity, physical place has been replaced by a new social space driven by the new technologies of highly differentiated and dynamic markets, but cultural orientations have yet to catch up to this pattern of time and space compression.

As with most economic postmodernists, Harvey emphasizes that markets now distribute services as much as they deliver commodities or "hard goods", and many of the commodities and services that are distributed concern the formation of an image of self and identity. Cultural images are now market driven, emphasizing fashion and corporate logos as well as other markers of culture, lifestyle, group membership, taste, status, and virtually anything that individuals can see as relevant to their identity. As boredom, saturation, and imitation create demands for new images with which to define self, cultural images constantly shift—being limited only by the imagination of people, advertisers, and profit-seeking producers. As a result, the pace and volatility of products to be consumed accelerates, and producers for markets as well as agents in markets (such as advertisers, bankers, investors) search for new images to market as commodities or services.

Given a culture that values instant gratification and easy disposability of commodities, people generally react with sensory block, denial, a blasé attitude, myopic specialization, increased nostalgia (for stable old ways), and an increased search for eternal but simplified truths and collective or personal identity. To the extent that these reactions are the mark of postmodernity, Harvey argues that they represent the lag between cultural responses to new patterns of capital displacement over time and place. Eventually, culture and people's perceptions will catch up to these new mechanisms for overcoming the latest incarnation of capital overaccumulation.

Scott Lash and John Urry

Like David Harvey, Scott Lash and John Urry argue that a *postmodern disposition* occurs with changes in advanced capitalism that shift time and space boundaries.[23] In their view, shifting conceptualizations of time and space are associated with changes in the distribution of capital. Moreover, like most postmodern theorists, they stress that postmodern culture is heavily influenced by the mass-media and advertising. Yet, revealing their Marxian roots, they add that the postmodern disposition is particularly dependent on the fragmentation of class experience and the rise of the service class. With their emphasis on capital and the class structure, the focus of Lash and Urry's analysis is thus Marxian, but their method is Weberian: They disavow causal sequences, preferring to speak in terms of preconditions and ideal-types. Thus, they see the postmodernism forms outlined by Lyotard, Jameson, and, as we will see, Baudrillard as an ideal-type against which different systems of culture can be compared.

Like Harvey, Lash and Urry do not see postmodern culture as entirely new, but unlike Harvey, they are less sure that it is a temporary phase waiting for culture to catch up to changed material conditions. Lash an Urry believe that postmodern culture will always appeal to certain audiences with "postmodern dispositions." These dispositions emerge in response to three forces: First, the boundary between reality and image must become blurred as the media, and especially advertising, present ready-made rather than socially constructed cultural images. Second, the traditional working class must be fractured and fragmented; at the same time, a new service class oriented to the consumption of commodities for their symbolic power to produce, mark, and proclaim distinctions in group memberships, taste, lifestyle, preferences, gender-orientation, ethnicity, and many other distinctions must become prominent. And third, the construction of personal and subjective identities must increasingly be built from cultural symbols detached from physical space and location, such as neighborhood, town, or region; as this detachment occurs, images of self become ever more transitory. As these three forces intensify, a postmodern disposition becomes more likely, and these dispositions can come to support a broader postmodern culture where symbols marking difference, identity, and location are purchased by the expanding service class.

Although Lash and Urry are reluctant to speak of causation, it appears that at least four deciding factors bring about these postmodern conditions. The first factor involves the shift from Taylorist or regimented forms of production, such as the old factory assembly line, to more flexible forms of organizing and controlling labor, such as production teams, "flex-time" working hours, reduced hierarchies of authority, and deconcentration of work extending to computer terminals at home. Like Harvey, Lash and Urry believe that these shifts cause, and reflect, decreased spatial concentration of capital and expanded communication and transportation technologies, spatial dispersion, deconcentration of capital, and rapid movement of information, people, and resources are the principle dynamics of change. The second factor concerns large-scale economic changes—the globalization of a market economy, the expansion of industry and banking across national boundaries, and the spread of capitalism into less developed countries. A third factor is increased distributive capacities that accelerate and extend the flow of commodities from the local and national to international markets. This increased scope and speed of circulation can empty many commodities of their ethnic, local, national, and other traditional anchors of symbolic and affective meaning. This rapid circulation of commodities increases the likelihood that many other commodities will be made and purchased for what they communicate aesthetically and cognitively about ever-shifting tastes, preferences, lifestyles, personal statements, and new boundaries of prestige and status group membership. And, a fourth factor is really a set of forces that follows from the other factors: (a) the commodification of leisure as yet one more purchased symbolic statement; (b) the breakdown of, and merger among, previously distinct and coherent cultural forms (revolving around music, art, literature, class, ethnic, or gender identity, and other cultural distinctions in modernism); (c) the general collapse of social space, designated physical locations, and temporal frames within which activities are conducted and personal identifications are sustained; and (d) the undermining of politics as tied to traditional constituencies (a time dimension) located in physical places like neighborhoods and social spaces such as classes and ethnic groups.

Together these factors create a spatially fragmented division of labor, a less clear-cut working class, a larger service class, a shift to symbolic rather than material or coercive domination, a use of cultural more than material resources for resistance, and a level of cultural fragmentation and pluralism that erodes nationalism. But Lash and Urry argue, in contrast with Jameson, this emptying out process is not as de-regulated as it might appear. They posit that new forms of distribution, communication, and transportation all create networks in time, social spaces, and physical places. Economic governance occurs where the networks are dense, with communications having an increasingly important impact on the difference between core and peripheral sites. Core sites are heavily networked communication sites that function as a "wired village of noncontiguous communities."[24]

All these economic postmodernists clearly have roots in Marxian analysis, both the critical forms that emerged in the early decades of the century (see Chapter 41) and the world-system forms of analysis that arose in the 1970s and continue to the present day (see Chapter 14). Early critical theorists had to come to terms with the Weberian specter of coercive and rational-legal authority as crushing emancipatory class activity, but this generation of postmodern critics has had to reconcile their rather muted emancipatory goals to the spread of world capitalism as the preferred economic system; the prosperity generated by capitalism; the breakdown of the proletariat as a coherent class (much less a vanguard of emancipation); the commodification of everything in fluid and dynamic markets; the production and consumption of symbols more than hard goods (as commodities are bought for their symbolic value); the destruction of social, physical, and temporal boundaries as restrictions of space and time are changed by technologies; the purchase of personal and subjective identities by consumer-driven actors; and the importance of symbolic and cultural "superstructures" as driving forces in world markets glutted with mass-media and advertising images. Given these forced adaptations of the Marxian perspective, it is not surprising that many postmodernists have shifted their focus from the economic base to culture.

CULTURAL POSTMODERNISM

All postmodern theories emphasize the fragmenting character of culture and the blurring of differences marked by symbols. Individuals are seen as caught in these transformations, participating in, and defining self from, an increasing array of social categories, such as race, class, gender, ethnicity, or status, while being exposed to ever increasing varieties of cultural images as potential markers of self. At the same time, individuals lose their sense of being located in stable places and time frames. Many of the forces examined by economic postmodernists can account for this fragmentation of culture, decline in the salience of markers of differences, and loss of identity in time, place, and social space, but cultural postmodernists place particular emphasis on mass-media and advertising because these are driven by markets and information technologies.

Jean Baudrillard

The strongest postmodern statement concerning the effects of the media on culture comes from Jean Baudrillard,[25] who sees the task before the social sciences today as challenging the "meaning that comes from the media and its fascination."[26] In contrast with philosophical postmodernism, Baudrillard's theory is based on the assumption that there is a potential equivalence or correspondence between the sign and its object, and based on this proposition, Baudrillard posits four historical phases of the sign. In the first phase, the sign represented a profound reality, with the correlation and correspondence between the sign and the obdurate reality it signified being very high. In the next two phases, signs dissimulated or hid reality in some way: In the second phase, signs masked or counterfeited reality, as when art elaborated or commented on life, whereas in the third phase, signs masked the absence of any profound reality, as when mass commodification produced a plethora of signs that have no real basis in group identity but have the appearance of originating in group interaction. The second phase roughly corresponds to the period of time from the Renaissance to the Industrial Revolution, whereas the third phase came with the Industrial Age, as production and new market forces created commodities whose *sign values* marking tastes, style, status, and other symbolic representations of individuals began to rival the *use value* (for some practical purpose) or *exchange value* (for some other commodity or resource like money) of commodities. In Baudrillard's view, then, the evolution of signs has involved decreasing, if not obfuscating, their connection to real objects in the actual world.

The fourth stage in the evolution of the sign is the present postmodern era. In this age, the sign "has no relation to any reality whatsoever: It is its own pure simulacrum."[27] Signs are about themselves and, hence, are simulations or *simulacrums* of other signs with little connection to the basic nature of the social or material world. Baudrillard's prime example of simulacrum is Disneyland. Disneyland presents itself as a representation of Americana, embodying the values and joys of American life. Disneyland is offered as imagery—a place to symbolically celebrate and enjoy all that is good in the real world. But Baudrillard argues that Disneyland is presented as imagery to hide the fact that it is American reality itself. Life in the surrounding "real" communities, for example, Los Angeles and Anaheim, consists simply of emulations of past realities: People no longer walk as a mode of transportation, rather, they jog or power walk; people no longer touch one another in daily interaction, rather, they go to contact-therapy groups. The essence of life in postmodernity is imagery; behavior is determined by image potential and is thus simply image. Baudrillard depicts Los Angeles as "no longer anything but an immense scenario and a perpetual pan shot."[28] Thus, when Disneyland is presented as a symbolic representation of life in America, when life in America is itself an image or simulation of a past reality, then Disneyland becomes a simulation of a simulation with no relationship to any reality whatsoever, and it hides the nonreality of daily life.

Baudrillard argues that the presentation of information by the media destroys information. This destruction occurs because there is a natural entropy within the information process; any information about a social event is a degraded form of that event and, hence, represents a dissolving of the social. The media is nothing more than a constant barrage of bits of image and sign that have been removed an infinite number of times from actual social events. Thus, the media does not present a surplus of information, but, on the contrary, what is communicated represents total entropy of information and, hence, of the social world that is supposedly denoted by signs organized into information. The media also destroys information because it stages the presentation of information, presenting it in a pre-packaged meaning form. As information is staged, the subject is told what constitutes their particular relationship to that information, thereby simulating for individuals their place and location in a universe of signs about signs.

Baudrillard argues that the break between reality and the sign was facilitated by advertising. Advertising eventually reduces objects from their use-value to their sign-value; the symbols of advertisements become commodities in and of themselves, image more than information about the commodity is communicated. Thus, advertisements typically juxtapose a commodity with a desirable image—for example, a watch showing one young male and two young females with their naked bodies overlapping one another—rather than providing information about the quality and durability of the commodity. So that what is being sold and purchased is the image rather than the commodity itself. But, further, advertising itself can become the commodity sought after by the consuming public rather than the image of the advertisement. In the postmodern era, the form of the advertisement rather than the advertisement itself becomes paramount. For example, a currently popular form of television commercials is what could be called the "MTV-style." Certain groups of people respond to these commercials not because of the product and not simply because of the images contained within the advertisements, but because they respond to the overall form of the message and not to its content at all. Thus, in postmodernity, the medium is the message, and what people are faced with, according to Baudrillard, is simulations of simulations and an utter absence of any reality.

Further Elaborations of Cultural Postmodernism

Kenneth Gergen. The self is best understood, in Kenneth Gergen's view,[29] as the process through which individuals categorize their own behaviors. This process depends on the linguistic system used in the physical and social spaces that locate the individual at a given time. Because conceptualizations of self are situational, the self generally tends to be experienced by individuals as fragmented and sometimes contradictory. Yet, people are generally motivated to eliminate inconsistencies in conceptualizations, and though Gergen grants that other possible factors influence efforts to resolve inconsistencies, people in western societies try to create a consistent self-identity because they are socialized to dislike cognitive dissonance in much the same way they are taught to reason rationally. Gergen thus sees an intrinsic relationship between the individual's experience of a self and the culture within which that experience takes place, a cultural stand that he exploits in his understanding of the postmodern self.

Gergen argues that the culture of the self has gone through at least three distinct stages—the romantic, modern, and current postmodern phases. During the romantic period, the self as an autonomous individual and agent was stressed as individuals came out from the domination of various institutions including the church and manorial estate; during the modern period, the self was perceived as possessing essential or basic qualities, such as psychologically defined inherent personality traits. But the postmodern self consists only of images, revealing no inherent qualities and, most significantly, has lost the ability as well as desire to create self-consistency. Further, because knowledge and culture are fragmented in the postmodern era, the very concept of the individual self must be questioned and the distinction between the subject and the object dropped. According to Gergen, then, the very category of the self has been erased as a result of postmodern culture.

Thus, like Baudrillard, Gergen sees the self in postmodern culture as becoming saturated with images that are incoherent, communicating unrelated elements in different languages. And corresponding to Baudrillard's death of the subject, Gergen posits that the category of the self has been eradicated because efforts to formulate consistent and coherent definitions of who people are have been overwhelmed by images on images, couched in diverse languages that cannot order self-reflection.

Norman Denzin and Douglas Kellner. In contrast with Baudrillard's claim that television is simply a flow of incessant disjointed and empty images, both Norman Denzin[30] and Douglas Kellner[31] argue that television and other media have formed people's ideas and actions in much the same way as traditional myth and ritual: These media integrate individuals into a social fabric of values, norms, and roles. In the postmodern culture and economy, media images themselves are the basis from which people get their identities—in particular, their identities of race, class, and gender.

In addition, Denzin and Kellner advocate a method of social activism: a critical reading of the texts from media presentations to discover the underlying ideologies, discourses, and meanings that the political-economy produces. The purpose of these critical readings is to "give a voice to the voiceless, as it deconstructs those popular culture texts which reproduce stereotypes about the powerless."[32] Both theorists disavow any grand narrative, and like more radical postmodernists, neither advocates a centerpoint, ultimate hope, and grand or totalizing discourse. Yet, like critical theorists before them, they both hold out the hope of forming new solidarities and initiating emancipatory conflicts through the exposé of the political-economy of signs.

But Denzin, unlike Kellner, takes issue with Baudrillard's and Jameson's views on self as an incoherent mirage of signs and symbols and as incapable of ordering images into some coherence. Denzin argues that the "lived experience" itself has become the final commodity in the circulation of capital and that the producers of postmodern culture selectively choose which lived experiences will be commodified and marketed to members of a society. Postmodern culture only commodifies those cultures that present a particular aesthetic picture of race, class, and gender relations, but because culture has become centrally important in postmodernity, this process of commodification has a positive value for giving individuals a sense of identity and for enabling them to act in the material world on the basis of this identity.

Mark Gottdiener. Like Denzin and Kellner, Mark Gottdiener wants to maintain a critical postmodern stand but also to argue that an objective referent is behind the infinite regress of meaning. In contrast with Denzin and Kellner, Gottdiener sees the effects of media and markets as trivializing the culture of postmodernity.[33] According to Gottdiener, signs in technologically advanced societies can circulate between the levels of lived experience and the level where the sign is expropriated by some center of power, including producers and marketers of symbols.

Capitalizing on Baudrillard's notion of sign-value, Gottdiener argues that there are three separate phases of interaction wherein a sign can be endowed with meaning. In the first stage, economically motivated producers create objects of exchange-value for money and profit, an intent that is decidedly different from the goals of those who purchase the objects for their use-value. In the second stage, these objects become involved in the everyday life of the social groups that use them. During this stage users might "transfunctionalize" the object from its use-value into a sign-value to connect the object to their subgroup or culture (for example, a type of denim jacket is personalized to represent a group, such as the Hell's Angels). The third stage occurs if and when the economic producers and retailers adopt these personalized and transfunctionalized objects and commodify them (for instance, the Hell's Angels' style of jacket can now be bought by any suburban teenager in a shopping mall). This third stage involves a "symbolic leveling" or "trivialization" of the signed object.[34]

Thomas Luckmann. Although Thomas Luckmann[35] recognizes the importance of the media and advertising in creating a postmodern culture, he focuses on the process of *de-institutionalization* as it pushes people into the cultural markets found in the mass-media. The basic function of any institution, Luckmann argues, is to provide a set of predetermined meanings for the perceived world and, simultaneously, to provide legitimation for these meanings. Religion, in particular, provides a shield of solidarity against any doubts, fears, and questions about ultimate meaning by giving and legitimating an ultimate *meaning set.* Yet, modern structural differentiation and specialization has, Luckmann contends, made the ultimate meanings of religion structurally unstable because individuals must confront a diverse array of secular tasks and obligations that carry alternative meanings. This structural instability has, in turn, resulted in the privatization of religion. This privatization of religion is, however, more than a retreat from secular structural forces; it is also a response to forces of the sacralization of subjectivity found in mass culture.

Because of the effects of structural differentiation, markets, and mass culture, consciousness within individuals is one of immediate sensations and emotions. As a consequence, consciousness is unstable, making acceptance of general legitimating myths, symbols, and dogmas problematic—Lyotard's "incredulity toward grand narratives." Yet, capitalist markets have turned this challenge into profitable business. The individual is now faced with a highly competitive market for ultimate meanings created by mass-media, churches and sects, residual nineteenth-century secular ideologies, and substitute religious communities. The products of this market form a more or less systematically arranged meaning set that refers to minimal and intermediate meanings but rarely to ultimate meanings. Under these conditions, a meaning set can be taken up by an individual for a long or short period of time and combined with elements from other meaning sets. Thus, just as early capitalism and the structural forces that it unleased undermined the integrative power of religion, so advanced capitalism creates a new, more postmodern diversity of commodified meaning sets that can be mass produced and consumed by individuals in search of cultural coherence that can stave off their anxieties and fears in a structurally differentiated and culturally fragmented social world.

Zygmunt Bauman. Like Luckmann, Zygmunt Bauman[36] examines the effects of de-institutionalization on meanings about self in chaotic, often random, and highly differentiated systems. Within these kinds of systems, identity formation consists of self constitution with no reference point for evaluation or monitoring, no clear anchorage in place and time, and no lifelong and consistent project of self formation. People thus experience a high degree of uncertainty about their identity, and as a consequence, Bauman argues, the only visible vehicle for identity formation is the body. Thus, in postmodernity, body cultivation becomes an extremely important dynamic in the process of self constitution. Because the body plays such an important role in constituting the postmodern-self, uncertainty is highest around bodily concerns, such as health, physique, aging, and skin blemishes; these issues become causes of increased reflexivity, evaluation, and, thus, uncertainty.

Bauman, like Luckmann, argues that the absence of any firm and objective evaluative guide tends to create a demand for a substitute. These substitutes are symbolically created, as other people and groups are seen as "unguarded totemic poles which one can approach or abandon without applying for permission to enter or leave."[37] Individuals use these others as reference points and adopt the symbols of belonging to the other. The availability of the symbolic tokens depends on their visibility which, in turn, depends on the use of the symbolic token to produce a satisfactory self-construction. In the end, the efficacy of these symbols rests on either expertise in some task or mass following. Bauman also argues that accessibility of the tokens depends on an agent's resources and increasingly is understood as knowledge and information. So, for example, people might adopt the symbols associated with a specific professional athlete—wearing the same type of shoe or physically moving in the same defining manner—or, individuals might assume all the outward symbols and cultural capital associated with a perceived group of computer wizards. The important issue for Bauman is that these symbols of group membership can be taken up or cast off without any commitment or punitive action because the individuals using the symbols have never been an interactive part of these groups or celebrities' lives.

The need for these tokens results in "tribal politics," defined as self-constructing practices that are collectivized.[38] These tribes function as *imagined communities* and, unlike premodern communities, exist only in symbolic form through the shared commitments of their members.[39] For example, a girl in rural North Carolina might pierce various body parts, wear mismatched clothing three sizes too large, have the music of "Bio-Hazard" habitually running through her mind, and see herself as a member of the grunge or punk community but never once interact with group members. Or, an individual might develop a concern for the use of animals in laboratory experiments, talk about it to others, wear proclamations on tee-shirts and bumper stickers, and attend an occasional rally, and, thus, might perceive himself as a group member but not be part of any kind of social group or interaction network. These quasi-groups function without the powers of inclusion and exclusion that earlier groups possessed; indeed, these "neo-tribes" are created only through the repetitive performance of symbolic rituals and exist only as long as the members perform the rituals. Neo-tribes are thus formed through concepts rather than through face-to-face encounters in actual social groups. They exist as imagined communities through self-identification and persist solely because people use them as vehicles for self-definition and as "imaginary sediments." Because the persistence of these tribes depends on the affective allegiance of the members, self-identifying rituals become more extravagant and spectacular. Spectacular displays, such as body scarring or extreme or random violence, are necessary because in postmodernity the true scarce resource on which self and other is based is public attention.

CONCLUSION: ASSESSING THE POSTMODERN PROJECT

Any assessment of postmodernism that will not simply have the characteristics of a polemic, will have to be based on some common ground. That common ground is provided by the general properties of language, culture, and knowledge. Postmodernism is based on a critique of science against which postmodernism itself cannot stand firmly. All culture and language is distanced from the physical world. Such abstraction is a necessary condition of culture and language because, without some degree of removal from the physical world, there would only be the thing-in-itself and no human meaning as we understand it. Because language and meaning are not moored in the physical world but are, actually, representations of the world, they are inherently contingent and unstable and, thus, must be reified and stabilized in some way. In addition, because culture is by its very nature abstract and contingent, it is self-referential and is undergirded by incorrigible propositions or unchallenged beliefs about the world.[40] Reification of ideas into reality, stabilization, and the protection of incorrigible assumptions occur principally through (1) the structuring or institutionalization of collective activities and (2) the investment of emotions by individuals—both of which are tied to group processes and identity.

The function of all cultural knowledge, particularly language and theory, is to call attention to some elements in the world, both social and physical, while excluding others. The process of inclusion and exclusion is a fundamental way in which meaning is created. As Max Weber indicated, culture is the process of singling out from "the meaningless infinity of the world process" a finite portion that is in turn infused with meaning and significance. And the incorrigible propositions undergirding any knowledge system also function through inclusion and exclusion: A system cannot simultaneously be based on pragmatism and mysticism.

Thus, postmodernists are in a sense correct in their critique of science: Science, like any knowledge system, is based on incorrigible assumptions, is an abstraction from physical reality, is in need of reification and stabilization through the processes of institutionalization and emotional investment, and is bent on systematically subjugating other knowledge systems to assert its own reality. But, what postmodernists have missed—despite disclaimers about having no "privileged voice"—is that their own knowledge systems are subject to the same properties. In creating a system of knowledge, postmodernists must reify and stabilize their knowledge through the same processes of institutionalization, emotional investment, and exclusion—or be subject to the nihilism of endless regression.

Mark Gottdiener and early critical theorists are correct when they assert that knowledge, beliefs, and group interests are inseparable. What he and most other critical theorists fail to understand, however, is that their own knowledge systems function in the same way as science and are open to the same critique. The battle over the definition of science, knowledge, and theory is a cultural war for legitimation on which turns the allocation of institutional and material resources. Based on a generalized understanding of how culture functions within and between groups, the behavior of both postmodernists and social scientists is fairly predictable, especially because most of the protagonists are situated within academia.

Postmodernism is premised on a fundamental error that originated with structuralism (see Chapter 36): The structure of the sign system is posited to be the dynamic on which human action and interaction depends. This prejudicial favoring of culture over other properties and processes in social life might be one defining characteristic of postmodernism. Even those who appear to want to consider other factors, such as Denzin, Kellner, and Gottdiener, end up simply analyzing cultural artifacts such as film or billboards and then imputing their findings to the social actors who might or might not interpret the artifact in the same manner or use the culture in the way the researcher supposes. This error has resulted in a general overemphasis on culture, the signification system, and the problem of representation to the neglect of human agency and interaction. Even if culture is as fragmented and free-floating as postmodernists claim, it will have little effect on people until it becomes the focus of their interactions. And in micro-level interaction there are processes that tend to mitigate the problems of free-floating signifiers and emotionally flat symbols, as the theories of interaction presented in the chapters of Part VI document. People respond to the contingent nature of culture at the micro-level by producing a Goffmanian type of interaction equilibrium and natural rituals to emotionally infuse symbols (see Chapter 37).

This fundamental error has also produced some questionable assertions by postmodernists concerning the self. For the category of the self to be obliterated or to be fragmented, as postmodernists claim, culture must be exclusively determinative, and it is not. The creation and organization of the self is informed and constrained by culture, but it is not a direct function of the sign system. From a sociological point of view, the self is a process that is the joint work of individuals and groups in relation to their social environment. The self is an internalized structure of meanings that has as its source the process of role-taking in real groups and with real people in a person's particular biographical history. Media images can inform the interaction through which the self is constituted, but the interaction itself determines how those images will be used and what meanings will be attached to them.

It appears that postmodernism is moving toward a more moderated position. Each of the founders of postmodern thought posited a radical break with modernity and a universal problem of meaning and signification, but most subsequent postmodern thinkers have made attempts at grounding their analyses in the material world. Thus, the intellectual crisis is not as deep as was first supposed. The economic postmodernists, in particular, are using more generalized principles and processes to explain social phenomena. If postmodernism is to have a substantial voice beyond a critical stand against social science, it must move toward these more moderated positions.

NOTES

1. Zygmunt Bauman, *Intimations of Postmodernity* (London and New York: Routledge, 1992).

2. Jean-François Lyotard, *The Postmodern Condition* (Minneapolis: University of Minnesota Press, 1979, 1984).

3. Ludwig Wittgenstein, *Philosophical Investigations* (New York: Macmillan, 1936–1949, 1973).

4. Richard Rorty, "Philosophy as a Kind of Writing: An Essay on Derrida," *New Literary History* 10 (1978), pp. 141–160; *Philosophy and the Mirror of Nature* (Princeton: Princeton University Press, 1979); "Method, Social Science, and Social Hope," in *The Postmodern Turn; New Perspectives on Social Theory,* ed. Steven Seidman (Cambridge: Cambridge University Press, 1994).

5. Rorty, "Method, Social Science, and Social Hope" (cited in note 4), p. 57.

6. For the postmodernist, all cultural expressions are to be understood as language. But the use of linguistic systems as a basis for understanding all social phenomena is not new with postmodernism. For example, this linguistic equivalence model is central to the work of Claude Lévi-Strauss. But, postmodernism has, building on the poststructuralist work of Jacques Derrida and Michel Foucault, explicitly made signs, sign systems, and texts human reality *in toto.*

7. Richard Harvey Brown, *Society as Text; Essays on Rhetoric, Reason, and Reality* (Chicago: The University of Chicago Press, 1987); *Social Science as Civic Discourse; Essays on the Invention, Legitimation, and Uses of Social Theory* (Chicago: The University of Chicago Press, 1989); "Rhetoric, Textuality, and the Postmodern Turn in Sociological Theory," *Sociological Theory* 8 (1990), pp. 188–197.

8. Brown, *Social Science as Civic Discourse* (cited in note 7), pp. 49–54.

9. Ibid., p. 188.

10. In the postmodern literature, a distinction is often made between social and sociological theory. Social theory is generally understood to be a text that is self-consciously directed toward improving social conditions through entering the social discourse, whereas sociological theory is a denotative text that is abstracted from social concerns and involvement. In the view of postmodernism, social theory is preferred to sociological theory.

11. Charles C. Lemert, "The Uses of French Structuralisms in Sociology," *Frontiers of Social Theory; The New Syntheses,* ed. George Ritzer (New York: Columbia University Press, 1990), pp. 230–254. Charles C. Lemert, "General Social Theory, Irony, Postmodernism," *Postmodernism and Social Theory,* eds. Steven Seidman and David G. Wagner (Cambridge, MA: Blackwell, 1990), pp. 17–46.

12. Steven Seidman, "The End of Sociological Theory," in *The Postmodern Turn: New Perspectives on Social Theory,* ed. Steven Seidman (Cambridge: Cambridge University Press, 1994), pp. 84–96.

13. Mark Gottdiener, "The Logocentrism of the Classics," *American Sociological Review* 55 (June 1990), pp. 460–463; "Ideology, Foundationalism, and Sociological Theory," *Sociological Quarterly* 34 (1993), pp. 653–671.

14. Gottdiener, "Ideology, Foundationalism, and Sociological Theory" (cited in note 13), p. 667.

15. Seidman, "The End of Sociological Theory" (cited in note 12), p. 120.

16. For example, Lemert (*Sociology After the Crisis,* Boulder, CO: Westview, 1995, p. 78) makes a distinction between radical postmodernism and strategic postmodernism. Radical postmodernists disavow any possibility of truth or reality whereas strategic postmodernists attempt to undercut the authority that modernist knowledge claims while preserving the language and categories that modernist knowledge uses. Strategic postmodernists, according to Lemert, still maintain the modernist hope of emancipation.

17. See, for example, Stephen Crook, Jan Pakulski, and Malcolm Waters, *Postmodernization* (London: Sage, 1992).

18. Fredric Jameson, *The Postmodern Condition* (Minneapolis: University of Minnesota Press, 1984).

19. Ibid., p. 92.

20. Ibid., p. 72.

21. Ibid., p. 64.

22. David Harvey, *The Conditions of Postmodernity: An Inquiry into the Origins of Cultural Change* (Oxford: Blackwell, 1989).

23. Scott Lash and John Urry, *The End of Organized Capitalism* (Madison, WI: University of Wisconsin Press, 1987); *Economies of Signs and Space* (Newbury Park, CA: Sage, 1994).

24. Lash and Urry, *Economies of Signs and Space* (cited in note 23), p. 28.

25. Jean Baudrillard, *For a Critique of the Political Economy of the Sign* (St. Louis: Telos, 1972,1981); *The Mirror of Production* (St. Louis: Telos, 1973,1975); *Simulacra and Simulation* (Ann Arbor: University of Michigan Press, 1981, 1994); *Symbolic Exchange and Death* (Newbury Park, CA: Sage, 1993).

26. Baudrillard, *Simulacra and Simulation* (cited in note 25), p. 84.

27. Ibid., p. 6.

28. Ibid., p. 13.

29. Kenneth J. Gergen, *The Saturated Self* (New York: Basic Books, 1991); *The Concept of Self* (New York: Holt, Rinehart and Winston, 1971).

30. Norman K. Denzin, "Postmodern Social Theory," *Sociological Theory* 4 (1986), pp. 194–204; *Images of Postmodern Society; Social Theory and Contemporary Cinema* (London: Sage, 1991); *Symbolic Interactionism and Cultural Studies* (Oxford: Blackwell, 1992).

31. Douglas Kellner, "Popular Culture and the Construction of Postmodern Identities," in *Modernity and Identity,* eds. Scott Lash and Jonathan Friedman (Oxford: Blackwell, 1992); *Media Culture; Cultural Studies, Identity and Politics Between the Modern and the Postmodern* (London: Routledge, 1995).

32. Denzin, *Images of Postmodern Society* (cited in note 31), p. 153.

33. Mark Gottdiener, "Hegemony and Mass Culture: A Semiotic Approach," *American Journal of Sociology* 90 (1985), pp. 979–1001; *Postmodern Semiotics; Material Culture and the Forms of Postmodern Life* (Oxford: Blackwell, 1995).

34. Gottdiener, "Hegemony and Mass Culture" (cited in note 33, p. 996.

35. Thomas Luckmann, "The New and the Old in Religion," *Social Theory for a Changing Society,* eds. Pierre Bourdieu and James S. Coleman (Boulder, CO: Westview, 1991).

36. Zygmunt Bauman, *Modernity and Ambivalence* (Ithaca NY: Cornell University Press, 1991); *Intimations of Postmodernity* (London and New York: Routledge, 1992).

37. Bauman, *Intimations of Postmodernity* (cited in note 36), p. 195.

38. Ibid., pp. 198–199.

39. See Benedict Anderson, *Imagined Communities* (London: Verso, 1983).

40. See H. Mehan and H. Wood, *Reality of Ethnomethodology* (New York: John Wiley, 1975); and Niklas Luhmann, "Society, Meaning, Religion—Based on Self-Reference," *Sociological Analysis* 46 (1985), pp. 5–20.

Name Index

Subject Index